CLAY MINERAL CEMENTS IN SANDSTONES

SPECIAL PUBLICATION NUMBER 34 OF THE INTERNATIONAL ASSOCIATION OF SEDIMENTOLOGISTS

Clay Mineral Cements in Sandstones

EDITED BY

Richard H. Worden and Sadoon Morad

Blackwell Publishing

Blackwell Science Ltd
a Blackwell Publishing company

350 Main Street, Malden, MA 02148-5018, USA
108 Cowley Road, Oxford OX4 1JF, UK
550 Swanston Street, Carlton, Victoria 3053, Australia
Kurfürstendamm 57, 10707 Berlin, Germany

First published 2003

Library of Congress Cataloging-in-Publication Data

Clay cements in sandstones/edited by Richard H. Worden and Sadoon Morad.
p. cm. — (Special publication number 34 of the International Association of Sedimentologists)
Includes bibliographical references and index.
ISBN 1-40510-587-9 (pbk.: alk. paper)
1. Clay minerals. 2. Sandstone. I. Worden, Richard H. II. Morad, Sadoon. III. Special publication . . . of the International Association of Sedimentologists; no. 34.

QE389.625.C517 2002
549′.6—dc21

2002070932

A catalogue record for this title is available from the British Library.

Set in 9.5/12pt Melior
by Graphicraft Limited, Hong Kong
Printed and bound in the United Kingdom
by MPG Books Ltd., Bodmin, Cornwall

For further information on
Blackwell Publishing, visit our website:
http://www.blackwellpublishing.com

Contents

Introduction

Clays are one of the most important groups of minerals that destroy permeability in sandstones, but they also react with drilling and completion fluids and induce fine-particle migration during hydrocarbon production. They are a very complex family of minerals that commonly are mutually intergrown and contain a wide range of solid solutions and form by a wide range of processes. They form under a wide diversity of pressure and temperature conditions, as well as rock and fluid compositional conditions.

In this volume, clay minerals in sandstones are reviewed in terms of their mineralogy and general occurrence, their stable and radiogenic isotope geochemistry and their relationship to sequence stratigraphy and palaeoclimate. Their relationship to the petrophysical properties of sandstones and their use in drilling technologies are also covered. The controls on the various clay minerals are addressed and a variety of geochemical issues, including the importance of mass flux, links to carbonate mineral diagenesis and linked clay diagenesis in interbedded mudstone–sandstone are explored. The thorny issue of the quantification of clay minerals using XRD (X-ray diffraction) is tackled from a variety of technical angles. A number of case studies are included for kaolin, illite and chlorite clay cements, and the occurrence of smectite in sandstone also is reviewed. Clay minerals grow at rates that are only just becoming known; experimental data on clay cements in sandstones are thus reviewed and there are two model-based case studies that address the rates of growth of kaolinite and illite.

This volume follows a volume on *Carbonate Cementation in Sandstones* (IAS Special Publication 26, edited by Sadoon Morad) and a volume on *Quartz Cementation in Sandstones* (IAS Special Publication 29, edited by Richard Worden and Sadoon Morad). With the publication of this Special Publication, the majority of mineral cements in sandstones will have been addressed in books sponsored by the IAS.

Amongst its readership this volume will attract a wide range of scientists and technologists including: (i) sedimentologists and petrographers who deal with the occurrence, spatial and temporal distribution patterns and importance of clay cements in sandstones, (ii) geochemists who are involved in unravelling the factors that control clay cement formation in sandstones and (iii) petroleum geoscientists who need to predict areas of a play that may contain formation-damaging clay minerals. The book also will be of interest to geologists involved in palaeoclimate studies, basin analysis and basin modelling.

Acknowledgements

Crucial to all refereed publications are the referees themselves. For this service, they receive no real recognition and yet their efforts commonly lead to much improved final products. We would therefore like to acknowledge the following reviewers for their hard work, dedication and care in reviewing the 21 papers published in this volume.

Alastair Ruffell
Alessandro Amorosi
Andrew Hogg
Andrew Hurst
Anne Grau
Atilla Juhasz
Bruce Yardley
Cathal Dillon
Chris Rochelle
Christoph Spötl
Craig Smalley
Dave Doff
Dave Manning
David Awwiller
Dewey Moore
Earle McBride
Etienne Brosse
Ghazi Kraishan
Giles Berger
Jean-Francois Deconinck
Jeff Grigsby
Jim Boles
Kevin Taylor
Kitty Lou Milliken
Knut Bjørlykke
Lou Macchi
Lyndsay Kaye
Mark Wilkinson
Michael Wilson
Mike Mayall
Norbert Clauer
Norman Oxtoby
Patrick Corbett
Per Aagaard
Reinhard Gaupp
Richard Evans
Shelagh Baines
Shirley Dutton
Steve Franks
Stuart Haszeldine
Susanne Schmid
Tony Fallick
Warren Dickenson

Review papers

Int. Assoc. Sedimentol. Spec. Publ. (2003) **34**, 3–41

Clay minerals in sandstones: controls on formation, distribution and evolution

R.H. WORDEN[1] *and* S. MORAD[2]

[1] *Department of Earth Sciences, University of Liverpool, Brownlow Street, Liverpool L69 3GP, UK, e-mail: r.worden@liverpool.ac.uk*

[2] *Department of Earth Sciences, Uppsala University, Villa vägen 16, S-752 36 Uppsala, Sweden, e-mail: sadoon.morad@geo.uu.se*

ABSTRACT

This paper addresses the origin, distribution pattern and burial diagenetic evolution of clay minerals in sandstone: kaolin, smectite, illite, chlorite, berthierine, glauconite and mixed-layer illite–smectite and chlorite–smectite. Clay minerals may be co-deposited with sand grains as sand-sized argillaceous intra- and extra-clasts and as flocculated clays. These sand-sized argillaceous clasts are deformed by mechanical compaction into clay pseudomatrix. Detrital clay minerals may be incorporated into sandy deposits by bioturbation and infiltration of muddy waters. Diagenetic clay minerals form by alteration of unstable detrital silicates and by transformation of detrital and precursor diagenetic clay minerals. The most common eogenetic clay minerals are kaolinite, dioctahedral and trioctahedral smectite, berthierine, glauconite and, less commonly, Mg-rich clay minerals such as palygorskite. The distribution of eogenetic clay minerals is strongly related to depositional facies and sequence stratigraphic surfaces. Illite and chlorite dominate the mesogenetic clay minerals and usually grow at the expense of eogenetic clay minerals and detrital feldspars and lithic grains. Mesogenetic illite and chlorite can result from widely different reactants and processes. Clay minerals usually are assumed to be detrimental to sandstone reservoir quality because they can plug pore throats and some clay minerals promote chemical compaction. However, coats of chlorite on sand grains can preserve reservoir quality because they prevent quartz cementation. Adding oil to a sandstone stops clay diagenesis if the sandstone is oil-wet but probably only slows clay reactions if the sandstone is water-wet. Sandstones tend to be more oil-wet as the Fe-bearing clay content of the sand increases and as oil becomes more enriched in polar compounds.

INTRODUCTION

The amount, distribution pattern and morphology of clay minerals have significant effects on sandstone properties in terms of porosity, permeability, density, natural radioactivity, electrical conductivity, the water content of petroleum fields and reactivity to various enhanced oil recovery practices. Prior to the routine use of the scanning electron microscope (SEM) in petrographic examination of sandstones, clay minerals were often wrongly assumed to be detrital in origin, being co-deposited with the primary host sand.

However, laws of hydrodynamics tend to cause separation of the clay- and sand-sized fractions, implying that post-depositional processes must be responsible for the incorporation of most clay minerals into sandstones. The intention of this paper is to review the following:
1 composition and mineralogy of clay minerals in sandstones;
2 how clay minerals are incorporated into sands prior to diagenesis;
3 early diagenetic (eogenetic) origin of clay minerals in sandstones;
4 burial diagenetic (mesogenetic) origin of clay minerals in sandstones;
5 uplift-related diagenetic (telogenetic) origin of clay minerals in sandstones;
6 effects of clay minerals on sandstone permeability;
7 effect of petroleum emplacement on clay diagenesis in sandstones.

Definitions

The word 'clay mineral' refers to diverse groups of minerals that are members of the hydrous aluminous phyllosilicates, whereas the word 'clay' is strictly a grain-size term, classically for particle diameters less than 3.9 μm (Wentworth, 1922). Unfortunately, in sedimentary petrology, the term 'clay' is frequently used synonymously with 'clay mineral'.

Eodiagenesis includes all processes that occur at or near the sediment surface, where the geochemistry of the interstitial waters is controlled mainly by the depositional environment. Eodiagenesis also can be defined in terms of temperature, and depth, where the upper temperature limit is < 70°C, typically equivalent to about 2 km burial (Morad *et al.*, 2000).

Mesodiagenesis occurs during burial and includes all diagenetic processes following eodiagenesis and through to the earliest stages of low-grade metamorphism (as defined by Choquette & Pray, 1970). In many cases, this includes sediments buried to depths with equivalent temperatures of about 200 to 250°C. The main factors that influence mesogenetic changes include the time–temperature history, the primary mineralogy and fabric, local eogenetic modifications, extent of material loss and gain to neighbouring lithologies, geochemistry of the pore water and the presence of petroleum-related fluids.

Telodiagenesis occurs in inverted basins that have experienced an influx of surface (usually meteoric) waters. Such water has the capacity to cause significant geochemical changes, including feldspar dissolution and alteration to kaolinite.

There are eight main ways that clay minerals are incorporated into sandstones:
1 clay-rich rock fragments formed in the hinterland (extraclastic, allochthonous);
2 clay-rich clasts formed within the sedimentary basin (intraclastic, autochthonous);
3 flocculated mud particles and faecal pellets;
4 inherited clay rims on sands grains;
5 post-depositional incorporation of detrital mud into the sandstone by bioturbation and clay infiltration;
6 eogenetic reaction products in sandstone;
7 mesogenetic reactions in sandstones;
8 telogenetic reactions in sandstones.

CLAY MINERALS IN SANDSTONES: SUMMARY OF CHEMISTRY AND STRUCTURE

General aspects of clay mineral structure

Clay minerals are hydrous aluminosilicates that belong to the phyllosilicate group of minerals (Deer *et al.*, 1998). In addition to aluminium and silicon they also may contain other cations, including alkali, alkaline earth and transition metals. Clay minerals have a sheet-like structure in which the building blocks are either tetrahedra or octahedra linked to each other into planar layers by sharing oxygen ions between Si or Al ions of the adjacent tetrahedra or octahedra (Bailey, 1980; Fig. 1). The tetrahedra result from the close packing of four O ions, with the space between them occupied by a Si^{4+} ion or, to a lesser extent, an Al^{3+} ion. The octahedra result from the close packing of six

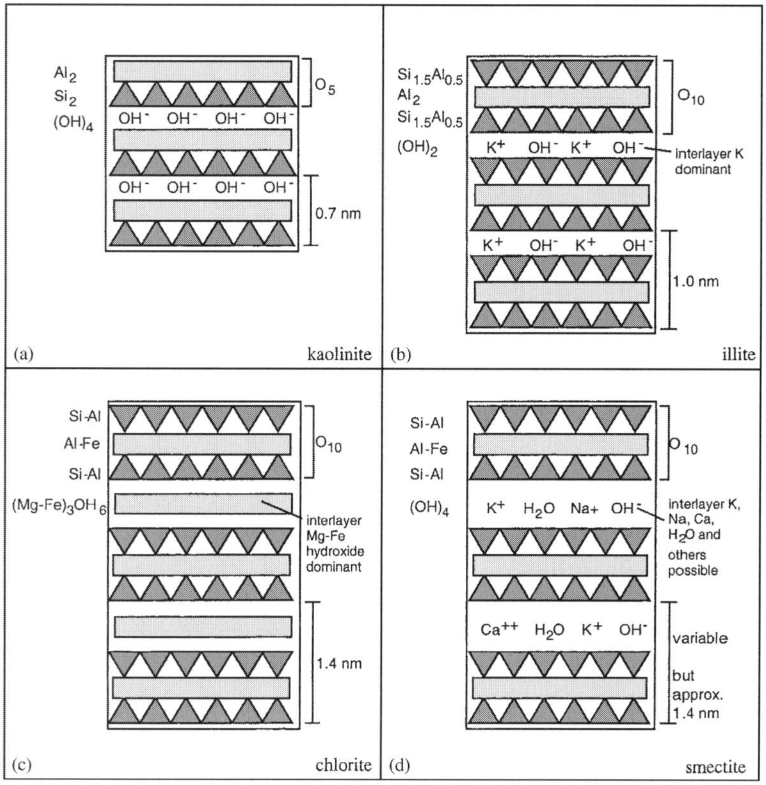

Fig. 1 Schematic diagram showing the structures of the common clay minerals (a) kaolin, (b) illite (c) chlorite and (d) dioctahedral smectite. The triangular motif represents tetrahedral layers. The solid grey bars represent octahedral layers.

anions that are dominantly oxygen but also can include some hydroxyl (OH) ions. The Si and Al ions mainly occupy the space between the oxygen octahedra and tetrahedra but other cations, such as iron, calcium, magnesium and potassium, are required in the clay structure to ensure charge balance. Tetrahedral and octahedral sheets are bound to each other in layers that extend for tens to thousands of nanometres (nm) in the *a* and *b* crystallographic directions. The layers are stacked on top of each other in the *c*-axis direction. Figure 2 is a schematic phase diagram that incorporates most of the important clay minerals. This figure is subdivided between K–Al-rich, Al- (and ferric iron) rich and Mg–Fe^{2+}-rich clay minerals.

Clay minerals can be classified based on the types of ions occupying the octahedral sites. If the ions are trivalent (Al, Fe^{3+}), the clay minerals are said to be dioctahedral because only two ions are needed to provide six positive charges. If the ions are divalent (Mg, Fe^{2+}), they are said to be trioctahedral because three ions are needed to provide six positive charges. Limited substitution of trivalent ions in dioctahedral clay minerals and divalent ions in trioctahedral clay minerals is possible. Thus typically Mg- and Fe^{2+}-rich clay minerals are trioctahedral whereas Al- and Fe^{3+}-rich clay minerals are dioctahedral. Interlayer cations are dominated by potassium. Ammonia (NH_4^+) can be present in small quantities in the interlayer site in illite (Williams *et al.*, 1992).

There are five *dominant* groups of clay minerals in sandstones: kaolin, illite, chlorite, smectite and mixed-layer varieties. A less common clay mineral in sandstones is palygorskite. Polytypes, or polymorphs, of clay mineral have the same composition but different crystal structures. The crystallographic differences typically occur in response to different temperature conditions.

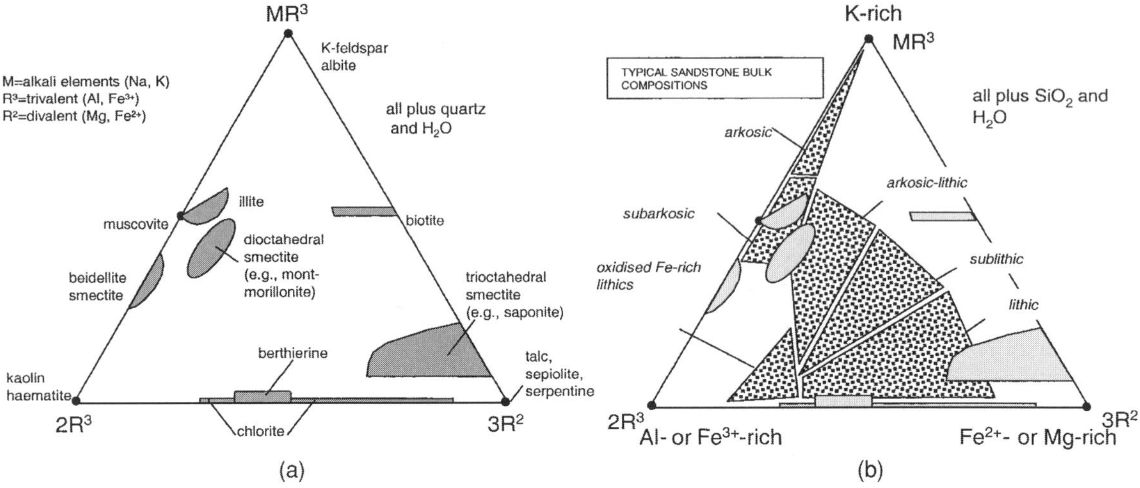

Fig. 2 Phase diagram showing the main clay minerals in terms of divalent, trivalent and alkali element (monovalent) ion proportions. The predominant alkali earth is potassium although smectites can contain measurable sodium. The main divalent ions are magnesium and iron so that a very important range of solid solutions (Fe–Mg) cannot be represented on this diagram. (a) Main minerals represented with areas of solid solution indicated. (b) Schematic representation of typical compositions of a variety of different sandstones from clean arenites through to highly lithic sandstones. Note that thermodynamic equilibrium cannot be assumed in low temperature sediments and it is possible to have many clay minerals together in one sample.

Kaolin–serpentine series clay minerals, including berthierine

Kaolin–serpentine series clay minerals are comprised of one tetrahedral layer linked to one octahedral layer with no interlayer cations and are termed 1 : 1 layer structures connected by O–H–O bonds. The chemical formula of kaolin is $Al_2Si_2O_5(OH)_4$, whereas the Mg end member serpentine has the formula $Mg_3Si_2O_5(OH)_4$. Serpentine can have Fe^{2+} substitution for Mg. Serpentine has not been reported in sandstones whereas berthierine $[Fe^{2+}_2Al(Si,Al)_2O_5(OH)_4]$ is a common clay mineral of the solid solution series between Fe-rich serpentinite and kaolin.

Kaolinite is the low temperature form, whereas dickite and nacrite are thought to be the high temperature forms of kaolin. Kaolinite has a unit cell of one octahedral–tetrahedral–octahedral package (unit cell of about 0.7 nm), whereas dickite has a unit cell made up of two of these packages (thus with a unit cell of about 1.4 nm) and nacrite has a unit cell made of six of these packages (unit cell of 4.3 nm). Kaolinite tends to form pseudohexagonal plates that commonly are stacked, in a book- or worm-like vermicular habit, whereas dickite tends to form small rhombic crystals (Fig. 3).

X-ray diffraction (XRD) analysis can be used to discriminate between dickite and kaolinite. However, more accurate distinction between the kaolin polymorphs can be made by determining the position and relative intensity of the OH-stretching bands in the 3600–3700 cm^{-1} region of infra-red spectra (Ruiz-Cruz, 1996; Hassouta *et al.*, 1999). Differential thermal analysis also can be used to discriminate between kaolin polymorphs, because they have considerably different dehydration temperatures (e.g. Beaufort *et al.*, 1998). Xia (1985) claimed that a variation in the relative solubilities of the kaolin polymorphs towards hydrofluoric acid allows a quantitative analysis of their abundances.

Illite and glauconite

Illite and glauconite are K-rich dioctahedral clay minerals comprised of one octahedral

(a) (b) (c) (d)

Fig. 3 Scanning electron microscope (SEM) micrographs illustrating the progressive burial diagenetic transformation of kaolinite into dickite. Note that dickite and kaolinite may abut the quartz overgrowths, which often lead to a misinterpretation of the paragenetic relationship between these two minerals. (a) Disordered vermicular kaolinite. (b) Well-ordered kaolinite in discrete booklets. (c) Euhedral dickite crystals: D, dickite; Q, quartz cement. (d) Illitized kaolinite (indicated by white arrow) with euhedral dickite preserved intact.

layer sandwiched between two tetrahedral layers and so are termed 2 : 1 structures. O–K–O bonds connect two opposing tetrahedral layers. The interlayer K^+ is required for charge balance accompanying the partial substitution of Al^{3+} for Si^{4+} in the tetrahedra and the substitution of divalent cations for Al^{3+} in the octahedra (Bailey, 1984). The O–K–O bonding is strong and prevents swelling behaviour in illite and glauconite mica. Illite has octahedral sites dominated by Al, whereas glauconite has octahedral sites with abundant Fe^{3+}.

The general chemical formula for illite is $K_yAl_4(Si_{8-y},Al_y)O_{20}(OH)_4$ (Velde, 1985), where *y* is typically significantly less than 2. Illite can occur as flakes, filaments or hair-like crystals (Fig. 4). Illite occurs as polytypes that reflect different ways in which layers are stacked.

(a) (b)

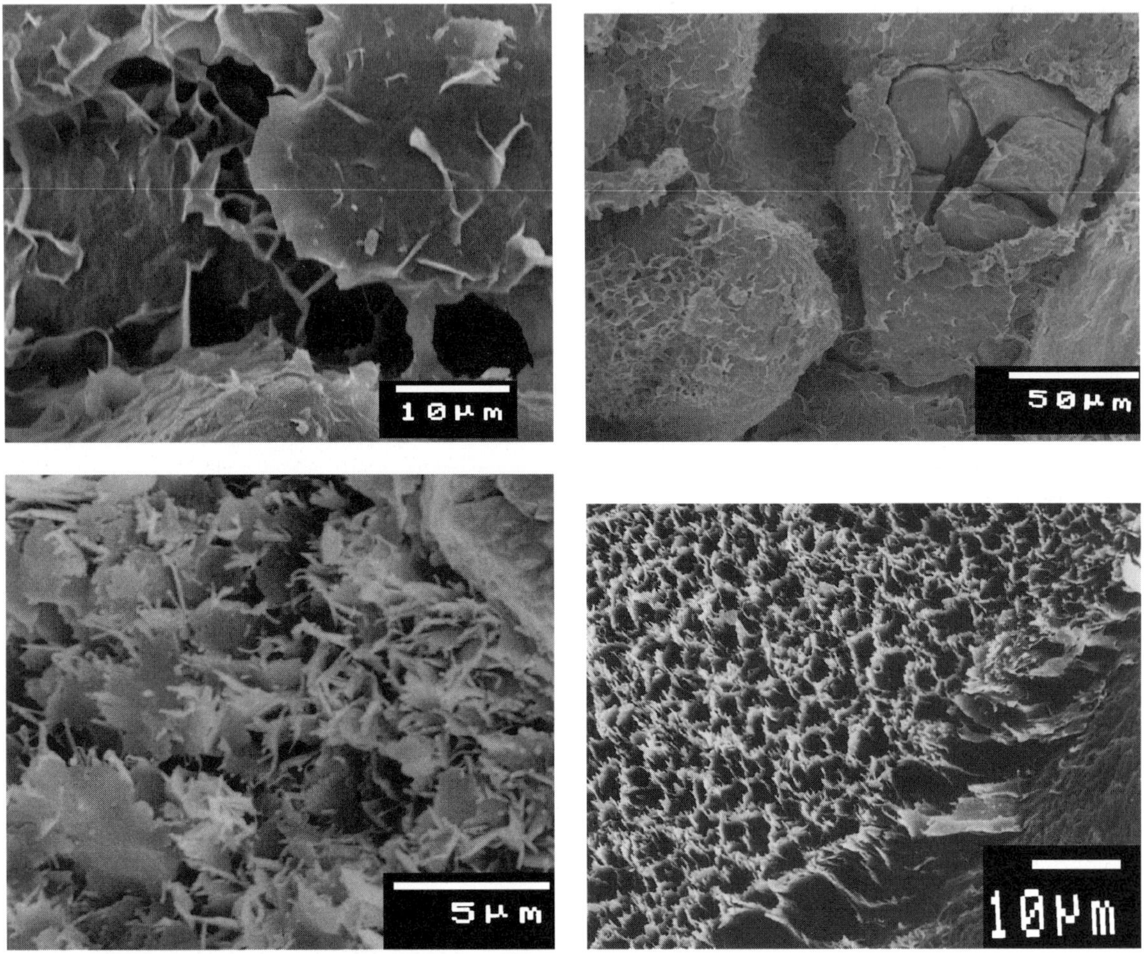

(c) (d)

(e)

Fig. 4 Scanning electron microscope (SEM) micrographs illustrating the progressive burial diagenetic transformation of grain-coating, infiltrated dioctahedral smectite into illite, via mixed-layer illite–smectite (I/S). (a) Initial stages of I/S, curling clay mineral flakes tangential to the detrital grain with 90% dioctahedral smectite, 10% illite. (b) Similar to (a) but with 80% smectite: (c) 60% smectite, (d) 20% smectite, (e) < 10% smectite. Note the change from tangential flakes to perpendicular fibres.

1M and $1M_d$ polytypes are prevalent for low temperature (i.e. diagenetic) illites and also for glauconites. The one common at lower diagenetic temperatures is the $1M_d$ polytype, in which the crystallography repeats for every single package of octahedral–tetrahedral–octahedral sheets (plus interlayer cations) and is disordered. The 1M polytype tends to form at higher diagenetic temperatures, and is more ordered than the $1M_d$ polytype. At high grade diagenetic to low grade metamorphic temperatures (> 200–250°C) illite is typically the 2M polytype, with a unit cell of about 2.0 nm comprised of a pair of tetrahedral–octahedral–tetrahedral layers.

Glauconite is a term that Odin & Matter (1981) have suggested be restricted to occurrences of dark green, Fe-rich, mica-type clay minerals of marine origin and with $K_2O > 6\%$. Glauconite has the formula $(K,Na,Ca)_{1.2-2.0}(Al,Mg,Fe)_4(Si_{7-7.6}Al_{1-0.4}O_{20})(OH)_4 \cdot nH_2O$. The term glaucony is thus recommended as a facies term that typically includes Fe-rich marine clay minerals that range in composition from glauconitic smectite to glauconitic mica.

Smectite

Smectite is a group of 2 : 1 clay minerals with one octahedral layer sandwiched between two tetrahedral layers. Smectite has the general formula $(0.5Ca,Na)_{0.7}(Al,Mg,Fe)_4(Si,Al)_8O_{20}(OH)_4 \cdot nH_2O$. Trioctahedral smectite has octahedral sites dominated by divalent metals (Fe^{2+}, Mg, Ca), whereas dioctahedral smectite has octahedral sites dominated by trivalent metals (Fe^{3+}, Al). There is less binding of opposing tetrahedral layers by K^+ than in illite, with interlayer water bound by weak van der Waal's forces. Cations present between layers are exchangeable and reflect the chemistry of the aqueous medium with which the smectite was last in contact. Interlayer cations are variably hydrated, resulting in the swelling characteristic of smectitic clay minerals. Smectites are defined by their tendency to swell when exposed to organic solvents, which can be absorbed between interlayers. Smectite usually occurs as flakes curling up from an attachment zone on the detrital sand grain surface (Fig. 4).

Chlorite

Chlorite has a 2 : 1 : 1 structure comprised of a negatively charged 2 : 1 tetrahedral–octahedral–tetrahedral layered structure interlayered with an additional octahedral layer that is positively charged and comprised of cations and hydroxyl ions (e.g. brucite layers; $Mg_3(OH)_6$). A general formula for chlorite is $(Mg,Al,Fe)_{12}[(Si,Al)_8O_{20}](OH)_{16}$. Solid solution is possible on all sites, leading to a very complex mineral group. Chlorite can exist as different polytype including the 1b polytype and the 2b polytypes with 1.4 and 2.8 nm basal spacings, respectively. Fe-rich diagenetic chlorite (e.g. chamosite) is typically the 1b polytype, whereas the more Mg-rich diagenetic varieties (e.g. clinochlore) are typically the 2b polytype. It has been proposed that the type 1 polytype may be prevalent at lower diagenetic temperatures, with the type 2 polytype forming as diagenetic temperatures approach low grade metamorphic conditions (Bailey & Brown, 1962). However, recent work has concluded that there is no well-defined link between temperature and chlorite polytypes (e.g. De Caritat *et al.*, 1993; Walker, 1993). Chlorite occurs in a variety of morphologies although classic chlorite occurs as a grain coating boxwork, with the chlorite crystals attached perpendicular to the grain surface (Fig. 5).

Mixed-layer clay minerals

Mixed-layer clay minerals result from the interstratification of different mineral layers in a single structure (Srodón, 1999). Most mixed-layer clay minerals contain smectite as a swelling component, and include illite–smectite and chlorite–smectite (abbreviated to I/S and C/S respectively). During progressive burial diagenesis I/S becomes more illite-rich and C/S becomes more chlorite-rich.

The stacking of layers in I/S usually is disordered (randomly interstratified) at the time of deposition and during eodiagenesis.

(a) (b)

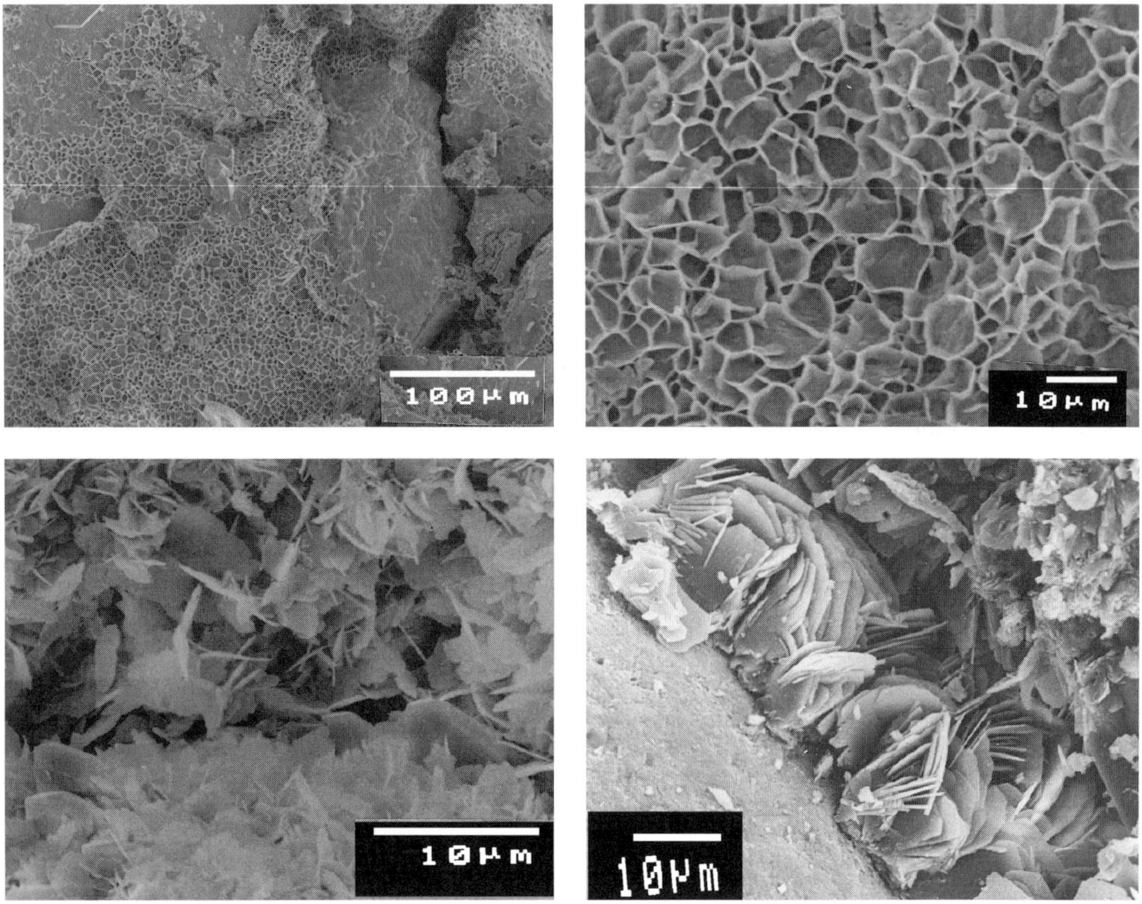

(c) (d)

Fig. 5 Scanning electron microscope (SEM) micrographs illustrating the progressive burial diagenetic transformation of grain-coating, infiltrated trioctahedral smectite into chlorite via mixed-layer chlorite–smectite. (a) and (b) represent a predominance of trioctahedral smectite present as a fine perpendicular honeycomb fabric. (c) Mixed chlorite–smectite showing a mixture of flakes and crystals. (d) Perfect Fe-chlorite occurring as perpendicular rosettes of well-formed crystals on detrital grain surfaces.

Randomly interstratified mixed-layer clay minerals are labelled according to the types of layers involved, with the most abundant layer type listed first (Reynolds, 1980). The term 'Reichweite', denoted by R, is used to describe ordering types. R = 0 describes totally random interstratification of smectite and other clay minerals. R = 1, R = 2 and R = 3 describe progressively more ordered intercalations (Reynolds, 1980; Wilson, 1999). The degree of disorder in I/S decreases and the proportion of illite increases in a semi-predictable manner during heating and burial diagenesis. Two types of ordered (R = 1, 1 : 1) mixed-layer clay minerals have been identified and given discrete names: corrensite (chlorite–smectite) and allevardite (illite–smectite).

Palygorskite

Palygorskite is comprised of laterally continuous two-dimensional trioctahedral sheets (dominated by Mg with OH ions) but does not have continuous SiO_4 tetrahedral sheets between the octahedral sheets. The tetrahedral part of the structure occurs as ribbons, two silica tetrahedra wide, with infinite length, between which there are systematic tunnel-like gaps. The tunnels have a width equivalent to two tetrahedra. Interlayer cations, required to keep charge balance, are exchangeable with organic ligands, resulting in an expandable structure.

DETRITAL CLAY MINERALS IN SANDSTONES

Detrital mud matrix

Laws of hydrodynamics control sediment transportation and deposition. The main variables are the velocity of flow and the grain diameter, shape and density (Allen, 1997). The fine grain size (< 0.004 mm) dictates that clay minerals are transported more easily than sand grade material and that during waning flow, primary deposition of sand is followed by clay deposition only at very low flow rates (Blatt *et al.*, 1980). Laws of hydrodynamics suggest that discrete clay grains should *not* be deposited at the same time as sand grade material, although incorporation of mud into turbidite, glacial and debris-flow sediments may occur during deposition if flow velocity decreases very rapidly.

Mud intraclasts

Intraclasts are autochthonous grains composed of aggregated clay minerals derived from the erosion of floodplain mud during fluvial channel migration, and typically are deposited as a lag in channel bases. In marine deposits they form at ravinement surfaces marking transgression and the initiation of sea-level rise. Intraclasts are thus typically associated with fluvial and deltaic depositional environments. Shelf incision by valleys at sequence boundaries also can result in mud intraclasts from sediments that originally were deposited as marine sediments.

Flocculated mud

Most suspended clay particles carried by rivers pass out of estuaries and into the open marine environment. Estuaries are areally restricted and tend not to accumulate suspended clay minerals, especially during times of sea-level fall. However, coalescence and deposition of suspended clay minerals into sand-sized particles can occur by two mechanisms: (i) flocculation of riverborne, suspended clay-sized particles upon mixing with seawater (van Olphen, 1977), and (ii) biogenic agglomeration. As the clay aggregates are weak, they tend to be broken up in high velocity currents (Krone, 1978). Flocculated particles settle out in low energy environments or at times in the tidal cycle when there is little particle motion.

Clay minerals carried in colloidal suspension in rivers have a negative surface charge that is neutralized by a layer of positively charged cations, known as the Gouy layer. This results in a positively charged cation concentration around each clay particle relative to the river water. The repulsive power created between clay particles is greater than attractive van der Waal's forces. Thus, colliding clay particles repel one another and tend not to coalesce in freshwater. However, when riverborne clay particles encounter seawater, the concentration of cations in the water increases. This leads to a collapse of the Gouy layer, reducing the repulsion between clay grains. Thus, in brackish water cohesive van der Waal's forces become stronger than the cation-layer-induced repulsion, and cohesion, or flocculation, of clay particles can occur. Flocculation of clay is enhanced by very high velocity gradients encountered in estuaries, which promotes a great degree of collision between the suspended clay particles (Krone, 1978).

Flocculation and coalescence of riverborne clay particles is controlled not only by the increase in aqueous salinity owing to mixing with seawater but also by the type of clay mineral involved. The sequence in which clay minerals have a decreasing tendency to flocculate is: kaolinite, illite and, finally, smectite (Edzwald *et al.*, 1974; Krone, 1978). Thus the order of deposition of clay aggregates will be kaolinite closest to the shore, followed by illite and then smectite furthest from the shore. Differential flocculation becomes less important when the clay particles have organic (particularly humic acid) or iron oxyhydroxide coatings that modify the surface properties of the clay minerals (Gibbs, 1977).

In salt-wedge estuaries, where river flow dominates over tidal forces, little mixing occurs between the seaward flowing river water in the upper layer, and the inward-flowing salt wedge below. This results in transfer of suspended, riverborne clay out to the sea to be deposited in deltas or offshore, in submarine canyons (Meade, 1972; Eisma *et al.*, 1978).

Biogenic agglomeration occurs when suspension feeders, such as bivalves, form faecal pellets. Deposited faecal pellets may be incorporated subsequently as grains in the sediment or disaggregated and resuspended (Pryor, 1975).

Clay-rich rock fragments (extraclasts)

Allochthonous clay-rich sand grains include sedimentary and low-grade metamorphic rock fragments. Additionally, high temperature minerals and rocks can be converted into low temperature clay minerals (i.e. argillized) without complete physical disaggregation into component clay crystals. In these cases, erosion and transport of such material will lead to lithic grains, dominated by clay minerals, being co-deposited with other sand grains. In some circumstances, detrital lithic fragments become clay-mineral rich only during subsequent diagenesis (e.g. Worden *et al.*, 1997, 2000) making it difficult to determine whether the clay was depositional (as clay-rich lithic fragments) or diagenetic (Whetton & Hawkins, 1970) in origin. Sediment source terrains dominated by metapelite and metabasites will be especially likely to lead to clay-rich rock fragments.

Inherited clay coats

Clay coats on sand grains can be detrital in origin, and occur most commonly within embayed surfaces of sand particles (Pittman *et al.*, 1992). In arid environments, clay coats form when wind-blown clay material adheres to moist sand grains (Krinsley, 1998). Inherited clay coats can be discriminated from diagenetic clay coats by: presence at point contacts between detrital sand grains, widely varying rim thickness, absence on diagenetic mineral surfaces and preferential occurrence in sediments resulting from lower energy sedimentary environments (Wilson, 1992).

POST-DEPOSITIONAL INCORPORATION OF DETRITAL CLAY MINERALS IN SANDSTONES

Bioturbation

Bioturbation operates as a post-depositional mechanism that mixes clay minerals from mud-rich layers with sand. Organisms can burrow into sand (seeking food or shelter) enabling the physical mixing of under- or overlying mud-rich layers with sand-rich layers. Faunal bioturbation is prolific in shallow-marine sandstones, where the supply of nutrients is high and there is plenty of light.

Infiltration of clay minerals

If water rich in suspended clay percolates through the vadose zone of a sandy aquifer, then clay will be filtered out of the water as coatings on sand grains (e.g. Moraes & De Ros, 1990). It is likely to occur in unconfined aquifers with substantial vadose zones that typify semi-arid climates and/or very thick sand sequences. Mud-rich surface waters are prevalent in floodplain and deltaic environments (Dunn, 1992). Multiple episodes of clay infiltration into coarse-grained, braided river sand deposits result in thick infiltrated grain coatings. Infiltrated clay coats are common in sandy meandering river deposits, but tend to be thinner than in braided river deposits.

EOGENETIC CLAY MINERALS

On deposition, the primary sand comprises a mixture of minerals that were formed under a wide range of conditions (e.g. temperature, pressure, oxidation state, water composition). Much freshly deposited sand contains unstable grains that survived weathering, erosion and transportation. Therefore, the detrital mineral assemblage may be inherently unstable and thus will tend to react with the ambient water during eodiagenesis.

The main eogenetic clay minerals are kaolinite, glauconite, berthierine, verdine, di- and trioctahedral smectite, I/S, C/S and Mg-clay minerals (palygorskite), which are formed by: (i) precipitation from pore waters, (ii) replacement of framework sand grains and (iii) replacement of precursor detrital or diagenetic clay minerals. Illite and chlorite do not form in eogenetic environments and they are depositional, rather than diagenetic, in origin where found in soils and sediments that have not suffered deep-burial diagenesis (Wilson, 1999). The formation of diagenetic clay minerals in sands at near-surface conditions and during shallow burial is controlled strongly by depositional facies, detrital composition of the sandstones and climatic conditions.

Smectite, mixed-layer clay minerals and palygorskite

Smectite and I/S form as grain-hugging flakes in sands under semi-arid climate. Mg-smectites (e.g. saponite, sepiolite) and palygorskite form as fibres and fibre bundles during near-surface eodiagenesis of lacustrine, fluvial and aeolian sediments, and, less commonly, coastal sabkha (Hover *et al.*, 1999; Pozo & Casas, 1999), which is subjected to strong evaporation conditions. Evaporitic conditions lead to the formation of hypersaline pore waters enriched with Mg^{2+} and dissolved SiO_2, but depleted in HCO_3^-, SO_4^{2-} and Cl^-, through the precipitation of calcite, aragonite, gypsum, anhydrite and halite. Trioctahedral smectite and palygorskite form either by precipitation from hypersaline pore waters or through the transformation of a clay mineral precursor or Mg–Si-rich gels (Mayayo *et al.*, 1998). They can be associated closely with pedogenic deposits developed on dolomitic and basaltic bedrocks, which act as a local source of Mg^{2+} (Karakas & Kadir, 1998). Dioctahedral smectite and I/S form in less evaporitic environments compared with trioctahedral smectites.

The most typical Fe-rich smectite is nontronite, which commonly forms on the abyssal plain of deep ocean basins, and in the vicinity of mid-oceanic ridges. An elevated Si content, owing to the presence of biogenic silica, and Fe content, owing to the presence of Fe-oxyhydroxides, in conjunction with low organic matter content in these sediments account for this Al-poor and Fe^{3+}-rich smectite. Kohler *et al.* (1994) suggested that Fe oxidation in nontronite from submarine hydrothermal chimneys of the Galapagos Rift and Mariana Trough is bacterially mediated.

Sepiolite, palygorskite and atapulgite are closely associated with Fe-oxides/oxyhydroxides, being typically formed by the replacement of the detrital ferroan silicates (biotite, pyroxene and amphibole) and volcanic rock fragments (Walker *et al.*, 1978; Surdam & Boles, 1979). The Mg-rich clay minerals are rare in ancient sediments, as these are sensitive to chloritization during subsequent mesodiagenesis (Stein *et al.*, 1990).

Green clay minerals: berthierines glauconite and verdine

Berthierine is an aluminous Fe^{2+}-rich 1 : 1 clay belonging to the kaolinite–serpentine series of minerals. Verdine represents a group of greenish, metastable aluminosilicates with a wide variety of crystal structures and chemical compositions. The most common minerals in this group include: (i) phyllite C, which is a 1.4–1.5 nm, Fe^{3+}- and Mg-rich clay mineral (Odin, 1990), and (ii) phyllite V (odinite), which is a 0.72 nm, Fe^{3+}- and Mg-rich clay mineral, with a 1 : 1 serpentine type structure intermediate between di- and trioctahedral structures (Bailey, 1988). Glauconite is an Fe^{3+}-rich dioctahedral clay, which upon primary formation has smectite-like swelling behaviour but which adopts mica-like (non-swelling) characteristics upon ageing (Odin & Matter, 1981).

Berthierine and verdine occur as small (< 5 μm), lath-shaped grain coatings and crystals, as coats (fringes or tangentially arranged), pellets, ooids and void-fillings, or form by the replacement of detrital grains (e.g. silicates and carbonate bioclasts). They form during diagenesis below the sediment–water interface in deltaic–estuarine deposits (Hornibrook & Longstaffe, 1996; De Hon *et al.*, 1999), primarily in tropical to subtropical seas (Odin, 1990; Kronen & Glenn, 2000; Thamban & Purnachandra, 2000). Berthierine, verdine and glauconite are often closely associated with authigenic apatite (Morad & Al-Aasm, 1994; Purnachandra *et al.*, 1995), suggesting growth in nutrient-rich coastal waters.

Berthierine authigenesis is favoured in volcanogenic sediments deposited in estuarine–coastal-plain environments (Jeans *et al.*, 2000). Verdine formation is favoured on shelves off river mouths, at water depths of < 200 m, under conditions characterized by low sedimentation rates (Kronen & Glenn, 2000; Vaz, 2000). Owing to the domination of Fe^{3+} over Fe^{2+} in verdine, its formation is anticipated to be at depths of centimetres or decimetres below the seafloor, being favoured by suboxic conditions. These are mildly reducing conditions that correspond to depths below the seafloor where nitrate or manganese reduction, or initial iron reduction occur. Conversely, the predominance of ferrous iron in berthierine suggests that it formed under strongly reducing conditions (i.e. pore waters relatively depleted in dissolved oxygen) compared with verdine, such as during iron-reducing suboxic conditions. Authigenesis of both verdine and berthierine occurs prior to the burial depths in sediments where bacterial sulphate reduction dominates, and where Fe^{2+} becomes incorporated preferentially in sulphide minerals.

Upon ageing and shallow burial to a few hundred metres, phyllite C is transformed into odinite (Purnachandra *et al.*, 1993), which in turn is transformed into berthierine (Odin, 1988). Odinite thus occurs only in Recent sediments, whereas berthierine occurs in ancient sedimentary rocks.

Berthierine and verdine typically form in ancient and Recent marine-shelf settings. However, a few fresh- and brackish water occurrences of berthierine have been reported, such as in Wealden (Early Cretaceous) sediments of southeast England (Taylor, 1990). However, even aeolian and fluvial deposits that become flooded by marine water during a transgression may be subjected to diagenetic alterations that result in growth of berthierine and glauconite (Ketzer *et al.*, this volume, pp. 43–61).

Berthierine occurs in estuarine and coastal-plain sediments whereas glauconite forms exclusively in open marine sediments (Odin & Matter, 1981), decimetres or metres below the seafloor. Berthierine forms in shallower waters and under more strongly reducing conditions, i.e. from pore waters richer in Fe^{2+}, relative to glauconite.

Kaolin

Typically, eogenetic kaolinite has a vermicular and book-like habit (Figs 3 & 6; Ketzer *et al.*, this volume, pp. 361–382). Kaolinite forms under humid climatic conditions in continental sediments by the action of low-pH groundwaters on detrital aluminosilicate minerals such as feldspars, mica, rock fragments, mud intraclasts and heavy minerals (Emery *et al.*, 1990). During forced regression and lowstand, large areas of marine sediment are subaerially exposed on the shelf, leading to an enlargement of the area of meteoric recharge. The basinward migration of the meteoric zone promotes the flushing of shallow-marine sediments and even deep-water turbidites in some cases.

The amount and distribution pattern of kaolinite is influenced by the amount of unstable detrital silicates, annual precipitation, hydraulic conductivity and rate of fluid flow in the sand body. Eogenetic grain dissolution is most prevalent in permeable sediments, such as channel sand deposits. Humid conditions result in the availability of greater amounts of meteoric waters, and hence promote eogenetic kaolinite.

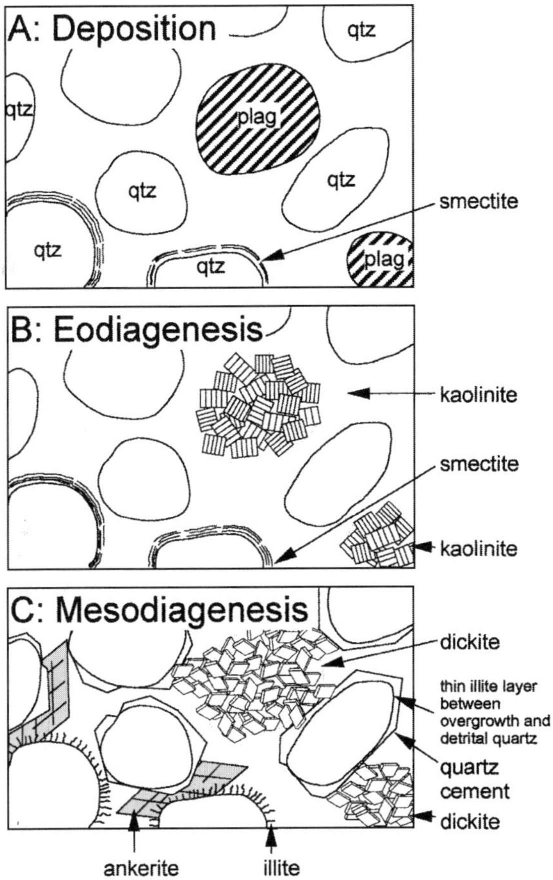

Fig. 6 Schematic representation of the development of eogenetic kaolinite and the illitization of detrital smectite. On deposition, the model arkosic sediment has a smectite grain coating. During eodiagenesis, feldspars are converted to kaolinite. Mesodiagenesis at temperatures > 110°C sees dickite growth from kaolinite, quartz overgrowths on detrital sand grains, conversion of grain-coating smectite into illite and concomitant growth of Fe dolomite.

Some of the earliest reactions to occur are the replacement of albite and calcic plagioclase by kaolinite.

$$\underset{\text{albite}}{2NaAlSi_3O_8} + 2H_2O + 2H^+ \Rightarrow \underset{\text{kaolinite}}{Al_2Si_2O_5(OH)_4} + 4SiO_2 + 2Na^+ \quad \text{(R1)}$$

$$\underset{\text{anorthite}}{CaAl_2Si_2O_8} + H_2O + 2H^+ \Rightarrow \underset{\text{kaolinite}}{Al_2Si_2O_5(OH)_4} + Ca^{2+} \quad \text{(R2)}$$

Calcic plagioclase and albite tend to be more susceptible to kaolinitization than K-feldspar. It is possible that the protons are supplied by partial dissociation of carbonic acid so that reactions may be accompanied by carbonate precipitation.

ROLE OF CLIMATE AND DEPOSITIONAL ENVIRONMENT ON EOGENETIC CLAY MINERALS

Depositional environment is a master control on eodiagenesis because it controls the type and amount of water present in sediment, water influx versus evaporation rate, temperature, exposure to atmospheric oxygen, plant-derived CO_2 and organic matter content. The role of depositional environment on clay mineral patterns in sandstones will be divided here between subaerial and marine systems. Continental environments will be split between warm and wet versus arid.

Warm and wet continental environments

Pore waters in humid, warm (subtropical to temperate) environments are dilute (less than a few hundred parts per million) and dominated by Ca^{2+} and HCO_3^- and slightly acidic (Fig. 7). Warm, wet, typically verdant, eogenetic environments also have an abundance of organic matter that undergoes bacterially mediated decay (Berner, 1980). Fe-bearing minerals in the sediment are readily reduced to aqueous Fe^{2+} by redox processes, which typically is available for siderite growth because very low concentrations of SO_4^{2-} leads to the absence of Fe-sulphides. The clay minerals in this environment are typified by kaolinite because its formation requires low ionic concentrations in pore waters.

Arid continental environments

Dry, typically but not always hot, continental environments (such as braid plains, playa margins and deserts) often have a low organic

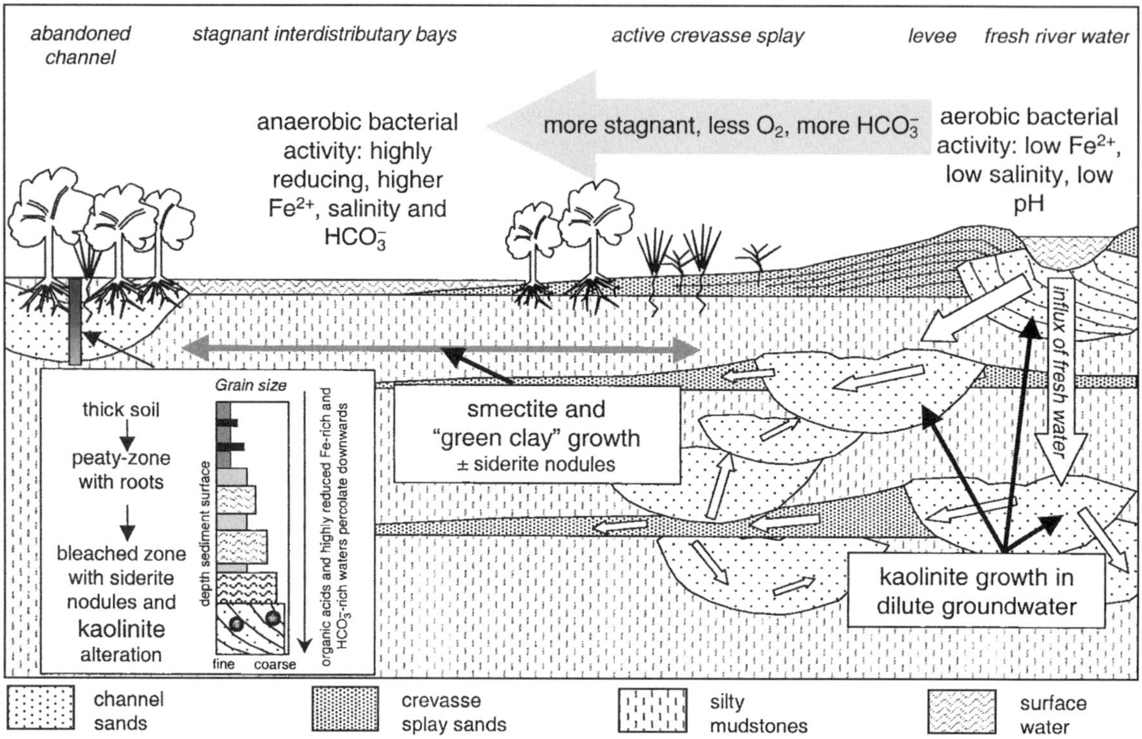

Fig. 7 Clay eodiagenesis in warm wet eogenetic environments. Kaolinite grows in areas actively flushed by river water as the ion activities are never given the chance to increase, thus leaving the water in the kaolinite stability field. Areas more remote from the main channel axis are able to stagnate and thus increase in ion concentration, taking groundwaters into the stability fields of smectite minerals. The relative absence of sulphur species minimizes growth of pyrite and so may allow available iron to enter clay minerals (mainly smectites). However, local availability of organic matter leads to increased alkalinity (following decay and oxidation) and thus Fe-rich carbonate growth (Curtis *et al.*, 1986). Thick soil horizons in abandoned channels will allow organic (humic) acids to build up, encouraging extreme feldspar weathering and leading to bleaching and growth of kaolinite deep in the soil profile.

matter content, deep water tables and fully oxidized sediments (Fig. 8). In sediment below topographic depressions, brines often evolve under strongly evaporitic conditions, such as in rift lakes. The pore waters are concentrated and tend to be dominated by Na^+, Ca^+, Mg^{2+} and HCO_3^- and, to a lesser extent, SO_4^{2-} (Eugster & Hardie, 1978). Iron tends to be fully oxidized (ferric) and often coats minerals as a hydroxide or sesquioxide. Evaporation often exceeds meteoric influx, leading to an upward flux of groundwater, evaporation and the consequent development of various smectite (e.g. montmorillonite, saponite) and Mg-rich clay minerals.

Marine environments

Marine environments are characterized by slightly alkaline waters (seawater pH is 8.3) and Na^+–Cl^- dominated water (with subordinate SO_4^{2-}, HCO_3^-, Ca^{2+} and Mg^{2+}), with a salinity of about 35 000 ppm (Fig. 9).

Bacterially catalysed processes characterize marine eodiagenesis. The interaction of organic matter and oxidizing inorganic solutes (e.g. SO_4^{2-}) and minerals (e.g. Fe^{3+} minerals) causes rapid eogenetic alteration of shallow buried sediments. The bacteria and a potent mix of oxidizing and reducing material leads to an

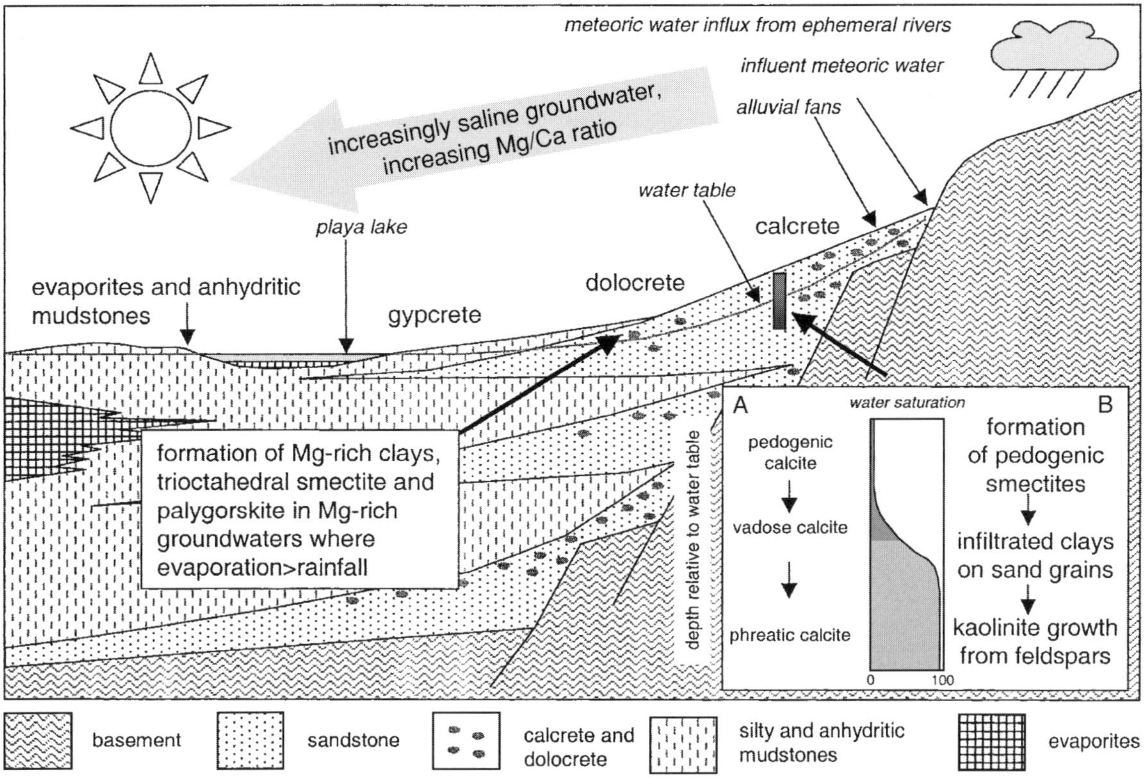

Fig. 8 Clay eodiagenesis in arid eogenetic environments. Pedogenesis will lead to smectite growth in soils on sands in the upper and middle reaches of hydrological basins. In the unsaturated zone of aquifers, clay infiltration will occur if percolating waters contain suspended clay minerals (Moraes & de Ros, 1990). Clay infiltration will occur when groundwater evaporation exceeds the rate of percolation, leaving the suspended clay minerals coated on sand grains. Kaolinite may form from detrital feldspars in the aquifer if the groundwater is flowing and being recharged. In the lower reaches of the hydrological basin, where the Mg/Ca ratio has increased following calcrete formation, Mg-rich clay minerals, as well as dolocrete, may form in the sediment.

authigenic suite of minerals, including Fe sulphides, various carbonate cements and a range of typically green sheet-silicates ('glaucony') including glauconite, berthierine and smectite (Stonecipher, 2000). Elevated concentrations of iron, responsible for berthierine or glauconite formation, are thought to be the result of enrichment of the detrital sediments in metabolizable organic matter and/or brackish pore waters that have lower concentrations of dissolved sulphate than marine pore waters. The low organic-matter content in abyssal, deep-sea sediments prevents the reduction of Fe-oxyhydroxides derived primarily from hydrothermal vents about mid-oceanic ridges, and hence favours the formation of Fe^{3+} smectite (typically nontronite).

SEQUENCE STRATIGRAPHY AND CLAY MINERALS IN SANDSTONES

Relative sea-level changes and the rate of sediment progradation influence strongly the position of the strand line, and hence the degree and patterns of mixing between marine and continental waters, and eodiagenesis of coastal and shallow-marine deposits (Morad *et al.*, 2000).

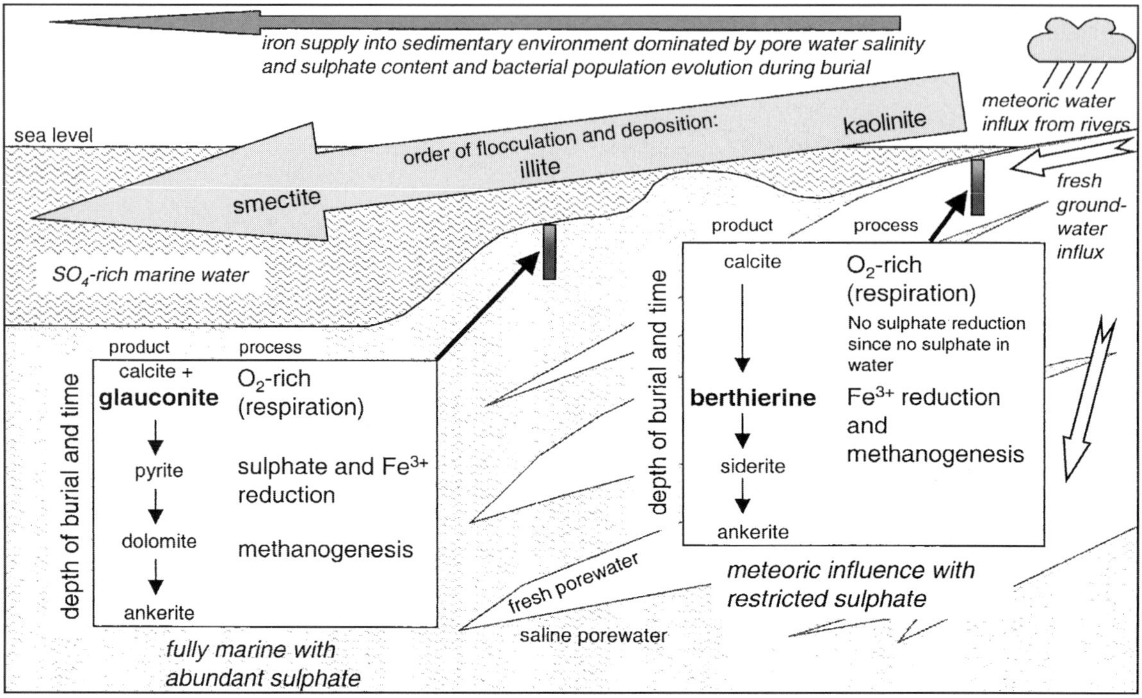

Fig. 9 Clay eodiagenesis in marine eogenetic environments. Marine sediments with a significant freshwater influx will have limited sulphate, and thus sulphide, and limited growth of pyrite. Berthierine forms close to the sediment surface but requires isolation from oxidized water to prevent re-oxidation (Odin & Matter, 1981). As primary Fe^{3+} is reduced, the subsequent Fe^{2+} is free to form the clay berthierine in the methanogenic zone and then siderite in the deeper decarboxylation zone (Pye *et al.*, 1990). Fully marine sediments have abundant pyrite and thus minimal berthierine in the methanogenic and sulphate reduction zones. Glauconite forms on, or near to, the sediment surface in condensed sequences that have time enough for biological alteration of detrital clay minerals (Pryor, 1975). In the deeper decarboxylation zone Fe-rich carbonates can form, as in the case of brackish pore waters. Flocculation leads to distinct zonation of geochemically aggregated clay minerals, with kaolinite floccules forming closest to shore, smectite floccules in the ocean basin and with illite intermediate (Edzwald *et al.*, 1974).

A schematic representation of the link between relative sea-level change and clay eodiagenesis is depicted in Fig. 10.

Relative sea-level changes are controlled by a combination of eustatic changes, subsidence/uplift of basin floor and/or substantial variations in the rate of sediment supply and result in three major stratigraphic surfaces that subdivide the depositional sequence into genetic packages known as systems tracts (Van Wagoner *et al.*, 1988). These surfaces include:

1 the transgressive surface (TS), which is the first significant marine-flooding surface across the shelf—this surface forms the boundary between the lowstand systems tract and overlying transgressive systems tracts;

2 the maximum flooding surface (MFS), which corresponds to the highest relative sea-level reached—this surface separates the transgressive systems tract from the overlying highstand systems tract;

3 the sequence boundary (SB), which forms as a response to relative sea-level fall. A sequence boundary is characterized by subaerial exposure and erosion associated with stream rejuvenation and truncation of the underlying strata as well as a basinward shift in facies. The SB, marked by unconformities and

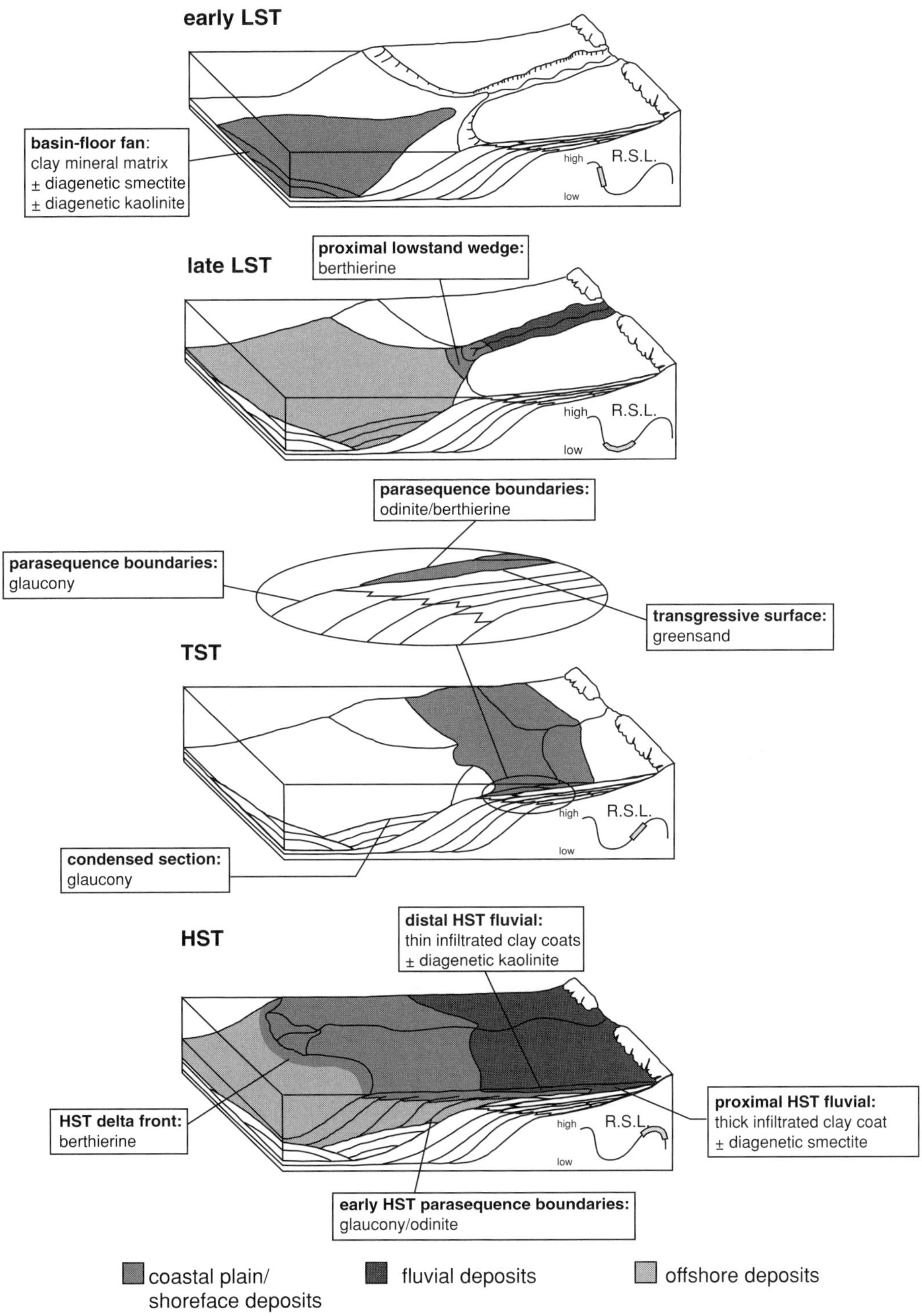

Fig. 10 Illustration of the location of clay minerals in sandstones in terms of the sequence stratigraphic organization of sands. The systems tracts developed during a cycle of sea-level change is adapted from Posamentier & Vail (1988). HST, highstand systems tract; LST, lowstand systems tract; TST, transgressive systems tract.

their correlative surfaces, bounds the depositional sequence.

Each cycle of relative sea-level changes results in lowstand, transgressive and then highstand systems tracts as defined by Van Wagoner *et al.* (1988). It should be noted that the original sequence model of Van Wagoner *et al.* (1988) and Posamentier & Vail (1988) has undergone substantial re-evaluation since its publication, although discussion of its evolution is beyond the scope of this paper. The distribution of clay minerals in sandstones can be constrained within these three types of systems tracts (for details see Ketzer *et al.*, this volume, pp. 43–61), as follows (Fig. 10).

Lowstand systems tract (LST)

Deposition of a LST occurs as a response to a fall and slow rise in the relative sea-level. The LST can be divided into two: early (lowstand fan) and late (lowstand wedge and valley infill) deposits (Van Wagoner *et al.*, 1988). The lowstand basin-floor fan is comprised of sediments that bypassed the shelf through incised valleys and is deposited on the slope and in the basin in the form of submarine fans. The clay minerals are predominantly detrital interstitial minerals deposited by turbidity currents and as sand-sized, mud- and glauconite intraclasts eroded by valley incision of highstand systems tract sediments exposed on the shelf. Burial and mechanical compaction of these ductile intraclasts result in the formation of pseudomatrix.

As valley incision is superseded by valley filling, the quantity of coarse-grained sediments delivered to the shelf decreases, resulting in the deposition of a lowstand wedge on the slope by a levee-channel complex with rhythmic turbidites, and later by delta progradation. These deposits are characterized by detrital clay minerals that are similar to those in the lowstand fan.

Clay minerals formed in sediments exposed on the shelf (i.e. valley-filling and exposed older strata) are related to the prevailing climatic conditions. Kaolinite forms under semi-humid to humid conditions. Clay infiltration occurs in semi-arid conditions. Strongly arid, evaporitic climates result in the formation of Mg-rich clay minerals (e.g. palygorskite, saponite). The interfluve sediments will be subject to pedogenesis. Kaolinite formation owing to meteoric water incursion commonly extends to include the sandy facies of the preceding highstand systems tract.

Transgressive systems tract (TST)

Deposition of a TST represents an abrupt landward shift of facies (i.e. transgression), which occurs as a response to a rapid rise in relative sea-level. A transgressive surface (TS) is developed on top of the incised valleys and their interfluves, and also at the upper boundary of the proximal lowstand-wedge. The valleys filled initially with fluvial deposits during late LST, begin to be filled with estuarine and, subsequently, shallow-marine sediments. Hence, berthierine may form in the upper parts of the incised valley sediment fill.

The TS is characterized by elevated concentrations of glauconite and verdine intraclasts that locally may result in the formation of greensand deposits on the coastal plain environments. A rise in relative sea-level is accompanied by a decrease in sedimentation rate on the shelf, because most of the coarse-grained sediments are entrapped landward. Shelf sediments are thus dominantly fine-grained and show a progressive upward increase in the amounts of authigenic glauconite and verdine, with maximum concentrations occurring along the MFS. The MFS is characterized by sediment starvation (i.e. condensed section) in the outer and middle shelf. In some cases, the formation of glauconite and verdine is favoured within bioturbation structures. Within the inner shelf, berthierine is formed in estuarine and shallow-marine environments.

Highstand systems tract (HST)

Deposition of the HST occurs in response to gradual slowing of relative sea-level rise and later during the initial stages of sea-level fall.

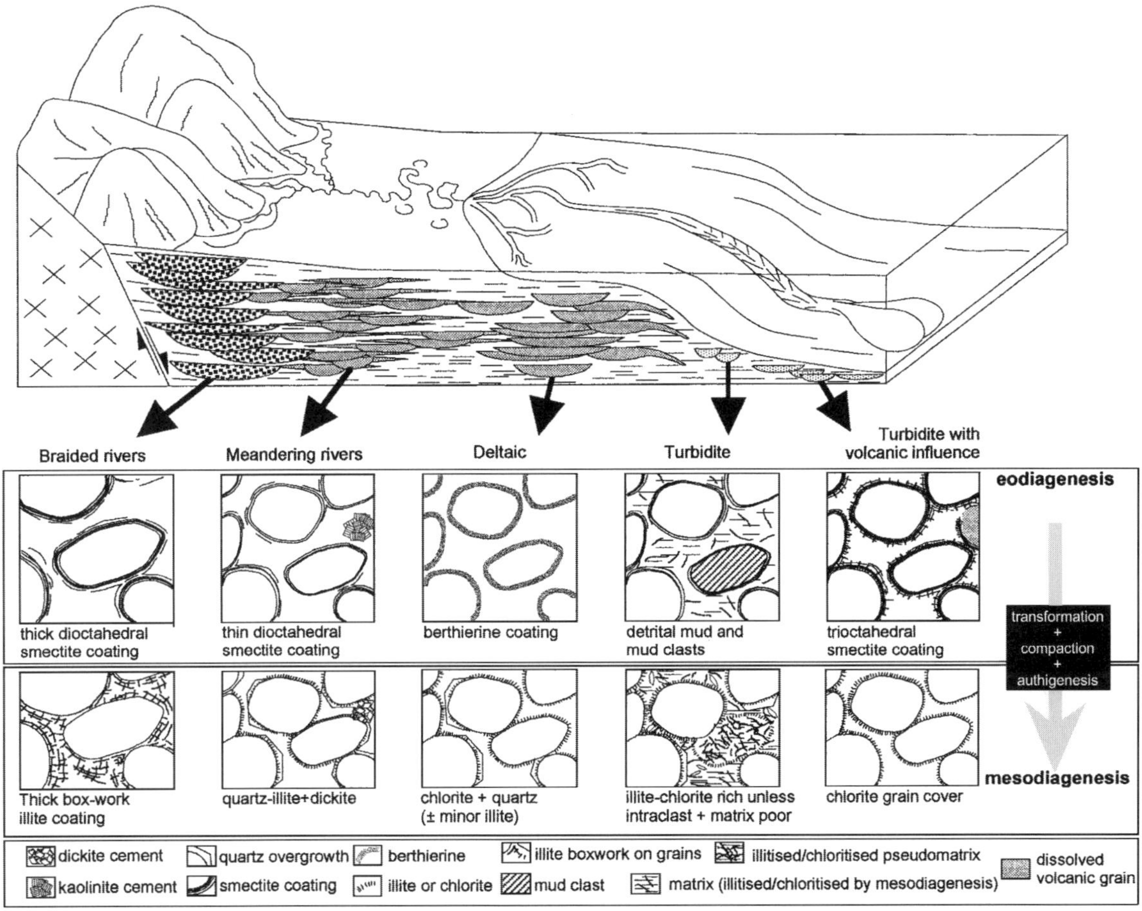

Fig. 11 Schematic representation of the spatial, facies-related eogenetic and subsequent mesogenetic evolution pathways of clay minerals in sandstones.

Hence, under conditions of high sediment flux, shoreline transgression gives way to regression and an increasingly progradational stacking pattern. The sand : mud ratio increases up the HST (i.e. away from MFS), being accompanied by a concomitant decrease in the amounts of authigenic glauconite and verdine, but an increase in the amounts of authigenic berthierine. Berthierine occurs mainly in coastal plain and shallow-marine sediments with a brackish pore water composition. Landward, the top of HST is associated with widespread fluvial deposition (Posamentier & Vail, 1988) that, depending on the climatic conditions, may be subjected to kaolinitization, clay infiltration, pedogenesis or the formation of Mg-rich clay minerals.

BURIAL DIAGENETIC (MESOGENETIC) CLAY MINERALS

Following eodiagenesis, sediment comprises a mixture of stable and metastable detrital components and eogenetic minerals. Further changes in pressure, temperature and porewater chemistry induce dynamic mesogenetic reactions (Fig. 11).

Temperature and the integrated thermal history are master controls on the clay minerals in

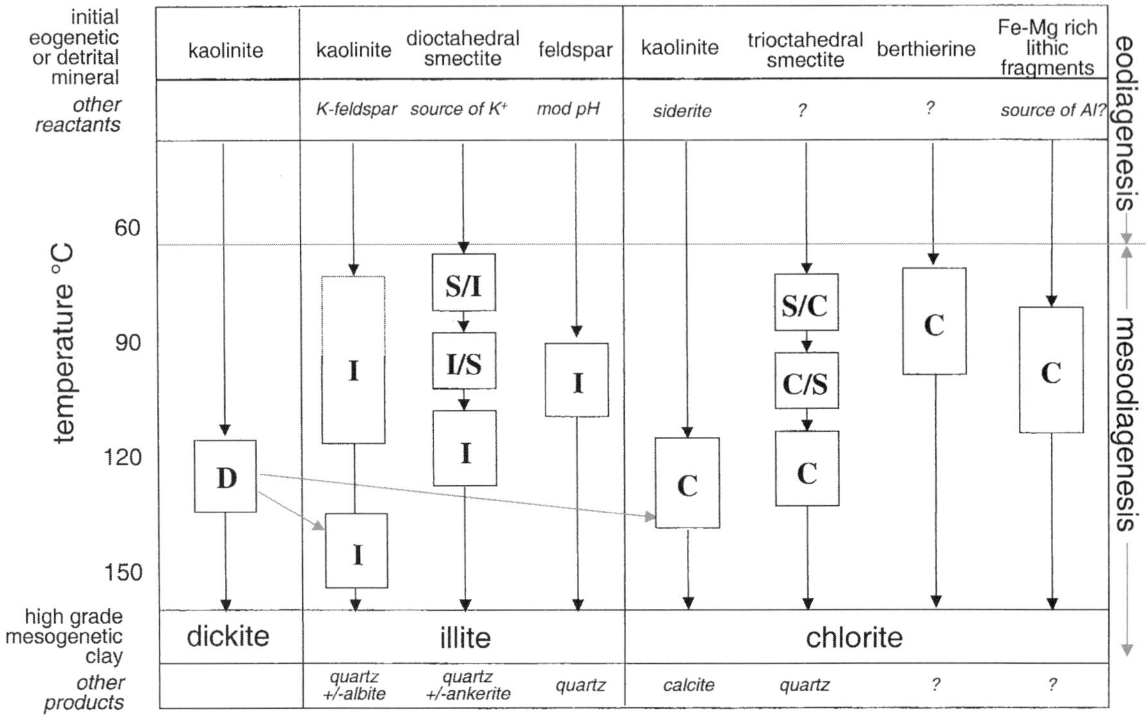

Fig. 12 Common mesogenetic pathways for clay minerals in sandstones, where D is dickite, S is smectite, I is illite and C is chlorite. Randomly interstratified mixed-layer clay minerals are named according to the types of layers involved, with the most abundant layer type listed first (Reynolds, 1980): S/I is mixed-layer smectite–illite dominated by smectite; I/S is the same mineral mixture dominated by illite. The same naming rules apply for interlayered smectite–chlorite. High-grade diagenesis leads to dickite, illite and chlorite clay minerals in sandstones. Kaolinite forms predominantly during eodiagenesis and can be cannibalized to form dickite, illite or even chlorite during mesodiagenesis. Illite forms by at least three main routes during mesodiagenesis. Although it can be a detrital clay (following incomplete weathering) it does not form during eodiagenesis. Chlorite also seems to be unlikely to form during eodiagenesis and forms by at least four main pathways.

sandstones. Clay-mineral assemblages change with increasing temperature, leading to a series of prograde diagenetic reactions. Eogenetic kaolinite, berthierine and smectite are replaced by mesogenetic dickite, illite and chlorite (Fig. 12).

The following sections consider the origins of the three main burial diagenetic clay minerals: dickite, illite and chlorite.

Kaolinite recrystallization and transformation into dickite

With progressive increase in burial depth and temperature (2–3 km; $T = 70–90°C$), thin vermicularly stacked and booklet-like, pseudohexagonal eogenetic kaolinite is progressively replaced by thick, well-developed crystals (Fig. 12). The evolution of kaolinite to dickite can be traced in the sequence of micrographs in Fig. 3. The thick crystals are believed to be well-ordered kaolinite and/or poorly disordered dickite that replaces poorly ordered, eogenetic kaolinite (Cassagnabere, 1998). This transformation, which seldom goes to completion, causes little or no disruption to the stacking pattern of the eogenetic kaolinite (Fig. 3). At burial depths between about 3.0 and 4.5 km (90–130°C), the remaining eogenetic kaolinite is subjected to dissolution and reprecipitation as blocky crystals of dickite (Beaufort *et al.*, 1998). This process increases the amounts of microporosity in the kaolin cement and a disruption of the stacking pattern of the eogenetic kaolinite. At burial depths greater than 4.5 km (> 130°C), kaolin is dominated by blocky, well-ordered dickite (Cassagnabere, 1998) and vermicular stacking patterns are not preserved.

The pervasive dissolution of eogenetic kaolinite and its replacement by dickite has been used to suggest that dickite forms by dissolution and reprecipitation (Ehrenberg *et al.*, 1993; Morad *et al.*, 1994). The greater degree of crystal disorder in kaolinite relative to dickite (Brindley & Brown, 1980) may account for the differential stability (Morad *et al.*, 1994). It is possible that the transformation of kaolinite to dickite is controlled kinetically. The transformation usually is more pervasive in high permeability sandstones than in low permeability sandstones (Cassagnabere, 1998). Furthermore, the transformation is expected to be retarded in oil-saturated sandstones and in sandstones that are severely compacted and/or enriched in clay matrix or by eogenetic, intergranular cement (Beaufort *et al.*, 1998). As depositional porosity and permeability varies within a single sequence, the extent of kaolin transformation probably is facies-related, being most extensive in coarse-grained, permeable sand facies. Accordingly, the kaolinite to dickite transformation should not be used as a simple palaeogeothermometer, as suggested previously (e.g. Ehrenberg, 1993).

The transformation of kaolinite into dickite is possibly aided by an increase in acidity of formation waters or a decrease in the aK^+/aH^+ ratio. Evidence supporting this includes the concomitant dissolution of framework silicates (e.g. feldspars) and carbonate cement precipitation. Considering the substantial depths at which dickite forms, it is unlikely that these modifications are a result of meteoric-water incursion, and it is instead attributed to the invasion of the reservoirs by organic acids or source-rock-derived CO_2 (Morad *et al.*, 1994; Clauer *et al.*, 1999).

Nacrite forms authigenically, in pores and fractures, often in association with hydrothermal barite and dolomite (Buatier *et al.*, 2000). Nacrite also may form by transformation of dickite (Xia, 1985). Although nacrite is considered to be formed at the highest temperature among the three kaolin polytypes, several authors have argued against its use as a qualitative geothermometric indicator of formation at relatively high temperatures (e.g. Buatier *et al.*, 2000).

Illitization of smectite

Smectitic clay minerals in siliciclastic rocks can be of both eogenetic and detrital origin (McKinley *et al.*, this volume, 109–128). With burial and heating, dioctahedral smectite transforms to illite, passing through interlayered forms. The evolution of smectite to illite can be traced in the sequence of micrographs in Fig. 4. Dioctahedral smectite can have a range of compositions but typically has a relatively low Al/Si ratio and contains Ca, Fe and Mg. Transformation of dioctahedral smectite to illite can occur by two routes. The first approximately conserves silica and thus requires a supply of aluminium to achieve a transformation into illite (R3). The second conserves aluminium and produces excess silica (probably as quartz cement, R4).

$$\begin{aligned}&0.45H^+ + 0.45K^+ + 0.4Al_2O_3 + \underset{\text{dioctahedral smectite}}{K_{0.1}Na_{0.1}Ca_{0.2}Mg_{0.4}Fe_{0.4}Al_{1.4}Si_{3.8}O_{10}(OH)_2 \cdot H_2O}\\ &\Rightarrow \underset{\text{illite}}{K_{0.55}Mg_{0.2}Fe_{0.15}Al_{2.2}Si_{3.5}O_{10}(OH)_2} + 0.1Na^+ + 0.2Ca^{2+} + 0.2Mg^{2+} + 0.125Fe_2O_3 + \underset{\text{quartz}}{0.3SiO_2}\\ &+ 1.22H_2O \qquad \text{(R3)}\end{aligned}$$

$$\begin{aligned}&1.242H^+ + 0.393K^+ + \underset{\text{dioctahedral smectite}}{1.58K_{0.1}Na_{0.1}Ca_{0.2}Mg_{0.4}Fe_{0.4}Al_{1.4}Si_{3.8}O_{10}(OH)_2 \cdot H_2O}\\ &\Rightarrow \underset{\text{illite}}{K_{0.55}Mg_{0.2}Fe_{0.15}Al_{2.2}Si_{3.5}O_{10}(OH)_2} + 0.16Na^+ + 0.31Ca^{2+} + 0.43Mg^{2+} + 0.24Fe_2O_3 + 2.47SiO_2\\ &+ 2.86H_2O \qquad \text{(R4)}\end{aligned}$$

By whatever reaction illitization of smectite occurs, it seems to be a progressive process whereby precursor smectite is replaced by illite on a layer-by-layer basis (Fig. 12). The evolution from dioctahedral smectite to illite requires loss of silicon at the expense of aluminium in the tetrahedral sites and the gain of potassium at the expense of water in the interlayer sites to maintain charge balance. The interlayering of smectite and the replacive illite is disordered in the first instance but seems to become progressively more ordered as temperature increases. Diagenetic illites formed at moderate temperatures (< 90°C) in sandstones have at least a minor component of smectite within them, possibly as a remnant of the original clay from which they formed (Burley & Macquaker, 1992). Conversely, there are very few reported cases of pure dioctahedral smectite in sandstones (most contain at least some illite). Although there are kinetic models for the conversion of smectite to illite in mudstones (e.g. Hower *et al.*, 1976; Huang *et al.*, 1993), it is not clear that these are applicable to sandstones (McKinley *et al.*, this volume, pp. 109–128).

Although potassium can be supplied locally from the concomitant dissolution of potassium feldspars (closed-system diagenesis), an increasing number of studies call upon an open-system diagenesis for relatively soluble metals such as alkali metals (e.g. K, Na) (e.g. Furlan *et al.*, 1996; Land *et al.*, 1997). Illitization potentially releases considerable amounts of Mg^{2+}, Fe^{2+}, Na^+ and Ca^{2+} ions that could be used in diagenetic reactions within the same rock or in adjacent lithologies. The common coeval mesogenetic formation of illite, quartz and ankerite may be, at least partially, the result of reactions such as R4 and R5.

Illitization of kaolin and K-feldspar

The transformation of kaolinite into illite is prevalent at temperatures greater than about 70°C but becomes pervasive at temperatures greater than about 130°C (Fig. 12). According to equilibrium thermodynamic models, K-feldspar and kaolinite seem to be universally unstable together and their coexistence at lower diagenetic temperatures has been assigned to the slow rate of their reaction (Bjørlykke & Aagaard, 1992). The most common reaction has been described as a pH-neutral, isochemical process (Bjørlykke, 1980; Bjørkum & Gjelsvik, 1988; Ehrenberg & Nadeau, 1989; Bjørlykke *et al.*, 1995; Ramm & Ryseth, 1996):

$$\underset{\text{kaolinite}}{Al_2Si_2O_5(OH)_4} + \underset{\text{K-feldspar}}{KAlSi_3O_8} \Rightarrow \underset{\text{illite}}{KAl_3Si_3O_{10}(OH)_2} + \underset{\text{quartz}}{2SiO_2} + H_2O \quad \text{(R5)}$$

The quartz by-product is thought to precipitate as discrete crystals or as overgrowths on quartz grains or on earlier generations of quartz overgrowths.

The illitization of kaolinite also may occur through a flux of K^+ from external sources, including the dissolution of adjacent evaporite deposits (Gaupp *et al.*, 1993):

$$\underset{\text{kaolinite}}{3Al_2Si_2O_5(OH)_4} + 2K^+ \Rightarrow \underset{\text{illite}}{2KAl_3Si_3O_{10}(OH)_2} + 2H^+ + 3H_2O \quad \text{(R6)}$$

This, and the previous reaction, are most important in fluvial sandstones that contain eogenetic kaolinite.

K-feldspar reactions to illite possibly can proceed in the absence of kaolinite (McAulay *et al.*, 1993; Platt, 1993) if there is a source of acidity as follows:

$$\underset{\text{K-feldspar}}{3KAlSi_3O_8} + H_2O + 2H^+ \Rightarrow \underset{\text{illite}}{KAl_3Si_3O_{10}(OH)_2} + \underset{\text{quartz}}{6SiO_2} + 2K^+ \quad \text{(R7)}$$

The source of acidity during mesodiagenesis is a controversial topic, but commonly has been ascribed to an influx of low pH, organic acid-bearing formation water (e.g. Surdam *et al.*, 1989). Another way to achieve low pH water is for gas phase CO_2 to migrate into the structure, partition into formation water, dissociate and thus produce low pH waters (Barclay & Worden, 2000). For dissociation to produce a constant supply of protons, the aqueous bicarbonate must be removed from solution by the concurrent precipitation of carbonate minerals.

The albitization of potassium feldspar coincides with, and provides K^+ needed for, illitization of kaolinite in sandstones. Indeed, the dissolution and albitization of potassium feldspar is thought to be controlled kinetically by the consumption of K^+ in the illitization of kaolinite (Aagaard *et al.*, 1989). The reaction can be written as follows (Morad, 1986):

$$\underset{\text{K-feldspar}}{2KAlSi_3O_8} + \underset{\text{kaolinite}}{2.5Al_2Si_2O_5(OH)_4} + Na^+$$

$$\Rightarrow \underset{\text{albite}}{NaAlSi_3O_8} + \underset{\text{illite}}{2KAl_3Si_3O_{10}(OH)_2} + \underset{\text{quartz}}{2SiO_2}$$

$$+ 2.5H_2O + H^+ \qquad \text{(R8)}$$

Albitization is thus controlled by the ionic activity of Na^+ and by pH and is accomplished with formation waters that vary widely in Na^+ activity and total salinity (Aagaard *et al.*, 1989). Pervasive or complete illitization of kaolinite and concomitant albitization of K-feldspar in sandstones occur at temperatures of about 130°C (Morad *et al.*, 1990).

Illite interlayer sites are dominated normally by potassium with little sodium tolerated. Small amounts of ammonia (NH_4^+) can occupy the interlayer sites because it has a molecular size similar to the atomic size of potassium and has the same charge. Ammonia-rich illite has been reported in several studies and may be more common than typically is appreciated. The paragenesis of ammonia-rich illite is not particularly well understood but is thought to be related to the thermal decomposition of nitrogen-rich kerogen (e.g. from coals at high grade) or other organic matter (Everlein & Hoffman, 1991; Williams *et al.*, 1992).

Dickite may be less susceptible to illitization than kaolinite, owing to its better ordered crystal structure relative to kaolinite (Morad *et al.*, 1994). This is demonstrated by the selective illitization of kaolinite remnants in partly dickitized, kaolinite vermicules. Therefore, sandstones that have suffered from pervasive dickitization of kaolinite prior to entering the illitization window, contain kaolin plus K-feldspar even at 100°C or higher temperatures. The factors controlling variations in the extents of kaolinite dickitization in sandstones have been discussed previously. Although there are no precise data demonstrating the pattern, temperature and timing of the transformation of dickite into illite, the reaction presumably occurs at temperatures greater than 130°C and becomes pervasive at temperatures > 150°C. This is manifested by the near absence of kaolin polymorphs in sandstones subjected to such elevated temperatures.

Grain replacive chlorite

Authigenic chlorite in sandstones may occur during mesodiagenesis as a result of the breakdown of volcaniclastic grains and Fe–Mg-rich detrital minerals such as garnet, biotite or amphibole (Fig. 12). This type of chlorite ultimately occurs as grain replacements, (Galloway, 1979; Surdam & Boles, 1979; Thomson, 1979; Chang *et al.*, 1986; Pirrie *et al.*, 1994; Remy, 1994) which, when deformed by compaction, appear as pseudomatrix (e.g. Burns & Ethridge, 1979; Seeman & Scherer, 1984; Chan, 1985; Mathisen & McPherson, 1991). Chlorite forms from Mg and Fe supplied from the dissolution of detrital ferromagnesian grains (De Ros *et al.*, 1994), or eogenetic haematite in red beds (Dixon *et al.*, 1989), from clay transformation reactions occurring in associated mudrocks (Boles & Franks, 1979; Moncure *et al.*, 1984), or from the destabilization of organometallic complexes (Surdam *et al.*, 1989).

Chlorite formation thus can be controlled strongly by detrital mineralogy of the sandstones (e.g. Buatier *et al.*, 2000). However, isotopic analyses of chlorite-rich and chlorite-poor Jurassic sandstones from the Norwegian continental shelf do not seem to support a model of localized chlorite formation as a result of localized influx of a unique Fe-rich provenance component (Ehrenberg *et al.*, 1998). These authors interpreted the Nd isotopic data to be consistent with authigenesis of Fe-clay on the sea-floor, localized by fluvial discharge into a nearshore marine setting, supporting the berthierine precursor model for the origin and distribution of mesogenetic chlorite cement.

Chloritization of odinite, berthierine and Mg-rich clay minerals

As chlorite is not formed directly during eodiagenesis, all the precompactional 'early chlorite coatings' described in the literature (e.g. Land & Dutton, 1978; Dutton & Land, 1985; Longstaffe, 1986, 1993) are most likely to be the product of chloritization of grain coating, eogenetic berthierine–odinite or smectite (Morad *et al.*, 2000) (Fig. 12). Chlorite occurs as coatings comprising small pseudo-hexagonal crystals arranged perpendicular to the sand grains (Anjos *et al.*, this volume, pp. 291–316). Transformation of precursor Fe-rich clay minerals to chlorite occurs at burial depths greater than about 2–3 km and temperatures greater than 60–100°C (Jahren & Aagaard, 1989; Burley & MacQuaker, 1992; Ehrenberg, 1993; Grigsby, 1999; Aagaard *et al.*, 2000). The precursor clay mineral was described by Aagaard *et al.* (2000) as an amorphous, Fe-bearing berthierine coating sand-grain surfaces. Odinite has been suggested as a precursor clay mineral for chlorite formation by Ryan & Reynolds (1996). The chloritization of odinite is believed to occur via an intermediate, early diagenetic mixed-layer serpentine–chlorite (Sp/C) with preservation of some of the original odinite textures (Ryan & Reynolds, 1996). These authors observed that with increasing burial depth of sandstones of the subsurface Tuscaloosa Formation between 1702 m and 6216 m, the proportion of serpentinite layers in Sp/C decreases from 21% to 1% (Ryan & Reynolds, 1997). As berthierine forms in estuarine and marine–pro-deltaic environments (Ehrenberg, 1993; De Hon *et al.*, 1999; Grigsby, 1999; Baker *et al.*, 2000), Fe-rich chlorite coatings are expected to occur in sandstones originally deposited in these environments. However, mesogenetic grain-coating chlorite also occurs in sandstones from deltaic–lacustrine environments (Luo *et al.*, in press), and fluvial incised-valley sandstones (S. Morad, unpublished data).

Mg-rich chlorite has been proposed to form in association with magnesium-rich brines resulting from the mesodiagenesis of stratigraphically, or areally, proximal evaporitic–carbonate sequences (Rossel, 1982; Goodchild & Whitaker, 1986; Janks *et al.*, 1992; Purvis, 1992; Gaupp *et al.*, 1993; Searl, 1994). However, magnesium-rich chlorites in many of these cases are possibly products of the undetected transformation of eogenetic magnesian smectitic clay precursors via transient corrensite (Hillier, 1994).

Chloritization of smectite

An alternative precursor to grain-coating chlorite is smectite (Hillier, 1994; Humphreys *et al.*, 1994), particularly the trioctahedral varieties, such as saponite (Fig. 12). The progressive transformation (Fig. 5) may require a source of aluminium (Chang *et al.*, 1986):

$$\underset{\text{trioctahedral smectite (saponite)}}{Ca_{0.1}Na_{0.2}Fe_{1.1}MgAlSi_{3.6}O_{10}(OH)_2} + 1.5Fe^{2+} + 1.2Mg^{2+} + 1.4Al^{3+} + 8.6H_2O$$

$$\Rightarrow \underset{\text{chlorite}}{Fe_{2.6}Mg_{2.2}Al_{2.4}Si_{2.8}O_{10}(OH)_8} + 0.1Ca^{2+} + 0.2Na^+ + \underset{\text{quartz}}{0.8SiO_2} + 9.2H^+ \qquad (R9)$$

An alternative process for the replacement of trioctahedral smectite by chlorite is conservative with respect to aluminium, requiring loss of excess silica (Chang *et al.*, 1986):

$$\underset{\text{trioctahedral smectite (saponite)}}{2.4Ca_{0.1}Na_{0.2}Fe_{1.1}MgAlSi_{3.6}O_{10}(OH)_2} + 0.88H_2O + 1.44H^+$$

$$\Rightarrow \underset{\text{chlorite}}{Fe_{2.6}Mg_{2.2}Al_{2.4}Si_{2.8}O_{10}(OH)_8} + 0.24Ca^{2+} + 0.48Na^+ + 0.04Fe^{2+} + 0.20Mg^{2+} + \underset{\text{quartz}}{5.84SiO_2} \qquad (R10)$$

In cases where reaction R10 operates, the liberated Fe and Mg may help to form ankerite, whereas the silica precipitates as quartz cement (Niu *et al.*, 2000). In cases where R10 operates, the sources of Fe, Mg and Al may be the dissolution of basic volcanic fragments, biotite and Fe-oxyhydroxide.

The chloritization of smectite is a dissolution–reprecipitation process (Ryan *et al.*, 1998) that occurs progressively and involves an intermediate mixed-layer chlorite–smectite phase, initially random but at greater depths with regular (corrensite) C/S stacking (Hillier, 1994). The chloritization of corrensite is accompanied by a successive decrease in Mg and a concomitant increase in Fe and Al (Biron *et al.*, 1999), which results in the formation of Fe-rich chlorite.

Chloritization of kaolinite

In a study of the active high-temperature diagenesis and metamorphism of upper Cenozoic sediments in the Salton Sea geothermal field and the Salton Trough (southern California), Muffler & White (1969) suggested that chloritization of kaolinite occurs according to the following outline reaction:

$$\text{kaolinite} + \text{dolomite} + \text{ankerite} \Rightarrow \text{chlorite} + \text{calcite} + CO_2 \qquad \text{(R11)}$$

Boles & Franks (1979) recorded that the chloritization of kaolinite occurs at burial depths between about 3500 and 4500 m (165–200°C). The reaction was envisaged to be:

$$\underset{\text{kaolinite}}{3Al_2Si_2O_5(OH)_4} + 3.5Fe^{2} + 3.5Mg^{2+} + 9H_2O \Rightarrow \underset{\text{chlorite}}{Fe_{3.5}Mg_{3.5}Al_{6.0}Si_{6.0}O_{20}(OH)_{16}} + 14H^+ \qquad \text{(R12)}$$

Iijima & Matsumoto (1982) suggested that siderite could be the source of Fe during the chloritization of kaolinite:

$$\underset{\text{kaolinite}}{3Al_2Si_2O_5(OH)_4} + 9FeCO_3 + 2H_2O \Rightarrow \underset{\text{chlorite}}{Fe_{9.0}Al_{6.0}Si_{6.0}O_{20}(OH)_{16}} + \underset{\text{quartz}}{SiO_2} + 9CO_2 \qquad \text{(R13)}$$

Chloritization of kaolinite in Triassic red-bed sandstone of the Lunde Formation, offshore Norway (S. Morad, unpublished data) occurs close to the oil–water contact at depths of about 2.5 km (100°C). The Fe^{2+} thus might be derived from the reduction of iron oxides (Curtis *et al.*, 1985), as envisaged below:

$$\underset{\text{kaolinite}}{2Al_2Si_2O_5(OH)_4} + \underset{\text{haematite}}{5Fe_2O_3} + \underset{\text{quartz}}{2SiO_2} + 4H_2O + \underset{\text{organic matter}}{2.5C} \Rightarrow \underset{\text{chlorite}}{Fe_{10.0}Al_{4.0}Si_{6.0}O_{20}(OH)_{16}} + 2.5CO_2 \qquad \text{(R14)}$$

This reaction therefore may be controlled by CO_2 fugacity. Analogous to the case of illite, dickite apparently is less sensitive to chloritization than kaolinite. Mixed-layer chlorite–kaolinite in reservoir sandstones from offshore Norway (Hillier & Velde, 1992) may be an intermediate phase involved in such reactions.

THE ROAD TO METAMORPHISM

The precise boundary between metamorphism and diagenesis is imperfectly defined. The attainment of metamorphic grade is characterized by thermodynamic equilibrium, by the near total reduction of porosity and by textural equilibrium (Frey & Robinson, 1999).

The probable evolutionary succession of clay minerals in sandstones is represented schematically in Fig. 13 in which the changing clay minerals for a given sandstone bulk composition can be traced. There are several distinct evolutionary routes that depend primarily on the bulk rock geochemistry. It is reasonable to assume an isochemical system because the occlusion of porosity will inhibit mass flux. The following analysis is based largely on the study of diagenetic to metamorphic mudstones (e.g. Dunoyer de Segonzac, 1970; Yardley, 1989; Frey & Robinson, 1999), as there has been relatively little study of the diagenesis to metamorphism transition in sandstones. Mudstones can be used as a satisfactory proxy because they have similar minerals, although naturally in different proportions.

Sandstones rich in Al but poor in K and Fe–Mg are dominated by kaolin, with quartz and minor albite. A low-grade metamorphic

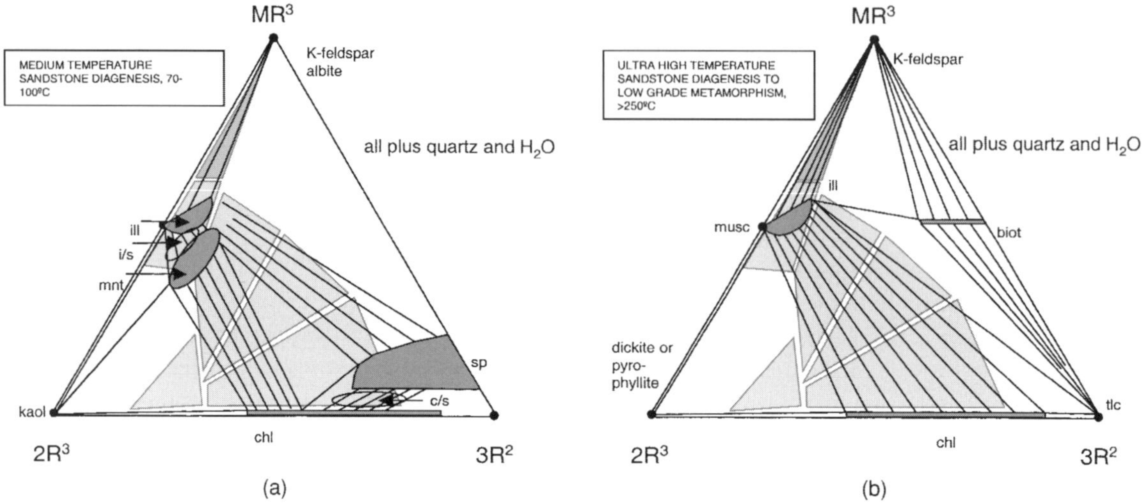

Fig. 13 Phase diagrams showing the main clay mineral assemblages and main sandstone compositional types (see Fig. 2) in terms of divalent, trivalent and alkali element (monovalent) ion proportions. (a) Typical lower mesogenetic temperature (below 100°C) assemblages. (b) Typical high grade (> 200°C) mesogenetic assemblages following the decomposition of smectites.

assemblage typical of quartz arenites and plagioclase-bearing sandstones would be pyrophyllite, quartz and albite, as kaolinite would be replaced by pyrophyllite at low metamorphic grades.

Sandstones rich in Al and K, but poor in Fe–Mg might evolve via a smectite to illite reaction or a kaolinite to illite reaction. Illite should evolve through a series of polytypes from disordered one-layer crystals, to ordered two-layer crystals during progressive diagenesis. Illite gains potassium progressively during the evolution of the polytype series, becoming true muscovite in low-grade metamorphic rocks. By the time a rock reaches metamorphic grade, it may not be clear whether the sandstone was enriched initially in kaolinite or montmorillonite. A low-grade metamorphic assembly typical of K- and Al-rich sandstones thus would be muscovite and quartz and possibly minor albite.

Sandstones rich in Al and Fe–Mg, but poor in K probably contained eogenetic trioctahedral smectite clay minerals. These evolve during heating with progressive replacement of the smectite by chlorite (as opposed to illite). Chlorite changes composition and polytypes during burial, although such patterns are poorly constrained from case studies. Low-grade metamorphic, lithic rich, arkosic poor sandstones will be dominated by chlorite and quartz, ±albite.

Sandstones rich in Al, K and Fe–Mg probably will contain K-feldspar and both illite and chlorite up to moderate diagenetic grades. At true metamorphic grades, any remaining K-feldspar probably undergoes reaction with chlorite to produce muscovite and biotite. Such reactions are characteristic of metapelite rocks, although they have not been well characterized for metasandstones. Low-grade metamorphic, lithic rich, arkosic rich sandstones probably are dominated by muscovite, biotite, albite and quartz and either chlorite or K-feldspar depending on the exact bulk-rock composition.

TELODIAGENESIS

The telogenetic regime is where formation waters resulting from early diagenetic processes or, more commonly from mesogenetic processes, are displaced by meteoric water during inversion, uplift, erosion and the formation of an unconformity. A pressure head associated with rainfall on upland regions can drive an artesian system and the flux of meteoric water. Influx of water usually is both permitted and driven by tectonic uplift. Telogenetic processes are more prevalent along basin margins and along fault blocks than in deep basin centres. The coincidence of conditions that permit meteoric influx is likely to be transient on a geological time-scale so that widespread influx of meteoric water is likely to be exceptional rather than routine.

Meteoric waters are commonly very dilute, oxidizing, saturated with CO_2 and thus acidic. Telogenetic changes include the alteration of feldspars to clay minerals, oxidation of reduced iron-bearing cements (ankerite, siderite and chlorite), and dissolution of calcite, dolomite and sulphate cements. Telogenetic processes have been interpreted to be responsible for the development of kaolinite in Brent Group and Kimmeridge Clay Formation sandstones during the Cimmerian uplift and development of an unconformity (e.g. Emery *et al.*, 1990; Ketzer *et al.*, this volume, pp. 43–61). However, Bjørkum *et al.* (1990) argued that the rates of uplift and erosion are higher than the rate of kaolinitization, precluding the preservation of telogenetic kaolinite in sandstones.

Most telogenetic processes occur within the first few metres of the surface. Although large volumes of meteoric water can pass deep into aquifers over protracted periods, the consequence may be muted because waters tend rapidly to reach saturation with the major aqueous species during meteoric influx. This limits the geochemical consequences of water flow into the deeper subsurface, and usually it is geochemically inappropriate to ascribe deep (hundreds of metres or more) processes to meteoric water influx (Shanmugam, 1990).

OPEN AND CLOSED SYSTEMS DURING MESODIAGENESIS: A QUESTION OF PERSPECTIVE

Many conceptual models of mesogenetic mineral dissolution and cementation have been built upon the notion that sandstones and other permeable rocks are effectively open to mass and heat flow. Such models envisage and require large volumes of water to pass through each pore volume, with the fluid driven by processes such as (i) faulting-induced seismic pumping and valving (e.g. Burley, 1993), (ii) artesian style meteoric water flushing (Stewart *et al.*, 2000) and (iii) overpressure release through mass exit of formation water from fractured top-seals (Wilkinson *et al.*, 1997). It is probable that each of these processes do indeed occur, to one degree or another. However, it is difficult to conceive of these processes acting over large areas, leading to widespread and consistent patterns of diagenesis under similar thermal regimes. It also is difficult to reconcile an external source of silica and alumina, given their extremely low solubility, with the vast quantities of quartz and various clay minerals found in sandstones. However, the issue of aluminium mobility during clay and related diagenesis remains unresolved, as there are petrographic (Milliken, this volume, pp. 343–360), mineralogical (Haszeldine *et al.*, 2000) and bulk-rock geochemical (Wilkinson *et al.*, this volume, pp. 147–160) data that seem to require movement of aluminium to provide a rational explanation for empirical patterns.

It was thought that organic acids might enhance aluminium solubility and affect clay diagenesis to such an extent that much smaller volumes of water could be responsible for clay cementation (e.g. Surdam & Macgowan, 1987). There was a huge research drive to first quantify organic acids in basinal waters and then to assess the complexing capability of a variety of organic acids (e.g. references in Pittman & Lewan, 1995). Organic acids are found routinely in formation waters; indeed in deep formation waters ethanoic acid commonly has more impact than bicarbonate on alkalinity. It also was discovered that trifunctional organic acids such as citric acid could enhance Al solubility. However, ethanoic acid is the only organic acid commonly found in formation waters (e.g. Hanor, 1994). Multifunctional organic acids are unusual or rare. It also was discovered that common divalent cations

(Ca^{2+}, Mg^{2+}) tend to complex preferentially with organic acids, thus minimizing their ultimate impact on Al solubility and mobility (Fein, 1991; Fein & Hestrin, 1994). The final verdict thus may be that organic acids cannot be held responsible for mass Al flux in formation waters. It is more likely that the Al in clay minerals is sourced locally within the sandstone (or in immediately adjacent mudstones). Such a conclusion has the notable advantage that it is modelled more easily.

Treating sandstone diagenesis as being either open or closed is to miss a crucial point. Sandstone geochemistry and clay diagenesis involve multiple components and not just alumina and silica. The oxides of iron, calcium, magnesium and potassium are also crucial, as is CO_2. A source or sink of protons also must be found for some reactions and many require loss or gain of oxygen. There is substantial evidence that some species undergo mass flux during sandstone diagenesis. For example, CO_2 must be added to sandstones in considerable volumes to lead to extensive mesogenetic carbonate cementation in sandstones and to cause the evolution of carbon isotopes to progressively more ^{12}C-rich values (e.g. Worden & Barclay, 2000). The presence of CO_2 can lead to enhanced formation water acidity, which in turn can lead to feldspar decay and clay mineral growth. Thus clay authigenesis may be initiated by influx of exotic fluids, even though the building blocks of the clay minerals themselves (e.g. Al- and Si-oxide) are supplied from the sandstone. Water passing through the Zechstein in the Southern North Sea gas basin can acquire elevated potassium concentrations and these are thought to be at least partly responsible for the growth of potassium-rich illite in the Rotliegend sandstones (Gaupp *et al.*, 1993).

Thus, it is not useful to imagine that sandstones are either open or closed during mesodiagenesis. Rather they will be relatively closed to low solubility refractory oxides, potentially open with respect to soluble alkali metals and open, to a lesser extent, to alkaline earth metals. Moreover, highly labile exotic species (e.g. CO_2) may be actively involved in initiating clay reactions even though they are not locked up in the clay minerals themselves.

EFFECTS OF CLAY MINERALS ON POROSITY AND PERMEABILITY IN SANDSTONES

Permeability is a function of many factors. Total porosity is clearly an important control on permeability but the nature of the porosity is equally significant. Microporosity tends to be ineffective, and isolated moldic pores may contribute nothing to the fluid-flow properties of a rock. Different clay minerals tend to have variable amounts of microporosity within them. For example clay-rich matrix in sandstones contains substantial void space, although the pores are micrometre scale or smaller, and tend to be poorly connected so that clay matrix acts as a baffle to fluid flow (Hurst & Nadeau, 1995).

Different clay-mineral cements can have different effects on permeability because they occupy different positions within the pore network (Stalder, 1973; Howard, 1992; Fig. 14). Clay minerals that are arranged tangentially to the grain surfaces have less of an effect on permeability than perpendicular clay minerals or clay minerals that sit within pores and pore throats (e.g. Pallatt *et al.*, 1984; Kantorowicz, 1990). Thin clay coats on grain surfaces may have little effect on permeability unless they become interwoven at pore throats. If thick coatings of illite or chlorite occur, permeability can again be seriously diminished, particularly in fine-grained sandstones, even though porosity is relatively unaffected.

Discrete aggregates of clay (e.g. altered extraclastic rock fragments or clay-rich intraclasts) will have little direct influence on permeability as long as the clay minerals are not spread throughout the pore network as greywacke-style matrix (Whetton & Hawkins, 1970). Abundant pseudomatrix between competent sand grains (see above, and Worden *et al.*, 1997, 2000) behaves like a plug or baffle in pore throats and can rapidly reduce permeability

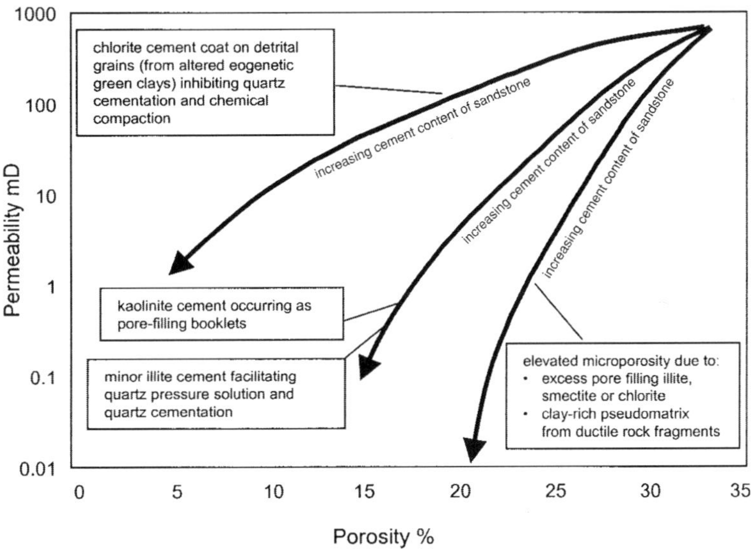

Fig. 14 Porosity–permeability diagram demonstrating the main end-member effects of the different types and morphologies of clay minerals on sandstones.

during progressive burial. Hence, compaction of ductile, clay-rich fragments can lead to catastrophic sandstone permeability loss.

Clay mineral type may affect permeability in other ways. The common presence of clay rims around detrital sand grains means that grain–grain interfaces often contain thin selvages of clay minerals. It is thought that mica-type minerals (e.g. illite) exacerbate pressure dissolution (chemical compaction) between quartz grains and that chlorite inhibits pressure dissolution (e.g. Fisher *et al.*, 2000). The possible active induction of quartz pressure dissolution and silica supply by thin grain-coating illite serves to reduce both sandstone porosity and permeability by assisting compaction and cementation. Conversely, grain-rimming chlorite (recrystallized from eogenetic berthierine in weakly compacted sandstones) can inhibit quartz cementation and pressure dissolution (chemical compaction). Thus, chlorite can be doubly beneficial to permeability by ameliorating pressure solution and preventing quartz overgrowth formation even with a thin layer on quartz surfaces. Note that sandstones containing deep-mesogenetic chlorite rims do not seem to show the same ability to preserve porosity.

THE EFFECT OF PETROLEUM ON CLAY MINERAL DIAGENESIS IN SANDSTONES

The study of clay diagenesis in sandstones has been advanced by examination of deeply buried sandstones from oil fields. One aspect of the study of clay diagenesis in such sandstones is the effect of oil-filling on the processes involved in clay diagenesis. Put simply, does oil emplacement stop, slow or have no affect on clay diagenesis in sandstones? This section addresses how oil-filling affects water in sandstones and the effect that this has on clay diagenesis.

Before petroleum enters a reservoir, pores are filled with formation water, which is displaced during petroleum emplacement. Even when a porous rock is full of petroleum, some of the original formation water remains, known as the irreducible water saturation (S_{wi}). The quantity of the residual water depends upon a variety of factors, including permeability and wettability. Low permeability encourages high residual water contents because the petroleum finds it harder to flush water from the system. Wettability describes the preference of a rock to be in contact with one type of fluid rather than another. The wetting state of a rock containing two immiscible fluids dictates which fluid is in contact with grain surfaces. For example, a water-wet petroleum reservoir is one where sand grains are coated with water, and where petroleum finds it difficult to come into contact with the grain surfaces, and vice versa. A continuum of behaviour exists in nature between oil- and water-wet reservoirs.

Physical influences of petroleum on clay mineral diagenesis

Clay mineral diagenesis is controlled by dissolution, transport and precipitation. Each of these can be influenced variably by addition of petroleum to a sandstone.

The rate of *advective* flux is proportional to the permeability of the carrier bed (according to the Darcy equation). Adding petroleum to the rock–water system means that the scaling factor, relative permeability, becomes important. In two-phase fluid–rock systems, it is the relative permeability of the rock to water that controls the rate of movement of reactants. Increasing petroleum saturation has a logarithmic effect on the relative permeability to water. At high petroleum saturations (70–80%), water advection is effectively stopped. Filling a reservoir with petroleum thus effectively stops the advective import of reactants.

The relative rate of the *diffusive* movement of material is inversely proportional to the tortuosity and proportional to the fraction of the pore network filled with water. The rate of diffusive flux in the petroleum-leg relative to the water-leg has been shown to be (Worden *et al.*, 1998; Worden & Heasley, 2000):

$$\text{diffusive flux}_{\text{oil-leg}}/\text{diffusive flux}_{\text{water-leg}} = S_{\text{w}}^{n} \quad \text{(R15)}$$

where S_w is the fractional water saturation and n is the saturation exponent (a constant for a particular rock fabric, mineralogy and petroleum type). This exponent is typically close to 2 in water-wet reservoirs but increases as oil-wetting increases (reaching up to 25 in extreme cases). Thus the relative inhibition of diffusive flux in petroleum fields (relative to the aquifer) increases as the system changes from water-wet to oil-wet. Diffusion is also slowed by the reduction in volume of the aqueous medium through which diffusion occurs, leading to a much higher tortuosity of the diffusion path.

The effect on the rate of mineral dissolution of adding petroleum to the water-filled sandstone may be negligible if the system is water-wet. If water coats all mineral surfaces, then the exposed (effective) surface area and dissolution rate constant will be the same as in the water-leg. If mineral surfaces are partially coated with petroleum then their exposed surface area (visible to the reactive aqueous medium) will be reduced. In fully oil-wet systems, all mineral surfaces are isolated from the residual water and the rate of supply by dissolution will be zero.

In petroleum-bearing sandstones both precipitation and dissolution are controlled by the amount of mineral surface area exposed to water. Water-wet reservoirs thus will have precipitation rates unaffected by oil emplacement. Conversely, in oil-wet reservoirs, precipitation will be halted entirely because the reactive fluid medium (water) cannot gain access to mineral surfaces.

Wettability is thus a key to understanding the effect of petroleum emplacement on clay diagenetic processes in sandstones. Water-wet systems will have unaffected rates of dissolution and precipitation but transport will be slowed or stopped. Diffusive transport rates will be reduced by several orders of magnitude at maximum oil saturation and advective transport rates will become negligible. Where diagenetic clay reactions entail movement of material, they will be slowed or stopped but if they are isochemical, they may continue, albeit at a reduced rate (e.g. Worden & Barclay, this volume, pp. 453–469). Oil-wet systems will have totally impeded dissolution and precipitation and diffusion and advection of species in water will be reduced to negligible rates.

Wettability, clay mineralogy and diagenesis

Wettability is a function of the characteristics of rock, water and petroleum (Treiber *et al.*,

1972; Barclay & Worden, 1997, 2000; Worden & Heasley, 2000). Petroleum rich in polar compounds (those bearing N-, S- and O-functional groups) tend to be oil-wet. Polar compounds commonly have a negative dipole and can be attracted to positively charged mineral surfaces. A simple corollary of this is that gas fields (dominated by polar-free short-chain alkanes) are likely to be water-wet.

The pH of formation water can control indirectly the wettability of a reservoir because the exact surface charge of a mineral is sensitive to pH. For example, the surface charge of calcite is positive at very low pH, and negative at high pH. The critical pH known as the zero point charge (ZPC) occurs over the pH range of 8.0 to 9.5 (Somasundaran & Agar, 1967). Thus, under typical petroleum reservoir conditions (pH range 4 to 7; Hanor, 1994), the surface of calcite has a positive surface charge and is preferentially oil-wet owing to dipole attraction to negatively charged molecules in the petroleum. Conversely, quartz typically has a negative surface charge under these pH conditions because the change from negative to positive occurs at a pH of about 3.

Some minerals display anisotropic ZPC, with different crystal faces having different ZPC values. Kaolinite is an example of this, with a lower ZPC on the surface of (001) sheets than on the faces perpendicular to the layers. Thus booklets of kaolinite tend to have the (010) and (110) faces oil-wet whereas the (001) surfaces of the crystals tend to be water-wet (Robin *et al.*, 1995; Barclay & Worden, 2000).

Formation water and minerals can contain a variety of transition metals in the form of either aqueous complexes, organometallic complexes or metallo-porphyrins (Collins, 1975; Saunders & Swann, 1990; Filby, 1994; Hanor, 1994). Trace quantities of aqueous transition metals can alter the wettability state of a reservoir (Wang & Guidry, 1994). Positively charged transition metal ions from any source may become adsorbed on to negatively charged mineral surfaces (Anderson, 1986), thus acting as a bridge between the mineral surface and negatively charged polar compounds.

Iron-rich clay minerals (such as Fe-chlorite or smectite) are typically oil-wet (Cuiec, 1987). Wang & Guidry (1994) suggested that catalytic oxidation of crude petroleum components by highly charged metal cations may result in the generation of polar compounds that then alter wettability. Sandstone reservoirs tend to display increasing oil-wetting as their bulk iron oxide quantity increases, and as the proportion of Fe-rich clay increases. However, even a thin coating of Fe-rich clay minerals on grain surfaces, representing a small fraction of the overall rock, may result in a rock being oil-wet. This may be one of the reasons that chlorite-cemented sandstones can have high porosity and negligible quantities of quartz cement. The Fe-rich chlorite may attract the polar compounds in the petroleum, encouraging oil-wetting and preventing access of water to grain surfaces, thus stopping aqueous geochemical processes such as quartz cementation. A contrast may be illite-cemented sandstones because illite is water-wet (Barclay & Worden, 2000), and will not exclude water from grain surfaces, thus allowing aqueous diagenesis to potentially continue even in the presence of petroleum.

Effects of gas versus oil on clay diagenesis

Gas differs from oil because it is dominated by short-chain alkanes (such as methane) and non-hydrocarbons (such as CO_2). It tends to contain few, if any, polar or aromatic compounds. It thus lacks the family of organic compounds that give rise to oil-wetting behaviour. There are few corroborated reports of gas-wet reservoirs and therefore it is reasonable to assume that structures that have only ever been filled with gas will be water-wet, even if the sandstone has abundant Fe-clay minerals that could result in oil-wetting behaviour. However, note that it is possible for oil-wet behaviour to lead to an asphaltene or polar compound coating on clay or other mineral surfaces that remains even if the oil is flushed out by a later gas charge or by oil-to-gas cracking.

Gas has much lower viscosity than oil and finds it easier to flush out the primary pore water from low permeability clay-rich sandstones. Thus even in clay-cemented beds with low permeability in sandstones, a gas charge could flush out most of the initial formation water in situations that an oil charge would not be able to influence. Gas-filled sandstones thus tend to have lower irreducible water saturations (S_{wi}) than oil-filled sandstones.

These two factors thus lead to an apparent paradox that water-wet, gas-filled sandstones can have very low water saturations.

Oil-filled sandstones cemented with Fe-rich clays encourage oil-wet behaviour but retain high water saturations. In contrast, the same sandstone filled with gas would have low S_{wi} but have water-coated grain surfaces. If a former oil charge had been oil-wetting and had left an organic coating on grains and clay cements, the ultimate gas charge could lead to negligible water remaining within the reservoir.

The very low water saturations in gas reservoirs would hugely inhibit clay mineral reactions that require diffusion, and stop any that require advective transport of material. The small amount of residual water probably would stop isochemical clay transformations, because even they involve aqueous dissolution and reprecipitation.

SOME FUTURE DIRECTIONS IN SANDSTONE CLAY MINERAL DIAGENESIS RESEARCH

An understanding of clay mineral distribution is beginning to be possible in a sequence stratigraphic context (Morad *et al.*, 2000; Ketzer *et al.*, this volume, pp. 43–61). Rather than being a mystery, the early diagenetic distribution of clay minerals can be related, in theory, to relative sea-level changes. This is helpful to clay mineral diagenesis because it brings what may seem to be an abstract branch of research into the main stream of sedimentology. Many more case studies are needed to prove the conceptual connections between sequence stratigraphy and clay mineral distributions in sandstones.

The prediction of the evolution of clay minerals seems to be possible during burial if the thermal history, initial rock composition and eogenetic clay assemblage are known. Influx or loss of material may be locally important in some basins, although it may be best to assume isochemical reactions in the cases of many sandstones. The initial rock composition can be related to fundamental sedimentological parameters such as source terrain, degree of weathering and length of sediment transfer system. The eogenetic clay assemblage seems to be predictable in a sequence context. Temperature history is predictable given knowledge of basin type, heat flow history and all the other thermal variables. The one significant unknown is whether it is safe to assume isochemical reactions; this can be reduced to a question of the size of the system under consideration. Mass transfer between mud and sand is poorly understood. Mudstones and sandstones contain similar minerals in many basins and it is not easy to understand why mass flux should occur. However, modern concepts of geochemical self-organization are beginning to suggest that progressive changes in clastic rocks during diagenesis may lead to movement (e.g. Ortoleva, 1987). Several studies seem to show that aluminium must leave sandstones during mesodiagenesis (e.g. Milliken, this volume, pp. 343–360; Wilkinson *et al.*, this volume, pp. 147–160). A pair of important issues thus remain to be solved: is the loss of aluminium from sandstones real and, if it is, then how can aluminium, so sparingly soluble in aqueous solution, be involved in mass flux? For prediction of clay diagenesis, the issue of aluminium flux, and the movement of other elements such as potassium, must be resolved.

Kinetics are very important for the burial diagenetic evolution of clay minerals. It is probable that disequilibrium is the norm during the earlier stages of burial, although high-grade diagenesis may be characterized by equilibrium as reaction rates increase with increasing temperature. However, the rates of clay reactions during sandstone diagenesis are largely unknown. Experiments typically have been performed under non-realistic conditions (Manning, this volume, pp. 177–190) and often they have given contradictory and thus unhelpful answers. The rate at which reactions occur is difficult to predict at present, although it seems likely that many sandstone assemblages are metastable during mesodiagenesis. It is not known whether the rates of smectite illitization deduced for mudstones are applicable to sandstones (e.g. McKinley *et al.*, this volume, pp. 109–128).

Bacteria are now known to be influential in sedimentary depositional settings (e.g. Konhauser *et al.*, 1998; Konhauser & Urrutia, 1999) and probably influence the growth of clay rims on detrital sand grains. Bacterial processes are known to be important for non-silicates (sulphides and carbonates) during earliest burial (Coleman, 1985) although their importance for silicate (clay) eodiagenesis has only been hinted at thus far (Pryor, 1975). If the full importance of bacteria during eodiagenesis is not yet fully appreciated then their role during mesodiagenesis is totally unknown. Bacteria, and other organisms, have been shown recently to be active in many sedimentary basins at far higher temperatures than previously thought possible (e.g. Spark *et al.*, 2000). Questions now exist as to whether bacteria are important in aluminosilicate reactions during burial diagenesis. Establishing and understanding the roles of bacteria in the origins of clay minerals in sandstones will be a challenging but exciting new development in sandstone clay diagenesis research.

ACKNOWLEDGEMENTS

We are grateful to Susanne Schmid, Attila Juhasz and Anne Grau for their comments, which significantly improved the final manuscript.

REFERENCES

Aagaard, P., Egeberg, P.K., Saigal, G.C., Morad, S. & Bjørlykke, K. (1989) Diagenetic albitization of detrital K-feldspars in Jurassic, Lower Cretaceous and Tertiary reservoir rocks from offshore Norway, II. Formation water chemistry and kinetic considerations. *J. Sed. Petrol.*, **60**, 575–581.

Aagaard, P., Jahren, J., Harstad, A.O., Nilsen, O. & Ramm, M. (2000) Formation of grain-coating chlorite in sandstones; laboratory synthesized vs. natural occurrences. *Clay Mineral.*, **35**, 261–269.

Allen, P.A. (1997) *Earth Surface Processes*. Blackwell Science, Oxford, 404 pp.

Anderson, W.G. (1986) Wettability literature survey: Part 1: rock/oil/brine interactions and the effects of core handling on wettability. *J. Petrol. Technol.*, **38**, 1125–1149.

Anjos, S.M.C., De Ros, L.F. & Silva, C.M.A. (2003) Chlorite authigenesis and porosity preservation in the Upper Cretaceous marine sandstones of the Santos Basin, offshore eastern Brazil. In: *Clay Mineral Cements in Sandstones*, International Association of Sedimentologists Special Publication 34 (Ed. by R.H. Worden and S. Morad), Blackwell Publishing, Oxford, pp. 291–316.

Bailey, S.W. (1980) Structures of layer silicates. In: *Crystal Structures of Clay Minerals and their X-ray Identification* (Ed. by G.W. Brindley and G. Brown), Mineralogical Society of London, London, pp. 1–123.

Bailey, S.W. (Ed.) (1984) *Micas*. Reviews in Mineralogy, Vol. 13, Mineralogical Society of America, Washington, DC, 584 pp.

Bailey, S.W. (Ed.) (1988) *Hydrous Phyllosilicates (Exclusive of Micas)*. Reviews in Mineralogy, Vol. 19, Mineralogical Society of America, Washington, DC, 725 pp.

Bailey, S.W. & Brown, B.E. (1962) Chlorite polytypism: I. Regular and semi-random one layer structures. *Am. Mineral.*, **47**, 819–850.

Baker, J.C., Havord, P.J., Martin, K.R. & Ghori, K.A.R. (2000) Diagenesis and petrophysics of the early Permian Moogooloo Sandstone, Southern Carnarvon basin, Western Australia. *AAPG Bull.*, **84**, 250–265.

Barclay, S.A. & Worden, R.H. (1997) Reservoir wettability and its effect upon cementation in oil fields. In: *Extended Abstract Volume of the Geofluids II '97 Conference* (Ed. by J. Hendry, P. Carey, J. Parnell, A. Ruffell and R.H. Worden), 10–14 March, Belfast, pp. 264–267.

Barclay, S.A. & Worden, R.H. (2000) Effects of reservoir wettability on quartz cementation in oil fields. In: *Quartz Cementation in Sandstones*, International Association of Sedimentologists Special Publication 29 (Ed. by R.H. Worden and S. Morad), Blackwell Science, Oxford, pp. 103–117.

Beaufort, D., Cassagnabere, A., Petit, S., *et al.* (1998) Kaolinite-to-dickite reaction in sandstone reservoirs. *Clay Mineral.*, **33**, 297–316.

Berner, R.A. (1980) *Early Diagenesis: a Theoretical Approach*. Princeton University Press, Princeton, 241 pp.

Biron, A., Sotak, J. & Bebej, J. (1999) Diagenetic trioctahedral phyllosilicates from sediments of the Sambron Zone (Eastern Slovakia): XRD, SEM, and EMPA study. *Geol. Carpath.*, **50**, 257–272.

Bjørkum, P.A. & Gjelsvik, N. (1988) An isochemical model for formation of authigenic kaolinite, K-feldspar and illite in sandstones. *J. Sed. Petrol.*, **58**, 506–511.

Bjørkum, P.A., Mjos, R., Walderhaug, O. & Hurst, A. (1990) The role of the late Cimmerian unconformity for the distribution of kaolinite in the Gullfaks Field, northern North Sea. *Sedimentology*, **37**, 395–406.

Bjørlykke, K. (1980) Clastic diagenesis and basin evolution. *Rev. Inst. Inv. Geol. Diput. Barcelona*, **34**, 21–44.

Bjørlykke, K. & Aagaard, P. (1992) Clay minerals in the North Sea. In: *Origin, Diagenesis, and Petrophysics of Clay Minerals in Sandstones*, Society of Economic Paleontologists and Mineralogists Special Publication 47 (Ed. by D.W. Houseknecht and E.D. Pittman), Tulsa, OK, pp. 65–80.

Bjørlykke, K., Aagaard, P., Egeberg, P.K. & Simmons, S.P. (1995) Geochemical constraints from formation water analyses from North Sea and the Gulf Coast basins on quartz, feldspar and illite precipitation in reservoir rocks. In: *The Geochemistry of Reservoirs*, Geological Society Special Publication 86 (Ed. by J.M. Cubbitt and W.A. England), Geological Society Publishing House, Bath, pp. 33–50.

Blatt, H., Middleton, G. & Murray, R. (1980) *Origin of Sedimentary Rocks*. Prentice Hall, New Jersey, 782 pp.

Boles, J.R. & Franks, S.G. (1979) Clay diagenesis in Wilcox sandstones of southwest Texas: implications of smectite diagenesis on sandstone cementation. *J. Sed. Petrol.*, **49**, 55–70.

Brindley, G.W. & Brown, G. (1980) *Crystal Structures of Clay Minerals and their X-ray Identification*. Monograph 5, Mineralogical Society, London, 495 pp.

Buatier, M.D., Deneele, D., Dubois, M., Potdevin, J.L. & Lopez, M. (2000) Nacrite in the Lodeve Permian Basin: a TEM and fluid-inclusion study. *Eur. J. Mineral.*, **12**, 329–340.

Burley, S.D. (1993) Models of burial diagenesis for deep exploration plays in Jurassic fault traps of the Central and Northern North Sea. In: *Petroleum Geology of Northwest Europe: Proceedings of the 4th Conference* (Ed. by J.R. Parker), Geological Society, London, pp. 1353–1375.

Burley, S.D. & Macquaker, J.H.S. (1992) Authigenic clays, diagenetic sequences and conceptual diagenetic models in contrasting basin-margin and basin center North Sea Jurassic sandstones and mudstones. In: *Origin, Diagenesis and Petrophysics of Clay Minerals in Sandstones*, Society of Economic Paleontologists and Mineralogists Special Publication 47 (Ed. by D.W. Houseknecht and E.D. Pittman), Tulsa, OK, pp. 81–110.

Burns, L.K. & Ethridge, F.G. (1979) Petrology and diagenetic effects of lithic sandstones: Paleocene and Eocene Umpqua Formation, southwest Oregon. In: *Aspects of Diagenesis*, Society of Economic Paleontologists and Mineralogists Special Publication 26 (Ed. by P.A. Scholle and P.R. Schluger), Tulsa, OK, pp. 307–317.

Cassagnabere, A. (1998) *Caractérisation et interprétation de la transition kaolinite-dickite dans les réservoirs à hydrocarbures de Froy et Rind (Mer du Nord), Norvège*. PhD thesis, University of Poitiers, 238 pp.

Chan, M.A. (1985) Correlations of diagenesis with sedimentary facies in Eocene sandstones, western Oregon. *J. Sed. Petrol.*, **55**, 322–333.

Chang, H.K., Mackenzie, F.T. & Schoonmaker, J. (1986) Comparisons between the diagenesis of dioctahedral and trioctahedral smectite, Brazilian offshore basins. *Clay. Clay Mineral.*, **34**(4), 407–423.

Choquette, P.W. & Pray, L. (1970) Geologic nomenclature and classification of porosity in sedimentary carbonates. *AAPG Bull.*, **54**, 207–250.

Clauer, N., Rinckenbach, T., Weber, F., Sommer, F., Chaudhuri, S. & O'Neil, J.R. (1999) Diagenetic evolution of clay minerals in oil-bearing Neogene sandstones and associated shales, Mahakam Delta Basin, Kalimantan, Indonesia. *AAPG Bull.*, **83**, 62–87.

Coleman, M.L. (1985) Geochemistry of non-silicate minerals: kinetic considerations. *Phil. Trans. Roy. Soc. London*, **315**, 39–56.

Collins, A.G. (1975) *Geochemistry of Oilfield Waters*. Elsevier, Amsterdam, 496 pp.

Cuiec, L. (1987) Wettability and oil reservoirs. In: *North Sea Oil and Gas Reservoirs* (Ed. by J. Kleppe), Graham and Trotman, London, pp. 193–207.

Curtis, C.D., Hughes, C.R., Whiteman, J.A. & Whittle, C.K. (1985) Compositional variation within some sedimentary chlorites and some comments on their origin. *Mineral. Mag.*, **49**, 375–386.

Curtis, C.D., Coleman, M.L. & Love, L.G. (1986) Controls on the precipitation of early diagenetic calcite, dolomite and siderite concretions in complex depositional sequences. *Geochim. Cosmochim. Acta*, **50**, 2321–2334.

De Caritat, P., Hutcheon, I. & Walshe, J.L. (1993) Chlorite geothermometry: a review. *Clay. Clay Mineral.*, **41**, 219–239.

De Hon, R.A., Clawe, L.N., Washington, P.A. & Young, L.M. (1999) Formation of northern Louisiana ironstones. *Geol. Soc. Am. Abstr. Programs*, **31**, 159–160.

De Ros, L.F., Anjos, S.M.C. & Morad, S. (1994) Authigenesis of amphibole and its relationship to the diagenetic evolution of Lower Cretaceous sandstones of the Potiguar rift basin, northeastern Brazil. *Sed. Geol.*, **88**, 253–266.

Deer, W.A., Howie, R.A. & Zussman, J. (1998) *An Introduction the Rock-forming Minerals*. Longman, London, 528 pp.

Dixon, S.A., Summers, D.M. & Surdam, R.C. (1989) Diagenesis and preservation of porosity in Norphlet Formation (Upper Jurassic), southern Alabama. *AAPG Bull.*, **73**, 707–728.

Dunn, T.L. (1992) Infiltrated material in Cretaceous volcanogenic sandstones, San Jorge Basin, Argentina. In:

Origin, Diagenesis, and Petrophysics of Clay Minerals in Sandstones, Special Publication of the Society of Economic Palaeontologists and Mineralogists 47 (Ed. by D.W. Houseknecht and E.D. Pittman), Tulsa, OK, pp. 159–174.

Dunoyer de Segonzac, G. (1970) The transformation of clay minerals during diagenesis and low grade metamorphism: a review. *Sedimentology*, **15**, 281–346.

Dutton, S.P. & Land, L.S. (1985) Meteoric burial diagenesis of Pennsylvanian arkosic sandstones, southwestern Anadarko Basin, Texas. *AAPG Bull.*, **69**, 22–38.

Edzwald, J.K., Upchurch, T.B. & O'Mella, C.R. (1974) Coagulation in estuaries. *Environ. Sci. Technol.*, **8**, 58–63.

Ehrenberg, S.N. (1993) Preservation of anomalously high porosity in deeply buried sandstones by grain-coating chlorite: examples from the Norwegian continental shelf. *AAPG Bull.*, **77**, 1260–1286.

Ehrenberg, S.N. & Nadeau, P.H. (1989) Formation of diagenetic illite in sandstones of the Garn Formation, Haltenbanken area, mid-Norwegian continental shelf. *Clay Mineral.*, **24**, 233–253.

Ehrenberg, S.N., Aagaard, P., Wilson, M.J., Fraser, A.R. & Duthie, D.M.L. (1993) Depth-dependent transformation of kaolinite to dickite in sandstones of the Norwegian continental shelf. *Clay Mineral.*, **28**, 325–352.

Ehrenberg, S.N., Dalland, A., Nadeau, P.H., Mearns, E.W. & Amundsen, H.E.F. (1998) Origin of chlorite enrichment and neodymium isotopic anomalies in Haltenbanken sandstones. *Mar. Petrol. Geol.*, **15**, 403–425.

Eisma, D., Kalf, J. & van der Gaast, S.J. (1978) Suspended matter in the Zaire estuary and adjacent Atlantic Ocean. *Neth. J. Sea Res.*, **12**, 382–406.

Emery, D., R.J. Myers, R.J. & Young, R. (1990) Ancient subaerial exposure and freshwater leaching in sandstones. *Geology*, **18**, 1178–1181.

Eugster, H.P. & Hardie, L.A. (1978) Saline lakes. In: *Lakes, Chemistry, Geology, Physics* (Ed. by A. Lehman), Springer-Verlag, Berlin, pp. 237–293.

Everlien, G. & Hoffmann, U. (1991) Nitrogen in natural gas. *Erdol und Kohle – Erdgas – Petrochemie vereinigt mit Brennstoff-Chemie*, **Bd. 44**, 4th April edition.

Fein, J. (1991) Experimental stability of aluminium-, calcium-, and magnesium-acetate complexeing at 80°C. *Geochim. Cosmochim. Acta*, **55**, 955–964.

Fein, J. & Hestrin, J.E. (1994) Experimental studies of oxalate complexation at 80°C: gibbsite, amorphous silica and quartz solubilities in oxalate-bearing fluids. *Geochim. Cosmochim. Acta*, **58**, 4817–4829.

Filby, R.H. (1994) Origin and nature of trace element species in crude oils, bitumens and kerogens: implications for correlation and other geochemical studies. In: *Geofluids: Origin, Migration and Evolution of Fluids in Sedimentary Basins*, Geological Society Special Publication 78 (Ed. by J. Parnell), Geological Society Publishing House, Bath, pp. 203–219.

Fisher, Q.J., Knipe, R. & Worden, R.H. (2000) Microstructures of deformed and non-deformed sandstones from the North Sea: implications for the origins of quartz cement in sandstones. In: *Quartz Cementation in Sandstones*, Special Publication of the International Association of Sedimentologists 29 (Ed. by R.H. Worden and S. Morad), Blackwell Science, Oxford, pp. 129–146.

Frey, M. & Robinson, D. (Eds) (1999) *Low-grade Metamorphism*. Blackwell Science, Oxford, 313 pp.

Furlan, S., Clauer, N., Chaudhuri, S. & Sommer, F. (1996) K transfer during burial diagenesis in the Mahakam Delta Basin (Kalimantan, Indonesia). *Clay. Clay Mineral.*, **44**, 157–169.

Galloway, W.E. (1979) Diagenetic control of reservoir quality in arc-derived sandstones: implications for petroleum exploration. In: *Aspects of Diagenesis*, Society of Economic Paleontologists and Mineralogists Special Publication 26 (Ed. by P.A. Scholle and P.R. Schluger), Tulsa, OK, pp. 251–262.

Gaupp, R., Matter, A., Platt, J., Ramseyer, K. & Walzebuck, J. (1993) Diagenesis and fluid evolution of deeply buried Permian (Rotliegende) gas reservoirs, northwest Germany. *AAPG Bull.*, **77**, 1111–1128.

Gibbs, R.J. (1977) Amazon River sediment transport in the Atlantic Ocean. *Geology*, **4**, 45–58.

Goodchild, M.W. & Whitaker, J.H.M. (1986) A petrographic study of the Rotliegendes sandstone reservoir (lower Permian) in the Rough gas field. *Clay Mineral.*, **21**, 459–477.

Grigsby, J.D. (1999) Origin of chlorite grain coats in sandstones of the lower Vicksburg Formation, South Texas. In: *American Association of Petroleum Geologists Annual Meeting, Extended Abstract*, Tulsa, OK, p. A51.

Hanor, J.S. (1994) Origin of saline fluids in sedimentary basins. In: *Geofluids: Origin, Migration and Evolution of Fluids in Sedimentary Basins*, Geological Society Special Publication 78 (Ed. by J. Parnell), Geological Society Publishing House, Bath, pp. 151–174.

Hassouta, L., Buatier, M., Potdevin, J.L. & Liewig, N. (1999) Clay diagenesis in the sandstone reservoir of the Ellon Field (Alwyn, North Sea). *Clay. Clay Mineral.*, **47**, 269–285.

Haszeldine, R.S., Macaulay, C.I., Marchand, A., *et al.* (2000) Sandstone cementation and fluids in hydrocarbon basins. *J. Geochem. Explor.*, **69–70**, 195–200.

Hillier, S. (1994) Pore-lining chlorites in siliciclastic reservoir sandstones: electron microprobe SEM and XRD data, and implications for their origin. *Clay Mineral.*, **29**, 665–679.

Hillier, S. & Velde, B. (1992) Chlorite interstratified with a 7 Angstrom mineral: an example from offshore Norway and possible implications for the interpretation of the composition of diagenetic chlorites. *Clay Mineral.*, **27**, 475–486.

Hornibrook, E.R.C. & Longstaffe, F.J. (1996) Berthierine

from the Lower Cretaceous Clearwater Formation, Alberta, Canada. *Clay. Clay Mineral.*, **44**, 1–21.

Hover, V.C., Walter, L.M., Peacor, D.R. & Martini, A.M. (1999) Mg-smectite authigenesis in a marine evaporative environment, Salina Ometepec, Baja California. *Clay. Clay Mineral.*, **47**, 252–268.

Howard, J.J. (1992) Influence of authigenic-clay minerals on permeability. In: *Origin, Diagenesis, and Petrophysics of Clay Minerals in Sandstones*, Special Publication of the Society of Economic Paleontologists and Mineralogists 47 (Ed. by D.W. Houseknecht and E.D. Pittman), Tulsa, OK, pp. 257–264.

Hower, J., Eslinger, E.V., Hower, M.E. & Perry, E.A. (1976) Mechanism of burial and metamorphism of argillaceous sediment: 1. Mineralogical and chemical evidence. *Geol. Soc. Am. Bull.*, **87**, 725–737.

Huang, W.-L., Longo, J.M. & Pevear, D.R. (1993) An experimentally derived kinetic model for smectite-to-illite conversion and its use as a geothermometer. *Clay. Clay Mineral.*, **41**, 162–177.

Humphreys, B., Kemp, S.J., Lott, G.K., Bermanto, F., Dharmayanti, D.A. & Samori, I. (1994) Origin of grain-coating chlorite by smectite transformation: an example from Miocene sandstones, North Sumatra back-arc basin, Indonesia. *Clay Mineral.*, **29**, 681–692.

Hurst, A. & Nadeau, P.H. (1995) Clay microporosity in reservoir sandstones: an application of quantitiative electron microscopy in petrophysical evaluation. *AAPG Bull.*, **79**, 563–573.

Iijima, A. & Matsumato, S.R. (1982) Berthierine and chamosite in coal measures of Japan. *Clay. Clay Mineral.*, **30**, 264–274.

Jahren, J.S. & Aagaard, P. (1989) Compositional variations in diagenetic chlorites and illites, and relationships with formation water chemistry. *Clay Mineral.*, **24**, 209–213.

Janks, J.S., Yusas, M.R. & Hall, C.M. (1992) Clay mineralogy of an interbedded sandstone, dolomite and anhydrite: the Permian Yates Formation, Winkler County, Texas. In: *Origin, Diagenesis, and Petrophysics of Clay Minerals in Sandstones*, Society of Economic Palaeontologists and Mineralogists Special Publication 47 (Ed. by D.W. Houseknecht and E.D. Pittman), Tulsa, OK, pp. 145–157.

Jeans, C.V., Wray, D.S., Merriman, R.J. & Fisher, M.J. (2000) Volcanogenic clays in Jurassic and Cretaceous strata of England and the North Sea basin. In: *Mineral Diagenesis and Reservoir Quality, the Way Forward* (Ed. by D.C. Bain, P.L. Hall, H.F. Shaw and D.A. Spears). *Clay Mineral.*, **35**, 25–55.

Kantorowicz, J.D. (1990) The influence of variations in illite morphology on the permeability of Middle Jurassic Brent Group sandstones, Cormorant Field, UK North Sea. *Mar. Petrol. Geol.*, **7**, 66–74.

Karakas, Z. & Kadir, S. (1998) Mineralogical and genetic relationships between carbonate and sepiolite–palygorskite formations in the neogene lacustrine Konya Basin, Turkey. *Carbonates Evaporites*, **13**, 198–206.

Ketzer, J.M., Morad, S. & Amorosi, A. (2003) Predictive diagenetic clay-mineral distribution in siliciclastic rocks within a sequence stratigraphic framework. In: *Clay Mineral Cements in Sandstones*, International Association of Sedimentologists Special Publication 34 (Ed. by R.H. Worden and S. Morad), Blackwell Publishing, Oxford, pp. 43–61.

Ketzer, J.M., Morad, S., Nystuen, J.P. & de Ros, L.F. (2003) The role of the Cimmerian Unconformity (Early Cretaceous) in the kaolinitization and related reservoir-quality evolution in Triassic sandstones of the Snorre Field, North Sea. In: *Clay Mineral Cements in Sandstones*, International Association of Sedimentologists Special Publication 34 (Ed. by R.H. Worden and S. Morad), Blackwell Publishing, Oxford, pp. 361–382.

Kohler, B., Singer, A. & Stoffers, P. (1994) Biogenic nontronite from marine white smoker chimneys. *Clay. Clay Mineral.*, **42**, 689–701.

Konhauser, K.O. & Urrutia, M.M. (1999) Bacterial clay authigenesis: a common biogeochemical process. *Chemical Geology*, **161**, 399–413.

Konhauser, K.O., Fisher, Q.J., Fyfe, W.S., Longstaffe, F.J. & Powell, M.A. (1998) Authigenic mineralization and detrital clay binding by freshwater biofilms: the Brahmani River, India. *Geomicrobiology*, **15**, 209–222.

Krinsley, D. (1998) Models of rock varnish formation constrained by high resolution transmission electron microscopy. *Sedimentology*, **45**, 711–725.

Krone, R.B. (1978) Aggregation of suspended particles in estuaries. In: *Estuarine Transport Processes* (Ed. by B. Kjerfve), University of South Carolina Press, Columbia, SC, pp. 177–190.

Kronen, J.D. & Glenn, C.R. (2000) Pristine to reworked verdine; keys to sequence stratigraphy in mixed carbonate–siliciclastic forereef sediments (Great Barrier Reef). In: *Marine Authigenesis; from Global to Microbial*, Society for Sedimentary Geology Special Publication 66 (Ed. by C.R. Glenn, L.L. Prevot and J. Lucas), Tulsa, OK, pp. 387–403.

Land, L.S. & Dutton, S.P. (1978) Cementation of a Pennsylvanian deltaic sandstone: Isotopic data. *J. Sed. Petrol.*, **48**, 1167–1176.

Land, L.S., Mack, L.E., Milliken, K.L. & Lynch, F.L. (1997) Burial diagenesis of argillaceous sediments, south Texas Gulf of Mexico sedimentary basin: a reexamination. *Geol. Soc. Am. Bull.*, **109**, 2–15.

Longstaffe, F.J. (1986) Oxygen isotope studies of diagenesis in the Basal Belly River Sandstone, Pembina I-Pool, Alberta. *J. Sed. Petrol.*, **56**, 77–88.

Luo, J., Morad, S., Ketzer, J.M., Yan, S., Zhang, X. & Xue, J. (in press) The diagenetic and related reservoir-quality evolution of fluvial and lacustrine-deltaic sandstones: evidence from Jurassic and Triassic sandstones of the Ordos basin, northwestern China. *Sedimentology.*

Manning, D.A.C. (2003) Experimental studies of clay mineral occurrences. In: *Clay Mineral Cements in Sandstones*, International Association of Sedimentologists Special Publication 34 (Ed. by R.H. Worden and S. Morad), Blackwell Publishing, Oxford, pp. 177–190.

Mathisen, M.E. & McPherson, J.G. (1991) Volcaniclastic deposits: implications for hydrocarbon exploration. In: *Sedimentation in Volcanic Settings*, Society of Economic Paleontologists and Mineralogists Special Publication 45 (Ed. by R.V. Fisher and G.A. Smith), Tulsa, OK, pp. 27–36.

Mayayo, M.J., Torres-Ruiz, J., Gonzalez-Lopez, J.M., *et al.* (1998) Mineralogical and chemical characterization of the sepiolite Mg-smectite deposit at Mara (Calatayud Basin, Spain). *Eur. J. Mineral.*, **10**, 367–383.

McAulay, G.E., Burley, S.D. & Johnes, L.H. (1993) Silicate mineral authogenesis in the Hutton and NW Hutton fields: implications for subsurface porosity development. In: *Petroleum Geology of Northwest Europe: Proceedings of the 4th Conference* (Ed. by J.R. Parker), Geological Society, London, pp. 1377–1394.

McKinley, J.M., Worden, R.H. & Ruffell, A.H. (2003) Smectite in sandstones: a review of the controls on occurrence and behaviour during diagenesis. In: *Clay Mineral Cements in Sandstones*, International Association of Sedimentologists Special Publication 34 (Ed. by R.H. Worden and S. Morad), Blackwell Publishing, Oxford, pp. 109–128.

Meade, R.H. (1972) Transport and deposition of sediment in estuaries. *Geol. Soc. Am. Mem.*, **133**, 91–120.

Milliken, K.L. (2003) Microscale distribution of kaolinite in Breathitt Formation sandstones (middle Pennsylvanian): implications for mass balance. In: *Clay Mineral Cements in Sandstones*, International Association of Sedimentologists Special Publication 34 (Ed. by R.H. Worden and S. Morad), Blackwell Publishing, Oxford, pp. 343–360.

Moncure, G.K., Lahann, R.W. & Siebert, R.M. (1984) Origin of secondary porosity and cement distribution in a sandstone/shale sequence from the Frio Formation (Oligocene). In: *Clastic Diagenesis*, American Association of Petroleum Geologists Memoir 37 (Ed. by D.A. McDonald and R.C. Surdam), Tulsa, OK, pp. 151–161.

Morad, S. (1986) Albitization of K-feldspar grains in Proterozoic arkoses and greywackes from southern Sweden. *Neues Jb. Mineral. Monat.*, **4**, 145–156.

Morad, S. & Al-Aasm, I.S. (1994) Origin and diagenetic evolution of phosphate nodules in Upper Proterozoic mudstones of the Visingsö Group (southern Sweden): evidence from petrology, mineral chemistry and isotopes. *Sed. Geol.*, **88**, 267–282.

Morad, S., Bergan, M., Knarud, R. & Nystuen, J.P. (1990) Albitization of detrital plagioclase in Triassic reservoir sandstones from the Snorre Field, Norwegian North Sea. *J. Sed. Petrol.*, **60**, 411–425.

Morad, S., Ben Ismail, H., Al-Aasm, I.S. & De Ros, L.F. (1994) Diagenesis and formation-water chemistry of Triassic reservoir sandstones from southern Tunisia. *Sedimentology*, **41**, 1253–1272.

Morad, S., Ketzer, J.M. & De Ros, L.F. (2000) Spatial and temporal distribution of diagenetic alterations in siliciclastic rocks: implications for mass transfer in sedimentary basins. *Sedimentology*, **47** (Millenium Reviews), 95–120.

Moraes, M.A.S. & De Ros, L.F. (1990) Infiltrated clays in fluvial Jurassic sandstones of Recôncavo Basin, northeastern Brazil. *J. Sed. Petrol.*, **60**, 809–819.

Muffler, L.J.P. & White, D.E. (1969) Active metamorphism of Upper Cenozoic sediments in the salton Sea geothermal field and the Salton Trough, Southeastern California. *Geol. Soc. Am. Bull.*, **94**, 161–180.

Niu, B., Yushimura, T. & Hirai, A. (2000) Smectite diagenesis in neogene marine sandstone and mudstone of the Niigata basin, Japan. *Clay. Clay Mineral.*, **48**, 26–42.

Odin, G.S. (1988) The verdine facies from the lagoon off New Caledonia. In: *Green Marine Clays*, Developments in Sedimentology 45 (Ed. by G.S. Odin), Elsevier, Amsterdam, pp. 57–82.

Odin, G.S. (1990) Clay mineral formation at the continental–ocean boundary: the verdine facies. *Clay Mineral.*, **25**, 477–483.

Odin, G.S. & Matter, A. (1981) De glauconiarum origine. *Sedimentology*, **28**, 611–641.

Ortoleva, P. (1987) Geochemical self-organisation: I. Reaction transport feedbacks and modeling approach. *Am. J. Sci.*, **296**, 420–452.

Pallat, N., Wilson, J. & McHardy, B. (1984) The relationship between permeability and the morphology of diagenetic illite in reservoir rocks. *J. Petrol. Technol.*, **36**, 2225–2227.

Pirrie, D., Ditchfield, P.W. & Marshall, J.D. (1994) Burial diagenesis and pore-fluid evolution in a Mesozoic back-arc basin: the Marambio Group, Vega Island, Antarctica. *J. Sed. Res.*, **64**, 541–552.

Pittman, E.D. & Lewan, M.D. (Eds) (1995) *Organic Acids in Geological Processes*. Springer-Verlag, New York.

Pittman, E.D., Larese, R.E. & Heald, M.T. (1992) Clay coats: occurrence and relevance to preservation of porosity in sandstones. In: *Origin, Diagenesis, and Petrophysics of Clay Minerals in Sandstones*, Special Publication of the Society of Economic Paleontologists and Mineralogists 47 (Ed. by D.W. Houseknecht and E.D. Pittman), Tulsa, OK, pp. 241–255.

Platt, J.D. (1993) Controls on clay mineral distribution and chemistry in the early Permian Rotliegend of Germany. *Clay Mineral.*, **28**, 393–416.

Posamentier, H.W. & Vail, P.R. (1988) Eustatic controls on clastic deposition II. Sequence and systems tract models. In: *Sea-level Changes: an Integrated Approach*, Society of Economic Paleontologists and Mineralogists Special Publication 42 (Ed. by C.K. Wilgus, B.S. Hastings, C.G.St.C. Kendall, H.W.

Posamentier, C.A. Ross and J.C. Van Wagoner), Tulsa, OK, pp. 125–154.

Pozo, M. & Casas, J. (1999) Origin of kerolite and associated Mg clays in palustrine-lacustrine environments. The Esquivias deposit (Neogene Madrid Basin, Spain). *Clay Mineral.*, **34**, 395–418.

Pryor, W.A. (1975) Biogenic sedimentation and alteration of argillaceous sediments in shallow marine environments. *Geol. Soc. Am. Bull.*, **86**, 1244–1254.

Purnachandra, R.V., Lamboy, M. & Dupueble, P.A. (1995) Verdine and other associated authigenic (glaucony, phosphate) facies from the surficial sediments of the southwestern continental margin of India. *Mar. Geol.*, **111**, 133–158.

Purvis, K. (1992) Lower Permian Rotliegend sandstones, southern North Sea: a case study of sandstone diagenesis in evaporite-associated sequences. *Sed. Geol.*, **77**, 155–171.

Pye, K., Dickson, J.A.D., Schiavon, N., Coleman, M.L. & Cox, M. (1990) Formation of siderite–Mg–calcite–iron concretions in intertidal marsh and sandflat sediments, north Norfolk, England. *Sedimentology*, **37**, 325–343.

Ramm, M. & Ryseth, A.E. (1996) Reservoir quality and burial diagenesis in the Statfjord Formation. *Petrol. Geosci.*, **2**, 313–324.

Remy, R.R. (1994) Porosity reduction and major controls on diagenesis of Cretaceous-Paleocene volcaniclastic and arkosic sandstone, Middle Park Basin, Colorado. *J. Sed. Petrol.*, **64**, 797–806.

Reynolds, R.C. (1980) Interstratified clay minerals. In: *Crystal Structures of Clay Minerals and their X-ray Identification* (Ed. by G.W. Brindley and G. Brown), Mineralogical Society, London, pp. 249–303.

Robin, M., Rosenburg, E. & Fassi-Fihri, O. (1995) Wettability studies at the pore level: a new approach by use of the cryo-SEM. *Soc. Petrol. Eng. Formation Eval.*, **10**, 75–89.

Rossel, N.C. (1982) Clay mineral diagenesis in Rotliegend aeolian sandstones of the southern North Sea. *Clay Mineral.*, **17**, 69–77.

Ruiz-Cruz, M.D. (1996) Dickite, nacrite and possible dickite/nacrite mixed-layers from the Betic Cordilleras (Spain). *Clay. Clay Mineral.*, **44**, 357–369.

Ryan, P.C. & Reynolds, R.C. Jr. (1996) The origin and diagenesis of grain-coating serpentine-chlorite in Tuscaloosa Formation sandstone, U.S. Gulf Coast. *Am. Mineral.*, **81**, 213–225.

Ryan, P.C. & Reynolds, R.C. Jr. (1997) The chemical composition of serpentine/chlorite in the Tuscaloosa Formation, U.S. Gulf Coast: EDX vs. XRD determinations, implications for mineralogic reactions, and the origin of anatase. *Clay. Clay Mineral.*, **45**, 339–352.

Ryan, P.C., Conrad, M.E., Brown, K., Chamberlain, C.P. & Reynolds, R.C. Jr. (1998) Oxygen isotope compositions of mixed layer serpentine–chlorite and illite–smectite in the Tuscaloosa Formation (U.S. Gulf Coast): implications for pore fluids and mineralogic reactions. *Clay. Clay Mineral.*, **46**, 357–368.

Saunders, J.A. & Swann, C.T. (1990) Trace-metal content of Mississippi oil field brines. *J. Geochem. Explor.*, **37**, 171–183.

Searl, A. (1994) Diagenetic destruction of reservoir potential in shallow marine sandstones of the Broadford Beds (Lower Jurassic), north-west Scotland: depositional versus burial and thermal history controls on porosity destruction. *Mar. Petrol. Geol.*, **11**, 131–147.

Seeman, U. & Scherer, M. (1984) Volcaniclastics as potential hydrocarbon reservoirs. *Clay Mineral.*, **9**, 457–470.

Shanmugam, G. (1990) Porosity prediction in sandstones using erosional unconformities. In: *Prediction of Reservoir Quality through Chemical Modelling* (Ed. by I.D. Meshri and P.J. Ortoleva). *AAPG Mem.*, **49**, 1–23.

Somasundaran, P. & Agar, G.E. (1967) The zero point of charge of calcite. *J. Colloid Interf. Sci.*, **24**, 433–440.

Spark, I., Patey, I., Duncan, B., Hamilton, A., Devine, C. & McGovern-Traa, C. (2000) The effects of indigenous and introduced microbes on deeply buried hydrocarbon reservoirs, North Sea. *Clay Mineral.*, **35**, 5–12.

Srodón, J. (1999) Use of clay minerals in constructing geological processes: recent advances and some perspectives. *Clay Mineral.*, **34**, 27–37.

Stalder, P.J. (1973) Influence of crystalline habit and aggregate structure of authigenic clay minerals on sandstone permeability. *Geol. Mijnbouw*, **52**, 217–220.

Stein, C.L., Krumhansl, J.L. & Kimball, K.M. (1990) Silicate diagenesis in the Salado Formation, southeastern New Mexico. *J. Sed. Res.*, **61**, 174–187.

Stewart, R.N.T., Haszeldine, R.S., Fallick, A.E., Wilkinson, M. & Macaulay, C.I. (2000) Regional distribution of diagenetic carbonate cement in Palaeocene deepwater sandstones: North Sea. *Clay Mineral.*, **35**, 119–133.

Stonecipher, S.A. (2000) *Applied Sandstone Diagenesis—Practical Petrographic Solutions for a Variety of Common Exploration, Development and Production Problems.* Society of Economic Mineralogists and Paleontologists Short Course Notes 50, Tulsa, OK.

Surdam, R.C. & Boles, J.R. (1979) Diagenesis of volcanic sandstones. In: *Aspects of Diagenesis*, Society of Economic Paleontologists and Mineralogists Special Publication 26 (Ed. by P.A. Scholle and P.R. Schluger), Tulsa, OK, pp. 227–242.

Surdam, R.C. & Macgowan, D.B. (1987) Oilfield waters and sandstone diagenesis. *Appl. Geochem.*, **2**, 613–619.

Surdam, R.C., Dunn, T.L., Heasler, H.P. & MacGowan, D.B. (1989) Porosity evolution in sandstone/shale systems. In: *Short Course on Burial Diagenesis* (Ed. by I.E. Hutcheon), Mineralogical Association of Canada, Montreal, pp. 61–133.

Taylor, K.G. (1990) Berthierine from the nonmarine Wealden (Early Cretaceous) sediments of south-east England. *Clay Mineral.*, **25**, 391–399.

Thamban, M. & Purnachandra, R.V. (2000) Distribution and composition of verdine and glaucony facies from the sediments of the western continental margin of India. In: *Marine Authigenesis; from Global to Microbial*, Society for Sedimentary Geology Special Publication 66 (Ed. by C.R. Glenn, L.L. Prevot and J. Lucas), Tulsa, OK, pp. 233–244.

Thomson, A. (1979) Preservation of porosity in the deep Woodbine/Tuscaloosa trend, Louisiana. *Trans. Gulf Coast Assoc. Geol. Soc.*, **29**, 396–403.

Treiber, L.E., Archer, D.L. & Owens, W.W. (1972) Laboratory evaluation of the wettability of 50 oil producing reservoirs. *Soc. Petrol. Eng. J.*, **12**, 531–540.

Van Olphen, H. (1977) *An Introduction to Clay Colloid Chemistry*. Wiley, New York, 318 pp.

Van Wagoner, J.C., Posamentier, H.W., Mitchum, R.M., Vail, P.R., Sarg, J.F., Loutit, T.S. & Hardenbol, J. (1988) An overview of the fundamentals of sequence stratigraphy and key definitions. In: *Sea-level Changes: an Integrated Approach*, Society of Economic Paleontologists and Mineralogists Special Publication 42 (Ed. by C.K. Wilgus, B.S. Hastings, C.G.St.C. Kendall, H.W. Posamentier, C.A. Ross and J.C. Van Wagoner), Tulsa, OK, pp. 39–46.

Vaz, G.G. (2000) Verdine and glaucony facies from continental margin off Chennai, bay of Bengal. *J. Geol. Soc. India*, **55**, 297–306.

Velde, B. (1985) *Clay Minerals. A Physico-chemical Explanation of Their Occurrence*. Developments in Sedimentology 40, Elsevier, Amsterdam, 427 pp.

Walker, J.R. (1993) Chlorite polytype geothermometry. *Clay Clay Mineral.*, **41**, 260–267.

Walker, T.R., Waugh, B. & Crone, A.J. (1978) Diagenesis in first-cycle desert alluvium of Cenozoic age, southwestern United States and northwestern Mexico. *Geol. Soc. Am. Bull.*, **89**, 19–32.

Wang, F.H.L. & Guidry, L.J. (1994) Effect of oxidation-reduction condition on wettability alteration. *Soc. Petrol. Eng. Formation Eval.*, **9**, 140–148.

Wentworth, C.K. (1922) A scale of grade class terms for clastic sediments. *J. Geol.*, **30**, 377–392.

Whetten J.T. & Hawkins, J.W. (1970) Diagenetic origin of graywacke matrix minerals. *Sedimentology*, **15**, 347–361.

Wilkinson, M., Darby, D., Haszeldine, R.S. & Couples, G.D. (1997) Secondary porosity generation during deep burial associated with overpressure leak-off: Fulmar Formation, United Kingdom Central Graben. *AAPG Bull.*, **81**, 803–813.

Wilkinson, M., Haszeldine, R.S. & Milliken, K.L. (2003) Cross-formational flux of aluminium and potassium in Gulf Coast (USA) sediments. In: *Clay Mineral Cements in Sandstones*, International Association of Sedimentologists Special Publication 34 (Ed. by R.H. Worden and S. Morad), Blackwell Publishing, Oxford, pp. 147–160.

Williams, L.B., Wicoxin, B.R., Ferrell, R.E. & Sassen, R. (1992) Diagenesis of ammonium during hydrocabon maturation and migration, Wilcox Group, Louisiana, USA. *Appl. Geochem.*, **7**, 123–134.

Wilson, M.J. (1992) Inherited grain-rimming clays in sandstones from eolian and shelf environments: their origin and control on reservoir properties. In: *Origin, Diagenesis, and Petrophysics of Clay Minerals in Sandstones*, Special Publication of the Society of Economic Palaeontologists and Mineralogists 47 (Ed. by D.W. Houseknecht and E.D. Pittman), Tulsa, OK, pp. 209–225.

Wilson, M.J. (1999) The origin and formation of clay minerals in soils: past, present and future perspectives. *Clay Mineral.*, **34**, 7–25.

Worden, R.H. & Barclay, S.A. (2000) Internally-sourced quartz cement due to externally-derived CO_2 in subarkosic sandstones, Northern North Sea, UKCS. *J. Geochem. Explor.*, **69–70**, 645–649.

Worden, R.H. & Barclay, S.A. (2003) The effect of oil emplacement on diagenetic clay mineralogy: the Upper Jurassic Magnus Sandstone Member, North Sea. In: *Clay Mineral Cements in Sandstones*, International Association of Sedimentologists Special Publication 34 (Ed. by R.H. Worden and S. Morad), Blackwell Publishing, Oxford, pp. 453–469.

Worden, R.H. & Heasley, E.C. (2000) Effects of petroleum emplacement on cementation in carbonate reservoirs. *Bull. Soc. Geol. Fr.*, **171**, 607–620.

Worden, R.H., Mayall, M.J. & Evans, I.J. (1997) Predicting reservoir quality during exploration: lithic grains, porosity and permeability in Tertiary clastics of the South China Sea basin. In: *Petroleum Geology of S E Asia*, Geological Society Special Publication 126 (Ed. by A.J. Fraser, A.J. Matthews and R.W. Murphy), Geological Society Publishing House, Bath, pp. 107–115.

Worden, R.H., Oxtoby, N.H. & Smalley, P.C. (1998) Can oil emplacement prevent quartz cementation in sandstones? *Petrol. Geosci.*, **4**, 129–138.

Worden, R.H., Mayall, M. & Evans, I.J. (2000) The effect of ductile compaction and quartz cementation on porosity and permeability in Tertiary clastics, South China Sea: prediction of reservoir quality. *AAPG Bull.*, **84**, 345–359.

Xia, Z. (1985) Petrological and mineralogical characteristics of Carboniferous–Permian kaolinitic claystone in northern China. *Acta Petrol. Sin.*, **4**, 70–78.

Yardley, B.W.D. (1989) *An Introduction to Metamorphic Petrology*. Longman, New York, 248 pp.

Int. Assoc. Sedimentol. Spec. Publ. (2003) **34**, 43–61

Predictive diagenetic clay-mineral distribution in siliciclastic rocks within a sequence stratigraphic framework

J.M. KETZER[1], S. MORAD[1] *and* A. AMOROSI[2]

[1]*Department of Earth Sciences, Uppsala University, Villa vägen 16, S-752 36, Uppsala, Sweden, e-mail: marcelo.ketzer@geo.uu.se, sadoon.morad@geo.uu.se*

[2]*Dipartimento di Scienze della Terra e Geologico-Ambientali, Università di Bologna, Via Zamboni 67, 40127, Bologna, Italy, e-mail: amorosi@geomin.unibo.it*

ABSTRACT

The distribution pattern of early diagenetic clay minerals such as kaolinite, smectite, palygorskite, glaucony and berthierine, as well as of mechanically infiltrated clays and mud intraclasts, is here predicted in siliciclastic rocks using a sequence stratigraphic approach. Changes in relative sea-level and in sediment supply/sedimentation rate, together with the climatic conditions prevalent during, and immediately after deposition of sediments control the type, abundance and spatial distribution of clay minerals by influencing the pore-water chemistry and the duration over which the sediments are submitted to a certain set of geochemical conditions. Diagenetic clay-mineral distribution is constrained along sequence and parasequence boundaries, and within parasequences, parasequence sets and systems tracts.

INTRODUCTION: AIM AND CONCEPTS

Sequence stratigraphy is a powerful tool for the prediction of facies distribution, and thus of depositional-related reservoir properties (Posamentier & Allen, 1993). These properties are also strongly controlled by facies-related and early diagenetic alterations (Bloch & McGowen, 1994). Recently, it was demonstrated that the overall spatial distribution of early diagenetic alterations and their impact on reservoir-quality evolution of siliciclastic sediments can be constrained within the context of sequence stratigraphy (Morad *et al.*, 2000). Such an approach is possible because changes in relative sea-level and sediment supply, which influence the near-surface pore-water chemistry, detrital composition and the 'residence time' of the sediments under a certain set of geochemical conditions, occur at predictable intervals within a sequence. Additionally, eogenetic pore-water chemistry is strongly influenced by the climatic conditions prevalent during and immediately after deposition and thus play an important role in controlling the type of clay mineral formed (Dutta & Sutner, 1986).

Diagenetic clay minerals occur in sandstones by direct precipitation from pore fluids (authigenesis), alteration of detrital silicates, mechanical clay infiltration and compaction of ductile argillaceous grains (i.e. pseudomatrix). We aim here to delineate a predictive model for the distribution pattern of early diagenetic clay minerals in siliciclastic sediments within a sequence stratigraphic framework, based upon a review of published material and on our ongoing research.

The large number of sequence stratigraphic models available in the literature (e.g. Vail *et al.*, 1977; Posamentier *et al.*, 1988; Galloway, 1989; Perlmutter & Matthews, 1989; Van Wagoner *et al.*, 1990; Einsele *et al.*, 1991; Schwarzacher, 1993; Helland-Hansen & Gjelberg, 1994; Embry, 1995) reflects, in part, the complexity of geological parameters controlling the stratigraphic record. Owing to the obvious difficulties in considering all stratigraphic terminology and models in this review, we constrained the distribution of diagenetic clay minerals according to the widely accepted sequence stratigraphic approaches of Posamentier *et al.* (1988), Posamentier & Vail (1988) and Van Wagoner *et al.* (1990), which are all influenced by the model of Vail *et al.* (1977), and developed for marine and transitional deposits. According to these models, facies distribution within a sequence is controlled by the interplay between sediment supply, basin-floor physiography and changes in relative sea-level, which refers to changes in the elevation of sea-level by a combination of eustatic sea-level fluctuation and basin-floor subsidence/uplift. The stratigraphic record is divided into sequences, which are bounded by unconformities or their correlative conformities. The sequence boundaries referred to in this review are formed by an abrupt fall in relative sea-level, which leads to a basinward shift of facies, subaerial exposure and erosion of the shelf (i.e. type 1 sequence boundary of Van Wagoner *et al.*, 1990). If the gradient of the exposed shelf is steeper than the gradient of the ancient alluvial profile, relative sea-level fall results in increasing gradient of the alluvial profile, valley incision and pronounced erosion of the shelf, and thus increasing sediment supply to the basin.

Sequences are divided into systems tracts, which are defined as a linkage of contemporaneous depositional systems (Brown & Fisher, 1977). Three types of systems tracts can be recognized and related to specific periods of a relative sea-level change cycle (Posamentier & Vail, 1988): the lowstand systems tract (LST), the transgressive systems tract (TST) and the highstand systems tract (HST). Systems tracts are composed of parasequences, which are genetically related strata bounded by marine flooding surfaces (Van Wagoner *et al.*, 1990). Parasequences can be grouped into parasequence sets depending on their stacking pattern (aggradational, progradational or retrogradational), which in turn, is controlled by the ratio of depositional rates and accommodation rates (Van Wagoner *et al.*, 1990).

Sediments of the LST are deposited between the point of maximum rate of relative sea-level fall and the beginning of relative sea-level rise, i.e. between the sequence boundary and the transgressive surface (TS). The TS is defined as the first significant marine flooding surface across the shelf (Van Wagoner *et al.*, 1990). The LST is composed of three subunits: the basin-floor fan, the slope fan, and the incised valley/lowstand wedge deposits, in which the latter deposits form a progradational parasequence set. The TST sediments occur between the TS and the maximum flooding surface (MFS) and are characterized by a retrogradational parasequence set (Van Wagoner *et al.*, 1990). The TST shelf deposits usually are fine-grained owing to the trapping of coarse-grained sediments in the coastal areas as a result of relative sea-level rise. When the rate of relative sea-level rise is at a maximum, little clastic sediment can reach the shelf, leading to the accumulation of hemipelagic and pelagic deposits at very slow rates (e.g. 1 cm kyr^{-1}), named the condensed section (Loutit *et al.*, 1988). Other periods of stratal condensation that are not related to maximum transgression may occur, such as along the basal part of parasequences, being referred to as condensed intervals. The HST deposits accumulate between the MFS and the next sequence boundary, being characterized by early aggradational to late progradational parasequence sets, formed between the maximum rate of rise of relative sea-level and the beginning of the subsequent fall (Van Wagoner *et al.*, 1990).

For sequences developed in continental strata, where the impact of changes in the relative sea-level are not as obvious as in transitional or marine settings, we refer to the

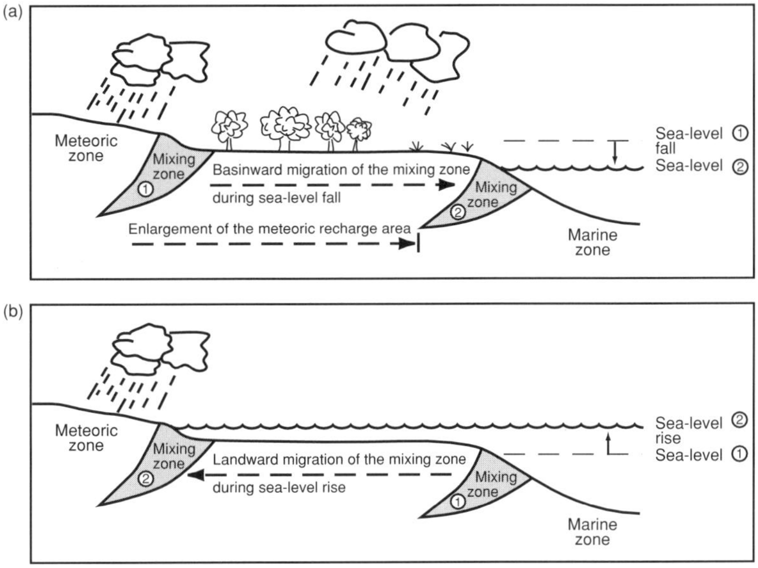

Fig. 1 A cartoon showing two hypothetical situations in which relative sea-level changes control the position of meteoric, mixed meteoric–marine and marine pore-water zones in transitional and shallow-marine deposits under a stable climatic condition. During sea-level fall (a) the mixing zone moves basinward enlarging the meteoric recharge area. During sea-level rise (b) the mixing zone moves landward and the meteoric zone evolves to mixing and marine zones.

sequence stratigraphic approach summarized in Shanley & McCabe (1994). According to this approach, sea-level changes control the depositional base-level and thus the architecture of alluvial deposits. During lowstand and early transgressive times, multilateral and multistorey braided river deposits are formed owing to a low rate of generation of accommodation. Such deposits fill the valleys that are formed during sea-level fall. When accommodation increases owing to progressive sea- and base-level rise, aggradation and formation of thick floodplain deposits take place and channel deposits are isolated from each other, forming single-storey sand bodies enclosed by fine-grained floodplain deposits (Wright & Marriott, 1993). During times of maximum transgression of the shoreline, fluvial-channel deposits may be reworked by tidal processes (Shanley & McCabe, 1993). Highstand times are characterized by isolated fluvial channels which tend to be amalgamated and interconnected towards the top of HST by lateral avulsion and accretion in response to the decreasing rate of addition of accommodation (Wright & Marriott, 1993).

CLAY-MINERAL DIAGENESIS AT SEQUENCE BOUNDARIES

A relative fall in sea-level and the formation of a sequence boundary can lead to subaerial exposure of the shelf, and hence erosion and pedogenesis of marine sediments. The preservation of pedogenic alterations is poorest in areas cut by incised valleys owing to the intense erosion by fluvial channels. Conversely, in the interfluve area, where the sequence boundary coincides with the transgressive surface (SB/TS), the preservation potential is higher, but dependent on the degree of erosion by wave action during the following transgressive event (Van Wagoner *et al.*, 1990).

A relative fall in sea-level results in a basinward shift of the meteoric water zone and in an increase in the hydraulic head and meteoric recharge area (Meisler *et al.*, 1984; Fig. 1). Under a humid climate, meteoric-water percolation is pervasive along subaerially exposed, older HST deposits, but may extend into the TST or even LST deposits (Carvalho *et al.*, 1995), if sufficient hydraulic head is created during sea-level fall. The Holocene sea-level fall related to the latest glacial maximum (16 kyr BP) and dislocated the freshwater–seawater interface several kilometres offshore along the northern Atlantic coast of USA, and hundreds of metres deeper than its present position, promoting meteoric water flushing of Upper Cretaceous sediments (Meisler *et al.*, 1984).

The clay-mineral types formed at subaerially exposed sequence boundaries are strongly related to the prevailing climatic conditions. Under warm, semi-humid to humid climatic conditions (precipitation > 400 mm yr^{-1}), extensive meteoric water flushing results in the dissolution of detrital silicates (e.g. feldspars) and formation of kaolinite and secondary intragranular porosity. The amount of kaolinite formed is thus dependent on the volumes of unstable silicate grains (e.g. micas and feldspars) and meteoric water involved, which in turn depends on: (i) meteoric water supply, which is a function of climate; (ii) permeability of the sediments; (iii) amplitude and frequency of sea-level change, and hence duration of subaerial exposure; and (iv) physiography and gradient of the basin floor exposed during relative sea-level fall, which determines the extent of subaerially exposed area and the meteoric recharge area.

The thickness of a kaolinitized zone below a sequence boundary depends on climate and duration of the subaerial exposure. In humid climates and during a subaerial exposure of 100–500 kyr (e.g. approximating a third-order sea-level cycle; cf. Vail *et al.*, 1977), the thickness of the kaolinitized zone developed in arkosic sands below a sequence boundary, assuming propagation of the meteoric alteration front to be ≈ 10 m Myr^{-1} (cf. Bjørkum *et al.*, 1990), would be around 1–5 m. Such kaolinitized horizons are useful for the recognition of an interfluve sequence boundary (McCarthy & Plint, 1998), and are thus potential stratigraphic markers for regional correlation within the basin. Extensively kaolinitized horizons possibly can be recognized in wireline logs by their high Th/K ratios in the spectral gamma-ray logs, as a result of K leaching during alteration of feldspars and micas (Emery *et al.*, 1990; Ruffell *et al.*, this volume, pp. 93–108).

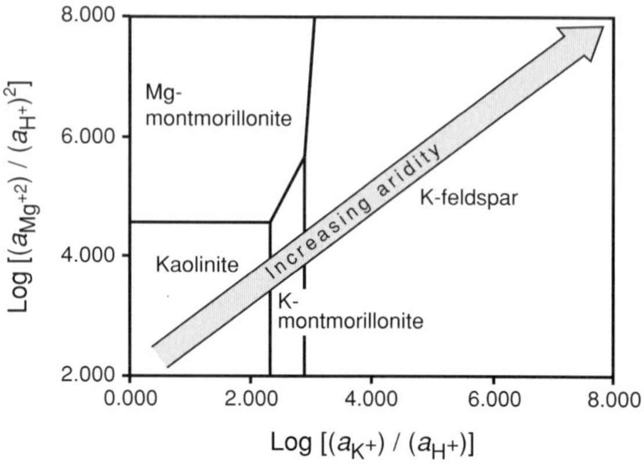

Fig. 2 Activity diagram of log [$a(Mg^{2+})/a(H^+)^2$] versus log [$a(K^+)/a(H^+)$] showing the stability fields of kaolinite, Mg- and K-smectite, and K-feldspar at 25°C and 1 atm (modified from Helgeson *et al.*, 1969). Humid climate results in effective leaching of ions and thus pore-waters are characterized by relatively high aH^+, leading to the formation of kaolinite. Conversely, under an arid climate, ionic leaching is inefficient and pore-waters are often characterized by high aMg^{2+}/aH^+ and high aK^+/aH^+ ratios, leading to the formation of smectitic clay minerals.

Under semi-arid to arid climatic conditions (precipitation < 400 mm yr^{-1}) the volume of meteoric water is strongly limited, and the removal of ions such as K^+, Mg^{2+}, Na^+ and Ca^{2+} is inefficient. Hence, pore-waters are often characterized by relatively high a^+_K/a^+_H, $a^{2+}_{Mg}/(a^+_H)^2$, a^+_{Na}/a^+_H and $a^{2+}_{Ca}/(a^+_H)^2$ ratios, which lead to the formation of smectite and mixed-layer illite–smectite (I/S) and chlorite–smectite (C/S; Dutta & Suttner, 1986; Fig. 2). Commonly, I/S and C/S occur as pore-lining or grain-replacive clays together with other early diagenetic minerals, such as K-feldspar, calcite, dolomite, sulphates and Fe-oxides (Walker *et al.*, 1978; Malicse & Mazzullo, 1996). Under strongly arid conditions, Mg-rich clays such as palygorskite and sepiolite are formed below unconformities developed under Mg-rich substrates such as dolostones (Rossi & Cañaveras, 1999).

Semi-arid to arid climatic conditions also favour the mechanical infiltration of clay below subaerially exposed sequence boundaries (Malicse & Mazzullo, 1996), owing to the deep position of the groundwater table (Walker,

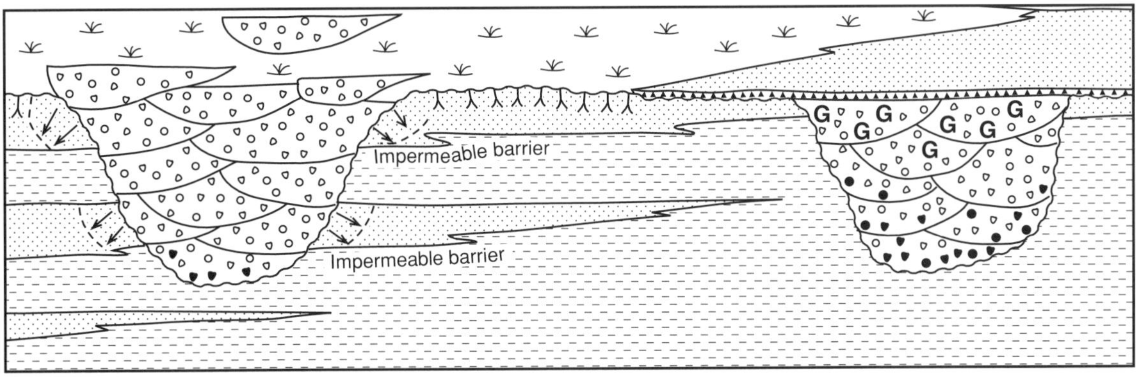

Coastal and flood plain fine-grained deposits

Fluvial channel sandstone deposits (e.g. incised valley deposits)

Coastal plain and shallow marine sandstone deposits

Shelf mudstone deposits

Transgressive surface and lag deposits formed by wave reworking during transgression

Sequence boundary

Infiltrated clays

G Glauconitization of micas and detrital smectites

• Mud intraclast and rip-up clast

Interfluve sequence boundary soils (abundant clay cutans)

Fig. 3 A cartoon illustrating the results of mechanical clay infiltration, which coats framework grains below a sequence boundary, and the effects of marine flooding over valley fill deposits. Mechanically infiltrated clays are abundant below valleys (which focus the flow of mud-rich river water), into the older, permeable HST sands, which are underlain by impermeable barriers, such as the HST offshore muds. In the interfluve areas the preservation potential of the infiltrated clays depends on the extent of wave reworking during the following transgression. Fluvial deposits in valleys that are flooded by sea water during transgression display glauconitization of detrital micas, especially close to the transgressive surface.

1976). Under more humid climatic conditions, an elevated groundwater table inhibits deep infiltration of muddy surface waters and thus of mechanical clay infiltration. Infiltration and accumulation of clays within sandstones occurs in the vadose zone by periodic flooding of permeable sediments by surface waters or in the phreatic zone by a fluctuating groundwater table or above impermeable layers (Moraes & De Ros, 1992).

Incised valleys are particularly favourable sites for clay infiltration because they focus the flow of muddy river water (Moraes & De Ros, 1992; Fig. 3). Accumulation of infiltrated clays might be expected to be important where incised valleys cut permeable HST sands (e.g. shoreface deposits), which are underlain by impermeable barriers, such as offshore mud deposits (Fig. 3). Sand deposits containing mechanically infiltrated clay minerals can be recognized in wireline logs by their characteristic low resistivity values owing to the high volumes of irreducible water (Moraes & De Ros, 1990; Seyler, 1997). Soils developed at interfluve sequence boundaries also may contain abundant clay cutans (McCarthy *et al.*, 1999; Fig. 3), and are characteristically highly mature as a result of prolonged exposure time (Wright & Marriott, 1993).

The infiltrated clay minerals, which often are smectitic in composition when related to weathering in semi-arid environments (Keller, 1970; Moraes & De Ros, 1992; De Ros *et al.*, 1994), exert a profound influence on porosity–permeability evolution of sandstones during burial diagenesis. Illitization of infiltrated clays induces serious deterioration in the reservoir quality by decreasing the permeability (Jiao &

Surdam, 1997) and increasing the ineffective microporosity and water saturation of sandstones (Hurst & Nadeau, 1995). Chloritization of the infiltrated clays enhances the preservation of anomalously high permeability and porosity in deeply buried sandstones by preventing the precipitation of quartz overgrowths (Dewers & Ortoleva, 1991; Ehrenberg, 1993). Although illite may have the same effect by inhibiting the formation of quartz overgrowths (Al-Jallal *et al.*, 2000), it commonly causes serious deterioration to permeability owing to the longer, fibrous illite crystals, which tend to block pore throats more efficiently than chlorite crystals. It is difficult to predict whether a smectitic clay will be illitized or chloritized, but it is believed that detrital composition of the sandstone and the chemical composition of the parent smectite play an important role. Magnesium-rich smectites, particularly those in sandstones with abundant biotite and mafic volcanic fragments or containing haematite coats, tend to be chloritized, whereas K- and Na-rich smectites in sandstones with abundant K-feldspar and muscovite tend to be illitized (Morad *et al.*, 2000).

CLAY-MINERAL DIAGENESIS AT PARASEQUENCE BOUNDARIES, IN PARASEQUENCES AND CONDENSED INTERVALS

As most parasequences represent upward-shoaling cycles, the top surface can be subaerially exposed, and subjected to meteoric water diagenesis and pedogenesis. Under warm, subhumid to humid climatic conditions, meteoric water percolation causes silicate-grain dissolution and formation of kaolinite (Fig. 4). As the duration of exposure is typically much shorter than for sequence boundaries, kaolinitization is likely to be less extensive below parasequence boundaries. Correlation of kaolinitized sediments at parasequence boundaries is probably more restricted and equivocal than correlation of the areally more extensive kaolinitized sequence boundaries (Van Wagoner *et al.*, 1990).

Subaerially exposed parasequence boundaries and the upper part of parasequences in coastal deposits may contain peat deposits (Van Wagoner *et al.*, 1990), which are microbially degraded into organic acids and CO_2 during early diagenesis. Thus, meteoric waters percolating through such deposits become acidic, with higher activity of H^+ relative to other ions (e.g. K^+), which favours the dissolution of silicates (e.g. feldspars) and formation of kaolinite in the first few decimetres of underlying sediments (Staub & Cohen, 1978). As the amount of kaolinite formed is believed to be proportional to the thickness of the peat (Staub & Cohen, 1978), most extensively kaolinitized sands are expected below thick peat accumulations, which in coastal environments are formed during periods of maximum accommodation creation (Ryer, 1984; Cross, 1988). In such cases, formation of kaolinite along parasequence boundaries might be much more extensive than along sequence boundaries, regardless of the subaerial exposure times. In extreme cases, extensive leaching of sandstones immediately below peat/coal deposits may lead to the formation of diagenetic quartzarenites or ganisters (Percival, 1986).

Coal measures and associated deposits also may contain lenses and concretions of berthierine (Iijima & Matsumoto, 1982). Berthierine formation is favoured by the strongly reducing conditions provided during organic matter degradation and the low SO_4^{2-}, brackish characteristic of pore waters, which inhibit sulphate reduction and incorporation of iron into sulphides. During deep burial diagenesis, thermal alteration of peat/lignite releases organic acids and CO_2, which contributes to the dissolution of silicates and formation of kaolin in the adjacent, over- and underlying beds (Surdam *et al.*, 1984; Cowan, 1989; Platt, 1993; Van Keer *et al.*, 1998). Like peat/lignite deposits, organic-rich, fine-grained marine sediments alter thermally during deep burial diagenesis to produce CO_2 and organic acids, which can dissolve carbonates and silicates and form kaolin in adjacent sandstones (Moncure *et al.*, 1984; Surdam *et al.*, 1984; Surdam & Yin, 1995; Van Keer *et*

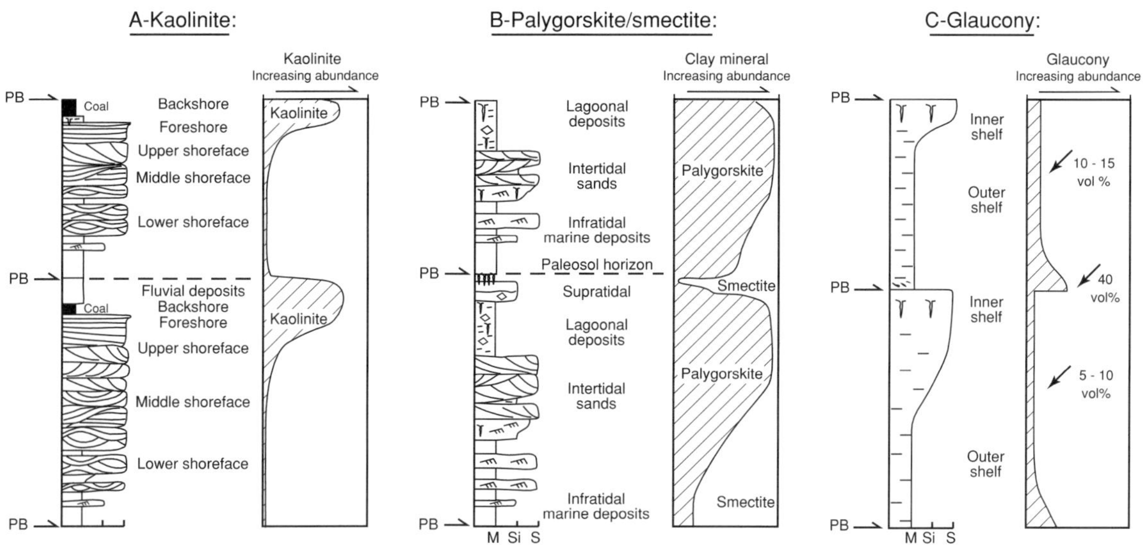

Fig. 4 Schematic sedimentary logs showing the distribution of (A) kaolinite, (B) palygorskite/smectite, and (C) glaucony in an aggradational parasequence set. In (A) kaolinite content increases towards the top of parasequences where continental facies can be exposed to extensive meteoric water flushing under semi-humid to humid climate. Kaolinite content is enhanced by the presence of unstable silicates and peat/coal deposits owing to the formation of acidic fluids during organic matter degradation. In (B) palygorskite is the dominant clay mineral in the top of parasequences capped by lagoonal/evaporitic deposits under a strongly arid climate, and tend to gradually decrease in abundance towards the bottom of the parasequences. In (C) maximum abundance of autochthonous glaucony occurs at the parasequence boundary (PB) and tends to decrease progressively towards the top of the parasequence.

al., 1998). Such organic-rich, marine sediments are related to periods of sediment starvation and anoxia in the basin, possibly related to the MFS and condensed section deposits (Loutit *et al.*, 1988) or other condensed intervals within a sequence (Curiale *et al.*, 1991; Palsey *et al.*, 1991). Condensed intervals can contain bentonite beds, which are smectite-rich clays formed by the alteration of volcanic ash material (Grim & Güven, 1978). During periods of high sedimentation rate, bentonite beds may not form because volcanic ash material is diluted within siliciclastic sediments (Leithold, 1993).

Under semi-arid to arid climatic conditions, palygorskite and sepiolite occur together with dolomite and evaporitic cements at the tops of parasequences that are capped by lagoonal/evaporitic deposits (Weaver, 1984; Fig. 4). Palygorskite formation in such environments is enhanced by the presence of Mg-rich sediments (e.g. dolomite, Mg-smectite) and strong evaporative conditions (Inglès & Anadón, 1991). If marine deposits occur in the lower part of the parasequence, aragonitic/high-Mg calcite shells can be replaced by palygorskite and dolomite (Weaver, 1984).

Various patterns of glaucony (*sensu* Odin & Matter, 1981) distribution occur within parasequences. *In situ*, autochthonous glaucony tends to be most abundant within the fine-grained, condensed intervals at parasequence boundaries or in the lower part of parasequences (Mitchum & Van Wagoner, 1991; Amorosi, 1995; Ghibaudo *et al.*, 1996; Dreyer *et al.*, 1999), followed by a marked upward decrease within the same parasequence (Amorosi & Centineo, 2000; Fig. 4). This distribution pattern is attributed to the cyclic alternation of favourable conditions for glaucony formation, related to changes in sedimentation rates (Amorosi & Centineo, 2000). Favourable conditions for glaucony formation include: (i) low sedimentation rates as a result of low siliciclastic input to the shelf, and (ii) relatively small amounts of organic matter accumulating in water depths > 100 m (Porrenga, 1967; Odin & Matter, 1981). Small amounts of organic matter inhibit rapid bacterial sulphate reduction and incorporation of iron into sulphides, and help establish mildly reducing conditions, i.e. $Fe^{2+} \approx Fe^{3+}$ (nitrate-reducing, suboxic condition; Berner, 1981). Conversely, elevated organic-matter content and warm waters, which occur for instance in the inner shelf sediments (e.g. water depths < 100 m), create a strongly reduced condition (i.e. $Fe^{2+} > Fe^{3+}$), which leads to the formation of Fe-sulphides, siderite or berthierine rather than glaucony. Odinite, a 7 Å Fe-rich clay mineral, forms in similar geochemical conditions to those required for berthierine, but tends to evolve to berthierine during early diagenesis (Odin *et al.*, 1988). Berthierine (and chamosite) also form oolitic ironstones, which may be closely associated with condensed intervals within a sequence characterized by intense sediment reworking on the sea floor (Burkhalter, 1995), such as at parasequence boundaries (Chan, 1992).

CLAY-MINERAL DIAGENESIS RELATED TO SYSTEMS TRACTS

Lowstand systems tract (LST)

A relative sea-level fall may result in valley incision and destabilization of the shelf, resulting in the delivery of large amounts of sediments to the basin, which are deposited as sand-rich, basin-floor fan turbidites. When the sea-level is stable at lowstand and begins to rise, during deposition of late LST, deep-water deposition results in the formation of channel–levee complexes (Van Wagoner *et al.*, 1990). In this case, channels potentially contain abundant mud intraclasts owing to the erosion of fine-grained, slope sediments (Carvalho *et al.*, 1995). Such mud intraclasts rapidly decrease permeability during burial owing to their mechanical compaction and formation of pseudomatrix.

In the late stages of the LST, accommodation is progressively created and valleys start to be filled by low sinuosity, braided river deposits (Shanley & McCabe, 1994), which may evolve to estuarine deposits and meandering rivers during the subsequent transgression. Fluvial deposits in valleys are prone to develop multiple and thick infiltrated clay coats owing to their typically high depositional permeability (Moraes & De Ros, 1992). If semi-arid to arid climatic conditions prevail, episodic floods produce ideal conditions for clay infiltration and accumulation (Walker, 1976; Moraes & De Ros, 1992). Infiltrated clays are, however, absent or scarce within the lag deposits along the transgressive surface above the valley fill as a result of intense wave reworking, which tends to produce clay-free sands (Matlack *et al.*, 1989; Fig. 3). Incorporation of large quantities of mud intraclasts into valley deposits (e.g. rip-up clasts) is favoured when valley incision occurs into offshore marine mud (Van Wagoner *et al.*, 1990). During burial and compaction, mud intraclasts form pseudomatrix, diminishing the initially high sandstone permeability.

The permeable LST fluvial deposits within valleys are also subjected to meteoric water flushing under humid to subhumid climatic conditions, which leads to the formation of kaolinite and improvement of reservoir quality owing to the formation of secondary intragranular porosity. The upper part of the LST is expected to contain progressively less kaolinite towards the TS owing to a progressive rise in relative sea-level, which leads to a landward migration of the meteoric pore-water zone (Meisler *et al.*, 1984), a decrease in the hydraulic head and thus, progressively less meteoric influence.

The LST coastal and marine deposits may contain extra- and intrabasinal glaucony derived from the erosion and redeposition of older deposits within the basin (i.e. detrital and parautochthonous glaucony *sensu* Amorosi, 1995). Such glaucony occurs in a variety of facies, such as tide- and storm-influenced coastal deposits as well as in deep-water turbidites (Baum & Vail, 1988; Glenn & Arthur, 1990; Mancini & Tew, 1993; Amorosi *et al.*, 1997). The glaucony typically displays textures indicative of reworking, such as deformed lamellar structure and lack of microcracks (Amorosi & Centineo, 2000). Considerable amounts of detrital and parautochthonous glaucony also occur in regressive shoreline sands (Walker & Bergman, 1993, e.g. green sands), and at sequence boundaries (Baum *et al.*, 1994; McCracken *et al.*, 1996). In contrast to detrital and parautochthonous glaucony, autochthonous glaucony is relatively rare in LST deposits, being restricted to more basinal areas, such as along parasequence boundaries of the lowstand wedge (Fig. 5).

The LST deposits of hydrologically closed, lacustrine environments, under strong evaporative conditions, are characterized by the formation of palygorskite and smectite, together with dolomite (Bellanca *et al.*, 1992; Colson & Cojan, 1996; Colson *et al.*, 1998). The formation of such Mg-rich clays is attributed to the mixing of fresh meteoric waters with hypersaline lake brines during lowstand times as a result of falling or rising lake-level, i.e. during deposition of the very early or late LST deposits, respectively (Colson *et al.*, 1998). Conversely, the formation of palygorskite and smectite is unlikely during lake-level highstand owing to the dilution of lake brine by fresh water (Bellanca *et al.*, 1992). However, minor amounts of palygorskite might be associated with isolated, ephemeral ponds in the lake margin during the HST (Colson *et al.*, 1998).

Transgressive systems tract (TST)

A relative sea-level rise results in an increase in accommodation and deposition of the TST. In transitional to marine settings, deposition of the TST involves retrogradational stacking of parasequences (Van Wagoner *et al.*, 1990), whereas in continental settings, deposition occurs through a rapid aggradation of the fluvial system (Wright & Marriott, 1993; Shanley & McCabe, 1994).

The continuous rise in relative sea-level during deposition of the TST causes a progressive decrease in the subaerially exposed portions of parasequence boundaries, a decrease in the hydraulic head, and a landward migration of the marine pore-water zone (Meisler *et al.*, 1984; Fig. 1). Consequently, one can expect that kaolinite formation along subaerially exposed parasequence boundaries decreases in importance towards the MFS if no major climatic changes occur (Fig. 5). In contrast to kaolinite, autochthonous glaucony is common within the TST deposits (Ghibaudo *et al.*, 1996; Amorosi, 1997; Breyer, 1997; Sugarman & Miller, 1997; Floquet, 1998; Ruffell & Wach, 1998; Kelly & Webb, 1999; Schulz-Rojahn *et al.*, this volume, pp. 473–488), and tends to increase in abundance towards the MFS. The global marine transgression during the Holocene following the latest glacial maximum has favoured the present-day formation of autochthonous glaucony on the outer shelf and slope environments (Bornhold & Giresse, 1985; Odin, 1988; Odin & Fullagar, 1988; Rao *et al.*, 1993; Corselli *et al.*, 1994; Gensous & Tesson, 1996). The flooding of fluvial LST sediments by marine water during transgression (e.g. flooded incised valley deposits) might be accompanied by the replacement of detrital micas by glaucony, such as observed in Triassic sandstones from France (Durand, 1977; Figs 3 & 6). Detrital smectites formed in continental settings and present in such fluvial sediments (e.g. infiltrated clays), which commonly are Ca-rich, may exchange Ca for K, Mg and Na from sea water (Sayles & Mangelsdorf, 1979). Incorporation of K from sea water also may lead to the glauconization of detrital smectites (Odin & Matter, 1981).

Autochthonous glaucony increases in abundance and maturity (i.e. increase in potassium content; cf. Odin & Matter, 1981) upwards through the condensed interval of the TST, especially towards the MFS (Loutit *et al.*, 1988; Baum *et al.*, 1994; Amorosi, 1995; McCracken *et al.*, 1996; Amorosi & Centineo, 1997; Duprat, 1997; Kitamura, 1998; Malartre *et al.*, 1998; Myrow, 1998; Pittet & Strasser, 1998; Vandenberghe *et al.*, 1998; Kim & Lee, 2000). The

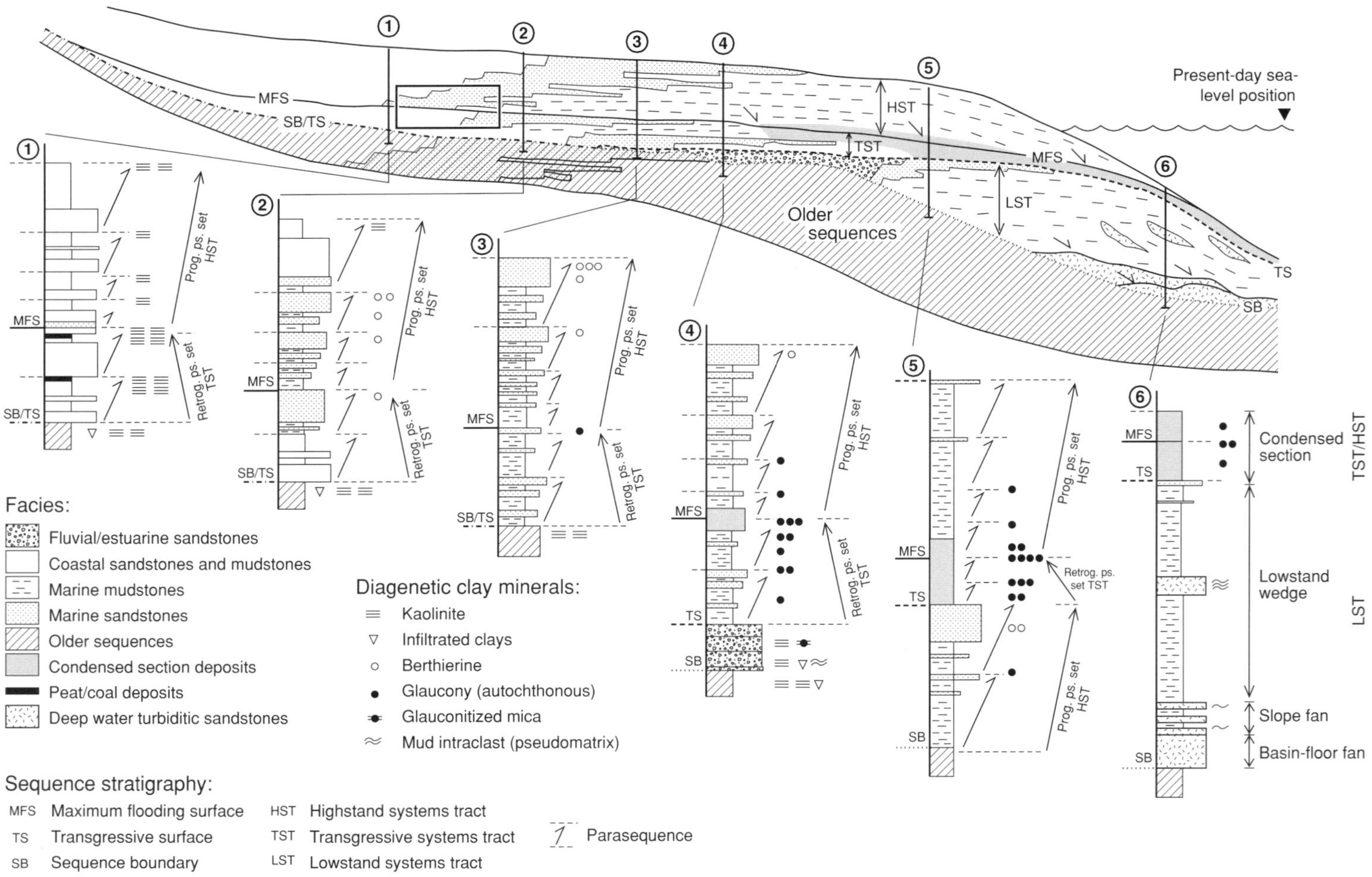

Fig. 5 A cartoon showing the distribution of diagenetic clay-minerals within a sequence. Sequence stratigraphic framework modified from Van Wagoner *et al.* (1990).

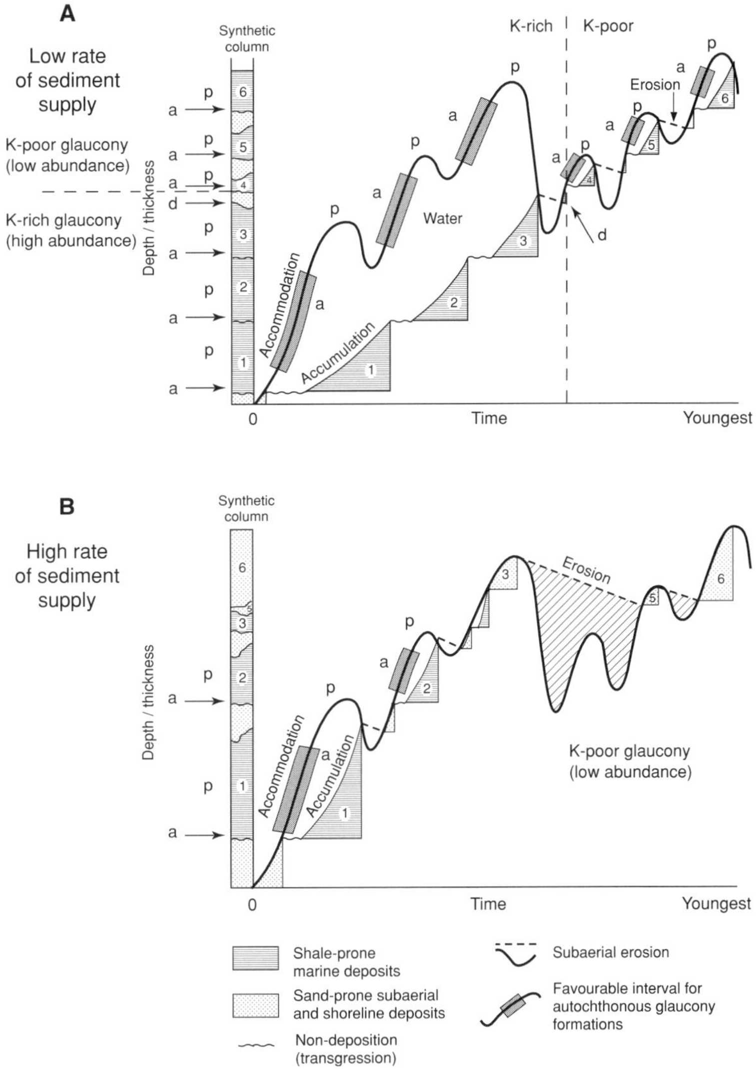

Fig. 6 Diagram showing the relationship between sediment accumulation, accommodation and the distribution of autochthonous (a), parautochthonous (p) and detrital (d) glaucony. The curve represents the evolution of accommodation through time. Cycles of deltaic progradation (1–6) are separated by non-depositional transgressive phases. Periods of subaerial exposure and erosion occur when the accommodation curve remains below the depositional base-level. In (A) sediment supply is low and the formation of autochthonous glaucony is more abundant than in (B), where the rate of sediment supply is higher. Low rates of sediment supply and high accommodation result in the formation of abundant K-rich (i.e. mature) glaucony (cycles 1, 2 and 3 in A). Reworked, detrital glaucony (e.g. green sands) may form between cycles 3 and 4. Conversely, high rates of sediment supply and low accommodation result in the formation of K-poor (i.e. immature) glaucony (cycles 4, 5 and 6 in A and in B). Synthetic column and accommodation curve adapted from Jervey (1988).

increase in maturity of glaucony occurs by diffusion and incorporation of K from sea water, owing to prolonged 'residence time' of glaucony at the sediment–water interface, caused by the low sedimentation rate (Odin & Matter, 1981). This pattern of widespread formation of mature (K-rich) autochthonous glaucony during transgressive events results in fairly reliable regional stratigraphic markers, such as those found in the Cretaceous to Oligocene glaucony-rich successions of northern central Europe (Robaszynski *et al.*, 1998; Vandenberghe *et al.*, 1998). Because the maturity, origin (autochthonous, parautochthonous and detrital) and abundance of glaucony possibly are related to sedimentation rate, these characteristics also may reflect the sediment influx to the basin (Amorosi & Centineo, 2000). Two hypothetical settings with different rates of sediment supply will have distinct distribution patterns of glaucony and glaucony maturity during the same relative sea-level cycle. Highly mature (K-rich) autochthonous glaucony, for instance, is much more common in siliciclastic starved periods than during periods of higher sediment supply (Fig. 6).

In deep-water marine settings, the TST is deposited under conditions of drastically reduced sediment supply. Such conditions, which are most prevalent during development of the MFS and condensed section, prolong the 'residence time' of the deep-water lowstand sediments on the sea floor and their reaction with sea water, leading to the formation of smectitic clays in deep-sea (> 1000 m) sediments (Thiry & Jacquin, 1993), especially those with abundant volcanic fragments.

In continental settings, relative sea-level rise and the consequent increase in accommodation during deposition of the TST result in an increase in the floodplain aggradation rate, which leads to an upward decrease in soil maturity (Wright & Marriott, 1993). Because aggradation of the floodplain during TST occurs by crevassing of channels (Kirschbaum & McCabe, 1992), one might expect that mud-rich waters from channels would promote mechanical clay infiltration in sandy substrates in the floodplain, such as in crevasse-splay deposits. Thus crevasse-splay deposits of the TST are potentially richer in infiltrated clays than comparable HST deposits (Fig. 7). However, only thin clay coats are developed owing to the rapid sealing of the deposits by floodplain fines and formation of soil profiles.

Highstand systems tract (HST)

The HST is characterized by an aggradational phase that tends to evolve to a progradational phase, leading to progressive basinward migration of the shoreline, and of the meteoric water zone. Therefore, it is expected that kaolinite content tends to increase towards the top of HST, owing to enhanced meteoric water incursion during shoreline progradation.

As accommodation decreases towards the top of HST, lateral rather than vertical accretion of the alluvial plain takes place, resulting in more laterally interconnected and amalgamated fluvial channels (Wright & Marriott, 1993; Shanley & McCabe, 1994), and a decrease in the importance of crevasse-splay deposits. Extensive lateral channel migration during deposition of the HST prevents the development of multiple infiltrated clay coats (De Ros *et al.*, 1994) and promotes floodplain erosion, decreasing the preservation potential of these deposits and of soil profiles (Fig. 7). These soils are more mature than those in TST deposits owing to the slower aggradation rates of the HST floodplain, and thus more prolonged subaerial exposure (Wright & Marriott, 1993). Increasing floodplain erosion results in the incorporation of more mud intraclasts into channel sands in the late HST (Fig. 7), which leads to rapid deterioration of reservoir quality as a result of their mechanical compaction and formation of pseudomatrix during burial.

Glaucony is less common in the HST compared with the underlying TST (Amorosi, 1995). Autochthonous glaucony in the HST is best developed in areas away from the influence of clastic input and occurs at the base of parasequences in the early HST and the condensed section, and less commonly in the late HST (Fig. 5). This upward decrease in abundance of glaucony results from the progressive increase in the rate of sediment supply to the shelf (Amorosi & Centineo, 1997). Reworking of autochthonous glaucony occurs in the HST within shoreface sands, tide-dominated shelf sand bodies and in deep-marine sands (Cant, 1996, 1998; McCracken *et al.*, 1996).

In contrast to glaucony, pellets and grain-coating or pore-lining berthierine form in shallower and warmer waters and in organic-rich marine sediments during periods of higher sedimentation rates and progradation of the shoreline, adjacent to river-mouths (Stonecipher & May, 1990; Hornibrook & Longstaffe, 1996; Kronen & Glenn, 2000), such as during late stages of the HST or lowstand wedge of the LST (Fig. 5). The relatively high burial rates of such sediments below the sea floor promote rapid establishment of post-oxic, Fe-reducing geochemical conditions that favour berthierine formation. The formation of berthierine in such deposits is possibly enhanced by an abundant supply of iron-bearing particles (Odin & Matter, 1981) and a low sulphate concentration in pore-waters (i.e. less Fe^{2+} is incorporated in

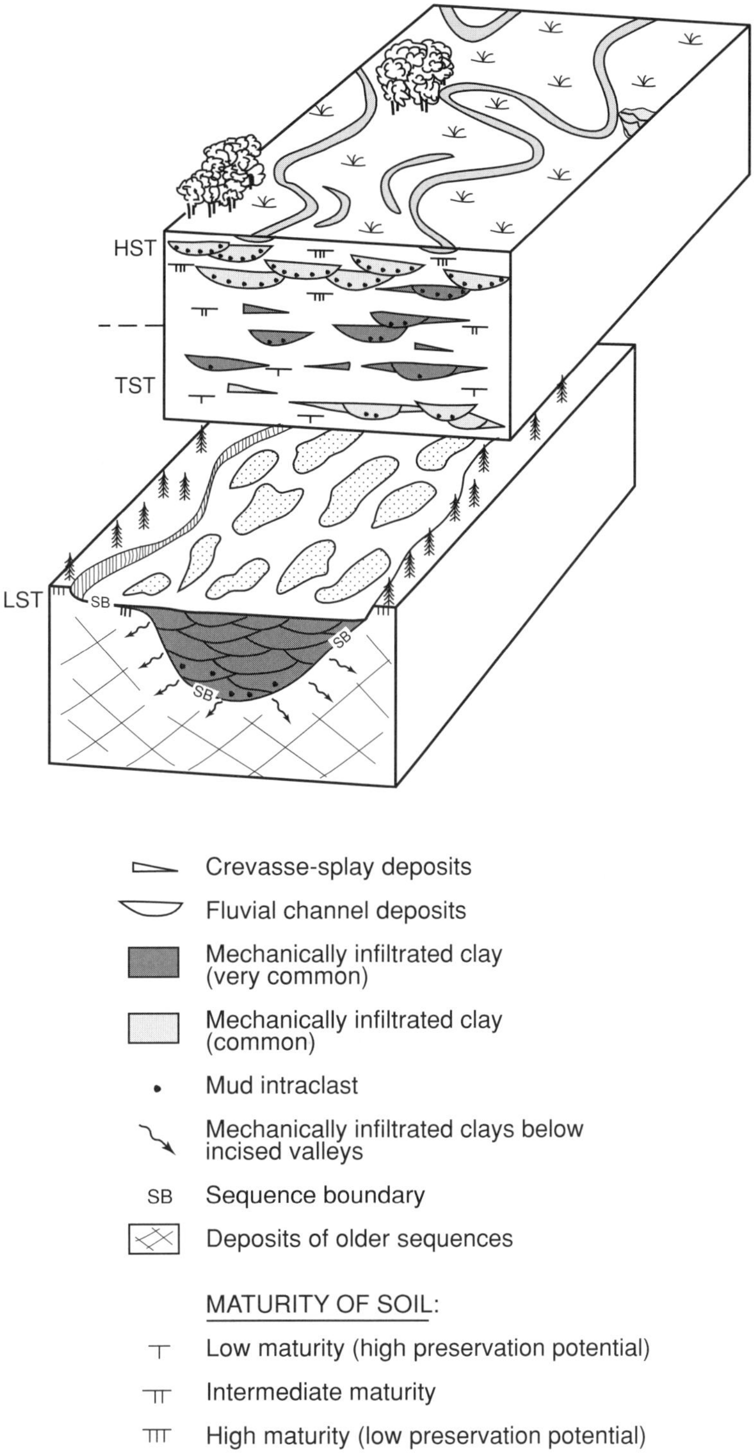

Fig. 7 Cartoon illustrating the distribution pattern of mechanically infiltrated clays, mud intraclasts and palaeosols in fluvial deposits within a sequence stratigraphic framework. Lowstand systems tract (LST) fluvial deposits within an incised valley are prone to mechanical clay infiltration because they focus the flow of mud-rich waters. Sediments containing mechanically infiltrated clays also occur in crevasse-splay deposits, which tend to be more important during the transgressive systems tract (TST), where the rate of vertical aggradation of the floodplain is highest. Highstand systems tract (HST) channels potentially contain the highest amounts of mud intraclasts owing to extensive lateral migration, and thus erosion and incorporation of floodplain sediments into channels. The HST palaeosols are much more mature and thicker than TST palaeosols owing to lower rates of floodplain aggradation and thus more prolonged subaerial exposure. Highly mature palaeosols are also developed on the interfluve sequence boundaries. Sequence stratigraphic framework based on Wright & Marriott (1993).

Fe-sulphides) owing to mixing of marine with meteoric waters as a result of shoreline progradation. Chloritization of pore-lining berthierine during burial diagenesis may preserve anomalously high porosity and permeability at great depths by preventing formation of quartz overgrowths (Hornibrook & Longstaffe, 1996).

CONCLUDING REMARKS

Despite the complexity of diagenetic systems, this study has demonstrated that the development of a predictive model of clay-mineral distribution patterns in siliciclastic sequences within a sequence stratigraphic framework is possible (Fig. 5). Distinct patterns of clay-mineral formation have been recognized along important sequence stratigraphic surfaces, such as sequence and parasequence boundaries, and trends in the extent of clay-mineral formation can be recognized along single parasequences, parasequence sets and systems tracts.

Changes in pore-water chemistry and sedimentation rates owing to relative sea-level change control the position of the meteoric, mixed meteoric–marine and marine pore-water zones, and the residence time of sediments under the same geochemical conditions during shallow burial diagenesis. Thus, relative sea-level changes help determine the type and distribution of neoformed clay minerals. During relative sea-level rise and transgression, and formation of the TS, marine flooding of subaerially exposed sediments promotes glauconization of detrital micas and smectitic clays in the underlying deposits. Autochthonous glaucony forms along condensed intervals such as the MFS, parasequence boundaries of the TST and early HST, and throughout the condensed section. As the extent of glaucony formation depends on the residence time at the sediment–water interface, the maxima in glaucony formation occurs along the MFS. Glaucony in the LST and HST usually is reworked, parautochthonous or detrital. Unlike glaucony, berthierine typically forms in progradational intervals with high sedimentation rates, such as during deposition of the late HST or of the late LST (lowstand wedge). Oolitic ironstones composed of Fe-rich clay minerals (e.g. berthierine and chamosite) are best developed in condensed intervals within a sequence, such as at parasequence boundaries.

The type and abundance of clay minerals formed along subaerially exposed sequence and parasequence boundaries are strongly dependent on climate. During humid to semi-humid climatic conditions, meteoric water flushing leads to the dissolution of silicates and precipitation of kaolinite. Formation of kaolinite is enhanced at the top of parasequences containing peat/coal deposits. Semi-arid to arid climatic conditions lead to the formation of smectitic clays and palygorskite. As the extent of clay mineral formation also depends on the duration of subaerial exposure, clay-mineral formation commonly is more pronounced along sequence boundaries than, for instance, along subaerially exposed parasequence boundaries.

In addition to clay-mineral formation by precipitation from pore-fluids and by replacement of unstable grains, the presence of mechanically infiltrated clays and mud intraclasts (which may evolve to pseudomatrix by mechanical compaction during diagenesis) also can be predicted within the context of sequence stratigraphy. Mechanically infiltrated clays are most important below sequence boundaries under semi-arid to arid climatic conditions and especially below and within valleys filled with fluvial deposits (especially braided rivers). Clay infiltration in meandering systems might be common during TST as a result of extensive channel crevassing. However, multiple and thick infiltrated clay coats are inhibited by rapid floodplain aggradation. Additionally, extensive channel migration during HST prevents development of multiple and thick infiltrated clay coats. Mud intraclasts are important in the late LST channel levee deep-water turbiditic complex, which erodes fine-grained slope deposits. In continental settings, LST fluvial deposits may contain abundant mud intraclasts when valleys cut into offshore mud. Mud intraclasts also may be important in the late HST, owing to the enhanced erosion of floodplain fines by channel avulsion.

ACKNOWLEDGEMENTS

J.M. Ketzer thanks the Brazilian Research Council (CNPq—grant no. 200059/98-0) and S. Morad thanks the Swedish Natural Science Research Council (NFR) for supporting research activities. We thank R. Worden, A. Ruffell and M. Mayall for critical and constructive reviews of the manuscript.

REFERENCES

Al-Jallal, I., Al-Hassan, S.A., Coker, J. *et al.* (2000) Jauf Formation subsurface mapping guided by seismic and gravity, its sand distribution, and reservoir quality: a regional integrated overview to help exploration in the northern Ghawar Area. In: *4th Middle East Geosciences Conference and Exhibition, March 25–27, Bahrain, Abstracts. GeoArabia*, **5**, 24–25.

Amorosi, A. (1995) Glaucony and sequence stratigraphy: conceptual framework of distribution in siliciclastic sequences. *J. Sed. Res.*, **B65**, 419–425.

Amorosi, A. (1997) Detecting compositional, spatial, and temporal attributes of glaucony: a tool for provenance research. *Sed. Geol.*, **109**, 135–153.

Amorosi, A. & Centineo, M.C. (1997) Glaucony from the Eocene of the Isle of Wight (southern UK): implications for basin analysis and sequence-stratigraphic interpretation. *J. Geol. Soc. London*, **154**, 887–896.

Amorosi, A. & Centineo, M.C. (2000) Anatomy of a condensed section: the Lower Cenomanian glaucony-rich deposits of Cap Blanc-Nez (Boulonnais, Northern France). In: *Marine Authigenesis: from Global to Microbial*, Society of Economic Paleontologists and Mineralogists Special Publication 65 (Ed. by C.R. Glenn, J. Lucas and L. Lucas), Tulsa, OK, pp. 405–413.

Amorosi, A., Centineo, M.C. & D'Atri, A. (1997) Lower Miocene glaucony-bearing deposits in the SE Tertiary Piedmont Basin (northern Italy). *Riv. Ital. Paleontol. Stratigr.*, **103**, 101–110.

Baum, G.R. & Vail, P.R. (1988) Sequence stratigraphic concepts applied to paleogene outcrops, Gulf and Atlantic basins. In: *Sea-level Changes: an Integrated Approach*, Society of Economic Paleontologists and Sedimentologists Special Publication 42 (Ed. by C.K. Wilgus, B.S. Hastings, C.G.S.C. Kendall, H.W. Posamentier, C.A. Ross and J.C. Van Wagoner), Tulsa, OK, pp. 309–327.

Baum, J.S., Baum, G.R., Thompson, P.R. & Humphrey, J.D. (1994) Stable isotopic evidence for relative and eustatic sea-level changes in Eocene to Oligocene carbonates, Baldwin County, Alabama. *Geol. Soc. Am. Bull.*, **106**, 824–839.

Bellanca, A., Calvo, J.P., Censi, P., Neri, R. & Pozo, M. (1992) Recognition of lake-level changes in Miocene lacustrine units, Madrid Basin, Spain. Evidence from facies analysis, isotope geochemistry and clay mineralogy. *Sed. Geol.*, **76**, 135–153.

Berner, R.A. (1981) A new geochemical classification of sedimentary environments. *J. Sed. Petrol.*, **51**, 359–365.

Bjørkum, P.A., Mjos, R., Walderhaug, O. & Hurst, A. (1990) The role of the late Cimmerian unconformity for the distribution of kaolinite in the Gullfaks Field, northern North Sea. *Sedimentology*, **37**, 395–406.

Bloch, S. & McGowen, J.H. (1994) Influence of depositional environment on reservoir quality prediction. In: *Reservoir Quality Assessment and Prediction in Clastic Rocks*, Society of Economic Paleontologists and Mineralogists Short Course 30 (Ed. by M.D. Wilson), Tulsa, OK, pp. 41–57.

Bornhold, B.D. & Giresse, P. (1985) Glauconitic sediments on the continental shelf off Vancouver Island, British Columbia, Canada. *J. Sed. Petrol.*, **55**, 653–664.

Breyer, J.A. (1997) Sequence stratigraphy of Gulf Coast lignite, Wilcox Group (Paleogene), south Texas. *J. Sed. Res.*, **67**, 1018–1029.

Brown, L.F. & Fisher, W.L. (1977) Seismic-stratigraphic interpretation of depositional systems: examples from Brazil rift and pull-apart basins. In: *Seismic stratigraphy—Applications to Hydrocarbon Exploration* (Ed. by C.E. Payton). *AAPG Mem.*, **26**, 213–248.

Burkhalter, R.M. (1995) Ooidal ironstones and ferruginous microbialites: origin and relation to sequence stratigraphy (Aalenian and Bajocian, Swiss Jura mountains). *Sedimentology*, **42**, 57–74.

Cant, D.J. (1996) Sedimentological and sequence stratigraphic organization of a foreland clastic wedge, Mannville Group, western Canada basin. *J. Sed. Res.*, **66**, 1137–1147.

Cant, D.J. (1998) Sequence stratigraphy, subsidence rates, and alluvial facies, Mannville Group, Alberta foreland basin. In: *Relative Role of Eustasy, Climate, and Tectonism in Continental Rocks*, Society of Economic Paleontologists and Sedimentologists Publication 59 (Ed. by K.W. Shanley and P.J. McCabe), Tulsa, OK, pp. 49–63.

Carvalho, M.V.F., De Ros, L.F. & Gomes, N.S. (1995) Carbonate cementation patterns and diagenetic reservoir facies in the Campos Basin Cretaceous turbidites, offshore eastern Brazil. *Mar. Petrol. Geol.*, **12**, 741–758.

Chan, M.A. (1992) Oolitic ironstone of the Cretaceous Western Interior Seaway, east-central Utah. *J. Sed. Petrol.*, **62**, 693–705.

Colson, J. & Cojan, I. (1996) Groundwater dolocretes in a lake-marginal environment: an alternative model for dolocrete formation in continental settings (Danian of the Provence Basin, France). *Sedimentology*, **43**, 175–188.

Colson, J., Cojan, I. & Thiry, M. (1998) A hydrogeological model for palygorskite formation in the Danian continental facies of the Provence Basin (France). *Clay Mineral.*, **33**, 333–347.

Corselli, C., Basso, D. & Garzanti, E. (1994) Paleobiological and sedimentological evidence of Pleistocene/Holocene hiatuses and ironstone formation at the Pontian islands, shelf break (Italy). *Mar. Geol.*, **117**, 317–328.

Cowan, G. (1989) Diagenesis of Upper Carboniferous sandstones: southern North Sea Basin. In: *Deltas: Sites and Traps for Fossil Fuels* (Ed. by M.K.G. Whateley and K.T. Pickering). *Geol. Soc. London Spec. Publ.*, **41**, 57–73.

Cross, T.A. (1988) Controls on coal distribution on trangressive–regressive cycles, Upper Createaceous, Western Interior, U.S.A. In: *Sea-level Changes: an Integrated Approach*, Society of Economic Paleontologists and Mineralogists Special Publication 42 (Ed. by C.K. Wilgus, B.S. Hastings, C.G.S.C. Kendall, H.W. Posamentier, C.A. Ross and J.C. Van Wagoner), Tulsa, OK, pp. 371–380.

Curiale, J.A., Cole, R.D. & Witmer, R.J. (1991) Application of organic geochemistry to sequence stratigraphic analysis: Four Corners area, New Mexico, U.S.A. *Org. Geochem.*, **19**, 53–75.

De Ros, L.F., Morad, S. & Paim, P.S.G. (1994) The role of detrital composition and climate on the diagenetic evolution of continental molasses: evidence from the Cambrio-Ordovician Guaritas sequence, southern Brazil. *Sed. Geol.*, **92**, 197–228.

Dewers, T. & Ortoleva, P. (1991) Influences of clay minerals on sandstone cementation and pressure solution. *Geology*, **19**, 1045–1048.

Dreyer, T., Corregidor, J., Arbues, P. & Puigdefabregas, C. (1999) Architecture of tectonically influenced Sobrarbe deltaic complex in the Ainsa Basin, northern Spain. *Sed. Geol.*, **127**, 127–169.

Duprat, M. (1997) Modéle tectono-sédimentaire des dépÙts Paléogènes dans le Nord-Est du Bassin de Paris: conséquences sur la géométrie du toit de la Craie. *Ann. Soc. Géol. Nord*, **5** (2éme série), 269–287.

Durand, M.M. (1977) Nature des colorations violettes et vertes de certains grès triasiques. *CR Acad. Sci. Paris*, **280D**, 2737–2740.

Dutta, P.K. & Sutner, L.J. (1986) Alluvial sandstone composition and paleoclimate, II Authigenic mineralogy. *J. Sed. Petrol.*, **56**, 346–358.

Ehrenberg, S.N. (1993) Preservation of anomalously high porosity in deeply buried sandstones by grain-coating chlorite: examples from the norwegian continental shelf. *AAPG Bull.*, **77**, 1260–1286.

Einsele, G., Ricken, W. & Seilacher, A. (1991) Cycles and events in stratigraphy—basic concepts and terms. In: *Cycles and Events in Stratigraphy* (Ed. by G. Einsele, W. Ricken and A. Seilacher), Springer-Verlag, Berlin, pp. 1–19.

Embry, A.F. (1995) Sequence boundaries and sequence hierarchies: problems and proposals. In: *Sequence Stratigraphy on the Northwest European Margin*, Norwegian Petroleum Society Special Publication 5, (Ed. by R.J. Steel, V.L. Felt, E.P. Johannessen and C. Marthieu), Elsevier, Amsterdam, pp. 1–11.

Emery, D., Myers, R.J. & Young, R. (1990) Ancient subaerial exposure and freshwater leaching in sandstones. *Geology*, **18**, 1178–1181.

Floquet, M. (1998) Outcrop cycle stratigraphy of shallow ramp deposits: the Late Cretaceous series on the Castilian ramp (northern Spain). In: *Mesozoic and Cenozoic Sequence Stratigraphy of European Basins*, Society of Economic Paleontologists and Mineralogists Special Publication 60 (Ed. by P.C. de Graciansky, J. Hardenbol, T. Jacquin and P.R. Vail), Tulsa, OK, pp. 343–361.

Galloway, W.E. (1989) Genetic stratigraphic sequences in basin analysis I: architecture and genesis of flooding-surface bounded depositional units. *AAPG Bull.*, **73**, 125–142.

Gensous, B. & Tesson, M. (1996) Sequence stratigraphy, seismic profiles, and cores of Pleistocene deposits on the Rhône continental shelf. *Sed. Geol.*, **105**, 183–190.

Ghibaudo, G., Grandesso, P., Massari, F. & Uchman, A. (1996) Use of trace fossils in delineating sequence stratigraphic surfaces (Tertiary Venetian Basin, northeastern Italy). *Palaeogeogr. Palaeoclimatol. Palaeoecol.*, **120**, 261–279.

Glenn, C.R. & Arthur, M.A. (1990) Anatomy and origin of a Cretaceous phosphorite greensand giant, Egypt. *Sedimentology*, **37**, 123–154.

Grim, R.E. & Güven, N. (1978) *Bentonites*. Elsevier, Amsterdam, 256 pp.

Helgeson, H.C., Brown, T.H. & Leeper, R.H. (1969) *Handbook of Theoretical Activity Diagrams Depicting Chemical Equilibra in Geological Systems Involving an Aqueous Phase at One atm and 0° to 300°C.* Freeman, Cooper & Co., San Francisco, 253 pp.

Helland-Hansen, W. & Gjelberg, J.G. (1994) Conceptual basis and variability in sequence stratigraphy: a different perspective. *Sed. Geol.*, **92**, 31–52.

Hornibrook, E.R.C. & Longstaffe, F.J. (1996) Berthierine from the Lower Cretaceous Clearwater Formation, Alberta, Canada. *Clay. Clay Mineral.*, **44**, 1–21.

Hurst, A. & Nadeau, P.H. (1995) Clay microporosity in reservoir sandstones: an application of quantitative electron microscopy in petrophysical evaluation. *AAPG Bull.*, **79**, 563–573.

Iijima, A. & Matsumoto, R. (1982) Berthierine and chamosite in coal measures of Japan. *Clay. Clay Mineral.*, **30**, 264–274.

Inglès, M. & Anadón, P. (1991) Relationship of clay minerals to depositional environment in the non-marine Eocene Pontils Group, SE Ebro Basin (Spain). *J. Sed. Petrol.*, **61**, 926–939.

Jervey, M.T. (1988) Quantitative geologic modeling of siliciclastic rock sequences and their seismic expression. In: *Sea-level Changes—an Integrated Approach*, Society of Economic Paleontologists and Mineralogists Special Publication 42 (Ed. by C.K. Wilgus, B.S. Hastings, C.G. Kendall, H.W. Posamentier, C.A. Ross and J.C. Van Wagoner), Tulsa, OK, pp. 47–69.

Jiao, Z.S. & Surdam, R.C. (1997) Clay diagenesis and sealing capacities of sequence boundaries in the Cretaceous section in Laramide basins. In: *American Association of Petroleum Geologists and Society for Sedimentary Geology Annual Meeting Abstracts* 6, Tulsa, OK, p. 56.

Keller, W.D. (1970) Environment aspects of clay minerals. *J. Sed. Petrol.*, **40**, 788–813.

Kelly, J.C. & Webb, J.A. (1999) The genesis of glaucony in the Oligo-Miocene Torquay Group, southeastern Australia: petrographic and geochemical evidence. *Sed. Geol.*, **125**, 99–114.

Kim, Y. & Lee, Y.I. (2000) Ironstones and green marine clays in the Dongjeom Formation (Early Ordovician) of Korea. *Sed. Geol.*, **130**, 65–80.

Kirschbaum, M.A. & McCabe, P.J. (1992) Controls on the accumulation of coal and on the development of anastomosed fluvial systems in the Cretaceous Dakota Formation of southern Utah. *Sedimentology*, **34**, 581–598.

Kitamura, A. (1998) Glaucony and carbonate grains as indicators of the condensed section: Omma Formation, Japan. *Sed. Geol.*, **122**, 151–163.

Kronen, J.D. Jr. & Glenn, C.R. (2000) Pristine to reworked verdine: keys to sequence stratigraphy in mixed carbonate–siliciclastic forereef sediments (Great Barrier Reef). In: *Marine Authigenesis: from Global to Microbial*, Society of Economic Paleontologists and Mineralogists Special Publication 65 (Ed. by C.R. Glenn, J. Lucas and L. Lucas), Tulsa, OK, pp. 387–403.

Leithold, E.L. (1993) Preservation of laminated shale in ancient clinoforms; comparison to modern subaqueous deltas. *Geology*, **21**, 359–362.

Loutit, T.S., Hardenbol, J., Vail, P.R. & Baum, G.R. (1988) Condensed sections: the key to age dating and correlation of continental margin sequences. In: *Sea-level Changes: an Integrated Approach*, Society of Economic Paleontologists and Mineralogists Special Publication 42 (Ed. by C.K. Wilgus, B.S. Hastings, C.G.S.C. Kendall, H.W. Posamentier, C.A. Ross and J.C. Van Wagoner), Tulsa, OK, pp. 183–216.

Malartre, F., Ferry, S. & Rubino, J.L. (1998) Interactions climat-eustatisme-tectonique. Les enseignements et perspectives du Crétacé supérieur (Cénomanien-Coniacen). *Geodin. Acta*, **11**, 253–270.

Malicse, A. & Mazzullo, J. (1996) Early diagenesis and paleosol features of ancient desert sediments: examples from the Permian Basin. In: *Siliciclastic Diagenesis and Fluid Flow: Concepts and Applications*, Society of Economic Paleontologists and Mineralogists Special Publication 55 (Ed. by L.J. Crossey, R. Loucks and M.W. Totten), Tulsa, OK, pp. 151–161.

Mancini, E.A. & Tew, B.H. (1993) Eustasy versus subsidence: Lower Paleocene depositional sequences from southern Alabama, eastern Gulf Coastal Plain. *Geol. Soc. Am. Bull.*, **105**, 3–17.

Matlack, K.S., Houseknecht, D.W. & Applin, K.R. (1989) Emplacement of clay into sand by infiltration. *J. Sed. Petrol.*, **59**, 77–87.

McCarthy, P.J. & Plint, A.G. (1998) Recognition of interfluve sequence boundaries: integrating paleopedology and sequence stratigraphy. *Geology*, **26**, 387–390.

McCarthy, P.J., Faccini, U.F. & Plint, A.G. (1999) Evolution of an ancient coastal plain: paleosols, interfluves and alluvial architecture in a sequence stratigraphic framework, Cenomanian Dunvegan Formation, NE British Columbia, Canada. *Sedimentology*, **46**, 861–891.

McCracken, S.R., Compton, J. & Hicks, K. (1996) Sequence-stratigraphic significance of glaucony-rich lithofacies. In: *Proceedings of the Ocean Drilling Program: Scientific Results*, Vol. 150 (Ed. by K.G. Miller, G.S. Mountain, P. Blum, C.W. Poag and D.C. Twitchell), College Station, TX, pp. 171–184.

Meisler, H., Leahy, P.P. & Knobel, L.L. (1984) Effect of eustatic sea-level changes on saltwater–freshwater in the Northern Atlantic Coast Plain. *US Geol. Surv. Water-supply Pap.*, **2255**, 27 pp.

Mitchum, R.M. & Van Wagoner, J.C. (1991) High-frequency sequences and their stacking pattern: sequence-stratigraphic evidence of high-frequency eustatic cycles. *Sed. Geol.*, **70**, 131–160.

Moncure, G.K., Lahann, R.W. & Siebert, R.M. (1984) Origin of secondary porosity and cement distribution in a sandstone/shale sequence from the Frio Formation (Oligocene). In: *Clastic Diagenesis* (Ed. by D.A. McDonald and R.C. Surdam). *AAPG Mem.*, **37**, 151–161.

Morad, S., Ketzer, J.M. & De Ros, L.F. (2000) Spatial and temporal distribution of diagenetic alterations in siliciclastic rocks: implications for mass transfer in sedimentary basins. *Sedimentology*, **47** (Supplement 1 Millennium Reviews), 95–120.

Moraes, M.A.S. & De Ros, L.F. (1990) Infiltrated clays in Jurassic fluvial sandstones of Recôncavo Basin, Northeastern Brazil. *J. Sed. Petrol.*, **60**, 809–819.

Moraes, M.A.S. & De Ros, L.F. (1992) Depositional, infiltrated and authigenic clays in fluvial sandstones of the Jurassic Sergi Formation, Recôncavo Basin, Northeastern Brazil. In: *Origin, Diagenesis, and Petrophysics of Clay Minerals in Sandstones*, Society of Economic Paleontologists and Mineralogists

Special Publication 47 (Ed. by D.W. Houseknecht and E.D. Pittman), Tulsa, OK, pp. 197–208.

Myrow, P. (1998) Transgressive stratigraphy and depositional framework of Cambrian tidal dune deposits, Peerless Formation, Central Colorado, U.S.A. In: *Tidalites: Processes and Products*, Society of Economic Paleontologists and Mineralogists Special Publication 61 (Ed. by C.R. Alexander, R.A. Davis and V.J. Henry), Tulsa, OK, pp. 143–154.

Odin, G.S. (1988) Glaucony from the Gulf of Guinea. In: *Green Marine Clays* (Ed. by G.S. Odin), Elsevier, Amsterdam, pp. 225–247.

Odin, G.S. & Fullagar, P.D. (1988) Geochemical significance of the glaucony facies. In: *Green Marine Clays* (Ed. by G.S. Odin), Elsevier, Amsterdam, pp. 295–332.

Odin, G.S. & Matter, A. (1981) De glauconiarum origine. *Sedimentology*, **28**, 611–641.

Odin, G.S., Bailey, S.W., Amouric, M., Frohlich, F. & Waychunas, G.A. (1988) Mineralogy of the facies verdine. In: *Green Marine Clays* (Ed. by G.S. Odin), Elsevier, Amsterdam, pp. 159–206.

Palsey, M.A., Gregory, W.A. & Hart, G.F. (1991) Organic matter variations in transgressive and regressive shales. *Org. Geochem.*, **17**, 483–501.

Percival, C.J. (1986) Paleosols containing an albic horizon: examples from the Upper Carboniferous of northern Britain. In: *Paleosols* (Ed. by V.P. Wright), Blackwell Scientific Publications, Oxford, pp. 87–110.

Perlmutter, M.A. & Matthews, M.D. (1989) Global cyclostratigraphy—a model. In: *Quantitative Dynamic Stratigraphy* (Ed. by T.A. Cross), Prentice Hall, New Jersey, pp. 233–260.

Pittet, B. & Strasser, A. (1998) Depositional sequences in deep-shelf environments formed through carbonate-mud import from the shallow platform (Late Oxfordian, German Swabian Alb and eastern Swiss Jura). *Eclogae Geol. Helv.*, **91**, 149–169.

Platt, J.D. (1993) Controls on clay mineral distribution and chemistry in the Early Permian Rotliegend of Germany. *Clay Mineral.*, **28**, 393–416.

Porrenga, D.H. (1967) Glauconite and chamosite as depth indicators in the marine environment. *Mar. Geol.*, **5**, 495–501.

Posamentier, H.W. & Allen, G.P. (1993) Variability of the sequence stratigraphic model: effects of local basin factors. *Sed. Geol.*, **86**, 91–109.

Posamentier, H.W. & Vail, P.R. (1988) Eustatic controls on clastic deposition II—sequence and systems tract models. In: *Sea-level Changes: an Integrated Approach*, Society of Economic Paleontologists and Mineralogists Special Publication 42 (Ed. by C.K. Wilgus, B.S. Hastings, C.G.S.C. Kendall, H.W. Posamentier, C.A. Ross and J.C. Van Wagoner), Tulsa, OK, pp. 125–154.

Posamentier, H.W., Jervey, M.T. & Vail, P.R. (1988) Eustatic controls on clastic deposition I—conceptual framework. In: *Sea-level Changes: an Integrated Approach*, Society of Economic Paleontologists and Mineralogists Special Publication 42 (Ed. by C.K. Wilgus, B.S. Hastings, C.G.S.C. Kendall, H.W. Posamentier, C.A. Ross and J.C. Van Wagoner), Tulsa, OK, pp. 109–124.

Rao, V.P., Lamboy, M. & Dupeuble, P.A. (1993) Verdine and other associated authigenic (glaucony, phosphate) facies from the surficial sediments from Krishna to Ganges river mouth, east coast of India. *Mar. Geol.*, **111**, 133–158.

Robaszynski, F., Gale, A., Juignet, P., Amédro, P. & Hardenbol, J. (1998) Sequence stratigraphy in the Upper Cretaceous series of the Anglo-Paris Basin: exemplified by the Cenomanian stage. In: *Sequence Stratigraphy of European Basins: Mesozoic and Cenozoic*, Society of Economic Paleontologists and Mineralogists Special Publication 60 (Ed. by P.C. de Graciansky, J. Hardenbol, T. Jacquin and P.R. Vail), Tulsa, OK, pp. 363–386.

Rossi, C. & Cañaveras, J.C. (1999) Pseudospherulitic fibrous calcite in paleo-groundwater, unconformity-related diagenetic carbonates (Paleocene of the Áger Basin and Miocene of the Madrid Basin, Spain). *J. Sed. Res.*, **69**, 224–238.

Ruffell, A. & Wach, G. (1998) Firmgrounds—key surfaces in the recognition of parasequences in the Aptian Lower Greensand Group, Isle of Wight (southern England). *Sedimentology*, **45**, 91–107.

Ruffell, A.H., Worden, R.H. & Evans, R. (2003) Palaeoclimate controls on spectral gamma-ray radiation from sandstones. In: *Clay Mineral Cements in Sandstones*, International Association of Sedimentologists Special Publication 34 (Ed. by R.H. Worden and S. Morad), Blackwell Publishing, Oxford, pp. 93–108.

Ryer, T. (1984) Transgressive–regressive cycles and the occurrence of coal in some Upper Cretaceous strata of Utah, U.S.A. In: *Sedimentology of Coal and Coal-bearing Sequences*, International Association of Sedimentologists Special Publication 7 (Ed. by R.A. Rahmani and R.M. Flores), Blackwell Scientific Publications, Oxford, pp. 217–227.

Sayles, F.L. & Mangelsdorf, P.C. (1979) Cation-exchange characteristics of Amazon River suspended sediment and its reaction with seawater. *Geochim. Cosmochim. Acta*, **43**, 767–779.

Schulz-Rojahn, J.P., Seeburger, D.A. & Beacher, G.J. (2003) Application of glauconite morphology in geosteering and for on-site reservoir quality assessment in very fine-grained sandstones: Carnarvon Basin, Australia. In: *Clay Mineral Cements in Sandstones*, International Association of Sedimentologists Special Publication 34 (Ed. by R.H. Worden and S. Morad), Blackwell Publishing, Oxford, pp. 473–488.

Schwarzacher, W. (1993) *Cyclostratigraphy and the*

Milankovitch Theory. Developments in Sedimentology 52, Elsevier, Amsterdam, 240 pp.

Seyler, B. (1997) Low resistivity in Aux Vases sandstone reservoirs in the Illinois Basin linked to clay minerals coating sand grains. In: *American Association of Petroleum Geologists Eastern Section and the Society for Organic Petrology Joint Meeting, Abstracts. AAPG Bull.*, **81**, 1564.

Shanley, K.W. & McCabe, P.J. (1993) Alluvial architecture in a sequence stratigraphic framework: a case history from the Upper Cretaceous of southern Utah, USA. In: *Quantitative Modelling of Clastic Hydrocarbon Reservoirs and Outcrop Analogues*, International Association of Sedimentologists Special Publication 15 (Ed. by S. Flint and I. Bryant), Blackwell Scientific Publications, Oxford, pp. 21–56.

Shanley, K.W. & McCabe, P.J. (1994) Perspectives on the sequence stratigraphy of continental strata. *AAPG Bull.*, **78**, 544–568.

Staub, J.R. & Cohen, A.D. (1978) Kaolinite-enrichment beneath coals: a modern analog, Snuggedy swamp, South Caroline. *J. Sed. Petrol.*, **48**, 203–210.

Stonecipher, S.A. & May, J.A. (1990) Facies controls on early diagenesis: Wilcox Group, Texas Gulf Coast. In: *Prediction of Reservoir Quality Through Chemical Modeling* (Ed. by I.D. Meshri and P.J. Ortoleva). *AAPG Mem.*, **49**, 25–44.

Sugarman, P.J. & Miller, K.G. (1997) Correlation of Miocene sequences and hydrogeologic units, New Jersey coastal plain. *Sed. Geol.*, **108**, 3–18.

Surdam, R.C. & Yin, P. (1995) Organic acids and carbonate stability, the key to predicting positive porosity anomalies. In: *Organic Acids in Geological Processes* (Ed. by E.D. Pittman and M.D. Lewan), Springer-Verlag, Berlin, pp. 398–448.

Surdam, R.C., Boese, S.W. & Crossey, L.J. (1984) The chemistry of secondary porosity. In: *Clastic Diagenesis* (Ed. by D.A. McDonald and R.C. Surdam). *AAPG Mem.*, **37**, 127–149.

Thiry, M. & Jacquin, T. (1993) Clay mineral distribution related to rift activity, sea-level changes and paleoceanography in the Cretaceous of the Atlantic Ocean. *Clay Mineral.*, **28**, 61–84.

Vail, P.R., Mitchum, R.M. & Thompson, S. (1977) Seismic stratigraphy and global changes of sea level, part 3: relative changes of sea level from coastal onlap. In: *Seismic Stratigraphy—Applications to Hydrocarbon Exploration* (Ed. by C.E. Payton). *AAPG Mem.*, **26**, 63–81.

Van Keer, I., Muchez, P.H. & Viaene, W. (1998) Clay mineralogical variations and evolutions in sandstone sequences near a coal seam and shales in the Westphalian of the Campine Basin (NE Belgium). *Clay Mineral.*, **33**, 159–169.

Van Wagoner, J.C., Mitchum, R.M., Campion, K.M. & Rahmanian, V.D. (1990) Siliciclastic sequence stratigraphy in well logs, cores, and outcrops: concepts for high-resolution correlation of time and facies. *AAPG Methods Explor. Ser.*, **7**, 55 pp.

Vandenberghe, N., Laga, P., Steurbaut, E., Hardenbol, J. & Vail, P.R. (1998) Tertiary sequence stratigraphy at the southern border of the North Sea Basin in Belgium. In: *Mesozoic and Cenozoic Sequence Stratigraphy of European Basins,* Society of Economic Paleontologists and Mineralogists Special Publication 60 (Ed. by P.C. de Graciansky, J. Hardenbol, T. Jacquin and P.R. Vail), Tulsa, OK, pp. 119–154.

Walker, R.G. & Bergman, K.M. (1993) Shannon Sandstone in Wyoming: a shelf–ridge complex reinterpreted as lowstand shoreface deposits. *J. Sed. Petrol.*, **63**, 839–851.

Walker, T.R. (1976) Diagenetic origin of continental red beds. In: *The Continental Permian in Central, West and South Europe* (Ed. by H. Falke), D. Reidel, Dordrecht, pp. 240–482.

Walker, T.R., Waugh, B. & Grone, A.J. (1978) Diagenesis in first-cycle desert alluvium of Cenozoic age, southwestern United States and northwestern Mexico. *Geol. Soc. Am. Bull.*, **89**, 19–32.

Weaver, C.E. (1984) Origin and geologic implications of the palygorskite of the SE United States. In: *Palygorskite–Sepiolite* (Ed. by A. Singer and E. Galen), Elsevier, Amsterdam, pp. 39–58,

Wright, V.P. & Marriott, S.B. (1993) The sequence stratigraphy of fluvial depositional systems: the role of floodplain sediment storage. *Sed. Geol.*, **86**, 203–210.

Int. Assoc. Sedimentol. Spec. Publ. (2003) **34**, 63–91

Oxygen and hydrogen isotopic composition of diagenetic clay minerals in sandstones: a review of the data and controls

S. MORAD[1], R.H. WORDEN[2] *and* J.M. KETZER[1]

[1] *Department of Earth Sciences, Uppsala University, Villa vägen 16, S-752 36 Uppsala, Sweden, e-mail: sadoon.morad@geo.uu.se, marcelo.ketzer@geo.uu.se*

[2] *Department of Earth Sciences, University of Liverpool, Brownlow Street, Liverpool L69 3GP, UK, e-mail: r.worden@liverpool.ac.uk*

ABSTRACT

Analyses of O and H isotopes ($\delta^{18}O$ and δ^2H) in diagenetic clay minerals of sandstones have been used increasingly to decipher primarily: (i) the palaeoclimatic conditions that prevailed during near-surface diagenesis (eo- and telodiagenesis), (ii) precipitation temperature of the clay minerals, and (iii) the isotopic composition, origin and geochemical evolution of formation waters. However, achieving these goals is fraught with a number of uncertainties, including the level of accuracy of the temperature-dependent, isotopic fractionation equations between clay minerals and water at temperatures encountered during diagenesis, degree of isotopic resetting subsequent to crystallization, and difficulties in obtaining pure, monophase clay minerals.

In this paper, we address these potential uses and uncertainties based on isotopic data on clay minerals (mainly kaolin, illite, chlorite and mixed-layer types) in sandstones compiled from the literature, and decipher the links between the isotopic evolution and equilibrium state between clay minerals and formation waters during basin evolution.

There is evidence indicating that, once formed, diagenetic clay minerals may preserve their original isotopic composition, unless subjected to dissolution–reprecipitation reactions (i.e. transformation into other clay minerals and recrystallization). Each group of diagenetic clay minerals has relatively wide O and H isotopic values both on global and basinal scales, implying wide variations of crystallization temperatures and/or in the isotopic composition of fluids involved. Kaolin has $\delta^{18}O$ values that range from +5‰ to +28.5‰ (75% between +12.5‰ and +18.5‰) and δ^2H (δD) values from −140‰ to −30‰ (50% between −70‰ and −40‰). The illitic clay minerals have $\delta^{18}O$ values that range from +5‰ to +26‰ (75% between +11.0‰ and +17.5‰) and δ^2H values from −110‰ to −30‰ (65% between −70‰ and −45‰). The chloritic clay minerals have $\delta^{18}O$ values that range between +0.0‰ and +20.0‰ (75% between +7‰ and +15‰) and δ^2H values between −110‰ and −10‰ (40% between −80‰ and −60‰).

Gulf Coast sandstones have clay $\delta^{18}O$ values that decrease and formation water $\delta^{18}O$ values that increase with depth, approximating to the maintenance of isotopic equilibrium during burial. In contrast, North Sea basin sandstones have widely variable clay $\delta^{18}O$ values despite having formation water $\delta^{18}O$ values that, like the Gulf Coast basin, increase with depth. The variability of North Sea clay $\delta^{18}O$ values is possibly related to (i) localized meteoric influx, inducing clay growth in exotic but transient water and (ii) initial equilibrium between water and clay that grew at moderate depth but failed to maintain equilibrium with the ambient water during continued burial. North Sea basin sandstones and waters have widely variable clay δ^2H values and do not show any simple patterns. Extremely low clay δ^2H values may be the result of localized petroleum–clay interactions.

INTRODUCTION

Clay minerals in sandstones form under all diagenetic regimes and have a substantial impact on reservoir-quality evolution. It therefore is of profound importance to unravel the conditions and timing of their formation. The oxygen and hydrogen isotopic compositions of diagenetic clay minerals in sandstones have been used widely over the past 20 yr as a tool for helping to unravel the physical and geochemical conditions encountered during the diagenetic evolution of clastic sequences. The fractionation of O and H isotopes between a clay mineral and water is controlled primarily by temperature. An increase in temperature induces an enrichment of the clay minerals in ^{16}O and ^2H relative to the aqueous medium. Stable isotopes of clay minerals in sandstones are also controlled by kinetics and mass balance considerations related to parameters such as permeability and the water/rock ratio (e.g. Longstaffe, 1989).

The solid-state exchange of structurally bonded O and H isotopes between clay minerals and pore-water at the low temperatures encountered during diagenesis generally is believed to be insignificant relative to the control by temperature and water $\delta^{18}O$ (Yeh & Savin, 1977; Longstaffe, 1983; Yeh & Eslinger, 1986; O'Neil, 1987). Thus, the isotopic composition of clay minerals has been used to constrain the origin of the formation water from which clay minerals grew, the precipitation temperature and palaeoclimatic conditions.

Water isotopic facies in sedimentary basins

There are different types of water, each of which may plausibly contribute to sedimentary formation water and thus influence clay mineral stable isotopes (Fig. 1).

1 Sea water is the standard used for stable isotopes of oxygen and hydrogen (see later), which presently has $\delta^{18}O$ and δ^2H values of 0‰ V-SMOW (Vienna Standard Mean Ocean Water), but has varied over geological time.

2 Rain water is generated primarily through the evaporation of sea water. The condensation of clouds into rain during movement to higher latitudes or altitudes causes the progressive enrichment of meteoric waters in ^{16}O relative to ^{18}O (i.e. decrease in $\delta^{18}O$), and in ^1H relative to ^2H (i.e. a decrease in δ^2H; Craig, 1961; Dansgaard, 1964). There is a global meteoric water line on which the vast majority of present-day rainwater plots (described by the equation $\delta^2H = 8 \cdot \delta^{18}O + 10$) and where the highest latitude meteoric water plots furthest from the origin (V-SMOW).

3 Sea water in enclosed basins, or even ground water, influenced by evaporation under arid conditions becomes enriched in ^{18}O and ^2H isotopes relative to the lighter ^{16}O and ^1H isotopes, respectively. Thus, evaporated sea water achieves elevated $\delta^{18}O$ and δ^2H values up to the point where the water has about four times the original salinity. At higher salinities, gypsum precipitation occurs, and the residual water gradually acquires lower $\delta^{18}O$ and δ^2H values.

4 The two common water sources of isotopically depleted hydrogen include high-latitude (or altitude) meteoric water and organic waters (Fig. 1). The latter waters are produced from a variety of organic-related processes including normal kerogen maturation, decomposition of oxidized compounds associated with petroleum (such as organic acids) and redox processes including petroleum fluids and oxidized

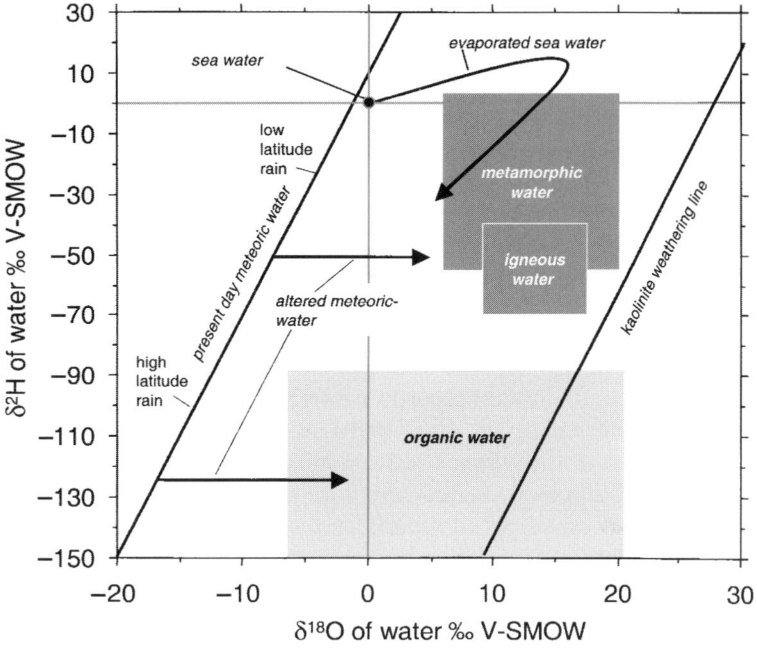

Fig. 1 Stable isotope ($\delta^{18}O$ and $\delta^{2}H$) characteristics of different types of water. Diagram modified from Sheppard (1986) and Longstaffe (1989). The kaolinite weathering line has been added to show where kaolinite, formed at Earth's surface, should plot assuming isotopic equilibrium with meteoric water (derived using Sheppard & Gilg's (1996) oxygen and hydrogen isotope fractionation equations). Evaporated sea water can have a range of isotopic compositions depending upon the degree of evaporation. Organic water has an arbitrary upper $\delta^{2}H$ limit of −90‰ V-SMOW and plausibly could reach much more depleted $\delta^{2}H$ values than shown. Sedimentary formation waters typically lie between the meteoric water line and the kaolinite weathering line.

species in minerals (e.g. SO_4^{2-}, Fe^{3+}; e.g. Surdam *et al.*, 1993; Worden *et al.*, 1996). Isotopically depleted hydrogen isotopes also can be derived during microbial fermentation of organic matter. Conversely, water derived from redox processes will probably have normal oxygen isotope values because the oxygen is typically derived from inorganic sources (Sheppard, 1986).

5 Metamorphic and igneous waters, which tend to be enriched in ^{18}O ($\delta^{18}O$ V-SMOW *c.* +6‰) relative to meteoric waters are unlikely to contribute volumetrically significant quantities of water to formation waters.

6 Connate waters, those present within a sediment at the time of deposition, are largely a theoretical concept, especially for permeable formations. It is highly unlikely that the original water could remain *in situ* given the typically complex history of relative sea-level changes during eodiagenesis, compaction and fluid expulsion, petroleum generation, and shale dewatering, to say nothing of localized uplift, exposure and telodiagenesis prior to reburial.

Sedimentary formation waters (diagenetic waters) have $\delta^{2}H$ values that plot to the right of the meteoric water line but rarely have values of more than about 30‰ greater than the meteoric water $\delta^{18}O$ for a given $\delta^{2}H$ value. Sedimentary formation waters can result from mixing of end-members (discussed above) but also are affected by water–rock interaction whereby low-temperature water–mineral assemblages re-equilibrate upon heating.

Clay mineral growth and diagenetic water types

Some clay minerals are formed shortly after deposition (eodiagenesis). Eogenetic clay minerals are formed by precipitation from near-surface pore waters (marine, meteoric or mixed) or through the replacement of framework grains (e.g. feldspar, mica, volcanic glass) or detrital clay-mineral precursors. Eogenetic clay minerals in continental sandstones usually form at higher surface temperatures from low $\delta^{18}O$ meteoric waters, and thus have lower $\delta^{18}O_{SMOW}$ values (typically +15‰ to +20‰) than eogenetic marine clay minerals ($\delta^{18}O$ = +26‰ to +31‰;

Savin & Epstein, 1970; Yeh & Savin, 1977; Clauer *et al.*, 1990). Telogenetic clay minerals typically form by incursion of meteoric waters below unconformity surfaces subsequent to uplift and subaerial exposure of the sequence. The stable isotopic composition of eogenetic and telogenetic clay minerals helps to constrain the porewater composition and the palaeoclimatic conditions.

For clay minerals formed at considerable burial depths (> 2 km) and elevated temperatures (> 70°C; mesodiagenesis, *sensu* Morad *et al.*, 2000), the stable isotope ratios can be used to help constrain the origin and geochemical evolution of formation waters, precipitation temperatures, degree of water–mineral equilibria, as well as the extent and scale of mass transfer in the subsurface. Mesogenetic clay minerals form mainly through the progressive transformation of eogenetic clay minerals, such as of kaolinite into dickite or illite, and of smectite into illite or chlorite and of berthierine into chlorite. These transformation reactions occur typically through small-scale dissolution and reprecipitation (Worden & Morad, this volume, pp. 3–41).

Temperatures of clay mineral growth

The precipitation temperatures or isotopic composition of fluids can be determined or constrained by using the temperature-dependent, equilibrium isotope fractionation equations between clay minerals and the reacting water. Caution should be used when interpreting calculated temperatures that are based on the $\delta^{18}O$ values of clay minerals, simply because the oxygen isotopic composition of the fluids from which the minerals were formed is typically poorly defined, and can be, at best, only inferred or broadly constrained. Caution also should be exercised because it is always difficult to prove the occurrence of isotopic equilibrium between clay *minerals* and their ambient water. Thus, the interpretation of temperatures that are calculated based on such premises should be performed with appropriate circumspection and within the framework of a sound paragenetic sequence, absolute dating (which can work for K-bearing minerals such as illite and glauconite) and microthermometric studies of fluid inclusion in suitable associated minerals, such as quartz and carbonate cements.

Oxygen isotopes, rather than hydrogen isotopes, typically are used to determine temperature for two reasons. First, the mineral–water oxygen isotope fractionation equations are much more sensitive to temperature than are hydrogen isotopes. Second, the temperature-dependent isotope fractionations are much less well known for hydrogen than for oxygen isotopes.

Isotopes and palaeoclimatic conditions

Eogenetic clay minerals formed under low latitude and/or dry climatic conditions generally have higher $\delta^{18}O$ and $\delta^{2}H$ values than those formed at high latitudes and altitudes (Savin & Epstein, 1970; Sheppard, 1977; Hassanipak & Eslinger, 1985). Clay minerals formed from meteoric waters in coastal areas or in oceanic islands have higher $\delta^{18}O$ and $\delta^{2}H$ values than those formed at similar latitude and evaporation/precipitation rates in continental interiors owing to more extensive isotopic fractionation between the water vapour (i.e. cloud) and liquid (i.e. rain) phases. If petrographic or other criteria suggest that clay minerals are eogenetic, then their stable isotopes could be used to define the $\delta^{18}O$ and $\delta^{2}H$ values of the meteoric water that induced their growth. If the criteria suggest that the clay minerals were subjected to recrystallization then obviously their stable isotopes must not be used for palaeoclimate reconstruction.

AIMS

Following this brief overview of the controls on, and uses of, isotopic composition of clay minerals, the aims of this paper are to:

1 discuss the derivation and values of isotopic equilibrium fractionation factors between different clay minerals and formation water;

2 present a compilation of published stable O and H isotopic data from various diagenetic clay minerals in sandstones;
3 provide a global constraint on the parameters controlling the isotopic compositions of clay minerals during basin evolution by highlighting clay-mineral and formation-water stable isotope data over wide depth ranges in the North Sea and Gulf Coast basins.

STABLE ISOTOPE FRACTIONATION FACTORS

The isotope ratios of ^{18}O to ^{16}O are expressed using the normal delta notation:

$$\delta^{18}O\ (‰\ \text{V-SMOW}) = [(R_{sample}/R_{SMOW}) - 1] \times 10^3 \quad (1)$$

where R is $^{18}O/^{16}O$ and V-SMOW is Vienna Standard Mean Ocean Water—the global oxygen isotope standard. The isotope ratio of a mineral is controlled by the temperature and the isotope ratio of the ambient fluid with which it is in (isotopic) equilibrium. Thus, it is conventional to consider the $\delta^{18}O$ of both a (clay) mineral and the $\delta^{18}O$ of the associated water. The fractionation, or partitioning, of isotopes between a mineral and the associated water is defined by the term 'α' where:

$$\alpha_{clay\text{-}water} = R_{clay}/R_{water} \quad (2)$$

The difference in the stable isotope ratios of two associated phases (e.g. clay and water) can be represented by:

$$\Delta_{clay\text{-}water} = \delta^{18}O_{clay} - \delta^{18}O_{water} \quad (3)$$

Now the fractionation factor $\alpha_{clay\text{-}water}$ also can be defined:

$$\alpha_{clay\text{-}water} = (1000 + \delta^{18}O_{clay})/(1000 + \delta^{18}O_{water}) \quad (4)$$

As $10^3 \ln 1.00X$ is very similar to the value of X, it is possible to use the approximation:

$$10^3 \ln \alpha \approx \delta^{18}O_{clay} - \delta^{18}O_{water} \quad (5)$$

The temperature dependence of the partition, or fractionation, of isotopes between mineral and water is often expressed by relations of the type:

$$10^3 \ln \alpha = A \cdot (10^6 T^{-2}) + B \quad (6)$$

where A and B are constants particular to the mineral in question. Thus, by relating equations (5) and (6), it is possible to conclude that differences in oxygen stable isotopes between clay and water vary as a function of temperature according to relationships of the type:

$$\delta^{18}O_{clay} - \delta^{18}O_{water} \approx A \cdot (10^6 T^{-2}) + B \quad (7)$$

The same approach can be taken for hydrogen stable isotopes. Hydrogen also uses standard mean ocean water (V-SMOW) as the stable isotopic standard.

Experimental and empirical fractionation equations (in the form of equation 7) have been derived for different clay minerals (Table 1). Oxygen isotope fractionation equations have been derived for all the major clay minerals. Hydrogen isotope fractionation relations are less well known for some of the clay minerals (smectite and chlorite) and water because empirical relationships have proved difficult to define and because existing analyses of the data have shown hydrogen isotopes to be very sensitive to the specific composition of the clay mineral (Sheppard & Gilg, 1996).

Oxygen isotope fractionation equations between clay minerals and water have been derived for kaolinite, smectite, illite and chlorite (Fig. 2). A noteworthy feature of these clay-mineral–water oxygen isotope relations is that they are very similar for the different clay minerals in that they largely have the same form and vary by only a few parts per thousand at any given temperature. The clay minerals are enriched in ^{18}O with respect to the water. At close to 0°C, the fractionation ($\Delta_{clay\text{-}water}$) is about +30‰. At 100°C, the fractionation is about +13‰ to +16‰. This similarity in clay-mineral–water oxygen isotope relations implies that: (i) $\delta^{18}O$ values of co-precipitated clay minerals cannot be used as a palaeothermometer, (ii) the oxygen isotope evolution of the formation water is not much affected by the specific *type* of clay mineral, although the amount of clay mineral

Table 1 Fractionation factors for different isotopes in different clay minerals.

Stable isotope	Mineral	Fractionation equation: $1000 \ln \alpha_{(\text{mineral-water})}$	Reference
Hydrogen	Kaolinite	$-4.53 \times 10^6T^{-2} + 19.4$	Lambert & Epstein, 1980
Hydrogen	Kaolinite	$-2.2 \times 10^6T^{-2} - 7.7$	Sheppard & Gilg, 1996
Hydrogen	Illite ± smectite	$-19.6 \times 10^3T^{-1} + 25$	Yeh, 1980
Hydrogen	Illite ± smectite	$-45.3 \times 10^3T^{-1} + 94.7$	Capuano, 1992
Oxygen	Kaolinite	$2.50 \times 10^6T^{-2} - 2.87$	Land & Dutton, 1978
Oxygen	Kaolinite	$10.6 \times 10^3T^{-1} + 0.42 \times 10^6T^{-2} - 15.337$	Savin & Lee, 1988
Oxygen	Kaolinite	$2.76 \times 10^6T^{-2} - 6.75$	Sheppard & Gilg, 1996
Oxygen	Smectite	$2.58 \times 10^6T^{-2} - 4.91$	Savin & Lee, 1988
Oxygen	Smectite	$2.55 \times 10^6T^{-2} - 4.05$	Sheppard & Gilg, 1996
Oxygen	Illite	$-2.87 + 1.83 \times 10^6T^{-2} + 0.0614 \times (10^6T^{-2})^2 - 0.00115 \times (10^6T^{-2})^3$	Lee, 1984
Oxygen	Illite	$2.39 \times 10^6T^{-2} - 4.19$	Savin & Lee, 1988
Oxygen	Illite	$2.39 \times 10^6T^{-2} - 3.76$	Sheppard & Gilg, 1996
Oxygen	Chlorite	$1.56 \times 10^6T^{-2} - 4.70$	Wenner & Taylor, 1971
Oxygen	Chlorite (>150°C)	$-8.38 \times 10^3T^{-1} + 4.81 \times 10^6T^{-2}$	Cole, 1985
Oxygen	Chlorite (Mg–Fe in hydroxide, Al–Fe in octahedral sheet, < 150°C)	$3.72 \times 10^3T^{-1} + 2.5 \times 10^6T^{-2} - 0.312 \times 10^9T^{-3} + 0.028 \times 10^{12}T^{-4} - 12.62$	Savin & Lee, 1988
Oxygen	Chlorite (Al–Mg–Fe in hydroxide, Mg in octahedral sheet, < 150°C)	$2.56 \times 10^3T^{-1} + 3.39 \times 10^6T^{-2} - 0.623 \times 10^9T^{-3} + 0.056 \times 10^{12}T^{-4} - 11.86$	Savin & Lee, 1988
Oxygen	Chlorite (Mg–Al in hydroxide, Al in octahedral sheet, < 150°C)	$6.78 \times 10^3T^{-1} + 1.19 \times 10^6T^{-2} - 13.68$	Savin & Lee, 1988

produced will, of course, be important, and (iii) the stable isotope composition of formation water is not indicative of any specific style of clay mineral diagenesis. For example, water oxygen isotope composition from a deeply buried sandstone cannot be used to say anything about the clay mineral present in the sandstone.

Kaolinite, illite and smectite have similar fractionation equations that result in similar suites of curves (for $\delta_{\text{mineral}} - \delta_{\text{water}}$ versus temperature). The curves seem to be unaffected by composition of the clay mineral (e.g. smectite/illite ratios). In contrast, chlorite seems to be strongly affected by the composition on both the octahedral and interlayer (brucite) sites. Chlorite has a very wide range of solid solutions including two different hydroxyl sites. Chlorite–water oxygen isotope fractionation systematics have been investigated as a function of composition (Fig. 2D) and have been shown to vary widely depending on mineral chemistry. Most diagenetic chlorites are Fe-rich, with the hydroxyl cations dominated by Fe and Mg. It is thus most likely that the curves labelled 1 and 3 on Fig. 2D are most representative of diagenetic chlorite–water isotope fractionation. Chlorite shows different mineral–water oxygen isotope fractionation compared with the other clay minerals in Fig. 2; the differences for the various types of diagenetic chlorite are substantial at low temperatures. At close to 0°C, the fractionation is between +25‰ and +35‰ V-SMOW. At 100°C, the fractionation is about +15‰ to +17‰ V-SMOW.

Hydrogen isotope fractionation equations between clay minerals and water have been derived for illite and kaolinite (Fig. 3). The most noteworthy feature of these fractionations is that the clay minerals are depleted in ^{2}H with respect to the water and the depletion is reduced as temperature increases. Two hydrogen isotope fractionation equations have been produced for kaolinite (Lambert & Epstein, 1980; Sheppard & Gilg, 1996) and for illite–smectite

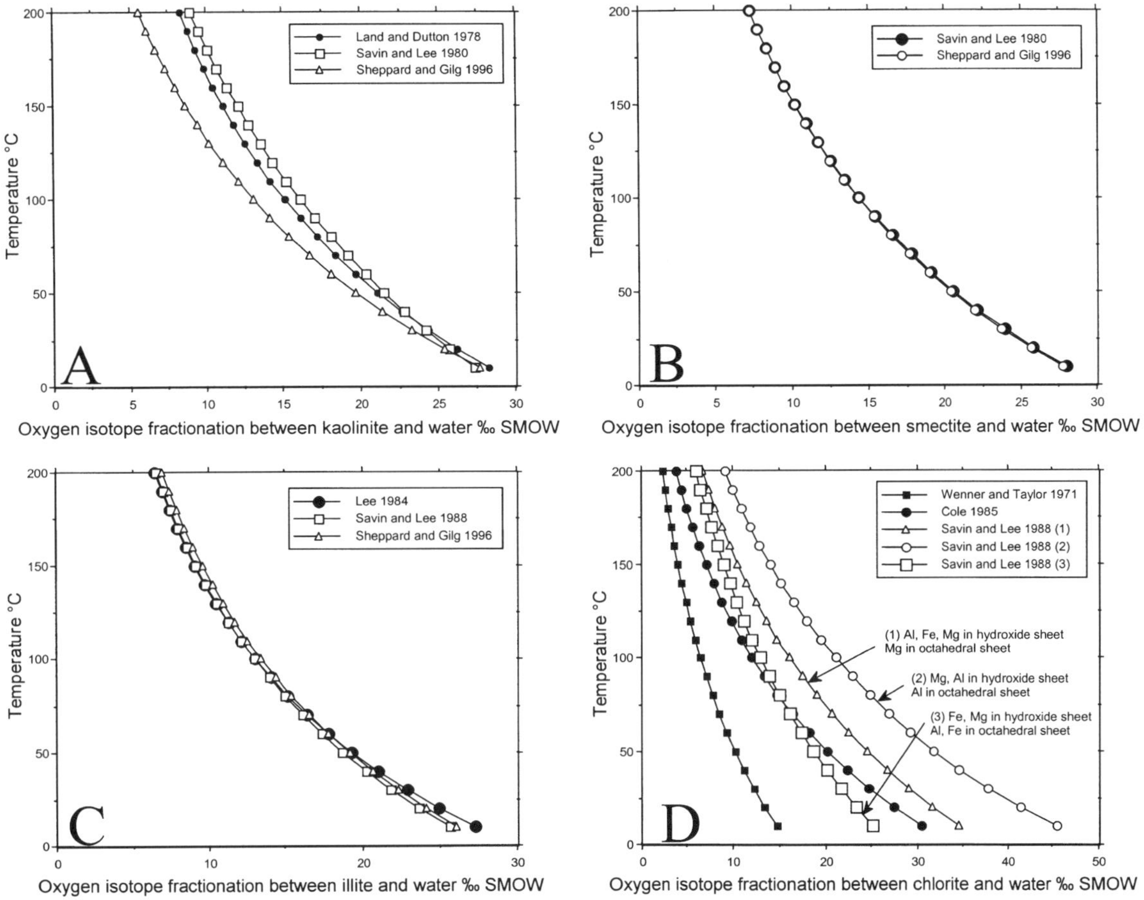

Fig. 2 Experimentally derived clay-mineral–water oxygen isotope fractionation factors: (A) kaolinite, (B) smectite, (C) illite and (D) chlorite.

(Yeh, 1980; Capuano, 1992). The illite–smectite fractionation equation includes illite–smectite mixtures as well as end-member illite. The two kaolinite fractionation equations disagree in detail and the disparity increases with increasing temperature (Fig. 3). The Sheppard & Gilg (1996) kaolinite–water hydrogen isotope fractionation equation is broadly similar to the Yeh (1980) illite–water hydrogen isotope fractionation equation.

Chlorite–water hydrogen isotope fractionation systematics are not yet fully understood. An experimental analysis of this clay mineral yielded imperfect results and an analysis of pre-existing experimental and empirical data did not yield a simple mathematical fractionation relationship (Graham *et al.*, 1987). The chlorite–water hydrogen isotope fractionation factor may be about −40 to −50‰ V-SMOW at 200°C, although values as low as −13.3‰ V-SMOW have been reported for temperatures of 130°C (Marumo *et al.*, 1980). If these two data points are correct then the form of the chlorite–water hydrogen isotope fractionation curve will be quite different from those of kaolinite and illite. Graham *et al.* (1987) concluded that the mineral chemistry of chlorite might affect the hydrogen isotope fractionation. Likewise, smectite has been shown to have (low temperature) hydrogen isotope fractionation factors varying by up to 60‰ V-SMOW as a function of the ferric iron content (Sheppard &

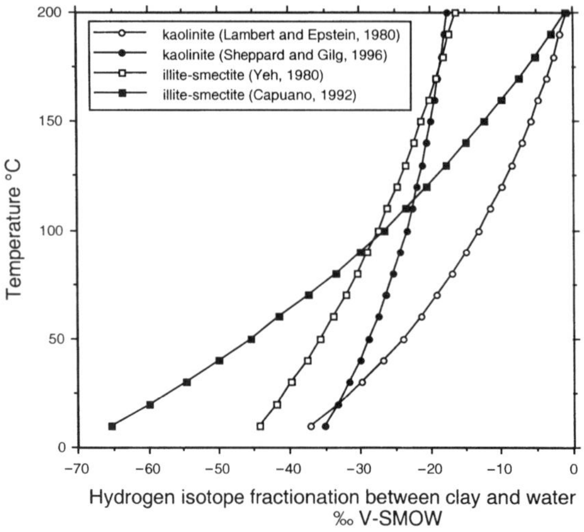

Fig. 3 Experimentally derived clay-mineral–water hydrogen isotope fractionation factors for kaolinite and illite. Data sources given in Table 1.

Gilg, 1996). There is no usable smectite–water hydrogen isotope fractionation equation but if there were, it would be essential to know the smectite mineral chemistry.

SOURCE AND SHORTCOMINGS OF PUBLISHED ISOTOPIC DATA

The oxygen and hydrogen isotopic data from kaolin minerals (kaolinite and dickite), illitic clays (illite and mixed-layer illite–smectite), chlorite and berthierine have been collated from a wide range of publications (Tables 2–4). Shortcomings of some of the published stable isotopic data include poor quantitative identification and polytype distinction of the clay minerals (e.g. the kaolin polytypes), the availability only of oxygen isotope data for some

Table 2 Sources of stable isotope data for the kaolin group minerals.

Reference	Kaolin mineral	Basin	Formation	Sandstone age
Longstaffe & Ayalon, 1987	Dickite	Alberta	Viking Formation	Lower Cretaceous
Maliva *et al.*, 1999	Dickite	Isle of Man	Derbyhaven beds	Lower Carboniferous
McAulay *et al.*, 1994	Dickite	North Sea	Brent Group	Jurassic
Parnell *et al.*, 2000	Dickite	Fault zone Scotland and Ireland		
Savin & Epstein, 1970	Dickite			
Tilley & Longstaffe, 1989	Dickite	Alberta Deep Basin	Falher Member	Lower Cretaceous
Ayalon & Longstaffe, 1988	Kaolinite	Western Canada Sedimentary Basin	Basal Belly River Sandstone	Upper Cretaceous
Brint, 1989	Kaolinite	North Sea	Brent Group	Jurassic
Franks *et al.*, 1997	Kaolinite			
Glasmann *et al.*, 1989a,b	Kaolinite	North Sea	Brent Group	Jurassic
Hogg *et al.*, 1995	Kaolinite	North Sea	Brent Group	Jurassic
Ketzer *et al.*, this volume, 361–382	Kaolinite	North Sea	Lunde Formation	Triassic
Longstaffe, 1989	Kaolinite			
Macaulay *et al.*, 1993	Kaolinite	North Sea	Magnus Sandstone	Upper Jurassic
Mátyás & Matter, 1997	Kaolinite	Pannonian Basin (Hungary)	Szolnok Formation	Neogene
McAulay *et al.*, 1994	Kaolinite	North Sea	Brent Group	Jurassic
Milliken *et al.*, 1981	Kaolinite	Frio Formation (Texas)		Oligocene
Mizota & Longstaffe, 1996	Kaolinite	Northeastern Japan	Namairi, Komatsu, Yokomichi Formations	Late Cretaceous + Oligocene
Osborne *et al.*, 1994	Kaolinite	North Sea	Brent Group	Jurassic
Parnell *et al.*, 2000	Kaolinite	Fault zone Scotland and Ireland		
Savin & Epstein, 1970	Kaolinite			
Shieh & Suttner, 1979	Kaolinite	Illinois Coal Basin		Pennsylvanian
Stewart *et al.*, 1994	Kaolinite	North Sea	Montrose Group	Lower Paleocene
Tilley & Longstaffe, 1989	Kaolinite	Alberta Basin	Cadomin Formation	Lower Cretaceous

Table 3 Sources of stable isotope data for illite and illite–smectite group minerals.

Reference	Basin	Formation	Depositional age
Ayalon & Longstaffe, 1988	Western Canada Sedimentary Basin	Basal Belly River Sandstone	Upper Cretaceous
Clauer *et al.*, 1995	Paris Basin		Uppermost Triassic
Fallick *et al.*, 1993	North Sea	Magnus Sandstone	Upper Jurassic
Girard & Barnes, 1995	Central Michigan Basin	St Peter Sandstone	Middle Ordovician
Glasmann *et al.*, 1989a,b	North Sea	Brent Group	Jurassic
Greenwood *et al.*, 1994	North Sea	Heather Formation	Middle Jurassic
Hogg *et al.*, 1995	North Sea	Brent Group	Jurassic
Longstaffe & Ayalon, 1987	Alberta	Viking Formation	Lower Cretaceous
Lynch *et al.*, 1997	Gulf Coast	Frio Formation	Tertiary
Mora *et al.*, 1998	Appalachian Basin	Mauch Chunk, Hinton, Pennington Formations	Upper Mississippian
Robinson *et al.*, 1993	North Sea	Leman Sandstone Formation	Lower Permian
Ryan *et al.*, 1998	Gulf Coast	Tuscaloosa Formation	Lower Cretaceous
Tilley & Longstaffe, 1989	Alberta Deep Basin	Falher Member	Lower Cretaceous
Uysal *et al.*, 2000	Bowen Basin (Australia)	Moranbah, Fort Cooper, Rangal coal measures	Late Permian
Wilkinson *et al.*, 1994	North Sea	Fulmar Formation	Jurassic
Ziegler & Longstaffe, 2000 and Harper *et al.*, 1995	Appalachian Basin	PreCambrian–Palaeozoic Unconformity	Uppermost Precambrian of southwest Ontario
Zwingmann *et al.*, 1999	Northern German Basin–Niedersächsische Rift		Permian

Table 4 Sources of stable isotope data for chlorite.

Reference	Basin	Formation	Depositional age
Ayalon & Longstaffe, 1988	Western Canada Sedimentary Basin	Basal Belly River Sandstone	Upper Cretaceous
Grigsby, 2001	Gulf Coast	Vicksburg	Tertiary
Hillier *et al.*, 1996		Hannover Formation (mainly) + Slochteren and Schneverdingen Formations	Permian
Longstaffe & Ayalon, 1987	Alberta	Viking Formation	Lower Cretaceous
Platt, 1993	Southern Permian Basin (Germany)	Rotliegend	Lower Permian
Ryan *et al.*, 1998	Gulf Coast	Tuscaloosa Formation	Lower Cretaceous
Spötl *et al.*, 1994	Arkoma Foreland Basin (USA)	Spiro Sandstone	Middle Pennsylvanian
Tilley & Longstaffe, 1989	Alberta Deep Basin	Falher Member	Lower Cretaceous
Whitney & Northrop, 1987	San Juan Basin (North Mexico)	Morrison Formation	Upper Jurassic
Ziegler & Longstaffe, 2000 and Harper *et al.*, 1995	Appalachian Basin	PreCambrian–Palaeozoic Unconformity	Uppermost Precambrian of southwest Ontario
Zwingmann *et al.*, 1999	Northern German Basin–Niedersächsische Rift		Permian

of the samples, the absence of present-day or palaeotemperature data, and tectonic uplift leading to potential underestimation of the maximum burial depths (and thus temperatures) reached by the host sandstone.

An additional problem in interpreting the isotopic data is uncertainty about whether or not clay minerals retain their isotopic composition from the time of crystallization. Very significant unrecognized problems with clay-mineral isotope data may arise through difficulties in obtaining pure single clay-mineral separates (Sheppard & Gilg, 1996). Size separation is often thought to be the best way to split similar clay minerals of different generations, including detrital clays (Girard & Barnes, 1995).

CLAY-MINERAL STABLE ISOTOPIC DATA

Kaolin group minerals

There are three polytypes of diagenetic kaolin in sandstones; kaolinite, dickite and, less commonly, nacrite. Identification and quantification of mixed kaolin polytypes using X-ray diffraction analyses are problematic, and thus should be aided by infrared spectroscopy (Worden & Morad, this volume, pp. 3–41). Kaolinite is the low-temperature polytype and forms owing to the incursion of meteoric waters into continental and paralic sediments during near-surface to shallow-burial eodiagenesis or during telodiagenesis. Meteoric water flushing is most efficient under semi-humid to humid climatic conditions, which tend to lead to the establishment of low concentrations of aqueous SiO_2 and low ionic activities of Ca, Na, Mg and K relative to H^+ activity, which favour the formation of abundant kaolinite. Kaolinite occurs mainly as vermicules and booklets consisting of thin (< 1 μm thick) pseudohexagonal crystals that grow at the expense of detrital feldspars, mica, felsic igneous rock fragments and mud intraclasts. Kaolinite formed at the expense of micas occurs as thin sheets that typically are interlaminated with sheet-like mica relicts. Kaolinite may have a poorly or well-ordered crystal structure. Increase in crystal structural ordering is believed to occur during burial to moderate depths (< 2 km; Worden & Morad, this volume, pp. 3–41).

Upon further increase of burial depth (> 2 km), kaolinite commences transformation into dickite, probably via an intermediate, mixed-layer kaolinite–dickite phase (Worden & Morad, this volume, pp. 3–41). Dickite occurs as aggregates that typically are comprised of blocky crystals. Pervasive to complete transformation of kaolinite into dickite occurs at burial depths greater than about 5 km and temperatures greater than about 150°C (Ehrenberg *et al.*, 1993). The depth range of dickite formation is similar to that of kaolinite illitization. Illite, however, usually grows at the expense of kaolinite, rather than dickite, in arkosic sandstones (Morad *et al.*, 1994). Hence, pervasive dickitization of kaolinite retards the conversion of kaolin into illite. Incomplete transformation of kaolinite into dickite results in serious problems to be faced in obtaining pure fractions of kaolinite or dickite for purposes of stable isotope analyses. Accordingly, most published O and H isotopic data from kaolin samples in the literature are likely to be for mixed kaolinite and dickite polytypes. We shall, therefore, treat our data on dickite and kaolinite under the umbrella term of kaolin. Nevertheless, the stable isotope data published for well-defined dickite and kaolinite are outlined and compared with the bulk kaolin data. The aggregation of kaolinite and dickite data may not be problematic as they are thought to have the same mineral–water stable isotope fractionation relationships (Sheppard & Gilg, 1996). Our compilation of data supports this because the two polytypes fall within the same range of oxygen and hydrogen isotopes.

A total of 319 published analyses reveal that kaolin has $\delta^{18}O$ values that range from +5‰ to +28.5‰ and δ^2H (δD) values from −140‰ to −30‰, both of which show a slight bimodality (Fig. 4A & B). However, 75% of the kaolin samples have $\delta^{18}O$ values ranging between +12.5‰ and +18.5‰ (Fig. 4A) and 50% of the samples have δ^2H values ranging between −70‰ and −40‰ (Fig. 4B). Conversely, samples specifically reported as dickite have $\delta^{18}O$ values ($n = 54$) mostly (90%) ranging between +10‰ and +16‰ and δ^2H values ($n = 29$) mostly (85%) ranging between −50‰ and −30‰ (Fig. 4A & B). The $\delta^{18}O$ and δ^2H values display a positive correlation, which is stronger for dickite ($r = +0.67$; Fig. 5), than for kaolinite ($r = +0.47$; Fig. 5). Neither $\delta^{18}O$ nor δ^2H values show a simple global pattern with increasing depth (Fig. 6).

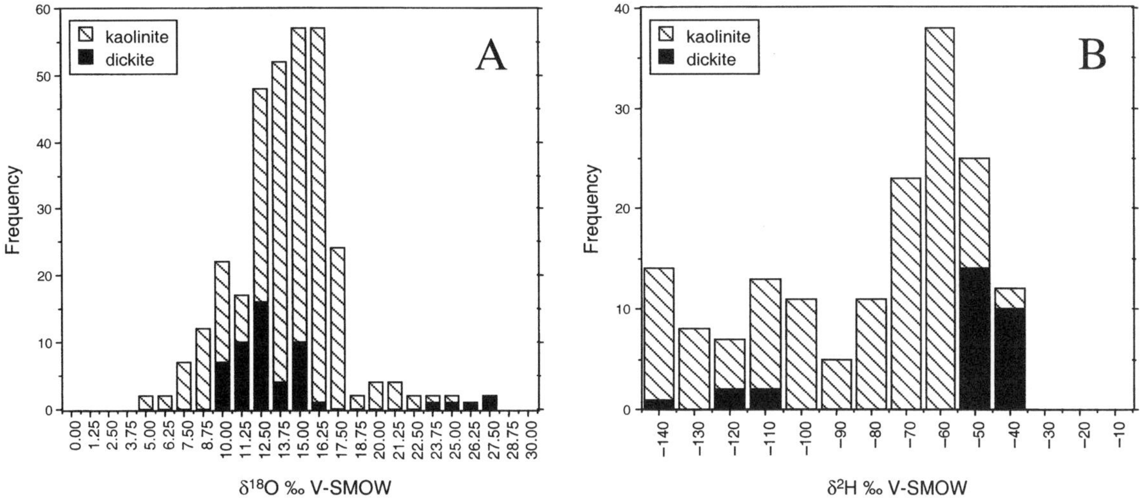

Fig. 4 Kaolin stable isotope data: (A) histogram of kaolin $\delta^{18}O$ values; (B) histogram of kaolin $\delta^{2}H$ values.

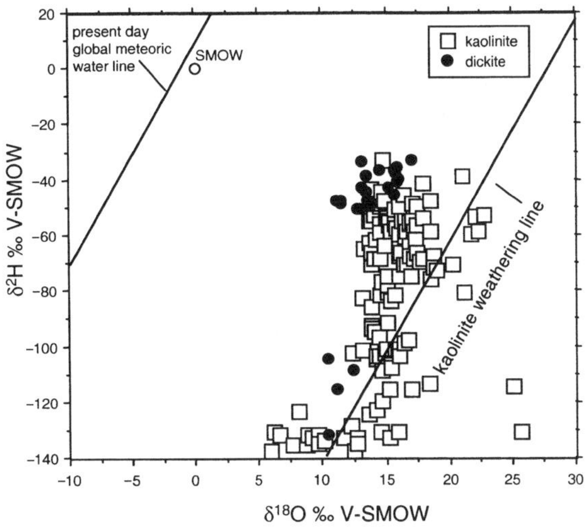

Fig. 5 Kaolin stable isotope $\delta^{18}O$ and $\delta^{2}H$ data.

Pure, initial meteoric-water signatures from unrecrystallized eogenetic kaolinite would be aligned along a line with very strong positive correlation between H and O isotopes (kaolinite weathering line; $\delta^{18}O$; Savin & Epstein, 1970) that runs subparallel to the Global Meteoric Water Line (GMWL; Figs 1 & 5). The difference in $\delta^{18}O$ values between the meteoric water and the eogenetic kaolinite should be equal to the equilibrium fraction factor at the temperature of the Earth's surface (28–30‰; Fig. 2A). There is no evidence to indicate that this relationship in ancient meteoric water isotopic data was different from that of present-day meteoric water (Sheppard, 1986). Oxygen isotopic exchange between clay minerals and pore-waters occurs only when the cation-oxygen bonds are broken and reformed, which is very slow at sedimentary temperatures (Savin & Epstein, 1970; O'Neil & Kharaka, 1976; Yeh & Savin, 1977).

Deviation of $\delta^{18}O$ and $\delta^{2}H$ values of kaolin from the kaolinite weathering line (KWL) can be attributed to: (i) variations in the isotopic re-equilibration rates of hydrogen and oxygen isotopes between stable solid clay minerals and formation water at a wide range of temperatures or (ii) the typical occurrence of kaolinite recrystallization and dickitization (i.e. dissolution and reprecipitation) processes that allow isotopic re-equilibration to occur. If the first option is correct, the hydrogen isotopes are more readily exchanged than oxygen isotopes during kaolinite recrystallization and dickitization. Several authors have argued that hydrogen isotope exchange between kaolin and pore-water may occur even at low temperatures (< 40°C) without a concomitant alteration of the oxygen isotopes (e.g. O'Neil & Kharaka, 1976; Bird & Chivas, 1988; Longstaffe & Ayalon, 1990). The

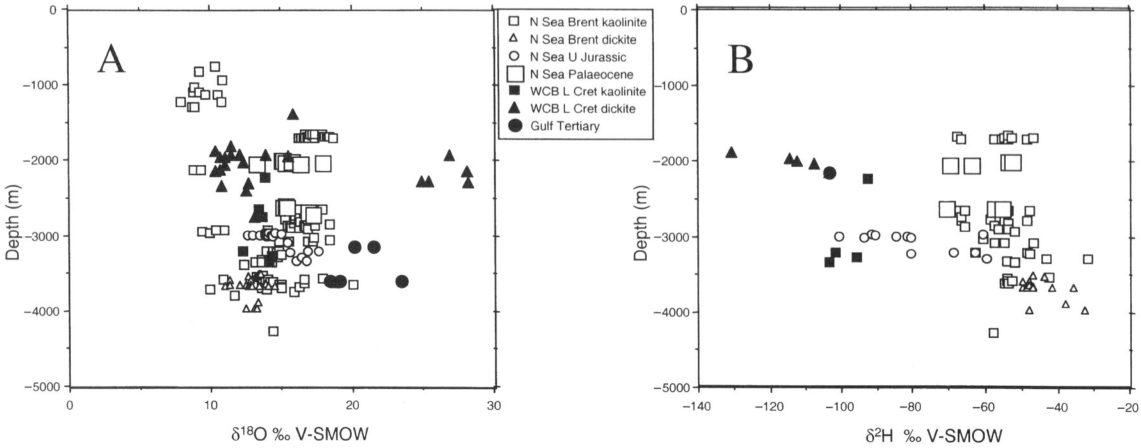

Fig. 6 Kaolin stable isotope data: (A) $\delta^{18}O$ values versus depth; (B) $\delta^{2}H$ values versus depth. WCB = Western Canada Basin.

difference between the rate of exchange of oxygen and hydrogen isotopes is thought to be due to the hydrogen isotopic exchange with porewater at temperatures lower than 200°C affecting H^+ rather than OH^- groups within the clay minerals (O'Neil & Kharaka, 1976; Kharaka & Carothers, 1986; Savin & Lee, 1988). The extent of hydrogen isotopic resetting in clay minerals during burial diagenesis is thought to increase with increasing water/rock ratios (Mizota & Longstaffe, 1996). The solid-state resetting of hydrogen isotopes (i.e. subsequent to crystallization and stabilization of clay minerals) has, however, been rejected by several authors (e.g. Osborne *et al.*, 1994; Stewart *et al.*, 1994) so that, once formed and stabilized, kaolin minerals may record the isotopic composition of fluids from which they formed. In other words, kaolin minerals do not undergo isotopic re-equilibration with formation waters (Osborne *et al.*, 1994), unless they are subjected to partial or pervasive recrystallization or transformation. Hence, kaolin can be used to reconstruct the spatial and temporal geochemical evolution of formation waters. Franks *et al.* (1997) claimed that the oxygen isotopic composition of kaolinite does not appear to be reset during transformation into dickite.

If the second option is correct, then the implications from Fig. 5 are either that most kaolin found in sandstones was either not eogenetic or that if it was eogenetic then it has undergone recrystallization. Another point worth making is that hydrogen isotopes have been affected by diagenesis. The existence of 'organic waters' (Sheppard, 1986; Fig. 1) means that recrystallization or formation from such waters would lead to kaolinite with lower $\delta^{2}H$ values than for kaolinite formed from meteoric water. Thus, kaolin with the lowest $\delta^{2}H$ values on Fig. 5 may have formed during mesodiagenesis (with commensurately lower $\delta^{18}O$ values) but adopted low $\delta^{2}H$ values from the organic water. Plausibly, this could move mesogenetic kaolin back on to the kaolinite weathering line. Clearly, care must be taken in interpreting kaolin stable isotope values: simple assumptions may be incorrect.

Illite and illite–smectite

Illite typically begins to form during mesodiagenesis at temperatures that exceed about 70°C but this transformation becomes volumetrically substantial at temperatures > 100°C (Worden & Morad, this volume, pp. 3–41). Illite forms by replacement of kaolin (initially mainly kaolinite) and smectite, by alteration of micas and feldspars and by precipitation from pore fluids. The illitization of eogenetic smectite involves

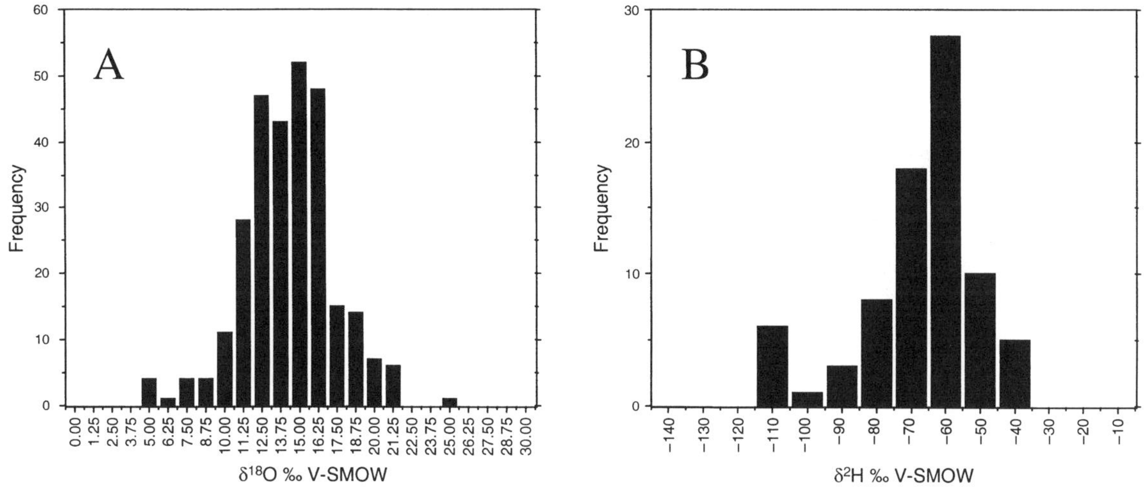

Fig. 7 Illite stable isotope data: (A) histogram of $\delta^{18}O$ values; (B) histogram of $\delta^{2}H$ values. Data sources given in Table 3.

an intermediate phase of mixed-layer illite–smectite (I/S) in which the proportion of illite increases progressively with increased burial and temperature (Worden & Morad, this volume, pp. 3–41). In the presence of suitable geochemical conditions (i.e. elevated aK^+/aH^+, for example, resulting from K-feldspar dissolution and albitization), the illitization reactions of kaolinite and smectite are completed at temperatures between 130°C and 150°C. At these temperatures, illite usually has < 5% interstratified smectite.

A total of 286 published analyses reveal that illitic clay minerals have $\delta^{18}O$ values that range from +5‰ to +26‰ and $\delta^{2}H$ values from −110‰ to −30‰ (Fig. 7A & B). However, most of the samples (75%) have $\delta^{18}O$ values ranging between about +11.0‰ and +17.5‰ (Fig. 7A) and 65% of the samples have $\delta^{2}H$ values ranging between −70‰ and −45‰ (Fig. 7B). The $\delta^{18}O$ and $\delta^{2}H$ values display a moderate positive covariation ($r = +0.49$; Fig. 8).

Analyses of illite from sandstones and associated mudstones from North Sea clastic reservoirs show similar oxygen isotope values (e.g. Greenwood *et al.*, 1994), which were used as evidence for the diagenetic (as opposed to detrital) origin of illite, even in mudstones. In contrast to oxygen, Greenwood *et al.* (1994) found

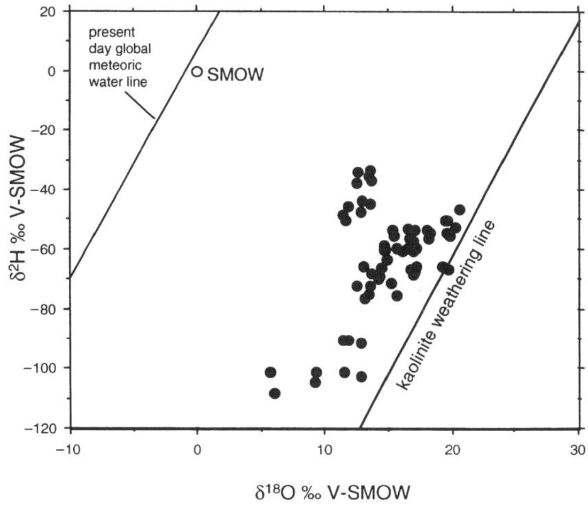

Fig. 8 Illite stable isotope data: cross-plot of $\delta^{18}O$ values versus $\delta^{2}H$.

that the hydrogen isotopic values for illite from sandstones are lower ($\delta^{2}H = -55$‰ to −51‰) than those from mudstones ($\delta^{2}H = -49$‰ to −36‰). This difference possibly precludes the anticipated, more profound influence of 'organic waters' in these mudstones compared with the neighbouring sandstones. Conversely, Uysal *et al.* (2000) found that the hydrogen isotopic signature in associated sandstones, mudstones and bentonites (from the Bowen Basin,

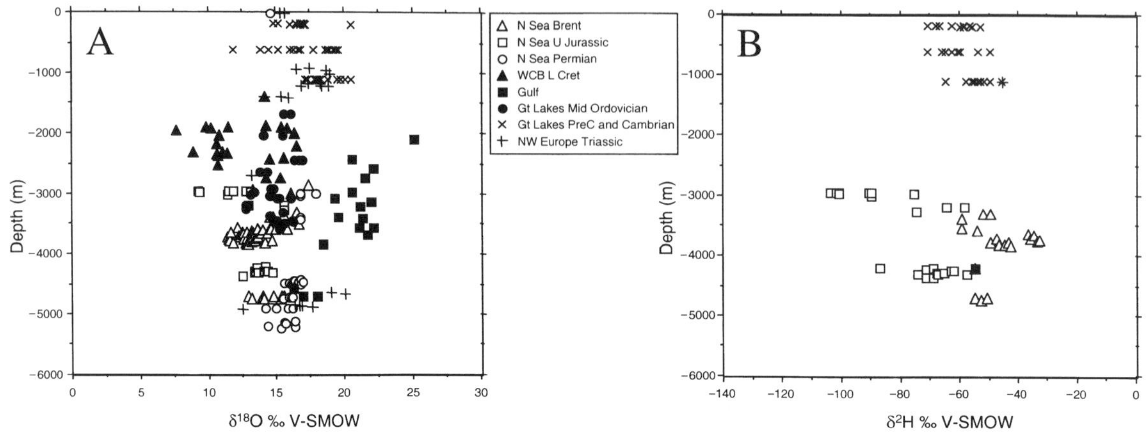

Fig. 9 Illite stable isotope data: (A) $\delta^{18}O$ versus depth; (B) $\delta^{2}H$ versus depth. WCB = Western Canada Basin.

Australia) subjected to 100°C burial temperature to be virtually constant, whereas the oxygen isotopic values vary considerably. The latter was attributed to variations in the water/rock and illite/smectite ratios (see below). Additionally, the hydrogen isotopic composition of illitic clay minerals in these sediments is similar to that of associated kaolin, which was attributed by these authors to re-equilibration of hydrogen isotopes of these clay minerals with fluids related to younger, more recent thermal or fluid flow events.

Whitney & Northrop (1987) concluded from a study of the isotopic composition of clay minerals in sandstones of the San Juan Basin, New Mexico that illite, formed by progressive replacement of smectite, tends to concentrate deuterium. Illitic clay minerals presently buried at depths of 0 to 4.8 km do not reveal simple global trends of $\delta^{18}O$ or $\delta^{2}H$ with depth and thus temperature (Fig. 9A & B).

The $\delta^{18}O$ values of illitic clay minerals decrease with increasing K_2O ($r = +0.45$; Fig. 10A) and with an increase in illite/smectite ratio ($r = 0.64$; Fig. 10B), suggesting that the progressive burial diagenetic transformation of smectite into illite involves an oxygen isotopic re-equilibration resulting from the dissolution–reprecipitation nature of the reaction. Such relationships have been described from a number of basins, such as the Gulf Coast (Lynch *et al.*, 1997), the Central Michigan Basin (Girard & Barnes, 1995) and the Bowen Basin, Australia (Uysal *et al.*, 2000). The rate of decrease in $\delta^{18}O$ values of illitic clay minerals with increase in illite (and K_2O) content, has been concluded by Uysal *et al.* (2000) to be higher at higher water/rock ratios, which allows a more intensive illitization reaction as well as isotopic reequilibration. There is no significant global correlation between $\delta^{2}H$ with K_2O content or with the illite/smectite ratio, suggesting that there are variable pathways of O and H isotope fractionation between clay minerals and formation waters with increasing temperature.

The solid-state exchange of structurally bonded O and H isotopes in illite and smectite with those in pore-waters is thought to be insignificant at the low temperatures encountered during diagenesis (Yeh & Savin, 1977; Longstaffe, 1983; Yeh & Eslinger, 1986; O'Neil, 1987). Conversely, the O and H isotopes of interlayer water of smectite are thought to re-equilibrate rapidly with pore-waters even at near-surface conditions (Savin & Epstein, 1970; O'Neil & Kharaka, 1976).

Chlorite

Chlorite forms through the transformation of smectite and berthierine during progressive burial and increase in temperature and does not

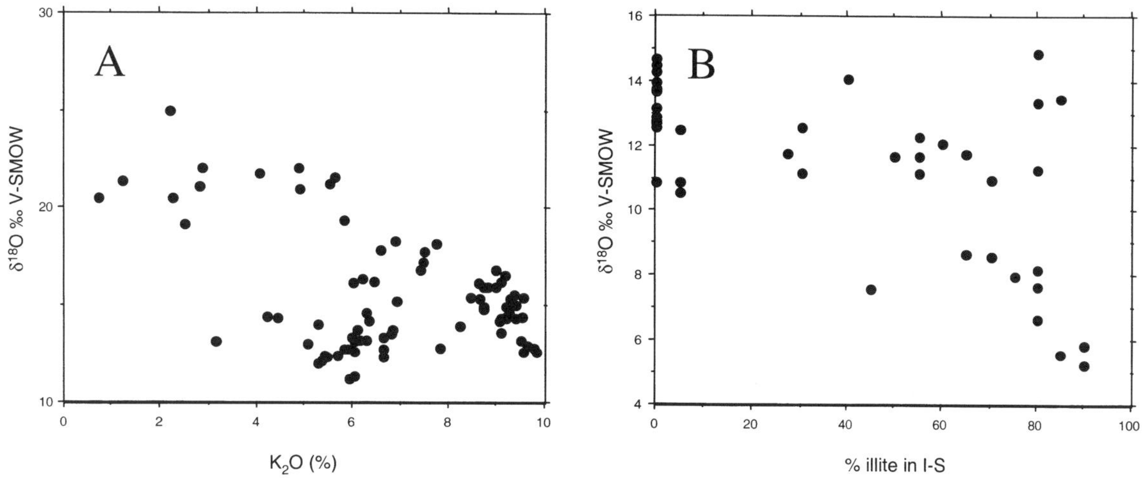

Fig. 10 $\delta^{18}O$ versus (a) K_2O content and (b) I/S ratio in illite.

form during eodiagenesis (Worden & Morad, this volume, pp. 3–41). The chloritization of smectite occurs through intermediate phases of mixed-layer chlorite–smectite (C/S). The C/S initially has an irregular and then a regular (called corrensite) chlorite–smectite stacking pattern. Pure chlorite typically is Fe-rich at lower diagenetic grades ($Mg_{0.41}Fe_{3.52}Al_{1.51}Al_{0.58}Si_{3.42}(O)_{10}(OH)_8$; Aagaard *et al.*, 2000) whereas the C/S, particularly corrensite, may contain elevated quantities of Mg. The precise mechanism for the transformation of berthierine into chlorite is less well known, but it may occur via an intermediate, mixed-layer chlorite–berthierine phase (S. Hillier, personal communication, 2001). Pervasive to complete chloritization of smectite and berthierine occurs at depths greater than 3 km. Hence, despite the apparently early diagenetic, pore-lining or particle-coating texture, chlorite is a mesogenetic mineral. Although chlorite forms most prominently by the replacement of grain-coating smectite and berthierine, it also forms by the replacement of volcanic fragments, biotite, feldspar and mud intraclasts (Anjos *et al.*, this volume, pp. 291–316).

Published stable isotope analyses reveal that chloritic clay minerals have an almost unimodal distribution of $\delta^{18}O$ values ($n = 98$) that ranges between +0.0‰ and +20.0‰ and bimodally distributed δ^2H values ($n = 20$) between −110‰ and −10‰ (Fig. 11A & B). However, 75% of the chloritic clay samples have $\delta^{18}O$ values ranging between +7‰ and +15‰ (Fig. 11A) and 40% of the samples have δ^2H values ranging between −80‰ and −60‰ (Fig. 11B). The number of samples for which there are both oxygen and hydrogen isotopic data is limited ($n = 20$). Like kaolin and illitic clays, the oxygen isotopic compositions of chloritic clays do not reveal any specific co-variation with depth (Fig. 12). Excluding the data from Hillier *et al.* (1996), which deviate considerably from values published by other authors, the isotopic data on chlorite display a weak correlation between $\delta^{18}O$ and δ^2H values (Fig. 13), similar to that observed for kaolin and illitic clay minerals (Figs 5 & 8). Ryan *et al.* (1998) concluded that different rates of oxygen isotopic exchange between clay minerals and formation water are encountered during simultaneous, successive burial diagenetic chloritization and illitization reactions. These authors observed that: (i) an increase in the chlorite at the expense of serpentine in mixed-layer serpentine–chlorite during increase in burial temperature from 1937 m to 5470 m was accompanied by a non-systematic small increase in $\delta^{18}O$ values,

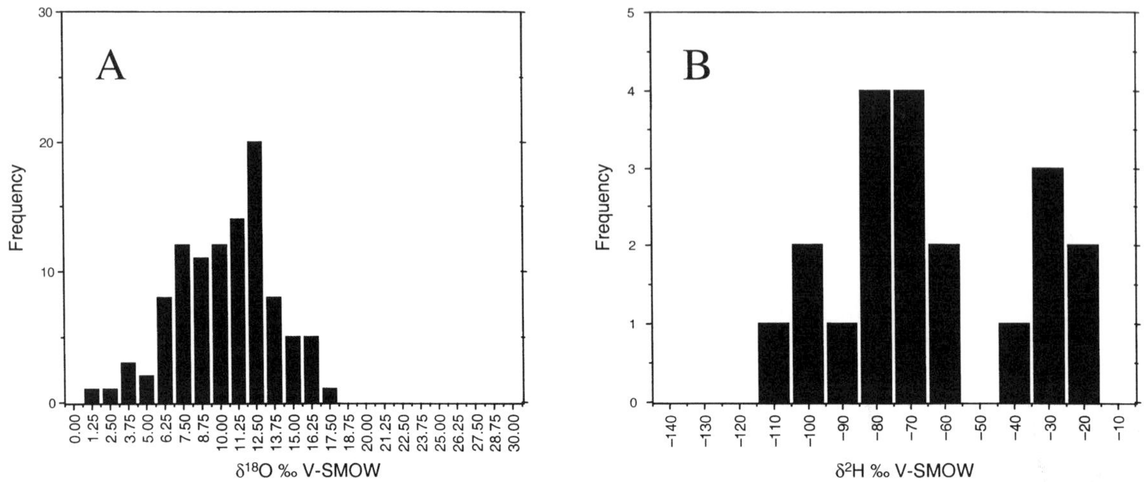

Fig. 11 Chlorite stable isotope data: (A) histogram of $\delta^{18}O$ values; (B) histogram of $\delta^{2}H$ values. Data sources given in Table 4.

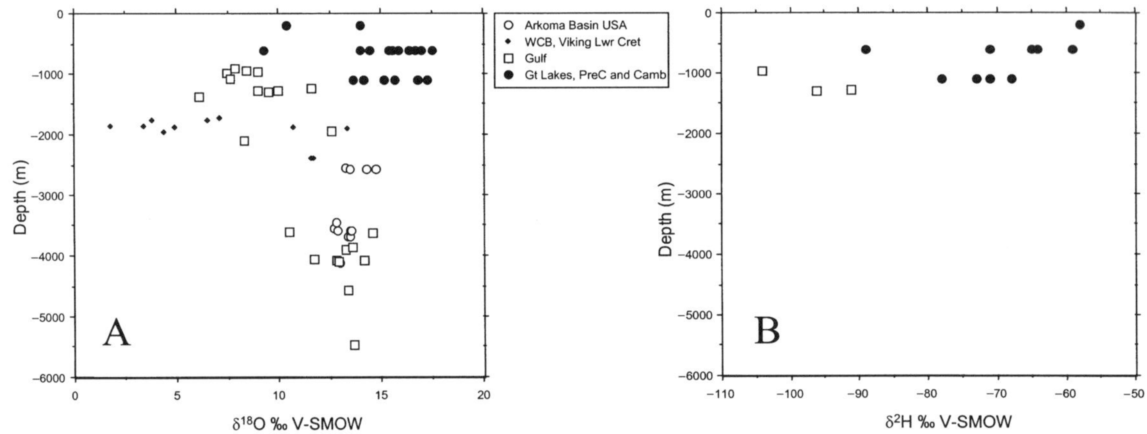

Fig. 12 Chlorite stable isotope data: (A) $\delta^{18}O$ values versus depth; (B) $\delta^{2}H$ values versus depth. WCB = Western Canada Basin.

and (ii) an increase in illite at the expense of smectite in I/S over this depth range was accompanied by a slight increase in $\delta^{18}O$ values. The increase in $\delta^{18}O$ values with increase in burial depth and temperature was poorly constrained by these authors and thus remains enigmatic. Conversely, Spötl *et al.* (1994) concluded that a decrease in $\delta^{18}O$ values of diagenetic chlorite owing to increase in temperature is accompanied by an increase of IIb polytype at the expense of Ib polytype chlorite.

Berthierine and glaucony

Berthierine is a 1 : 1 clay mineral and thus structurally similar to kaolin, but chemically similar to Fe-rich chlorite. Berthierine forms at near-surface temperatures by precipitation as coatings around framework grains, as pellets, and by replacement of some framework grains such as biotite (Morad & Al-Aasm, 1994) and of earlier formed, diagenetic odinite (Worden & Morad, this volume, pp. 3–41). Berthierine is

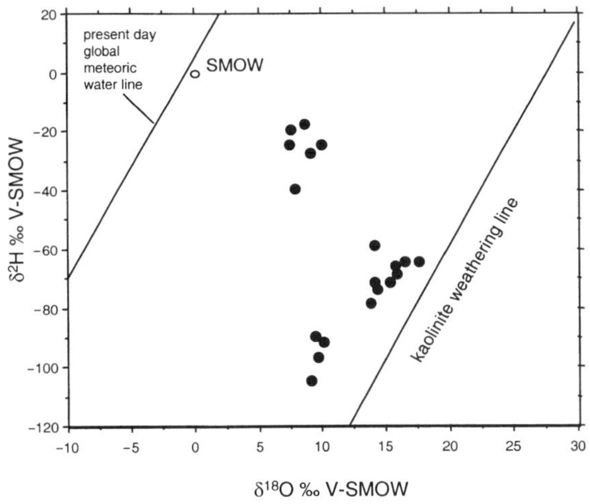

Fig. 13 Cross-plot of $\delta^{18}O$ values versus $\delta^{2}H$ for chlorite.

formed in marine and paralic sediments either from brackish (e.g. in estuaries) or marine pore-waters that undergo suboxic and anoxic, Fe-reduction diagenesis (Froelich *et al.*, 1979; Berner, 1980). Upon burial, berthierine is transformed progressively into chlorite (Worden & Morad, this volume, pp. 3–41).

Five isotopic analyses have been published on berthierine (Hornibrook & Longstaffe, 1996), which show $\delta^{18}O$ values that range between +6.6‰ and +11.4‰ and $\delta^{2}H$ values between −119‰ and −116.5‰. These authors suggested that the $\delta^{18}O$ values indicate that the low-temperature (15–45°C) berthierine formation occurred from pore-waters that are substantially depleted in ^{18}O ($\delta^{18}O_{SMOW}$ = −14‰ to −6‰). However, meteoric waters with $\delta^{18}O_{SMOW}$ = −14‰ to −6‰ should have a wide range of $\delta^{2}H$ values of −98‰ to −54‰, which are higher than the values obtained. One possible source of light hydrogen, suggested by these authors, is from microbial fermentation of organic matter (which would also enhance the concentration of Fe^{2+} in pore-waters, and hence favour berthierine formation). In sediments where microbial methanogenesis is accompanied by the formation of methane hydrates (enriched in ^{18}O and ^{2}H), the remaining pore-waters become enriched in ^{16}O and ^{1}H (Hesse *et al.*, 1985).

Glaucony includes a group of potassium- and iron-rich aluminum phyllosilicates, that form from mildly reducing pore-waters (i.e. suboxic, nitrate and manganese reduction diagenesis; Froelich *et al.*, 1979), immediately below the sediment–water interface in marine environments characterized by low sedimentation rates and oxic bottom conditions. Glaucony occurs in sandstones mainly as pellets that form through the replacement of argillaceous clasts, carbonate bioclasts, mica and volcanic fragments. The chemical composition of glaucony is often related to its state of maturity; smectitic immature glaucony can evolve into micaceous mature glaucony (glauconite) upon incorporation of potassium and iron.

Chemical evolution occurring below the sea floor is also accompanied by a successive enrichment of the glauconitic mineral in ^{16}O isotope (Kelly *et al.*, 2001). These authors found that the precursor smectite fraction in Oligo-Miocene marine sediments from Australia has a $\delta^{18}O$ value of +26.4‰, whereas the least and most evolved glauconitic fractions have $\delta^{18}O$ values of +22.3‰ to +23.4‰ and 20.3‰, respectively. These values, as well as those reported by other authors (Friedrichsen, 1984; Keppens & O'Neil, 1984; Clauer *et al.*, 1992) are lower than what would be expected if the formation of glauconitic minerals occurred in equilibrium with pristine marine pore-water at low temperatures. Thus, Kelly *et al.* (2001) concluded that glauconitic minerals developed in marine pore waters that were depleted in ^{18}O owing to the removal and incorporation of ^{18}O in low-temperature, diagenetic clay minerals, such as smectite (e.g. Lawrence *et al.*, 1975). Kelly *et al.* (2001) thus suggested that the degree of depletion in ^{18}O upon evolution of glauconitic minerals would be greatest in low-porosity sediments with considerable amounts of diagenetic clay minerals. Low porosity enhances the development of a closed pore-water system with respect to the overlying sea water.

STABLE ISOTOPES OF FORMATION WATER AND CLAY MINERALS IN SANDSTONES

North Sea

North Sea basin oxygen isotopes in formation waters and clay minerals in sandstone

Formation water oxygen and hydrogen isotope data from clastic reservoirs of the North Sea basin have been reported from most reservoir ages and for most geographical areas (Table 5). The $\delta^{18}O$ values of formation waters from sandstones range between −5‰ and +5‰ (average +0.3‰) and show a general pattern of increase as a function of depth (Fig. 14), also noted by Aplin *et al.* (1993) and Aplin & Warren (1994). The $\delta^{18}O$ values of formation waters from carbonate reservoirs (between +5‰ and +12‰; average +6.3‰; Fig. 14) are higher than those from clastic reservoirs. The more positive $\delta^{18}O$ values from formation water from carbonate reservoirs suggest that whatever is controlling the $\delta^{18}O$ values of formation waters in porous hydrocarbon reservoirs, it is local to the reservoir itself and is not a basin-scale phenomenon because the carbonate and sandstone reservoirs lie within the same basinal context.

Table 5 Sources of stable isotope data for formation waters from the North Sea reservoirs.

Reference	North Sea province
Aplin *et al.*, 1993	Range of formations
Burley & Macquaker, 1993	Jurassic
Egeberg & Aagaard, 1989	Range of formations
Gluyas *et al.*, 1997	Rotliegend
Macaulay *et al.*, 1992	Upper Jurassic
Osborne *et al.*, 1994	Brent
Warren & Smalley, 1994	Range of formations
Wilkinson *et al.*, 1994	Upper Jurassic

The loose pattern of increasing $\delta^{18}O$ with depth for the sandstone formation waters may be the result of progessive sediment–water interaction (increasing fractionation of ^{16}O into diagenetic minerals and simultaneous, or consequent, enrichment of formation water in ^{18}O)

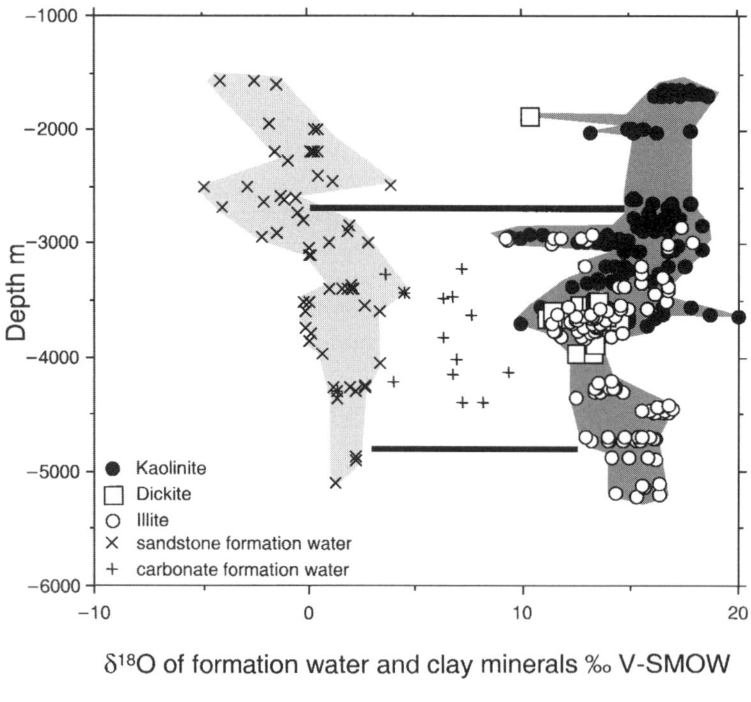

Fig. 14 North Sea $\delta^{18}O$ data from formation waters and clay minerals plotted as a function of present-day depths. Formation water data come from both clastic and carbonate reservoirs. The illite data probably include mixed-layer illite–smectite although the differences in the fractionation factors for $\delta^{18}O$ between smectite and illite with water are trivial (Fig. 2). The heavy horizontal bars represent the approximate equilibrium fractionation (using fractionation factors derived using equations in Table 1 at given depths assuming a general geothermal gradient of 30°C km^{-1} and a surface temperature of 10°C). Data sources given in Tables 1–5.

to increase in temperature. Importantly, the increase in $\delta^{18}O$ with depth seems to reduce the possibility of involvement of extensive post-burial meteoric water deep in the basin. If there had been a major and sustained influx of meteoric water deep into the basin, then the formation water would not have evolved with depth but would have retained a pristine meteoric signal (i.e. negative $\delta^{18}O$ values—see the meteoric water line on Fig. 1). However, there are still notably negative water $\delta^{18}O$ values (approaching −5‰) at 2500 m burial, leaving open the possibility of meteoric water influx even at that depth of burial.

The $\delta^{18}O$ values of clay minerals from the North Sea reservoirs have been plotted on the North Sea formation water $\delta^{18}O$ versus depth diagram (Fig. 14). The clay minerals have more positive $\delta^{18}O$ values than formation water from the same depth. There is a lot of scatter in the clay mineral $\delta^{18}O$ values but they seem to display a general decrease with increasing depth down to about 3800 m. As the North Sea formation waters tend to achieve increasing $\delta^{18}O$ values with increasing depth there is an overall decrease in the difference between the clay minerals and water $\delta^{18}O$ values with increasing depth down to about 3800 m (Fig. 14). Heat flow and conductivity vary locally in the North Sea but it is feasible to generalize and use a geothermal gradient of 30°C km^{-1} and a surface temperature of 10°C, which means that 3000 m burial equates to a temperature of about 100°C. Thus, there should be a difference in the $\delta^{18}O$ values of clay minerals and formation waters of about +15‰ at 3000 m (Fig. 2) if an isotopic equilibrium has been achieved between the clay minerals and formation waters. At that depth, the broad difference between clay minerals and water oxygen isotopes ($\Delta_{clay\text{-}water}$) is indeed approximately +15‰ (Fig. 14) so that it is possible that the formation waters and clay minerals are in isotopic equilibrium at that depth. However, there are obvious outliers to the trends above, where the clay minerals tend to have less positive $\delta^{18}O$ values than required for isotopic equilibrium with water from that depth. Such outliers could result from:

1 uplift and cooling of a mineral–water pair;
2 growth of clay minerals at the current depth (*c.* 3000 m) in the presence of water with much lower $\delta^{18}O$ values than suggested by the pattern on Fig. 14.

Option 2 above may result most plausibly from an influx of meteoric water.

If the diagenetic system was dominated by extrinsic clay minerals in mudstones (i.e. in the whole basin) then the carbonate reservoirs should have $\delta^{18}O$ values identical to, and not heavier than, those in formation waters in the sandstone reservoirs (Fig. 14). This suggests that the host mineralogy of the reservoir, rather than the basin-filling mudstones, dominates the formation water $\delta^{18}O$ values.

It is possible that the reservoir clay minerals and formation water represent a reasonably closed system in many cases (but see exception above) and are mutually locked into a system of isotopic equilibrium. However, another exception to this may be in the deepest parts of basins. At about 5000 m, the reservoirs have temperatures > 150°C and the equilibrium difference in $\delta^{18}O$ values between clay and water should be ≤ +9‰. The clay $\delta^{18}O$ data from these depths average at about +16‰, implying formation water should have an average $\delta^{18}O$ value of about +7‰ rather than the observed values of about +2 to +4‰ (Fig. 14). There are different ways to explain this disparity:

1 clay minerals at depths of 5000 m were formed at shallower depths (about 3000 m) and underwent progressive burial without further re-equilibration with the formation water;
2 the formation water isotope data base is simply not extensive enough—the apparent difference may be the result of poor statistical coverage of the formation water at that depth;
3 isotopically depleted waters entered the deeper reservoirs since the clay minerals formed, although the new waters did not lead to diagenetic reactions that could have facilitated the attainment of isotopic equilibrium—the most commonly cited source of isotopically depleted waters in sedimentary basins is meteoric water.

Explanation **3** is improbable because (i) it would be highly coincidental that all deep

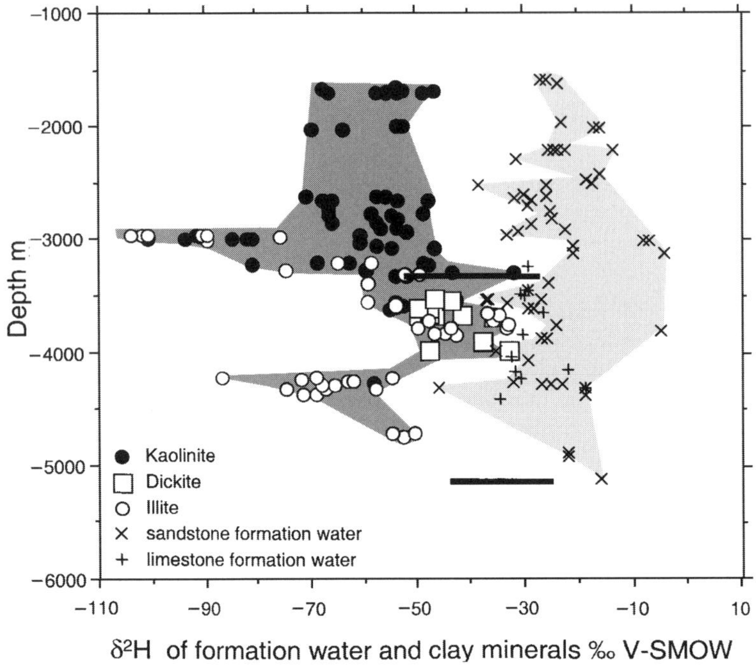

Fig. 15 North Sea δ^2H data on formation waters and clay minerals data plotted as a function of present-day depths. Formation water data come from both sandstone and carbonate reservoirs. The heavy horizontal bars represent the approximate equilibrium fractionation (using fractionation factors derived using equations in Table 1 at given depths assuming a general geothermal gradient of 30°C km^{-1} and a surface temperature of 10°C).

reservoirs received meteoric water, and (ii) it is unlikely that meteoric water would preferentially affect the deepest reservoirs. Explanation **1** is quite feasible because much clay diagenesis occurs between about 80 and 120°C (Worden & Morad, this volume, pp. 3–41), dominated by the growth of illite from a wide variety of sources, whereas relatively less clay diagenesis occurs between 120 and 180°C. Explanation **2** is, of course, an option although the existing spread of the data seems to suggest coverage is good enough and the pattern is real.

North Sea basin hydrogen isotopes in formation waters and clay minerals in sandstone

The δ^2H values of formation waters from North Sea clastic and carbonate reservoirs range between −5 and −50‰ (clastic average −25‰; carbonate average −29‰) and show no simple pattern of variation as a function of depth (Fig. 15). Most formation waters lie within 10‰ of −20‰ and seem to vary little with depth. There are no obvious formation water candidates for the category of 'organic water' (i.e. very negative δ^2H values). Clastic and carbonate formation waters seem to have similar δ^2H values in the North Sea basin. This may be coincidental or a consequence of an identical set of controls that operate on a basin scale. Carbonates do not routinely contain hydrogen-bearing minerals and thus cannot influence the hydrogen isotopes. This might be taken to imply that the controls on the hydrogen isotopes in carbonate reservoir formation waters are largely controlled by basinal phenomena (as opposed to local phenomena). As sandstone and carbonate formation waters seem to have similar δ^2H values then it is possible that sandstone hydrogen isotopes also are controlled by basinal phenomena. Alternatively, the clays in the sandstones may evolve identically to, but independently of, the basinal mudstones (the coincidence option).

There is a wide degree of scatter in the clay-mineral–hydrogen-isotope data (Fig. 15). The highest values (about −40‰) occur in all minerals over a wide depth range. Fast hydrogen-isotope exchange through protonation and/or

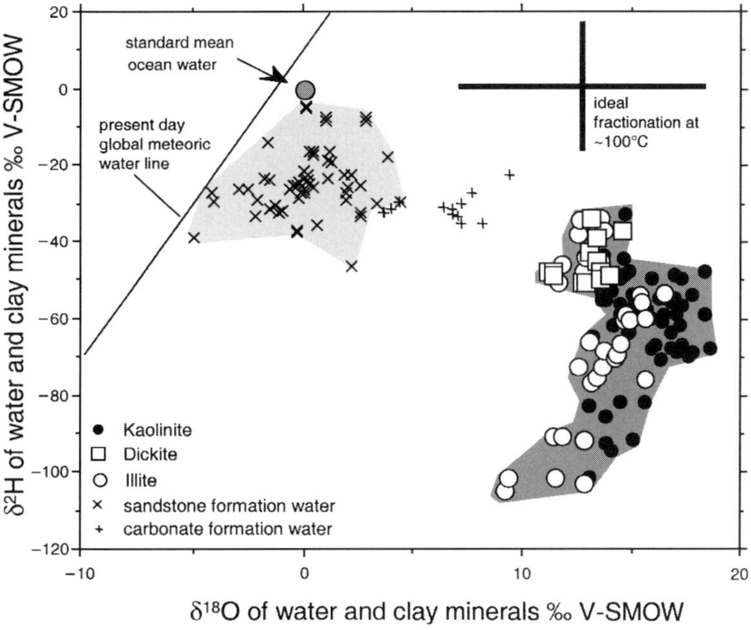

Fig. 16 North Sea $\delta^{18}O$ and δ^2H cross-plotted data from formation waters and clay minerals. Formation water data come from both sandstone and carbonate reservoirs. The cross represents the approximate equilibrium fractionations for $\delta^{18}O$ and δ^2H between water and clay (kaolinite and illite) at 100°C calculated using the fractionation equations in Table 1.

dissolution–precipitation reactions (Kharaka & Carothers, 1986) has been proposed as a mechanism for allowing the maintenance of hydrogen isotopic equilibrium between clay minerals and water. In the deeper parts of sedimentary basins, where compaction causes a decrease in the rock/water ratio, the δ^2H composition of clay minerals, particularly in mudstones, has been proposed as a strong control on the hydrogen isotopic composition of formation waters (Kharaka & Carothers, 1986).

The greatest difference between the oxygen isotope and the hydrogen isotope patterns is the occurrence of clay minerals with anomalously low δ^2H values. The negative δ^2H anomalies could result from:

1 an influx of meteoric water from very high latitudes or from high-altitude regions;

2 an influx of organic water (Sheppard, 1986).

It is not easy to discriminate between these options, although it may be noteworthy that the lowest clay mineral $\delta^{18}O$ and δ^2H values seem to occur at about the same depth (3000 m; Figs 14 & 15). The lack of negative clay mineral δ^2H values at shallow depth may be a result of the absence of meteoric water penetration into shallow buried formations. Alternatively, if the negative clay-mineral δ^2H values are the result of organic water, the absence of such a signal in shallow formations may indicate an origin related to source-rock maturation.

There are also positive clay-mineral δ^2H anomalies shown on Fig. 15 that tend to occur in the deepest parts of the basin. These positive anomalies may result from:

1 uplift of formations previously equilibrated at greater depth;

2 influx of water associated with evaporitic rocks.

It is not possible to discriminate between these possibilities.

Comparison of North Sea basin oxygen and hydrogen isotopes in formation waters and clay minerals in sandstone

A comparison of δ^2H and $\delta^{18}O$ values of clay minerals and water values (Fig. 16) confirms the occurrence of extremely low δ^2H anomalies. The heavy cross (Fig. 16) portrays the ideal fractionation at an illustrative 100°C. There is a broad agreement between the $\delta^{18}O$ of the clay minerals but there is a distinct tail to the δ^2H of the clay minerals that cannot be in equilibrium with any of the present formation water samples in the data base. To reiterate, of the two common water sources of isotopically depleted hydrogen (high-latitude, or altitude, meteoric water and organic) it is not easy to conceive of a situation that allowed the influx of meteoric water to depths of 4000 m. Thus, the remaining source of the low δ^2H in the clay minerals is organic water. Anomalously low δ^2H values in clay minerals have been noted previously and assigned to interaction between petroleum-related fluids and clay minerals (e.g. Fallick *et al.*, 1993; Ziegler *et al.*, 1994). Direct interaction between clay minerals and petroleum may occur, especially if it is rich in aromatic compounds. This is because aromatic compounds are more soluble in water than alkanes (Hunt, 1996). Short-chain organic acids are relatively common in formation waters and these contain abundant hydrogen in a form that also may be accessible to the clay minerals (Pittman & Lewan, 1994). A corollary of this should be that the δ^2H composition of petroleum-related fluids should be *more* affected by fluid–rock interaction in clay-mineral-bearing sandstone reservoirs than in clay-mineral-free carbonate reservoirs. Published hydrogen isotope data from, for example, organic acids are not available to sustain or contradict this assertion.

Gulf Coast

Formation water data from the USA Gulf Coast basin display an increasing $\delta^{18}O$ with increasing depth from less than 0‰ at shallow depths to +9.3‰ at 5000 m (average +2.8‰: Fig. 17 & Table 6). The pattern is similar to, if better defined than, North Sea reservoir formation waters (Fig. 15), but the Gulf Coast formation waters are more enriched in ^{18}O. This trend has been noted even on the scale of a single formation within the Gulf Coast basin (e.g. Land & Fisher, 1987).

The clay-mineral stable isotope data from sandstone reservoirs of the Gulf Coast are dominated by illite and, less extensively, chlorite $\delta^{18}O$ data from a wide range of depths (Fig. 17). With increasing depth, illite $\delta^{18}O$ values tend to decrease. The water and illite $\delta^{18}O$ data appear to be converging, suggesting that they may be in isotopic equilibrium over a wide range of depths of burial. The difference between the $\delta^{18}O$ of formation waters and illite approximately obey illite–water equilibrium fractionation rules such that at 3000 m the difference is about +12‰ and at depths greater than 5000 m the difference is about +7‰. This pattern suggests that illite in sandstone reservoirs and the associated formation water remain in isotopic equilibrium during burial. It has been concluded previously (e.g. Land & Fisher, 1987) that mudstones dominate the formation water $\delta^{18}O$ values in Gulf Coast sandstones. The sandstone illite–water plot (Fig. 17) suggests that this may not be necessarily true: the formation water may be in equilibrium with illite in the sandstones (rather than, or as well as clays in mudstones). It is also possible that the mudstone–water and sandstone–water isotopic systems independently change in the same way by parallel evolution.

Gulf Coast chlorite stable isotope data seem at first sight to represent a significant deviation from the equilibrium status apparent for illite and formation water. The chlorite data appear to require formation waters that are more depleted in ^{18}O than the current formation waters. However, the strong mineral chemical

Table 6 Sources of stable isotope data for formation waters from the Gulf Coast reservoirs.

Reference	Gulf Coast province
Clayton *et al.*, 1966	Various
Fisher & Kreitler, 1987	Texas, Permian
Kharaka *et al.*, 1977	Various
Kharaka *et al.*, 1987	Central Mississippi, various
Land & Macpherson, 1989	Offshore Louisiana, Plio-Pleistocene
Land & Prezbindowski, 1981	South-central Texas, Lower Cretaceous carbonates
Macpherson, 1992	Texas, Oligocene
Milliken *et al.*, 1981	Texas, Oligocene
Moldovanyi *et al.*, 1993	Gulf Coast
Morton & Land, 1987	Texas, Oligocene
Taylor & Land, 1996	Offshore Texas, Miocene

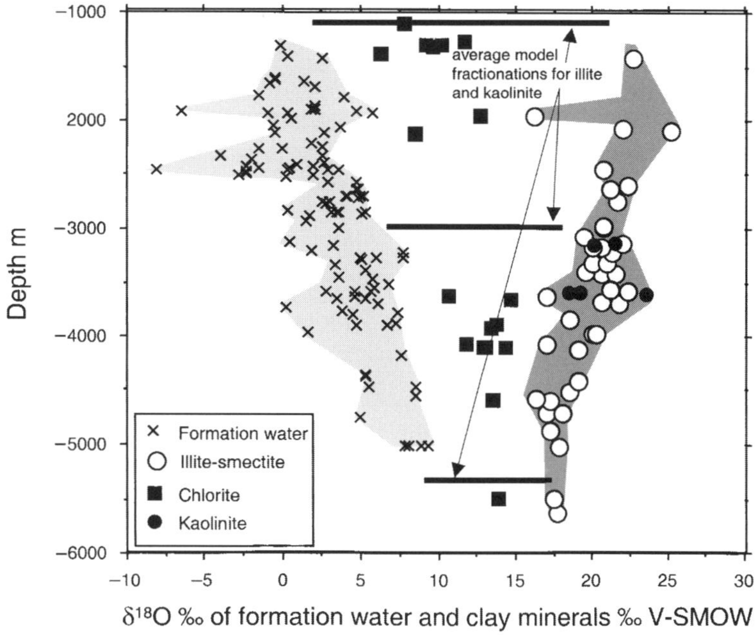

Fig. 17 Gulf Coast $\delta^{18}O$ data from formation water and clay minerals. Formation water data come predominantly from sandstone reservoirs. Clay minerals include kaolinite, chlorite and illite. The illite data probably include mixed-layer illite–smectites, although the differences in the fractionation factors for $\delta^{18}O$ between smectite and illite with water are trivial (Fig. 2). The heavy horizontal bars represent the approximate equilibrium fractionation at given depths assuming a general geothermal gradient (of 31°C km^{-1} and a surface temperature of 12°C). Data sources given in Tables 1–5.

dependence of chlorite–water fractionation has been noted already (Fig. 2D). It is possible that chlorites from the Gulf Coast reservoirs can be represented by the Wenner & Taylor (1971) fractionation equation, in which case equilibrium may have been reached and preserved.

CONCLUDING REMARKS

Once formed and stabilized, the diagenetic clay minerals may preserve their original, particularly oxygen, isotopic signatures, unless subjected to recrystallization and transformation (i.e. dissolution–reprecipitation) reactions. Hence clay minerals can be used to track the origin, as well as the spatial and temporal geochemical evolution of formation waters. However, hydrogen isotopes have been suggested by some authors (but dismissed by others) to undergo re-equilibration with formation waters without involving dissolution and reprecipitation reactions.

Statistical data for all kaolinite, dickite, illite and chlorite oxygen and hydrogen stable isotope data (in Figs 3–13) are presented in Table 7. The data include mode values (primary and secondary) and simple averages. The averages of $\delta^{18}O$ for kaolinite, dickite and illite are very similar, although the dickite mode is conspicuous for being slightly lower. This could be taken to suggest that all minerals formed from similar water at similar temperatures (acknowledging that their fractionation factors are similar over the relevant range of temperatures; Fig. 2). Alternatively the similarity may be merely coincidental such that differences in actual fractionation factors (controlled by different temperatures of growth) cancel out owing to growth from differentially evolved water. The $\delta^{18}O$ average for chlorite is noticeably different to the other clays, being 3.5‰ lighter. This may be a useful signpost to the systematically different oxygen isotope fractionation between chlorite and water, relative to the other common clay minerals (hinted at in Fig. 2). As for oxygen, the hydrogen isotope modes and averages are similar for kaolin, dickite and illite but different for chlorite. Chlorite does not seem to have the low $\delta^{2}H$ mode. This simply may be a result of the diminutive chlorite $\delta^{2}H$ data set.

Table 7 Summary modes and averages of $\delta^{18}O$ and $\delta^{2}H$ for kaolinite, dickite, illite and chlorite for all collated data from all basins.

	$\delta^{18}O$ per mil SMOW				$\delta^{2}H$ per mil SMOW			
Mineral	Primary mode	Secondary mode	Average	Number of samples	Primary mode	Secondary mode	Average	Number of samples
Kaolinite	+15.5	+10.0	+15.0	265	−60	−140	−57	139
Dickite	+12.5	+15.0	+14.4	54	−50	−120	−55	29
Illite	+15.0	+12.5	+14.5	293	−60	−110	−79	79
Chlorite	+12.5	+7.5	+11.0	103	−75	−30	−60	21

The relatively wide variation in the $\delta^{18}O$ and $\delta^{2}H$ values for all the clay-mineral types, both on global and basinal or sequence scale, implies that each of these minerals form from fluids with spatially and temporally variable isotopic (and probably chemical compositions), and/or at a wide range of temperatures.

Studies of the stable isotope ratios of clay minerals in sandstones should be coupled with careful petrographic reconstruction of the paragenetic sequence or even absolute dating (e.g. of illitic clay minerals; see Hamilton, this volume, pp. 253–287) within a well-constrained burial–thermal history of the basin, as well as fluid inclusion thermometry of closely associated minerals, such as quartz and carbonate cements.

North Sea basin sandstone formation waters acquire broadly increasing $\delta^{18}O$ values with increasing depth of burial. Clay minerals in sandstones have a wide range of $\delta^{18}O$ values. Some of the clay mineral data suggest that isotopic equilibrium has been achieved between the clay minerals in the sandstones and the present formation water. However, there is *no* simple pattern because there are at least two anomalous groups of data. Some clay minerals have anomalously low $\delta^{18}O$ values, suggesting either that there has been influx of meteoric water or that the clay minerals grew at significantly higher temperatures than currently found in the host sandstone. Conversely, some clay minerals in the more deeply buried sandstone have anomalously high $\delta^{18}O$ values, suggesting that the clay minerals grew at significantly lower temperatures than currently found in the host sandstone.

The marked systematic difference between $\delta^{18}O$ values in sandstone and carbonate formation waters in the same North Sea basin suggests that waters in permeable rocks are controlled by local phenomena and *not* by basin-scale processes that are capable of overprinting local mineral–water phenomena.

North Sea formation waters have relatively uniform $\delta^{2}H$ values with increased depth, whereas clay minerals in sandstone tend to acquire increased $\delta^{2}H$ values with increasing depth of burial, suggesting the development and maintainence of hydrogen isotopic equilibrium between the formation water and clay minerals. There are pronounced negative clay-mineral $\delta^{2}H$ anomalies, especially at about 3000 m (but also at about 4000 m). The negative anomalies may be the result of meteoric water incursion at these depths. Alternatively the anomalies may indicate that there has been a localized influx of distinct 'organic water', but only at depths at which organic-rich source rocks can generate petroleum (implying a link between the two).

Gulf Coast basin formation waters and clays in sandstones behave in a simpler way than North Sea basin sandstones because there seem to be few anomalies. Isotopic equilibrium seems to be developed and maintained at least between illite and kaolinite and formation water. Chlorite isotopes seem to be anomalously depleted in ^{18}O, although that may be the result of ill-defined chlorite–water oxygen isotope fractionation. The simpler data patterns in the Gulf Coast $\delta^{18}O$ data relative to the North Sea may be a function of the former being a basin with a simpler thermal and burial history.

SUGGESTIONS FOR FUTURE RESEARCH

This review has highlighted some significant shortfalls in our state of knowledge of clay minerals and stable isotopes. The following are possible directions for future research.

1 The clay-mineral–water oxygen isotope fractionation equations for kaolinite, illite and smectite seem to be reasonably robust but the selection of an oxygen isotope fractionation equation for chlorite seems to be unclear. Understanding the compositional dependence of chlorite–water oxygen isotope fractionation, and a protocol of how to use the suite of fractionation equations (especially when details of chlorite mineralogy are not known) are sorely needed.

2 Clay-mineral–water hydrogen isotope fractionation equations are comparatively poorly known. The kaolinite–water and illite–water hydrogen isotope systems have been defined (although there is a lack of agreement between published fractionation equations). For smectite and chlorite, the state of knowledge is relatively poor. The mineral compositional dependence of smectite–water and chlorite–water hydrogen isotope fractionation is particularly poorly known.

3 The effect of the specific origin of diagenetic clay-mineral's stable isotopic composition is not known. Thus, the isotopic compositions of illite from transformations of precursor kaolinite and dioctahedral smectite, and from feldspar dissolution may be different. Understanding of this could be improved by careful experimental work and empirical case studies.

4 The nature of isotopic modifications during progressive transformation of kaolinite into mesogenetic dickite and of smectite (and berthierine) into mesogenetic chlorite are poorly known. As for illite, understanding of the influence of the isotopic signature of the precursor clay on the ultimate mesogenetic clay minerals could be improved by careful experimental work and empirical case studies.

5 The specific influence of the pore-water/rock ratio on the extent of isotopic re-equilibration during clay-mineral reactions is poorly known. Empirical oxygen and hydrogen isotope data from clay minerals, reviewed here, suggest that mass flux of allochthonous water into deeply buried sandstones is potentially important, even in sandstones buried to 3000 m. This conclusion has important implications for the debate on open- versus closed-system diagenesis. Carefully chosen case studies of contrasting situations, previously and independently characterized, of both no-mass-flux and then open-system (e.g. artesian basins) conditions could help define what the clay mineral isotopic signal of high water/rock ratios should look like.

6 More (probably experimental) work is required to resolve issues of solid state stable isotopic resetting during continued diagenesis of clay minerals versus preservation of O and H isotopes during burial.

7 Ion (laser) probe isotope analyses of clay minerals may be needed to avoid the problems of contamination of one clay mineral by another and contamination by other minerals and organic matter.

ACKNOWLEDGEMENTS

The authors would like to acknowledge the careful and thoughtful reviews of Tony Fallick, Steve Franks, Reinhard Gaupp and Michael Wilson. Their questions, comments and suggestions led to many improvements.

REFERENCES

Aagaard, P., Jahren, J., Harstad, A.O., Nilsen, O. & Ramm, M. (2000) Formation of grain-coating chlorite in sandstones; laboratory synthesized vs. natural occurrences. *Clay Mineral.*, **35**, 261–269.

Anjos, S.M.C., De Ros, L.F. & Silva, C.M.A. (2003) Chlorite authigenesis and porosity preservation in the Upper

Cretaceous marine sandstones of the Santos Basin, offshore eastern Brazil. In: *Clay Mineral Cements in Sandstones*, International Association of Sedimentologists Special Publication 34 (Ed. by R.H. Worden and S. Morad), Blackwell Publishing, Oxford, pp. 291–316.

Aplin, A.C. & Warren, E.A. (1994) Oxygen isotopic indications of the mechanisms of silica transport and quartz cementation in deeply buried sandstones. *Geology*, **22**, 847–850.

Aplin, A.C., Warren, E.A., Grant, S.M. & Robinson, A.G. (1993) Mechanisms of quartz cementation in North Sea reservoir sandstones: constraints from fluid compositions. In: *Diagenesis and Basin Development* (Ed. by A. Horbury and A.G. Robinson). *AAPG Stud. Geol.*, **36**, 7–23.

Ayalon, A. & Longstaffe, F.J. (1988) Oxygen isotope studies of diagenesis and pore-water evolution in the Western Canada Sedimentary Basin: evidence from the upper Cretaceous basal Belly River Sandstone, Alberta. *J. Sed. Petrol.*, **58**, 489–505.

Berner, R.A. (1980) *Early Diagenesis, a Theoretical Approach*. Princeton University Press, Princeton.

Bird, M.I. & Chivas, A.R. (1988) Stable-isotopes evidence for low-temperature kaolinite weathering and post-formational hydrogen-isotope exchange in Permian kaolinites. *Chem. Geol. (Isotope Geosci. Sect.)*, **77**, 249–265.

Brint, J.F. (1989) *Isotope diagenesis and palaeofluid movement: Middle Jurassic Brent sandstones, North Sea*. D. Phil. thesis, University of Strathclyde, 288 pp.

Burley, S.D. & Macquaker, J.H.S. (1993) Authigenic clays, diagenetic sequences and conceptual diagenetic models in contrasting basin-margin and basin center North Sea Jurassic sandstones and mudstones. In: *Origin, Diagenesis and Petrophysics of Clay Minerals in Sandstones*, Society of Economic Paleontologists and Mineralogists Special Publication 47 (Ed. by D.W. Houseknecht and E.D. Pittman), Tulsa, OK, pp. 81–110.

Capuano, R.M. (1992) The temperature dependence of hydrogen isotope fractionation between clay minerals and water: evidence from geopressured systems. *Geochim. Cosmochim. Acta*, **56**, 2547–2554.

Clauer, N., O'Neil, J.R., Bonnot-Courtois, C. & Holtzapffel, T. (1990) Morphological, chemical and isotopic evidence for an early diagenetic evolution of detrital smectite in marine sediments. *Clays Mineral.*, **38**, 33–46.

Clauer, N., Stille, P., Keppens, E. & O'Neil, J.R. (1992) The glauconitization process: constraints from the strontium, neodymium and oxygen isotope geochemistry of Recent glauconites. *CR Acad. Sci. Paris (Ser. II)*, **315**, 321–327.

Clauer, N., O'Neil, J.R. & Furlan, S. (1995) Clay minerals as records of temperature conditions and duration of thermal anomalies in the Paris Basin, France. *Clay Mineral.*, **30**, 1–13.

Clayton, R.N., Friedman, I., Graf, D.L., Mayeda, T.K., Meents, W.F. & Shimp, N.F. (1966) The origin of saline formation waters 1. Isotopic composition. *J. Geophys. Res.*, **71**, 3869–3882.

Cole, D.R. (1985) A preliminary evaluation of oxygen isotopic exchange between chlorite and water. *Geol. Soci. Am. Abstr. Programs*, **17**, 550.

Craig, H. (1961) Isotope variations in meteoric waters. *Science*, **133**, 1702–1703.

Dansgaard, W. (1964) Stable isotopes in precipitation. *Tellus*, **16**, 436–468.

Egeberg, P.A. & Aagaard, P. (1989) Origin and evolution of formation waters from oil fields on the Norwegian shelf. *Appl. Geochem.*, **4**, 131–142.

Ehrenberg, S.N., Aagaard, P., Wilson, M.J., Fraser, A.R. & Duthie, D.M.L. (1993) Depth-dependent transformation of kaolinite to dickite in sandstones of the Norwegian continental shelf. *Clay Mineral.*, **28**, 325–352.

Fallick, A.E., Macaulay, C.I. & Haszeldine, R.S. (1993) Implications for linearly correlated oxygen and hydrogen isotopic compositions for kaolinite and illite in the Magnus sandstone, North Sea. *Clay. Clay Mineral.*, **41**, 184–190.

Fisher, R.S. & Kreitler, C.W. (1987) Geochemistry and hydrodynamics of deep-basin brines, Palo Duro Basin, Texas, USA. *Appl. Geochem.*, **2**, 459–476.

Franks, S.G., Hameg, A. & Liang, L. (1997) Diagenetic evolution of Cambrian sandstones, Hassi Bir Rekaiz area, Algeria. In: *American Association of Petroleum Geologists 1997 Annual Convention, Annual Meeting Abstracts*, 6, Tulsa, OK, p. 37.

Friedrichsen, H. (1984) Oxygen and hydrogen isotope studies on glauconites. *Terra Cognita*, **4**, 218.

Froelich, P.N., Klinkhammer, G.P., Bender, M.L. *et al.* (1979) Early oxidation of organic matter in pelagic sediments of the eastern equatorial Atlantic: suboxic diagenesis. *Geochim. Cosmochim. Acta*, **43**, 1075–1090.

Girard, J.P. & Barnes, D.A. (1995) Illitisation and paleothermal regimes in the Middle Ordovician St. Peter Sandstone, central Michigan Basin: K–Ar, oxygen isotopes, and fluid inclusion data. *AAPG Bull.*, **79**, 49–69.

Glasmann, J.R., Clark, R.A., Larter, S., Briedis, N.A. & Lundegard, P.D. (1989a) Diagenesis and hydrocarbon accumulation; Brent Sandstone (Jurassic), Bergen High area, North Sea. *AAPG Bull.*, **73**, 1341–1360.

Glasmann, J.R., Larter, S., Briedis, N.A. & Lundegard, P.D. (1989b) Shale diagenesis in the Bergen High area, North Sea. *Clay. Clay Mineral.*, **37**, 97–112.

Gluyas, J., Robinson, A. & Primmer, T. (1997) Rotliegend sandstone diagenesis: a tale of two waters. In: *Geofluids II '97: Extended Abstract Volume* (Ed. by J. Hendry, P. Carey, J. Parnell, A. Ruffell and R.H. Worden), 10–14 March, Belfast, pp. 291–294.

Graham, C.M., Viglino, J.A. & Harmon, R.S. (1987) Experimental study of hydrogen-isotope exchange between aluminous chlorite and water and of hydrogen diffusion in chlorite. *Am. Mineral.*, **72**, 566–579.

Greenwood, P.J., Shaw, H.F. & Fallick, A.E. (1994) Petrographic and isotopic evidence for diagenetic processes in Middle Jurassic sandstones and mudrocks from the Brae area, North Sea. *Clay Mineral.*, **29**, 637–650.

Grigsby, J.D. (2001) Origin and growth mechanism of authigenic chlorite in sandstones of the Lower Vicksburg Formation, South Texas. *J. Sed. Res.*, **71**, 27–36.

Hamilton, P.J. (2003) A review of radiometric dating techniques for clay mineral cements in sandstones. In: *Clay Mineral Cements in Sandstones*, International Association of Sedimentologists Special Publication 34 (Ed. by R.H. Worden and S. Morad), Blackwell Publishing, Oxford, pp. 253–287.

Harper, D.A., Longstaffe, F.J., Wadleigh, M.A. & McNutt, R.H. (1995) Secondary K-feldspar at the Pre-Cambrian–Paleozoic unconformity, southwestern Ontario. *Can. J. Earth Sci.*, **32**, 1432–1450.

Hassanipak, A.A. & Eslinger, E.V. (1985) Mineralogy, crystallinity, $^{18}O/^{16}O$ and D/H of Georgia kaolins. *Clay. Clay Mineral.*, **33**, 99–106.

Hesse, R., Lebel, J. & Gieskes, J.M. (1985) Interstitial water chemistry of gas hydrate bearing sections on the Middle America Trench Slope Deep Sea Drilling Project, Leg 84. *Init. Rep. Deep Sea Drilling Proj.*, **84**, 727–737.

Hillier, S., Fallick, A.E. & Matter, A. (1996) Origin of pore-lining chlorite in the aeolian Rotliegend of northern Germany. *Clay Mineral.*, **31**, 153–171.

Hogg, A.J.C., Pearson, M.J., Fallick, A.E. & Hamilton, P.J. (1995) An integrated thermal and isotopic study of the diagenesis of the Brent Group, Alwyn South, U.K. North Sea. *Appl. Geochem.*, **10**, 531–546.

Hornibrook, E.R.C. & Longstaffe, F.J. (1996) Berthierine from the Lower Cretaceous Clearwater Formation, Alberta, Canada. *Clay. Clay Mineral.*, **44**, 1–21.

Hunt, J.M. (1996) *Petroleum Geochemistry and Geology*, 2nd ed. W.H. Freeman, New York.

Kelly, J.C., Webb, J.A. & Maas, R. (2001) Isotopic constraints on the genesis and age of autochthonous glaucony in the Oligo-Miocene Torquay Group, south-eastern Australia. *Sedimentology*, **48**, 325–338.

Keppens, E. & O'Neil, J.R. (1984) Oxygen isotope variations in glauconites. *Terra Cognita*, **4**, 224.

Ketzer, J.M., Morad, S., Nystuen, J.P. & De Ros, L.F. (2003) The role of the Cimmerian Unconformity (Early Cretaceous) in the kaolinitization and related reservoir-quality evolution in Triassic sandstones of the Snorre Field, North Sea. In: *Clay Mineral Cements in Sandstones*, International Association of Sedimentologists Special Publication 34 (Ed. by R.H. Worden and S. Morad), Blackwell Publishing, Oxford, pp. 361–382.

Kharaka, Y.K. & Carothers, W.W. (1986) Oxygen and hydrogen isotope geochemistry of deep basin brines. In: *Handbook of Environmental Isotope Geochemistry*, Vol. 2 (Ed. by P. Fritz and J. Ch. Fontes), Elsevier, Amsterdam, pp. 305–360.

Kharaka, Y.K., Callender, E. & Carothers, W.W. (1977) Geochemistry of geopressured geothermal waters from the Texas Gulf Coast. In: *Proceedings of the 3rd Conference on Geopressured-Geothermal Energy*, US Gulf Coast, Vol. 1, University of Southwestern Louisiana, pp. 121–165.

Kharaka, Y.K., Maest, A.S., Carothers, W.W., Law, L.M., Lamothe, P.J. & Fries, T.L. (1987) Geochemistry of metal rich brines from Central Mississippi Salt Dome Basin, USA. *Appl. Geochem.*, **2**, 543–561.

Lambert, S.J. & Epstein, S. (1980) Stable isotope investigations of an active geothermal system in Valles Caldera, Jemez Mountains, New Mexico. *J. Volcanol. Geoth. Res.*, **8**, 111–129.

Land, L.S. & Dutton, S.P. (1978) Cementation of a Pennsylvanian deltaic sandstone: isotopic data. *J. Sed. Petrol.*, **48**, 1167–1176.

Land, L.S. & Fisher, R.S. (1987) Wilcox sandstone diagenesis, Texas Gulf Coast: a regional isotopic comparison with the Frio Formation. In: *Diagenesis of Sedimentary Sequences* (Ed. by J.D. Marshall). *Geol. Soc. London Spec. Publ.*, **36**, 219–235.

Land, L.S. & Macpherson, G.L. (1989) Geochemistry of formation water, Plio-Pleistocene reservoirs, offshore Louisiana. *Trans. Gulf Coast Assoc. Geol. Soc.*, **39**, 421–430.

Land, L.S. & Prezbindowski, D.R. (1981) The origin and evolution of saline formation water, Lower Cretaceous carbonates, South-Central Texas. *J. Hydrol.*, **54**, 51–74.

Lawrence, J.R., Gieskes, J.M. & Broecker, W.S. (1975) Oxygen isotope and cation composition of DSDP pore waters and the alteration of layer II basalt. *Earth Planet. Sci. Lett.*, **27**, 1–10.

Lee, M. (1984) *Diagenesis of the Permian Rotliegend sandstone, North Sea: K–Ar, $^{18}O/^{16}O$ and petrographic evidence.* PhD thesis, Case Western Reserve University, Ohio, 346 pp.

Longstaffe, F.J. (1983) Diagenesis, IV. Stable isotope studies of diagenesis in clastic rocks. *Geosci. Can.*, **10**, 44–58.

Longstaffe, F.J. (1989) Stable isotopes as tracers in clastic diagenesis. In: *Short Course in Burial Ddiagenesis* (Ed. by I.E. Hutcheon), Mineralogical Association of Canada, Montreal, pp. 201–277.

Longstaffe, F.J. & Ayalon, A. (1987) Oxygen-isotope studies of clastic diagenesis in the Lower Cretaceous Viking Formation, Alberta: implications for the role of meteoric water. In: *Diagenesis of Sedimentary Sequences* (Ed. by J.D. Marshall). *Geol. Soc. London Spec. Publ.*, **36**, 277–296.

Longstaffe, F.J. & Ayalon, A. (1990) Hydrogen-isotope geochemistry of diagenetic clay minerals from Cretaceous sandstones, Alberta, Canada: evidence for exchange. *Appl. Geochem.*, **5**, 657–668.

Lynch, F.L., Mack, L.E. & Land, L.S. (1997) Burial diagenesis of illite/smectite in shales and the origins of authigenic quartz and secondary porosity in sandstones. *Geochim. Cosmochim. Acta*, **61**, 1995–2006.

Macaulay, C.I., Haszeldine, R.S. & Fallick, A.E. (1992) Diagenetic pore waters stratified for at least 35 million years: Magnus oil field, North Sea. *AAPG Bull.*, **76**, 1625–1634.

Macaulay, C.I., Fallick, A.E. & Haszeldine, R.S. (1993) Textural and isotopic variations in diagenetic kaolinite from the Magnus Oilfield sandstones. *Clay Mineral.*, **28**, 625–639.

Macpherson, G.L. (1992) Regional variations in formation water chemistry: major and minor elements, Frio Formation fluids, Texas. *AAPG Bull.*, **76**, 740–757.

Maliva, R.G., Dickson, J.A.D. & Fallick, A.E. (1999) Kaolin cements in limestones: potential indicators of organic-rich pore waters during diagenesis. *J. Sed. Res.*, **69**, 158–163.

Marumo, K., Nagasawa, K. & Kuroda, Y. (1980) Mineralogy and hydrogen isotope geochemistry of clay minerals in the Ohnuma geothermal area, northeastern Japan. *Earth Planet. Sci. Lett.*, **47**, 255–262.

Mátyás, J. & Matter, A. (1997) Diagenetic indicators of meteoric flow in the Pannonian Basin, Southeast Hungary. In: *Basin-wide Diagenetic Patterns: Integrated Petrologic, Geochemical and Hydrological Considerations*, Society of Economic Paleontologists and Mineralogists Special Publication 57 (Ed. by I.P. Montañez, J.M. Gregg and K.L. Shelton), Tulsa, OK, pp. 281–296.

McAulay, G.E., Burley, S.D., Fallick, A.E. & Kusznir, N.J. (1994) Paleohydrodynamic fluid flow regimes during diagenesis of the Brent Group in the Hutton–NW Hutton reservoirs: constraints from oxygen isotopic studies of authigenic kaolin and reverse flexural modelling. *Clay Mineral.*, **29**, 609–626.

Milliken, K.L., Land, L.S. & Loucks, R.G. (1981) History of burial diagenesis determined from isotopic geochemistry, Frio Formation, Brazoria County, Texas. *AAPG Bull.*, **65**, 1397–1413.

Mizota, C. & Longstaffe, F.J. (1996) Origin of Cretaceous and Oligocene kaolinites from the Iwaizumi clay deposit, Iwate, northeastern Japan. *Clay. Clay Mineral.*, **44**, 408–416.

Moldovanyi, E.P., Walter, L.M. & Land, L.S. (1993) Strontium, boron, oxygen and hydrogen isotope geochemistry of brines in basal strata of the Gulf Coast sedimentary basin, USA. *Geochim. Cosmochim. Acta*, **57**, 2083–2099.

Mora, C.I., Sheldon, B.T., Elliott, W.C. & Driese, S.G. (1998) An oxygen isotope study of illite and calcite in three Appalachian Paleozoic vertic paleosols. *J. Sed. Res.*, **68**, 456–464.

Morad, S. & Al-Aasm, I.S. (1994) Origin and diagenetic evolution of phosphate nodules in Upper Proterozoic mudstones of the Visingsö Group (southern Sweden): evidence from petrology, mineral chemistry and isotopes. *Sed. Geol.*, **88**, 267–282.

Morad, S., Ben Ismail, H., Al-Aasm, I.S. & de Ros, L.F. (1994) Diagenesis and formation-water chemistry of Triassic reservoir sandstones from southern Tunisia. *Sedimentology*, **41**, 1253–1272.

Morad, S., Ketzer, M. & de Ros, L.F. (2000) Spatial and temporal distributions of diagenetic alterations in siliciclastic rocks. *Sedimentology*, **47**, 95–120.

Morton, R.A. & Land, L.S. (1987) Regional variations in formation water chemistry, Frio Formation (Oligocene), Texas Gulf Coast. *AAPG Bull.*, **71**, 191–206.

O'Neil, J.R. (1987) Preservation of H, C and O isotopic ratios in the low temperature environment. In: *Stable Isotope Geochemistry of Low Temperature Fluids* (Ed. by T.K. Kyser), Short Course 13, Mineralogical Association of Canada, Montreal, pp. 85–128.

O'Neil, J.R. & Kharaka, Y.K. (1976) Hydrogen and oxygen isotope exchange reactions between clay minerals and water. *Geochim. Cosmochim. Acta*, **40**, 241–246.

Osborne, M., Haszeldine, R.S. & Fallick, A.E. (1994) Variations in kaolinite morphology with growth temperature in isotopically mixed pore-fluids, Brent Group, UK North Sea. *Clay Mineral.*, **29**, 591–608.

Parnell, J., Baron, M. & Boyce, A. (2000) Controls on kaolinite and dickite distribution, Highland Boundary Fault Zone, Scotland and Northern Ireland. *J. Geol. Soc. London*, **157**, 635–640.

Pittman, E.D. & Lewan, M.D. (1994) *Organic Acids in Geological Processes*. Springer-Verlag, New York.

Platt, J.D. (1993) Controls on clay mineral distribution and chemistry in the Early Permian Rotliegend of Germany. *Clay Mineral.*, **28**, 393–416.

Robinson, A.G., Coleman, M.L. & Gluyas, J.G. (1993) The age of illite cement growth, Village Fields area, southern North Sea: evidence from K–Ar ages and $^{18}O/^{16}O$ ratios. *AAPG Bull.*, **77**, 68–80.

Ryan, P.C., Conrad, M.E., Brown, K., Chamberlain, C.P. & Reynolds, R.C. (1998) Oxygen isotope compositions of mixed-layer serpentine–chlorite and illite–smectite in the Tuscaloosa Formation (U.S. Gulf Coast): implications for pore fluids and mineralogic reactions. *Clay. Clay Mineral.*, **46**, 357–368.

Savin, S.M. & Epstein, S. (1970) The oxygen and hydrogen isotope geochemistry of clay minerals. *Geochim. Cosmochim. Acta*, **34**, 25–42.

Savin, S.M. & Lee, M. (1988) Isotopic studies of phyllosilicates. In: *Hydrous Phyllosilicates (Exclusive of Micas)*, Reviews in Mineralogy 19 (Ed. by S.W. Bailey), Bookcrafter Inc., Michigan, MI, pp. 189–223.

Sheppard, S.M.F. (1977) The Cornubian batholith, SW England: D/H and $^{18}O/^{16}O$ studies of kaolinite and other alteration minerals. *J. Geol. Soc. London*, **133**, 573–591.

Sheppard, S.M.F. (1986) Characterization of isotopic variations in natural waters. In: *Stable Isotopes in High Temperature Geological Processes* (Ed. by J.W.

Valley, H.P. Taylor Jr. and J.R. O'Neil). *Rev. Mineral.*, **16**, 165–183.

Sheppard, S.M.F. & Gilg, H.A. (1996) Stable isotope geochemistry of clays. *Clay Mineral.*, **31**, 1–24.

Shieh, Y.N. & Sutter, T.G. (1979) Formation conditions of authigenic kaolinite and calcite in coals by stable isotope determinations. *Clay. Clay Mineral.*, **27**, 154–156.

Spötl, C., Houseknecht, D.W. & Longstaffe, F.J. (1994) Authigenic chlorites in sandstones as indicators of high-temperature diagenesis, Arkoma foreland basin, USA. *J. Sed. Res.*, **A64**, 553–566.

Stewart, R.N.T., Fallick, A.E. & Haszeldine, R.S. (1994) Kaolinite growth during pore-water mixing: isotopic data from Palaeocene sands, North Sea, UK. *Clay Mineral.*, **29**, 627–636.

Surdam, R.C., Jiao, Z.S. & MacGowan, D.B. (1993) Redox reactions involving hydrocarbons and mineral oxidants: a mechanism for significant porosity enhancement in sandstones. *AAPG Bull.*, **77**, 1509–1518.

Taylor, T.R. & Land, L.S. (1996) Association of allochthonous waters and reservoir enhancement in deeply buried Miocene sandstones: Picaroon Field, Corsair Trend, offshore Texas. In: *Siliciclastic Diagenesis and Fluid Flow: Concepts and Applications*, Society of Economic Paleontologists and Mineralogists Special Publication 55 (Ed. by L.J. Crossey, R. Loucks and M.W. Totten), Tulsa, OK, pp. 37–48.

Tilley, B.J. & Longstaffe, F.J. (1989) Diagenesis and isotopic evolution of porewaters in the Alberta Deep Basin: the Falher Member and Cadomin Formation. *Geochem. Cosmochim. Acta*, **53**, 2529–2546.

Uysal, I.T., Golding, S.D. & Baublys, K. (2000) Stable isotope geochemistry of authigenic clay minerals from Late Permian coal measures, Queensland, Australia: implications for the evolution of the Bowen Basin. *Earth Planet. Sci. Lett.*, **180**, 149–162.

Warren, E.A. & Smalley, P.C. (1994) *North Sea Formation Water Atlas*. Geological Society Memoir 15, Geological Society Publishing House, Bath.

Wenner, D.B. & Taylor, H.P. Jr. (1971) Temperatures of serpentinization of ultramafic rocks based upon O^{18}/O^{16} fractionation between coexisting serpentine and magnetite. *Contrib. Mineral. Petrol.*, **32**, 165–185.

Whitney, G. & Northrop, H.R. (1987) Diagenesis and fluid flow in the San Juan Basin, New Mexico—regional zonation in the mineralogy and stable isotope composition of clay minerals in sandstone. *Am. J. Sci.*, **287**, 353–382.

Wilkinson, M., Fallick, A.E., Keaney, G.M.J., Haszeldine, R.S. & McHardy, W.J. (1994) Stable isotopes in illite: the case for meteoric water flushing within the Upper Jurassic Fulmar Formation sandstones, UK North Sea. *Clay Mineral.*, **29**, 567–574.

Worden, R.H. & Morad, S. (2003) Clay minerals in sandstones: controls on formation, evolution and distribution. In: *Clay Mineral Cements in Sandstones*, International Association of Sedimentologists Special Publication 34 (Ed. by R.H. Worden and S. Morad), Blackwell Publishing, Oxford, pp. 3–41.

Worden, R.H., Smalley, P.C. & Oxtoby, N.H. (1996) The effects of thermochemical sulfate reduction upon formation water salinity and oxygen isotopes in carbonate gas reservoirs. *Geochim. Cosmochim. Acta*, **60**, 3925–3931.

Yeh, H.-W. (1980) D/H ratios and late stage dehydration of shales during burial. *Geochim. Cosmochim. Acta*, **44**, 341–352.

Yeh, H.-W. & Eslinger, E.V. (1986) Oxygen isotopes and extent of diagenesis of clay minerals during sedimentation and burial in the sea. *Clay. Clay Mineral.*, **34**, 403–406.

Yeh, H.-W. & Savin, S.M. (1977) Mechanism of burial metamorphism of argillaceous sediments: 3. O-isotope evidence. *Geol. Soc. Am. Bull.*, **88**, 1321–1330.

Ziegler, K. & Longstaffe, F.J. (2000) Multiple episodes of clay alteration at the Precambrian/Paleozoic unconformity, Appalachian Basin: isotopic evidence for long-distance and local fluid migrations. *Clay. Clay Mineral.*, **8**, 474–493.

Ziegler, K., Sellwood, B.W. & Fallick, A.E. (1994) Radiogenic and stable isotope evidence for age and origin of authigenic illites in the Rotliegend, Southern North Sea. *Clay Mineral.*, **29**, 555–565.

Zwingmann, H., Clauer, N. & Gaupp, R. (1999) Structure-related geochemical (REE) and isotopic (K–Ar, Rb–Sr, $\delta^{18}O$) characteristics of clay minerals from Rotliegend sandstone reservoirs (Permian, northern Germany). *Geochim. Cosmochim. Acta*, **63**, 2805–2823.

Int. Assoc. Sedimentol. Spec. Publ. (2003) **34**, 93–108

Palaeoclimate controls on spectral gamma-ray radiation from sandstones

A.H. RUFFELL[1], R.H. WORDEN[2] *and* R. EVANS[1]*

[1] *School of Geography, Queen's University, Belfast BT7 1NN, UK, e-mail: a.ruffell@qub.ac.uk*

[2] *Department of Earth Sciences, University of Liverpool, Brownlow Street, Liverpool L69 3BX, UK, e-mail: r.worden@liverpool.ac.uk*

ABSTRACT

Spectral gamma-ray data from borehole logs or outcrop measurements are cheap and quick to access, providing useful information for reservoir geologists, sequence stratigraphers, petrographers and now palaeoclimatologists. Potassium (K), uranium (U) and thorium (Th) are the three most abundant radioactive elements found in rocks and are thus considered to be the likely sources of gamma-ray variation detected both at outcrop and in the subsurface. The spectral gamma-ray (SGR) variation shown by these three elements in rocks has been considered to reflect: grain sorting at flooding surfaces; U concentration during anoxia; Th dilution in carbonates and changes in the weathering regime of sediment source-lands. Potassium is routinely assumed to reflect clay-mineral, feldspar, glauconite and evaporite abundance during wireline log interpretation. This study makes an assessment of the interpretation of spectral gamma-ray data in sandstones in terms of palaeoclimate influence. The different hosts and mobility of K, U and Th can provide interpretative data on palaeoclimates. The content and weathering of K-feldspars, heavy minerals and clays or micas, when observed in conjunction with facies analysis and petrography, all provide palaeoclimatic information. We document three case studies where SGR data in clay-bearing sandstones have yielded valuable palaeoclimatic and reservoir quality information. These are: Cretaceous greensands (southern England); Carboniferous feldspathic ganisters (north of Ireland); and Jurassic feldspathic reservoir sandstones (North Sea). We also document and discuss the reasons why one case study using SGR data failed to reveal significant palaeoclimatic information.

INTRODUCTION

K, U and Th hosts and mobility in sandstones

Potassium is common in many sediments that bear evaporite minerals, feldspar, mica or clays. Potassium is known to be leached from feldspars and muscovite during kaolinite formation in hot and humid climates, whereas Th is considered to be insoluble in most weak alkali liquids and concentrated during weathering (Parkinson, 1996). Consequently, Th/K ratios in mudrocks have been cited as a palaeoclimate indicator (Myers, 1987; Ruffell & Worden, 1999), where elevated Th contents (compared with K) may reflect a humid hinter-

*Present address: Reservoir Management Ltd, 7 Bon Accord Square, Aberdeen AB11 6DJ, UK, r.evans@rml.co.uk land climate in which K was leached from the clays; a hypothesis which has yet to be tested. Uranium and thorium have a number of host minerals in sedimentary rocks including clays, feldspars, heavy minerals, phosphates and organic matter. Uranium and Th may thus have a number of hosts in sandstones, ranging from detrital grains (e.g. micas and heavy minerals) through to diagenetic cements (e.g. clays). There are no hard data on the U and Th contents of different clay species: the following suppositions are based on inferred observations and statistical correlations. Although the various clay species have different sites for U and Th, there is evidence to support a relationship between detrital kaolinite and U or detrital illite with Th (Pliler & Adams, 1962; Chamberlain, 1984; Myers, 1987; Myers & Bristow, 1989; Slatt *et al.*, 1992). In addition, specific heavy minerals have high Th or U contents, creating a potential problem in interpreting the gamma-ray emission of sedimentary rocks containing zircon, uraninite or thorite (Hurst, 1990; Myers & Bristow, 1989). Uranium is also thought to be more soluble in aqueous solution than Th and thus prone to mobilization during leaching (Frondel, 1958). Thus the Th/K ratio, although undoubtedly reflecting the detrital K content of sedimentary rocks, occasionally may be influenced by concentrations of thoriferous heavy minerals (for instance at transgressive surfaces), or K-rich feldspars, especially in sandstones (Hurst & Milodowski, 1996). Caution must be taken as the dilution effect of carbonate on Th concentration can produce the same Th/U ratios as anoxia (McRoberts *et al.*, 1998).

Structure of this paper

This paper presents opportunities for the interpretation of the spectral gamma-ray response from clay-bearing sandstones in terms of palaeoclimatic influence. Similar previous works have concentrated on mudstones (Parkinson, 1996; Ruffell & Worden, 1999), soils (Osmond & Ivanovich, 1992; Roberts & Plater, 1999) and limestones (McRoberts *et al.*, 1998). We will discuss initially the spectral gamma-ray responses that have a palaeoclimatic significance and possible problems in interpretation, followed by some illustrative case studies in which methods, results and interpretations will be presented.

SPECTRAL GAMMA-RAY RADIATION OF SANDSTONES

Clays and micas

Illites, white micas (e.g. muscovite) and biotite may contain (typically) 5–10% K_2O; chlorite, smectite, vermiculite and palygorskite contain between zero and 1.0% (Myers & Bristow, 1989) and kaolinite (and the other more rare kandites) contains no K (Pye & Krinsley, 1986; Chamley, 1989). Uranium solubility is pH dependent; U may be highly mobile and is often unevenly distributed in clays and micas. Definitive statements on the hosts and amounts of U and Th in clays are hard to find. Frondel (1958), Myers & Wignall (1987), Myers & Bristow (1989), van Buchem *et al.* (1992) and Davies & McLean (1996) all give elemental analysis of kaolinite-rich and illite-rich sediments that show higher concentrations of U (10 to 100 ppm) than in smectitic sediments such as fuller's earths (below 20 ppm). Beyond this, the mobility of U during both weathering and deep burial diagenesis makes consideration of its location in clays and micas difficult as single mineral analysis is required, which in clays is hard to achieve for two reasons. Firstly, in detrital clays, the size of individual crystals makes identification, and thus probe analysis difficult. Secondly, in the case of mixed-layer (or interstratified) clay minerals (such as illite–smectite) we are not too sure which phase is being measured. The sensitivity of U to weathering does make it a useful proxy for palaeoclimate studies so long as some constraints on the timing of U migration are made. Many micas contain Th from 30 ppm to 100 ppm. Thorium-rich minerals are also known from micro-inclusions in kaolinite (and other clay minerals: see Hurst, 1999), although the insoluble nature of Th in aqueous solution (Osmond & Ivanovich, 1992) means that strong acids are often required to mobilize this element.

Glaucony

The various K-rich and K-poor expandable and non-expandable forms of glauconite (potassium iron silicates) are grouped together as 'glaucony', which in its pelletal or oolitic form is indicative of a chemical environment as much as a mineralogical entity. Many consider glauconite *sensu stricto* as a clay mineral with a lattice akin to the white micas (Odin & Matter, 1981). The glaucony environment usually is mildly alkaline (i.e., sea water, pH *c.* 8) and weathering of glauconite is usually accelerated by oxidizing and mildly acidic conditions. Such weathering causes a loss of K and Fe and the crystallization of smectite. Thus the weathering of glaucony minerals can yield significant palaeoclimatic data. Glaucony-rich sandstones ('greensands') also contain significant U and Th in associated phosphates and organic matter, making the comparative use of Th/K and U/K ratios useful in assessing element mobility. The enrichment of K, U and Th in often permeable greensand successions (that will be more susceptible to leaching than mudrocks) suggests that glaucony may have considerable use as a highly sensitive indicator of weathering.

Feldspars

Feldspathic sandstones (e.g. arkoses) can contribute significantly to the total count gamma-ray emission as well as the K-sourced gamma-rays in sandstones. The level of gamma-ray emission depends largely on the K content of the detrital feldspars. Feldspars can also host Th to a level of 2 ppm (Myers & Wignall, 1987; McRoberts *et al.*, 1998, both after previous workers). The K content of alkali feldspars makes this group the most important source of gamma-ray radiation: the 5–10% elemental K content (Deer *et al.*, 1992) of alkali feldspars is a major contribution to the gamma-ray flux. The detrital feldspar content of sandstones is dependent on abundance in the hinterlands (provenance), weathering (during erosion and transport) and sorting. High feldspar contents indicate mineralogically immature sediment yields typical of rapid erosion and deposition. Identification of feldspar as the source of high gamma-ray emission, ideally using spectral data, requires preservation of this mineral group in sediments, which in turn may suggest certain palaeoclimatic conditions. The original K content of feldspars may be reduced through leaching. This may occur during weathering of the parent rock, transportation and deposition, or diagenesis.

Lithic clasts

Lithic sandstones present a problem for the correct interpretation of gamma-ray logs as the K, U or Th host of the individual derived rock fragments may be very variable. In a perfect world, we would prefer to see elemental analysis of individual grains plotted cumulatively, which could then be compared with gamma-ray logs. However, as gamma-ray logs are quick and cheap to obtain, many studies are made in the reverse order (like this one) where the gamma-ray signature is known and the most likely mineralogical response worked out thereafter. Down-hole spectral gamma-ray logs are rarely run through minerals and the mixture of minerals that rocks represent are no better exemplified than by lithic sandstones, which contain a vast admixture of minerals and thus chemistries. The conjunctive use of spectral gamma-ray analysis with sediment petrography is thus required. Potassium may be hosted in feldspathic, clay-rich or micaceous clasts. Thorium and U may occur in granite, pegmatite or gneiss clasts (Kearey & Brooks, 1991). Differentiation of K, U or Th may well indicate that high-K feldspars or high-U clays constitute a major component of the lithic clasts yet differentiating a feldspathic greywacke from a lithic sandstone may become difficult. Many lithic sandstones do result in higher gamma-ray emission than pure quartz sandstones by virtue of the derived igneous, metamorphic and sedimentary clasts that they contain (Chamberlain, 1984). The exception to this rule (lithic clasts from quartzites or polycrystalline quartz), rarely presents a problem for palaeoclimate analysis because it is the stability of the included clasts that is important. Excepting quartzite clasts, lithic material usually is derived in the same manner as feldspar: the inherently unstable nature of many rock fragments requires rapid erosion, transportation and deposition in order to preserve such material in subsequent sediment. The original K, U or Th content of a lithic sandstone can be compared with its weathered counterpart. Changes in radioelement content or ratios may reflect selective removal under specific climatic regimes.

Heavy minerals

Although volumetrically unimportant, the high gamma-ray emission and naturally high specific gravities of many heavy minerals makes them susceptible to hydraulic concentration and thus significant in terms of the gamma-ray response from discrete layers (Myers & Bristow, 1989; Hurst, 1990; Hurst & Milodowski, 1996). Potassium is not a significant component of the common heavy minerals. Thorium is found in abundance in apatite, monazite, thorianite, thorite and uranothorite; zircon, epidote and titanite also contain Th. Uraninite is actually an oxide of U, Pb, Ra and Th and with apatite may constitute the main mineralogical source of U in heavy minerals. The identification and effect on gamma-ray logs of heavy minerals is discussed in Dresser-Atlas (1983) and Hurst (1990). If a positive identification can be made then the susceptibility of many heavy minerals to weathering suggests that they could be used in palaeoclimate studies. Many provenance and correlation studies avoid sampling heavy mineral concentrates: down-hole spectral gamma-ray logs are not so selective and thus we might actually look closely at such concentrates to see what the weathering and palaeoweathering effects have been on such minerals.

CASE STUDIES

Spectral gamma-ray methods

Little use has been made of SGR data in palaeoclimate studies. In the present work, analyses of gamma-ray spectra have been carried out using two types of field-operated device. Measurements were made using appropriate sampling geometries with a Scintrex GIS-5 machine, calibrated to a Th standard and used initially at a 30 to 50 cm sample distance. The method of Slatt *et al.* (1992) was followed, wherein five readings were taken at each site, the lowest and highest were discarded and the three remaining values averaged. The Th/K ratio is not generated automatically by this device and must be derived by extraction of the U- and Th-based gamma-rays from the combined K + U + Th measurement. As a cross-check, further analyses have been undertaken using an Exploranium GR256 Spectrometer, allowing automatic derivation of Th/K and Th/U. The operational limitations and detection limits of all these machines are described by Slatt *et al.* (1992) and Davies & Elliot (1996). We also use and discuss a previously documented case study using a Schumberger down-hole SGR tool, for which the usual limitations on interpretation apply.

The K, U or Th hosts in any given sedimentary rock are rarely understood ahead of a full laboratory geochemical analysis (e.g. ICP–MS (inductively coupled plasma–mass spectrometry) or ICP–AES (atomic emission spectroscopy), XRF (X-ray fluorescence)). It seems fair to suppose that the close monitoring of the SGR technique by other methods and the use of the gamma-ray tool in sequence stratigraphic analysis suggests that, at the very least, major changes in clastic sedimentation are undoubtedly detected and thus the possibility is open for further investigation and other applications (like this one). The ease and speed with which measurements can be made has allowed integration of Th/K or Th/U ratios from the outcrop sections studied with mineralogical and isotope analyses. When attempting to detect a palaeoclimate signal in Phanerozoic marine successions, as many techniques as are available should be utilized. The possible overinterpretation of inappropriate data suggests that the conjunctive use of a wide range of information is recommended. This will diminish the chances of a non-climatically controlled signal (sea-level change, tectonism, diagenesis) being overemphasized. Thus we either use or discuss as many of the available indicators that may have palaeoclimatic significance. Sedimentological techniques that have proven valuable in previously published accounts of palaeoclimate analysis include the sediment type and facies, sediment mineralogy (especially clay mineralogy), total organic carbon contents and stable isotopes (usually oxygen and carbon). Relevant macro- and micropalaeontological analyses are discussed in this text. Some of the analyses of spectral gamma-ray characteristics in this study required a higher resolution sample density than the recommended 30 cm. To achieve this, some analyses were taken at a 15 cm vertical and/or horizontal spacing and the presumed overlap of analysed rock volume noted.

An example of K, U and Th measurement from clays in Cretaceous glauconitic sandstones from southern England

The Aptian (Lower Cretaceous: paralic and shallow marine) succession exposed in the sea-cliff sections of the Isle of Wight comprises the topmost non-marine Wealden Group (Wessex and Vectis Formations) and marine Lower Greensand Group (Ruffell & Wach, 1998; Fig. 1). Ruffell & Wach (1998) consider the lowest marine unit (Atherfield Clay Formation) to be a shallow-marine (inner shelf) mudrock, overlain by similar (outer) shelf sands/silty mudstones of the fossiliferous Ferruginous Sands Formation. This study concentrates on the Ferruginous Sands Formation, where retrogradational and progradational parasequences and parasequence sets (Ruffell & Wach, 1998) are usually detectable by the firmground faunas developed at flooding surfaces. The beds are all siliciclastic-dominated, being fine-grained silts and clays in the lower Aptian strata, passing into coarser glauconite- and limonite-rich sands with thin conglomerates in the upper Aptian and lower Albian strata. The most noticeable change from clay to mixed clay and fine-sand grain-size occurs at the Atherfield Clay–Ferruginous Sands transition (Fig. 1). Glauconite, iron oxides (limonite, goethite, haematite) and phosphates are all common in the sand-rich horizons, which become common in the lower Aptian Zone and recur throughout the section into the lower Albian.

The Isle of Wight section has been sampled and logged extensively using the instruments and techniques described above. Over 580 horizons (Fig. 1) were sampled at densities of 15 cm to 50 cm (30 cm average) using the devices described above. This work was conducted to derive comparative palaeoclimate data (Ruffell & Worden, 1999) and for sequence stratigraphic correlation to offshore hydrocarbon borehole data. To overcome the problem of possible misinterpretation of the derivation of K, U and Th, SGR data were taken specifically at clay-poor or clay-rich beds, to avoid the most obvious sandstone host. This still leaves a problem in how to interpret each flooding surface because hydraulic sorting will have caused variable mineral concentration at parasequence, parasequence set and systems tract boundaries. Thus we cannot regard all flooding surfaces as the same. Conversely, by selectively sampling clay beds, we have at least diminished the importance of sand-grade heavy minerals and short-term sea-level rise in influencing the results. Fifteen clay samples were also separated and the $< 2\ \mu m$ fraction collected and analysed by XRF and ICP–MS or ICP–AES. The resulting clay residue contained between 2 and 12 ppm of Th, and similar Th/K or Th/U ratios to the spectral gamma-ray data.

One hundred and sixteen XRD (X-ray diffraction), 56 SEM (scanning electron microscope) and 23 electron microprobe analyses were conducted (Fig. 1). Sandstones in the succession are quartzose with abundant limonite/goethite and glauconite ooids. Feldspar is rare in the Lower Greensands and apparently absent from mudstones. Heavy minerals are very well-known from this succession (Garden, 1991; Wach, 1991) and include rutile- and/or zircon-rich layers 2–3 mm thick. These commonly are present in sandstone and siltstone beds, causing problems for interpretation of Th and U ratios. Kaolinite/illite ratios, obtained from the

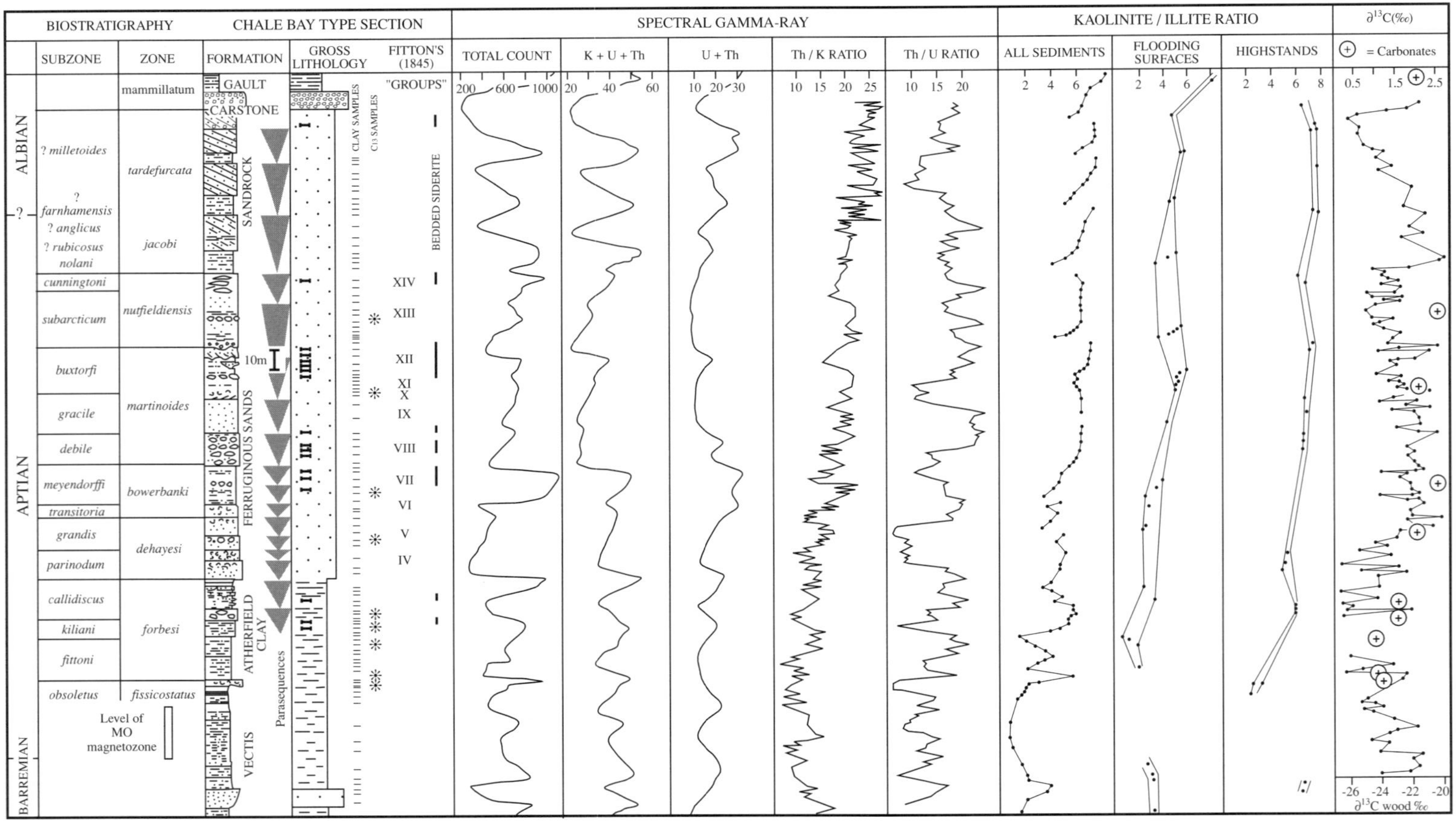

Fig. 1 Selective stratigraphic analysis of spectral gamma-ray, kaolinite/illite and $\delta^{13}C$ in order to compare stratigraphic lithology to palaeoclimate proxies. The Aptian–Albian Lower Greensand Group reference section at Chale Bay (Isle of Wight, southern England). Stratigraphy after Ruffell & Wach (1998). $\delta^{13}C$ of wood from Grocke *et al.* (1999). All other data from this study. Gross lithology is separated from formations to emphasize the early Aptian change from clay to sand.

clay fraction of the sediments show a strong relationship to both the grain size of the host sediment and thus to the parasequences of Ruffell & Wach (1998). The rise in kaolinite contents through each parasequence is thought to be caused by both increases in the detrital clay content (from SEM) and kaolinite diagenesis. Together, a bathymetric control on the detrital clay suite as well as a potential diagenetic control on secondary kaolinite makes a palaeoclimate interpretation very difficult. To negate the effect of short-term changes in bathymetry, we compared only the SGR response and kaolinite/illite ratio from the thicker clays developed at flooding surfaces. This simplification assumes that the flooding surfaces caused similar landward shifts of facies (and thus the nearshore settling of kaolinite) as the entire succession shows no evidence of deep marine (pelagic) or fluvial environments. However, even within the shelf environment, great changes in the flocculation state are possible (Gibbs, 1977), thus the comparison of flooding surfaces is taken as a guide only, rather than an absolute measurement. As an internal standard, the measurements from highstands are also compared. The similarity in trends suggests that either all lithologies were subjected to homogeneous diagenesis (for which there is no evidence) or that the diagenesis that undoubtedly did occur had a minor effect on the kaolinite/illite ratio. One notable change in the kaolinite/illite ratio is the sharp rise in kaolinite observed in the *Forbesi* Zone 15 m below and thus before the sand influx of the basal Ferruginous Sands.

The pattern of Th/K and Th/U variation is similar to the C-isotope stratigraphy (derived from fossil wood) by Grocke *et al.* (1999: Fig. 1 of this study) and the kaolinite/illite ratio. This especially is true of the SGR readings from flooding surfaces, although the overall trend is evident in the non-biased readings as well, suggesting that the whole succession has retained a spectral gamma-ray signal that can be positively compared both to palaeoclimate indicators (such as the kaolinite/illite ratio) and to grain size. This may be an effect of shallow burial (and thus limited fluid movement during diagenesis) or the result of internal barriers to fluid movement (mudstone flooding surfaces). The SGR curve from clays at flooding surfaces is similar to kaolinite/illite ratios and to the carbon isotope curve of Grocke *et al.* (1999), which suggest that the controls on detrital clay mineralogy are similar to Th/K and Th/U ratios and thus possible partitioning in the Aptian hinterlands. To summarize this section, the selective SGR outcrop analysis of mudstones intercalated with sandstones can be used in a multidisciplinary analysis of palaeoclimates.

An example of K mobility in Carboniferous feldspathic sandstones from northeast Ireland

The Dinantian–Namurian succession exposed in the sea-cliff sections north-east of Ballycastle (Co. Antrim, Northern Ireland) comprise over 600 m of cyclic coals, shales, limestones and variably feldspathic/quartzose sandstones. Such cyclically arranged lithofacies occur throughout the Dinantian and are thus mostly older than the classic Yoredale Cycles, and Millstone Grit successions of northern England, which are very similar. The sedimentology of this succession is summarized by Evans *et al.* (1998) who record six facies associations, which cover the transitional subenvironments between fluvio-deltaic and offshore sediments. A typical cycle consists of fining upward fluvio-deltaic and/or nearshore marine sandstones 6–8 m thick, which form a multistorey stack. These cycles are separated by coals, seatearths, shales and limestones, which record episodes of abandonment or transgression. Moreover, sediments show increasing evidence of marine or tidal influence towards the top of the succession. These rocks lie abruptly on basaltic lavas of Carboniferous age, and are overlain by Cretaceous limestones and Tertiary basalts. Tertiary sills, such as the 30-m-thick Fair Head Sill intrude some of the thicker shale horizons. The horizon selected was several kilometres from the nearest Tertiary sill or dyke, negating any significant contact metamorphic effect. The

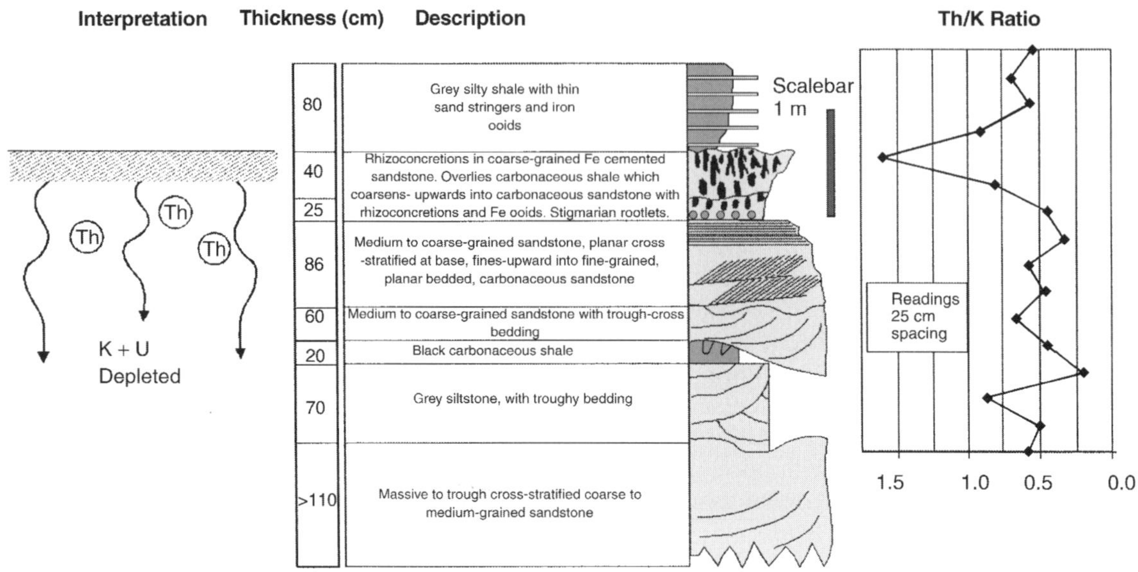

Fig. 2 Sedimentary log, facies description, Th/K ratio (from both Scintrex GIS-5 and Exploranium GR-256 run at the same time) and interpretation through a ganister or seatearth, Carboniferous, Ballycastle, Co. Antrim, Northern Ireland.

stratigraphy and sedimentology of the Ballycastle succession is covered in greater detail by Evans *et al.* (1998).

This study concentrates on a seatearth horizon developed just below the highest marine band (McGildoney's Marine Band) in the succession (Fig. 2). The marine band is one of the four found throughout the stratigraphy, which contain shales and marine shelly fauna (Fig. 2). The McGildoney's Marine Band (or McGildoney's Member, Evans *et al.*, 1998) is approximately 65 m above the Main Limestone and is dominated by two (1–1.5 m) shale layers, which are separated by thick, lenticular sandstones (Fig. 2). The shales contain ironstone bands and the bivalve *Lingula*, and the top shale contains the Namurian E2 zone fossil *Schellwienella rotundata* (Wilson & Robbie, 1966). The underlying seatearth consists of very fine-grained, quartzose sandstone containing ironstone rhizoconcretions (Fig. 2). This horizon was selected because it is probably the best-exposed example of a seatearth in the succession. We also chose a location where we were sure to be studying a true palaeosol or ganister (our interpretation is supported by Dr. J.O. Buckman, ichnologist, personal communication, 1994, who suggested that this succession was ideal). Furthermore, we were careful to select a ganister profile that had a thick carbonaceous shale above it in order to diminish the possibility of subsequent meteoric flushing of K, U or Th. Thinner silts and shales also occur below the horizon studied (Fig. 2). At outcrop the ganisters are 60 cm to 2 m in thickness, passing into coals or erosive surfaces above and massive or trough cross-stratified sandstones below. Lithological logs through the seatearths themselves are uninformative, being massive, pink to pure white variably feldspathic to quartzose sandstones. Both Scintrex GIS-5 and Exploranium GR-256 spectrometers were used in our study at a 25 cm interval across the horizon of interest. The data were collected (using a 30 s count rate) following the published method of Slatt *et al.* (1992). Measurements were calculated for total counts, K, U, Th, Th/U and Th/K ratios; the latter are plotted against a lithological log (Fig. 2).

The depletion of K (relative to Th) below the top surface of the two seatearths is of significance to this study because it demonstrates how

SGR data may be affected by palaeoclimate. The presence of coals, abundant fluvial sediments, positive $\delta^{13}C$ from shelly fauna relative to the rest of the succession (Evans *et al.*, 1998) all suggest that a warm and wet climate prevailed at this time in the Carboniferous of northeast Ireland. As a result, soils formed on feldspathic sandstones experienced progressive removal of K and U from feldspars and clays. This may have been through the action of meteoric water (natural acid rain) from above. The mineralogy, and especially the clay mineralogy of this succession, have several implications. In the ganisters, the low Th/K recorded at the outcrop, together with the clean, quartz-rich nature of the bed could be interpreted to be the result of severe leaching. The low Th/K values in the softer, clay-rich sandstones below the ganisters reflect preservation of K relative to Th, relative to the sandstones above. The heterolithic sandstones, siltstones and shales above and below the ganister show the highest Th/K, suggesting minimal removal of K in comparison to the ganisters. This interpretation does not preclude the removal of K from feldspars in the hinterland: moreover it suggests that when attempting to reconstruct hinterland weathering conditions (as in Fig. 2), we may wish to avoid ganisters for fear of measuring the K content of clays and feldspars that have been weathered both in the hinterland and during early diagenesis. The loss of K from feldspars during kaolinitization is proportional to the degree of weathering: complete removal of feldspars and kaolinite in the most severely leached seatearths has resulted in a very low gamma-ray flux from these units.

Unconformity leaching of K in Jurassic arkosic reservoir sandstones

Emery *et al.* (1990) document low levels of K (from down-hole SGR data) in the Upper Jurassic Magnus Field sandstones of the northern North Sea, which they ascribe to leaching during subsequent subaerial exposure. The kaolinitization of feldspars was also thought by Emery *et al.* (1990) to control the distribution of clay cements below the late Cimmerian Unconformity in the Gullfaks Field (northern North Sea). In this study we infer a strong palaeoclimatic control on the K loss.

The Late Jurassic environment of the northern North Sea (including the Magnus Field area) was a period of extensional tectonics that coincided with a period of raised sea-levels (Faerseth, 1996). During this time, the crests of tilted fault blocks were actively eroded, depositing turbiditic subarkosic sandstones, which now form the main reservoir target for the 600 million barrel oil field of Magnus. A further period of tectonism, coupled with a regional sea-level fall in latest Jurassic times, caused transient subaerial exposure of the Magnus Sandstones. During early Cretaceous times, thermal relaxation, subsidence and deposition of the regional topseal occurred. Emery *et al.* (1990) used depth-converted seismic data to generate a backstripped plot of the burial history for the Magnus Field. They concluded that at the time the early Cretaceous unconformity was formed, there was 250–750 m of uplift, making freshwater leaching of the Magnus Sandstones a distinct possibility.

Emery *et al.* (1990) then examined SGR logs through the unconformity and noted a zone of decreased K content extending for about 10 ft (3 m) into the Magnus Sandstone (Fig. 3). They compared this K depletion with petrographic analysis, which showed 7–9 vol% less K-feldspar and 3–5 vol% more kaolinite, which was not thought to reflect a regional change in facies. The partial or complete dissolution of feldspars was thought to have created a 5% increase in porosity, making the topmost Magnus Sandstone the main reservoir target. Emery *et al.* (1990) concern themselves mainly with the implications of their study for reservoir quality, uplift estimates and SGR log interpretation. An additional consideration is the palaeoclimatic implication for Emery *et al.*'s (1990) interpretation. During deposition of the Magnus Sandstone an arid or semi-arid climate prevailed over most of northern Europe (Hallam *et al.*, 1991). During early Cretaceous times, the peak of this aridity was reached, followed by

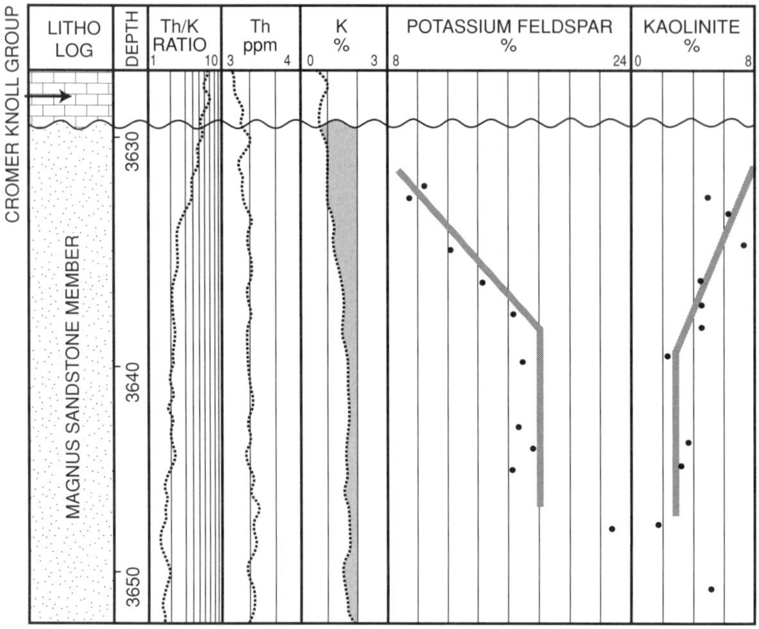

Fig. 3 Comparison of natural gamma-ray spectrometry log, petrographic and core analysis data through a borehole in the Magnus Field, northern North Sea. The upwards decline in K approaching the sub-Cretaceous unconformity corresponds to decreasing K-feldspar contents in the sandstone. Kaolinite and porosity both increase with progressive loss of K and K-feldspar. From Emery *et al.* (1990).

a period of more humid climatic conditions (Hallam *et al.*, 1991). Sladen & Batten (1984) made a palaeohydrological reconstruction of this time period from clay mineralogy, microflora and macroflora and suggested that arid or semi-arid periods were represented by annual rainfall rates of less than 500 mm. More humid periods, such as those that might have prevailed during leaching of the Magnus Sandstone, could have been as high as 1000 mm yr^{-1}. This inferred palaeoclimate history suggests that the eroded Magnus sands probably were feldspathic at the time of deposition (as shown by petrographic examination of sandstones distant from the unconformity; Barclay & Worden, 1998) as a result of rapid erosion and transport from the uplifted rift shoulders. These same feldspathic sandstones were then actively leached, with commensurate growth of kaolinite in the sandstone during transient Cretaceous uplift in a more humid climatic regime. The distribution of kaolinite (recorded by Emery *et al.*, 1990) from a similar tectonic–stratigraphic position, adds weight to the linked sea-level, tectonic and palaeoclimate hypothesis developed here through combining these disparate works. In addition, Emery *et al.* (1990) suggested that burial diagenesis alone could not be responsible for the widespread absence of kaolinite from beneath the unconformity. The implication of this is that subunconformity weathering and Th/K variation would not have been recorded had the climate been arid. Palaeoclimate information (in terms of Th/K, K-feldspar dissolution and clay mineralogy) often may be accessible from unconformities above sandstones.

Where is the palaeoclimate signal in Triassic Sherwood Sandstone Group reservoirs of the Wytch Farm Oilfield?

The Triassic Sherwood Sandstone Group is the main reservoir in the Wytch Farm Oilfield (365 million barrels) in southern England. The sandstones were deposited in a braided river system with interfluves and soil horizons (Holloway *et al.*, 1989), now represented by calcrete horizons. Calcretes attest to the semi-arid and arid nature of the palaeoclimate (Holloway *et al.*,

1989; Hogg *et al.*, 1996), which is somewhat at odds with the fluvial nature of the sandstones themselves. The explanation for this apparent contradiction comes from climate modelling: Parrish *et al.* (1982) interpreted the Triassic climate of Pangaea to be 'mega-monsoonal', suggesting that highly seasonal rainfall and drought occurred. This accounts for the nature of the Triassic deposits in Wytch Farm, which were predominantly fluvially transported, but then may have lain in an arid climate in which calcretes formed. The evidence of emergence (palaeosols, calcretes) and the extreme climate of the time makes these deposits attractive to palaeoclimatologists.

We examined numerous spectral gamma-ray logs (from boreholes) through the succession with a view to finding evidence for the actions of the prevailing climate or palaeoclimatic change. However, we found no evidence for any systematic depletion of K, U or Th, neither through the whole Sherwood Sandstone Group reservoir succession, nor beneath or above juvenile palaeosols or calcrete (Fig. 4). Initially, this was disappointing, given the good results from the other three case studies. Further consideration of the Sherwood Sandstone leads us to believe that although the hinterland probably experienced a high degree of seasonality, the lowland plain that was to become the Wytch Farm area was more or less arid, with brief and intense periods of flooding (Ruffell & Shelton, 1999). This apparent lack of consistent or significant groundwater flux, probably in an area of low vegetation and hence less humic acid generation, may have diminished the likelihood of K, U or Th removal. Thus, spectral gamma-ray evidence of palaeoclimate may not be preserved in soils formed under arid conditions.

The formation of soil profiles, calcretes and weathering profiles in sandstones laid down under arid conditions is thus unlikely to be associated with major changes of K, U or Th because the degree of chemical weathering of the detrital silicate minerals (e.g. K-feldspars) and the commensurate formation of pedogenic clays (e.g. kaolinite or mixed-layer illite–smectites) will have been minimal. Regional

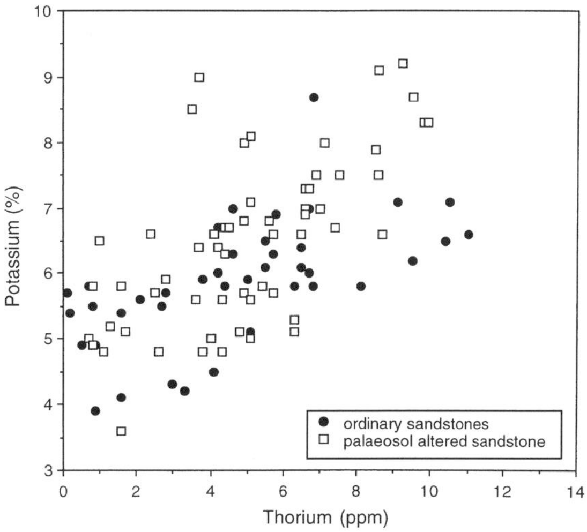

Fig. 4 Spectral gamma-ray data from the Sherwood Sandstone Group (Triassic) reservoir in Wytch Farm, Dorset. Cross-plot of potassium against thorium from both sandstones and palaeosol sandstones. The prediction of low-K sandstones from the leached (palaeosols) is not supported by these data. See text for discussion. (Data courtesy of Jonathon Evans, British Petroleum plc)

evidence for this is provided by the typically fresh appearance of detrital feldspars in many Triassic sandstones across northwest Europe (e.g. Worden *et al.*, 1999). Furthermore, we should consider the common occurrence of smectite in Triassic sandstones across northern Europe. Regardless of source lithology, smectites are almost ubiquitous in the reservoir sandstones of the North Sea, southern England, Paris Basin and northern Germany. The widespread occurrence of smectite, together with the presence of palygorskite and sepiolite suggests that these minerals formed during early diagenesis as a result of extreme wetting and drying cycles.

CONCLUSIONS

The general conclusions that can be drawn from the review section of this paper are summarized in terms of the K, U, or Th mineral hosts discussed above.

Clays and micas

Ratios of K, U and Th may serve as proxies for changing clay mineral populations when part of a multidisciplinary study. Thus low K and U, and high Th clays that form under hot and humid palaeoclimatic conditions in sandstones during eogenesis may provide a characteristic spectral gamma-ray response. Similarly, the *in situ* leaching of K, U or Th from clay in sandstones also may provide evidence of palaeoclimatic conditions (e.g. Th leaching may require humic acid migration from organic-rich sediments; in other acidic pore waters, Th is stable). Such leaching commonly accompanies clay diagenesis, which, if early and responding to meteoric influence, also may be useful in palaeoclimate studies (Tang *et al.*, 1994). With progressively deeper burial diagenesis, the interpretation of any palaeoclimate influence becomes harder to discern, being riven with too many variables. A far more effective approach to obtaining palaeoclimatic information from clay in sandstones is the selective sampling technique outlined here. Obtaining spectral gamma-ray data from thick clay beds within sandstones overcomes some of the problems of diagenesis because, although clay neoformation may have occurred, element transfer through the rock unit may not. Conversely, a clay sorting effect may be generated by measuring flooding surfaces (Davies & McLean, 1996). The varying magnitude of sea-level rise associated with each flooding surface may then produce a consistent pattern unconnected with palaeoclimate, unless such changes can be compared positively with other climate indicators.

Feldspars and glaucony

These mineral groups are considered together in light of their high K contents and similar reaction to weathering, albeit with very different end-products. The susceptibility of feldspars in sandstones to eogenetic dissolution and replacement by clay minerals makes this mineral group both useful and somewhat risky (owing to potential mesogenetic leaching/authigenesis) in palaeoclimate studies. The high K content of detrital K-feldspar results in usually obvious gamma-ray signatures. When borehole spectral gamma-ray devices are passed through sandstones, K-feldspars, high-K micas or glauconites, they will all provide elevated K-sourced gamma-ray emissions. Distinguishing the source of this K is dependant on Th/K or U/K ratios and the mineralogy of cuttings. If linked to facies analysis, the presence of such feldspar-rich sediment is suggestive of rapid physical erosion and deposition and is most common in areas with good sediment transfer mechanisms, such as glaciers and rivers (Reading, 1996). Again, when constrained through dating, the rates of deposition of such feldspar-bearing sandstones lead us to consider whether rapid erosion and deposition or lack of chemical weathering is the reason the feldspar has survived. Chronostratigraphically constrained horizons of K-depletion can be used as a proxy for *in situ* weathering. Care must be taken to ensure that non-depositional surfaces do not occur within the weathered horizons, especially if such a surface separates K-depleted from K-enriched sands. Conversely, when compared with other palaeoclimate proxies such as coals, bauxites, clay minerals, isotopes, etc., the presence of K-leaching and pedogenic clay cementation in sandstones may add significantly to our palaeoclimate data base.

Lithic clasts

The presence of clasts of rocks and minerals that are susceptible to weathering can be interpreted in a similar manner to feldspathic sandstones in that a suitable parent rock type must have been rapidly eroded, transported and deposited. This has obvious connotations for tectonics and climate, with implications of rapid basin subsidence and efficient transport mechanisms. The variable K, U and Th content of such lithic sandstones generally reflects hinterland lithologies, modified somewhat by sorting, weathering and diagenesis. Identification of the source terrain and its K, U and Th distribution can be compared with the resultant lithic sandstones, providing a monitor of elemental changes between source and sedimentary basin. Furthermore, the eogenetic removal of lithic clasts also can be monitored by comparing fresh to weathered sandstones. This comparative analysis must allow for changes in sediment input that are unrelated to, or coincident with, post-depositional diagenesis. The best opportunity for such study comes from the identification of hiatal surfaces that show little evidence of erosion yet abundant evidence of palaeoweathering.

Heavy minerals

The use of heavy minerals in palaeoclimatic studies is different to many of the other minerals outlined in these conclusions in that many such mineral grains have low solubility and are thus less susceptible to chemical weathering. The K, U and Th content of the heavy minerals in the source terrain can be compared with the weathered product, thus allowing some analysis of leaching. This is best achieved when constrained by stratigraphy or better still, chronostratigraphy. The variation in K, U and Th content and heavy mineral types makes further comparison difficult, especially when element mobility in these sediments is poorly understood.

Case studies

The more specific conclusions from our case studies show how selective analysis of K, U and Th from clays in sandstones may yield useful conjunctive palaeoclimate data and that K leaching in feldspathic sandstones was driven by a humid palaeoclimate. Together, our general observations, reinforced in three instances by case studies, lead us to consider the general mobility of K, U and Th in different climatic conditions. These are summarized in Fig. 5,

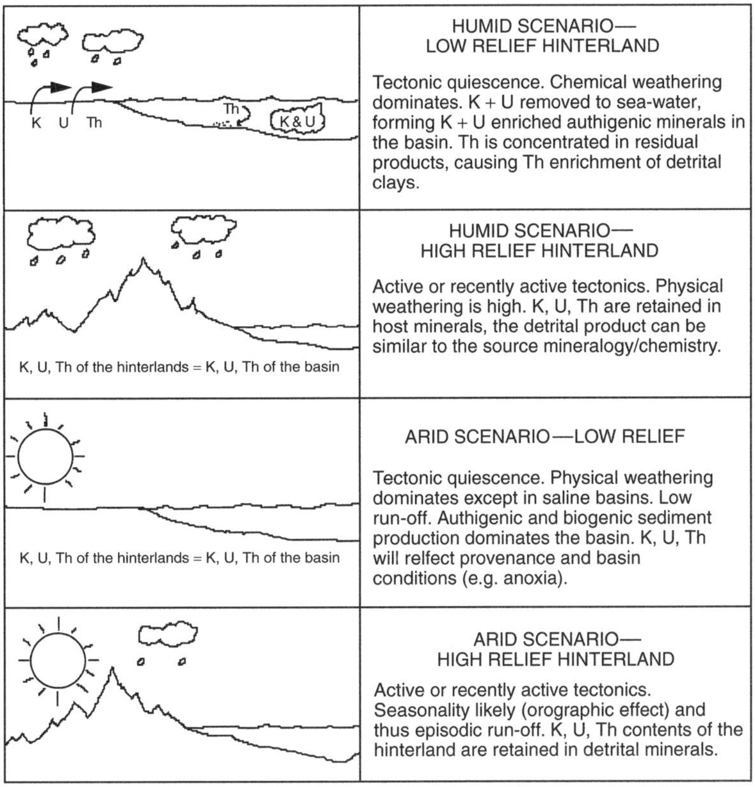

Fig. 5 Cartoon representation of the K, U and Th flux in different weathering systems at an idealized basin-wide scale. Fluxes of K, U and Th may differ at both finer and grosser scale (e.g. the movement of Th in soil profiles or the global cycling of K through sea water and authigenesis).

where we show what happens to K, U and Th in residual or weathered soils (and thence their detrital product, as in Parkinson (1996) or Ruffell & Worden (1999)) and thereafter in aqueous solution. If K and U are leached preferentially by non-humic acid-containing waters in the hinterland, these elements must go into ground water or river water. In the latter case, this may lead to a K or U flux into the adjacent basins. Although K and U fixing in glauconites, phosphates and black shales has often been discussed (Odin & Matter, 1981), possible changes in their abundance from source have not. The global geochemical cycles of elements such as K or U are determined partially by weathering. Changes to this weathering regime may be closely monitored by isotopes of C or Sr or clay minerals and possibly the K, U and Th variation in clays (Parkinson, 1996; Ruffell & Worden, 1999) and in sandstones.

ACKNOWLEDGEMENTS

We would like to thank John Parnell for his help in supervision during research by RE, and Grant Wach for his ideas on the Isle of Wight succession. Thanks also to our reviewers Andy Hurst and Jean-Francois Deconinck who made numerous improvements. Ian Meighan clarified the effect of apatite on radioactive emission. This work was funded partly by the British Council/DAAD and by the Royal Society (London). Jon Evans (BP-Amoco) is graciously thanked for allowing us to use the spectral gamma-ray data and sedimentological logs from the Wytch Farm example.

REFERENCES

Barclay, S.A. & Worden, R.H. (1998) Quartz cement volumes across oil–water contacts using core petrography and wireline logs: preliminary results from the Magnus Field in the northern North Sea. In: *Core-log Integration* (Ed. by P.K. Harvey and M.A. Lovell). *Geol. Soc. London Spec. Publ.*, **136**, 327–339.

Chamberlain, A.K. (1984) Surface gamma-ray logs: a correlation tool for frontier areas. *AAPG Bull.*, **68**, 1040–1043.

Chamley, H. (1989) *Clay Sedimentology.* Springer-Verlag, Berlin, 354 pp.

Davies, S.J. & Elliot, T. (1996) Spectral gamma-ray characterisation of high-resolution sequence stratigraphy: examples from Upper Carboniferous fluvio-deltaic systems, Co. Clare, Ireland. In: *High Resolution Sequence Stratigraphy: Innovations and Applications* (Ed. by J.A. Howell and J.F. Aitken). *Geol. Soc. London Spec. Publ.*, **104**, 25–35.

Davies, S.J. & McLean, D. (1996) Spectral gamma-ray and palynological characterisation of Kinderscoutian marine bands in the Namurian of the Pennine Basin. *Proc. Yorks. Geol. Soc.*, **51**, 103–114.

Deer, W.A., Howie, R.A. & Zussman, J. (1992) *An Introduction to the Rock Forming Minerals,* 2nd edn. Longman, London, 720 pp.

Dresser-Atlas (1983) *Log Interpretation Charts.* Dresser-Atlas Publication, Houston, TX.

Emery, D., Myers, K.J. & Young, R. (1990) Ancient subaerial exposure and freshwater leaching in sandstones. *Geology*, **18**, 1178–1181.

Evans, R., Hendry, J.P., Parnell, J. & Kalin, R.M. (1998) Origin and significance of fracture-related dolomite in porous sandstones: an example from the Carboniferous of County Antrim, Northern Ireland. In: *Carbonate Cementation in Sandstones*, International Association of Sedimentologists Special Publication 26 (Ed. by S. Morad), Blackwell Science, Oxford, pp. 409–435.

Faerseth, R.B. (1996) Interaction of Permo-Triassic and Jurassic extensional fault-blocks during the development of the northern North Sea. *J. Geol. Soc. London*, **153**, 931–944.

Fitton, W.H. (1845) Comparative remarks on the sections below the Chalk on the coast near Hythe and Atherfield, in the Isle of Wight. *J. Geol. Soc. London*, **1**, 179–189.

Frondel, C. (1958) Systematic mineralogy of uranium and thorium. *Bull. US Geol. Surv.*, **1964**, 162 pp.

Garden, I.R. (1991) Changes in the provenance of pebbly detritus in southern Britain and northern France associated with basin rifting. In: *Developments in Sedimentary Provenance Studies* (Ed. by A.C. Morton, S.P. Todd and P.D. Haughton). *Geol. Soc. London Spec. Publ.*, **57**, 273–289.

Gibbs, R.J. (1977) Clay mineral segregation in the marine environment. *J. Sed. Petrol.*, **47**, 237–243.

Grocke, D., Hesselbo, S.P. & Jenkyns, H.C. (1999) Carbon-isotope stratigraphy of Lower Cretaceous fossil wood: ocean–atmosphere chemistry and relation to sea-level change. *Geology*, **91**, 1108–1112.

Hallam, A., Grose, J.A. & Ruffell, A.H. (1991) Palaeoclimatic significance of changes in clay mineralogy across the Jurassic–Cretaceous boundary in England and France. *Palaeogeogr. Palaeoclimatol. Palaeoecol.*, **81**, 173–187.

Hogg, A.J.C., Mitchell, A.W. & Young, S. (1996) Predicting well productivity from grain size analysis and logging while drilling. *Petrol. Geosci.*, **2**, 1–15.

Holloway, S., Milodowski, A.E., Strong, G.E. & Warrington, G. (1989) The Sherwood Sandstone Group (Triassic) of the Wessex Basin, southern England. *Proc. Geol. Assoc.*, **100**, 383–394.

Hurst, A. (1990) Natural gamma-ray spectrometry in hydrocarbon-bearing sandstones from the Norwegian continental shelf. In: *Geological Applications of Wireline Logs* (Ed. by A. Hurst, M. Lovell and A. Morton). *Geol. Soc. London Spec. Publ.*, **48**, 211–222.

Hurst, A. (1999) Textural and geochemical micro-analysis in the interpretation of clay mineral characteristics: lessons from sandstone hydrocarbon reservoirs. *Clay Mineral.*, **34**, 137–150.

Hurst, A. & Milodowski, A. (1996) Thorium distribution in some North Sea sandstones: implications for petrophysical evaluation. *Petrol. Geosci.*, **2**, 59–68.

Kearey, P. & Brooks, M. 1991. *An Introduction to Geophysical Exploration.* Blackwell Scientific Publication, Oxford, 254 pp.

McRoberts, C.A., Furrer, H. & Jones, D.S. (1998) Palaeoenvironmental interpretation of a Triassic–Jurassic boundary section from Western Australia based on palaeoecological and geochemical data. *Palaeogeogr. Palaeoclimatol. Palaeoecol.*, **136**, 79–95.

Myers, K.J. (1987) *Onshore outcrop gamma-ray spectroscopy as a tool in sedimentological studies.* Unpublished PhD thesis, University of London, 100 pp.

Myers, K.J. & Bristow, C.S. (1989) Detailed sedimentology and gamma-ray log characteristics of a Namurian deltaic succession II: gamma-ray logging. In: *Deltas: Sites and Traps for Fossil Fuels* (Ed. by M.K.G. Whateley and K.T. Pickering). *Geol. Soc. London Spec. Publ.*, **41**, 81–88.

Myers, K.J. & Wignall, P.B. (1987) Understanding Jurassic organic-rich mudrocks—new concepts using gamma-ray spectrometry and palaeoecology: examples from the Kimmeridge Clay of Dorset and the Jet Rock of Yorkshire. In: *Marine Clastic Environments: Concepts and Case Studies* (Ed. by J.K. Legget, and G.G. Zuffa), Graham and Trotman, London, pp. 175–192.

Odin, G.S. & Matter, A. (1981) De Glauconarium Origine. *Sedimentology*, **28**, 611–641.

Osmond, J.K. & Ivanovich, M. (1992) Uranium-series mobilisation and surface hydrology. In: *Uranium-series Disequilibrium: Applications to Earth, Marine and Environmental Sciences* (Ed. by M. Ivanovich and R.S. Harmon), Clarendon Press, Oxford, pp. 259–289.

Parkinson, D.N. (1996) Gamma-ray spectrometry as a tool for stratigraphical interpretation: examples from the western European Lower Jurassic. In: *Sequence Stratigraphy in British Geology* (Ed. by S.P. Hesselbo and D.N. Parkinson). *Geol. Soc. London Spec. Publ.*, **103**, 231–255.

Parrish, J.T., Ziegler, A.M. & Scotese, C.R. (1982) Rainfall patterns and the distribution of coals and evaporites in the Mesozoic and Cenozoic. *Palaeogeogr. Palaeoclimatol. Palaeoecol.*, **40**, 67–102.

Pliler, R. & Adams, J.A.S. (1962) The distribution of thorium, uranium and potassium in the Mancos Shale. *Geochim. Cosmochim. Acta*, **26**, 1115–1135.

Pye, K. & Krinsley, D.H. (1986) Microfabric, mineralogy and early diagenetic history of the Whitby Mudstone Formation (Toarcian), Cleveland Basin, UK. *Geol. Mag.*, **123**, 191–203.

Reading, H. (ed.) (1996) *Sedimentary Environments: Processes, Facies and Stratigraphy*, 3rd edn. Blackwell Science, Oxford, 688 pp.

Roberts, H.M. & Plater, A.J. (1999) U- and Th-series disequilibria in coastal infill sediments from Praia da Rocha (Algarve Region, Portugal): a contribution to the study of late Quaternary weathering and erosion. *Geomorphology*, **26**, 223–238.

Ruffell, A. & Shelton, R.H. (1999) The control of sedimentary facies by climate during phases of crustal extension: examples from the Triassic of onshore and offshore England and Northern ireland. *J. Geol. Soc. London*, **156**, 779–789.

Ruffell, A.H. & Wach, G.D. (1998) Firmgrounds—key surfaces in the recognition of parasequence sets in the Lower Greensand Group (Aptian-Albian) of the Isle of Wight, southern England. *Sedimentology*, **45**, 91–107.

Ruffell, A. & Worden, R.H. (1999) Palaeoclimate analysis using spectral gamma-ray data from the Aptian (Cretaceous) of southern England and southern France. *Palaeoecol., Palaeogeogr., Palaeoclimatol.*, **154**, 313–325.

Sladen, C.P. & Batten, D.J. (1984) Source-area environments of Late Jurassic and Early Cretaceous sediments in Southeast England. *Proc. Geol. Assoc.*, **95**, 149–163.

Slatt, R.M., Jordan, D.W., D'Agostino, A. & Gillespie, R.H. (1992) Outcrop gamma-ray logging to improve understanding of subsurface well log correlations. In: *Geological Applications of Wireline Logs II* (Ed. by A. Hurst, C.M. Griffiths and P.F. Worthington). *Geol. Soc. London Spec. Publ.*, **65**, 3–19.

Tang, Z., Parnell, J. & Ruffell, A. (1994) Deposition and diagenesis of the lacustrine–fluvial Canfanggou Group (uppermost Permian to Lower Triassic) southern Junggar Basin, NW China: a contribution from sequence stratigraphy. *J. Palaeolimnol.*, **11**, 67–90.

Van Buchem, F.S.P., Melnyk, D.H. & McCave, I.N. (1992) Chemical cyclicity and correlation of Lower Lias mudstones using gamma-ray logs, Yorkshire, UK. *J. Geol. Soc.*, **149**, 991–1002.

Wach, G.D. (1991) *Sedimentology and stratigraphy of the Lower Cretaceous of the Channel Basin*. PhD thesis, University of Oxford, 350 pp.

Wilson, H.E. & Robbie, J.A. (1966) *Geology of the Country Around Ballycastle*. Memoirs of the Geological Survey of Northern Ireland, Sheet 8, Belfast, 125 pp.

Worden, R.H., Coleman, M.L. & Matray, J.M. (1999) Basin scale evolution of formation waters: a diagenetic and formation water study of the Triassic Chaunoy Formation, Paris Basin. *Geochim. Cosmochim. Acta*, **63**, 2513–2528.

Int. Assoc. Sedimentol. Spec. Publ. (2003) **34**, 109–128

Smectite in sandstones: a review of the controls on occurrence and behaviour during diagenesis

J.M. McKINLEY[1], R.H. WORDEN[2] *and* A.H. RUFFELL[3]

[1] *Department of Geology, Queen's University, Belfast BT7 1NN, UK, e-mail: j.mckinley@qub.ac.uk*

[2] *Department of Earth Sciences, University of Liverpool, Brownlow Street, Liverpool L69 3GP, UK, e-mail: r.worden@liverpool.ac.uk*

[3] *School of Geography, Queen's University, Belfast BT7 1NN, UK, e-mail: a.ruffell@qub.ac.uk*

ABSTRACT

Smectite comprises a chemically complex group of clay minerals and is common in sandstones, although there have been relatively few studies of their origin and distribution in comparison to both other clays in sandstones and smectites in mudstones. The occurrence of smectite in sandstones is a function of both depositional and diagenetic factors. It can occur as a component of primary sediment when the source terrain lithology, climate, duration of weathering and groundwater conditions are favourable. Smectite is characteristic of inhibited chemical weathering, for example, in arid, high-relief terrain. It thus is most common in aeolian, fluvial, lacustrine and turbidite sandstones, where the opportunity for advanced degrees of weathering and smectite breakdown (to other clays) is minimized. Smectite may be more abundant in deep-water settings because it tends to be finer grained than other clays and thus can be carried further into deep waters before settling. Smectites in sandstones are not always co-deposited with sand grade sediment but can be incorporated into the sand by bioturbation, soft-sediment deformation and mechanical infiltration. Smectite cements also can develop as authigenic minerals in sandstones owing to the presence and breakdown of unstable lithic grains and the formation of pseudomatrix resulting from compacted clay-rich clasts.

Dioctahedral smectites in sandstones transform to illite and trioctahedral smectites transform to chlorite via mixed-clay intermediates during progressive diagenesis because they become unstable as the temperature and compositional environment changes during burial. The processes of smectite illitization and chloritization are not isochemical because typically they require movement of a wide range of elements (K, Fe, Mg, and also Al, and Si, etc.). Typically, the processes involve a net relative increase of the Al/Si ratio in the clay, frequently resulting in concomitant quartz cementation. The occurrence and rates of smectite illitization and chloritization in sandstones can be both similar and dissimilar to equivalent processes in neighbouring mudstones. If smectite in a sandstone is authigenic it may be relatively more stable during burial than detrital smectite in mudstones. Conversely, if smectite breakdown requires exotic components, the typically higher permeability of sandstones may facilitate faster transformation of the sandstone smectite than the mudstone smectite. The key controls on smectite transformation are the rock mineralogy, the pore-water geochemistry and the higher permeability of sandstones than mudstones.

INTRODUCTION

Smectites are a chemically complex group of clay minerals that, in comparison to other clays, have been relatively poorly documented in sandstones. Smectites are important in sandstones for a number of reasons. They represent about 25% of clay-cemented sandstones and thus are a reservoir quality issue (Primmer *et al.*, 1997). They potentially supply material for other diagenetic processes such as illitization, quartz cementation and zeolite growth (Boles & Franks, 1979). They more readily adsorb and incorporate organic material than many other rock-forming minerals (Chamley, 1994), and thus can alter the wetting state of a sandstone, potentially rendering it oil-wet. Smectites contain a relatively large quantity of structural water so that breakdown and growth of new clays can also lead to *in situ* generation of water, potentially leading to overpressuring (Burst, 1969). Smectite clays in the pore network can have a highly detrimental effect on reservoir quality because they typically contain much ineffective microporosity.

The importance of smectite minerals in mudstone diagenesis has been recognized for a long time (e.g. see review by Dunoyer de Segonzac, 1970). Progressive depth- and temperature-related changes in smectite mineralogy and crystal structure in mudstones have been studied widely but smectite diagenesis in sandstones has been relatively poorly addressed. It is wrong to assume simply that smectite in the two different rock types will have identical patterns of behaviour because sandstones and mudstones have different mineralogy, water–rock ratios, permeability, rates of diffusive flux and oxidation–reduction buffers. The origin and significance of smectites in sandstones and the part they play in sandstone diagenesis is the subject of this review.

Clay minerals in sandstones may be either detrital or authigenic in origin. Detrital (allogenic) smectite refers to clay minerals that originate from a source external to the host rock. In contrast authigenic clay minerals are formed or regenerated *in situ* (Wilson & Pittman, 1977). Both detrital and authigenic smectites occur in sandstones (Whitney & Northrop, 1987; Humphreys *et al.*, 1994; Huggett, 1996; Weibel, 1999). Three main origins have been proposed for smectites (Chamley, 1994; Deconinck & Chamley, 1995). The first is detrital smectite from the reworking of weathering products in surficial soils and transient sediment deposits. These are also known as terrigenous smectites. The second is volcanogenic smectite, formed either from the marine alteration of volcanic material or from only slightly weathered volcaniclastic sedimentary material. The third is early diagenetic smectite resulting from a variety of authigenic processes during eodiagenesis.

There are studies that focus on smectite in mudstone successions. The primary (sedimentary) origin of smectite in mudstones is likely to be similar to the origin of primary smectite in sandstones (e.g. hinterland lithology, relief, climate and possibly depositonal environment). For example, a source terrain that ultimately produced smectite-rich mudstones probably would lead to the primary clay fraction of related sandstones also being smectite-rich. For this reason a review of the wider literature (including mudstone-focused papers, because the genesis of detrital clay in mudstone is likely to be similar to the genesis of detrital clay in sandstones) on the subject of smectite was undertaken. One of the main problems concerning smectite literature is that the type of smectite and the mineral chemistry are seldom described. It will be shown sub-sequently that the two main types of smectite (dioctohedral and trioctohedral) have different primary sources and behave differently during burial, heating and diagenesis. Thus, we will show that it is inadequate to blandly describe a mineral as 'smectite' in studies of sandstone sediment provenance, properties or reservoir quality.

The paper will be divided into a brief review of the structure and mineral chemistry of common smectites in sandstones followed by an assessment of the sedimentary (depositional) controls on the occurrence and types of smectite in sands and then an assessment of the burial diagenetic controls on the occurrence and types of smectite in sandstones.

SMECTITE CHEMISTRY AND STRUCTURE

Smectites, like many clay minerals, are 2 : 1 layered silicates in which two tetrahedral sheets (dominated by Si and Al) are sandwiched either side of one octahedral sheet of metal hydroxides (e.g. $Al_2(OH)_6$, $Mg_3(OH)_6$ and $Fe_3(OH)_6$; Brindley & Brown, 1980).

Smectites are classified as dioctahedral when only two out of three sites in the three fundamental octahedral units are occupied. This occurs when the octahedral site is occupied mainly by trivalent cations such as Al^{3+} or Fe^{3+}. They are classified as trioctahedral when all or most octahedral units are occupied by divalent cations such as Fe^{2+} and Mg^{2+} (Güven, 1988). Thus dioctahedral smectites tend to be Al-rich and trioctrahedral smectites tend to be Fe–Mg rich. Both groups of smectite are important in sandstones. Smectites can have various cation substitution sites on both the tetrahedral and octahedral positions. Substitutions of ions of the same charge, e.g. Fe^{2+} for Mg^{2+} and Fe^{3+} for Al^{3+}, is common in octahedral positions. Substitutions by ions of lower charge (e.g. Al^{3+} for Si^{4+} in tetrahedral positions and Mg^{2+} for Al^{3+} or Fe^{2+} for Fe^{3+} in octahedral positions) produces a net negative charge for the tetra–octa–tetrahedral sandwich. This is redressed by the presence of a limited quantity of interlayer cations, such as Na^{+} or Ca^{2+} or Mg^{2+} ions, although other cations, including organic ions, also can be introduced by exchange reactions (Brindley & Brown, 1980).

The smectite group of clay minerals is characterized by properties of cation exchange and expandability. They have the distinctive ability to transiently take up and release water, or organic molecules, between their layers and thus expand or contract parallel to the layers (Deer *et al.*, 1998). Smectites include any clay with a basal spacing (the thickness of the 2 : 1 sheet plus interlayer space) that expands to 1.7 nm on treatment with ethylene glycol (Brindley & Brown, 1980). Expansion by ethylene glycol is indicative of a structure in which the number of positive charges (i.e. positively charged metal ions) needed in the interlayer positions corresponds to < 0.2 to 0.5 unit per formula unit $[O_{10}(OH)_2]$. In the Al-rich smectite, montmorillonite, the need for interlayer cations is the result of substitution of Mg^{2+} for Al^{3+} and the ensuing charge imbalance in the octahedral layers. In contrast, in beidellite (also Al-rich), the need for interlayer cations is the result of substitution of Al^{3+} for Si^{4+} in the tetrahedral layers. Smectites usually are very fine grained even by clay mineral standards (i.e. usually less than 1 μm; Drever, 1982). The smectite group (Table 1) includes the dioctohedral smectites (e.g. montmorillonite, beidellite, nontronite) and the trioctahedral smectites (e.g. saponite, hectorite and sauconite).

Smectite minerals can form an enormous range of solid solutions. It is often thought that smectites, of ill-defined mineral chemistry at the microprobe scale (tens of micrometres), can incorporate intergrowths of different types/compositions of discrete smectite and other clay minerals. Smectite is often interlayered with other clay minerals, mainly illite and chlorite (Deer *et al.*, 1998). The stacking of layers in mixed-layer smectite usually is disordered at the time of deposition. Randomly interstratified mixed-layer clays are labelled according to the types of layers involved, with the most abundant layer type listed first (Reynolds, 1980). The degree of disorder in mixed-layer smectite evolves in a semi-predictable manner during burial and heating. The most common variety of ordering involves a regular alternation of two-layer types (Reynolds, 1980). The term 'Reichweite', denoted by S or R, is used to describe ordering types: R = 0 describes totally random interstratification of smectite and other clay minerals; R = 1, R = 2 and R = 3 describe progressively more ordered intercalations up to four-layer structures (Reynolds, 1980; Wilson, 1999). Illite–smectite and chlorite–smectite are generic names for interlayered smectite (abbreviated to I/S and C/S respectively). The most

Table 1 Chemical composition and occupancy of the different sites in typical sedimentary smectites (after Deer *et al.*, 1998). Formulae calculated on the basis of 20 oxygen atoms per unit cell (ignoring hydroxyls).

Smectites are subdivided according to substitutions involved	Z tetrahedral sites	Y octahedral sites	X interlayer exchange cations	Burial diagenesis—increasing temperature and depth
Substitutions in dioctahedral sites:		Al-rich (lesser amounts of Mg, Fe)		Dioctahedral smectite → illite/smectite → illite
montmorillonite	Si_8	$Al_{3.3}Mg_{0.7}$	$(^1/_2Ca,Na)_{0.7}$	
beidellite	$Si_{7.3}Al_{0.7}$	Al_4	$(^1/_2Ca,Na)_{0.7}$	
nontronite	$Si_{7.3}Al_{0.7}$	Fe^{3+}_4	$(^1/_2Ca,Na)_{0.7}$	
Substitutions in trioctrahedral sites:		Mg-rich (lesser amounts of Al, Fe)		Trioctrahedral smectite → chlorite/smectite → chlorite
saponite	$Si_{7.4}Al_{0.8}$	Mg_6	$(^1/_2Ca,Na)_{0.8}$	
hectorite				
sauconite				

common dioctohedral and trioctohedral smectites in clastic sediments are montmorillonite and saponite respectively (Table 1). The chemistry of the smectite group is also illustrated in Table 1, with approximate formulae taken from Deer *et al.* (1998).

SEDIMENTARY (DEPOSITIONAL) CONTROLS ON THE OCCURRENCE AND TYPES OF SMECTITE IN SANDS

The role of sediment provenance in smectite occurrence in sands

Smectite derived during continental weathering

Smectite can form as the weathering product of most igneous rocks (Table 2). The weathering of acid (granite) and intermediate (diorite) igneous rocks typically leads to the growth of smectite when the local weathering microenvironments provide the necessary conditions (Christidis & Dunham, 1993, 1997). Weathering and replacement of different minerals in an igneous rock can lead to subtle differences in the resulting smectite. The weathering of plagioclase and alkali feldspars commonly leads to the formation of pseudomorphic dioctahedral smectite (e.g. montmorillonite; Aoudjit *et al.*, 1995). Conversely the weathering of Fe–Mg-rich minerals (i.e. biotite or pyroxene) results in them being pseudomorphed by trioctahedral smectite such as saponite (Hill *et al.*, 2000). The weathering of acid and intermediate igneous rocks and more silicic metamorphic rocks produces a predominance of dioctahedral smectites, whereas more mafic-rich basaltic volcanics and volcaniclastics and metabasites tend to produce mainly trioctahedral smectites (Chang *et al.*, 1986). Weathered sandstone and psammitic sediment sources do not typically lead to smectite-bearing sedimentary rocks. The weathering of limestones and their metamorphosed equivalent (marble) also may produce trioctahedral smectite in the form of saponite, but only if the original rock was dolomitic (Post, 1984).

Weathering occurs by a series of steps. The source materials usually are high temperature, anhydrous minerals (e.g. feldspars) that might evolve (i) initially through zeolite minerals, (ii) then through various types of smectite clay minerals, (iii) to kaolinite and (iv) then to oxides and hydroxide minerals such as gibbsite during extreme weathering (Fig. 1). Smectite is

Table 2 Effect of provenance (parent rock) on the type and likelihood of smectite in sandstone. Sources of information: (1) Aoudjit *et al.* (1995), (2) Chang *et al.* (1986), (3) Christidis & Dunham (1997), (4) Christidis & Dunham (1993), (5) Robinson & Santana De Zamora (1999), (6) Ryu & Niem (1999), (7) Moberly & Jenkyns (1981), (8) Jin Hwan Noh & Boles (1993), (9) Stern *et al.* (1997), (10) Deconinck & Chamley (1995), (11) Post (1984).

Common sandstone parent rocks	Probability of smectite in sandstone (references)	Smectite type
Granite	**High** (if plagioclase/biotite provide basic microsystems) (1) **High** (if drainage is poor, pH is high) (1)	Plagioclase → montmorillonite Biotite →biotite/smectite (beidellite) Biotite → montmorillonite Muscovite → muscovite/smectite
Weathering of granitic and gneissic terrains	**Low** (if good drainage) (1) **High** (2)	No smectite Dioctahedral smectite
Rhyolite/volcanic glass	**High** (smectite forms only when system is well flushed) (3)	Al-rich beidellite and Tatatilla-type montmorillonite
Andesite	**High** (4) (5)	Mg-rich smectite (4) Smectite with minor chlorite (composition between a di- and trismectite) (5)
Basalt (mafic volcanic, pyroclastic, hyaloclastic)	**High** (6)(7)	
Volcanic-glass zeolites	**High** (8)	
Metamorphosed volcano-sedimentary	**High** (2)	Trioctahedral smectite (saponite)
Interbedded sandstone and mudstone	**High** (9)	Ca-saturated smectite
Chalk/limestone	**Medium** (10)	
Meta-dolomitic limestones	**Medium** (11)	Saponite

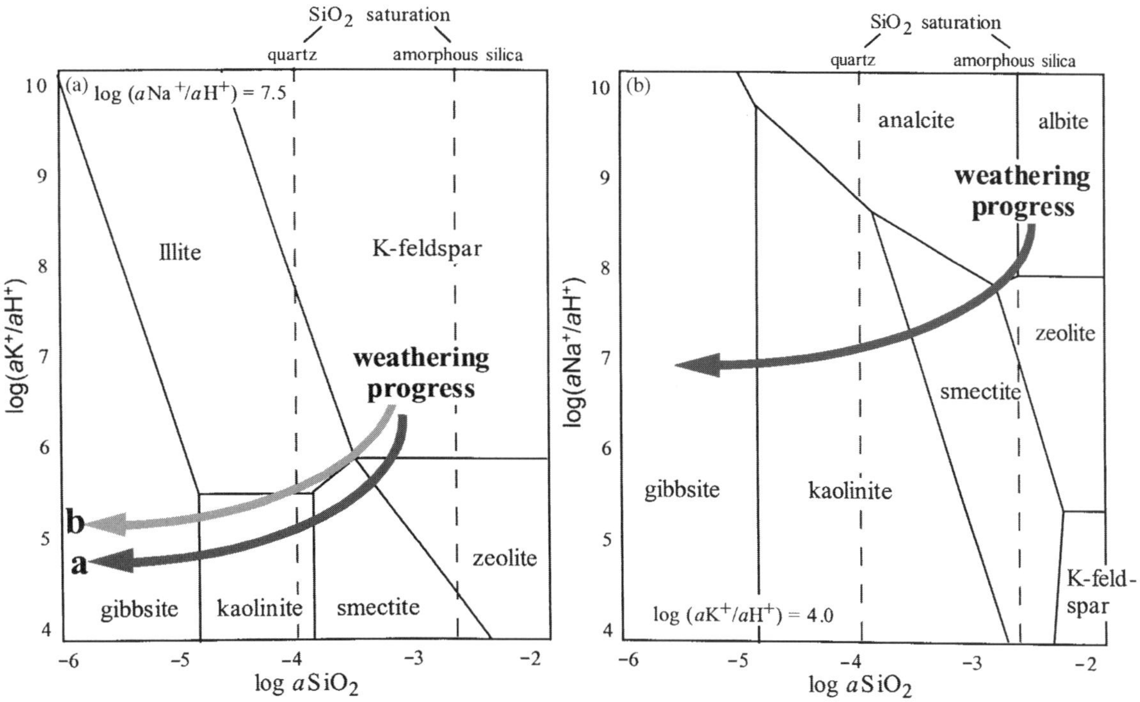

Fig. 1 Activity diagrams for the K_2O–Na_2O–Al_2O_3–SiO_2–H_2O system (after Drever, 1982). (a) Diagram of (aK^+/aH^+) and $aSiO_2$ at a fixed value of (aNa^+/aH^+) with stability fields of different minerals indicated. (b) Diagram of (aNa^+/aH^+) and $aSiO_2$ at a fixed value of (aK^+/aH^+). Water geochemical evolution paths during weathering are shown on both diagrams. In (a), the path labelled 'a' implies high pH (i.e. not acidic) or low potassium concentration conditions leads to smectite growth. The path labelled 'b', occurs at either higher pH or higher aqueous metal concentrations. Smectite will thus grow (or survive) if either weathering is inhibited (limited movement down the trend), pH and aqueous metal concentration are low or if the SiO_2 solubility is controlled by amorphous silica.

dominant during the early and intermediate stages of rock weathering (Chesworth, 1977). However, the growth of smectite is only a transient step in the weathering cycle. Whether the middle or extreme stages of weathering are achieved and smectite preservation minimized depends on climate (especially humidity) and the duration of weathering prior to sediment being deposited in its final resting place. Warm and humid climates will promote more rapid chemical weathering and may reduce the opportunity for the preservation of smectite clays. The duration of the total weathering process itself depends on the terrain (steep terrain results in rapid transport and sedimentation) and position in the depositional systems tract (with fluvial sediments typically having had less time to weather than, for example, shallow-marine sediments). The fixation of potassium at surface temperatures decreases both the ability of smectite to swell, and its cation exchange capacity. Wetting and drying experiments performed with several different smectites and four cations (K, Na, Mg, Ca) indicated that only potassium is fixed and the original cation exchange capacity of the smectite is decreased by as much as 52% by potassium fixation (Eberl *et al.*, 1986). Evidence that wetting and drying is important in the smectite to illite transition in natural settings is provided in the Swiss and French Jura Mountains. Al–Fe-rich smectites deposited on the shallow-water Purbeckian platform were probably subjected to repetitive wetting by K-rich marine waters and drying under a hot climate leading to the development of Fe-illite (Deconinck & Strasser, 1987; Deconinck *et al.*, 1988).

Smectites in the Atlantic sediments of late Mesozoic to early Cenozoic age are mostly the Al-rich dioctahedral smectite, beidellite. Aluminium is thus present in both the tetrahedral and octahedral sites but there is also partial substitution of ferric iron and magnesium for octahedral aluminium (Debrabant *et al.*, 1985). This contrasts with the smectites derived from the alteration of volcanic glass or basalt in hydrothermal environments, which are enriched in magnesium or iron. The strong comparison to the Al–Fe dioctohedral smectite of terrigeneous origin is evidence, according to Chamley (1994), for the continental parent-rock origin of smectite in Atlantic sediments.

Volcanogenic smectite

Volcanogenic smectite forms either from the marine alteration of volcanic material or from only slightly weathered volcaniclastic sedimentary material (Chamley, 1994; Deconinck & Chamley, 1995). The chemical composition of smectites derived from submarine volcanic rocks is dependent upon the initial composition of the original volcanic materials, although there are broad differences relative to typical terrigeneous smectites (Chamley, 1994). The presence of non-altered volcanic debris (volcanic glass, pumiceous ash, etc.) usually still exists in deposits that are unquestionably rich in volcanogenic smectite (Moberly & Jenkyns, 1981; Jeans *et al.*, 1982; Chamley, 1994). Sediments studied from the South and West Pacific Ocean and the eastern Mediterranean Sea display a clear correlation between the development of smectite and the occurrence of volcanic glass (Chamley, 1994). A study of Miocene sandstones from the North Sumatra back-arc basin, Indonesia shows the potential relationship between smectite diagenesis and an environment of simultaneous sedimentation and volcanism (Humphreys *et al.*, 1994). The development of grain-coating cements in these Miocene sandstones is probably related to abundant volcanic material in sediments (comprising porphyritic andesite to dacitic tuffs, calc-alkaline basalts, agglomerates and tuffs) at the time of deposition.

Smectite-rich mudstones derived from volcanic ash-falls are termed bentonites. Such sediments are commonly found adjacent to explosive volcanoes and may occur in any depositional environment. However, owing to preservation conditions, they are most common in shallow- and deep-water successions with moderate to high rates of sedimentation, where they may be admixed with sand-grade material through bioturbation or turbidity. Many bentonite-bearing sandstones are dominated by smectite-rich early cements (see references in Jeans *et al.*, 2000). Shallow-marine examples include the laterally persistent Cretaceous bentonites of the western interior of North America (e.g. the Black Hills of Montana, Wyoming) and the Jurassic (Na montmorillonite) and Cretaceous (Ca montmorillonite) fuller's earths of southern England (Jeans *et al.*, 2000).

Early diagenetic smectite

The authigenic clay-mineral assemblage produced by early diagenesis of framework grains, particularly in volcanic sediment provinces, often closely matches the clay assemblage produced by *in situ* weathering of these rocks in the source terrain (Wilson & Pittman, 1977). Grain-coating smectite, absent at grain contacts, is a common early diagenetic cement in volcaniclastic litharenites (Carrigy & Mellon, 1964; Galloway, 1974; Davies & Ethridge, 1975). Local dissolution of detrital framework grains can result in replacive smectite growth at the expense of plagioclase and glassy volcanic grains in plagioclase-rich arkoses and litharenites (Ryu & Niem, 1999). Thiry & Jacquin (1993) conclude *in situ* transformation and recrystallization of terrigenous clay components to be the process most likely to explain the systematic evolution of clay minerals resulting in smectite enrichment of ocean basin sediments.

The effect of hinterland terrain on weathering and smectite occurrence in sands

Weathering will result in smectite rather than kaolinite, illite, etc., only if the excess metal cations and silica cannot be lost from the aqueous geochemical system. In stagnant groundwater conditions, loss of the excess material is impeded, leading to minimal chemical weathering and the opportunity for smectite minerals to be preserved. In flowing and active ground waters, such as found in high-relief regions or where there are well-connected aquifers, loss of excess material is easily achieved, resulting

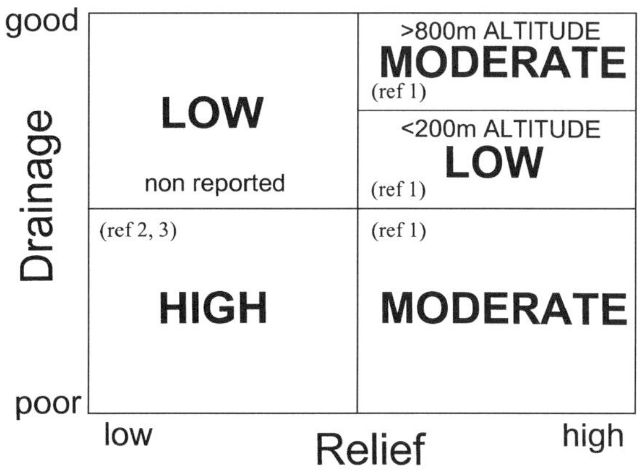

Fig. 2 Effect of hinterland relief on likelihood of smectite deposition in sandstone. Sources of information: (1) Aoudjit *et al.* (1995), (2) Hartmann *et al.* (1999), (3) Henderson (1998).

in the possibility of more advanced chemical weathering and reduced preservation potential of smectite minerals (Fig. 2). Low-lying topography, poor drainage and Ca–Mg-rich parent material lead to favourable chemical conditions for the formation of smectite minerals (high pH, high silica activity and an abundance of alkaline earth cations: Wilson, 1999). In weathered granitic rocks of the French American Massif, poorly drained and relatively stagnant groundwater conditions in a downslope position (with the water enriched in silica and basic cations), led to the preferential formation of dioctahedral smectite (montmorillonite; Aoudjit *et al.*, 1995). However, smectite was not formed during weathering of the same parent material in an upslope, well-drained position (Fig. 2).

The effect of climate on weathering and smectite occurrence in sands

Smectite minerals preferentially form during weathering under arid climatic conditions (Fig. 3). Note that smectite is reported to be the dominant clay-mineral weathering product in modern desert environments (Chamley, 1989). All geochemical processes involved in chemical weathering require water to (at least) catalyse the dissolution, transport and reprecipitation steps. In order to achieve advanced stages of

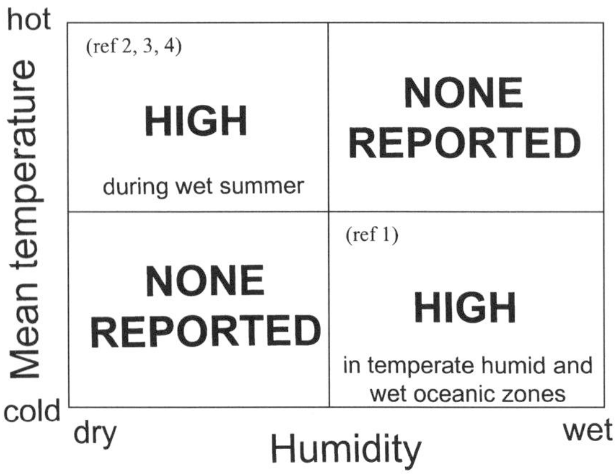

Fig. 3 Effect of climate on likelihood of smectite deposition in sandstone. Sources of information: (1) Aoudjit *et al.* (1995), (2) Stern *et al.* (1997), (3) Bjørlykke & Aagaard (1992), (4) Hartmann *et al.* (1999).

weathering (kaolinite, hydroxides), the regolith must lose a substantial proportion of cations by percolating or flowing fresh water. Under arid conditions, there is minimal throughput of water and thus limited ability to lose cations (Ca, Mg, Na, K, etc.). Arid climates lead to mineralogically immature sediment that increases the chance of retaining smectite minerals in the resulting rock.

The various reservoir sandstones of the North Sea Basin (Fig. 4) potentially highlight the importance of climate in controlling clay mineralogy because smectite is most important in rocks formed under the most arid conditions (Permo-Triassic aeolian and fluvial sandstones; Bjørlykke & Aagaard, 1992). Intraformational climate change, from a cold and temperate climate to a warmer semi-arid/arid climate, has been used to explain the relative increase in abundance of smectite layers in the initial composition of mixed-layer illite–smectites from base to top of the (Permo-Carboniferous) Haushi Group, in the Interior Oman Sedimentary Basin (Hartmann *et al.*, 1999; Table 3).

Table 3 Effect of sandstone depositional environment on likelihood of smectite deposition or authigenesis in sandstones. Sources of information: (1) Ramseyer & Boles (1986), (2) Hillier (1994), (3) Surdam & Boles (1979), (4) Bühmann (1992), (5) Bjørlykke & Aagaard (1992), (6) Zhao *et al.* (1999), (7) Burley (1984), (8) Boles & Franks (1979), (9) Ko & Hesse (1998), (10) Huggett (1996), (11) Dypvik (1983), (12) Ryu & Riem (1999), (13) Gibbs (1977). Please note the numbers of citations are skewed by the abundance of sandstone type and the likelihood of finding water, oil or gas resources in sandstones from different environments.

Sandstone depositional environment	Reported occurrence of smectite (references)
Continental:	
alluvial	(1)
lacustrine	(2)(3)
fluvial	(4)(5)(6)(7)
aeolian	(5)
Coastal/shelf:	
deltaic	(8)(9)
shallow marine	(4)(5)(8)(10)
beach	None reported
Deep marine	(5)(11)(12)(13)

Age	Climate	Depositional Environment	Interpreted relative initial smectite content and type
Lower Tertiary	Humid	Deep marine	I/S
Upper Jurassic	Humid	Deep marine	I/S
Mid Jurassic	Humid	Shallow marine (±deltaic, fluvial)	
Lower Jurassic	Humid	Fluvial (±shallow marine)	I/S
Permo-Triassic	Arid-semi arid	Aeolian and fluvial	S, I/S and C/S

Fig. 4 Relative distribution of smectite clays in sandstones from the North Sea (adapted from Bjørlykke & Aagaard, 1992). I/S = illite–smectite; S = smectite; C/S = chlorite–smectite.

The effect of sand depositional environment on smectite occurrence

The abundance of primary smectite (and other clays) in sands is affected by the depositional environment for two main reasons. First, environments closer to the initial source of the sediment (the weathering zone) are more likely to be mineralogically immature because the sediment supply will have had less time to allow rock fragments to weather fully. The lack of mineralogical maturity will lead to the increased potential for smectite growth during diagenesis (see below). Thus, fluvial and aeolian sand may be more prone to smectite growth than time-equivalent sands deposited in marine environments. Second, environments with wildly fluctuating energy conditions (turbulent systems, especially) lead to less well-sorted sediments at the point of deposition, potentially allowing intimate mixing of both clays and sand-grade material.

Smectite, like other syndepositional clay matrix, is typically absent or minor in continually high-energy environments, such as beaches, where the action of waves and ebb and flow tidal currents winnow out the fine-grained clays from the sand- and silt-grade material (Davies & Ethridge, 1975; Lynch, 1996). Smectite, and other clays, tend to be deposited with sand in environments that experience large fluctuations in energy in which sand is transported and deposited under transient high-energy conditions, with the clay minerals settling out during quiescent periods (e.g. in fluvial, aeolian and turbidite sands). In consequence, marine sands contain little detrital clay in their matrix, although various types of bioturbation and soft-sediment deformation can cause mixing of the discretely deposited sand and clay layers (Wilson & Pittman, 1977; Boles & Franks, 1979).

Fluvial and lacustrine sands tend to contain more smectite than those formed in environments further basinward (e.g. shallow-marine environments) because these sands and the correlative clay minerals in the mud layers tend to be more mineralogically immature, reflecting the lesser degree of reworking and chemical weathering (Primmer *et al.*, 1997). Many aeolian, fluvial and lacustrine sediments have not had the opportunity to achieve the advanced stages of weathering that lead to kaolinite growth or the total breakdown of reactive lithic grains and clay minerals (Fig. 1; and see Surdam & Boles, 1979). Lacustrine sandstones seem to be especially prone to (trioctahedral) smectite, where periodic evaporation leads to concentration of alkaline earth elements and Fe in ground and lake waters (e.g. Hillier, 1994).

Detrital smectite can occur in almost any depositional environment, although it is rarely found in beach sands (Table 3). However, it is most abundant in sands deposited in fluvial environments, especially those with an aeolian influence, and also in deep marine environments (Fig. 5; Bjørlykke & Aagaard, 1992).

Smectite tends to be finer grained (smaller particle size) than the other clay types, ranging from 0.9 μm to below 0.1 μm, with a mean about 0.4 μm (Gibbs, 1977). Smectite thus can be carried further than other clays down the depositional slope (into deeper water) before settling out of suspension. From the mouths of major river systems, illite, chlorite and kaolinite preferentially tend to be deposited close to the shoreline, whereas smectite is transported further into the ocean. The dominant mechanism responsible for the lateral change in clay type along a depositional slope is physical sorting of the sediments by size. The mixed-layer I/S clays tend to become more smectite-rich along the flow path; pure smectite tends to occur instead of mixed-layer clays only when fully open-marine conditions are well established (Gutiérrez-Mas *et al.*, 1997). Flocculation is thought to be important in fractionating clays during transport from rivers into oceans. Flocculation depends on the surface charge of particles and the chemistry of the water in which they are suspended. Water of low cation concentration (fresh water) allows fine-grained material to remain dispersed (i.e. particles repel each other). With increasing cation concentration (freshwater to marine transition), the forces of mass attraction come into play and multi-particle associations are formed

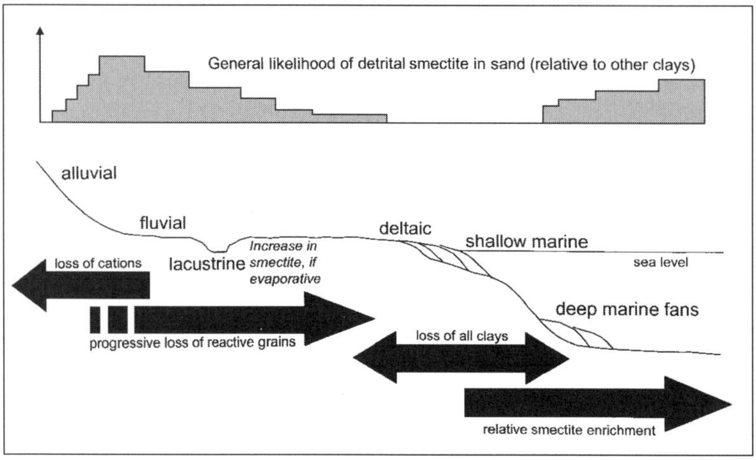

Fig. 5 Relative abundance of smectite in sandstones from different depositional environments (all other factors, e.g. supply, being equal). Smectite supply may be controlled by hinterland provenance, the palaeoclimate of the hinterland and site of deposition and depositional environment. This diagram highlights depositional environment and presumes that no significant changes in sediment sourcelands or palaeoclimate regime have occurred during deposition.

(floccules). Smectites form loose floccules of low bulk specific gravity and are easily torn apart in turbulent water, whereas illite, chlorite and kaolinite form dense, tight floccules of relatively high specific density and thus have greater effective settling diameters (Jeans, 1989). However, differential flocculation does not always appear to be of significance. For example, in the Amazon River and Altantic Ocean area, where turbulence or organic and metallic particle coatings interfere with the process (Gibbs, 1977), and in the Upper Jurassic sediments of northern France, where weathering conditions and relief in the source terrain seem to have been much more important controls (Chamley *et al.*, 1990; El Albani *et al.*, 1993).

DIAGENETIC CONTROLS ON THE OCCURRENCE AND TYPES OF SMECTITE IN SANDSTONES

Smectite evolution during burial and heating in sandstones

Thermal effects

During simple heating experiments, the primary effect on smectites is the loss of interlayer water, mostly between 100 and 250°C, although some remains to 300°C. Although smectite commonly is found in low-temperature geochemical environments, it is not just a very low-temperature mineral. Dioctahedral smectite may react to form mixed-layer clays at relatively low temperatures, whereas trioctahedral smectites may persist to higher temperatures (Eberl *et al.*, 1978). The stability of smectite depends on the geochemical environment (mineralogy and water chemistry) as much as the ambient temperature.

The smectite to illite transformation

Smectite does not convert to illite by simple cation exchange processes such as the substitution of interlayer cations by K^+ (Weaver, 1958). The conversion also involves an increase in the net negative layer charge of the tetrahedral and/or octahedral layers (Hower *et al.*, 1976). The charge increase must result either from substitution of Al for Si in the tetrahedral layer or the substitution of divalent cations (Fe^{2+}, Mg^{2+}) for Al^{3+} in the octahedral layer (Dunoyer de Segonzac, 1970):

$$\text{smectite} + Al^{3+} + K^+ \Rightarrow \text{illite} + Si^{4+}$$
(with loss of Fe and Mg; Hower *et al.*, 1976) (R1)

It is theoretically possible that Al^{3+} may behave either as a mobile component (reaction 2; also adapted from Boles & Franks, 1979) or as a conservative component (reaction 3; adapted from Boles & Franks, 1979):

$$0.45H^+ + 0.45K^+ + 0.4Al_2O_3 + K_{0.1}Na_{0.1}Ca_{0.2}Mg_{0.4}Fe_{0.4}Al_{1.4}Si_{3.8}O_{10}(OH)_2 \cdot H_2O$$

dioctahedral smectite

$$\Rightarrow K_{0.55}Mg_{0.2}Fe_{0.15}Al_{2.2}Si_{3.5}O_{10}(OH)_2 + 0.1Na^+$$

illite

$$+ 0.2Ca^{2+} + 0.2Mg^{2+} + 0.125Fe_2O_3 + 0.3SiO_2 + 1.225H_2O \quad (R2)$$

$$1.242H^+ + 0.393K^+ + 1.57K_{0.1}Na_{0.1}Ca_{0.2}Mg_{0.4}Fe_{0.4}Al_{1.4}Si_{3.8}O_{10}(OH)_2$$

dioctahedral smectite

$$\cdot H_2O \Rightarrow K_{0.55}Mg_{0.2}Fe_{0.15}Al_{2.2}Si_{3.5}O_{10}(OH)_2$$

illite

$$+ 0.157Na^+ + 0.314Ca^{2+} + 0.428Mg^{2+} + 0.239Fe_2O_3 + 2.466SiO_2 + 2.761H_2O \quad (R3)$$

Note that reactions 2 and 3 have been modified slightly from Boles & Franks (1979) to ensure that there is charge balance and material balance.

In reaction 2, the original 2 : 1 layers remain intact and the mineralogical change involves only ionic substitution within the structure (Chang *et al.*, 1986). The development of a sufficiently large negative charge deficiency and the subsequent fixation of potassium are two independent events, which may have their own rate-limiting steps (Howard, 1981). An association of highly charged expandable layers and a tetrahedral layer charge-deficiency, is found in the layers of mixed-layer I/S clays from the sand laminae of interlaminated sandstones and mudstones from the Gulf Coast, Texas. This suggests that the supply of K is not fixed immediately in interlayer spaces and stresses the importance of tetrahedral substitution in the illitization process (Howard, 1981). The I/S clays in sandstones show an increase in tetrahedral Al with progressive illitization, whereas octahedral Al remains essentially constant (Howard, 1981). The implication is that the available Al from the octahedral sheet does not simply diffuse to the tetrahedral sheet, where it would substitute for vacant Si. Instead an external source for Al in the sandstone seems to be required for illitization. Dissolution of feldspar (and mica), discrete illite or kaolinite, or the cannibalization of I/S may provide a source of Al and K (Perry & Hower, 1970; Hower *et al.*, 1976; Howard, 1981; Lynch, 1997). It is thus likely that a net reaction between K-feldspar and dioctahedral smectite could result in abundant illite and quartz cement (Robertson & Lahann, 1981).

The smectite to chlorite transformation

The smectite to chlorite transformation, as for smectite to illite, involves the change of an expandable mineral into a non-expandable one (Humphreys *et al.*, 1994). The differences between the smectite chloritization process and the smectite illitization process, described earlier, are (after Chang *et al.*, 1986):

1 chloritization requires addition of Fe, Mg and H_2O to preserve the original structure, whereas if Al is conserved and acts as an immobile element, only a supply of H_2O and H^+ is required;

2 the interlayer (Z) sites are progressively occupied by K^+ during illitization, but chlorite forms by fixation of (typically Fe or Mg) hydroxide in these sites.

In the smectite chloritization process, as for the smectite to illite transformation, Al may be either mobile (see reaction 4 below) or conservative (reaction 5). The following reaction stoichiometries are taken from Chang *et al.* (1986):

$$Ca_{0.1}Na_{0.2}Fe_{1.1}MgAlSi_{3.6}O_{10}(OH)_2 + 1.5Fe^{2+}$$

trioctahedral smectite (saponite)

$$+ 1.2Mg^{2+} + 1.4Al^{3+} + 7.6H_2O$$

$$\Rightarrow Fe_{2.6}Mg_{2.2}Al_{2.4}Si_{2.8}O_{10}(OH)_8 + 0.1Ca^{2+}$$

chlorite

$$+ 0.2Na^+ + 0.8SiO_2 + 9.2H^+ \quad (R4)$$

quartz

$$2.4Ca_{0.1}Na_{0.2}Fe_{1.1}MgAlSi_{3.6}O_{10}(OH)_2 + 0.88H_2O$$

trioctahedral smectite (saponite)

$$+ 1.44H^+ \Rightarrow Fe_{2.6}Mg_{2.2}Al_{2.4}Si_{2.8}O_{10}(OH)_8$$

chlorite

$$+ 0.24Ca^{2+} + 0.48Na^+ + 0.04Fe^{2+} + 0.20Mg^{2+} + 5.84SiO_2 \quad (R5)$$

quartz

The smectite to kaolinite transformation

In experimental work, Eberl & Hower (1976) produced both illite and kaolinite at the expense of smectite. Kaolinite appears to have formed as a reaction product in response to a low pH value. However, observations of natural mudstones and sandstones do not appear to show an increase of kaolinite with depth (Hower *et al.*, 1976; Boles & Franks, 1979; Lynch 1997). In experiments by Robertson & Lahann (1981), solutions were buffered by carbonate ions, which kept the pH at higher values. Thus the solubility of Al^{3+} was minimized and the production of kaolinite prevented. Although there is evidence to show that the transformation of smectite to kaolinite through interstratified kaolinite–smectite does occur (e.g. Thiry *et al.*, 1977; Amouric & Olives, 1998; Corti *et al.*, 1998), it does not appear to be of significance in sandstones.

General controls on smectite diagenesis in sandstones

The effects of burial

It has been long been noted that smectite minerals transform to illite or chlorite with increasing depth during progressive burial (e.g. Burst, 1969; Perry & Hower, 1970). However, it is unlikely that depth itself is important because the depth of transformation varies widely between different sedimentary basins. The depth at which transformation occurs is a function of many controls (see later), although a master variable seems to be temperature.

Temperature

Perry & Hower (1970) suggested that for mudstones, the occurrence of the smectite to illite reaction is directly related to temperature. However, the presence or absence of smectite in sandstone cannot simply be assumed to be an unequivocal indicator of the maximum temperature reached in diagenesis because, by analogy with mudstones, smectite potentially can be stable over the full range of temperatures encountered during diagenesis, through to low-grade metamorphism (Dunoyer de Segonzac, 1970). Despite this, the expandability and quantity of smectite in *mudstones* tends to be inversely correlated with maximum temperature (or at least the integrated time–temperature history; Środoń, 1999). However, clay-mineral diagenetic reactions in *sandstones* may be quite different to those documented in mudstones (Fig. 6). Note that during thermal history modelling and validation, only homogeneous mudstones are typically assumed to give consistent results for palaeotemperature estimation (Smart & Clayton, 1985). Geochemically exotic formation-water compositions may be involved

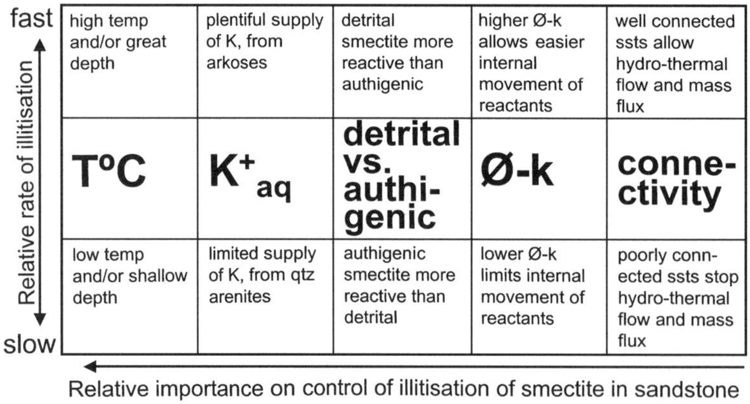

Fig. 6 A synthesis of the general controls on the transformation of smectite to illite in sandstones as described in this paper. ϕ–k = porosity and permeability; ssts = sandstones.

in sandstone diagenesis and the high permeability of sandstones may provide conduits for anomalously hot and geochemically active fluids (Pytte & Reynolds, 1989). For example, Whitney & Northrop (1987) concluded that the transformation of smectite might reflect the temperature of migrating fluids rather than the ambient burial temperatures.

It has been suggested that the temperature at which the illitization reaction occurs may influence the distance over which the products evolved will migrate (Lahann, 1980). Rapid sedimentation and the transiently reduced geothermal gradient (and the resulting lower temperatures) could cause clay units to function as isochemical systems incapable of adding material to sandstones (Louisana Gulf Coast; Hower *et al.*, 1976). In contrast, relatively slow sedimentation allows the maintenance of thermal equilibrium, where the higher temperatures (all other factors being equal) in sandstones could produce more extensive migration and cementation (southwest Texas; Boles & Franks, 1979).

Thus temperature is likely to be important in sandstones, although the effects of temperature and heating rate have not been quantified sufficiently to reveal the kinetics of smectite transformation to other clay minerals.

Pressure

The conversion of smectite to illite releases water and may lead to overpressure and variable pressure regimes (Burst, 1969; Bruce, 1984; Lynch, 1997). Although these studies have concentrated on mudstone intervals, the release of water in the smectite to illite reaction also occurs in sandstones, albeit transiently dependent upon gross permeability and connectivity of the sandstone. In principle, developing overpressure could inhibit reaction progress, although this is likely to be a minor or negligible control on the rate of smectite transformation. Ko & Hesse (1998) found that the smectite to illite reaction was advanced in geopressured zones and they related this to the higher water/rock ratios compared with the normally pressured zones. However, it is not clear whether smectite breakdown caused, or was caused by, the overpressure.

Availability of potassium

During burial diagenesis, the conversion of dioctahedral smectite to illite is thought generally to require the addition of aqueous potassium. An absence of available potassium in sandstones could increase the stability of smectite and inhibit the transformation process (Fig. 6). It is possible that smectite illitization is linked in time and space to K-feldspar dissolution during sandstone diagenesis, because this process typically may release potassium into the formation water. Similar links have been made for the equivalent processes in mudstones, whereby smectite and K-feldspar both decrease in abundance and illite simultaneously increases in abundance during progressively deeper burial (Hower *et al.*, 1976). The absolute concentration of aqueous potassium appears to be not as important as the amount of potassium that actually comes into contact with the highly charged interlayer (Ramseyer & Boles, 1986). The conversion of detrital smectite into random S/I at relatively shallow depths and low temperatures (47–68°C) in sandstones of the Triassic Skagerrak Formation, Denmark, emphasizes the importance of the availability of potassium for the smectite to illite process (Weibel, 1999). Associated salt structures and faults promoted fluid migration of the necessary potassium ions. It is possible that illitization of dioctahedral smectite in quartz arenites may be inhibited in comparison to arkoses. The absence of an internal source of potassium may be important, especially as mudstones tend to import rather than export potassium during burial diagenetic processes (e.g. Land & Milliken, 2000).

Porosity and permeability

The porosity and permeability of sandstones may play a significant role in the composition of an interstratified smectite clay mineral such that high permeability, especially, promotes the transformation from smectite to illite or chlorite (Fig. 6). For example, interlayered smectite–chlorite is more ordered (with less smectite) in sandstones at lower temperature and shallower depth in sandstones (*c.* 60°C at 1500 m) than in mudstones (*c.* 70°C at 2000 m) in the Cretaceous sediments of offshore Brazil (Chang *et al.*, 1986). This has been attributed to the higher permeability of the sandstones than the mudstones, which allowed more fluid–rock interaction and thus led to more chemical exchange between the solid and fluid phases. Early cementation and permeability degradation of a rock can retard the progress of the smectite to illite reaction, possibly by impeding the convective influx of aluminium or, more likely, potassium (Ramseyer & Boles, 1986). Furthermore, low permeability, mixed-layer smectite–illite cemented Tertiary sandstones from the San Joaquin Basin had a relatively greater proportion of smectite (greater expandability) than adjacent uncemented porous sandstones in the San Joaquin Valley (Ramseyer & Boles, 1986), presumably for the same reason.

Detrital smectite composition

The original detrital composition of a particular smectite may have a great impact on its behaviour during burial and heating (Fig. 6; Boles & Franks, 1979; Niu & Ishida, 2000). The slower rate of illitization in the sandstones compared with the associated mudstones from the Niigata Basin, Japan, was dependent upon the original detrital smectite composition and the pore-fluid composition (Niu & Ishida, 2000). The compositional differences of residual smectite layers at different depths of burial was used by Boles & Franks (1979) to show how Al-rich smectite underwent more rapid transformation to non-expandable clays than Fe- and Mg-rich smectites. Trioctahedral smectites react more slowly than dioctahedral smectites to develop increasingly illite-rich I/S, because trioctahedral smectites (with less intrinsic Al) do not have the correct Al/Si ratio necessary to fix potassium (Eberl *et al.*, 1978; Boles & Franks, 1979). The substitution of Mg^{2+} for Al^{3+} in the octahedral sheet is not possible in the trioctahedral smectites, as in this case the octahedral sheet already has its full complement of divalent or monovalent cations, enhancing its stability (Eberl *et al.*, 1978). The smectite to illite transformation is slowed by the presence of Ca^{2+} and Mg^{2+} by an order of magnitude relative to Na^{+}-rich smectite (Robertson & Lahann, 1981), as these probably increase the thermodynamic stability of smectite.

Comparative smectite diagenesis in sandstone and neighbouring mudstones

Sandstones often have relatively little detrital (i.e. primary depositional) clay owing to hydrodynamic sorting (see above) and yet can have high permeability, thus permitting import and export of material (rock-forming solutes). In contrast, mudstones commonly are dominated by aluminosilicate clay minerals and have poor permeability, rendering mass movement (especially influx) more difficult. Petrographic and sedimentological evidence has shown that in sandstones most smectite (and other clay) cements and pore linings are the result of: (i) bioturbation, (ii) infiltration, (iii) authigenesis in the ambient formation water (Jeans, 1989) or (iv) deformation of detrital mud clasts (e.g. intraclasts). In contrast, X-ray diffraction (XRD) and petrographic evidence from mudstones suggests that smectite (and other clay) minerals are inherited directly from the original sediment (Clauer *et al.*, 1999; Niu & Ishida, 2000). Two factors are important in determining the clay-mineral assemblages exhibited by alternating mudstone–sandstone facies if a single detrital source is assumed (after Jeans, 1989): (i) the action of differential hydrodynamic segregation at the point of sedimentation (i.e. fractionation of clay types during sedimentation) and (ii) variations in primary depositional water chemistry and diagenetic formation water chemistry.

If a uniform source of detrital sediment is broadly assumed for interlayered sand- and mud-dominated sedimentary layers, the action

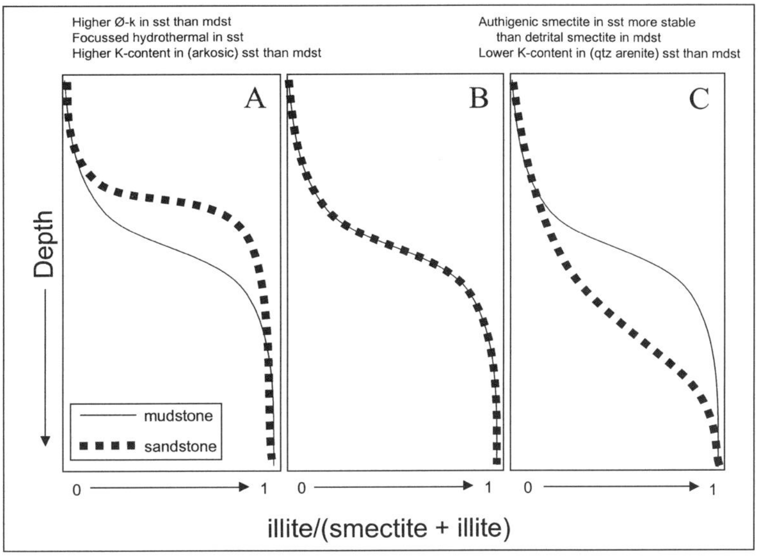

Fig. 7 Comparison of smectite to illite transformation rates in sandstones (sst) and interbedded mudstones (mdst). Depth is generalized and is dependent on the variables given in Fig. 4. 0 represents 100% smectite and 0% illite and 1 represents 0% smectite and 100% illite. (A) Where illitization occurs at lower depth/temperature in sandstones than mudstone. (B) Where the processes are the same in the two lithologies. (C) Where illitization occurs at higher depth/temperature in sandstones than mudstone. ϕ–k = porosity and permeability.

of differential segregation could produce a marked difference in clay mineralogy of sandstones and mudstones, as long as there has been relatively little subsequent interchange between pore-fluids of the two lithologies (Jeans, 1989). Neoformed components of the assemblage tend to be similar except where there has been an invasion of allochthonous formation water into the coarser, more permeable lithology (e.g. Hillier *et al.*, 1996).

The evolution of smectite in sandstones and interbedded mudstones during diagenesis can be divided into three types:

1 those in which the clay mineralogy is very similar to that in adjacent mudstones (in terms of clay type, specific composition and relative proportions);

2 those in which there is less smectite in sandstones than coeval mudstones;

3 those in which there is more smectite in sandstones than coeval mudstones (Fig. 7).

The detrital and authigenic mineralogy of the Tertiary sandstones of the Central North Sea are almost identical to those in interlayered and adjacent mudstone sequences, the principal difference being the relative proportions of the clays (Huggett, 1996). Detrital I/S clays in the matrix of the Wilcox sandstones of southwest Texas show similar diagenetic patterns similar to those of the interbedded mudstones, however, the illitization process did not proceed as far in the sandstones as in the mudstones (Boles & Franks, 1979).

The less deeply buried sandstones (< 800 m) from the Niigata Basin, Japan were found to have a different clay-mineral assemblage relative to the associated mudstones (Niu & Ishida, 2000). The illitization process at depth (> 3000 m) tended to produce similar diagenetic assemblages in the sandstones and mudstones. However, the rate of illitization was slower in the sandstones than in the mudstones. This is largely because the smectites in the mudstones were derived from partly weathered detrital illite (thus making back-conversion easier because they were far from equilibrium with the diagenetic environment, according to Niu & Ishida, 2000). Conversely, the smectite in the sandstone was generated by authigenesis during burial, implying that it grew in equilibrium with the pore-fluids. The implication is that smectite reactivity (i.e. susceptibility to illitization) may depend on whether the smectite was detrital or authigenic, the latter being intrinsically more stable than the former. Similarly, Howard (1981), Smart & Clayton (1985), Boles & Franks (1979) and Eberly & Crossey (1989) all observed that I/S in clays tended to be much more expandable (higher smectite content) in sandstones than in adjacent mudstones.

The clay-mineral diagenetic patterns in Miocene mudstones and sandstones from the Ulleung (Tsushima) backarc basin, Korea are quite distinct (Hillier *et al.*, 1996). Although the transformation from smectite to illite occurs at relatively shallow depths in both lithologies (owing to the elevated basal heat flow), the sandstone loses its smectite component at shallower depths than the mudstones. The sandstone has lost the smectite by 1500 m whereas the mudstone lost its smectite by 2000 m. Hydrothermal activity was thought to have been focused in the more permeable sandstones, allowing them to reach higher temperatures sooner than the less permeable mudstones. These data concur with those of Ramseyer & Boles (1986) who found that I/S in mudstones tended to show a slightly higher content of smectite layers (about 5–10%) than I/S in adjacent sandstones. For example, illitization of smectite is advanced in the sandy sections of the Beaufort–Mackenzie Basin relative to the muddy sections. This is thought to be a result of the greater porosity and permeability of the sandstone and thus higher chance of reaction between I/S and pore-water solutes rather than hydrothermal activity (Ko & Hesse, 1998).

POSSIBLE EFFECTS OF SMECTITE DIAGENESIS IN SANDSTONES ON OTHER DIAGENETIC PHENOMENA

It is very likely that the diagenetic evolution of smectite in sandstone has significant consequences for the overall diagenetic mineralogy and formation-water composition (by reference to the reactions R2–R5). Smectite transformation to other clays during burial leads to quartz cementation, possibly carbonate cementation and probably changes the cation ratios in the formation water (especially increasing the concentration of aqueous sodium and decreasing the concentration of aqueous potassium).

Because illitization increases the Al/Si ratio of the clay mineral, it is likely to lead to simultaneous quartz growth. The transformation of dioctahedral smectites involves consumption of potassium from the formation water and release of sodium to the formation water. Illitization may thus have an important control on the Na/K ratio of formation waters. Illitization also releases calcium and magnesium (and possibly iron). Although some formation waters do have unusually elevated alkaline earth elements, it is also likely that these will lead to the growth of dolomite (possibly ferroan dolomite) if the formation water contains any bicarbonate. Because illitization leads to complete loss of calcium and sodium from the clay to the host rock the ratio of these elements will not change unless subsequently calcium is lost during carbonate cement growth.

Because the Al/Si ratio of the clay mineral increases during the change from a trioctahedral smectite to chlorite, it is likely that chloritization also leads to simultaneous quartz growth. Chloritization may produce relatively more quartz than does illitization. The transformation of trioctahedral smectites involves loss of sodium from the clay to the formation water. Chloritization may thus have an important control on the Na^+ content of formation waters. Chloritization may also lead to release of calcium (and possibly magnesium) and thus may lead to carbonate cementation (if bicarbonate is present in the formation water or CO_2 is present in a gas phase). As chloritization leads to total loss of calcium and sodium to the formation, the ratio of these elements will not change unless subsequently calcium is lost during carbonate cement growth.

SYNTHESIS

The occurrence of smectite in sandstones is a function of both depositional and diagenetic factors. It occurs in the primary sediment where the source terrain can produce smectite in preference to other clays typical of more advanced weathering. These conditions occur in either arid or rapidly eroded, high-relief areas where weathering cannot reach advanced stages. They also occur when drainage is poor, because this prevents removal of aqueous weathering products and inhibits advanced weathering. Smectites are somewhat influenced by the lithology of the source terrain, although this has a greater impact on the type of smectite rather than its occurrence. Granitic, andesitic, metapsammite and arkosic sandstones may weather to produce dioctahedral smectites (such as montmorillonite). Basalt, metabasites, impure marbles, metapelites, mudstones and lithic sandstones may weather to trioctahedral smectites (such as saponite). Smectites are most common in aeolian, fluvial, lacustrine and turbidite sandstones. In aeolian, lacustrine and fluvial sandstones, the opportunity for advanced degrees of weathering and smectite breakdown (to other clays) are minimized by (i) the arid climate and (ii) limited time to weather relative to the time the sediments are in transport. In turbidite sandstones, smectites are relatively more abundant because they tend to be finer grained than other clays (and thus can be carried further into the ocean prior to settling).

Smectites, like other clays, usually are not codeposited with non-marine or shallow marine sand-grade sediment but they are incorporated into the sand by bioturbation, soft-sediment deformation and fines infiltration into the sand. Smectite minerals also can develop as authigenic minerals in sandstones, owing to continued breakdown of unstable lithic grains in the primary sand.

Smectites in sandstones transform to illite or chlorite via mixed clay intermediates during progressive diagenesis, although it should be noted that smectites are not intrinsically unstable at high diagenetic temperatures. Rather they become unstable as the geochemical (aqueous) environment changes during burial. Thus breakdown of K-feldspar at elevated temperature releases potassium, which then can react with (dioctahedral) smectite to produce illite. The processes of smectite illitization and chloritization are unlikely to be isochemical.

The occurrence and rates of smectite illitization and chloritization in sandstones can be both similar and dissimilar to equivalent processes in neighbouring mudstones. The key controls are the rock mineralogy, the local water geochemistry and the fact that sandstones have higher permeability than mudstones, which can permit hot water circulation (and thus promote thermally controlled reactions) and movement of components essential for smectite breakdown.

ACKNOWLEDGEMENTS

We would like to thank the reviewers of this paper, Jim Boles and Jean-Francois Deconinck. Their suggestions have led to a much better paper. The editorial work and suggestions by Sadoon Morad are also much appreciated.

REFERENCES

Amouric, M. & Olives, J. (1998) Transformation mechanisms and interstratification in conversion of smectite to kaolinite: an HRTEM study. *Clay. Clay Mineral.*, **46**, 521–527.

Aoudjit, H., Robert, M., Elsass, F. & Curmi, P. (1995) Detailed study of smectite genesis in granitic saprolites by analytical electron microsopy. *Clay Mineral.*, **30**, 135–147.

Bjørlykke, K. & Aagaard, P. (1992) Clay minerals in the North Sea sandstones. In: *Origin, Diagenesis and Petrophysics of Clay Minerals in Sandstones*, Society of Economic Paleontologists and Mineralogists Special Publication 47 (Ed. by D.W. Houseknecht and E.D. Pittman), Tulsa, Oklahoma, pp. 65–80.

Boles, J.R. & Franks, S.G. (1979) Clay diagenesis in Wilcox sandstones of southwest Texas: implications of smectite diagenesis on sandstone cementation. *J. Sed. Petrol.*, **49**, 55–70.

Brindley, G.W. & Brown, G. (1980) Crystal structures of clay minerals and their X-ray identification. *Mineral. Soc. Monogr.*, **5**, 495 pp.

Bruce, C.H. (1984) Smectite dehydration—its relation to structural development and hydrocarbon accumulation in northern Gulf of Mexico basin. *AAPG Bull.*, **68**, 673–683.

Bühmann, C. (1992) Smectite-to-illite conversion in a geothermally and lithogically complex Permian sedimentary sequence. *Clay. Clay Mineral.*, **40**, 53–64.

Burley, S.D. (1984) Patterns of diagenesis in the Sherwood Sandstone Group (Triassic), United Kingdom. *Clay Mineral.*, **19**, 403–440.

Burst, J.F. Jr. (1969) Diagenesis of Gulf Coast clayey sediments and its possible relation to petroleum migration. *AAPG Bull.*, **53**, 73–93.

Carrigy, M.A. & Mellon, G.B. (1964) Authigenic clay mineral cements in Cretaceous and Tertiary sandstones of Alberta. *J. Sed. Petrol.*, **34**, 461–472.

Chamley, H. (1989) *Clay Sedimentology.* Springer-Verlag, Berlin.

Chamley, H. (1994) Clay mineral diagenesis. In: *Quantitative Diagenesis: Recent Developments and Applications to Reservoir Geology* (Ed. by A. Parker and B.W. Sellwood), Kluwer Academic, Dordrecht, pp. 161–188.

Chamley, H., Deconinck, J.-F. & Millot, G. (1990) About the abundance of smectitic minerals in common marine sediments deposited during high sea level stages in late Jurassic–Paleogene times. *CR Acad. Sci. Paris*, **311**, 1529–1536.

Chang, H.K., MacKenzie, F.T. & Schoonmaker, J. (1986) Comparision between the diagenesis of dioctahedral and triochedral smectite, Brazilian offshore basins. *Clay. Clay Mineral.*, **34**, 407–423.

Chesworth, W. (1977) Weathering stages of the common igneous rocks, index minerals and mineral assemblages at the surface of the earth. *J. Soil Sci.*, **28**, 490–497.

Christidis, G. & Dunham, A.C. (1993) Compositional variations in smectites. Part 1: alteration of intermediate volcanic rocks. A case study from Milos Island, Greece. *Clay Mineral.*, **28**, 255–273.

Christidis, G. & Dunham, A.C. (1997) Compositional variations in smectites. Part 2: alteration of acidic precursors, a case study from Milos Island, Greece. *Clay Mineral.*, **32**, 253–270.

Clauer, N., Rinckenbach, T., Weber, F., Sommer, F., Chaudhuri, S. & O'Neil, J.R. (1999) Diagenetic evolution of clay minerals in oil-bearing eogene sandstones and associated shales, Mahakam delta basin, Kalimantan, Indonesia. *AAPG Bull.*, **83**, 62–87.

Corti, G., Sanjurjo, M.J.F. & Ugolini, F.C. (1998) Randomly interstratified kaolinite–smectite from Galicia (NW Spain): a new procedure for determination. *Clay. Clay Mineral.*, **46**, 705–711.

Davies, D.K. & Ethridge, F.G. (1975) Sandstone composition and depositional environment. *AAPG Bull.*, **59**, 239–264.

Debrabant, P., Delbart, S. & Lemaguer, D. (1985) Microanalyses géochimiques de minéraux argileux de sediments prélevés en Atlantique Nord (forages du DSDP). *Clay Mineral.*, **20**, 125–145.

Deconinck, J.-F. & Chamley, H. (1995) Diversity of smectite origins in late Cretaceous sediments: example of chalks from Northern France. *Clay Mineral.*, **30**, 365–379.

Deconinck, J.-F. & Strasser, A. (1987) Sedimentology, clay mineralogy and depositional environment of Purbeckian green marls (Swiss and French Jura). *Eclogae Geol.*, **80**, 753–772.

Deconinck, J.-F., Strasser, A. & Debrabant, P. (1988) Formation of illitic minerals at surface temperatures in Purbeckian sediments (Lower Berriasian, Swiss and French Jura). *Clay Mineral.*, **23**, 91–103.

Deer, W.A., Howie, R.A. & Zussman, J. (1998) *An Introduction the Rock-forming Minerals.* Longman, Harlow, 528 pp.

Drever, J.I. (1982) *The Geochemistry of Natural Waters.* Prentice Hall, Englewood Cliffs, NJ, 388 pp.

Dunoyer De Segonzac, G. (1970) The transformation of clay minerals during diagenesis and low-grade metamorphism: a review. *Sedimentotogy*, **15**, 281–346.

Dypvik, H. (1983) Clay mineral transformations in Tertiary and Mesozoic sediments from North Sea. *AAPG Bull.*, **67**, 160–165.

Eberl, D. & Hower, J. (1976) Kinetics of illite formation. *Geol. Soc. Am. Bull.*, **87**, 1326–1330.

Eberl, D., Whitney, G. & Khoury, H. (1978) Hydrothermal reactivity of smectite. *Am. Mineral.*, **63**, 401–409.

Eberl, D., Środoń, J. & Northrop, H.R. (1986) Potassium fixation in smectite by wetting and drying. In: *Geochemical Processes at Mineral Surfaces* (Ed. by J.A. Davies and K.F. Hayes), American Chemical Society, Washington, DC, pp. 296–326.

Eberly, P.O. & Crossey, L.J. (1989) Compositional variation in the clay mineral fraction of fine- and coarse grained units in Westwater Canyon Member (Morrison Formation San Juan, New Mexico). *AAPG Bull.*, **73**, 1154.

El Albani, A., Deconinck, J.-F., Herbin, J.-P. & Proust, J.-N. (1993) Geochemical characterisation of organic matter and clay mineralogy of the Kimmeridgian of the Boulonnais (Northern France). *Ann. Soc. Géol. Nord*, **2**, 113–120.

Galloway, W.E. (1974) Deposition and diagenesis alteration of sandstone in northeast Pacific arc-related basins: Implications for greywacke genesis. *Geol. Soc. Am. Bull.*, **85**, 379–390.

Gibbs, R.J. (1977) Clay mineral segregation in the marine environment. *J. Sed. Petrol.*, **47**, 237–243.

Gutiérrez-Mas, J.M., López-Galindo, A. & López-Aguayo, F. (1997) Clay minerals in recent sediments of the continental shelf and the Bay of Cádiz (SW Spain). *Clay Mineral.*, **32**, 507–515.

Güven, H. (1988) Smectites. In: *Hydrous Phyllosilicates*, Reviews in Mineralogy 19 (Ed. by S.W. Bailey), Mineralogical Society of America, Washington, DC, pp. 497–559.

Hartmann, B.H., Bodnár, K.J., Ramseyer, K. & Matter, A. (1999) Effect of Permo-Carboniferous climate on illite–smectite, Haushi Group, Sultanate of Oman. *Clay. Clay Mineral.*, **47**, 131–143.

Henderson, R.A. (1998) Eustatic and palaeoenvironmental assessment of the mid-Cretaceous Bathurst Island Group of the Money Shoals Platform, northern Australia Source. *Palaeogeogr. Palaeoclimatol. Palaeoecol.*, **138**, 115–138.

Hill, I., Worden, R.H. & Meighan, I.G. (2000) Geochemical evolution of a palaeolaterite: the interbasaltic formation, Northern Ireland. *Chem. Geol.*, **166**, 65–84.

Hillier, S. (1994) Pore lining chlorite in siliciclastic reservoir sandstones: electron microprobe, SEM and XRD data, and implications for their origin. *Clay Mineral.*, **29**, 665–679.

Hillier, S., Son, B.K. & Velde, B. (1996) Effects of hydrothermal activity on clay mineral diagenesis in Miocene shales and sandstones from the Ulleung (Tsushima) backarc basin, East Sea (Sea of Japan), Korea. *Clay Mineral.*, **31**, 113–126.

Howard, J.J. (1981) Lithium and potassium saturation of illite–smectite clays from interlaminated shales and sandstones. *Clay. Clay Mineral.*, **29**, 136–142.

Hower, J., Eslinger, E.V., Hower, M.E. & Perry, E.A. (1976) Mechanism of burial and metamorphism of argillaceous sediment: 1. Mineralogical and chemical evidence. *Geol. Soc. Am. Bull.*, **87**, 725–737.

Huggett, J.M. (1996) Aluminosilicate diagenesis in a Tertiary sandstone-mudstone sequence from the Central North Sea, UK. *Clay Mineral.*, **31**, 523–536.

Humphreys, B., Kemp, S.J., Lott, G.K., Bermanto, D.A., Dharmayanti, D.A. & Samsori, I. (1994) Origin of grain-coating chlorite by smectite transformation: an example from Miocene sandstones, North Sumatra Back-arc Basin, Indonesia. *Clay Mineral.*, **29**, 681–692.

Jeans, C.V. (1989) Clay diagenesis in sandstones and shales: an introduction. *Clay Mineral.*, **24**, 127–136.

Jeans, C.V., Merriman, R.J., Mithell, J.G. & Bland, D.J. (1982) Volcanic clays in the Cretaceous strata of Southern England and Northern Ireland. *Clay Mineral.*, **17**, 105–156.

Jeans, C.V., Wray, D.S., Merriman, R.J. & Fisher, M.J. (2000) Volcanogenic clays in Jurassic and Cretaceous strata of England and the North Sea Basin. *Clay Mineral.*, **35**, 25–55.

Jin Hwan Noh & Boles, J.R. (1993) Origin of zeolite cements in the Miocene sandstones, North Tejon oil fields, California Source. *J. Sed. Petrol.*, **63**, 248–260.

Ko, J. & Hesse, R. (1998) Illite/smectite diagenesis in the Beaufort–Mackenzie Basin, Arctic Canada: relation to hydrocarbon occurrence? *Bull. Can. Petrol. Geol.*, **46**, 74–88.

Lahann, R.W. (1980) Smectite diagenesis and sandstone cement: the effect of reaction temperature. *J. Sed. Petrol.*, **50**, 755–760.

Land, L.S. & Milliken, K.L. (2000) Regional loss of SiO_2 and $CaCO_3$ and gain of K_2O during burial diagenesis of Gulf Coast mudrocks, USA. In: *Quartz Cementation in Sandstones*, Special Publication of the International Association of Sedimentologists 29 (Ed. by R.H. Worden and S. Morad), Blackwell Science, Oxford, pp. 183–197.

Lynch, F.L. (1996) Mineral/water interaction, fluid flow, and Frio sandstone diagenesis: evidence from the rocks. *AAPG Bull.*, **80**, 486–504.

Lynch, F.L. (1997) Frio shale mineralogy and the stoichiometry of the smectite-to-illite reaction: the most important reaction in clastic sedimentary diagenesis. *Clay. Clay Mineral.*, **45**, 618–631.

Moberly, R. & Jenkyns, H.C. (1981) Cretaceous volcanogenic sediments of the Nauru Basin, Deep Sea Drilling Project, Leg 61. *Init. Rep. Deep Sea Drilling Proj.*, **61**, 533–548.

Niu, B. & Ishida, H. (2000) Different rates of smectite illitization in mudstones and sandstones from the Niigata basin, Japan. *Clay Mineral.*, **35**, 163–173.

Perry, E. & Hower, J. (1970) Burial diagenesis in Gulf Coast pelitic sediments. *Clay. Clay Mineral.*, **18**, 165–178.

Post, J.L. (1984) Saponite from near Ballarat, California. *Clay. Clay Mineral.*, **32**, 147–153.

Primmer, T.J., Cade, C.A., Evans, I.J., *et al.* (1997) Global patterns in sandstone diagenesis: application to reservoir quality prediction for petroleum exploration. *AAPG Mem.*, **69**, 61–78.

Pytte, A.M. & Reynolds, R.C. (1989) The thermal transformation of smectite to illite. In: *Thermal History of Sedimentary Basins: Methods and Case Histories* (Ed. by N.D. Naeser and T.H. McCulloh), Springer-Verlag, New York, pp. 133–140.

Ramseyer, K. & Boles, J.R. (1986) Mixed-layer illite/smectite minerals in Tertiary sandstones and shales, San Joaquin Basin, California. *Clay. Clay Mineral.*, **34**, 115–124.

Reynolds, R.C. (1980) Interstratified clay minerals. In: *Crystal Structures of Clay Minerals and their X-ray Identification* (Ed. by G.W. Brindley and G. Brown), Mineralogical Society, London, pp. 249–303.

Robertson, H.E. & Lahann, R. (1981) Smectite to illite conversion rates: effects of solution chemistry. *Clay. Clay Mineral.*, **29**, 129–135.

Robinson, D. & Santana De Zamora, A. (1999) The smectite to chlorite transition in the Chipilapa geothermal system, El Salvador. *Am. Mineral.*, **84**, 607–619.

Ryu, I. & Niem A.R. (1999) Sandstone diagenesis, reservoir potential, and sequence stratigraphy of the Eocene Tyee basin, Oregon. *J. Sed. Res.*, **69**, 384–393.

Smart, G. & Clayton, T. (1985) The progressive illitization of interstratified illite–smectite from Carboniferous sediments of northern England and its relationship to organic maturity indicators. *Clay Mineral.*, **20**, 455–466.

Środoń, J. (1999) Use of clay minerals in constructing geological processes: recent advances and some perspectives. *Clay Mineral.*, **34**, 27–37.

Stern, A.L., Chamberlain, C.P., Reynolds, R.C. & Johnson, G.D. (1997) Oxygen isotope evidence of climate change from pedogenic clay minerals in Himalayan molasse. *Geochim. Cosmochim. Acta*, **61**, 731–744.

Surdam, R.C. & Boles, J.R. (1979) Diagenesis of volcanic sandstones. In: *Aspects of Diagenesis*, Society of

Economic Palaeontologists and Mineralogists Special Publication 26 (Ed. by P.A. Scholle and P.R. Schluger), Tulsa, OK, pp. 227–242.

Thiry, M. & Jacquin, T. (1993) Clay mineral distribution related to rift activity, sea level changes and palaeoceanography in the Cretaceous of the Atlantic Ocean. *Clay Mineral.*, **28**, 61–84.

Thiry, M., Cavelier, C. & Trauth, T. (1977) Les sédiments de L'éocène inférieur du bassin de Paris et leurs relations avec la paléoaltération de la craie. *Sci. Géol. Bull.*, **30**, 113–128.

Weaver, C.E. (1958) The effects and geological significance of potassium 'fixation' by expandable clay minerals derived from muscovite, biotite, chlorite, and volcanic material. *Am. Mineral.*, **43**, 839–861.

Weibel, R. (1999) Effects of burial on the clay assemblages in the Triassic Skagerrak Formation, Denmark. *Clay Mineral.*, **34**, 619–635.

Whitney, G. & Northrop, H.R. (1987) Diagenesis and fluid flow in the San Juan Basin, New Mexico—regional zonation in the mineralogy and stable isotope composition of clay minerals in sandstone. *Am. J. Sci.*, **287**, 353–382.

Wilson, M.D. & Pittman, E.D. (1977) Authigenic clays in sandstones: recognition and influence on reservoir properties and paleoenvironmental analysis. *J. Sed. Petrol.*, **47**, 3–31.

Wilson, M.J. (1999) The origin and formation of clay minerals in soils: past, present and future perspectives. *Clay Mineral.*, **34**, 7–25.

Zhao, G., Donald, R. & Mc Dowell, S.D. (1999) 'Retrograde diagenesis' of clay minerals in the Precambrian Freda sandstone, Wisconsin. *Clay. Clay Mineral.*, **47**, 119–130.

Int. Assoc. Sedimentol. Spec. Publ. (2003) **34**, 129–145

Patterns of clay mineral diagenesis in interbedded mudrocks and sandstones: an example from the Palaeocene of the North Sea

H.F. SHAW *and* D.M. CONYBEARE*

Department of Earth Science and Engineering, Imperial College, Prince Consort Road, London SW7 2 BP, UK, e-mail: h.shaw@ic.ac.uk

ABSTRACT

Patterns of clay mineral (and non-clay mineral) diagenesis in a Palaeocene sequence of interbedded deep-marine-fan turbidite sandstones and mudrocks from the Everest Complex of the North Sea are described in order to compare and contrast the diagenetic processes in these different lithotypes. The study demonstrates broad similarities in the patterns of clay mineral and non-clay mineral diagenesis in the sandstones and mudrocks, except for quartz cementation and dissolution. This suggests generally similar diagenetic processes in which the sandstones and mudrocks behave, at least in part, as relatively 'open' systems involving fluid-mediated migration of chemical components within and across the interbedded mudrock–sandstone sequences. The mudrocks appear to display varying degrees of openness depending on their permeability, thickness and composition. This conclusion poses fundamental questions about the behaviour of mudrocks during burial diagenesis that have important consequences for understanding fluid migration and diagenetic processes in sedimentary basins.

INTRODUCTION

The literature on clay mineral diagenesis is predominantly represented by separate studies of either mudrock sequences or sandstones. In some sandstone studies diagenesis is interpreted in terms of the supply of chemical components from the adjacent shales (e.g. silica for quartz cements, organic acids to induce secondary porosity or kaolinitization). However, more often than not the associated mudrocks have not been examined, but are cited as a source of the components required with little direct evidence.

In this paper detailed petrographic and mineralogical data from a study of interbedded Palaeocene mudrocks and sandstones from the North Sea are reviewed. The extent to which the diagenetic processes in the mudrocks and sandstones are linked or separate are examined and an attempt is made to draw more general conclusions about how diagenetic processes operate in sandstones versus mudrocks.

GEOLOGICAL SETTING OF THE PALAEOCENE SEQUENCE

The samples for this study were taken from four wells located in the Everest Complex of the Central Graben of the northern North Sea (Fig. 1). The Palaeocene Everest Complex, on

* Current address: Baker Atlas Geoscience, Kettock Lodge, Campus 2, Aberdeen Science and Technology Park, Balgownie Drive, Bridge of Don, Aberdeen AB22 8GU, UK.

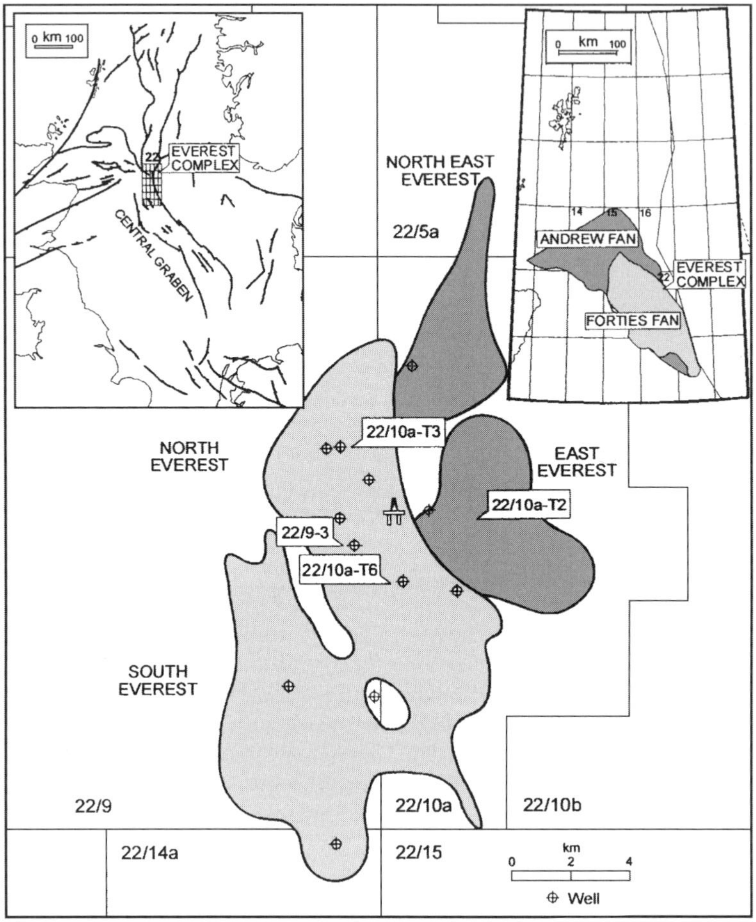

Fig. 1 Study area and well locations.

the eastern side of the Central Graben of the North Sea, consists of a sequence of interbedded deep-marine-fan turbidite sandstones and basinal mudrocks. The submarine fan sequence formed as a consequence of terrigenous clastic sediment input from the uplifted Shetland Platform at the beginning of the Tertiary. The turbidite reservoir sands, deposited in middle and outer fan settings, overlie the Cretaceous Danian chalks and form the eastward extension of the Forties and Andrew fans. The reservoir sands form part of the Maureen Formation, Andrew Sandstone Unit and Forties Sandstone Member; the Maureen Formation and Andrew Sandstone Unit are sealed by the Lista Formation mudrocks and the Forties Sandstone Member by the mudrock-dominated Sele Formation (Fig. 2) (Thompson & Butcher, 1991; Knox & Holloway, 1992; O'Connor & Walker, 1993).

The Maureen Formation consists of medium- to coarse-grained high-density turbidite sandstones interbedded with thick bioturbated, often calcareous, basinal mudrocks. The Andrew Sandstone Unit is subdivided into the Lower Andrew Sandstone and the Upper Andrew Sandstone, separated by the Mid-Andrew Shale (Fig. 2). Sedimentation in the Lower Andrew Sandstone can be described broadly as representing pulses of progradation of mid-fan, medium- to coarse-grained turbidite sands into basinal muds. The overlying Mid-Andrew Shale is dominated by basinal, often bioturbated mudrocks with some thin medium-

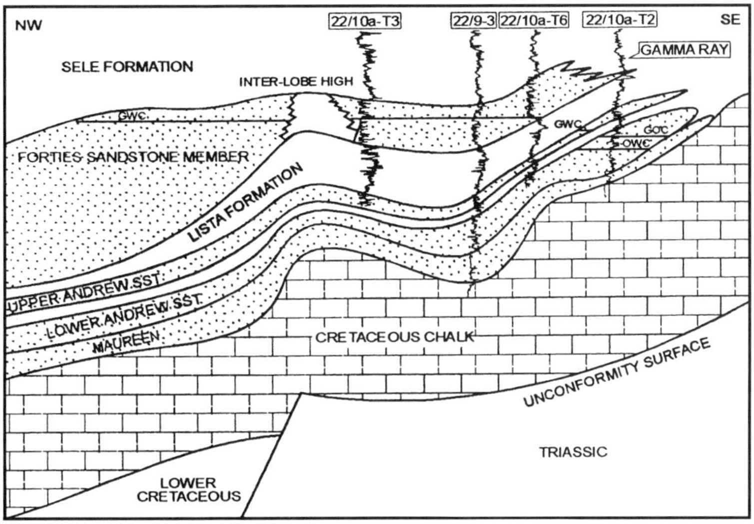

Fig. 2 Tertiary stratigraphy of the Everest Complex. OWC = oil–water contact; GWC = gas–water contact; GOC = gas–oil contact.

grained to very fine-grained sands representing deposition from dilute low-density turbidite flows. In the Upper Andrew Sandstone there is a return to sedimentation dominated by clastic fan deposition and fan progradation with lithologies similar to the Lower Andrew Sandstone.

The Lista Formation, incorporating the Andrew Sandstone Member, is dominated by basinal non-bioturbated and bioturbated mudrocks. The Forties Sandstone Member principally is composed of graded medium- to coarse-grained and fine- to medium-grained sandstones representing prograding mid- to outer fan and fan fringe deposits. The sandstones are interbedded with extensive mudrock horizons indicating periodic cessation of sand deposition.

Following deposition, some deformation, including injection of sandstone bodies, and fracturing accompanied burial compaction of the Palaeocene sequences in the Everest Complex. A burial history curve for one of the wells in the Everest Complex is shown in Fig. 3.

SAMPLING AND METHODOLOGICAL APPROACH

Samples were selected from four wells to provide a good coverage of the stratigraphy and the various fan and interfan facies. Conventional optical petrographic analyses were made on selected sandstone samples by the examination of thin-sections, stained with Alizarin Red-S and potassium ferricyanide to aid differentiation of carbonate phases (Dickson, 1966). Estimates of variations in the relative mineral compositions of the sandstones were based on point-count data (300 counts were made for each thin-section) and the sandstones classified using the classification of Pettijohn *et al.* (1973). Electron microscopy, in both secondary and backscattered electron imaging mode, together with electron microprobe analysis to aid mineral identification and monitor compositional variations, was used extensively to examine the petrography of mudrock and sandstone samples (Mackinnon, 1990). In addition to conventional secondary electron imaging, field emission scanning electron microscopy was used to provide high-resolution images, principally of mudrock samples (Huggett & Shaw, 1997).

The mineralogy of bulk sandstone and mudrock samples and clay (< 2 μm) fractions was determined using standard X-ray powder diffraction (XRD) analysis. The relative mineral proportions in the bulk samples were estimated based on corrected peak intensities using correction factors derived from X-ray diffraction

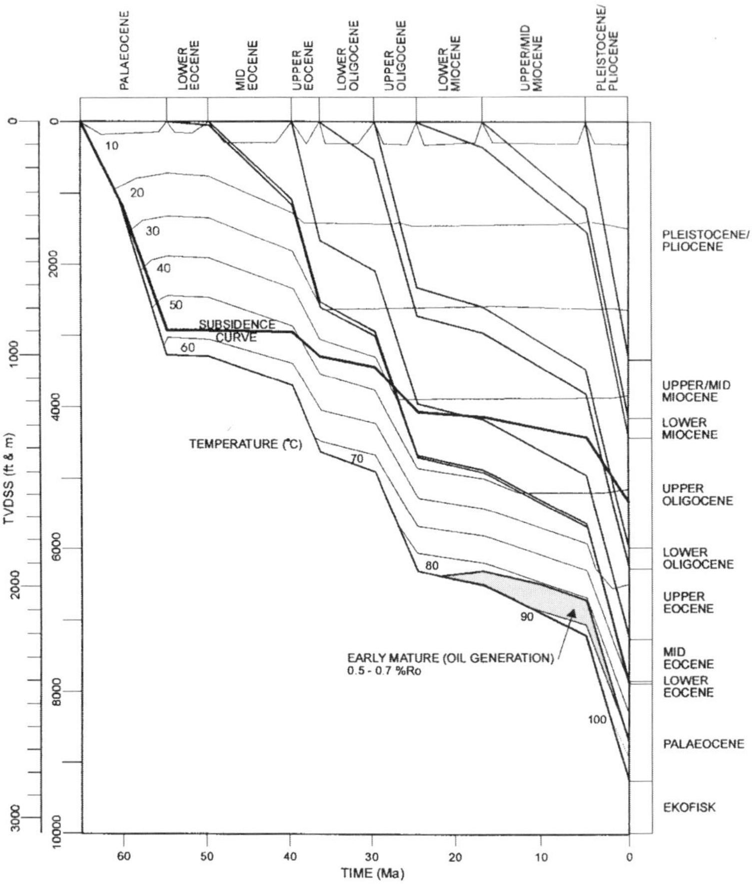

Fig. 3 Burial history curve for one well in the Everest Complex derived using BasinMod (Platte River Associates). TVDSS = total vertical depth subsea.

(XRD) analyses of pure mineral–quartz mixtures in known proportions. The detailed XRD analyses of the clay minerals were made on specimens of the orientated clay fractions prepared using the method of Shaw (1972) and analysed in the air-dry state and after glycolation and heating at 350°C and 550°C. Estimates of clay-mineral proportions from the XRD analyses were made, based on the method described by Schultz (1964).

The compositions of the mixed-layer illite–smectite were determined from the XRD analyses using the method described by Reynolds (1980) based on variations in the $002_{illite}/003_{smectite}$ reflection after glycolation of the orientated clay-fraction specimens.

Stable isotope analyses were made on separated authigenic kaolinites (oxygen and hydrogen) and Fe-calcite cements (oxygen and carbon) using standard stable isotope analytical methods. The authigenic kaolinites were separated using a combination of grain size and high-density magnetic separation to achieve > 95% purity. The Fe-calcite cements were physically extracted from their host sediments using a microdrill (Conybeare, 1996).

The total organic carbon contents of the mudrock samples were determined using a Perkin Elmer PE240C elemental analyser (Conybeare, 1996).

PARAGENESIS OF CLAY MINERALS IN MUDROCKS AND SANDSTONES

Based on conventional optical petrography plus backscatter scanning electron microscopy and secondary electron field emission scanning electron microscopy studies of mudrock and

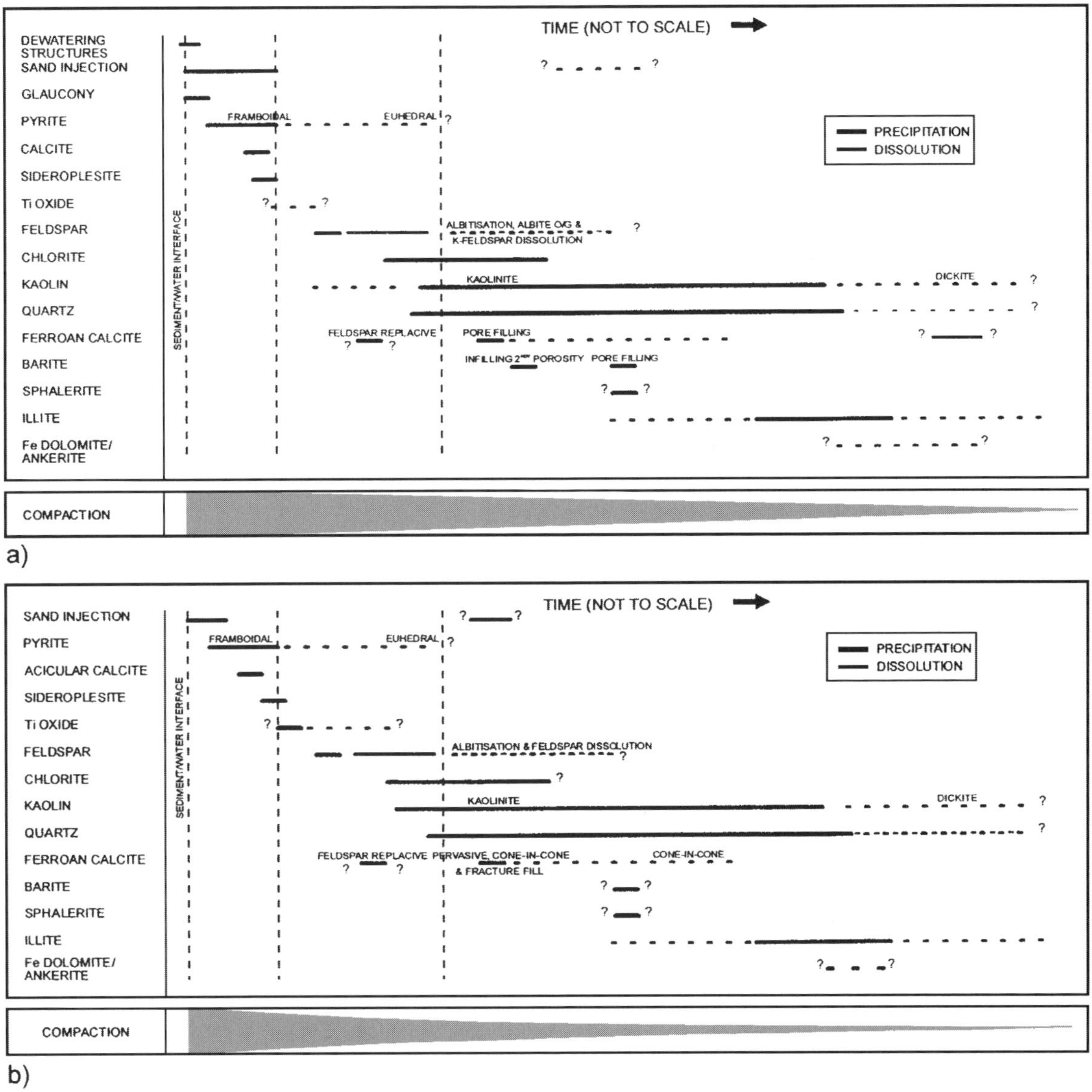

Fig. 4 Paragenetic sequences of the Everest Palaeocene sandstones (a) and mudrocks (b). O/G = overgrowth.

sandstone samples, it has been possible to produce paragenetic sequences of the diagenetic events in the mudrocks and sandstones (Fig. 4).

Mudrocks

The clay mineralogy of the mudrocks is dominated by the mixed-layer illite–smectites (I/S) which make up most of the mudrock matrix material. The mixed-layer I/S show a wide range of compositions ranging from random mixed-layer smectite-rich (I_5Sm_{95}) to ordered illitic I/S ($I_{80}Sm_{20}$); with a mean composition of $I_{60}Sm_{40}$ ordered mixed layer. The variations in I/S compositions with depth in the four wells sampled are shown in Fig. 5. The wide ranging differences in the I/S compositions show no recognizable depth-related pattern. The presence of smectite-rich I/S at greater depths than illite-rich ordered I/S and statistically significant

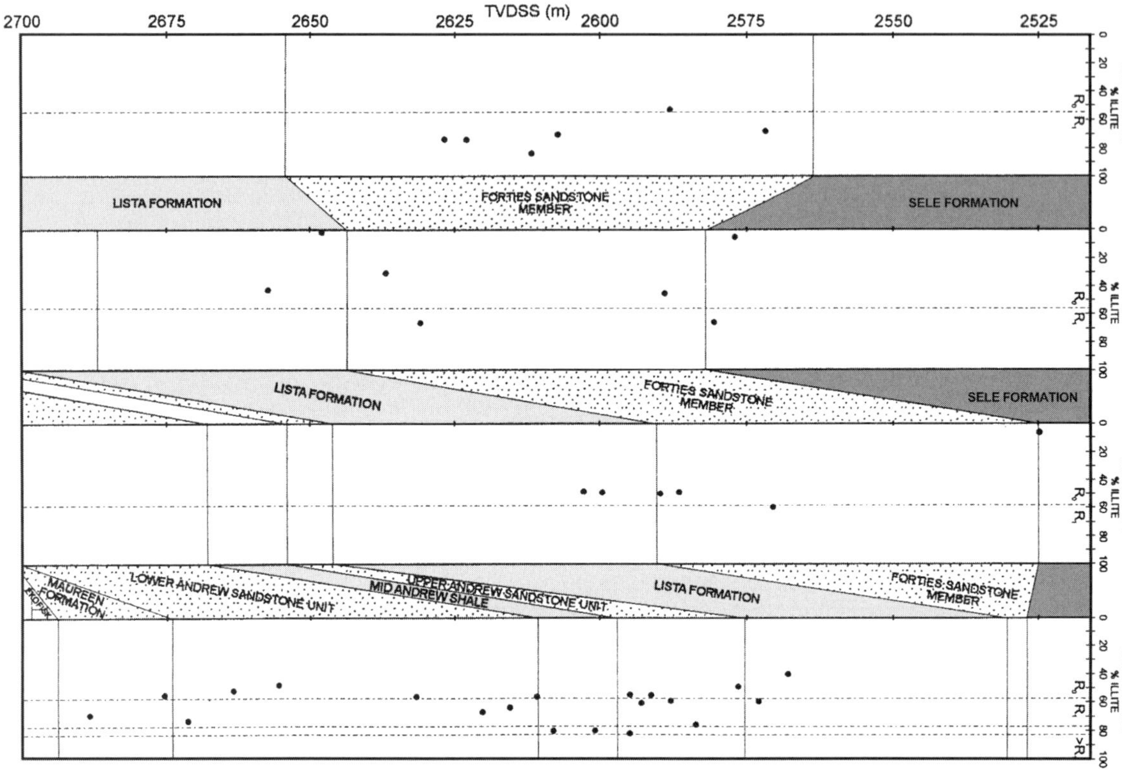

Fig. 5 Variations in I/S compositions of Everest Palaeocene mudrocks. The mixed-layer illite–smectites compositions are expressed as percentage illite in the mixed-layer illite–smectites. The degree of ordering in the mixed-layer illite–smectites is identified as follows: R0 = random mixed-layer illite–smectite; R1 = ordered mixed-layer illite–smectites; > R1 = ordered mixed-layer illite–smectites with more than 80% illite. TVDSS = total vertical depth subsea.

differences in I/S compositions within and between stratigraphical units together suggest that the I/S compositions predominantly reflect their detrital compositions rather than reflecting any general pattern of diagenetic illitization. This interpretation also is supported by the petrographic evidence showing platy I/S particles aligned parallel to bedding and compacted around detrital quartz and feldspar clasts and early diagenetic phases, consistent with a detrital origin.

This interpretation of the nature of the I/S phases is in agreement with other studies of Palaeocene mudrocks from the North Sea (Huggett, 1995). Huggett (1995) described the presence of detrital ordered I/S with compositions ranging from $I_{50}Sm_{50}$ to $I_{75}Sm_{25}$ in samples from the Palaeocene Sele Formation of the central North Sea.

The principal authigenic clay minerals in the Palaeocene mudrocks are kaolinite, illite and chlorite. Chlorite is the earliest formed of the authigenic clay cements, with slightly greater amounts observed in the Maureen Formation and Andrew Sandstone Unit (mean 3%) compared with the Lista Formation, Forties Sandstone Member and Sele Formation (mean 1%). Most of the authigenic chlorite is associated with, and may have formed from, altered biotites (Fig. 6), but smaller amounts are found infilling foram tests and secondary porosity in feldspar and quartz, and as aggregates or patches within the matrix. Microprobe analyses show that the chlorites are Fe-rich with a mean Fe/Mg ratio of 2.7 and that they have similar compositions in all their different occurrences.

The proportion of kaolinite in the mudrocks is significantly higher than in the sandstones

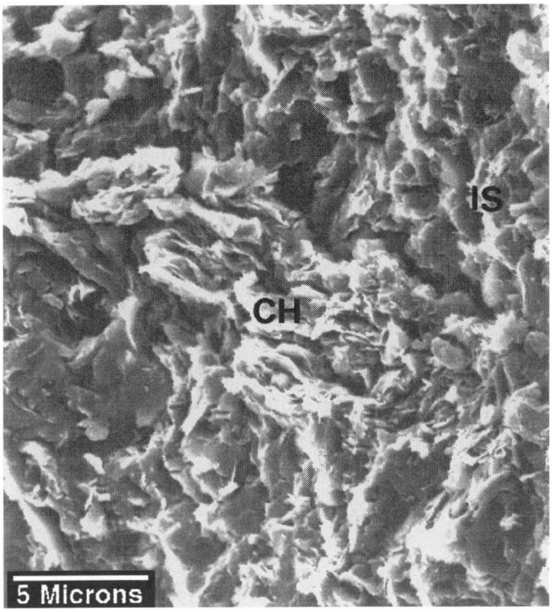

Fig. 6 Secondary electron image (SEI) photomicrograph of authigenic chlorite (CH) possibly resulting from biotite alteration in illite–smectite (IS) matrix in Everest Palaeocene mudrocks.

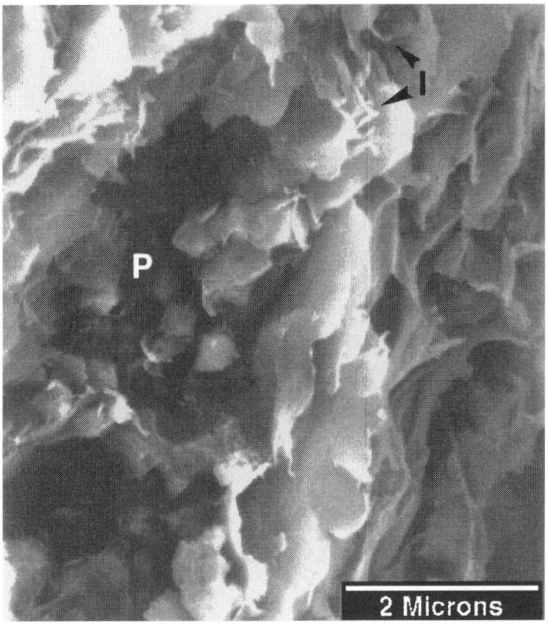

Fig. 8 Backscatter scanning electron microscope (BSEM) photomicrograph of fibrous illite (I) on platy detrital I/S in Everest Palaeocene mudrocks with some fine euhedral pyrite (P).

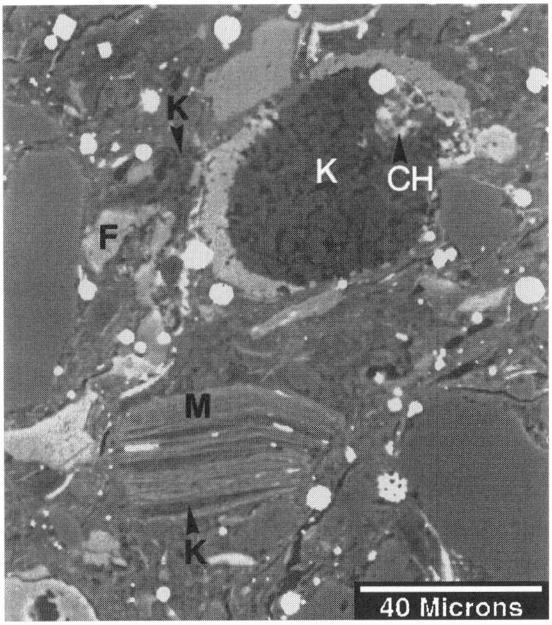

Fig. 7 Backscatter scanning electron microscope (BSEM) photomicrograph of authigenic kaolinite (K) and minor amounts of chlorite (CH) within a foram test and forming displacively between mica cleavage planes (M) in Everest Palaeocene mudrocks. F = feldspar.

and is significantly greater in the Forties Formation (mean 7% and maximum 30%) compared with the Maureen Formation, Andrew Sandstone Unit and Lista Formation (mean 2% and maximum 6%). Most, if not all, of this kaolinite appears to be diagenetic rather than detrital in origin. Authigenic kaolinite is observed most commonly forming displacively between mica cleavage planes and, less commonly, infilling foram tests or within intraparticle secondary porosity in partly dissolved feldspars (Fig. 7). There is some evidence from XRD analyses of very minor amounts of dickite in addition to the kaolinite in two samples of the Forties Formation mudrocks that have very high kaolin contents. The authigenic kaolinite appears largely to have formed after chlorite.

Authigenic illite is difficult to quantify because of problems in differentiating the abundance of authigenic illite from detrital illite/micas. The authigenic illite occurs as lath-like growths on platy detrital I/S (Fig. 8). This morphology is similar to that of authigenic

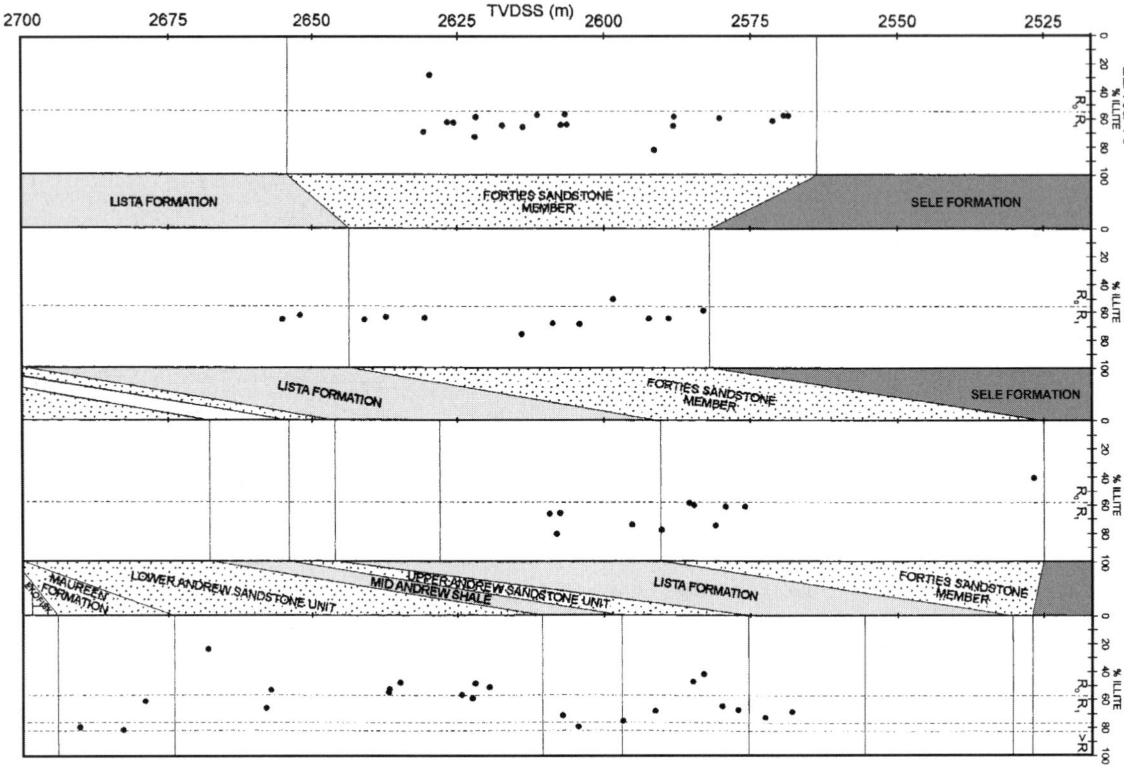

Fig. 9 Variations in I/S compositions of Everest Palaeocene sandstones. The mixed-layer illite–smectites compositions are expressed as percentage illite in the mixed-layer illite–smectites. The degree of ordering in the mixed-layer illite–smectites is identified as follows: R0 = random mixed-layer illite–smectite; R1 = ordered mixed-layer illite–smectites; > R1 = ordered mixed-layer illite–smectites with more than 80% illite. TVDSS = total vertical depth subsea.

illite described by Huggett (1995) in her studies of diagenesis in Sele Formation mudrocks.

Sandstones

According to the classification of Pettijohn *et al.* (1973), the sandstones of the Everest Complex are predominantly subarkoses with subordinate arkosic wackes (Conybeare, 1996).

Qualitatively the clay mineralogy of the sandstones is similar to the mudrocks but differs in relative abundances, microfabrics and distribution. The matrix of the sandstones is composed predominantly of ordered mixed-layer illite–smectites with a mean composition of $I_{60}Sm_{40}$, although the I/S compositions vary from randomly ordered smectite-rich to ordered illite-rich $I_{80}Sm_{20}$. There is no observable trend to the I/S compositions with depth or with stratigraphy (Fig. 9).

Authigenic chlorite is a common but minor (up to 1%) clay cement in all the sandstones and shows little stratigraphic variation in its distribution. As in the mudrocks the authigenic chlorite commonly is associated with altered biotites and, consequently, locally higher chlorite abundances are associated with sandstones that originally were enriched in detrital biotite (Fig. 10). Chlorite also occurs as clay rims around detrital grains and could have inhibited later quartz cementation, as reported in several previous studies (e.g. Imam & Shaw, 1987).

Authigenic kaolin, predominantly in the form of kaolinite but with some minor dickite, is ubiquitous in the Palaeocene sands of the Everest Complex, with abundances ranging

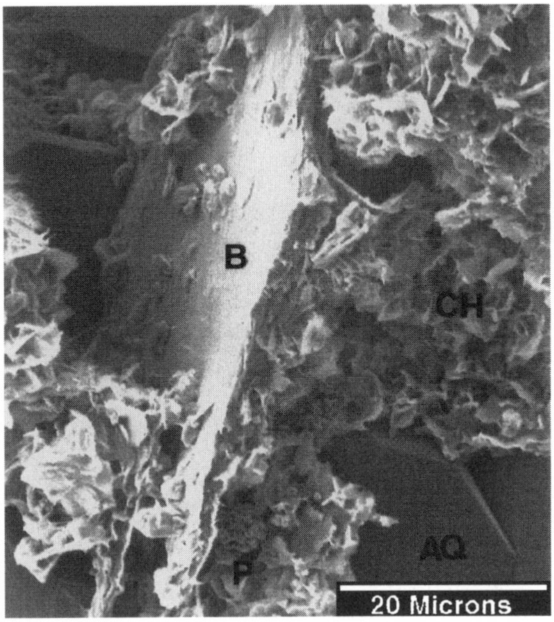

Fig. 10 Secondary electron image (SEI) photomicrograph of pore-filling chlorite (CH) associated with detrital biotite (B) in Everest Palaeocene sandstone. AQ = authegenic quartz; P = pyrite.

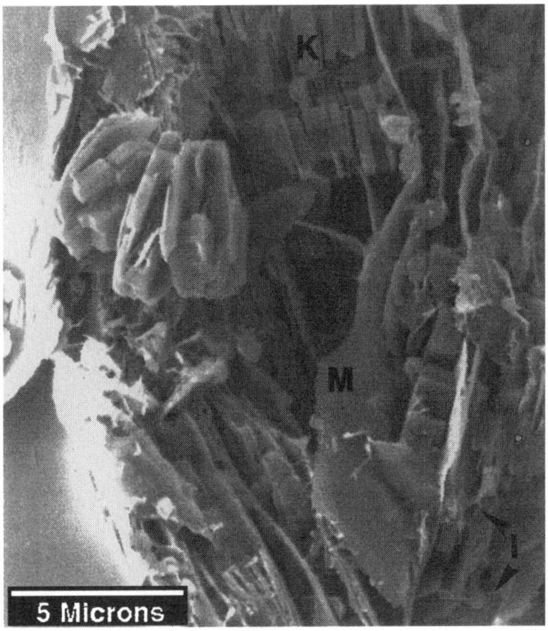

Fig. 12 Secondary electron image (SEI) photomicrograph of blocky kaolinite (K) formed between mica (M) cleavage sheets in Everest Palaeocene sandstone.

Fig. 11 Secondary electron image (SEI) photomicrograph of blocky pore-filling kaolinite in acetic-acid-etched sandstone. The area outlined by the white box in the lower part is shown in the upper part.

from < 1% to 11%. As observed for the mudrocks, the kaolins generally are more abundant in the Forties Sandstone Member than in the sandstones of the Maureen Formation and Andrew Sandstone Unit. Most of the kaolin forms as pore-filling and pore-lining phases with blocky or vermicular habits (Fig. 11). It also grows between cleavage planes of detrital muscovites in a manner similar to the chlorites within detrital biotites (Fig. 12).

Mass-balance calculations, based on concepts outlined by Helgeson *et al.* (1978) and Giles & de Boer (1990) and using quantitative petrological data from point-count analyses of the sandstones, indicate that dissolution of detrital feldspars in the sandstones generally would be insufficient to supply the components for formation of the authigenic kaolinite and an external source would be needed (Conybeare, 1996). There is an estimated maximum dissolution of up to 8% K-feldspar that could yield a maximum of about 3.6% kaolinite in a closed

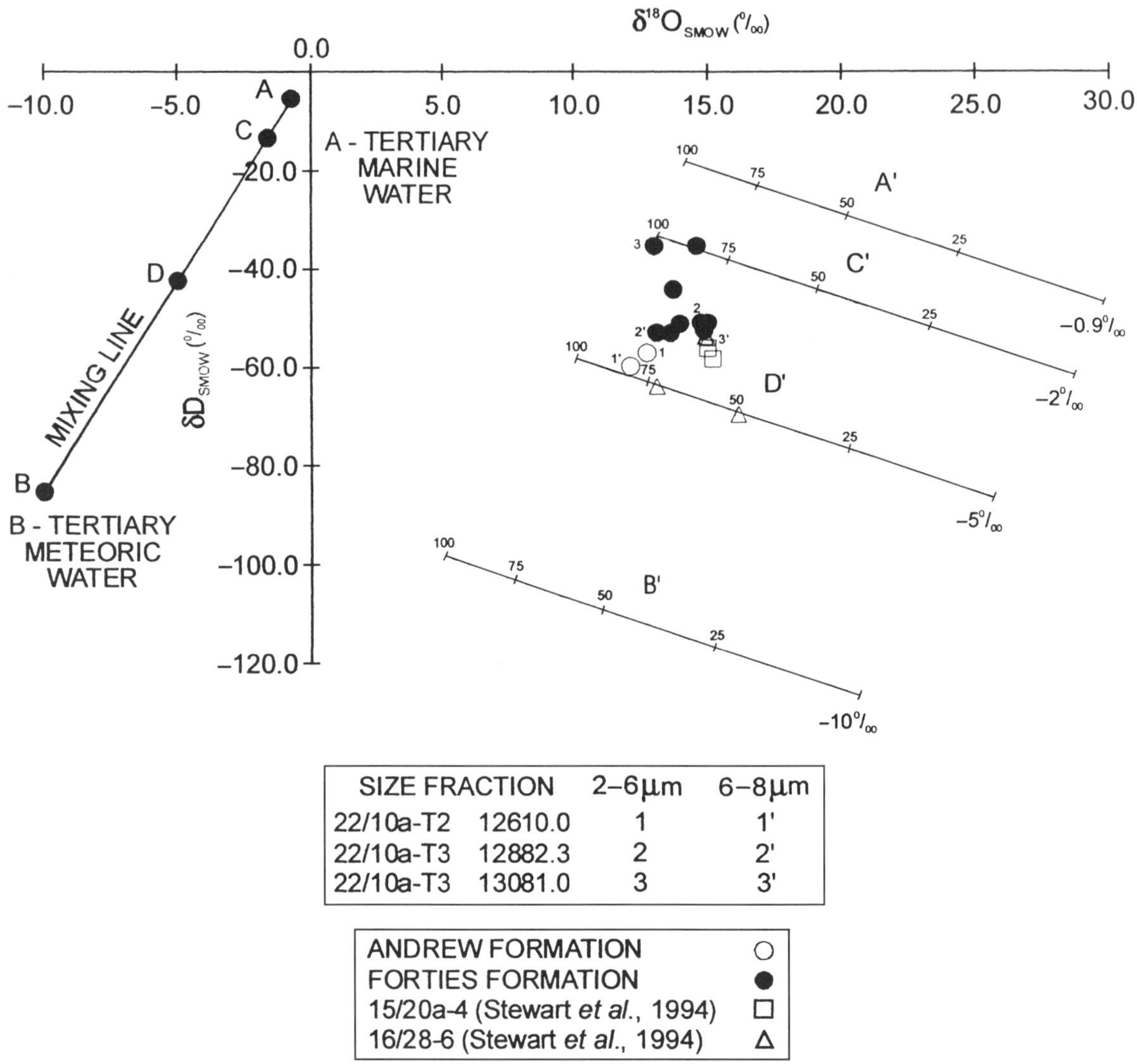

Fig. 13 Cross-plot of δD and $\delta^{18}O$ data for authigenic kaolins in Everest complex sandstones. A = Tertiary marine water composition, B = Tertiary meteoric water; C = formation water with $\delta^{18}O$ composition of −2‰, D = formation water with $\delta^{18}O$ composition of −5‰; line A–D represents a simple mixing line of formation water compositions.

system, well below the maximum observed 11% kaolinite. The estimated observed abundance of kaolinite may be overestimated owing to the presence of some microporosity, but it seems unlikely that this could account for the difference between the observed and calculated kaolinite values.

The authigenic kaolinites in the sandstones were of sufficient abundance that they could be separated for oxygen and hydrogen stable isotope analyses. The $\delta^{18}O$ (SMOW—Standard Mean Ocean Water) values vary from 12.2 to 14.9‰ (mean value 13.8‰) and δD (SMOW) from −35 to −60‰ (mean value −49‰) (Conybeare, 1996).

A cross-plot of δD versus $\delta^{18}O$ allows the temperature and fluid composition at the time of kaolin formation to be constrained (Fig. 13). This is derived from the application of the fractionation equations for oxygen and hydrogen

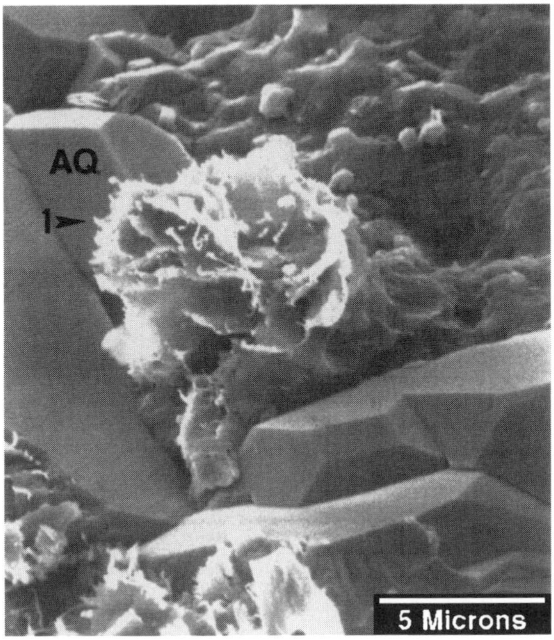

Fig. 14 Secondary electron image (SEI) photomicrograph of illite (I) fibres growing on the edges of detrital illite–smectite flakes. Note presence of authigenic quartz overgrowths (AQ).

isotopes between water and kaolin proposed respectively by Land & Dutton (1978) and Lambert & Epstein (1980). The plot suggests that the authigenic kaolin principally formed in the sandstones at about 75–85°C in a fluid of mixed meteoric–marine composition. Stewart *et al.* (1994) also proposed a similar origin for authigenic kaolins in North Sea Palaeocene sandstones, but at a lower formation temperature of 50–75°C and with a slightly greater influence of meteoric waters on the formation fluids. Stewart *et al.* (1994) related the influx of meteoric waters to the marine regression and uplift of the East Shetlands Platform at the end of the Palaeocene, thus creating a hydraulic head to drive the meteoric waters into the reservoir sands, where they mixed with the marine depositional fluids.

Late diagenetic illite is a minor clay cement phase and, as in the mudrocks, occurs as filamentous growths from detrital clay or mica plates or, less commonly, other detrital grains (Fig. 14). There is no evidence of the illite replacing kaolins as has been described in other studies, notably in the Jurassic reservoir sands of the North Sea. The relatively low abundance of the filamentous illite is unlikely to have a significantly deleterious effect on the permeability of the reservoir sands.

Green pelletal or pore-filling glauconitic clays also are often observed in the Palaeocene sands, but only in very small amounts. These are likely to have formed at or near the sediment–water interface during very early diagenesis, but it also is possible that the pelletal glaucony grains may be second-cycle and have been reworked after their formation.

INTERPRETATIONS

Apart from the glaucony, all the diagenetic clay minerals present in the sandstones are also found in the mudrocks. The sandstones and mudrocks have very similar paragenetic sequences for the clay and non-clay mineral diagenetic phases and show similar mineral associations and habits. This is well illustrated by the illite cements, which are the last significant diagenetic phase to form in both the sandstones and the mudrocks. In both lithotypes the illite grows as fibrous growths on clastic grains, principally platy I/S particles, micas, kaolins, quartz and feldspar. It is more abundant in the mudrocks, where the greater abundance of the detrital I/S could be a significant controlling factor. Small *et al.* (1992) showed experimentally that fibrous illite formation was favoured by rapid precipitation from fluids with high Si and K concentrations and organically complexed Al, heated to temperatures of 100–120°C, which could cause the breakdown of the Al complexes and allow the Al to be incorporated in the illite. However, there is no direct evidence for the presence of organic complexing of Al in oilfield formation waters (R.H. Worden, personal communication, 2001). The maximum burial temperature of the Everest Complex samples is 110°C and thus within the range for late-stage diagenetic formation in accordance with the experimental conditions described by Small *et al.* (1992). This formation process would require the presence of organic acids, which could be derived from the mudrocks, and thus indicate the need for the influx of fluids from the mudrocks to allow authigenic illite formation in the sandstones. The components for authigenic illite formation in the mudrocks could be wholly internally derived.

These observations are broadly in agreement with those of Huggett (1996) in her study of interbedded Palaeocene mudrock–sandstone sequences from the central North Sea. However, the general increase with burial depth in the abundances of authigenic illite and chlorite in the sandstones that Huggett (1996) observed has not been found in this study.

The pattern of clay-mineral diagenesis in the mudrocks is thus generally seen to be very similar to that in the sandstones in both style and abundance. This would further suggest that similar diagenetic processes affected the mudrocks and sandstones, requiring similar fluid–rock interactions to have occurred in both lithotypes. Huggett (1996) interpreted the broad similarities in the diagenesis of the Palaeocene sandstones and mudrocks she studied to be indicative of internally sourced components for the diagenetic reactions in the two lithotypes, with little influence of cross-lithology formation fluid flow on diagenesis. An alternative view could be that the mudrocks were as chemically open as the sandstones and thus underwent the same diagenetic reactions at similar times.

To test these hypotheses for internal sources versus 'open' chemical systems, we need to look in more detail at the non-clay mineral diagenesis in the mudrocks and sandstones. The most abundant of the non-clay cements is Fe-calcite, which forms pervasive poikilotopic cements in the sandstones and destroys reservoir potential. It is also common in the mudrocks as pore-filling cements, but also occurs infilling fractures, cone-in-cone structures and sandstone injection features. The similarities in the distribution of this cement suggest that similar cementation processes ocurred in both lithotypes. The stable isotope analyses of the Fe-calcite cements from the sandstones show $\delta^{18}O$ (PDB—a standard based on a Jurassic belemnite from the Pee Dee Formation) values ranging from −10 to −12‰ and $\delta^{13}C$ (PDB) values from 0 to −6‰. The isotopic compositions of the Fe-calcite cements hosted in the mudrocks show similar $\delta^{13}C$ (PDB) values (0.6 to −5.1‰) but greater variability in the $\delta^{18}O$ (PDB) values (−2.3 to −13‰). However, most of the Fe-calcite cements hosted in the mudrock samples have $\delta^{18}O$ (PDB) values in the range −8 to −13‰, indicating sources and timing of the calcite cements in the mudrocks similar to those in the sandstones. Interpretations of the stable isotope data indicate that the bicarbonate source for the Fe-calcite cements was from dissolution and recrystallization processes in the underlying Cretaceous limestones, Tertiary calcareous mudrocks and limestones, mixed with thermal decarboxylation of organic matter in the mudrocks, i.e. external to the sandstones and, in part, to the mudrocks (Conybeare & Shaw, 2000). This would imply migration of fluids from the mudrocks into the sandstones and also significant influx of fluids into the mudrocks along fractures and with the intrusion of fluidized sands. Conybeare & Shaw (2000) suggested that fracturing and sand injection was related to overpressure release, with an accompanying fluid flux from which calcite cements precipitated.

The calculated formation temperatures for the Fe-calcite cements range from 48 to 62°C (Conybeare & Shaw, 2000), which implies that Fe-calcite cementation occurs before kaolinite cementation. However, there is petrographic evidence that Fe-calcite also formed at the same time as the kaolinite (Fig. 3). It therefore is probable that kaolinite cement formation occurred in a similar fluid flux environment from which the Fe-calcite cements were generated, albeit under a different temperature regime.

In contrast to the similarity in cement types in the mudrocks and sandstones, the pattern of quartz cementation and detrital quartz grain dissolution shows distinct differences between the two lithotypes. In the sandstones, quartz overgrowth cements are most abundant in clean coarse sands, whereas they are rarely found in the mudrocks, where limited quartz cementation is most commonly in the form of finely crystalline silica. However, evidence of clastic quartz grain dissolution is observed to a greater extent in the mudrocks than the sandstones. This could suggest that the mudrocks are a source of silica for the quartz cements in the sandstones, although there is no statistical correlation between the degree of quartz cements in the sandstones and the sandstone : mudrock ratio.

In both the sandstones and mudrocks there is a significant lithofacies control on diagenetic processes. The amount of organic matter is a dominant control on the distribution of pyrite. In the mudrocks, the early diagenetic pyrite is most abundant in those mudrocks with the higher total organic carbon (TOC) values (> 2%) that were deposited under restricted bottom-water conditions with a high organic supply and lack of bioturbation favouring organic matter preservation. There also is evidence of greater clastic grain dissolution in such mudrocks, which suggests that the dissolution processes are mediated by the organic matter. In the sandstones, pyrite is preferentially hosted in the organic and clay-rich laminae and is absent from the clean coarse sands.

Authigenic kaolin and illite are more abundant in the finer grained micaceous sandstones, reflecting the association of the authigenic kaolins and illites with micas and illite–smectites.

The development of quartz overgrowth cements is controlled by primary porosity and permeability and the inhibiting influence of grain-coating and pore-filling matrix clays. Consequently, quartz overgrowths are best developed in the coarser clean sandstones and, within individual facies, the distribution of overgrowths correlates with the higher primary permeability zones.

The coarser mudrock facies contain higher proportions of detrital micas and feldspars and appear to be more permeable than the finer distal-facies mudrocks. In the proximal fan facies there is a greater sandstone : mudrock ratio and the mudrocks are coarser grained than in the distal facies associations. This provides the potential for more open diagenetic systems in the proximal facies associations and a greater degree of interchange of fluids between the coarser mudrocks and sandstones than in the distal facies. This manifests itself in the greater degree of cementation and grain dissolution in the coarser proximal mudrocks. For example, there is up to 30% authigenic kaolinite in the coarser mudrock facies that cannot have been sourced internally by processes such as dissolution of feldspars, judging from the petrological evidence available. There generally is less than 10% feldspar in the mudrocks and if the feldspars were the principal source of the kaolinite, this would require a maximum of up to 70% feldspar to originally have been present in the mudrocks, assuming a closed system (Giles & de Boer, 1990).

DISCUSSION

Based on the diagenetic processes described above, it is possible to reconstruct the pathways by which diagenetic fluids might have accessed the sandstones and mudrocks to produce the observed diagenetic reactions in the Tertiary Everest Complex sequences. These pathways are summarized in Fig. 15. Connectivity within the formations and across the stratigraphy is required to explain the similar diagenetic features observed in the sandstones and mudrocks, carbonate cemented fractures and the presence of hydrocarbon fluid inclusions in quartz cements throughout the Everest Complex Formations. Within the sandstones, sedimentary structures and textures exert important localized controls on poroperm properties and cause focusing of fluid flow. Similar facies controls are also observed in the mudrocks, with the coarser mudrocks providing potentially more permeable pathways than the finer grained mudrocks.

Quantitative mineralogy and mass-balance calculations for the sandstones indicate that they were not geochemically closed and that

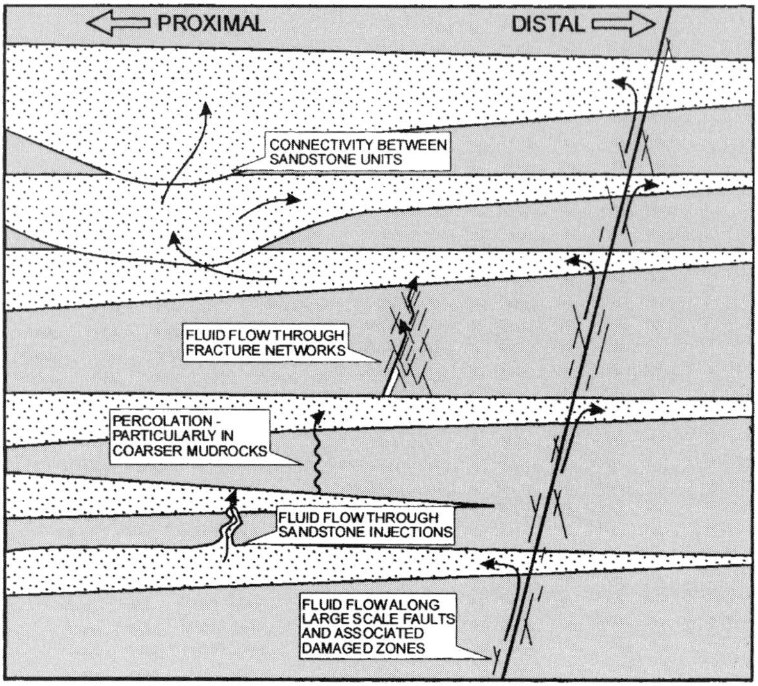

Fig. 15 Potential sources and pathways of formation fluids in the Everest Complex.

diagenetic components were, at least in part, externally sourced. However, there is no association between the diagenetic phases in the sandstones and proximity to mudrock horizons or correlations with the sandstone : mudrock ratio that might imply that the mudrocks are the principal source. Difficulties in quantifying the mineralogy and making mass-balance calculations for the mudrocks make it more difficult to interpret to what extent the mudrocks behaved as open or closed systems during diagenesis. In the coarser grained mudrocks, the relatively large quantities of authigenic kaolinite (estimated at up to 30%) would be difficult to source internally, suggesting that these mudrocks at least behaved to some extent as open systems. It is suggested that the mudrocks are likely to have displayed varying degrees of openness, depending on their permeability, thickness and composition.

Several other recent studies of interbedded sandstone–shale sequences (Milliken, 1992; Greenwood *et al.*, 1994; Jubril *et al.*, 1998) have demonstrated that extensive and varied products of diagenesis are present in the mudrocks, which would require significant pore-fluid–rock interactions similar to those found in the adjacent sandstones. The conventional view of shales being essentially impermeable thus appears to be challenged by these observations and poses fundamental questions concerning fluid movement in mudrocks.

In a detailed study of feldspars in sandstone–shale sequences from the Oligocene Frio Formation, Milliken (1992) concluded that the feldspars in the mudrocks underwent progressive alteration with depth via dissolution/precipitation mechanisms similar to feldspars in sandstones. The rate of feldspar alteration reactions was found to be somewhat slower in the mudrocks than the adjacent sandstones, but not dramatically so.

Milliken (1992) found evidence of textures in the mudrocks that might suggest diffusional grain boundary reactions that were absent in the sandstones. There was also evidence that the degree of 'open' system behaviour in the mudrocks was significantly greater in the siltstones than the claystones, with transport likely to be at least over hundreds of microns in the siltstones.

From the data of Milliken (1992) it is suggested that the reactions in the mudrocks were coincident in time and space with those in the sandstones, although they are slightly delayed by an indeterminate amount in the mudrocks. Similarities in the paragenetic sequences for the mudrocks and sandstones from several studies also might imply that the diagenetic events in the two lithotypes are essentially synchronous. However, measured ages of the authigenic phases (e.g. illites) in the mudrocks and sandstones are needed to confirm this potential coincidence of diagenetic events. Greenwood *et al.* (1994) did attempt to date authigenic illite in interbedded Middle Jurassic sandstones and mudrocks. They reported much older ages for the illites in the mudrocks (84–97 Ma) compared with the sandstones (48 Ma). Other measured ages of authigenic illites in Jurassic mudrocks from the North Sea have been made by Glasmann *et al.* (1989) and Burley & Flisch (1989), who reported dates of 106–31 Ma and 153–68 Ma, respectively, and generally much older than the comparable ages for the authigenic illites in the sandstones (55–30 Ma). Based on these ages, illites in the mudrocks appear to have formed earlier during the burial history than the sandstones, when the mudrocks are likely to have undergone less compaction and perhaps been relatively more permeable. However, there needs to be a cautious use of these age data as it is impossible to be certain that there is no contamination of the analysed illite separates by detrital illites and micas. Even a small amount of such contaminants would significantly affect the measured ages. For the Palaeocene Everest Complex samples it proved impossible to adequately separate out authigenic illites from the mudrocks or the sandstones without the risk of extensive detrital mica and illite contamination. An added complication is that detailed examination of filamentous illite cements in sandstones has demonstrated several generations of illite growth rather than one event (Berry, 1995). The question of the relative ages of similar diagenetic events in the mudrocks compared with the adjacent sandstones therefore must remain open until more reliable and extensive data are collected.

Even if similar diagenetic events occurred earlier in the mudrocks, many of them appear to have taken place after significant burial compaction. Consequently the matrix permeability of the mudrocks would be expected to be reduced significantly and thus limit the access of fluids through the mudrocks, and restrict fluid–rock diagenetic reactions. The petrological data would seem to require fluid-mediated migration of chemical components in the mudrocks that challenge conventional opinions about fluid flow in buried compacted mudrocks. How then are these observed extensive diagenetic reactions occurring in the mudrocks; how do the diagenetic fluids access the 'impermeable' mudrocks?

One answer in part would be diffusive grain boundary processes as suggested by Milliken (1992), but this alone would not appear to be sufficient to be responsible for the extensive and varied diagenetic process that are often observed in mudrocks. Capuano (1993) proposed that the flow of fluids in the mudrocks via microfracture permeability rather than matrix permeability could support flow comparable to that in sandstones. Flow through such microfractures in mudrocks previously has been cited to explain hydrocarbon migration from source rocks, quartz cementation in sandstones (Land, 1984), dissolution of feldspars and transfer of potassium to mudrocks for illitization reactions (Awwiller, 1993). The release of overpressures in mudrocks are considered to be a major trigger for microfracturing and could allow formation fluids to be released from the mudrocks (Cartwright, 1994a,b; Lonergan *et al.*, 1998a,b). Bjørlykke & Hoeg (1997) questioned whether the fluid flux generated by such overpressure release would make a significant contribution to the fluid flow required for diagenetic processes. However, overall there is considerable evidence that significant fluid flow can occur along fractures in brittle mudrocks that dilate during deformation and allow fracture permeability to develop (Dewhurst *et al.*, 1999; Ingram & Urai, 1999).

There remain considerable gaps in our knowledge of fluid flow within and across mudrocks that have important consequences for understanding fluid migration and diagenetic processes in sedimentary basins. Too often fluid migration and palaeohydrologic basin models have been based on studies of sandstones alone and have paid little or no detailed attention to the role of associated mudrocks. Thus there is a need for more integrated studies of mudrocks and sandstones, such as described here.

ACKNOWLEDGEMENTS

The authors would like to acknowledge the contribution of both data and funding from the Everest Group (operated at the time by Amoco (UK) Exploration Company) which made this study possible. Within the Everest Group particular thanks are due to Peter Thompson, Robin Crawford and Greig Cowan. Martin Gill and Dick Giddens of Imperial College are thanked for their assistance with XRD and SEM analyses, respectively. Tony Fallick and staff at SURRC, and David Mattey and staff at the University of London stable isotope facility provided guidance and assistance with the stable isotope analyses.

An earlier version of the paper benefited significantly from reviews by Kitty Milliken, Ghazi Kraishan and the editorial input of Richard Worden.

REFERENCES

Awwiller, D.N. (1993) Illite/smectite formation and potassium mass transfer during burial diagenesis of mudrocks: a study from the Texas Gulf Coast Palaeocene–Eocene. *J. Sed. Petrol.*, **63**, 501–512.

Berry, J.S. (1995) *Diagenesis of Westphalian sediments from the southern North Sea gas basin.* Unpublished PhD thesis, University of London, 359 pp.

Bjørlykke, K. & Hoeg, K. (1997) Effects of burial diagenesis on stress, compaction and fluid flow in sedimentary basins. *Mar. Petrol. Geol.*, **14**, 267–276.

Burley, S.D. & Flisch, M. (1989) K–Ar geochronology and the timing of detrital I/S clay illitisation and authigenic illite precipitation in the Piper and Tartan Fields, Outer Moray Firth. *Clay Mineral.*, **24**, 285–315.

Capuano, R.M. (1993) Evidence of fluid flow in microfractures in geopressured shales. *AAPG Bull.*, **77**, 1303–1314.

Cartwright, J.A. (1994a) Episodic basin wide fluid expulsion from geopressured shale sequences in the North Sea basin. *Geology*, **22**, 447–450.

Cartwright, J.A. (1994b) Episodic basin-wide hydrofracturing of overpressured Early Cenozoic mudrock sequences in the North Sea Basin. *Mar. Petrol. Geol.*, **11**, 587–607.

Conybeare, D.M. (1996) *Diagenesis and reservoir quality in the Everest Complex, North Sea.* Unpublished PhD thesis, University of London, 348 pp.

Conybeare, D.M. & Shaw, H.F. (2000) Fracturing, overpressure release and carbonate cementation in the Everest Complex, North Sea. *Clay Mineral.*, **35**, 139–153.

Dewhurst, D.N., Yang, Y. & Aplin, A.C. (1999) Permeability and fluid flow in natural mudstones. In: *Muds and Mudstones: Physical and Fluid Flow Properties* (Ed. by A.C. Aplin, A.J. Fleet and J.H.S. Macquaker). *Geol. Soc. London Spec. Publ.*, **158**, 23–43.

Dickson, J.A.D. (1966) Carbonate identification and genesis as revealed by staining. *J. Sed. Petrol.*, **36**, 491–505.

Giles, M.R. & de Boer, R.B. (1990) Origin and significance of redistributional secondary porosity. *Mar. Petrol. Geol.*, **7**, 378–397.

Glasmann, J.R., Larter, S., Breidis, N.A. & Lundegard, P.D. (1989) Shale diagenesis in the Bergen High Area, North Sea. *Clay. Clay Mineral.*, **37**, 97–112.

Greenwood, P.J., Shaw, H.F. & Fallick, A.E. (1994) Petrographic and isotopic evidence for diagenetic processes in Middle Jurassic sandstones and mudrocks from the Brae Area, North Sea. *Clay Mineral.*, **29**, 637–650.

Helgeson, H.C., Delaney, J.M., Nesbitt, H.W. & Bird, D.K. (1978) Summary and critique of the thermodynamic properties of rock forming minerals. *Am. J. Sci.*, **278A**, 1–229.

Huggett, J.M. (1995) Formation of authigenic illite in Palaeocene mudrocks from the central North Sea: a study by high resolution electron microscopy. *Clay. Clay Mineral.*, **41**, 682–692.

Huggett, J.M. (1996) Aluminosilicate diagenesis in a Tertiary sandstone–mudrock sequence from the central North Sea, UK. *Clay Mineral.*, **31**, 523–536.

Huggett, J.M. & Shaw, H.F. (1997) Field emission scanning electron microscopy—a high resolution technique for the study of clay minerals in sediments. *Clay Mineral.*, **32**, 197–204.

Imam, M.B. & Shaw, H.F. (1987) Diagenetic controls on the reservoir properties of gas bearing Neogene Surma

Group sandstones in the Bengal Basin, Bangladesh. *Mar. Petrol. Geol.*, **4**, 103–111.

Ingram, G.M. & Urai, J.L. (1999) Top seal leakage through faults and fractures: the role of mudrock properties. In: *Muds and Mudstones: Physical and Fluid Flow Properties* (Ed. by A.C. Aplin, A.J. Fleet and J.H.S. Macquaker). *Geol. Soc. London Spec. Publ.*, **158**, 125–136.

Knox, R.W.O'B. & Holloway, S. (1992) Palaeogene of the central and northern North Sea. In: *Lithostratigraphic Nomenclature of the UK North Sea* (Ed. by R.W.O'B. Knox and W.G. Cordey), British Geological Survey, Keyworth, 139 pp.

Jubril, M.A., Shaw, H.F. & Fallick, A.E. (1998) Stable isotope and geochemical evidence of formation pore fluid evolution during diagenesis of Tertiary sandstones and mudrucks of the Niger Delta. *J. Afr. Earth Sci.*, **27**, 417–435.

Lambert, S.J. & Epstein, S. (1980) Stable isotope investigation of an active geothermal system in Valles Caldera, Jemez Mountains, New Mexico. *J. Volcanol. Geotherm. Res.*, **8**, 111–129.

Land, L.S. (1984) Frio sandstone diagenesis, Texas Gulf Coast; a regional isotopic study. In: *Clastic Diagenesis* (Ed. by D.A. McDonald and R.C. Surdam). *AAPG Mem.*, **37**, 37–62.

Land, L.S. & Dutton, S.P. (1978) Cementation of a Pennsylvanian deltaic sandstone: isotopic data. *J. Sed. Petrol.*, **48**, 1167–1176.

Lonergan, L., Cartwright, J.A. & Jolly, R. (1998a) The geometry of polygonal fault systems in Tertiary mudrocks of the North Sea. *J. Struct. Geol.*, **20**, 529–548.

Lonergan, L., Cartwright, J.A., Laver, R. & Staffurth, J. (1998b) Polygonal faulting in the Tertiary of the central North Sea: implications for reservoir geology. In: *Structural Geology in Reservoir Characterisation* (Ed. by M.P. Coward, T.S. Daltaban and H.D. Johnson). *Geol. Soc. London Spec. Publ.*, **127**, 191–207.

Mackinnon, I.D.R. (1990) Introduction to electron beam techniques. In: *Electron Optical Methods in Clay Science* (Ed. by I.D.R. Mackinnon), Workshop Lectures 2, Clay Minerals Society, Boulder, CO, pp. 2–13.

Milliken, K.L. (1992) Chemical behaviour of detrital feldspars in mudrocks versus sandstones, Frio Formation (Oligocene), South Texas. *J. Sed. Petrol.*, **62**, 790–801.

O'Connor, S.J. & Walker, D. (1993) Palaeocene reservoirs of the Everest trend. In: *Proceedings 4th Conference on Petroleum Geology of North West Europe* (Ed. by J.R. Parker), Geological Society of London, pp. 145–160.

Pettijohn, F.J., Potter, P.E. & Siever, R. (1973) *Sand and Sandstone*. Springer-Verlag, Berlin, 618 pp.

Reynolds, R. (1980) Interstratified clay minerals In: *Crystal Structures of Clay Minerals and their X-ray Identification*, Mineralogical Society of Great Britain and Ireland Monograph 5 (Ed. by G.W. Brindley and G. Brown), London, pp. 249–304.

Schultz, L.G. (1964) Quantitative interpretation of mineralogical composition from X-ray and chemical data for the Pierre Shale. *US Geol. Surv. Prof. Pap.*, **391-C**, 31 pp.

Shaw, H.F. (1972) Preparation of oriented clay mineral specimens for X-ray diffraction analysis by a suction-onto-ceramic-tile method. *Clay Mineral.*, **3**, 349–350.

Small, J.S., Hamilton, D.L. & Habesch, S. (1992) Experimental simulation of clay precipitation within reservoir sands 2: mechanisms of illite formation and controls on morphology. *J. Sed. Petrol.*, **62**, 520–529.

Stewart, R.N.T., Fallick, A.E. & Haszeldine, R.S. (1994) Kaolinite growth during porewater mixing: isotopic data from Palaeocene sands, North Sea, UK. *Clay Mineral*, **29**, 627–636.

Thompson, P.J. & Butcher, P.D. (1991) The geology and geophysics of the Everest Complex. In: *Generation, Accumulation and Production of Europe's Hydrocarbons*, Special Publication of the European Association of Petroleum Geoscientists (Ed. by A.M. Spencer), Oxford, pp. 89–98.

Int. Assoc. Sedimentol. Spec. Publ. (2003) **34**, 147–160

Cross-formational flux of aluminium and potassium in Gulf Coast (USA) sediments

MARK WILKINSON[1], R. STUART HASZELDINE[1] *and* KITTY L. MILLIKEN[2]

[1] *Department of Geology and Geophysics, Grant Institute, Kings Buildings, University of Edinburgh, Edinburgh EH9 3JW, UK, e-mail: m.wilkinson@glg.ed.ac.uk*

[2] *Department of Geological Sciences, University of Texas at Austin, Austin, Texas 78712, USA, e-mail: kittym@mail.utexas.edu*

ABSTRACT

The diagenetic dissolution of minerals to form net porosity in oilfield sandstones requires export of major-element constituents. Mass-balance must exist at some size scale, however, this is difficult to identify owing to the depositional variability within sedimentary formations. We have taken sets of whole-rock geochemical data from the Frio Formation of the USA Gulf Coast, for both sandstones and enclosing shales. Graphs of 'mobile'/'immobile' element ratios show a statistically significant decrease of K (mobile) with depth for sandstones, and a corresponding relative increase in Ti, Zr and Hf (immobile) in shales. Data for Al are also suggestive of a net flux from sandstones to shales, although of marginal statistical validity. Assuming that the sediment had an initially uniform chemical composition, then K and possibly Al are exported from sandstones into shales, despite the common assumption of Al immobility. There is insufficient export of K from sandstones to account for the import of K into shales, so another source of K must exist. However, there is probably Al conservation within the Frio Formation, as the calculated outflux from the sands and the influx from the shales approximately balance. More data are required to conclusively settle the question of Al mobility. Aluminium export potentially can preserve permeability within sandstones by reducing the volume of authigenic clay minerals precipitated.

INTRODUCTION

Clay minerals are important authigenic cements within many sandstones, and are considered to originate by recrystallization of detrital clays, or by the precipitation of material derived from the dissolution of unstable detrital minerals. Many detrital minerals (feldspars, micas) are metastable under surface conditions or at diagenetic temperatures and pressures. Increasing temperature during progressive burial increases the rate of breakdown reactions, so that the detrital minerals dissolve and liberate their constituent ions, which react to form authigenic minerals. The products of these reactions are frequently clay minerals, although carbonates, quartz, feldspars and zeolites are also important. Although it is clear that at least some of the components of authigenic clay minerals are sourced within the sandstones themselves by the dissolution of detrital phases, there has been debate as to the extent of material flux between the sandstones and the surrounding shales. This is of importance, because shales are volumetrically dominant in the majority of sedimentary basins, and the dissolution of detrital smectite to form authigenic illite and possibly many other minerals could be the most important silicate reaction in some basins (Boles & Franks, 1979; Lynch, 1997).

The debate about the mobility of elements during sandstone diagenesis can be considered as two opposing points of view: open-system versus closed-system. For example, it has been reported that silicon is imported into sandstones from shales (e.g. Lynch, 1996) and also that this is impossible (e.g. Bjørlykke, 1997). Geochemical mass-balances of shale sequences can show either import or export of potassium and other major elements from different examples (Gluyas & Coleman, 1992; Wintsch & Kvale, 1994). There is similar discussion concerning the mobility, or otherwise, of aluminium. Advocates of aluminium transport present empirical evidence for feldspar dissolution as a key observation, and claim that, with respect to aluminium, some sandstones contain insufficient authigenic minerals to balance the estimated quantity of dissolved feldspars (e.g. Milliken *et al.*, 1989; Wilkinson & Haszeldine, 1996). Hence, the aluminium must have left the sandstones. Sceptics point to the very low solubility of aluminium under subsurface conditions, with limited volumes of compactional water, and deduce that transport of significant quantities of aluminium is impossible over even a thin-section scale (e.g. Bjørlykke, 1979; Giles & De Boer, 1990). Hence, it is sometimes argued that any supposed observations indicating aluminium transport must be reinterpreted to fit the theoretical predictions.

Advocates of aluminium transport admit the problems of low aluminium solubility, but appeal to organic acids or CO_2. These are claimed to be abundant in the subsurface and to increase Al solubility sufficiently to enable transport to take place (Pittman & Lewan, 1994, and references therein). The occurrence and role of organic acids has become as controversial as the original secondary porosity debate, with conflicting evidence and workers in entrenched positions (see references within Pittman & Lewan, 1994). It is not the aim of the present paper to discuss the solubility issue itself, rather we present geochemical evidence from the USA Gulf Coast, which we interpret in favour of aluminium transport, and leave speculation as to the solubility conditions and mechanism of transport to others.

In order to assess the quantity of feldspar dissolution that has occurred in a sediment, it is necessary to estimate the original sediment composition. This can be done either by measuring secondary porosity after feldspar dissolution by standard petrographic techniques, or by assuming an initially uniform sandstone composition within a formation and comparing the compositions of deeply buried samples with more shallowly buried equivalents. The two approaches frequently lead to radically different conclusions. A common finding is that the percentage of point-counted secondary porosity is small, and approximately constant over the depth range for which data are available (e.g. a constant 3% below 2 km in fig. 5 of Bjørlykke *et al.* (1992) for the Brent Group of the northern North Sea). This can be interpreted as showing that there is little feldspar dissolution at these depths. However, Wilkinson *et al.* (1997) proposed that small values of visible secondary porosity are not reliable indicators of the cumulative quantity of feldspar dissolved. They suggested that feldspar has continually dissolved during burial and that compaction has eliminated the diagnostic petrographic textures. Consequently estimates of the quantity of dissolved feldspar derived by comparing deep and shallow buried sandstones are frequently an order of magnitude greater than observed secondary porosity values (e.g. Milliken *et al.*, 1989, 1994; Wilkinson & Haszeldine, 1996). The greatest problem with this 'abundance with depth' method is that the assumption must be made that all the rocks sampled had an approximately uniform composition. Critics claim that changes in feldspar abundance with depth could be the result of changing provenance, rather than diagenesis (e.g. Oxtoby & Gluyas, 1997). However, a compilation of data from sandstones in several different basins shows that similar patterns of feldspar decline occur during burial, and thus that changing provenance is unlikely to be the answer (Wilkinson *et al.*, 2001).

The implication of Al transfer from sands to shales is that even though the dissolution of feldspar may occur during deep burial, the volume of clay minerals precipitated within the sand will be less than predicted on the basis of a closed-system model. This especially is important for 'hairy' illite, which significantly reduces sand permeability if present in even small quantities.

APPROACH

A debate exists concerned with the size scale of mass-balance within the diagenetic system. Is this on the scale of millimetres, metres, hundreds of metres or kilometres? This inherently is difficult to assess, because 'proving' export from a sandstone unit and import to a mudrock unit is confused by the large variability within the natural starting materials—influenced, for example, by grain size and hydrodynamic sorting between and within different facies. To tackle this problem, we have sought a location where analytical data exist for major elements and trace elements within both sandstones and mudrocks: the USA Gulf Coast. We specifically searched for evidence for mobility of major elements during burial, by selecting a suite of minor elements that generally are considered to be immobile (or less mobile) under diagenetic conditions. These are inferred to have remained constant in the host sandstone or mudstone. By contrast, mobile elements may have become enriched or depleted during burial, which is expressed as changes in element ratios relative to our chosen immobile elements.

GEOLOGICAL SETTING AND METHODS

Sediments of the Texas Gulf Coast form a thick, predominantly clastic succession, which has

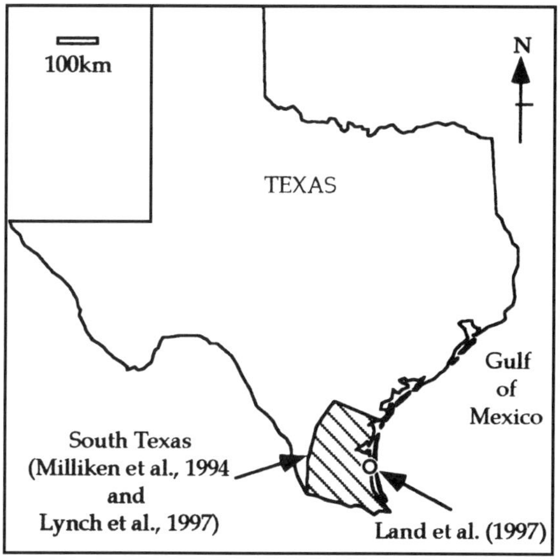

Fig. 1 Location map for geochemical data sets, Texas Gulf Coast, USA.

prograded into the Gulf of Mexico (Fig. 1). The area is a major oil province, so consequently the sediments have been studied extensively, and both the sedimentology and diagenesis are well documented (e.g. Land, 1984; Loucks *et al.*, 1984). According to some workers, the estimated quantity of mineral dissolution in sandstones exceeds the observed volumes of authigenic minerals (e.g. Milliken *et al.*, 1994). Therefore large-scale element transport during burial has been proposed. The extent of potassium feldspar dissolution within sandstones estimated by comparing sediments at different burial depths is significantly greater than the quantity of authigenic clays observed (Milliken *et al.*, 1989). Note that secondary porosity is not regarded as a reliable indicator of the total volume of feldspar dissolved owing to the effects of compaction (Wilkinson *et al.*, 1997).

There have been a number of unusually extensive geochemical studies of the Frio Formation (Table 1). Land *et al.* (1997) analysed shale cuttings from an upper Oligocene to Lower Miocene succession in Kennedy County, south Texas (Fig. 1). Their analysis included several 'immobile' elements, such that 'mobile'/

Table 1 Data sources and stratigraphy.

Reference	Age	Stratigraphy	Geographical location	Rock type	Sample type	Analytical method*
Milliken *et al.* (1994)	Oligocene	Frio Formation	South Texas Gulf Coast	Sandstone	Conventional core	ICP–AES
Land *et al.* (1997)	Upper Oligocene–Lower Miocene	Frio and ?Anahuac Formations	Kennedy County	Shale	Cuttings	XRD
Lynch *et al.* (1997)	Oligocene	Frio Formation	South Texas Gulf Coast (others omitted)	Shale (< 0.1 μm size fraction)	Conventional core	ICP (K by isotope dilution)

* ICP–AES, inductively coupled plasma–atomic emission spectroscopy; XRD, X-ray diffraction.

'immobile' ratios could be calculated as a function of depth. They concluded that shale diagenesis is an open-system process, and that potassium was imported into the shales. Land *et al.* (1997) found that Al_2O_3 correlated with TiO_2, and that the correlation line passed close to the origin, from which they concluded that aluminium was immobile. The Land *et al.* (1997) data are here re-examined specifically to test for the movement of aluminium during diagenesis.

Milliken *et al.* (1994) analysed samples from the Oligocene Frio sandstones of the southern Texas Gulf Coast (Fig. 1) by ICP–AES (inductively coupled plasma–atomic emission spectroscopy), to an estimated accuracy of ±5% for most elements. Samples are from conventional cores, prepared so as to eliminate contamination by drilling mud (Table 1). Sampling was carried out with the aim of minimizing variations in geographical area (and hence provenance), grain size and sorting, while obtaining a wide and even spread of sample depths. The analyses included a number of 'immobile' elements, of which Zr showed the least petrographic evidence of mobility. Milliken *et al.* (1994) concluded that substantial quantities of potassium and lesser quantities of aluminium had been lost from the sandstones during burial. Again, these data are re-examined below.

Chemical data on clay-mineral size fractions of Frio Formation shales are available, e.g. Hower *et al.* (1976) and Lynch *et al.* (1997). The Hower *et al.* (1976) data set shows clear trends in both Al_2O_3/Ti and K_2O/Ti with depth, but the accuracy of the analyses has been questioned by Lynch *et al.* (1997): the sample sizes available were too small for accurate analysis using the technology of the day. In addition, the Hower *et al.* (1976) data set is from north Texas, which has a substantially different provenance from the south Texas area sampled by Land *et al.* (1997) and Milliken *et al.* (1994). The data set of Lynch *et al.* (1997) hence is used here.

To assess the mobility of K and Al, these elements are ratioed to 'immobile' elements, The results are plotted as 'mobile'/'immobile' element ratios versus burial depth, and straight lines fitted using depth as the independent variable. All data are plotted as weight per cent oxide for major elements, and parts per million for trace elements. The two most shallow samples in the Milliken *et al.* data set are omitted from the correlations, as they are of significantly different facies from the remainder (Milliken *et al.*, 1994).

RESULTS

Common trends in element ratio versus depth are observed for both K and Al as 'mobile' elements, and Ti, Zr and Hf as 'immobile' elements (Table 2; Figs 3–12). The ratio of 'mobile' to 'immobile' elements systematically decreases with depth within the sandstones,

Table 2 'Mobile' to 'immobile' element correlations. Correlation coefficients (r^2) for mobile/immobile element (or oxide) ratios versus depth. Critical values at the 90% confidence limit are given in parentheses. Significant correlations are shown in italic.

		Immobile element or oxide		
	Mobile oxide	**TiO_2**	**Zr**	**Hf**
Sands (whole-rock)	K_2O	*0.57* (0.23)	*0.50* (0.23)	*0.48* (0.23)
(Milliken *et al.*, 1994)	Al_2O_3	0.11 (0.23)	*0.60** (0.23)	*0.52** (0.23)
Shales (whole-rock)	K_2O	*0.78* (0.14)	*0.70* (0.19)	*0.68* (0.19)
(Land *et al.*, 1997)	Al_2O_3	*0.25* (0.14)	*0.23* (0.19)	0.16 (0.19)
Shales (< 0.1 μm)	K_2O	*0.15* (0.15)		
(Lynch *et al.*, 1997)	Al_2O_3	0.10 (0.15)		

* Data from 2465 m omitted from correlation (see Figs 9 & 11).

and increases within the shales. The rate of change of the ratios with depth is greater for sands than for shales. Correlation coefficients generally are higher for shales than for sands (Table 2). Whereas the shale data correlations are mostly significant at the 90% confidence interval for both K_2O and Al_2O_3, owing to scatter in the sandstone data, the Al_2O_3/'immobile' trends are significant at the 90% confidence level only for Zr.

We have fitted straight lines through the data, in order to calculate correlation coefficients. This does not imply that the diagenetic fluxes of K and Al are linear with depth, but simply that this is the most convenient way of analysing the data. In reality, a plot of concentration of Al (or K) versus depth is most likely to be sigmoidal (Fig. 2) owing to the interplay of potential reaction rate and solute source. With the small number of available data points it is not possible to distinguish statistically between a sigmoidal and a linear trend with depth.

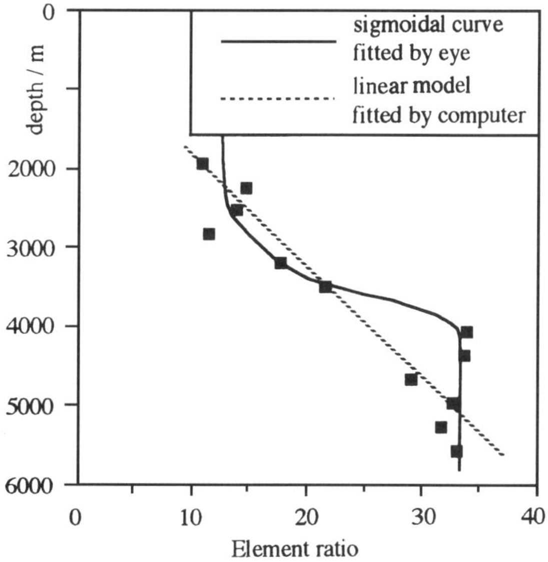

Fig. 2 Specimen data set showing a probable sigmoidal trend of element-ratio with depth (line fitted by eye) and also a linear model (i.e. a straight line). Although the sigmoidal trend is more realistic in terms of reaction kinetics, the small number of data points available do not justify fitting a sigmoidal curve by attempting to model element fluxes. Instead, we have used linear models to assess whether there are significant changes in element-ratios with depth.

DISCUSSION

Element 'immobility'

We have assumed that the elements Ti, Zr and Hf are 'immobile', following the literature on igneous and metamorphic geochemistry (Table 3). The geochemical conditions of sub-solidus cooling and metasomatism, and low-temperature metamorphism, are assumed to be sufficiently similar to those of diagenesis for this to be valid. For sedimentary systems, Land *et al.* (1997) studied element mobility during shale diagenesis within the Gulf Coast, and identified Ti, Zr and Hf as immobile, as well as Al and the heavy rare earth elements (REEs).

Table 3 Elements identified as immobile during alteration processes.

Reference	Elements
Floyd & Winchester (1975)	Ti, P, Zr, Y, Nb
Pearce (1975)	Ti, Zr, Y, Nb ± REE
Wood *et al.* (1979)	Zr, Nb, Ti, Y, P ± middle and heavy REEs
Merriman & Roberts (1990)	Hf, Nb, Ta, Th, Y, Zr + REEs
Van Baalen (1993)	Ti, Zr, Hf, Nb, Ta ± Th

This is discussed below. For Frio Formation sandstone diagenesis, Milliken *et al.* (1994) calculated a small loss in Al during burial, but concluded that the case for Al-mobility was unproven. They used Ti, Zr and REE as 'immobile'.

From Table 3 it is apparent that Ti, Zr and Hf are widely regarded as immobile by geochemists, despite petrographic evidence for mobility during diagenesis on a millimetre scale, i.e. the occurrence of authigenic anatase, titanite and sphene within deeply buried sandstones (e.g. van Panhuys-Sigler & Trewin, 1990). The apparent mobility of an element is dependant on the scale of observation. Clearly, all elements are mobile on an atomic scale, and are immobile on some larger scale. For the purposes of this paper, an 'immobile' element is one displaying closed-system behaviour on the scale of sampling (centimetres for sandstones, metres for shales), whereas open-system behaviour would define a 'mobile' element.

From Table 3 it is clear that the REEs are potentially useful as 'immobile' elements during diagenesis. However, whereas Ti, Zr and Hf show increasing concentration within the shales with depth, the pattern for the REEs are more complex. The heavy REEs show increasing concentrations with depth, although their ratios relative to each other (and the other 'immobiles') change subtly. The light REEs (LREEs) show decreasing concentrations with depth (Fig. 5). Awwiller & Mack (1991) described evidence for the possible mobility of Nd and Sm within the Frio and Wilcox formations of the Texas Gulf Coast, although in a later study, Awwiller (1994) concluded that there was little evidence for REE redistribution on anything more than a local scale. Mobility of REE during diagenesis also has been recorded within sandstones of the European continental shelf, associated with the growth of authigenic francolite (e.g. Bouch *et al.*, 1995), and there is some evidence that fractionation can occur between the heavy REEs (HREEs) and the LREEs during authigenic mineral growth (Bouch *et al.*, 1997). As the data from the Frio shales are inconsistent with REE conservation, we conclude that the REEs are lost from the shales during burial, with the LREEs being most mobile and the HREE the least mobile. This pattern of REE mobility is consistent with that found in some studies of subsolidus metasomatic alteration of igneous rocks (Table 3).

Potassium transfer

Figure 3 shows K_2O/TiO_2 versus burial depth for whole-rock sandstones (Milliken *et al.*, 1994) and shales (Land *et al.*, 1997). The sandstones show a decrease in K_2O/TiO_2 ratio with depth over the interval 2000–4200 m, whereas the shales have a small but significant increase in the ratio over the same depth interval ($r^2 = 0.78$, significant at the 99% confidence level). The decrease in K_2O/TiO_2 ratio for sands is almost entirely the result of a decrease in K_2O, as TiO_2 is almost constant with depth (Fig. 4). For shales, both K_2O and TiO_2 vary with depth, an increase in TiO_2 being the result a loss of calcite and Si (Fig. 5; Land *et al.*, 1997). Hence the increase in K_2O/TiO_2 ratio is the result of a very significant K_2O import.

Figure 6 shows chemical data for the < 0.1 μm size fraction of shales from the south Texas area, which consists predominantly of illite–smectite (90–100%, Lynch *et al.*, 1997). For this clay-mineral fraction, it is apparent that TiO_2 (assumed immobile) is approximately constant, such that the increase in K_2O represents import. It is also apparent that the TiO_2 data has a high degree of scatter, which may result from primary variation, diagenetic remobilization, or the

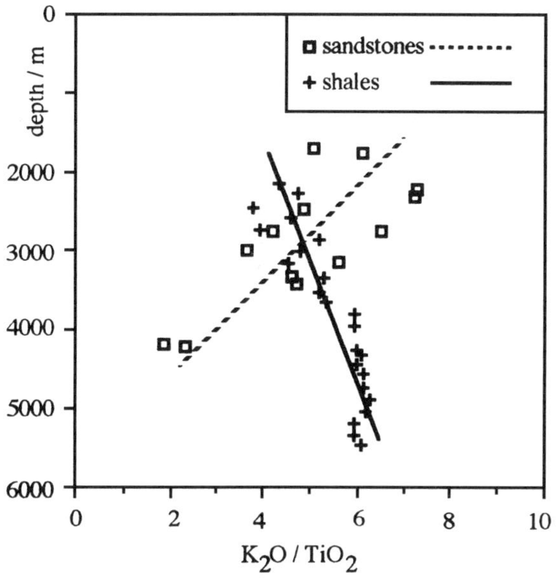

Fig. 3 K_2O/TiO_2 decreases over the interval 2000–4400 m for sandstones, which implies that potassium is being exported (assuming that Ti is immobile). Over the same depth interval, K_2O/TiO_2 increases slightly for whole-rock shales. Assuming that Ti is immobile, this indicates import of K from sandstones into the shales. We infer that potassium is transported from the sandstones into the shales.

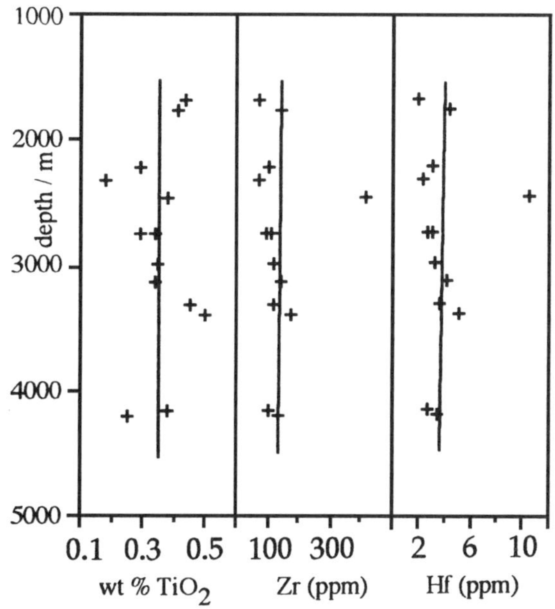

Fig. 4 Concentrations of immobile elements (TiO_2, Zr and Hf) with depth for sandstones, with correlation lines shown. All are approximately constant.

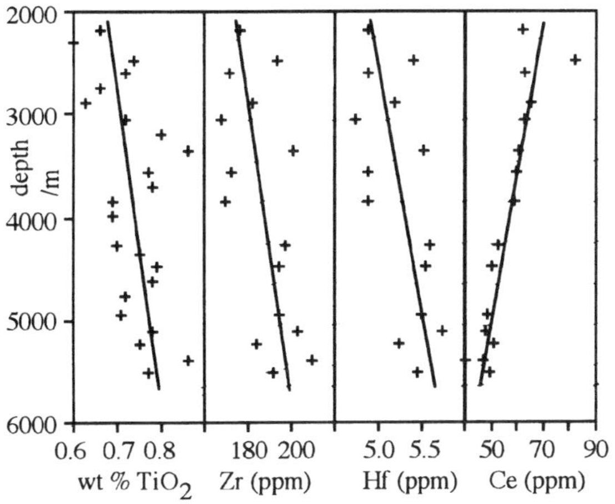

Fig. 5 Concentrations of immobile elements (TiO_2, Zr and Hf) with depth for shales, with correlation lines. All show an increase with depth owing to loss of calcite and other components from the rocks. Note that Ce shows a decrease with depth, and consequently must be considered to be mobile, as must the other REEs.

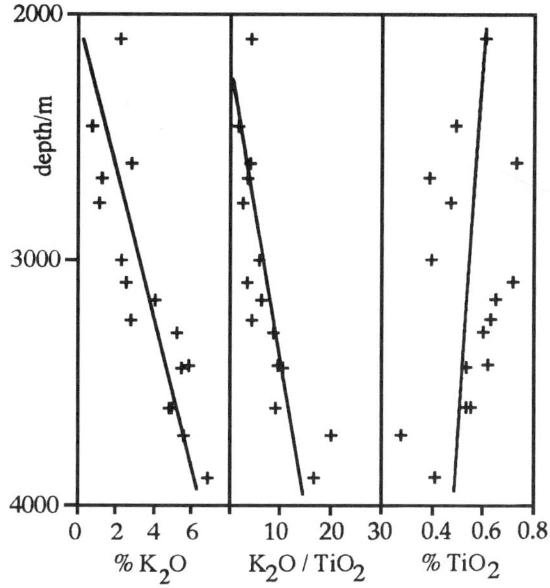

Fig. 6 K and Ti data for the clay-mineral fraction of south Texas shales. Both K_2O and K_2O/TiO_2 show an increase with depth, interpreted as K import and incorporation into illite–smectite during illitization.

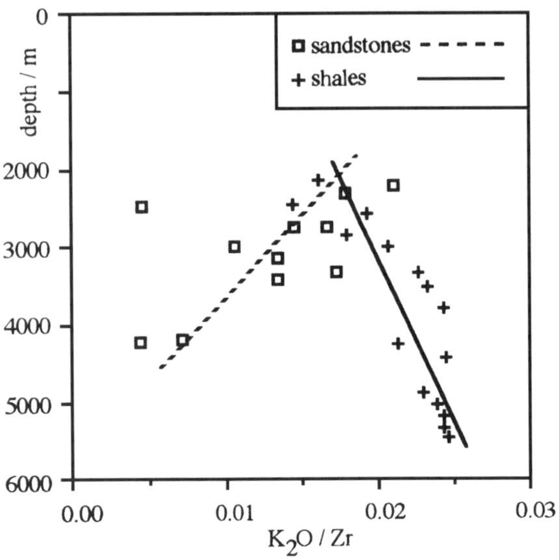

Fig. 7 Change in whole-rock K_2O/Zr for sandstones and shales with depth. The ratio K_2O/Zr decreases over the interval 2000–4400 m for sandstones, which implies that K is being exported (assuming that Zr is immobile). Over the same depth interval, K_2O/Zr increases for shales, implying K import. This supports the hypothesis that potassium is transported from the sandstones into the shales during burial.

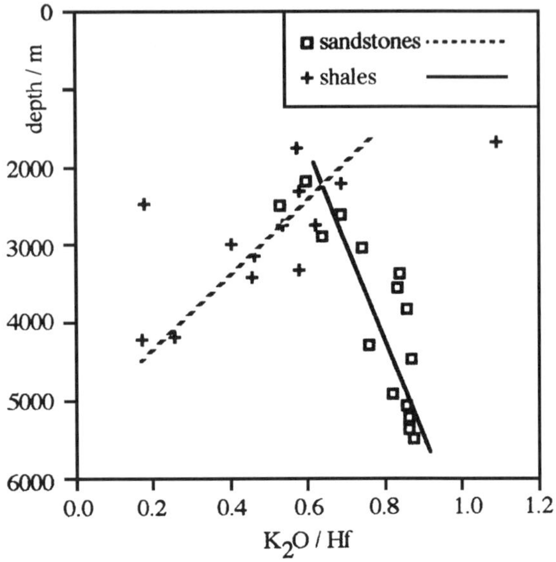

Fig. 8 Change in K_2O/Hf with depth for sandstones and shales. Assuming Hf is immobile, this supports the hypothesis of K transport from sandstones into shales.

procedure used to separate the clay-mineral fraction for analysis.

Figures 7 and 8 are whole-rock plots for K_2O/Zr and K_2O/Hf versus depth, which show the same trends as K_2O/TiO_2. Assuming that Ti, Zr and Hf are immobile, we interpret this to indicate that potassium is being transported from the sandstones into the surrounding shales over the depth interval 2000–4200 m. The volumetrically large reactions over the depth range 2–4 km are feldspar dissolution in sandstones and the conversion of smectite to illite in mudstones. We infer that the majority of the potassium released from feldspars in sandstones is being consumed within the clay-mineral fraction of the shales.

The conclusion that potassium is transported from sandstones into shales during diagenesis also has been reached by Furlan *et al.* (1996) who studied the Mahakam Delta Basin in Indonesia. An increasing number of studies of Gulf Coast shales have also concluded that the smectite to illite transformation is dependent upon an external source of potassium for at least some of the depth interval (Awwiller, 1993; Wintsch & Kvale, 1994; Land *et al.*, 1997; Berger *et al.*, 1999). Berger *et al.* (1999) found that the early stages of the smectite to illite transformation were sourced by potassium feldspar dissolution within the shale. However, below around 2 km burial depth the smectite transformation required an external source of potassium, the nature of which has been a source of speculation. None of the above studies considered data from both sandstones and shales. We have presented evidence above which shows that at least some of the influx of potassium into shales is sourced from interbedded sandstones, at least for the case of the Frio Formation. As the sandstone to shale ratio within this formation is approximately 1 : 8 (Sharp *et al.*, 1988, cited in Lynch *et al.*, 1997), the outflux of K from the sandstones is insufficient to account for the influx into the volumetrically dominant shales (Table 4).

Table 4 Change in sediment composition with depth, south Texas.

Lithology	Δ Al_2O_3 ± 2SE	ΔK_2O ± 2SE	Abundance relative to sandstone
Shale (Land *et al.*, 1997)	+0.10 ± 0.08	+0.12 ± 0.03	8
Sandstone (Milliken *et al.*, 1994)	−0.54 ± 0.22	−0.14 ± 0.06	1

Change in element content over depth range 2000–4500 m. A positive number is an import, values in moles kg^{-1} of original sediment, with ±2 standard error uncertainties. Although the Al export from sandstones may be sufficient to account for the import into shales, it is clear that the K export from sandstones is not sufficient to account for the import into shales. Based on immobility of Ti, Zr and Hf. Sand : shale ratio from Sharp *et al.* (1988).

Consequently there must be another source of K during diagenesis, such that whole-rock K_2O concentrations rise from *c.* 3 to *c.* 5 wt%.

Aluminium

Figure 9 shows Al_2O_3/Zr versus depth for both whole-rock sandstones (Milliken *et al.*, 1994) and shales (Land *et al.*, 1997). Similar geochemical trends with increasing depth are apparent for Al_2O_3 as were described above for K_2O: a decrease in Al_2O_3/Zr ratio within sandstones (Zr is almost constant, Fig. 4) and an increase in Al_2O_3/Zr ratio shales (although Zr increases, Fig. 5). Again, the increase in the Al_2O_3/Zr ratio for the whole-rock shale samples is small but significant ($r^2 = 0.23$, significant at the 95% level). We interpret this to indicate that aluminium is being transported from the sandstones into the surrounding shales. Figures 10 & 11 show Al_2O_3/TiO_2 and Al_2O_3/Hf. These results are interpreted to be consistent with the transport of aluminium from the sandstones to the shales, although the data are more scattered and the Ti-correlation is not significant at the 90% level. Nevertheless, it is still simpler to propose that Al has been transported from the sandstones to the shales, than to suggest that all of Ti, Zr and Hf have been transported from the shales into the sandstones. As Fig. 12 shows an increase in Al_2O_3/TiO_2 for the clay-mineral fraction of the rocks (< 0.1 μm, Lynch *et al.*, 1997), it is likely that the imported Al is absorbed by the clay-mineral fraction during the smectite to illite reaction. In terms of mass-balance, the calculated outflux of Al from the sandstones approximately balances the influx into the volumetrically dominant shales to within the resolution of the data (Table 4). The Frio Formation sands and muds probably are a

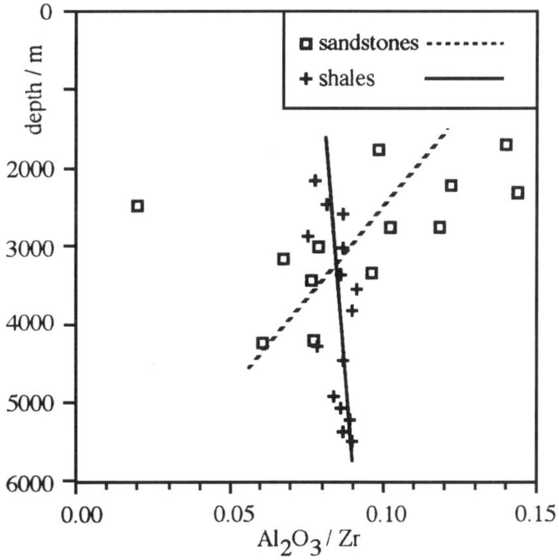

Fig. 9 Change in whole-rock Al_2O_3/Zr for sandstones and shales with depth. The trends are similar to those in Fig. 7 for K_2O, increasing Al in shales with depth, and a decrease of Al in sandstones. We infer that Al is transported from the sandstones into the shales during burial. This is supported by similar trends being apparent when aluminium is compared with other elements that normally are considered to be relatively immobile: Ti and Hf.

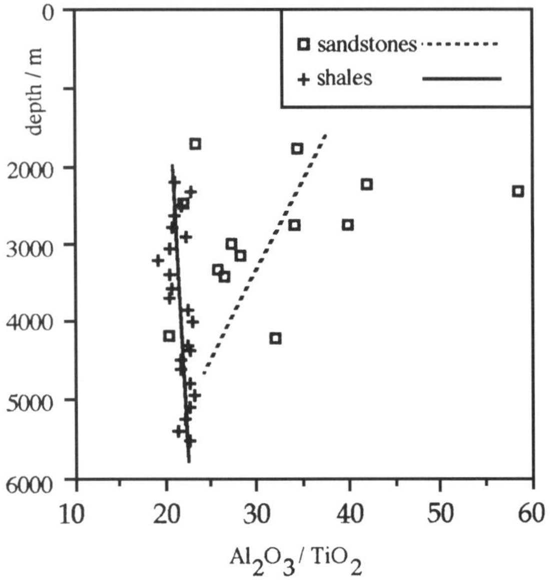

Fig. 10 Change in whole-rock Al_2O_3/TiO_2 for sandstones and shales with depth. A pronounced export from sandstones is mass-balanced by a small import in whole-rock shales (to within the resolution of the data). The data for Al_2O_3/TiO_2 shows similar patterns to K_2O/TiO_2.

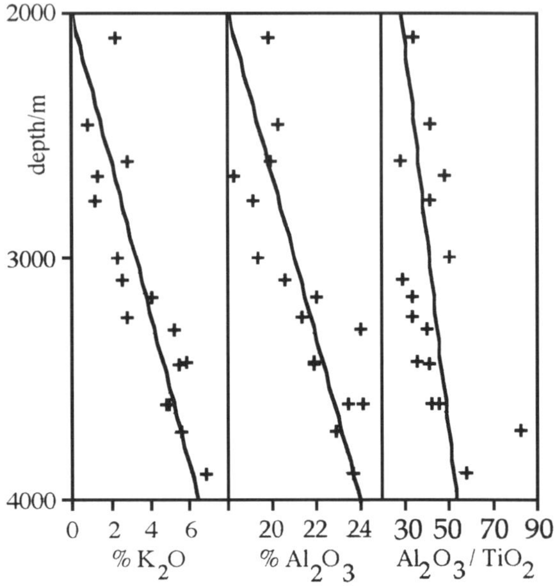

Fig. 12 K_2O, Al_2O_3 and Al_2O_3/TiO_2 data versus depth for clay mineral fraction (< 0.1 μm) of south Texas shales, with best-fit lines. Both Al_2O_3 and Al_2O_3/TiO_2 increase with depth, implying import of Al during the illitization of illite–smectite.

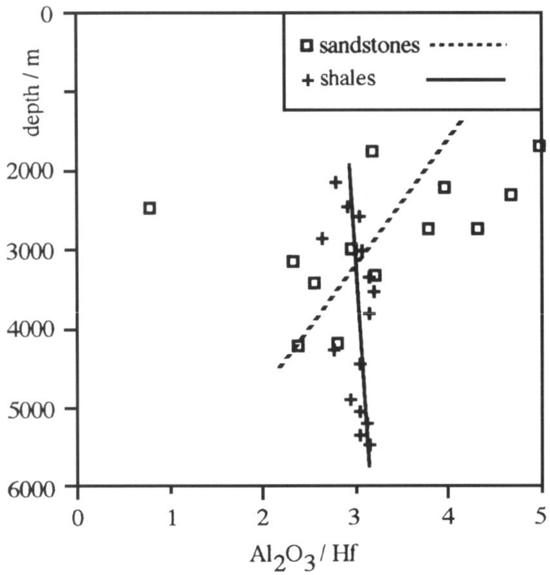

Fig. 11 Change in whole-rock Al_2O_3/Hf for sandstones and shales with depth. A pronounced export from sandstones is mass-balanced by a small import in whole-rock shales (to within the resolution of the data).

closed system with respect to Al during burial from 2000 to 4500 m.

Two previous studies have concluded that Al can be considered immobile during shale diagenesis within the Texas Gulf Coast (Land *et al.*, 1977; Wintsch & Kvale, 1994). In both cases plots of Al_2O_3 versus TiO_2 were cited to show a strong correlation, with the best-fit line passing close to the origin. Because the import of Al into shales is only small as a proportion of total Al, the deflection of the best-fit line from the origin is also small, and easily overlooked. However, in terms of Al transport within the sedimentary basin, subtle variations within the Al_2O_3 versus TiO_2 ratio are important, and can be interpreted to indicate Al mobility. In addition, Wintsch & Kvale (1994) combined data sets from 13 different authors covering a wide geographical area. Small analytical variations between these data sets might be expected to obscure any subtle variations in Al_2O_3/'immobile' ratios, as might geographical provenance differences.

Can we eliminate the possibility that the apparent depth-related changes in sediment chemistry are related to changing source areas during deposition, or to a change in other sedimentary factors, e.g. climate, transport distance, or degree of reworking? This is the same problem faced by studies of progressive diagenesis in any geological sequence in which the most deeply buried sediments are also the oldest. The best known example may be the data set of Hower *et al.* (1976), which shows a progressive change in the illite content of illite–smectite with depth within shales of the Texas Gulf Coast. Although Hower *et al.* (1976) interpreted this change as a diagenetic phenomenon, they were unable to disprove conclusively the alternative explanation that the illite content of illite–smectite was a primary sedimentary feature. Subsequent work in other geographical locations has confirmed the interpretation of Hower *et al.* (1976), e.g. Clauer & Chaudhuri (1996). Similarly, we are unable to offer independent evidence that the changes in sandstone and shale composition with depth within the study data sets are the result of diagenesis and not a primary sedimentary feature. However, we consider that it would require a high degree of coincidence for all the 'immobile' elements to show the same patterns with increasing burial depth, as in Figs 3–12, although see Bloch *et al.* (1998) for a different point of view. Recently reported independent evidence of aluminium mobility during diagenesis includes the occurrence of kaolin cements within limestones (Maliva *et al.*, 1999). As that kaolin was reported to be associated with hydrocarbons, it was suggested that complexing by organic compounds had increased the solubility of the aluminium.

From the data available to us it is not possible to draw any conclusions about the spatial distribution of aluminium and potassium sinks within shales. If elements are exported from the sandstones into the shales, then it might be predicted that these elements would be enriched within the shales close to the sandstone–shale contacts (e.g. Moncure *et al.*, 1984). However, the Land *et al.* (1997) data set is derived from cuttings. These have the advantage of homogenizing material over a relatively wide depth range (compared with a core sample), but the disadvantage that the exact location of the samples is not known. Some previous studies of diagenetic changes at sandstone–shale contacts have concluded that the flux of formation water, and hence permeability patterns, are more important in determining the location of authigenic minerals than the local availability of components (Sullivan & McBride, 1991).

The effect of sample type also can be seen in the quality of the 'mobile'/'immobile' element ratio–depth correlations. The correlation coefficients for the shales generally are much higher than required for significance at the 90% level, whereas those for sands are much closer to the critical value (Table 2). This partly is a result of the greater number of samples in the shale data set, but mostly results from the much lower scatter in these data. This can be attributed partly to the fine-grained nature of shales, such that a small sample contains many grains, but also must reflect how cuttings samples mix rocks from a depth interval (metres to tens of metres scale), and hence smooth out any local variations in shale chemistry. In contrast, the sandstone samples are from conventional cores, and each sample is from only a very small depth interval (centimetre scale). Hence small-scale variations in sandstone chemistry (especially heavy mineral bands) heavily influence the data quality.

IMPLICATIONS

If the dissolution of potassium feldspar within sandstones is providing aluminium that is being absorbed by the reaction of smectite to illite within shales, then the rate of the two reactions would be linked. As it is not possible to hold large volumes of the relevant ions in solution (when compared with the volumes of minerals undergoing reaction), then the source and sink reactions (in sandstones and shales, respectively) must progress at the same rate if the system is closed. The question then becomes: which of these reactions is rate limiting? It is even possible that the rates of both reactions are determined by the rate of transport of Al from the sandstones into the shales. This is not a trivial question, because numerous workers have attempted to determine the reaction rate of the smectite to illite transformation for use as a geothermometer (e.g. Velde & Vasseur, 1992; Berger *et al.*, 1997). It is possible that the source of potassium, and hence the rate control upon the smectite to illite reaction, varies during burial. Berger *et al.* (1999) found that the early stages of the smectite to illite transformation (R_0 structure) were sourced by potassium feldspar dissolution within the shales. However, at higher temperatures (R_1 structure) the smectite transformation required an external source of potassium. Eberl (1993) also noted a change in reaction style with depth. This could give rise to the change in reaction rate observed by Velde & Vasseur (1992). The activation energy calculated by Velde & Vaseur (1992) for the smectite to illite transformation (37–70 kJ mol^{-1}) is characteristic of mineral dissolution or ion exchange reactions. This again is consistent with the smectite to illite reaction being coupled to the dissolution of potassium feldspar, but does not help to determine which element(s) are common to both reactions, or where the K-feldspar is located.

The volumes of clay minerals precipitated within a sandstone may be controlled not only by reactions taking place within the sandstone, but also by reactions within the surrounding shales. As authigenic clay minerals, and especially illite, dramatically reduce sandstone permeability, this conclusion has obvious economic implications. If the transport of solutes from sandstones into shales is not a spatially uniform process, but occurs preferentially in certain areas (determined by fluid-flow patterns), then these areas become potential targets for the oil industry. As fluid flow, driven by overpressure, is concentrated into structurally favoured regions (termed leak-points, e.g. Wilkinson *et al.*, 1997), then these regions have less volume of reaction products (i.e. permeability-reducing authigenic clays) than would be predicted using a closed-system diagenetic model (Haszeldine *et al.*, 1999). Importantly, these areas will have higher porosity and permeability than the regional norm for their burial depth.

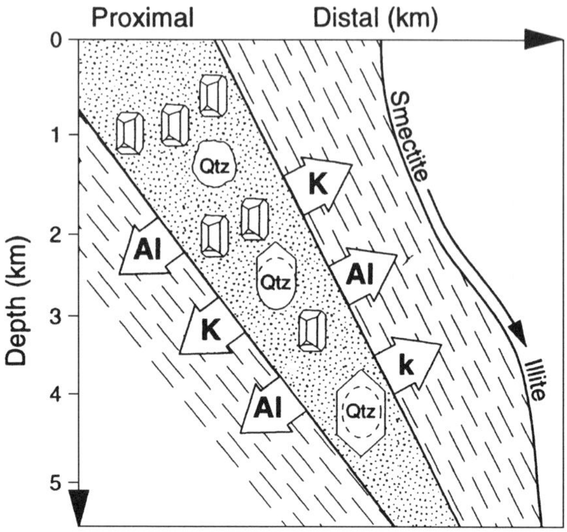

Fig. 13 Summary of solute fluxes during diagenesis of the Gulf Coast siliciclastic succession. Potassium and aluminium, sourced from dissolving alkali feldspars, are exported from the sandstones to the surrounding shales during burial from *c.* 2 to 5 km. These solutes are incorporated into the clay mineral fraction of the shales as the detrital smectite transforms to illite. There is a significant change in the bulk chemistry of both the shales and the sandstones during burial, with the sandstones becoming more silica-rich (more quartzitic) with depth.

CONCLUSIONS

1 Ratios of potassium to 'immobile' elements show systematic changes in depth for sediments of the south Texas Gulf Coast that are consistent with the movement of potassium from the sandstones into the shales during diagenesis from 2 to 4.5 km burial depth (Fig. 13). This is supported by recent studies in the Gulf Coast and Mahakam Delta. There is insufficient K released by the sandstones to supply the influx into the shales, so there must be another K source.

2 Aluminium/'immobile' element ratios show the same trends as for potassium, although with more scatter. This is consistent with, although does not prove conclusively, the movement of aluminium from the sandstones to the shales during diagenesis from 2 to 4.5 km burial depth. The sandstone data set has more scatter than that from the shales, suggesting that the problem of Al mobility could be better addressed by a larger sandstone data set. The outflux of Al from the sandstones approximately balances the influx into the shales, to within the resolution of the data. The Frio Formation sands and muds are together probably a closed system with respect to Al during burial from 2 to 4.5 km. The scale on which the system is closed must be at least tens of metres.

3 The volume of authigenic clay minerals developed within a sandstone will depend upon the breakdown of unstable clastic grains and the proportion of solutes exported from the sandstone as they are generated. These clays include kaolinite, which infills porosity, and illite, which is deleterious to the permeability of hydrocarbon reservoir sandstones. Useful (net) secondary porosity is only generated if the products of mineral dissolution are exported from sandstones and the rock can support the extra void space.

4 The rate of the smectite to illite within shales reaction may be determined by the rate of feldspar dissolution within sandstones, or even by the transport rate of aluminium or potassium from sandstones into shales. This has implications for the use of the smectite–illite geothermometer.

ACKNOWLEDGEMENTS

MW is funded by NERC (UK) on a RoPA grant (No. GR3/R9671). The paper was improved by reviews by Craig Smalley and Olav Walderhaug, and by editorial comments from Richard Worden.

REFERENCES

Awwiller, D.N. (1993) Illite/smectite formation and potassium mass transfer during burial diagenesis of mudrocks: a study from the Texas Gulf Coast Paleocene–Eocene. *J. Sed. Petrol.*, **63**, 501–512.

Awwiller, D.N. (1994) Geochronology and mass-transfer in Gulf Coast mudrocks (south-central Texas, U.S.A.): Rb–Sr, Sm–Nd and REE systematics. *Chem. Geol.*, **116**, 61–84.

Awwiller, D.N. & Mack, L.E. (1991) Diagenetic modification of Sm–Nd model ages in Tertiary sandstones and shales, Texas Gulf Coast. *Geology*, **19**, 311–314.

Berger, G., Lacharpagne, J.C., Velde, B., Beaufort, D. & Lanson, B. (1997) Kinetic constraints on illitization reactions and the effects of organic diagenesis in sandstone/shale sequences. *Appl. Geochem.*, **12**, 23–35.

Berger, G., Velde, B. & Aigouy, T. (1999) Potassium sources and illitization in Texas Gulf Coast shale diagenesis. *J. Sed. Res.*, **69**, 151–157.

Bjørlykke, K. (1979) Cementation of sandstones, discussion. *J. Sed. Petrol.*, **49**, 1358–1359.

Bjørlykke, K. (1997) Mineral/water interaction, fluid flow, and Frio sandstone diagenesis: evidence from the rocks: discussion. *AAPG Bull.*, **81**, 1536–1537.

Bjørlykke, K., Nedkvitne, T., Ramm, M. & Saigal, G.C. (1992) Diagenetic processes in the Brent Group (Middle Jurassic) reservoirs, of the North Sea: an overview. In: *Geology of the Brent Group* (Ed. by A.C. Morton, R.S. Haszeldine, M.R. Giles and S. Brown). *Geol. Soc. London Spec. Publ.*, **61**, 263–287.

Bloch, J., Hutcheon, I.E. & de Caritat, P. (1998) Tertiary volcanic rocks and the potassium content of the Gulf Coast shales—the smoking gun. *Geology*, **26**, 527–530.

Boles, J.R. & Franks, S.G. (1979) Clay diagenesis in Wilcox Sandstones of Southwest Texas: implications of smectite diagenesis on sandstone cementation. *J. Sed. Petrol.*, **49**, 55–70.

Bouch, J.E., Hole, M.J., Trewin, N.H. & Morton, A.C. (1995) Low temperature aqueous mobility of the rare earth elements during sandstone diagenesis. *J. Geol. Soc. London*, **152**, 895–898.

Bouch, J.E., Hole, M.J. & Trewin, N.H. (1997) Rare earth and high field strength element partitioning behaviour in diagenetically precipitated titanites. *Neues Jb. Mineral. Abh.*, **172**, 3–21.

Clauer, N. & Chaudhuri, S. (1996) Inter-basinal comparison of the diagenetic evolution of illite/smectite minerals in buried shales on the basis of K–Ar systematic. *Clays. Clay Mineral.*, **44**, 818–824.

Eberl, D.D. (1993) 3 Zones for illite formation during burial diagenesis and metamorphism. *Clay. Clay Mineral.*, **41**, 26–37.

Floyd, P.A. & Winchester, J.A. (1975) Magma type and tectonic setting discrimination using immobile elements. *Earth Planet. Sci. Lett.*, **27**, 211–218.

Furlan, S., Clauer, N., Chaudhuri, S. & Sommer, F. (1996) K transfer during burial diagenesis in the Mahakam Delta Basin (Kalimantan, Indonesia), *Clay. Clay Mineral.*, **44**, 157–169.

Giles, M.R. & De Boer, R.B. (1990) Origin and

significance of redistributional secondary porosity. *Mar. Petrol. Geol.*, **7**, 378–397.

Gluyas, J. & Coleman, M. (1992) Material flux and porosity changes during sediment diagenesis. *Nature*, **356**, 52–54.

Haszeldine, R.S., Wilkinson, M., Darby, D., *et al.* (1999) Porosity creation and destruction on a graben scale. In: *Petroleum Geology of NW Europe: Proceedings of the 5th conference* (Ed. by A. Fleet and S.A.R. Boldy), Geological Society Publishing House, Bath, pp. 1339–1350.

Hower, J., Eslinger, E.V., Hower, M.E. & Perry, E.A. (1976) Mechanism of burial metamorphism of argillaceous sediment: 1. Mineralogical and chemical evidence. *Geol. Soc. Am. Bull.*, **87**, 725–737.

Land, L.S. (1984) Frio Sandstone diagenesis, Texas Gulf Coast: a regional isotopic study. In: *Clastic Diagenesis* (Ed. by D.A. McDonald and R.C. Surdam). *AAPG Mem.*, **37**, 47–62.

Land, L.S., Mack, L.E., Milliken, K.L. & Lynch, F.L. (1997) Burial diagenesis of argillaceous sediment, south Texas Gulf of Mexico sedimentary basin: a reexamination. *Geol. Soc. Am. Bull.*, **109**, 2–15.

Loucks, R.G., Dodge, M.M. & Galloway, W.E. (1984) Regional controls on diagenesis and reservoir quality in Lower Tertiary sandstones along the Texas Gulf Coast. In: *Clastic Diagenesis* (Ed. by D.A. McDonald and R.C. Surdam). *AAPG Mem.*, **37**, 15–45.

Lynch, F.L. (1996) Mineral/water interaction, fluid-flow, and Frio sandstone diagenesis—evidence from the rocks. *AAPG Bull.*, **80**, 486–504.

Lynch, F.L. (1997) Frio shale mineralogy and the stoichiometry of the smectite-to-illite reaction: the most important reaction in clastic sedimentary diagenesis. *Clay. Clay Mineral.*, **45**, 618–631.

Lynch, F.L., Mack, L.E. & Land, L.S. (1997) Burial diagenesis of illite/smectite in shales and the origins of authigenic quartz and secondary porosity in sandstones. *Geochim. Cosmochim. Acta*, **61**, 1995–2006.

Maliva, R.G., Dickson, J.A.D. & Fallick, A.E. (1999) Kaolin cements in limestones: potential indicators of organic-rich porewaters during diagenesis. *J. Sed. Res.*, **69**, 158–163.

Merriman, R.J. & Roberts, B. (1990) Metabentonites in the Moffat Shale Group, Southern Uplands of Scotland: geochemical evidence of ensialic marginal basin volcanism. *Geol. Mag.*, **127**, 259–271.

Milliken, K.L., McBride, E.F. & Land, L.S. (1989) Numerical assessment of dissolution versus replacement in the subsurface destruction of detrital feldspars, Frio Formation, South Texas. *J. Sed. Petrol.*, **59**, 740–757.

Milliken, K.L., Mack, L.E. & Land, L.S. (1994) Elemental mobility in sandstones during burial: whole-rock chemical and isotopic data, Frio formation, south Texas. *J. Sed. Res.*, **A64**, 788–796.

Moncure, G.K., Lahann, R.W. & Siebert, R.M. (1984) Origin of secondary porosity and cement distribution in a sandstone/shale sequence from the Frio Formation (Oligocene). In: *Clastic Diagenesis* (Ed. by D.A. McDonald and R.C. Surdam). *AAPG Mem.*, **37**, 151–161.

Oxtoby, N. & Gluyas, J. (1997) Discussion on aluminium loss during sandstone diagenesis. *J. Geol. Soc.*, **154**, 747–751.

Pearce, J.A. (1975) Basalt geochemistry used to investigate past tectonic environments on Cyprus. *Tectonophysics*, **25**, 41–67.

Pittman, E.D. & Lewan, M.D. (1994) *Organic Acids in Geological Processes.* Springer-Verlag, Berlin. 482 pp.

Sharp, J.M., Galloway, W.E., Land, L.S., *et al.* (1988) Diagenetic processes in Northwestern Gulf of Mexico sediments. In: *Diagenesis II* (Ed. by G.V. Chilingarian and K.H. Wolfe), Elsevier, Amsterdam, pp. 43–113.

Sullivan, K.B. & McBride, E.F. (1991) Diagenesis of sandstones at shale contacts and diagenetic heterogeneity, Frio Formation, Texas. *AAPG Bull.*, **75**, 121–138.

Van Baalen, M.R. (1993) Titanium mobility in metamorphic systems—a review. *Chem. Geol.*, **110**, 233–249.

Van Panhuys-Sigler, M. & Trewin, N.H. (1990) Authigenic sphene cement in Permian sandstones from Arran. *Scot. J. Geol.*, **26**, 139–144.

Velde, B. & Vasseur, G. (1992) Estimation of the diagenetic smectite to illite transformation in time–temperature space. *Am. Mineral.*, **77**, 967–976.

Wilkinson, M. & Haszeldine, R.S. (1996) Aluminium loss from arkoses produces diagenetic quartzites and secondary porosity: Fulmar Formation, North Sea. *J. Geol. Soc.*, **153**, 657–660.

Wilkinson, M., Darby, D., Haszeldine, R.S. & Couples, G.D. (1997) Secondary-porosity generation during deep burial associated with overpressure leak-off: Fulmar Formation, UKCS. *AAPG Bull.*, **81**, 803–813.

Wilkinson, M., Milliken, K.L. & Haszeldine, R.S. (2001) Systematic destruction of K-feldspar in deeply buried rift and passive margin sandstones. *J. Geol. Soc. London*, **158**, 675–683.

Wintsch, R.P. & Kvale, C.M. (1994) Differential mobility of elements in burial diagenesis of siliciclastic rocks. *J. Sed. Res.*, **64**, 349–361.

Wood, D.A., Joron, J-L. & Treuil, M. (1979) A reappraisal of the use of trace elements to classify and discriminate between magma series erupted in different settings. *Earth Planet. Sci. Lett.*, **45**, 326–336.

Int. Assoc. Sedimentol. Spec. Publ. (2003) **34**, 161–176

Silicate–carbonate reactions in sedimentary systems: fluid composition control and potential for generation of overpressure

IAN HUTCHEON[1] *and* STEPHEN DESROCHER[2]

[1] *Geology and Geophysics, University of Calgary, 2500 University Drive NW, Calgary, Alberta T2N IN4, Canada, e-mail: ian@earth.geo.ucalgary.ca*

[2] *Golder and Associates, 2180 Meadowvale Blvd., Mississauga, Ontario L5N 5S3, Canada*

ABSTRACT

Silicate–carbonate reactions have been identified as probable controls on two trends of P_{CO_2} and temperature, one observed in sedimentary basins and the other in geothermal fluids associated with mafic volcanic rocks. These trends probably reflect equilibrium between the fluids and silicate–carbonate minerals, primarily clay minerals and feldspars in sediments, and clay minerals, feldspars, zeolites and zoisite in mafic rocks. Silicate–carbonate mineral reactions have been identified as pH buffering systems with the capacity to set pH in organic acid-bearing waters, but it is possible that hydrogen released by organic maturation reactions also could influence pH and lead to development of secondary porosity. Computer simulation of these processes suggests that hydrogen from such reactions would have to be highly focused by fluid flow to develop significant porosity in pH buffered silicate–carbonate systems or in unbuffered carbonate rock systems. Silicate–carbonate reactions can lead to the formation of a free vapour phase containing CO_2. Potentially, continued progress of such reactions could lead to pressures above hydrostatic (overpressure). Simulations based on silicate–carbonate reactions observed in sedimentary basins suggest that these reactions could account for as much as a 77% increase in hydrostatic pressure. The transfer of CO_2 from a carbonate mineral phase, when present, to the gas phase by silicate–carbonate reactions could be a previously unrecognized mechanism for the generation of overpressure during progressive burial of siliciclastic sand and shale.

INTRODUCTION

Silicate–carbonate reactions are known to occur in hydrothermal and sedimentary systems (Zen, 1959; Levinson & Vian, 1966; Muffler & White, 1969; Hutcheon *et al.*, 1980, 1993; Smith & Ehrenberg, 1989; Hutcheon, 1990; Coudrain-Ribstein *et al.*, 1998). Silicate–carbonate reactions have been suggested as a source of CO_2 in fluids of sedimentary basins (Hutcheon *et al.*, 1980, 1993; Lundegard *et al.*, 1984; Smith & Ehrenberg, 1989; Hutcheon, 1990), in hydrothermal systems, and as a buffer of pH, and therefore to influence the partial pressure of carbon dioxide (P_{CO_2}). Hutcheon *et al.* (1993) noted that there are two nearly linear trends of log P_{CO_2} versus temperature, a high P_{CO_2} trend (Smith & Ehrenberg, 1989; Lundegard & Trevena, 1990) for sedimentary rocks, and a low P_{CO_2} trend in hydrothermal systems dominated by mafic volcanic rocks (Arnórsson *et al.*, 1983). Hutcheon *et al.* (1993) examined the potential for silicate–carbonate reactions to buffer pH relative to reactions involving only carbonate minerals and organic acids. They concluded that pH in sedimentary and hydrothermal systems probably is set by silicate–carbonate reactions by recognizing that most of the waters in these systems appeared to reach metastable equilibrium with a carbonate–silicate mineral assemblage at approximately 100°C.

There are three issues that are raised by papers published on this topic since Hutcheon *et al.* (1990, 1993).

1 Coudrain-Ribstein *et al.* (1998) noted that the absence of a free gas phase in a confined aquifer renders the concept of P_{CO_2} to be confusing. They use the dissociation of carbonic acid to determine what reactions potentially buffer P_{CO_2}. This does not account for reactions that must occur in the aqueous phase involving H^+. This paper examines reactions in the aqueous phase and correlates those same reactions with the observed trends of T–P_{CO_2}.

2 Lewan (1997) notes that H_2, produced in hydrous pyrolysis experiments and by the reaction of hydrogen with water, should represent a source of acidity. The second objective of this paper is to consider the potential for hydrogen generated by organic reactions to influence the H^+ of formation fluids and the degree to which dissolution and the production of secondary porosity might be influenced by such processes in the presence of silicate–carbonate reactions.

3 Hutcheon *et al.* (1980) and Hutcheon (1990) note that the progress of silicate–carbonate reactions during burial is such that higher P_{CO_2} must be imposed at higher temperatures. Because a gas phase can result from the progress of silicate–carbonate reactions as temperature increases, there is the potential that these reactions could contribute to fluid pressures higher than hydrostatic (overpressure). The third objective of this paper is to evaluate the degree to which silicate–carbonate reactions might contribute to fluid pressures that are greater than hydrostatic in sedimentary basins.

THE ROLE OF SILICATE–CARBONATE REACTIONS IN SETTING pH AND P_{CO_2} IN FLUIDS ASSOCIATED WITH SEDIMENTARY AND VOLCANIC ROCKS

Zen (1959) is the first author who noted the potential of silicate–carbonate reactions to take place in sedimentary rocks. He observed kaolinite–calcite bearing altered volcanic rocks and concluded this probably was a stable assemblage with phase relations possibly affected by the presence of montmorillonite. Various authors have since documented the occurrence of silicate–carbonate reactions in sediments (Muffler & White, 1969; Hutcheon *et al.*, 1980; Hutcheon, 1990). The role of such reactions in the control of P_{CO_2} was reported by Smith & Ehrenberg (1989) and the mechanism of and control of pH and P_{CO_2} via buffering of fluid phase compositions was examined by Hutcheon *et al.* (1993). Hutcheon & Abercrombie (1990) noted the probable occurrence of two different trends of P_{CO_2} versus temperature related to associated rock types and this relationship was examined in detail by Coudrain-Ribstein *et al.* (1998). These last authors examined a number of reactions of the general form:

$$\text{silicate} + \text{calcite} = \text{Ca-silicate} + CO_2 \quad \text{(R1)}$$

and fit equilibrium constants to trends of temperature versus P_{CO_2}. They calculated trends of P_{CO_2} versus temperature in confined aquifers (lacking a free gas phase) from the relationship:

$$P_{CO_2} = \frac{[H^+][HCO_3^-]}{K_{CO_2}[H_2O]} \quad \text{(R2)}$$

Two nearly linear trends of log P_{CO_2} versus temperature were reported by Hutcheon *et al.* (1993). A low P_{CO_2} trend occurs in hydrothermal systems in Iceland dominated by mafic volcanic rocks (data reported by Arnórsson *et al.*, 1983) and a high P_{CO_2} trend (Smith & Ehrenberg, 1989; Lundegard & Trevena, 1990) for sedimentary rocks. The Icelandic geothermal systems generally are boiling and the hydrocarbon reservoirs reported by Smith & Ehrenberg (1989) and

Lundegard & Trevena (1990) contain a gas phase, thus both systems effectively have a separate gas phase.

Agreement of fit of equilibrium constants to the trends of P_{CO_2} is problematic if a separate gas phase is not present, because many reactions may fit the data. In a water–rock–gas system, any proposed mineral reaction must also fit the water chemistry data to be considered to be a potential buffer of P_{CO_2}. Data from Iceland (Arnórsson *et al.*, 1983) for both water composition and P_{CO_2} were examined to determine which reactions might be involved in determining the observed trends in P_{CO_2}.

The silicate reactions, of a parallel form to (R1), that do not include calcite or CO_2 have the general form:

$$\text{Al-silicate (hydrous)} + Ca^{2+} = \text{Ca–Al-silicate} + H^+ \qquad \text{(R3)}$$

the equilibrium constant for which (assuming pure solids) is:

$$K = \frac{aCa^{2+}}{a(H^+)^2} \qquad \text{(R4)}$$

The reactions examined by Coudrain-Ribstein *et al.* (1998), designated V1, V2, etc., were recast not to include calcite and CO_2, so as to be of the same form as (R4).

Many such reactions involve silica (SiO_2) and the activity of this component either must be defined from water analyses or set using a silica mineral phase. The choice of silica polymorph to set aqueous silica activity is problematic. It generally is accepted that waters in sedimentary water–rock systems are in equilibrium with quartz at, or above, about 100°C (Kharaka & Mariner, 1988). However, in volcanic hydrothermal systems, comparison of the silica activity at temperatures measured by various aqueous geothermometers with the stability of various silica minerals, suggests that up to about 180°C, chalcedony is close to equilibrium (Abercrombie, 1988). The silica activity for calculations (Fig. 1) involving the Iceland data (Arnórsson *et al.*, 1983) was set below 180°C using chalcedony. The activity of silica for reactions of interest above 180°C (V9 primarily) was set using quartz. The actual position of all reactions usually is affected only slightly by the choice of silica mineral species.

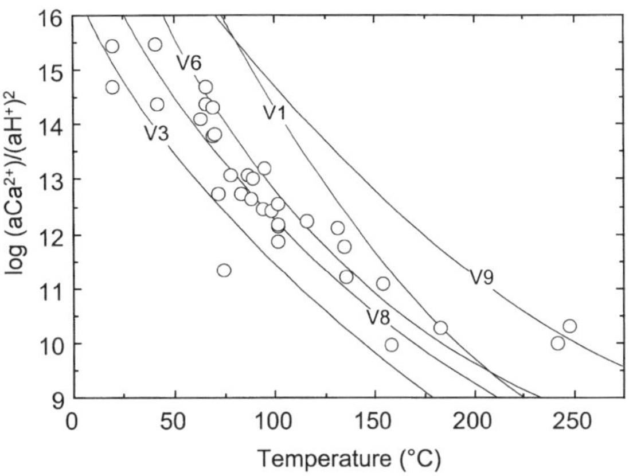

Fig. 1 The relative positions of reactions V1, V3, V6, V8 and V9 from Coudrain-Ribstein *et al.* (1998) and log ($a Ca^{2+}/a(H^+)^2$) determined from water samples in Arnórsson *et al.* (1983) versus temperature. Reactions calculated using Ge0-CALC (Brown *et al.*,1988) on the liquid–vapour curve at saturation with chalcedony except V9, where quartz was used. Circles represent values derived from water analyses as reported in the text.

Various reactions from Coudrain-Ribstein *et al.* (1998) were compared with the water chemistry reported by Arnórsson *et al.* (1983). A plot of reactions (Fig. 1) of the form of reaction (R4) versus temperature, was calculated using Ge0-CALC (Berman, 1988; Brown *et al.*, 1988). The circles represent measured water compositions, the lines the equilibrium constants for different reactions.

Only the reactions (numbers used are the same as those used by Coudrain-Ribstein *et al.*, 1988):

$$\text{kaolinite} + Ca^{2+} = \text{anorthite} + 2H^+ + H_2O \qquad \text{(V1)}$$

$$0.5\ \text{kaolinite} + 0.5\ \text{chalcedony} + Ca^{2+} = 0.5\ \text{prehnite} + 2H^+ + 0.5H_2O \qquad \text{(V3)}$$

$$0.75\ \text{muscovite} + 1.5\ \text{chalcedony} + Ca^{2+} + 0.5H_2O = 0.5\ \text{zoisite} + 0.75\ \text{K-feldspar} + 2H^+ \qquad \text{(V6)}$$

$$0.75\ \text{paragonite} + 1.5\ \text{chalcedony} + Ca^{2+} + 0.5H_2O = 0.5\ \text{zoisite} + 0.75\ \text{albite} + 2H^+ \qquad \text{(V8)}$$

$$16 \text{ zoisite} + 15 \text{ phlogopite} + 32Ca^{2+} + 28H_2O = 15 \text{ muscovite} + 21 \text{ quartz} + 9 \text{ chlorite} + 64H^+ \quad \text{(V9)}$$

are within the range of the measured activity data for log $(aCa^{2+}/a(H^+)^2)$ for the waters. A reaction between wairakite, K-feldspar and muscovite (V7, Coudrain-Ribstein *et al.*, 1998) falls to the lower side of the range of log $(aCa^{2+}/a(H^+)^2)$. None of the reactions above are close to the high temperature data at about 240°C. The closest reaction to these points is a reaction among zoisite, phlogopite, muscovite and chlorite (V9, Coudrain-Ribstein *et al.*, 1998). It is possible that this represents a reaction in the rocks from an assemblage of kaolinite–prehnite–muscovite–zoisite–albite to one containing chlorite and in which prehnite is no longer stable. The trend of T–P_{CO_2} seems to cross from V3 at lower temperatures to V8, V6, V1 and V9 as temperature increases, suggesting that different mineral assemblages are influencing fluid composition at different temperatures. Careful examination of the rocks, which is not possible in the context of this paper, would be required to demonstrate the validity of any change in mineral assemblage resulting from a reaction in the rocks.

Having established which reactions are potentially in equilibrium with the measured water composition, it is possible to determine if any of these reactions can reproduce the observed trends of T–P_{CO_2}. These reactions can be formed from those above by adding the reaction:

$$CaCO_3 + 2H^+ = Ca^{2+} + CO_2 + H_2O \quad \text{(R5)}$$

This gives reactions of exactly the form reported by Coudrain-Ribstein *et al.* (1998), for example, reaction (V1 above) would become:

$$\text{kaolinite} + \text{calcite} = \text{anorthite} + 2H_2O + CO_2 \quad \text{(V1)}$$

Figure 2 shows the position of a number of reactions calculated using Ge0-CALC (Brown *et al.*, 1988). Reaction (V1) does not fit the CO_2

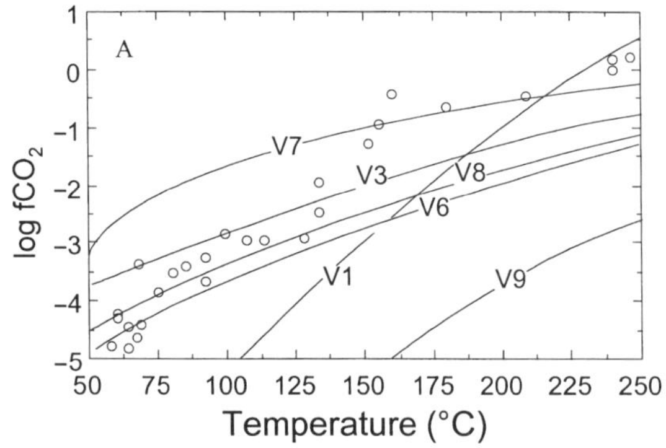

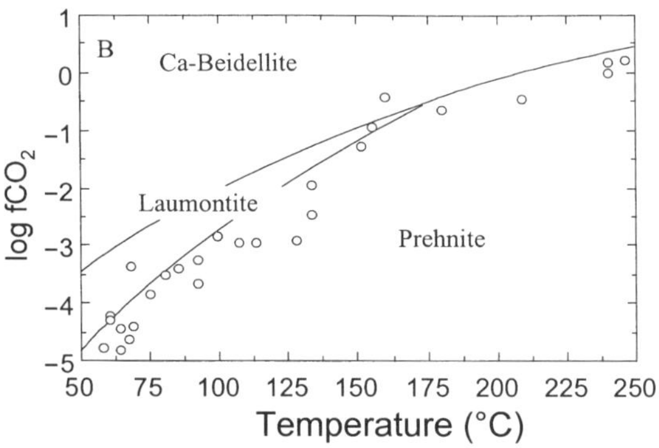

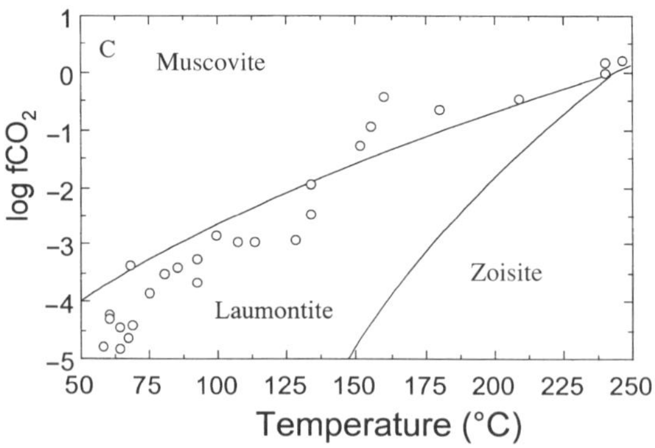

Fig. 2 (A) The relative positions of reactions V1, V3, V6, V7, V8 and V9 from Coudrain-Ribstein *et al.* (1998) and log P_{CO_2} reported by Arnórsson *et al.* (1983) versus temperature. (B) Phase diagram for reactions among Ca-beidellite–prehnite–laumontite relative to log P_{CO_2} reported by Arnórsson *et al.* (1983) versus temperature. (C) Phase diagram for reactions among muscovite–zoisite–laumontite relative to log P_{CO_2} determined from water samples reported by Arnórsson *et al.* (1983) versus temperature. All reactions calculated using Ge0-CALC (Brown *et al.*, 1988) on the liquid–vapour curve at saturation with chalcedony except V9, where quartz was used. Circles represent values derived from gas analyses as reported in the text.

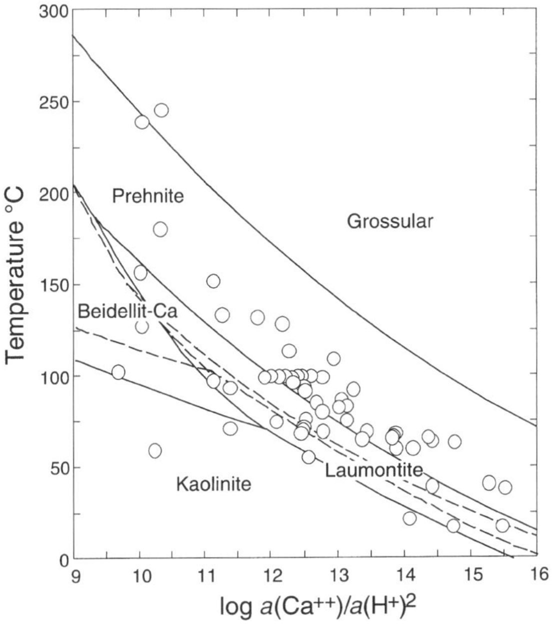

Fig. 3 Phase diagram for reactions among Ca-beidellite–prehnite–laumontite–kaolinite–grossular relative to log (aCa^{2+}/a(H$^+$)^2) determined from water samples in Arnórsson *et al.* (1983) versus temperature. Reactions were calculated using 'The Geochemist's Workbench® 3.0' (Bethke, 1998) on the liquid–vapour curve at saturation with chalcedony (solid lines) or quartz (dashed lines). Circles represent values derived from water analyses as reported in the text.

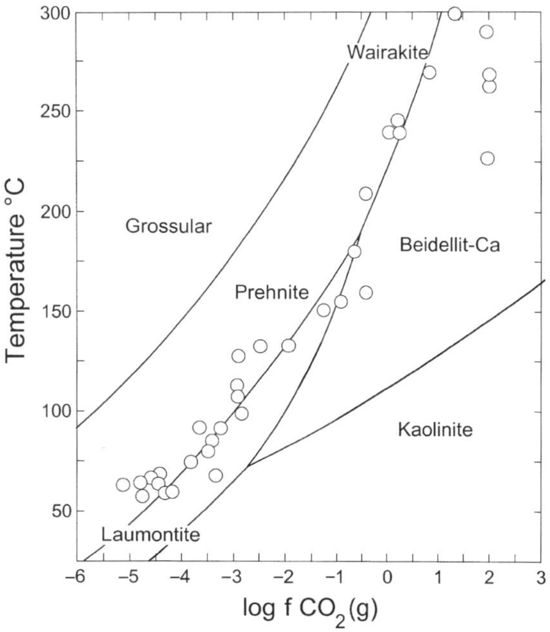

Fig. 4 Phase diagram for reactions among Ca-beidellite–prehnite–laumontite–kaolinite–wairakite–grossular relative to log P_{CO_2} reported by Arnórsson *et al.* (1983) versus temperature. All reactions calculated using 'The Geochemist's Workbench® 3.0' (Bethke, 1998) on the liquid–vapour curve at saturation with chalcedony. Circles represent values derived from gas analyses as reported in the text.

data at all, and reactions (V3), (V6) and (V8) only fit below 140°C. Reaction (V7) and (V9) fit the log (aCa^{2+}/a(H$^+$)^2) data at high temperatures, but (V7) only fits the CO_2 data above 150°C, and (V9) does not fit the CO_2 data at all. The high temperature fit suggests that different reactions control CO_2 at different temperatures, consistent with the changes in mineral assemblages suggested by Arnórsson *et al.* (1983).

The process of fitting individual reactions does not account for mineral phases that might be the most stable. The geochemical modelling software package 'The Geochemist's Workbench® 3.0' (Bethke, 1998) was used to examine the systems Ca–O–H–Al–Si and Ca–C–O–H–Al–Si. Reactions involving K and Na minerals were not considered. Figures 3 & 4 show the stable reactions in this system on the liquid–vapour curve (the Iceland waters are at, or close to, boiling at temperatures ranging from 20 to 250°C), with the silica activity set by chalcedony and aluminium activity set by kaolinite. The choice of quartz to set silica activity moves the kaolinite–laumontite and kaolinite–beidellite boundaries to temperatures about 15°C higher, reducing the size of the laumontite stability field (dashed lines, Fig. 3). However, as noted previously, chalcedony is considered to be a more appropriate choice to set silica activity below 180°C in this geothermal system. The stability of the zeolite phases clinoptilolite and heulandite was suppressed because they do not appear as minerals as reported by Arnórsson *et al.* (1983). The log (aCa^{2+}/a(H$^+$)^2) values (Fig. 3) cluster around reactions between prehnite–laumontite and kaolinite–laumontite at lower temperatures, with some data clustering around the kaolinite–beidellite and prehnite–beidellite reactions. The two high temperature points (247, 249°C) are coincident with the prehnite–grossular reaction. Figure 4 shows the carbonate–silicate reactions plotted as a function of log f_{CO_2}. The data cluster around the prehnite–beidellite and prehnite–laumontite reactions, as observed on the plot of log $(aCa^{2+}/a(H^+)^2)$ versus temperature. This suggests that these reactions are most likely to be responsible for the observed T–P_{CO_2} trend.

The data for sedimentary rocks are considered in a similar manner. Published water data with a known free gas phase are those from the Patani Basin in Thailand (Lundegard & Trevena, 1990) and the North Sea (Barth, 1991). The P_{CO_2} data for Patani were used to calculate the pH at reservoir conditions and were provided by Paul Lundegard (personal communication, 1995). Smith & Ehrenberg (1989) report the P_{CO_2} data used for the North Sea waters. Hutcheon *et al.* (1993) had little success in fitting log $(aCa^{2+}/a(H^+)^2)$ data for waters from a number of sedimentary basins to mineral reactions in the system Ca–O–H–Al–Si if aqueous species of variable activity were not included. They found that the reaction smectite–kaolinite–calcite best fit the observed trend of log f_{CO_2} with temperature. As noted by Coudrain-Ribstein *et al.* (1998), such reactions do not explicitly define the activity values of aqueous species. Accordingly, the log $(aCa^{2+}/a(H^+)^2)$ and log f_{CO_2} data from sedimentary basins (Fig. 5 & 6) are reconsidered using 'The Geochemist's Workbench® 3.0' (Bethke, 1998). The reactions that fit the Iceland log $(aCa^{2+}/a(H^+)^2)$ data generally fit the data from sedimentary basins (Fig. 5) but there are significant differences. In particular, the range of log $(aCa^{2+}/a(H^+)^2)$ for sedimentary waters is lower than for the Iceland data, and the majority of waters cluster around the kaolinite–beideillite reaction, with some waters on the laumontite–beidellite reaction. None of these reactions fit the position or the trend of P_{CO_2}, which has a different slope than the phase boundaries, suggesting that even though calcite is present, the reactions examined here are not appropriate as possible controls of fluid composition or P_{CO_2} (Fig. 6). The reaction of

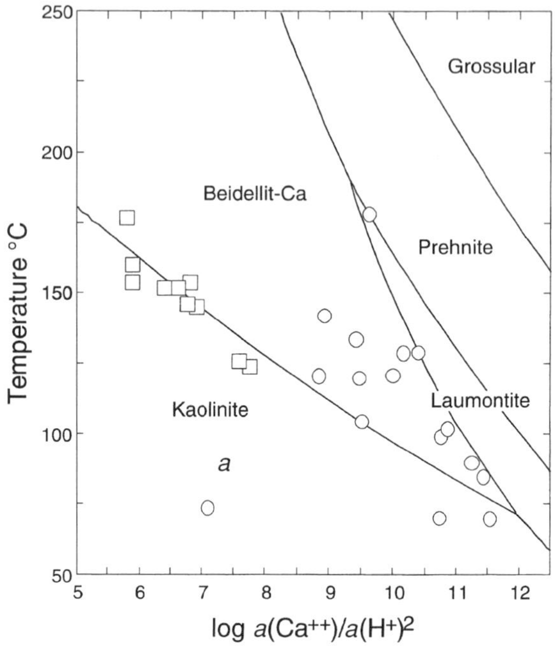

Fig. 5 Phase diagram for reactions among Ca-beidellite–prehnite–laumontite–kaolinite–grossular relative to log $(aCa^{2+}/a(H^+)^2)$ determined from water samples reported by Lundegard & Trevena (1990) for the Patani Basin (squares) and Barth (1991) for the North Sea (circles) versus temperature. All reactions calculated using 'The Geochemist's Workbench® 3.0' (Bethke, 1998) on the liquid–vapour curve at saturation with chalcedony. Circles represent values derived from water analyses as reported in the text.

kaolinite–dolomite–chlorite–calcite, originally proposed by Hutcheon *et al.* (1980), has a similar trend to the data (Fig. 7), but falls at temperatures approximately 50°C too high. This reaction, however, considers only the pure Mg end-member chlorite, clinochlore. Authigenic chlorite in sedimentary rocks typically ranges from iron-rich to 1 : 1 Mg : Fe, however, the compositions required to estimate the activity of the clinochlore end-member are rarely reported. Based on electron microprobe analyses, authigenic chlorite from the Venture Field in Canada (Hutcheon, 1990) has about 0.1 to 0.2 mol fraction of the Mg end-member. Aagaard & Helgeson (1983) consider activity models for clay compositions. Assuming a simplified ideal 'mixing on sites' solid solution model of the

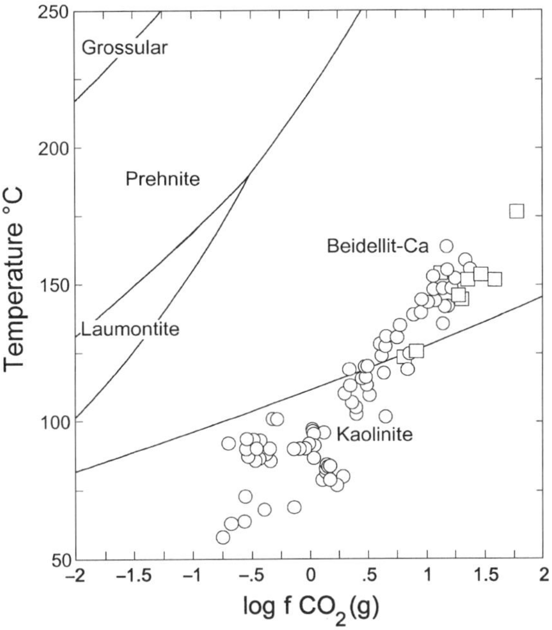

Fig. 6 Phase diagram for reactions among Ca-beidellite–prehnite–laumontite–kaolinite–grossular relative to log P_{CO_2} from Lundegard & Trevena (1990) for the Patani Basin (squares) and Smith & Ehrenberg (1989) for the North Sea (circles) versus temperature. All reactions calculated using 'The Geochemist's Workbench® 3.0' (Bethke, 1998) on the liquid–vapour curve at saturation with chalcedony. Circles represent values derived from gas analyses as reported in the text.

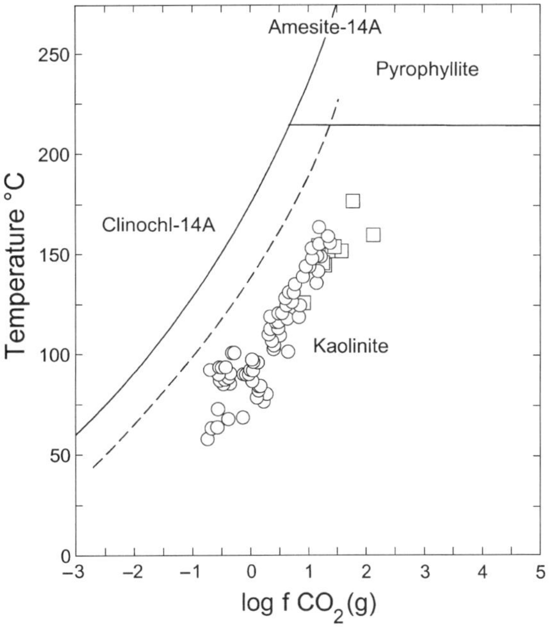

Fig. 7 Phase diagram for reactions among kaolinite–dolomite–clinochlore–calcite relative to log P_{CO_2} from Lundegard & Trevena (1990) for the Patani Basin (squares) and Smith & Ehrenberg (1989) for the North Sea (circles) versus temperature. All reactions calculated using 'The Geochemist's Workbench® 3.0' (Bethke, 1998) on the liquid–vapour curve at saturation with chalcedony. The dashed line represents the position of the reaction of kaolinite and dolomite to form clinochlore and calcite. The solid lines represent the reaction of kaolinite and dolomite to amesite and calcite (curved line) and the reaction of kaolinite to pyrophyllite (horizontal line). The dashed line is adjusted for the composition of chlorite as reported by Hutcheon (1990). Circles represent values derived from gas analyses as reported in the text.

form below and using 0.1 mol fraction Mg end-member:

$$a_{\text{Chlorite}}^{\text{Mg}_5\text{Al}_2\text{Si}_3\text{O}^{10}(\text{OH})_8} = (X^{\text{Mg}})^5$$

the new position of the kaolinite–dolomite–calcite–clinochlore reaction is closer to the P_{CO_2} data by about 30°C (dashed line on Fig. 7), but still does not fit the data very well. Increasing iron content will shift the curve to higher values of f_{CO_2} at any temperature. Coudrain-Ribstein *et al.* (1998) proposed reactions similar to the kaolinite–dolomite–chlorite–calcite but involving phlogopite, K-feldspar, muscovite and zoisite. Of these reactions, one between amesite (an aluminous Mg chlorite) and zoisite fits the data much better than the kaolinite–dolomite–calcite–clinochlore reaction. However, the authigenic occurrence of zoisite is not reported in sandstones from the Gulf Coast or North Sea, so the fit of this reaction to the data seems to be coincidental.

In summary, the log ($a\text{Ca}^{2+}/a(\text{H}^+)^2$) and log P_{CO_2} versus temperature trends for waters and gases from Iceland can be fitted by reactions that are consistent with minerals, including calcite, observed in those rocks. However, log ($a\text{Ca}^{2+}/a(\text{H}^+)^2$) and log P_{CO_2} versus temperature trends are difficult to fit for data from sedimentary basin fluids. The trends of log P_{CO_2} versus temperature argue strongly that pH and P_{CO_2} are set by mineral reactions involving carbonates and silicates, even if the exact form of these reactions may be elusive in sedimentary systems. The two trends of P_{CO_2} versus temperature observed for sedimentary basins and for volcanic settings reflect the $(a\mathrm{Ca}^{2+}/a(\mathrm{H}^+)^2)$ ratios in water and the reactions that set the activity ratio. Mineral reactions that impose low values of $(a\mathrm{Ca}^{2+}/a(\mathrm{H}^+)^2)$ on the water require high values of P_{CO_2} and vice versa.

The control, or buffering, of pH by mineral reactions in sediments has been examined for dissolved aqueous species and dissolved organic acids (Hutcheon *et al.*, 1993). However, Lewan (1997) proposed that hydrogen produced by organic reactions also could influence the pH of sedimentary fluids. The next section of this paper considers the potential for organic reactions producing H_2 to buffer pH.

Hydrogen from organic reactions and pH

Lewan (1997) suggested that hydrogen (H_2) produced by organic reactions and the concomitant reduction of water could influence the pH of waters in sedimentary basins where hydrocarbon maturation reactions are taking place. This is suggested to result from a series of reactions (Hostettler, 1984), which overall reduce to:

$$\mathrm{H}_{2(g)} = 2\mathrm{H}^+_{(aq)} + 2\mathrm{e}^-_{(aq)} \qquad \text{(R6)}$$

It is important to recognize that (R6), as written, is a half reaction that represents the oxidation of H_2 gas. There must be a corresponding reduction half reaction(s), in other words some part of the system must be reduced. A familiar example of redox reactions in sedimentary rocks is the common observation of red iron staining (oxidized) replaced by green staining (reduced). If a rock had available ferric (Fe^{3+}) iron, the resulting reaction would be:

$$\mathrm{Fe}^{3+}_{(aq)} + \mathrm{e}^-_{(aq)} = \mathrm{Fe}^{2+}_{(aq)} \qquad \text{(R7)}$$

and combining (R6) and (R7) would give:

$$2\mathrm{Fe}^{3+}_{(aq)} + \mathrm{H}_{2(g)} = 2\mathrm{Fe}^{2+}_{(aq)} + 2\mathrm{H}^+_{(aq)} \qquad \text{(R8)}$$

Reaction (R8) would still allow the production of H_2 to result in H^+ being added to the water. However, in a rock with no minerals that could be reduced, the water or some other constituent of the water, would be reduced. The stability of liquid water (the reduction of water to H_2 gas) represents the lower limit for such a process.

If a system of organic reactions that generates H_2 is in equilibrium for reaction (R6), then each additional mole of H_2 generated by an organic reaction produces two moles of H^+. Table 7 in Lewan (1997, p. 3692) shows the amounts of various gases generated from hydrous pyrolysis of Woodford Shale. The experiments range from 300°C to 350°C and the hydrous experiments produce from 5.06 to 6.07 mmol of H_2 per 400 g of shale with an organic carbon content of 22.39 wt%.

The REACT reaction path module in 'The Geochemist's Workbench® 3.0' (Bethke, 1998) was used to simulate the effect at 100°C of the release of all the hydrogen generated by a rich source rock on the pH of the pore fluid. It was assumed that there is an immediately adjacent sandstone reservoir rock with 10% porosity. There are 400 g of rock and 320 g of water in Lewan's (1997) experiments, so by mass the rock–water ratio is approximately 1.25. Porosity is a volume fraction and assuming a rock density of about 2.0 (these are very organic rich rocks) would give a rock–water ratio of 0.625. The modelling was carried out at 100°C and at the depths that would be required to produce these temperatures, it is not likely that a rock would have more than 10–20% porosity (Dixon *et al.*, 1989). REACT works with 1.0 kg of water, so to simulate sandstone with 10% porosity a rock density of 2.6 g cm^{-3} was assumed. Given a density of pure water of 1.0 g cm^{-3}, this requires 23.4 kg of rock to produce a rock–water system with 1 kg of water and 10% porosity. Using the ratio of about 5 mmol of H_2 produced from 400 g of rock in Lewan's experiments and taking the 23.4 kg of rock in the simulation requires that 292.5 mmol of H_2 or 585 mmol of H^+ be added to the 1 kg of water as a reactant. It should be noted that this is a huge amount of H^+ and also represents a very extreme set of conditions. About 36% of the H_2 produced is from the reduction of water (M.D.

Lewan, personal communication, 1999). The water–rock ratio in the experiments is much higher than in a rock with 10% porosity and the amount of H_2 produced from water reduction would be expected to be much less than is actually used in this simulation. Also, the rock in the experiments is very organic-rich shale with 22.39 wt% organic content. More typical organic shale would contain perhaps 3–5 wt% and would produce less hydrogen. To ensure the maximum effect from H_2 the amount of H^+ used in the simulation was doubled. The calculation assumes that all the H_2 released is immediately transferred into the pore-water of the sandstone and converted to H^+, giving the maximum possible effect. Although there is no easy way to place suitable constraints on the relative masses of shale, sandstone and H_2, the conditions assumed here would seem to provide for an abundance of H_2 and H^+ unlikely to be realized in nature. Doubling the sandstone porosity to 20% would not increase the total amount of hydrogen, but would increase the mass of water, decreasing the net effect of the hydrogen on dissolution.

Table 1 shows the rock composition and system conditions set in 'The Geochemist's Workbench® 3.0' (Bethke, 1998) and Fig. 8 shows the results of the simulation. More than a mole (1.17 mol) of H^+ is added to 1.0 kg of water and 23.4 kg of rock containing kaolinite, chlorite, calcite, dolomite, albite, illite, K-feldspar and quartz. This is a huge amount of H^+ to add to 1.0 kg of water, however, the pH decreases from approximately 5.925 to 5.9 or about 0.025 pH units. It is important to realize that the rock–water ratio is critical in such simulations and with 10% porosity the mass of rock clearly dominates the system. No attempt was made to calculate a buffer index (Hutcheon *et al.*, 1993) because reaction (R6) releases H^+ as an aqueous species and is thus equivalent to a completely dissociated (strong) acid such as HCl.

The main minerals dissolved during the simulation are clinochlore (35 cm^3) and calcite (46 cm^3) while dolomite (31 cm^3) and kaolinite (20 cm^3) precipitate. This is consistent with the reaction proposed by Hutcheon *et al.* (1980) being driven to the kaolinite–dolomite side by the addition of H^+. The total volume change is a 30 cm^3 decrease in the mineral volume, which would appear as porosity. This represents an increase in porosity of 0.3%, which would not be detectable. Other modelling scenarios could be considered. For example, higher hydrogen contents could be envisioned by focusing the H_2 from a large volume of shale into a small volume of sandstone. However, considering that the constraints on the simulation were established to produce a maximum effect, it seems unlikely that the release of H_2 from maturation of organic matter would be

Table 1 Run conditions for input file of 'The Geochemist's Workbench ® 3.0' for addition of 1.17 mol of H^+ to 23 400 g of rock of the composition shown. Silica activity is set by chalcedony and the stability of the minerals muscovite, magnesite, paragonite, saponite-Na, saponite-K, saponite-H, saponite-Mg, saponite-Ca, talc and phengite are suppressed.

Rock composition (wt%):
kaolinite = 5% = 1170 g
chalcedony = 75% = 17 550 g
illite = 1.5% = 351 g
albite = 12% = 2808 g
calcite = 5% = 1170 g
clinochlore = 1.5% = 351 g
Temperature = 100
Swap kaolinite for Al^{3+}
Swap chalcedony for SiO_2(aq)
Swap illite for K^+
Swap albite for H^+
Swap calcite for Ca^{2+}
Swap clinochlore-14A for Mg^{2+}
1 kg free H_2O
Free kg kaolinite = 1.17
Free kg chalcedony = 17.55
Free gram illite = 351
Free kg albite = 2.808
Free kg calcite = 1.17
Free gram clinochlore-14A = 351
Total molality of Na^+ = 3
Balance on Cl^-
Total molality of HCO_3^- = 0.2
React 1.17 mol of H^+

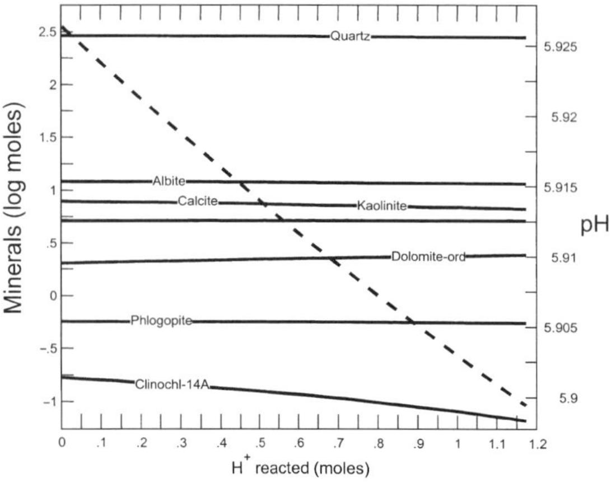

Fig. 8 Reaction path simulation using the REACT module in 'The Geochemist's Workbench® 3.0' (Bethke, 1998) of addition of H^+ to a sandstone with 10% porosity and containing the minerals albite–K-feldspar–illite–calcite–dolomite and quartz. The solid lines show the changes in the amounts of minerals (scale on left-hand side) produced by reaction with a rock containing the minerals noted above; the dashed line shows the change in pH (scale on right-hand side). Run conditions as in Table 1.

higher than the amounts simulated. Based on the simulation, it appears that H_2 is not an effective agent of dissolution in rocks that contain potential buffers of H^+, such as the silicate–carbonate reactions considered here.

In rocks that do not contain H^+-buffering assemblages, such as limestone, dolomite or carbonate-cemented quartz sandstone without significant amounts of feldspar or clay minerals, the addition of H^+ could result in more dissolution, increasing CO_2 in the gas phase and increased porosity. To examine this possibility the effect of releasing the same amount of H_2 into limestone was simulated. The simulation conditions are shown in Table 2 and the results are presented in Fig. 9. The conditions (10% porosity, 9 L of rock) are similar to those for the sandstone simulation, however, the density difference requires a slightly larger volume of limestone (calcite). Adding 1.2 mol of H^+ to 1.0 kg of water in the pore system results in a decrease in pH from 4.35 to 4.05. To place these low pH values in perspective, neutral pH at 100°C is 6.13. The rock volume decreases by about 30 cm^3 in 9000, or 0.3%. These amounts are very close to those for the sandstone simulation and the decrease in pH and rock volume is rather small.

Table 2 Run conditions for input file of 'The Geochemist's Workbench ® 3.0' for addition of 1.17 mol of H^+ to 24 480 g limestone containing only calcite.

Temperature = 100
Swap calcite for H^+
1 kg free H_2O
Free gram calcite = 24 480
Total molality of Ca^{2+} = 4
Total molality of Na^+ = 3
Total molality of Mg^{2+} = 0.565
Total molality of K^+ = 0.02448
Total molality of Al^{3+} = $4.95e^{-8}$
Total molality of $SiO_2(aq)$ = 0.000763
Total molality of SO_4^{2-} = 0.067
Total molality of HCO_3^- = 0.2
Balance on Cl^-
React 1.2 mol of H^+
Suppress all
Unsuppress calcite

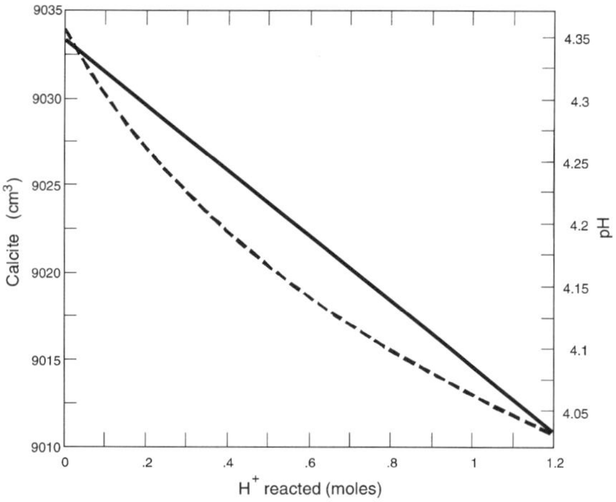

Fig. 9 Reaction path simulation using the REACT module in 'The Geochemist's Workbench® 3.0' (Bethke, 1998) of addition of H^+ to a limestone with 10% porosity and containing calcite. The solid line shows the changes in the volume of calcite (scale on left-hand side), the dashed line shows the change in pH (scale on right-hand side). Run conditions as in Table 2.

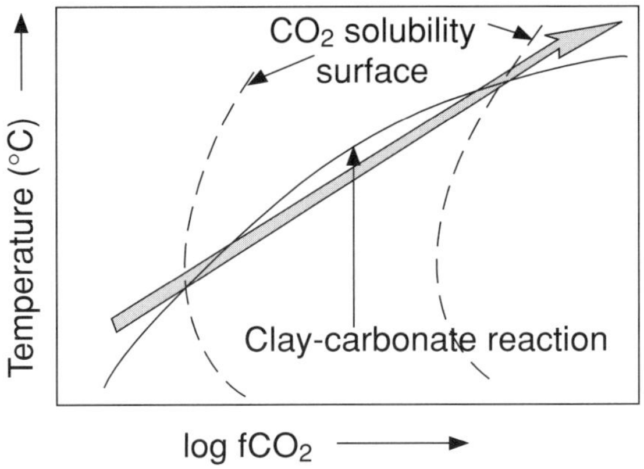

Fig. 10 Schematic diagram showing the intersection of the solubility of CO_2 in water (dashed curves), with the trace of a silicate–carbonate reaction (solid curve). Note that the position of the solubility surface is influenced both by pressure and salinity, as well as temperature, whereas the silicate–carbonate reaction is relatively insensitive to pressure or salinity. The stippled arrow shows the path that will produce a separate gas phase with the potential to generate overpressure.

SILICATE–CARBONATE REACTIONS AND OVERPRESSURE

In the previous sections, prograde silicate–carbonate reactions among common authigenic minerals in clastic rocks have been shown to impose increasing f_{CO_2} as temperature increases. This suggests that the transfer of CO_2 from a carbonate mineral phase, when present, to the gas phase by silicate–carbonate reactions could be a previously unrecognized mechanism for the generation of overpressure during progressive burial of siliciclastic sand and shale. The generation of overpressure by this mechanism requires that the solubility of CO_2 in the aqueous phase be exceeded so as to produce a free gas phase. Temperature, pressure and water salinity determine the solubility of CO_2 in the aqueous phase. Figure 10 shows schematically how the intersection of silicate–carbonate reactions (solid curves) with the CO_2 solubility surface produce an increasing path of fugacity (or partial pressure) of CO_2 with increasing temperature and result in the addition of CO_2 to a free gas phase, thus increasing pressure.

The position of the CO_2 solubility surface is a complex function of pressure, temperature and the salinity of the water. The partial pressure of CO_2 required to saturate water of a particular salinity can be computed from Barta & Bradley (1985) and the fugacity coefficients for CO_2 can be obtained from Angus *et al.* (1976). The position of the silicate–carbonate reaction, which then determines at what conditions CO_2 solubility in water is exceeded, is a function of the particular mineral reaction chosen and the temperature. Pressure has a less significant effect on the stability of silicate–carbonate reactions, but has an extreme effect on the solubility of CO_2 in water. In order to evaluate the effect of silicate–carbonate reactions on overpressure it is necessary to calculate the solubility of CO_2 and determine the T–f_{CO_2} trajectory of representative silicate–carbonate reactions. Recognizing that this is a mass-balance problem, and that the stoichiometry of individual silicate–carbonate reactions is poorly known, we chose to use the stoichiometry reported by various authors. The reactions

5 dolomite + kaolinite + quartz + $2H_2O$ = chlorite + 5 calcite + CO_2
(Hutcheon *et al.*, 1980)

4 dolomite + ankerite + 2.5 kaolinite = chlorite + 9 calcite + $9CO_2$
(Muffler & White, 1969)

illite + 12 quartz + calcite + KCl = 6 K-feldspar + $2H_2O$ + $2CaCl_2$ + $9CO_2$
(Smith & Ehrenberg, 1989)

were evaluated. Although the exact form of these reactions could be debated, for example, should $CaCl_2$ be used as a component of the fluid, the net effect of the exact choice of reaction mechanism on the value of f_{CO_2}, and thus the solubility in water, is relatively small.

Silicate–carbonate reactions can generate significant overpressure only if enough CO_2 is produced so as to exceed the solubility in water. Once the solubility is exceeded, CO_2 is added to the gas phase and generates pressure according to the relationship: $PV = ZnRT$

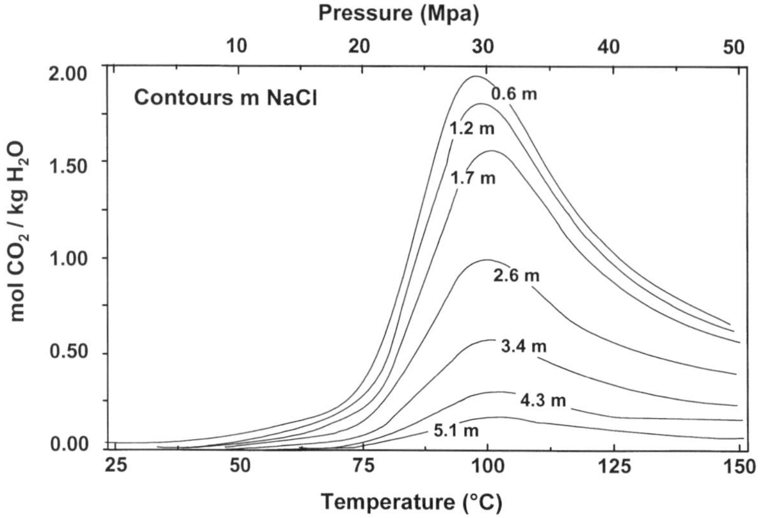

Fig. 11 The solubility of CO_2 calculated along a 31°C km^{-1} geotherm, from Barta & Bradley (1985). The contours represent the solubility at different concentrations (molal) of NaCl.

(where Z is compressibility and n is the number of moles of CO_2). If a gas phase were already present, for example, in a natural gas deposit, any CO_2 generated would result in a partial pressure of CO_2 in the gas and increase the total fluid pressure. The amount of overpressure that can be produced is related directly to the number of moles of CO_2 generated by a particular reaction, hence the influence of reaction stoichiometry. For the purposes of modelling the potential for silicate–carbonate reactions to increase overpressure, we assumed sandstone with 10% porosity and allowed 2 wt% of kaolinite or illite to react. Higher temperature and higher salinity will tend to increase the fractionation of CO_2 into the vapour phase and produce the greatest effect on calculated overpressure. The effect of different geothermal gradients was evaluated by choosing P–T paths constructed assuming two values for a geotherm, 20°C km^{-1} and 31°C km^{-1} at a hydrostatic pressure gradient of 10.6 MPa km^{-1}. The solubility of CO_2 in waters of varying salinity along a geotherm of 31°C km^{-1} at hydrostatic pressure is shown in Fig. 11. As expected, the solubility of CO_2 decreases as salinity increases. Note that there is a maximum in the solubility along this geotherm at about 100°C.

The solubility of CO_2 strongly influences the amount of overpressure that might be produced. Once the limit of solubility is reached, any additional CO_2 enters the gas phase, increasing calculated overpressure. Thus, factors that tend to decrease the solubility will tend to produce higher calculated overpressures. Figure 11 shows that high salinity decreases solubility and would be expected to increase overpressure. Solubility increases rapidly above 75°C, so high temperatures would decrease overpressures, but counteracting this effect is the higher f_{CO_2} imposed by the silicate–carbonate reactions. High pressure increases solubility, so in general, a low geothermal gradient in low-salinity fluids will tend to produce low calculated overpressures, whereas a high geothermal gradient in high-salinity fluids will tend to increase calculated overpressures.

For the lower geotherm scenario salinity was chosen as 0.6 M NaCl to simulate low potential for increased overpressure. A scenario that probably is more realistic uses the higher geotherm and 3.4 M NaCl salinity. Increasing the geothermal gradient and/or the fluid salinity would generate higher calculated overpressures at any particular depth. Figure 12 shows the effect on fluid pressure for illite and kaolinite silicate–carbonate reactions in a 'low overpressure' scenario. The kaolinite reaction tends to have a greater effect on overpressure than the illite reaction, but both have a significant effect.

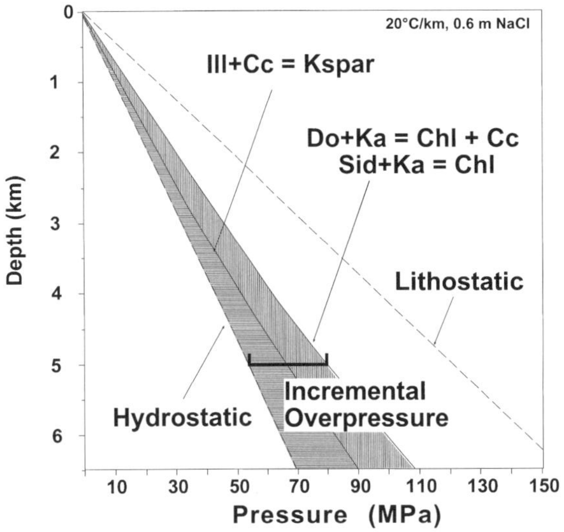

Fig. 12 Potential incremental overpressure generated in a 'low overpressure' scenario (20°C km^{-1} geothermal gradient and 0.6 M NaCl) by the illite–calcite–K-feldspar (reaction suggested by Smith & Ehrenberg, 1989) and dolomite–kaolinite and siderite–kaolinite reacting to chlorite (reaction suggested by Hutcheon, 1990).

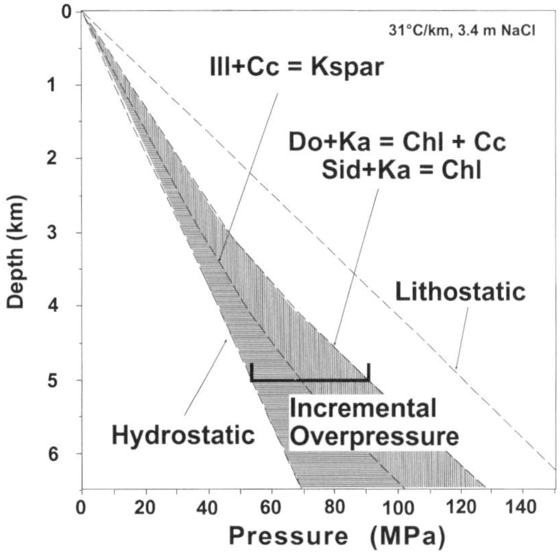

Fig. 13 Potential incremental overpressure generated in a 'high overpressure' scenario (31°C km^{-1} geothermal gradient and 3.4 M NaCl) by the illite–calcite–K-feldspar (reaction suggested by Smith & Ehrenberg, 1989) and dolomite–kaolinite and siderite–kaolinite reacting to chlorite (reaction suggested by Hutcheon, 1990).

At 5000 m depth the illite reaction produces pressures 13 MPa above hydrostatic (or about 24% increase over hydrostatic pressure) and the kaolinite reaction produces pressures 26 MPa above hydrostatic (or about 50% increase over hydrostatic pressure). The same reactions in the higher overpressure scenario produce greater pressures. Figure 13 shows that the illite reaction produces 16 MPa above hydrostatic (or about 31% increase over hydrostatic pressure) and the kaolinite reaction produces nearly 40 MPa above hydrostatic (or about 77% increase over hydrostatic pressure).

The calculations assume that all the CO_2 generated is retained within the rock and none is lost to reactions that precipitate carbonate minerals. Because the 'prograde' trend of all silicate–carbonate reactions dissolves calcite, this is a reasonable assumption. Silicate–carbonate reactions have been inferred to occur in sandstone and geothermal systems based on petrographic examination and/or fluid chemistry. The same minerals occur in shale and it is probable, although not documented in the literature to our knowledge, that such reactions occur in shale. It is interesting to speculate further that silicate–carbonate reactions could produce overpressure in shale, potentially leading to fracturing. The CO_2 produced may have an additional effect. It is well known from engineering oilfield practice that CO_2 causes a volume increase and viscosity decrease in oils. It is possible that CO_2 generated from silicate–carbonate reactions could be dissolved in oil formed in source rocks and may aid in the expulsion of oil from source rocks. At least two mechanisms could be envisaged:

1 build-up of overpressure could lead to fracturing of source rocks and associated transport of oil, as pressure is released;

2 the dissolution of CO_2 in oil would tend to decrease the viscosity and increase the volume of oil, making transport through the pore network of source rocks more effective.

It needs to be emphasized that we do not suggest that silicate–carbonate reactions are the only source of overpressure in sedimentary basins. Numerous other mechanisms have been proposed and we accept that some or all of these mechanisms are probably active in most basins. However, the results as calculated here suggest that silicate–carbonate reactions may well be an additional source that could augment overpressure. Evidence that silicate–carbonate reactions have contributed to overpressure might include carbonate-filled fractures in the caps of reservoir structures, shale with 'beef', isotopic composition of calcite and CO_2 in equilibrium at the reservoir temperature, and CO_2 partial pressures greater than solubility in the aqueous phase. Gathering of such evidence is difficult because:

- in high-pressure reservoirs water sampling is difficult, reducing the potential to include water composition in CO_2 solubility calculations;
- reservoir cap-rock typically is shale and cores are not often collected in shale because it does not normally produce hydrocarbon—thus, the expected location of carbonate-filled fractures is not sampled by coring;
- because the cap-rocks cannot be observed, it is not possible to compare the isotopic composition of carbonate cement in cap-rocks with that of CO_2 in the gas.

It is possible that careful examination of reservoir rocks, particularly cap-rocks, could provide data to give evidence for the contribution of silicate–carbonate reactions to overpressure, but no such studies are known to the authors.

CONCLUSIONS

The trend of increasing P_{CO_2} with temperature in sedimentary basins and in volcanic-dominated geothermal settings can be attributed to the 'buffering' of P_{CO_2} by silicate–carbonate reactions. Although increasing temperature requires such reactions to coexist with increasing P_{CO_2}, it is not necessary that they be the only source of CO_2 in fluids. However, if extraneous sources, such as organic maturation or decarboxylation of organic acids, do introduce CO_2, the partial pressure will regain values consistent with silicate–carbonate reactions. This requires that appropriate reactants are present in the rocks and precipitation of calcite or other carbonates would be observed. Silicate–carbonate reactions that produce low values of the $(a\mathrm{Ca}^{2+}/a(\mathrm{H}^+)^2)$ ratio in a fluid at any particular temperature will result in correspondingly higher P_{CO_2} values and vice versa. Sedimentary rocks containing clay minerals, feldspar and calcite tend to impose low $(a\mathrm{Ca}^{2+}/a(\mathrm{H}^+)^2)$ ratios in the fluid, thus producing high P_{CO_2}. Basaltic volcanic rocks, such as found in Iceland, contain zeolites, clay minerals and calcite and tend to impose high $(a\mathrm{Ca}^{2+}/a(\mathrm{H}^+)^2)$ ratios in the fluid, thus producing low P_{CO_2}.

The addition of H_2 from organic maturation reactions is an interesting potential source of H^+ that could result in dissolution of reservoir rocks and produce porosity. In rocks that contain H^+-buffering silicate–carbonate mineral assemblages the potential for dissolution of rock is low. However, in a rock that does not contain H^+ buffers there is the possibility that H_2 from organic maturation reactions could create porosity.

Silicate–carbonate reactions represent a potential source of overpressure. If silicate–carbonate reactions do produce CO_2 in sufficient quantities to induce overpressure, they also could be implicated in the generation of fractures and the migration of hydrocarbons in source-rock shale. It should be noted that P_{CO_2} values above those reported in the literature (Smith & Ehrenberg, 1989; Hutcheon *et al.*, 1993) are not a requirement of this process. In this model, exceeding the solubility of CO_2 in the aqueous phase *at equlibrium* with the silicate–carbonate assemblage drives overpressure by the transfer of CO_2 from the aqueous phase to the gas phase. Overpressured zones attributable to this process will not have higher P_{CO_2} than adjacent carbonate-containing clastic rocks. Although these are clearly speculative suggestions, the model calculations shown here, when combined with evidence that silicate–carbonate reactions do occur in many types of rocks, indicate that these speculations warrant further investigation.

ACKNOWLEDGEMENTS

Thanks are due to Glenn Wilson for numerous ideas and to Mike Lewan for many useful suggestions (and answering a lot of dumb questions from Hutcheon). Dirk Kirste read an early version of this manuscript and provided many thoughtful comments. Acknowledgement is made to the donors of the Petroleum Research Fund, administered by the American Chemical Society, for partial support of this research. Research funds to the senior author from NSERC and the Alberta Department of Innovation and Science also supported this work.

REFERENCES

Aagaard, P. & Helgeson, H.C. (1983) Activity/composition relations among silicates and aqueous solutions: II. Chemical and thermodynamic consequences of ideal mixing of atoms on homological sites in montmorillonites, illites, and mixed-layer clays. *Clay. Clay Mineral.*, **31**, 207–217.

Abercrombie, H.J. (1988) *Water–rock interaction during diagenesis and thermal recovery, Cold Lake Alberta.* PhD thesis, Department of Geology and Geophysics, The University of Calgary, Calgary, Alberta, 183 pp.

Angus, S., Armstrong, B. & Reuck, K.M. de (1976) *International Thermodynamic Tables of the Fluid State: Carbon Dioxide.* International Union of Pure and Applied Chemistry, Commission on Thermodynamics and Thermochemistry, Pergamon Press, Elmsford, New York, 385 pp.

Arnórsson, S., Gunnlaugsson, E. & Svavarsson, H. (1983) The chemistry of geothermal waters in Iceland. II. Mineral equilibria and independent variables controlling water compositions. *Geochim. Cosmochim. Acta*, **47**, 547–566.

Barta, L. & Bradley, D.J. (1985) Extension of the specific interaction model to include gas solubilities in high temperature brines. *Geochim. Cosmochim. Acta*, **49**, 195–203.

Barth, T. (1991) Organic acids and inorganic ions in waters from petroleum reservoirs, Norwegian continental shelf; a multivariate statistical analysis and comparison with American reservoir formation waters. *Appl. Geochem.*, **6**, 1–16.

Berman, R.G. (1988) Internally consistent thermodynamic data for minerals in the system Na_2O–K_2O–CaO–MgO–FeO–Fe_2O_3–Al_2O_3–SiO_2–TiO_2–H_2O–CO_2. *J. Petrol.*, **29**, 445–522.

Bethke, C. (1988) *The Geochemist's Workbench® 3.0, Users' Guide to Rxn, Act2, Tact and Gtplot.* Hydrogeology Program, University of Illinois, Urbana-Champaign, 184 pp.

Brown, T.H., Berman, R.G. & Perkins, E.H. (1988) Ge0-CALC: software package for calculation and display of pressure–temperature–composition phase diagrams using an IBM or compatible personal computer. *Comput. Geosci.*, **14**, 279–289.

Coudrain-Ribstein, A., Gouze, P. & de Marsily, G. (1998) Temperature–carbon dioxide partial pressure trends in confined aquifers. *Chem. Geol.*, **145**, 73–89.

Dixon, S.A., Summers, D.M. & Surdam, R.C. (1989) Diagenesis and preservation of porosity in Norphlet Formation (Upper Jurassic) Southern Alabama. *AAPG Bull.*, **73**, 707–728.

Hosttettler, J.D. (1984) Electrode electrons, aqueous electrons, and redox potentials in natural waters. *Am. J. Sci.*, **284**, 734–759.

Hutcheon, I. (1990) Clay–carbonate reactions in the Venture Field, Nova Scotia, Canada. In *Fluid–Mineral Interactions: a Tribute to H.P. Eugster*, Applied Geochemistry Special Publication 2 (Ed. by R.J. Spencer and I-Ming Chou), Geochemical Society, University Part, PA, pp. 199–212.

Hutcheon, I. & Abercrombie, H.J. (1990) Carbon dioxide in clastic rocks and silicate hydrolysis. *Geology*, **18**, 541–544.

Hutcheon, I., Oldershaw, A. & Ghent, E. (1980) Diagenesis of sandstones of the Kootenay Formation at Elk Valley (southeastern British Columbia) and Mount Allan (southwestern Alberta). *Geochim. Cosmochim. Acta*, **44**, 1425–1435.

Hutcheon, I., Shevalier, M. & Abercrombie, H.J. (1993) pH buffering by metastable mineral–fluid equilibria and evolution of carbon dioxide fugacity during burial diagenesis. *Geochim. Cosmochim. Acta*, **57**, 1017–1027.

Kharaka, Y.K. & Mariner, R.H. (1988) Chemical geothermometers and their application to formation waters from sedimentary basins. In: *Thermal History of Sedimentary Basins* (Ed. by N.D. Naeser and T.H. McCulloh), Springer-Verlag, New York, pp. 99–118.

Lewan, M.D. (1997) Experiments on the role of water in petroleum formation. *Geochim. Cosmochim. Acta*, **61**, 3691–3723.

Levinson, A.A. & Vian, R.W. (1966) The hydrothermal syntheses of montmorillonite group minerals from kaolinite, quartz and various carbonates. *Am. Mineral.*, **51**, 495–498.

Lundegard, P.D. & Trevena, A.S. (1990) Sandstone diagenesis in the Patani Basin (Gulf of Thailand): history of water–rock interaction and comparison with the Gulf of Mexico. *Appl. Geochem.*, **5**, 669–686.

Lundegard, P.D., Land, L.S. & Galloway, W.E. (1984) Problem of secondary porosity: Frio Formation (Oligocene), Texas Gulf Coast. *Geology*, **12**, 399–402.

Muffler, L.P.J. & White, D.E. (1969) Active metamorphism of Upper Cenozoic sediments in the Salton Sea geothermal field and the Salton Trough, Southeastern California. *Geol. Soc. Am. Bull.*, **80**, 157–182.

Smith, J.T. & Ehrenberg, S.N. (1989) Correlation of carbon dioxide abundance with temperature in clastic hydrocarbon reservoirs: relationship to inorganic chemical equilibrium. *Mar. Petrol. Geol.*, **6**, 129–135.

Zen, E-An (1959) Clay mineral–carbonate relations in sedimentary rocks. *Am. J. Sci.*, **257**, 29–43.

Int. Assoc. Sedimentol. Spec. Publ. (2003) **34**, 177–190

Experimental studies of clay mineral occurrence

D.A.C. MANNING

School of Civil Engineering and Geosciences, University of Newcastle upon Tyne, Newcastle upon Tyne NE1 7RU, UK, e-mail: david.manning@ncl.ac.uk

ABSTRACT

Fundamental controls on the occurrence and distribution of clay minerals in sedimentary rocks are determined by the application of experimental data relating to their thermodynamic stabilities and the kinetics of clay reactions. The thermodynamic properties of individual representatives of the major clay-mineral groups (kaolin, illite, smectite, chlorite) and polymorphs (kaolinite, dickite, nacrite, halloysite) have been determined mainly from synthesis experiments but also from solubility determinations and calorimetric investigations of natural and synthetic clays. In all cases, mineral starting materials need to be fully characterized to quantify impurities and heterogeneities at a molecular scale. In synthesis and solubility experiments, analysis of the fluid composition is integrated ideally with detailed high-resolution mineralogical analysis of solid experimental products. It is clear from many studies that kinetic factors control both the natural occurrence of clay minerals and the results of synthesis experiments, and this can lead to controversy in the interpretation of experimental data. Recent experimental work on clay-mineral reactions has been undertaken under conditions of pressure and temperature that either match the saturated vapour pressure curve for aqueous solutions, under autoclave conditions, or are at elevated pressure controlled independently of temperature. This review explains the principles involved in experimental design and their limitations, and provides signposts to key studies of clay stability.

INTRODUCTION

In clastic petroleum reservoirs, the clay-mineral assemblage plays a vital role in controlling key petrophysical and geochemical parameters. The morphology, wettability and stability of authigenic clay-mineral cements influence permeability to both water and hydrocarbon liquids, and clay reactions control many of the compositional characteristics of pore fluids in mudstones and sandstones (e.g. Hutcheon *et al.*, 1993). Detailed reviews of these aspects of clay-mineral–pore-fluid interaction are provided by many authors in this volume, as well as previous collections of papers (e.g. Manning *et al.*, 1993).

Almost all quantitative studies of clay-mineral occurrence are based ultimately on experimental studies, which provide information concerning the reactions shown by clay minerals in response to two important controls: changes in temperature and changes in water chemistry. Temperature changes may govern polymorphic transitions, such as the stabilities of the kaolin group polymorphs: kaolinite, dickite and nacrite. It is also observed from natural sequences that thermal history and clay-mineral assemblage correlate; for example, it is unusual to find smectites in deep, hot reservoirs, in which chlorite-group clays occur as authigenic minerals. The influence of water chemistry is also well known from studies of hydrothermal systems (Hemley, 1959; Hemley *et al.*, 1980) and is particularly relevant for artificial situations in which clays are used to contain potentially harmful aqueous liquids, as in waste containment systems.

This review focuses on some of the key issues that arise as a consequence of the technical limitations associated with experimental work. It acknowledges the fact that there are rather few centres with experimental facilities, and that their work is strongly influenced by the equipment at their disposal. First, it assumes ideal mineralogical compositions and summarizes typical experimental methods and procedures, examining their limitations as well as identifying their strengths. Second, examples of particular clay reactions are discussed, introducing problems that arise from the natural complexity and heterogeneity of the clays. In this way, the review sets out to illustrate principles and problems, and does not set out to provide an exhaustive list of all experimental work on clay reactions.

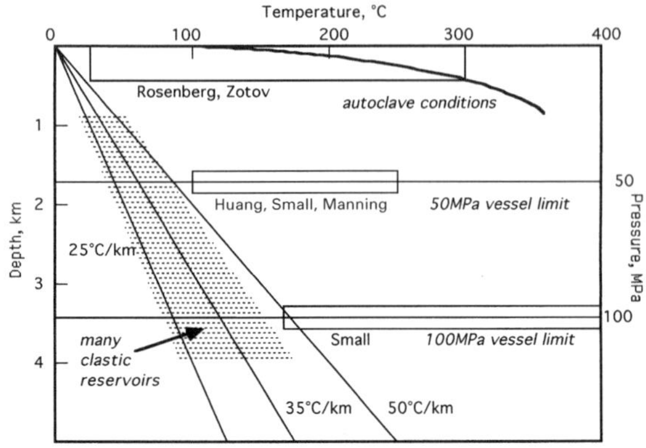

Fig. 1 General pressure–temperature relationships in petroleum reservoirs (shaded), along typical geothermal gradients in sedimentary basins and as used in experimental studies carried out in the laboratories of the named researchers.

GENERAL PRINCIPLES

Experimental conditions

The conditions required for an experimental study of clay reactions typically represent a compromise between those expected in nature and those that can be achieved in a human time-scale in the laboratory using the equipment that is currently available at an acceptable price. Typical pressures and temperatures for diagenetic systems in sedimentary environments that follow 'normal' burial curves with different temperature gradients are shown in Fig. 1. Typical reservoir *P–T* conditions are shown, using data from Warren & Smalley (1994) for the North Sea.

Figure 1 also shows the saturated vapour pressure curve for pure water in terms of pressure and temperature. This is the pressure achieved when water is taken to a particular temperature in a closed vessel, and corresponds to the conditions obtained using an autoclave (effectively a heavy-duty pressure cooker). It is quite clear that autoclave conditions are distant from those encountered on a burial curve. They correspond closely (and may be applicable directly) to conditions generated during steam flooding of oilfields during enhanced recovery, recovery from oil sands (e.g. Boon & Hitchon, 1983) or other engineering procedures when temperatures are increased artificially at low pressure. Using autoclaves, the experimental system is constrained by the saturated vapour pressure curve, and the fluid components are present in a two-phase system as an aqueous liquid and an aqueous vapour phase.

To duplicate reservoir pressures for a given temperature, an increased pressure in excess of that available using an autoclave needs to be generated. This is achieved by providing an external source of pressure. It means that the experimental system will contain a single aqueous phase, which is liquid at temperatures below the critical temperature for water (374°C) and which is the situation for all burial diagenetic systems. Figure 1 shows that experimental work at pressures in excess of autoclave conditions has focused at 50 MPa, equivalent to a depth of about 1.8 km, and at 100 MPa (3.5 km). Temperatures used in the experimental work are consistently in excess of those encountered with a 50°C km^{-1} gradient, in order to achieve results on a human time-scale.

The chemical composition of the aqueous phase also is a key variable to be considered in any experimental study. It not only influences the chemical stability of the mineral of interest, but the presence of solutes in an aqueous solution affects the pressure and temperature of the saturated vapour pressure curve. Thus with autoclaves, strictly it is impossible to carry out experiments under identical conditions of pressure and temperature unless aqueous fluid compositions are also identical. This cannot be achieved because in addition to any solute added prior to an experiment the minerals of interest have solubilities and contribute solutes as well. It is usually assumed (with good reason) that these effects are negligible, depending on the goal of a particular experimental study.

Experimental design

Most experimental studies consider clay-mineral reactions within a coupled system in which the aqueous phase is an essential constituent. 'Dry' reactions between clay minerals are rarely relevant for burial diagenetic systems, because of the presence of water under almost all circumstances. Natural variation in clay-mineral composition within the kaolin group is subtle and includes intercalation of illitic interlayers, particularly in vermiform kaolinite (Crowley, 1991; Robertson & Eggleton, 1991; Psyrillos *et al.*, 1999). Chemical variability arising from solid solution becomes increasingly substantial for the illite, smectite and chlorite groups, all of which also can show internal mineralogical heterogeneity. It is practically impossible to obtain a 'pure' natural mineral starting material that is chemically and crystallographically homogeneous (indeed, Rae & Manning (1996) found that mineralogically pure kaolinite from a widely used industrial source carried residual chemicals from the mineral processing stream). In an attempt to overcome these difficulties, minerals of known composition are chosen as experimental starting materials. In some cases these are end-members of solid solution series (e.g. pyrophyllite) and in other cases they are proxies that are not 'true' clay minerals (e.g. muscovite).

The framework for discussing clay stabilities depends upon the phase diagram for the system K_2O (or Na_2O)–Al_2O_3–SiO_2–H_2O, and this also influences selection of the starting material. This phase diagram permits kaolinite and illite to be described and has long formed the basis for the interpretation of clay stabilities from natural systems (e.g. Garrels, 1984). In the system K_2O–Al_2O_3–SiO_2–H_2O, muscovite ($KAl_3Si_3O_{10}(OH)_2$) is widely used as a proxy for illite, and pyrophyllite ($Al_2Si_4O_{10}(OH)_2$) for dioctahedral smectite. Both muscovite and pyrophyllite are relatively easily available as mineralogically pure phases, and consequently are well defined chemically and thermodynamically. Work with chlorite is complicated by the importance of iron as a constituent; experimentally, consideration of iron-bearing phases requires control of oxygen fugacities to ensure that Fe has an appropriate oxidation state. At high temperatures, this is achieved using mineral buffers (Ulmer & Barnes, 1987), but this technique is practically impossible below 400°C. To simplify matters, chlorite stabilities are considered using the iron-free system MgO–Al_2O_3–SiO_2–H_2O, directly corresponding to the clinochlore type, or by using low-iron chlorites (Aja & Small, 1999). The use of the two model systems (K_2O–Al_2O_3–SiO_2–H_2O and MgO–Al_2O_3–SiO_2–H_2O) allows simplified or end-member compositions to be studied, permitting eventual extrapolation to a wide range of natural compositions. In all cases, the low solubility of aluminium in natural waters means that this component is assumed to be saturated, and mineral stabilities are expressed in terms of K or Mg, Si and hydrogen ions (pH).

Figure 2 shows the phase diagram for the system K_2O–Al_2O_3–SiO_2–H_2O (150°C; 50 MPa) in the form that is very widely used for the design and interpretation of experimental studies of clays. The principle behind the construction of this diagram can be illustrated using the reaction of kaolinite to muscovite as an analogue for illite:

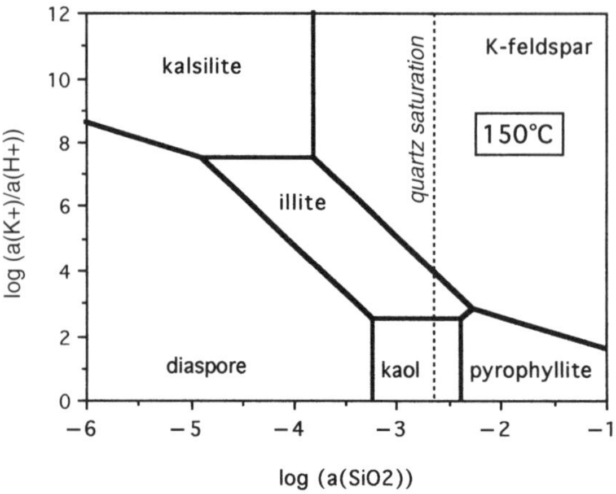

Fig. 2 Simplified activity phase diagram for the system $K_2O–Al_2O_3–SiO_2–H_2O$ at 150°C, 50 MPa, showing the topology of the stability fields for kalsilite, potassium feldspar, diaspore, illite (with the muscovite composition $KAl_3Si_3O_{10}(OH)_2$), kaolinite and pyrophyllite, and quartz saturation.

$$3Al_2Si_2O_5(OH)_4 + 2K^+ = 2KAl_3Si_3O_{10}(OH)_2 + 2H^+ + 3H_2O \quad \text{(R1)}$$

Assuming that the mineral phases are pure (unit activities for the solids), rearranging gives:

$$\log\left(\frac{a_{K^+}}{a_{H^+}}\right) = -\frac{1}{2}\log K$$

where log K is the logarithm of the equilibrium constant for the reaction. An equilibrium assemblage has a defined single value for the ratio of the activities of potassium and hydrogen ions in aqueous solution. This is independent of the solution silica content, giving a horizontal phase boundary in Fig. 2. If pH is known, the potassium concentration can be calculated, and vice versa.

The significance of the pressure used in an experimental investigation can be assessed by considering the volume changes associated with clay reactions. For reaction (R1), the molar volume change is approximately 37 cm^3, the left-hand side of the equation having a lower volume, indicating that the reaction will not be favoured by an increase in pressure.

Thus in any study of the stability of a clay mineral, in which it is desired to establish its thermodynamic properties at a given pressure or temperature, all that is required is to generate an equilibrium assemblage of the clay mineral in a known reaction with a second phase of known composition, and to measure the composition of the aqueous phase. The activities of solid end-member components must be known for all mineral phases, reactants and products. The experiment must acknowledge the importance of the Gibbs Phase Rule: for example, if quartz is present in the equilibrium assemblage one degree of freedom is lost and fluid silica contents are fixed. This approach represents the basic philosophy behind much of the experimental work that has been carried out on clay-bearing systems, and depends on the assumption that no changes take place in the chemical or mineralogical composition of the clay minerals.

In addition to use of natural clay minerals as starting materials, synthesis experiments have been used to generate clays with known compositions (determined partly from the reagent mix but most importantly by analysis of the products using high-resolution methods such as transmission electron microscopy (TEM) to assess heterogeneity). This approach requires knowledge of the thermodynamic stabilities of the clay of interest in pressure, temperature and composition (PTX) space, and so synthesis itself becomes an integral part of an experimental investigation of clay reactions (e.g. Small *et al.*, 1992b; Roberson *et al.*, 1999).

Key experimental methods

Technical constraints underpin the practical design of any experiment, and available options for the design of equipment are given in detail by Ulmer & Barnes (1987). Considerations that need to be taken into account include the following:

- chemical inertness of the reaction containment system;
- rates and mechanisms for applying and reducing heat and pressure;
- ease of access to reaction products.

The first of these can be addressed using a wide

Table 1 Materials widely used for reaction containment.

Material	Advantages	Disadvantages
PTFE (Teflon™)	Relatively cheap, unreactive to a wide range of solution chemistries	Relies on a mechanical seal; may be permeable; mechanical seals may be unreliable above about 200°C; breaks down chemically above 300°C. No strength; must be used within a steel or alloy vessel
Titanium	Available as autoclaves and in suitable form for a wide range of fabrications	Expensive; relies on mechanical seals (although welding is possible); limited strength so suitable mainly for autoclaves or liners of high-pressure vessels
Gold	Scrap has value; easy to weld and seal; highly flexible when annealed; impermeable to water and gases within temperature range of interest	Expensive; no strength and so must be used within a steel or alloy vessel. Can work-harden, leading to embrittlement and fracturing

variety of materials, for which advantages and disadvantages of typical materials are given in Table 1. This table by no means gives an exhaustive list of materials and their properties, but highlights some factors of immediate importance to those planning an experimental investigation.

Once sample containment has been chosen, consideration needs to be given to its incorporation into a piece of equipment that will perform as required. Table 1 identifies gold and PTFE as having negligible strength (certainly so if working under conditions for a burial curve). Thus these materials cannot be used on their own to make a reaction vessel. Titanium can be used to make an autoclave, but it is not strong enough to be used as a material for a pressure vessel for use at much higher pressures, without being prohibitively large in vessel wall thickness. However, if experiments are required to lie on the saturated vapour pressure curve for water, titanium autoclaves (with titanium closures and PTFE seals) can be very satisfactory (e.g. Aagaard *et al.*, 2000).

All three materials given in Table 1 can be used to line a potentially reactive steel (or alloy) vessel, in which the mechanical properties of the steel are used to design a vessel that can perform at high pressures and temperatures. Difficulties arise with designing a seal, given that potentially corrosive liquids and gases may attack materials outside the lining system if access to these is possible. One way around this problem is to use flexible gold (or PTFE) cells within a volume of water that is pressurized (Seyfried *et al.*, 1987); the gold acts as a membrane transmitting pressure from the water but preventing reaction with the contents of the cell. For work at high pressures, or where fluid compositional information can be obtained at the end of an experiment, sealed gold capsules can be used within cold-seal pressure vessels (Small *et al.*, 1992a; Huang, 1993).

An important aspect of the design of a reaction vessel is its thermal inertia. A large vessel, designed to accommodate high pressures encountered on a burial curve, can take 24 h to heat up to and stabilize at run temperature, and a similarly long time to cool so that it can be opened to gain access to the experimental products. This provides opportunities for reactions to take place, especially during cooling, and these may change the solid assemblage or the fluid composition. In circumstances where carbonate is present, reduction in pressure during cooling or sampling perturbs fluid compositions through the precipitation of calcite (or other carbonates) as a coating on grain or metal surfaces.

It is often the case that the experimental equipment has features that influence the design of the experiment. In these circumstances, the experiment has to be designed and interpreted with care.

Use of complexing

Aluminium and Si, essential constituents of clays, have low aqueous solubilities compared with cations such as K^+, Mg^{2+}, etc., and are known to be influenced by complexing with ligands such as chloride or the organic acid anions. The organic acid anions differ from chloride by virtue of their low thermal stability; they decompose to give carbon dioxide, methane and water (depending on their composition, temperature and overall fluid composition). It is highly debatable whether or not organic acid anions form complexes with Al or Si in nature (Crossey, 1991; Fein, 1994; Fein *et al.*, 1994, 1995). Irrespective of this, complexes with organic ligands provide an artificial experimental device that can be used to generate solutions supersaturated with respect to Al, influencing the crystallization of clays (e.g. Small & Manning, 1994a). Small (1993) has used this technique to influence the crystallization of illite and kaolinite, and Small (1994) did so to investigate the smectite-to-illite reaction. The approach involves generation of an experimental fluid that is strongly enriched in aluminium by allowing a readily soluble source, such as an artificial gel, to dissolve in the presence of oxalate, so that aluminium oxalate complexes are formed in solution. These complexes are unstable at run conditions of pressure and temperature, and decompose. Aluminium is liberated, supersaturating the fluid, and precipitates out as a clay mineral according to its stability field, such as shown in the phase diagram of Fig. 2 (stoichiometric phases) or similar diagrams for non-stoichiometric phases (Aagaard & Helgeson, 1983; Aja *et al.*, 1991a).

The use of complexing as an experimental device also has been reported for flow experiments (Small & Manning, 1994b) in which kaolinite was precipitated in a rock core within a hydrothermal flow cell. Aluminium entered the core as an aluminium malonate complex. When temperature was raised above 120°C, the aluminium malonate complex was destroyed, and kaolinite precipitation took place. On reduction of the core temperature to below 120°C, the complex was again stable and precipitation ceased. Although there can be no direct evidence of the occurrence of oxalate and malonate complexes in natural waters in diagenetic systems, the widespread presence of calcium oxalate minerals in a range of geological settings demonstrates that at least oxalate may occur much more widely than currently accepted (Manning, 2000). Any calcium oxalate mineral that is present will buffer solution oxalate contents, albeit sparingly, and so provide the ligands required for formation of aluminium oxalate complexes.

Clay reaction kinetics and experiment design

In all experiments of clay reactions, it is essential to consider reaction kinetics. Apart from the fact that the goal of an investigation might involve determination of precipitation or dissolution kinetics, or the rate of a mineral reaction, the experiment design has to ensure that an appropriate amount of time is spent at the *P–T* conditions of interest, and that the time taken for heating or cooling before or after the experiment will not affect the composition of the reaction products. It also is important to consider how the design of the experiment permits sampling for analysis. In sealed reaction vessels, such as autoclaves or welded precious metal capsules, generally neither the solid nor the fluid products can be investigated until the experiment is completed and the vessel or capsule has been opened at room temperature and atmospheric pressure. This gives a single point result, with no indication of the path taken by fluid or mineral composition during the time spent at temperature and pressure or during heating and cooling.

As an alternative, the use of fluid sampling pressure vessels or flow systems (Ulmer & Barnes, 1987) allows samples of solution to be taken periodically before the end of an experiment, without disturbing pressure or temperature of the reactants. Using these techniques, fluid compositions can be determined and plotted within a phase diagram to chart the progress of a mineral reaction, including the determination of reaction rates and assessment of changes that take place during heating and cooling. At the end of an experiment of this type, the residual solids can be recovered for investigation in the context of the observed changes in fluid chemistry (e.g. Huang, 1993; Small, 1993). Additionally, in some circumstances suspended solids also can be sampled in the course of an experiment without disturbing pressure or temperature to provide evidence of changes in mineralogy. This has been undertaken very effectively by Small (1994) in an investigation of the smectite-to-illite reaction (discussed in detail below).

The design of an experimental investigation of clay-mineral reaction kinetics also has to take into account the extent to which the composition of the reactant assemblage differs from that of the product (Oelkers, 1996). High reaction rates are obtained for experiments in which the reactants are far from equilibrium initially; in nature, most reacting systems are expected to be close to equilibrium and this situation should be reproduced experimentally. In practical terms, establishing a perturbation from equilibrium and then monitoring the response requires very careful experiment design coupled with detailed thermodynamic modelling and thorough investigation of the mineral phases that are used (see below—for example, the apparent conflict between Rosenberg's group (Sass *et al.*, 1987; Aja *et al.*, 1991a) and Essene & Peacor (1995, 1997)).

EXAMPLES OF CLAY REACTIONS

Having discussed some of the general features of experiment design and execution, a number of cases have been chosen to illustrate specific reactions and problems that have arisen. In some cases, these are a consequence of the design of the experiment, and demonstrate (for example) the need for additional characterization of the solids. In other cases examples are given of conflicts arising from differences in the interpretation of experimental data.

Kaolinite stability

Despite its widespread occurrence as a reservoir clay mineral, relatively few experimental studies have addressed the stability of kaolinite and its polymorphs under diagenetic conditions. The most recent determination of the thermodynamic properties of the kaolin polymorphs used drop-solution calorimetry (de Ligny & Navrotsky, 1999). This technique involves measurement of the heat energy liberated when the sample dissolves completely in molten lead borate at 975 K (702°C), and has been specially developed to measure the heat of formation of hydrous phases. It provides fundamental information and is independent of the method of synthesis of the clay or any reactions within the experimental system (natural samples were used in this study). De Ligny & Navrotsky (1999) determined the enthalpy of formation of kaolinite, dickite, nacrite and halloysite, and calculated values for the standard free energy of formation for kaolinite, dickite and halloysite, paying attention to the chemical impurities (Table 2). These data show that kaolinite is the most stable of the polymorphs, and suggest that dickite and halloysite (and nacrite—although the lack of entropy data prevents calculation of this polymorph's standard free energy of formation) are metastable. As metastable phases, their origin has to be related to kinetic effects, and so de Ligny & Navrotsky (1999) argue that the concept of a stability field for dickite (Ehrenberg *et al.*, 1993) is erroneous, because the dissolution of kaolinite with increasing temperature and subsequent crystallization of dickite is impossible thermodynamically in the light of their experimental results.

In contrast to de Ligny & Navrotsky (1999), Zotov *et al.* (1998) reported thermodynamic data for kaolinite and dickite, demonstrating that dickite is the stable phase to temperatures of up to at least 350°C, and that kaolinite is metastable relative to dickite. These experiments determined the solubilities of kaolinite and dickite, thus modelling experimentally what would be expected in nature if it is assumed that the two phases are produced by

Table 2 Standard enthalpy and standard free energy of formation at 25°C and 1 bar from the elements for kaolinite, dickite, nacrite and halloysite: (1) data from de Ligny & Navrotsky (1999); (2) data from Zotov *et al.* (1998); (3) data from Devidal *et al.* (1996).

	Standard enthalpy of formation (kJ/mol^{-1})			Standard free energy of formation (kJ/mol^{-1})		
Polymorph	**1**	**2**	**3**	**1**	**2**	**3**
Kaolinite	−4120.2	−4120.9	−4121.9	−3799.4	−3799.3	−3800.3
Dickite	−4107.6	−4127.1		−3785.1	−3803.1	
Nacrite	−4104.0					
Halloysite	−4097.5			−3776.8		

Table 3 Experimentally determined dissolution and precipitation rates for kaolinite and illite.

Study	Temperature (°C)	Kaolinite dissolution rate (moles m^{-2} s^{-1})	Kaolinite precipitation rate (moles m^{-2} s^{-1})	Illite precipitation rate (moles m^{-2} s^{-1})
Nagy *et al.*, 1991	80	1.2×10^{-12}		
Huang, 1993	200	2.2×10^{-12}		
	250	0.85×10^{-11}		
Small, 1993	160		1.0×10^{-8}	
	180		4.7×10^{-8}	
	200		1.9×10^{-7}	1.6×10^{-9}
	230			3.3×10^{-9}
	250			3.5×10^{-8}

dissolution of one and precipitation of the other. The experiments were carried out in PTFE-lined titanium autoclaves and titanium autoclaves, using hydrochloric acid solutions with concentrations of 0.1 and 0.25 M (approximate pH 1.1–1.25 at *P* and *T*). Careful analysis was carried out for the experimental solutions, extracted after cooling the autoclaves, or by sampling during the experiment without cooling the autoclaves. As de Ligny and Navrotsky (1999) point out, Zotov *et al.*'s (1998) experiments depend on robust knowledge of activity coefficients for the dissolved species, and these may not be known. The achievement of chemical equilibrium is also questioned, and incongruent dissolution is identified as an additional possible source of error. Clearly, further work is required to resolve this issue.

Kaolinite dissolution–precipitation kinetics

In addition to determination of the stability of kaolinite and its thermodynamic properties, a number of studies have addressed the kinetics of its dissolution and precipitation (Table 3). Dissolution rates are presented for kaolinite by Nagy *et al.* (1990, 1991), determined at 80°C and 1 atmosphere pressure, using a flow cell. In these experiments, kaolinite dissolution was undertaken. Huang (1993) carried out similar experiments, investigating reactions involving kaolinite and boehmite. In both cases, the calculated rates of reaction are close to equilibrium, and can be considered as estimates of both dissolution and precipitation rates. Strictly, however, Huang's (1993) kinetic data describe the forward and reverse rates for the reaction of kaolinite to boehmite. An alternative approach has been taken by Small (1993), who presented precipitation rates for kaolinite (and illite), determined by monitoring changes in fluid composition when kaolinite precipitates from a fluid supersaturated with respect to aluminium and silicon derived from dissolution of a gel. Table 3 shows that the precipitation rates determined by Small (1993) are greatly in excess of those determined by Huang (1993) at the same temperature. This can be explained in part by difference in experimental design: Small's experiments were far from equilibrium, whereas Huang's were derived for an equilibrium reaction between kaolinite and boehmite and so represent the kinetics of the kaolinite precipitation close to equilibrium (which is more likely to be the case in diagenetic *in situ* conditions).

Illite stability

Again, controversy enters the question of the stability and thermodynamic properties of illite. Substantial work has been carried out at Rosenberg's laboratory (Sass *et al.*, 1987; Aja, 1991; Aja *et al.*, 1991a,b; Aja & Rosenberg, 1992; Yates & Rosenberg, 1996, 1997), using PTFE-lined reaction vessels, which were cooled prior to extraction of both solid and liquid reaction products for analysis. The approach taken was to monitor fluid composition for KCl–HCl solutions of various concentrations, equilibrated with illite or muscovite, well characterized using X-ray diffraction and analytical transmission electron microscopy. Phase relationships for the system $K_2O–Al_2O_3–SiO_2–H_2O$ between 25 and 250°C are presented by Aja *et al.* (1991a), who publish phase diagrams analogous to that shown in Fig. 2. Aja *et al.* report that muscovite is stable at temperatures above 200°C. Between 100 and 200°C, an illite with 85% occupancy of the K site is stable, and below 100°C the stable phase is an illite with 69% occupancy. This phase diagram has been criticized by Essene & Peacor (1995), who claim that it violates Schreinemakers' rules of phase boundary construction, that it cannot be regarded as an equilibrium phase diagram, and consequently that an equilibrium mineral assemblage was not produced experimentally. Essene & Peacor (1995) present an alternative diagram that is consistent with fluid compositional data derived from the experiments and obeys Schreinemakers' rules. A consequence of this is that the relative thermal stabilities of illite and muscovite are reversed (muscovite becomes the stable phase at low temperature, illite at high temperature). A robust defence of Aja *et al.*'s (1991a) interpretation of their experimental data has been made by Aja & Rosenberg (1996), leading to a further retort by Essene & Peacor (1997). Undoubtedly both parties in this discussion raise valid points, but the problem of the relative stabilities of muscovite and illite are not yet resolved.

Kinetics of illite precipitation

As shown in Table 3, illite precipitation rates have been determined by Small (1993) in experiments under conditions that were far from equilibrium and in which illite precipitated from a supersaturated solution derived by dissolution of an amorphous gel. These experimental results have been compared with independent empirical evidence for illite growth rates that can be constrained by dates obtained for authigenic illites. Using the experimentally determined illite precipitation rates in thermal history modelling for specific reservoirs where the age of authigenic illite is known, the onset of precipitation can be predicted and compared with that age. Small (1993) showed that there is excellent correspondence between the predicted and the observed ages for authigenic illite within reservoirs from the North Sea. This indicates that the experimentally determined illite precipitation rates are valid. As these are derived from experiments with enhanced solution Al contents, it appears that there may be circumstances in petroleum reservoirs in which illite crystallization also takes place far from chemical equilibrium (this is supported by morphological studies; see 'influences on grain morphology' below). One possible mechanism that matches the experimental design could be thermal destabilization of complexes of aluminium with organic acid anions within the reservoir.

The smectite-to-illite reaction

Much attention has been paid to the distribution with depth of smectite and illite, with the definitive work from the Gulf of Mexico (Hower *et al.*, 1976) showing that illite takes the place of smectite at temperatures above approximately 95°C and depths below 3700 m. The nature of this reaction has received much attention (see Bloch *et al.* (1998) for a recent critical appraisal), and has been addressed experimentally by Small (1994). In a very elegant set of experiments, Small used fluid sampling pressure vessels to remove a composite sample of solution and suspended clay solids periodically during long experiments at 200°C and 50 MPa (500 bars). The role of organic acid anions (acetate and oxalate) played an important role in these experiments, particularly in controlling pH by enabling the aqueous phase to act as a buffer solution. Mass-balance calculations showed that mineral dissolution–precipitation rates in this complex experiment could not be determined directly from fluid data, as the amount of illite produced (determined by X-ray diffraction) was too great to be explained from the changes observed in fluid composition. Instead, Small (1994) proposed that local dissolution of smectite and precipitation of illite took place, with very limited solute residence in the aqueous phase, and concluded that organic acid anions associated with petroleum formation could accelerate the formation of illite.

Influences on grain morphology: the fibrous illite problem

One of the major causes of poor hydrocarbon production rate from a reservoir with apparently good porosity properties is the occurrence of fibrous illite as an authigenic phase (e.g. Stalder, 1973; Wilson & Pittman, 1977; Small & Manning, 1993). Fibrous illite causes problems by baffling pore throats without a major reduction in pore volume. Petrographic information gives no insight into the controls on the formation of fibrous illite, and so a number of experimental studies have been undertaken to determine the physical and/or chemical conditions that favour crystallization of illite with a fibrous morphology. Illite has been synthesized by Huang (1992), in the temperature range 200–350°C and at 50 and 100 MPa (500 and 1000 bars) pressure, and by Small *et al.* (1992a) under similar *P–T* conditions. These studies both identified fluid compositions from which illite could form, and with a fibrous morphology. Small *et al.* (1992b) and Small & Manning (1993) show clearly in the experiments that the formation of fibrous illite is associated with rapid crystallization from solutions designed to have high fluid Al from the decomposition of a thermally unstable aluminium oxalate complex ($Al(COO)_2^-$), and which have higher K^+/H^+ ratios than appropriate for the illite stability field in the phase diagram for the system $K_2O–Al_2O_3–SiO_2–H_2O$ (Fig. 3). For illite crystallization from other solutions with starting compositions that were close to equilibrium, plotting within the illite stability field in this

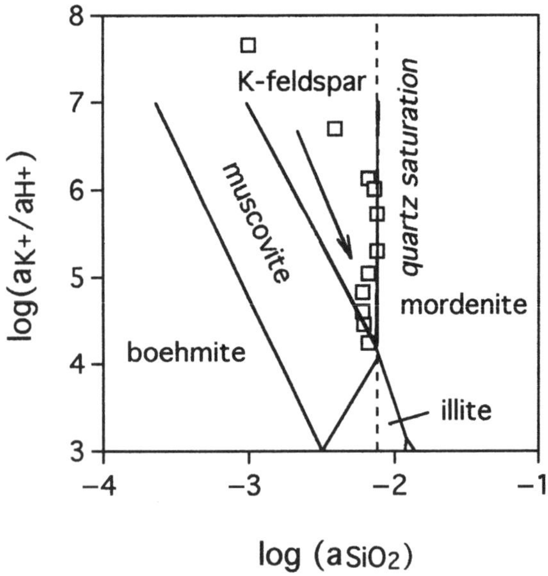

Fig. 3 Changes in fluid composition observed for experiments precipitating fibrous illite at 250°C and 50 MPa (Small & Manning, 1993).

phase diagram, a 'normal' platy morphology was obtained. Small & Manning's (1993) experiments yield information for the change in fluid composition as crystallization took place, and as shown in Fig. 3, fluid compositions coincide with the potassium feldspar–mordenite phase boundary. Mordenite is a zeolite, with a fibrous morphology, and these results are interpreted to show that the initial phase to crystallize is a zeolite-like precursor phase, which gives its fibrous morphology to illite that pseudomorphed the zeolite.

Behaviour of composite illite–kaolinite systems

Illite or kaolinite rarely occur naturally as a pure single mineral, and the influence of composite assemblages has to be taken into account (the matrix effect), particularly in experiments that are designed to assess aqueous solute yields from mudstones in response to thermal maturation. This work has significance with respect to the investigation of organic-rich mudstones as petroleum source rocks; to obtain an integrated assessment of their thermal maturation, mineral–fluid reactions are as important as the behaviour of their organic constituents.

As examples of the way in which experimental work has been used to investigate the thermal maturation of mudstones, two studies can be considered. Rae *et al.* (1993) report results for experimental investigation of the thermal maturation of two North Sea source rocks: the Kimmeridgian and the Permian Marl Slate Formation. Experiments were carried out between 150 and 310°C at 50 MPa (500 bars). In these experiments, the mineral assemblage in the two natural rocks included carbonate minerals as well as silicates. Using fluid sampling pressure vessels, a pressure quench takes place during sampling, and calcium carbonates precipitated in the equipment's sampling tubes. This experimental problem is directly analogous to the development of scale within petroleum production pipework. To overcome the problem of carbonate precipitation, Rae & Manning (1996) report results for experiments to determine the behaviour of kaolinite and natural kaolinite–illite (ball clay) assemblages that were carbonate-free. In this instance, fluid compositions equilibrated at values predicted using equilibrium thermodynamic geochemical solution modelling, demonstrating that fluid compositions are controlled by mineral assemblages (cf. Seewald *et al.*, 1990; Hutcheon *et al.*, 1993).

Chlorite synthesis and morphology

Experiments that synthesize chlorite have focused on metamorphic conditions of pressure and temperature (e.g. Saccocia & Seyfried, 1994). Roberson *et al.* (1999) carried out experiments designed to synthesize corrensite by reaction of the trioctahedral smectite saponite at 200 MPa and 350 or 500°C. Clearly these conditions are well in excess of those encountered in burial diagenesis, but they do demonstrate overall stability conditions and indicate those required for the formation of a chlorite. Closer to burial diagenesis, Small *et al.* (1992a) report the synthesis of an expandable 14 Å chlorite-like phase from hydrothermal recrystallization of a gel with composition $Si_6Al_2Fe_2Mg_2O_{20}$ at 250°C and 100 MPa (1 kbar). They were also able to synthesize a 7 Å berthierine and trioctahedral smectite as a product of reaction between siderite, magnesite, amorphous silica and alumina. Non-expandable 14 Å chlorite (chamosite, clinochlore) has been synthesized, but only at temperatures above 400°C (Nelson & Roy, 1958; Velde, 1973). Small *et al.* (1992a) suggest that the 7 Å phase is readily produced in the laboratory, and that the 14 Å phase requires a slow precipitation rate from fluids that are close to saturation. These conditions are difficult to achieve experimentally but may predominate in natural reservoirs. There are circumstances, however, in which authigenic 14 Å chlorites develop a platy boxwork texture (Humphreys *et al.*, 1989; Hillier, 1994) similar to those produced experimentally: these textures could result from the transformation of an initial 7 Å phase (Small *et al.*, 1992a). This is supported by synthesis work of Aagaard *et al.*

(2000), who produced 14 Å overgrowths on a reservoir sand originally coated with berthierine, at 200–250°C and saturated vapour pressure (*c.* 4 MPa). Finally, a lithium chlorite, cookeite, has been synthesized unintentionally in feldspar-dissolution experiments reported by Manning *et al.* (1991), in which lithium acetate solutions were used to buffer pH (the assumption being made that lithium would be conservative). These experiments were carried out at 150°C and 50 MPa, and yielded a 7 Å phase determined as cookeite when investigated using analytical transmission electron microscopy, confirming the relative ease with which a 7 Å phase can be synthesized.

CONCLUSIONS

This brief review has highlighted some of the problems that arise in the experimental investigation of clay-mineral reactions. Relatively few laboratories are active in providing the experimental information that is required to develop predictive theoretical models of clay reactions that are known to take place in sedimentary rocks. There are also technical restrictions on the design of experimental work, and these restrict the possible diversity of conditions that are available. Controversies surround experimental work on the relative stabilities of the kaolin polymorphs (kaolinite, dickite, nacrite and halloysite), and of illites with compositions that deviate from an ideal muscovite composition. Despite these problems, it is possible to synthesize most authigenic clay minerals.

Although it is important to obtain thermodynamic data relating to clay-mineral stability, it is quite clear that the kinetics of clay precipitation are fundamental to controlling the progress of an experimental study, and are highly likely to influence the distribution of specific clay-mineral assemblages in nature. As an experimental device, the limited thermal stability of organic complexes of Al (such as aluminium oxalate and aluminium malonate complexes) has been exploited to supersaturate a solution with respect to Al, driving clay crystallization and permitting the determination of precipitation rates from fluids with compositions far from equilibrium. There can be no direct evidence for the occurrence of complex species of this type in natural reservoir systems.

It is also clear from most published studies that solid experimental products (and reactants) need to be characterized as fully as possible in order to interpret fluid compositional data. In addition to X-ray diffraction, transmission electron microscopy is often essential to determine the chemical and mineralogical composition of clay particles, which often show heterogeneity at a very small scale (tens of nanometres). Without thorough investigation of the solids that participate in experimental reactions, the value of any experimental study is severely limited.

REFERENCES

Aagaard, P. & Helgeson, H.C. (1983) Activity/composition relations among silicates and aqueous solutions: II. Chemical and thermodynamic consequences of ideal mixing of atoms on homological sites in montmorillonites, illites and mixed-layer clays. *Clay. Clay Mineral.*, **31**, 207–217.

Aagaard, P., Jahren, J.S., Harstad, A.O., Nilsen, O. & Ramm, M. (2000) Formation of grain-coating chlorite in sandstones. Laboratory synthesized vs. natural occurrences. *Clay Mineral.*, **35**, 261–269.

Aja, S.U. (1991) Illite equilibria in solutions: III. A reinterpretation of the data of Sass *et al.* (1987). *Geochim. Cosmochim. Acta*, **55**, 3431–3453.

Aja, S.U. & Rosenberg, P.E. (1992) The thermodynamic status of compositionally-variable clay minerals: a discussion. *Clay. Clay Mineral.*, **40**, 292–299.

Aja, S.U. & Rosenberg, P.E. (1996) The thermodynamic status of compositionally-complex clay minerals: discussion of 'Clay mineral thermometry—a critical perspective'. *Clay. Clay Mineral.*, **44**, 560–568.

Aja, S.U. & Small, J.S. (1999) The solubility of a low-Fe clinochlore between 25 and 175°C and $P_v = P_{H_2O}$. *Eur. J. Mineral.*, **11**, 829–842.

Aja, S.U., Rosenberg, P.E. & Kittrick, J.A. (1991a) Illite equilibria in solutions: I. Phase relationships in the system $K_2O–Al_2O_3–SiO_2–H_2O$ between 25 and 250°C. *Geochim. Cosmochim. Acta*, **55**, 1353–1364.

Aja, S.U., Rosenberg, P.E. & Kittrick, J.A. (1991b) Illite equilibria in solutions: II. Phase relationships in the system $K_2O–MgO–Al_2O_3–SiO_2–H_2O$ between 25 and 250°C. *Geochim. Cosmochim. Acta*, **55**, 1365–1374.

Bloch, J., Hutcheon, I.E. & de Caritat, P. (1998) Tertiary volcanic rocks and the potassium content of Gulf Coast shales; the smoking gun. *Geology*, **26**, 527–530.

Boon, J.A. & Hitchon, B. (1983) Application of fluid–rock reaction studies to *in situ* recovery from oil sand deposits, Alberta, Canada—I. Aqueous phase results for an experimental–statistical study of water–bitumen–shale reactions. *Geochim. Cosmochim. Acta*, **47**, 235–248.

Crossey, L.J. (1991) Thermal degradation of aqueous oxalate species. *Geochim. Cosmochim. Acta*, **55**, 1515–1527.

Crowley, S.F. (1991) Diagenetic modification of detrital muscovite: an example from the Great Limestone cyclothem (Carboniferous) of Co. Durham, UK. *Clay Mineral.*, **26**, 91–103.

De Ligny, D. & Navrotsky, A. (1999) Energetics of kaolin polymorphs. *Am. Mineral.*, **84**, 506–516.

Devidal, J.-L., Dandurand, J.L. & Gout, R. (1996) Gibbs free energy of formation of kaolinite from solubility measurement in basic solution between 60 and 170°C. *Geochim. Cosmochim. Acta*, **60**, 553–564.

Ehrenberg, S.N., Aagaard, P., Wilson, M.J., Fraser, A.R. & Duthie, D.M.L. (1993) Depth-dependent transformation of kaolinite to dickite in sandstones of the Norwegian continental shelf. *Clay Mineral.*, **28**, 325–352.

Essene, E.J. & Peacor, D.R. (1995) Clay mineral thermometry: a critical perspective. *Clay. Clay Mineral.*, **43**, 540–553.

Essene, E.J. & Peacor, D.R. (1997) Illite and smectite: metastable, stable or unstable? Further discussion and a correction. *Clay. Clay Mineral.*, **45**, 116–122.

Fein, J.B. (1994) Porosity enhancement during clastic diagenesis as a result of aqueous metal–carboxylate complexation: experimental studies. *Chem. Geol.*, **115**, 263–279.

Fein, J.B., Yane, L. & Handa, T. (1994) The effect of aqueous complexation on the decarboxylation rates of oxalate. *Geochim. Cosmochim. Acta*, **58**, 3975–3981.

Fein, J.B., Yane, L., Jyoti, A. & Handa, T. (1995) Experimental study of aluminium and calcium malonate complexation at 25, 35 and 80 degrees C. *Geochim. Cosmochim. Acta*, **59**, 1053–1062.

Garrels, R.M. (1984) Montmorillonite/illite stability diagrams. *Clay. Clay Mineral.*, **32**, 161–166.

Hemley, J.J. (1959) Some mineral equilibria in the system K_2O–Al_2O_3–SiO_2–H_2O. *Am. J. Sci.*, **257**, 241–270.

Hemley, J.J., Montoya, J.W., Marineko, J.W. & Luce, R.W. (1980) Equilibria in the system Al_2O_3–SiO_2–H_2O and some general implications for alteration/mineralisation processes. *Econ. Geol.*, **25**, 210–228.

Hillier, S.D. (1994) Pore-lining chlorites in siliciclastic reservoir sandstones: electron microprobe, SEM and XRD data and implications for their origin. *Clay Mineral.*, **29**, 665–679.

Hower, J., Eslinger, E., Hower, M.E. & Perry, E.A. (1976) Mechanism of burial metamorphism of of argillaceous sediments—1. Mineralogical and chemical evidence. *Geol. Soc. Am. Bull.*, **87**, 725–737.

Huang, W.-L. (1992) Illitic-clay formation during experimental diagenesis of arkoses. In: *Origin, Diagenesis, and Petrophysics of Clay Minerals in Sandstones*, Society of Economic Paleontologists and Mineralogists Special Publication 47 (Ed. by D.W. Housenecht and E.D. Pittman), Tulsa, OK, pp. 49–63.

Huang, W.-L. (1993) Stability and kinetics of kaolinite to boehmite conversion under hydrothermal conditions. *Chem. Geol.*, **105**, 197–214.

Humphreys, B., Smith, S.A. & Strong, G.E. (1989) Authigenic chlorite in late Triassic sandstones from the Central Graben, North Sea. *Clay Mineral.*, **24**, 427–444.

Hutcheon, I., Shevalier, M. & Abercrombie, H.J. (1993) pH buffering by metastable mineral–fluid equilibria and evolution of carbon dioxide fugacity during burial diagenesis. *Geochim. Cosmochim. Acta*, **57**, 1017–1027.

Manning, D.A.C. (2000) Carbonates and oxalates in sediments and landfill: monitors of death and decay in natural and artificial systems. *J. Geol. Soc. London*, **157**, 229–238.

Manning, D.A.C., Rae, E.I.C. & Small, J.S. (1991) An exploratory study of acetate decomposition and dissolution of quartz and Pb-rich feldspar at 150°C, 50 MPa (500 bars). *Mineral. Mag.*, **55**, 183–195.

Manning. D.A.C., Hall, P.L. & Hughes, C.R. (1993) *Geochemistry of Clay–Pore Fluid Interactions*. Chapman and Hall, London, 427 pp.

Nagy, K.L., Steefel, C.I., Blum, A.E. & Lasaga, A.C. (1990) Dissolution and precipitation kinetics of kaolinite: initial results at 80°C with application to porosity evolution in a sandstone. In: *Prediction of Reservoir Quality through Chemical Modelling* (Ed. by D. Meshri and P.J. Ortoleva). *AAPG Mem.*, **49**, 85–101.

Nagy, K.L., Blum, A.E. & Lasaga, A.C. (1991) Dissolution and precipitation kinetics of kaolinite at 80°C and pH 3: the dependence on solution saturation state. *Am. J. Sci.*, **291**, 649–686.

Nelson, B.W. & Roy, R. (1958) Synthesis of the chlorites and their structural and chemical composition. *Am. Mineral.*, **43**, 707–725.

Oelkers, E.H. (1996) Physical and chemical properties of rocks and fluids for chemical mass transport calculations. In: *Reactive Transport in Porous Media*, Reviews in Mineralogy 34 (Ed. by P.C. Lichtner, C.I. Steefel and E.H. Oelkers), Mineralogical Society of America, Washington, DC, pp. 131–191.

Psyrillos, A., Howe, J.H., Manning, D.A.C. & Burley, S.D. (1999) Geological controls on kaolin particle shape and consequences for mineral processing. *Clay Mineral.*, **34**, 193–208.

Rae, E.I.C. & Manning, D.A.C. (1996) Experimentally determined solute yields from kaolinite–illite/muscovite assemblages under diagenetic conditions of pressure and temperature. *Clay Mineral.*, **31**, 537–547.

Rae, E.I.C., Manning, D.A.C. & Hughes, C.R. (1993) Experimental diagenesis of mudrocks. In: *Geochemistry of Clay–Pore Fluid Interactions* (Ed. by D.A.C. Manning, P.L. Hall and C.R. Hughes), Chapman and Hall, London, pp. 213–242.

Roberson, H.E., Reynolds, R.C. & Jenkins, D.M. (1999) Hydrothermal synthesis of corrensite: a study of the transformation of saponite to corrensite. *Clay. Clay Mineral.*, **47**, 212–218.

Robertson, I.D.M. & Eggleton, R.A. (1991) Weathering of granitic muscovite to kaolinite and halloysite and of plagioclase-derived kaolinite to halloysite. *Clay. Clay Mineral.*, **39**, 113–126.

Saccocia, P.J. & Seyfried, W.E. (1994) The solubility of chlorite solid solutions in 3.2 wt-percent NaCl fluids from 300–400 degrees C, 500 bars. *Geochim. Cosmochim. Acta*, **58**, 567–585.

Sass, B.M., Rosenberg, P.E. & Kittrick, J.A. (1987) The stability of illite/smectite during diagenesis: an experimental study. *Geochim. Cosmochim. Acta*, **51**, 2103–2115.

Seewald, J.S., Seyfried, W.E. & Thornton, E.C. (1990) Organic-rich sediment alteration: an experimental and theoretical study at elevated temperatures and pressures. *Appl. Geochem.*, **5**, 193–209.

Seyfried, W.E., Janecky, D.R. & Berndt, H.E. (1987) Rocking autoclaves for hydrothermal experiments II. The flexible reaction-cell system. In: *Hydrothermal Experimental Techniques* (Ed. by G.C. Ulmer and H.L. Barnes), John Wiley, New York, pp. 216–239.

Small, J.S. (1993) Experimental determination of the rates of precipitation of authigenic illite and kaolinite in the presence of aqueous oxalate and comparison to the K–Ar ages of authigenic illite in reservoir sandstones. *Clay. Clay Mineral.*, **41**, 191–208.

Small, J.S. (1994) Fluid composition, mineralogy and morphological changes associated with the smectite-to-illite reaction: an experimental study of the effect of organic acid anions. *Clay Mineral.*, **29**, 539–554.

Small, J.S. & Manning, D.A.C. (1993) Laboratory reproduction of morphological variation in petroleum reservoir clays: monitoring of fluid composition during illite precipitation. In: *Geochemistry of Clay–Pore Fluid Interactions* (Ed. by D.A.C. Manning, P.L. Hall and C.R. Hughes), Chapman and Hall, London, pp. 181–212.

Small, J.S. & Manning, D.A.C. (1994a) Competition between metal–carboxylate complexing, mineral precipitation, and carboxylate decomposition reactions. *Goldschmidt Conf. Abstr., Mineral. Mag.*, **58A**, 850–851.

Small, J.S. & Manning, D.A.C. (1994b) On-line monitoring of clay mineral precipitation in sandstone porespace under flow conditions. *Goldschmidt Conf. Abstr., Mineral. Mag.*, **58A**, 852–853.

Small, J.S., Hamilton, D.L. & Habesch, S. (1992a) Experimental simulation of clay mineral precipitation in reservoir sandstones 1: techniques and examples. *J. Sed. Petrol.*, **62**, 508–519.

Small, J.S., Hamilton, D.L. & Habesch, S. (1992b) Experimental simulation of clay mineral precipitation in reservoir sandstones 2: mechanisms of illite formation and controls on morphology. *J. Sed. Petrol.*, **62**, 520–529.

Stalder, P.J. (1973) Influence of crystallographic habit and aggregate structure of authigenic minerals on sandstone permeability. *Geol. Mijnbouw*, **52**, 217–219.

Ulmer, G.C. & Barnes, H.L. (1987) *Hydrothermal Experimental Techniques*. John Wiley, New York, 523 pp.

Velde, B. (1973) Phase equilibria in the system $MgO–Al_2O_3–SiO_2–H_2O$. *Mineral. Mag.*, **39**, 297–312.

Warren, E.A. & Smalley, P.C. (1994) *North Sea Formation Waters Atlas*. Memoir 15, Geological Society Publishing House, Bath, 208 pp.

Wilson, M.D. & Pittman, E.D. (1977) Authigenic clays in sandstones—recognition and influence on reservoir properties and palaeoenvironmental analysis. *J. Sed. Petrol.*, **47**, 3–31.

Yates, D.M. & Rosenberg, P.E. (1996) Formation and stability of endmember illite: I. Solution equilibration experiments at 100–250°C and $P_{v,soln}$. *Geochim. Cosmochim. Acta*, **60**, 1873–1883.

Yates, D.M. & Rosenberg, P.E. (1997) Formation and stability of endmember illite: II. Solid equilibration experiments at 100–250°C and $P_{v,soln}$. *Geochim. Cosmochim. Acta*, **61**, 3135–3144.

Zotov, A., Mukhamet-Galeev, A. & Schott, J. (1998) An experimental study of kaolinite and dickite relative stability at 150–300°C and the thermodynamic properties of dickite. *Am. Mineral.*, **83**, 516–524.

Int. Assoc. Sedimentol. Spec. Publ. (2003) **34**, 191–211

Effect of clay content upon some physical properties of sandstone reservoirs

PAUL F. WORTHINGTON
Gaffney, Cline & Associates, Bentley Hall, Blacknest, Alton, Hants GU34 4PU, UK,
e-mail: pworthington@gaffney-cline.com

ABSTRACT

The presence of clays and clay minerals within a petroleum reservoir has an impact upon reservoir properties such as porosity and permeability, upon those measured physical data that are used to evaluate reservoir properties, and upon the forms of the petrophysical relationships between subsets of these parameters. Under these conditions, deterministic petrophysical interpretation takes place within either the effective porosity system or the total porosity system, one of which usually is adopted exclusively for interpretation purposes according to contemporary petroleum industry culture. A collation of some pertinent physical properties of clay minerals demonstrates their variability and emphasizes that each sandstone reservoir must be studied separately and thoroughly. The physical manifestation of clay minerals in measured physical data primarily is the result of their composition and internal structure. On the other hand, the manifestation of clays and clay minerals in the context of reservoir properties is principally morphological, as exemplified by microporosity, capillarity and sand–shale laminations. All these manifestations have an impact upon the interpretative petrophysical process itself. Future investigations should be directed at the quantitative incorporation of clay and clay-mineral effects within an overall continuum of reservoir behavioural characteristics.

INTRODUCTION

The geoscience community's awareness of the influence of clays upon the petrophysical and hydraulic properties of siliciclastic reservoir rocks has evolved significantly in response to a growing ability to measure the concentration and physicochemical properties of intraformational and interstitial clays, both in the laboratory and down-hole. Thus, contemporary understanding has been facilitated classically, through a stronger measurement capability that has benefited from real-time quality assurance derived from advances in sensor design, digital data acquisition and computing power over the past 30 yr.

Petrophysical interpretation has responded to the increased measurement capability of core sensors and well logs by seeking to refine interpretative models that take account of clay effects on the evaluation of reservoir rocks. These effects can be attributed to two key sources: the intrinsic properties of clay minerals and the pore geometry associated with the occurrence of clays and clay minerals. These two sources have an impact on the measured physical properties of the reservoir (e.g. sonic velocity), the reservoir properties that are to be evaluated (e.g. porosity), and the relationships between them (e.g. the sonic time-average equation for the evaluation of porosity (Wyllie *et al.*, 1956)). Each of these sources will be considered in turn.

Before doing so, it is necessary to clarify the terminology that is adopted in petrophysics, the reservoir models that are used in conventional petrophysical evaluation, and how clay minerals are accommodated within them. Throughout, the context is one of clay minerals within the reservoir, rather than one of the intrinsic properties of a (pure) clay-mineral assemblage. The treatment is not intended to be exhaustive but rather to highlight some issues of topical interest at the interface of clay mineralogy and petrophysics.

PETROPHYSICAL TERMINOLOGY

Different subject areas maintain their own terminology in respect of clay geoscience. The following terminology is set within a petrophysical context and is included for clarity.

- Clay mineral. A member of the hydrous aluminous phyllosilicates that dominate the fine fractions of reservoir rocks. The term 'clay mineral' refers to composition, not grain size, and where clay minerals are present they should be accommodated petrophysically regardless of grain size.
- Clay. Clay is strictly a grain-size term, classically with particle diameter less than 3.9 μm (Wentworth, 1922). It does not describe composition. Yet, in petrophysics the term 'clay' is frequently used to mean 'clay mineral'.
- Silt. Silt is another grain-size term, classically with particle diameter between 3.9 and 62.5 μm. Although it does not describe composition, petrophysicists often specify silt to comprise exceedingly fine-grained quartz. Occasionally, silt is treated as a separate mineral (e.g. Kuttan *et al.*, 1980).
- Matrix. In a petrophysical sense the term 'matrix' does not include clay minerals. Petrophysics admits three types of rock matrix; sandstone, limestone and dolomite. No others are accommodated, although mixtures are allowed. Evaporites, carbonaceous beds and igneous rocks are dealt with separately by the interpretative literature (e.g. Dewan, 1983).
- Bound water. In its unqualified form, the term 'bound water' is the electrochemically bound molecular water that is adsorbed at mineral surfaces as the result of a negative surface charge, which attracts free positive ions into an electrical double layer. This effect is most pronounced in the presence of clay minerals, for which it also includes interlayer molecular water. Tradition has it that bound water can be removed by drying at 110°C, a practice that alters interface-sensitive clay minerals, but there are experimental indications that this temperature is clay-mineral specific and that a temperature as high as 200°C might be needed (Pallatt & Thornley, 1990). Thereafter the clay minerals are said to be in a 'dry' state. A dry clay mineral retains its structural water (see below). The complement of bound water is electrochemically 'free water', not to be confused with movable water, which is free of capillary retention forces and which forms part of the electrochemically free water.
- Structural water. Structural 'water' takes the form of non-liquid hydroxyl groups. It is taken to be an integral part of the dry clay mineral for petrophysical purposes. Structural water can be removed only by heating to temperatures above 300°C (Pallatt & Thornley, 1990). After removal of the structural water, or dehydroxylization, the clay minerals are said to be in a 'fired' state. They have been altered irreversibly.
- Shale. Shale is predominantly an assemblage of silt, clay and clay minerals. Although petrophysical analysis often assumes that shale comprises only clay minerals, in reality shale might contain more than 70% by volume of silt and larger detritus. A perfect shale is one that comprises 100% clay minerals, regardless of grain size. A perfect shale contains no electrochemically free water. An imperfect shale comprises a significant percentage of minerals other than clay minerals. It contains both free and bound waters.

The interrrelationship of the above definitions is illustrated in Fig. 1. They are frequently misused in petrophysical evaluation, where the term 'shale' is often used loosely to indicate clay minerals. Furthermore, this term makes no distinction between, for example, detrital clay

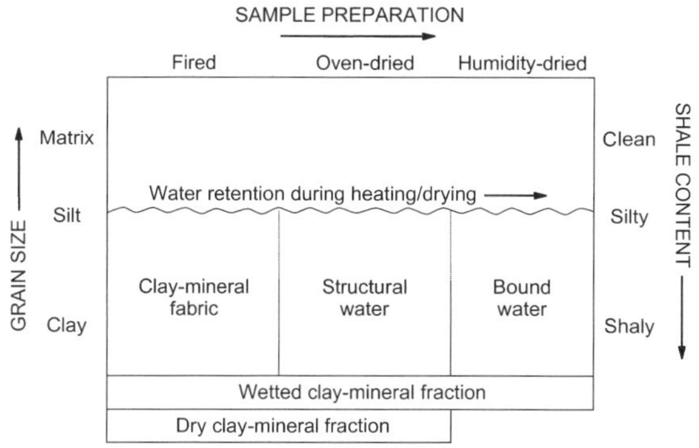

Fig. 1 Interrelationship of clay-related petrophysical terminology.

minerals within intraformational shale beds and authigenic clay minerals within larger arenaceous units, but instead it groups both types together as congenital. In a petrophysical sense, a 'shaly sand' is therefore one that contains clay minerals, which can be present as shale laminations, structural shale (load-bearing grains), and dispersed shale in the form of allogenic clay minerals (e.g. transported detritus) or authigenic clay minerals (e.g. cements or overgrowths).

A nomenclature of symbols is included as Table 1.

PETROPHYSICAL MODELS

There are two interpretative models in common use in the deterministic petrophysical evaluation of clay-mineral-bearing sands. They differ in the way in which electrochemically bound water is accommodated.

Effective porosity model

The effective porosity model incorporates the electrochemically bound water into the clay-mineral volume, which then becomes a wetted clay-mineral fraction. In petrophysical parlance, the wetted clay-mineral fraction is often described loosely as a wetted shale fraction. The effective porosity ϕ_e and water saturation S_{we} are defined in terms of the electrochemically free pore space (Fig. 2). The effective porosity of a perfect shale is zero.

Total porosity model

The total porosity model incorporates the electrochemically bound water within the pore space, so that the clay-mineral volume becomes a dry clay-mineral fraction (which includes structural 'water'). In petrophysical parlance, the dry clay-mineral fraction often is described loosely as a dry clay fraction. The total porosity ϕ_t and water saturation S_{wt} are defined in terms of the total pore space (Fig. 2). The total porosity of a perfect shale is greater than zero.

Conjunctive use of petrophysical models

Most organizations that are concerned with deterministic petrophysical analysis operate these two models exclusively. However, there

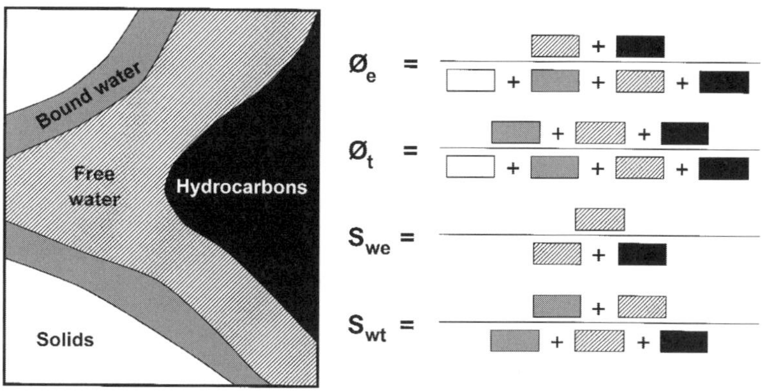

Fig. 2 Porosity (ϕ) and water saturation (S_w) in the effective (subscript e) and total (subscript t) porosity models.

Table 1 Nomenclature of symbols.

Symbol	Definition
B	Equivalent conductance of (sodium) clay-exchange cations (S m^{-1} eq^{-1} litre)
C_o	Conductivity of fully water-saturated reservoir rock (S m^{-1})
C_{sh}	Conductivity of intraformational shale (S m^{-1})
C_t	Conductivity of partially water-saturated reservoir rock (S m^{-1})
C_w	Conductivity of formation water (S m^{-1})
F	Formation resistivity factor
F_a	Apparent formation resistivity factor
F_e	Formation resistivity factor in the effective porosity system
F_t	Formation resistivity factor in the total porosity system
GR	Gamma-ray response (API units)
I_R	Resistivity index
I_{Ra}	Apparent resistivity index
K	Intergranular permeability (mD)
PEF	Photoelectric factor (barns/electron)
Q_v	Cation exchange capacity per unit pore volume (eq $litre^{-1}$)
S_w	Water saturation
S_{we}	Water saturation in the effective porosity system
S_{wirr}	Irreducible water saturation
S_{wt}	Water saturation in the total porosity system
T_2	NMR transverse relaxation time (ms)
U	Generic log response
U_{corr}	Generic log response corrected for shale effects
U_{ma}	Generic log response to matrix
U_{sand}	Generic log response to clean sand
U_{sh}	Generic log response to shale
U_{shale}	Generic log response to perfect shale
U_{wcm}	Generic log response to wetted clay-mineral content
V_p	Compressional wave velocity (km s^{-1})
V_s	Shear wave velocity (km s^{-1})
V_{sh}	Fractional volume of shale
V_{wcm}	Fractional volume of wetted clay mineral(s)
X	Generalized shale parameter (S m^{-1})
m	Intrinsic porosity exponent
m_a	Apparent porosity exponent
n	Intrinsic saturation exponent
n_a	Apparent saturation exponent
Δt	Sonic transit time (μs m^{-1})
$\Delta t(C)$	Compressional-wave transit time (μs m^{-1})
$\Delta t(S)$	Shear-wave transit time (μs m^{-1})
Φ_N	Neutron porosity log response (limestone units)
$\Phi_N(T)$	Thermal neutron porosity log response (limestone units)
$\Phi_N(E)$	Epithermal neutron porosity log response (limestone units)
ϕ	Porosity
ϕ_e	Effective porosity
ϕ_t	Total interconnected porosity
ρ_b	Log-derived bulk density (g cm^{-3})

are quality-assurance benefits to be gained by operating both models conjunctively. The association of the effective and total porosity models within a petrophysical interpretative context has been analysed by Worthington (1998).

PETROPHYSICAL INTERPRETATION

This treatment is intended to provide a schematic overview. More detailed accounts of log data acquisition and interpretation can be found in Theys (1999) and Dewan (1983), respectively. The measurement and petrophysical application of core data form the subject of Tiab & Donaldson (1996).

Three interpretative approaches are considered for clay-bearing sandstones; the deterministic effective porosity system, the deterministic total porosity system, and the statistical solution of log response equations. Deterministic methods are sequential and therefore they lend themselves to a structured quality-control procedure, an attribute that is central to the contemporary industry thrust of seeking to reduce uncertainty in integrated reservoir description. Statistical methods are simultaneous and therefore they are less appropriate to a structured quality-control scheme. The discussion will presuppose that density and neutron logs already have been corrected for any light hydrocarbon effects, so that these logs respond primarily to lithology and porosity.

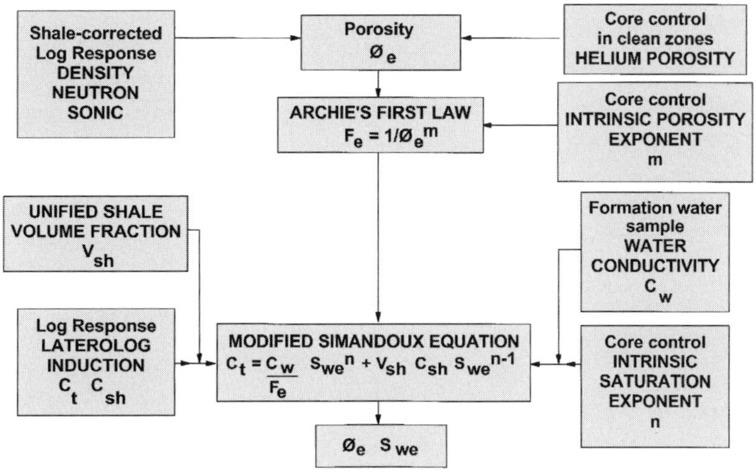

Fig. 3 Scheme of petrophysical evaluation in the effective porosity system.

Deterministic interpretation within the effective porosity system

Petrophysical interpretation in the effective porosity system follows the scheme of Fig. 3. The family of porosity logs (density, neutron and sonic) must be corrected for clay-mineral effects before porosity can be evaluated. The generic equation for making this correction is:

$$U_{corr} = U + V_{wcm}\,(U_{ma} - U_{wcm}) \qquad (1)$$

where U is the log response (corrected for environmental effects), U_{corr} is the log response corrected also for wetted clay-mineral effects, U_{ma} is the log response to pure matrix (i.e. a solid block of quartz, limestone or dolomite, or some mixture thereof), U_{wcm} is the log response to the wetted clay mineral(s), and V_{wcm} is the volume fraction of the wetted clay mineral(s). Equation (1) can be applied to the three porosity logs at each digital sampling level within the evaluation interval. The intention is to replace the wetted clay-mineral fraction by geometrically identical rock matrix. In doing so, the value of V_{wcm} used to correct a particular tool response should be physically compatible with the tool measurement. For example, values of V_{wcm} inferred from the gamma-ray log may not furnish a meaningful clay-mineral correction to the sonic log. A full discussion of this issue has been presented elsewhere (Worthington, 1996).

In practice, the petrophysical interpreter substitutes V_{wcm} by a term known as the (wetted) shale volume fraction V_{sh} and U_{wcm} by the log response to wetted shale U_{sh}. There are several ways of evaluating V_{sh}, the simplest method being the linear model for the response of a logging tool that is being used as a single shale indicator:

$$V_{sh} = (U - U_{sand})/(U_{shale} - U_{sand}) \qquad (2)$$

where U_{sand} is the (environmentally corrected) response of the logging tool to a clean (clay-mineral-free) sand and U_{shale} is the tool response to a (perfect) shale. The quantity U_{shale} usually is taken as the (environmentally corrected) log response opposite a sufficiently thick shale bed. This approach mixes the properties of, for example, detrital shale that comprises the intraformational beds and authigenic shale in the reservoir sands. An alternative approach that avoids mixing potentially incompatible data is to calculate an interim V_{sh} using, say, the gamma-ray log and to cross-plot this interim V_{sh} and the tool response U using data sampling points that lie within the reservoir sands. The data trend within the sands can be extrapolated to the end-point at interim $V_{sh} = 1$, where $U = U_{shale}$. Note that this extrapolation need not be linear. This method, too, has its shortcomings, not least those resulting from imperfect tool resolution. However, where data

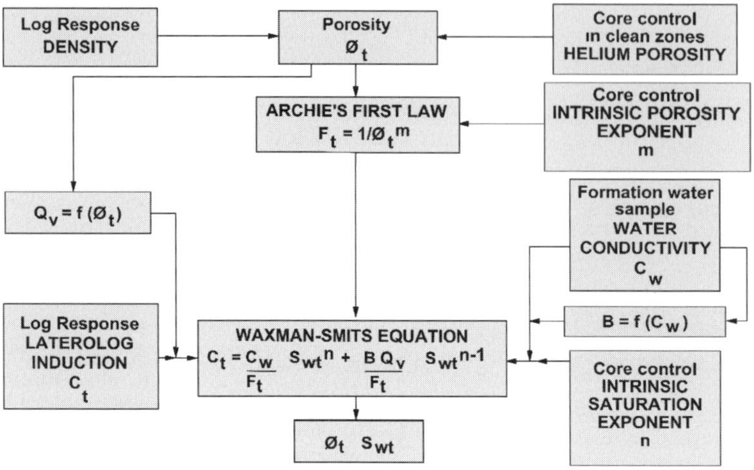

Fig. 4 Scheme of petrophysical evaluation in the total porosity system.

quality is sufficiently good, the method might allow a comparison of the potential errors associated with the use of a detrital shale bed as a reference end-point for U_{shale}, a practice that contributes to overestimating V_{sh} because a shale bed is not a perfect assemblage of clay minerals.

Once equation (1) has been applied to the family of porosity logs, effective porosity ϕ_e is evaluated through the method that would be used if the reservoir were clean. However, the issue of V_{sh} raises its head again when effective water saturation S_{we} is evaluated from resistivity logs (Fig. 3). Here, the resistivity log is not initially corrected for shale effects. Instead, the measured formation conductivity C_t (the reciprocal of resistivity) is substituted into an equation that expresses C_t (or a function of C_t) as the sum of two terms, a clean-sand term and a shale term. An example is the modified Simandoux equation (Bardon & Pied, 1969):

$$C_t = (C_w/F_e)S_{we}^n + (C_{sh}V_{sh})S_{we}^{n-1} \quad (3)$$

where C_w is the conductivity of the interstitial water, F_e is the intrinsic formation resistivity factor (i.e. corrected to be independent of water salinity), and n is the intrinsic saturation exponent (again corrected to be independent of water salinity). For a discussion of these parameters and their corrections see Worthington (1985). The shale term contains V_{sh} and the shale conductivity C_{sh}. The evaluation of C_{sh} broadly follows the discussion of the previous paragraph. However, because resistivity logs also respond to changes in fluid type, the intrareservoir determination of C_{sh} should be confined to the water zone. The choice of the most appropriate V_{sh} is less clear. The resistivity log cannot easily be used to infer V_{sh} on a level-by-level basis outside the water zone. For this reason an integrated V_{sh} derived from several log responses is recommended: see Worthington (1996) for a full discussion. A key message is that different (sets of) logs furnish different values of V_{sh} and it is important to use one that is fit for the purpose.

Deterministic interpretation within the total porosity system

Petrophysical interpretation in the total porosity system follows the scheme of Fig. 4. Total porosity ϕ_t is obtained by using only the density log, just as one would if the formation were clean. There is no shale correction. It is assumed that the dry clay-mineral fraction has the same density as the petrophysical rock matrix at every digital sampling level within the evaluation interval. If the matrix density varies with depth, the dry clay-mineral density is assumed to vary in exactly the same manner. The density log is the only one for which the assumption of equality between rock matrix and dry clay-mineral properties is approximately satisfied.

The evaluation of total water saturation S_{wt} from resistivity logs again uses an equation that expresses log-measured formation conductivity C_t as the sum of two terms, a clean-sand term and a shale term (Fig. 4). An example is the equation of Waxman & Smits (1968):

$$C_t = (C_w/F_t)S_{wt}^n + (BQ_v/F_t)S_{wt}^{n-1} \quad (4)$$

where the shale term contains the equivalent conductance of (sodium) clay-exchange cations B and the cation exchange capacity per unit pore volume of the reservoir rock Q_v, and the product BQ_v has units of conductivity. The quantity B is a published function of C_w and temperature (e.g. Juhasz, 1981). The quantity Q_v cannot be measured down-hole and has to be inferred through algorithms based on (generally weak) correlations with other physical properties, most commonly porosity (e.g. Austin & Ganley, 1991).

Statistical interpretation of well logs

These methods are derived from the so-called 'global' method (Mayer & Sibbitt, 1980). They develop a simultaneous solution to a set of log-response equations. In order to set up the equations it is necessary to assume the components of the reservoir. For example, an oil-bearing, calcareous shaly sand might have the following components; quartz, calcite, shale, water and oil. It also is necessary to specify the logs that will enter the calculation. Solutions usually are effected on a well-by-well basis. They take the form of computed volume fractions of each component at each digital sampling level. If the input logs are changed, the solutions will be different.

The statistical method as described above cannot be used in the total porosity system, because it is unrealistic to assume that the matrix and dry shale values of each admitted physical property are equal at every digital sampling level. Therefore the global method has to be set in the effective porosity system. This means that the computed V_{sh} values will be based on every input log and in this respect they represent an average of those V_{sh} values that could be computed separately using single logs, such as the gamma ray, or pairs of logs, such as the density–neutron combination.

More recent advances have extended statistical applications to take not just a composite account of the wetted shale fraction, but separate accounts of its dry clay-mineral constituents and the electrochemically bound water. These multimineral methods do not require the selection of an interpretative system beforehand, because the computed volume fractions can be combined subsequently according to the requirements of either system. For example, in the effective porosity system, the quantity V_{sh} would be the sum of the dry clay-mineral fractions and the saturation of the electrochemically bound water. However, these methods do require the responses of the input logging tools to the dry clay-mineral components and this information might not be available in unambiguous form.

In addition to their diagnostic capability for V_{sh}, statistical methods are especially useful as an intermediate facility in lithologically complex reservoirs, where they can be used to infer on a level-by-level basis the physical properties of a matrix of variable composition.

EFFECT OF CLAY MINERALS ON MEASURED PHYSICAL PROPERTIES

The effect of clay minerals upon petrophysical properties can be parameter-specific. For example, it is quite possible to have a situation in which the neutron porosity log is significantly affected by clay minerals but the electrical resistivity log is not. For this reason, each of the following parameters is considered separately. A collation of some physical properties of clay minerals taken from different sources is presented as Table 2. Although many of these data are average values, they expose ranges of variability for different clay minerals. Moreover, where they are averages, the data mask ranges of variability for individual clay minerals. Therefore the data of Table 2 should be taken

Table 2 Some reported physical properties of clay minerals.

Parameter	Mineral: Kaolinite	Illite	Smectite	Chlorite	Glauconite	Source
Gamma ray (API units)	104	160	168	56	150	Herron & Matteson (1993)
Density (g cm^{-3})	2.62	2.78	2.63	3.42	2.85	Herron & Matteson (1993)
Φ_N(T) (limestone %)	45.1	24.7	21.8	>48.2	>50	Herron & Matteson (1993)
Φ_N(E) (limestone %)	43.3	17.6	17.8	61.0	20.2	Herron & Matteson (1993)
PEF (barns/electron)	2.05	4.01	2.89	8.06	6.73	Herron & Matteson (1993)
Δt(C) (μs m^{-1})	694.6					Ellis *et al.* (1988)
Δt(S) (μs m^{-1})	1075.2					Ellis *et al.* (1988)

solely for illustrative purposes. Every reservoir that is under evaluation should be studied separately and thoroughly.

Natural radioactivity

This is the most established indicator of clay minerals. Total natural gamma counts can be attributed to potassium, thorium and uranium components. This last component is sometimes eliminated, with the sum of the potassium and thorium counts being described as a corrected gamma-ray response. Quantitative estimates of clay-mineral concentration in reservoir rocks are degraded because the reference shale is imperfect and because clay-mineral composition varies from place to place. Therefore the application of equation (2) can afford at best an uncalibrated overestimate of clay-mineral content. Conventionally, estimates of V_{sh} are taken to represent a wetted clay-mineral content as per the effective porosity system, because the reference shale at which V_{sh} is arbitrarily set to unity is in the wetted state. Qualitative evaluations of clay-mineral type have been attempted using the natural-gamma spectral log, but this tool is diagnostically inadequate when used without induced-gamma spectral data for identification of other elements. The principal application of the natural-gamma log has been semi-quantitative, with the computed values of V_{sh} being used to identify net sands. The net-sand cut-off is often computed as a limit beyond which the inclusion of additional rock does not materially alter the calculated hydrocarbons-in-place. By defining this limit in terms of hydrocarbons-in-place, many of the shortcomings of the natural-gamma log as an indicator of clay-mineral content do not have an adverse impact on the subsequent evaluation of net pay.

Density

In the effective porosity system log-derived bulk density (ρ_b) is nominally corrected for clay-mineral effects using equation (1). Again, this correction is imperfect, even when the value of V_{wcm}, or its substitute V_{sh}, is first evaluated using a method that incorporates the density log, e.g. the neutron-density combination. The correction requires a knowledge of the density of the wetted clay-mineral assemblage, which usually is inferred by extrapolating a cross-plotted trend of log-derived ρ_b versus V_{sh}. The effect of constituent clay minerals upon sandstone density can be positive or negative. In most cases, however, the wetted clay-mineral fraction is less dense than the rock matrix, so that the effect of clay minerals is to lower the density for a given effective porosity. Once this correction has been applied, the density log is used to evaluate effective porosity as if the sandstone were clean, by drawing upon a conceptual summation-of-masses equation. Porosity evaluation is often effected using the neutron–density log combination.

In the total porosity system no distinction is drawn between the densities of the rock matrix (as defined petrophysically) and of the constituent clay minerals in the dry state. This assumption is particularly questionable where either the matrix or the clay-mineral component shows significant variations in composition. The density log is used to evaluate total porosity as though the sandstone were clean, regardless of its clay-mineral content.

Neutron porosity

Neutron porosity (Φ_N) is related empirically to neutron logging tool response. It is recorded in limestone units, i.e. a given neutron tool response is reported in terms of the porosity indicated empirically by that response, assuming that the formation is a pure limestone. The neutron tool responds to hydrogen and therefore the bound and structural waters associated with clay minerals are manifested as an increased limestone porosity. Neutron porosity in limestone units can be corrected for clay-mineral effects using equation (1) with the matrix term set to zero. Again, this correction is imperfect, even when the value of V_{wcm}, or its substitute V_{sh}, is first evaluated using a method that incorporates the neutron log, e.g. the neutron-density combination. The correction requires a knowledge of the neutron limestone porosity of the wetted clay-mineral assemblage, which can be inferred by extrapolating a cross-plotted trend of log-derived Φ_N versus V_{sh}. When this correction has been applied, neutron porosity can be used in conjunction with the log-derived density to evaluate effective porosity using a computerized neutron-density cross-plot. Neutron porosity is rarely used on a stand-alone basis.

Photoelectric factor (PEF)

The photoelectric factor is a photoelectric absorption cross-section index. It is delivered as a product of the (spectral) density log. The photoelectric region is the lower region of gamma-ray energy. Low-energy gamma rays (< 0.2 MeV) are absorbed on impact with an atom, with the emission of a photoelectron. Photoelectric absorption is related to lithology. The photoelectric factor is used in conjunction with density and other lithology indicators for mineralogical analysis. It is especially useful as a parametric input to statistical well-log analysis (see 'Statistical interpretation of well logs' above).

Sonic velocity

Log-derived compressional wave velocity, or its reciprocal, sonic transit time Δt, is ostensibly corrected for clay-mineral effects using equation (1) for application within the effective porosity system. It is recognized that this correction is imperfect, even when the value of V_{wcm}, or its substitute V_{sh}, is first evaluated using a method that incorporates the sonic log, e.g. the neutron–sonic combination. The correction requires a knowledge of the sonic log compressional-wave response to wetted clay minerals, which usually is inferred by extrapolating a cross-plotted trend of log-derived Δt versus V_{sh}. Once this correction has been applied, sonic velocity is used to evaluate effective porosity as though the sandstone were clean, by drawing upon the conceptual time-average equation (Wyllie *et al.*, 1956) or the empirical relationship proposed by Raymer *et al.* (1980).

In sandstones with constant porosity, clay minerals cause a reduction in compressional-wave velocity, with the incremental effect being greater as small amounts of clay minerals are initially introduced and then declining as additional clay minerals are added (e.g. Eberhart-Phillips *et al.*, 1989). A similar effect is seen for shear-wave velocity (Fig. 5). These effects are demonstrably greater for clay-mineral cements than for dispersed clay minerals because acoustic measurements are governed by grain-to-grain coupling (Minear, 1982). The relationship between compressional-wave velocity and clay-mineral fraction is also influenced by the nature of the supporting rock fabric, i.e. matrix supported or clay-mineral supported. The results of this investigative work have not yet been incorporated into

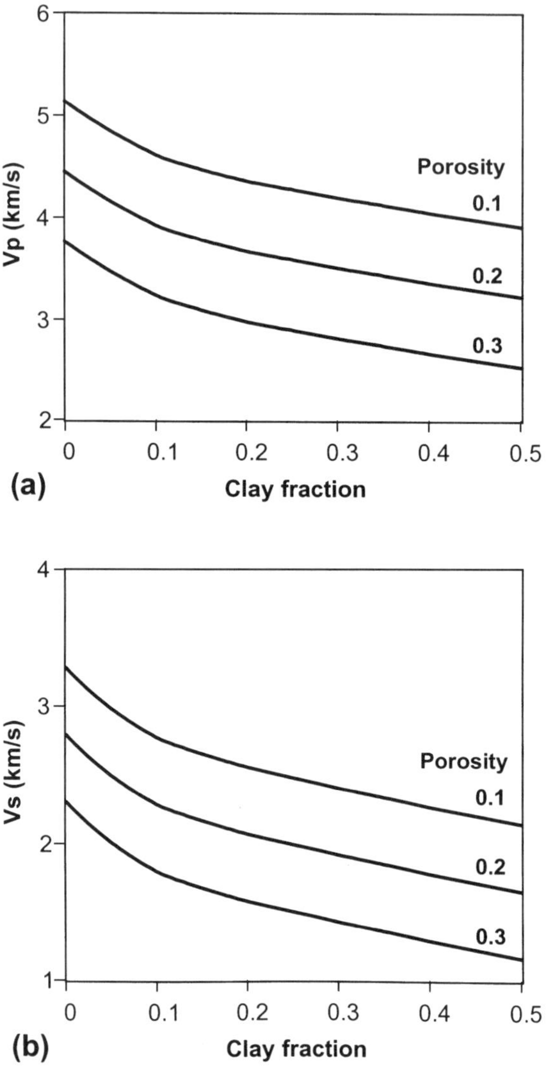

Fig. 5 Empirical relationships between sonic velocity and clay content for specified values of the porosity of a water-saturated sandstone at an effective stress of 0.2 kbar: (a) compressional-wave velocity; (b) shear-wave velocity. (Constructed from Eberhart-Phillips *et al.*, 1989.)

operational petrophysical practice, for example by modifying the form of equation (1) to reflect the non-linearity of the relationship between compressional-wave velocity and clay-mineral content for sandstones.

The relationship between compressional and shear velocities remains well-defined and quasi-linear in the presence of clay minerals (Castagna *et al.*, 1985). As more clay minerals are introduced the data trend is nudged to reflect a proportionately greater reduction in shear-wave velocity. However, the effects are small until fractional clay-mineral content exceeds about 0.35 (Vernik & Nur, 1992).

Electrical conductivity

For some 50 yr it has been recognized that sands containing clay minerals show an electrical conductivity beyond that which can be attributed to the saturating aqueous electrolyte alone (Patnode & Wyllie, 1950; Winsauer & McCardell, 1953). This extra conductivity has been associated with the ionic double layer that forms at charged mineral surfaces in the presence of an ionic solution (Waxman & Smits, 1968). Various models have been proposed to describe the excess conductivity effect, and many of these can be described through the following generic equation:

$$C_t = (C_w/F)S_w^n + XS_w^{n-1} \quad (5)$$

where X is a generalized composite clay term with units of conductivity. Indeed, equation (5) is a generalized form of equations (3) and (4), all of which can be seen as containing a clean-sand term and a clay term on the right-hand side. Where the clay term is small relative to the clean-sand term, the effect of clay on conduction is minimal. This situation can arise where clay(-mineral) content is low, so that X is low, or where formation-water salinity is high, so that C_w is high, or where both these conditions are satisfied.

The above comments emphasize the relationship between the effect of constituent clays on electrical conduction in a sandstone and the salinity of the saturating electrolyte. In this respect, a sandstone behaves as a coupled system and it is not appropriate to consider one element of this coupling in isolation from another. It also should be noted that the term 'clay content' may or may not include clay minerals. Very-fine-grained sandstones that contain no clay minerals can show electrical characteristics that mimic clay-mineral effects, especially if the saturating electrolyte is fresh. This happens because, although the surface charge density of these rocks is relatively low, the pore surface area is very large. Therefore the integral of charge over the pore surface area can itself be large, a property that can contribute to a significant electrical conduction capability over and above that resulting from ionic conduction through the saturating electrolyte. On the other hand, some rocks that do contain clay minerals show no clay-mineral effects on conduction, because the saturating electrolyte is sufficiently saline to suppress the effects of any extra conduction associated with the clay minerals. All this suggests that the conductivity of reservoir rocks can be described in terms of a continuum of electrical properties that include the conductivity of the saturating electrolyte. Worthington (1997) has described the use of this continuum concept to identify whether or not the conductivity of a given reservoir rock does include clay(-mineral) effects. The concept is illustrated for fully water-saturated conditions in Fig. 6.

Equation (5), or one of its special forms, is used to evaluate water saturation from electrical properties. If the clay term is omitted when it should have been included, S_w will be overestimated and thence the hydrocarbon saturation will be underestimated. If the clay term is included unnecessarily, the greater uncertainties associated with the quantification of this term might actually degrade the evaluation of water saturation rather than enhance the process. It is therefore important to understand the mechanism of electrical conduction in reservoirs that contain clays and/or clay minerals.

The electrical conductivity of clay-mineral assemblages is usually considered in the presence of an electrolyte: an exception is the investigation of electronic conduction (see below). Thick shale beds typically have an electrical resistivity within the range 3–10 Ωm. This is not necessarily the resistivity of wetted clay minerals for reasons stated earlier about the imperfect nature of shales.

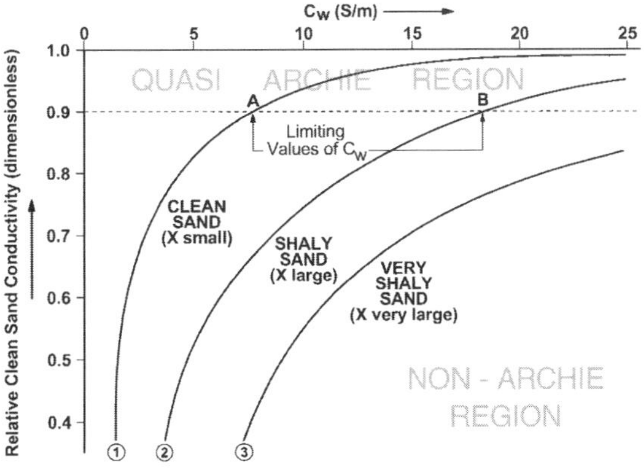

Fig. 6 Schematic identification of the minimum values of C_w necessary for clean-sand (Archie) electrical conditions in the cases of (1) lithologically clean sand; (2) shaly sand. The very shaly sand does not attain Archie conditions for any value of C_w. The ordinate is the Archie conductivity as a fraction of total conductivity: the limiting value of 0.9 for Archie conditions broadly follows Hoyer & Spann (1975).

Electronic conductivity

Electronic conductivity is over and above that due to ionic conduction. It takes place within solid matter. Electronic conduction is confined to iron-bearing minerals such as glauconite, which can in many respects be regarded as a clay mineral even though its structure is similar to that of mica. There is very little documented on the petrophysical implications of electronic conduction associated with glauconite and other iron-bearing silicates. The presence of electronic conduction can be confirmed by measuring the conductivity of oven-dried core plugs, which ought to be zero where this phenomenon is not active. However, although a finite conductivity measured under these conditions does indeed indicate electronic conduction within a rock, a zero conductivity measurement does not rule it out. The reason is that iron-bearing minerals can occur in clusters that are not in direct electrical contact with each other in the dry state. However, when the rock is saturated with an aqueous electrolyte, the clusters are in contact with the electrolyte itself and with other clusters through that electrolyte, so that they can increase the conductivity of the electrolyte-saturated reservoir rock.

Cook *et al.* (1990) have shown interesting examples of the effect of glauconite on the resistivity of the gas-bearing Stanley Sand within the Upper Cretaceous Eutaw play in the Mississippi Interior Salt Basin. The low resistivity of the gas leg (*c.* 1 Ωm) was at least partly attributed to conductivity enhancement by this iron-rich mineral. If this effect is overlooked, the resulting values of water saturation will be too high.

Nuclear magnetic resonance (NMR) properties

Petrophysical principles of NMR logging have been reviewed by Kenyon (1997). Hydrogen nuclei (protons) in water and hydrocarbons can be aligned under the influence of a steady magnetic field so that all nuclei precess at the same rate. An oscillating radiofrequency magnetic field is then applied at 90° to the steady field with a frequency that resonates with the (Lamor) frequency of precession of the nuclei in the steady field. A pulse of the oscillating field redirects the nuclear magnetization through 90°. When this pulse is switched off, the redirected (transverse) magnetization vector dephases by precessing around the direction of the steady field and inducing an alternating voltage in a receiver coil. In order to improve signal quality, a sequence of 180° pulses is then applied. These allow the dephasing process to be reversed with a subsequent rephasing and a signal known as a spin echo. The sequence of measured spin-echo amplitudes constitutes the output of NMR down-hole measurements. The decay of these amplitudes contains information about formation and fluid properties. It is used to compute a transverse relaxation time T_2, which is the key parameter for NMR interpretation.

Because rocks contain a range of pore sizes, and because smaller pores correspond to faster relaxation rates, a computed spectrum of T_2 values contains important information about pore size and thence capillary-bound water. Further, recent experiments suggest that T_2 values of less than about 3 ms in a T_2 spectrum correspond to clay-mineral-bound water, which is a subset of capillary-bound water (T_2 less than about 33 ms). A key issue is how to distinguish the manifestation of clay-mineral-bound water from the influence of very small pores. Again, is this effect predominantly the result of ionic bound water, occurring where clay minerals are present, or does it include all surface-affected water regardless of the mineralogy? These issues have to be borne in mind when considering the T_2 signatures associated with clay minerals, which are also a function of clay-mineral distribution and water content (Prammer *et al.*, 1996). Such considerations form part of ongoing efforts to extend the capability of NMR logs to furnish a total porosity in clay-mineral-bearing sands rather than the traditional free-fluid porosity, the fractional volume of fluids that are free of capillary retention effects.

CLAY GEOMETRY

The effects of clay geometry on measured petrophysical properties are essentially fourfold. First, the size and distribution of constituent clays and clay minerals have a profound influence on intergranular permeability. Second, the presence of microporosity within clay-mineral overgrowths on quartz and other detrital minerals lowers the formation resistivity and thereby increases the log-derived water saturation. Third, both the above can cause high capillarity. Finally, clays and clay minerals present as shale laminations within a subresolution sand–shale sequence can cause deep resistivity logging tools to underread in the sands, where the measured deep formation resistivity can be typical of water-bearing rock even though the sand laminae actually produce dry hydrocarbons.

Effect on permeability

It is well known that grain size exerts a strong influence upon the ability of a sedimentary rock to transmit interstitial fluids. The effect is

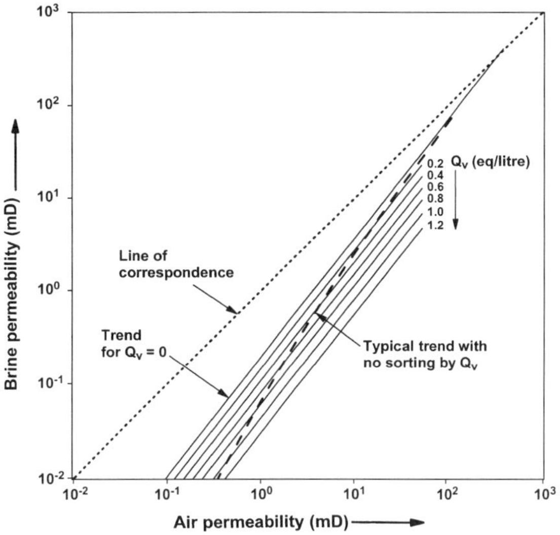

Fig. 7 Effect of clay-mineral content upon a validated relationship between ambient-condition air permeability and brine permeability to water of salinity 20 000 ppm NaCl equivalent at an effective stress of 3000 psi. Clay-mineral content is expressed electrochemically in terms of Q_v. Essentially, as Q_v increases, the relationship moves further away from the line of correspondence, owing to the greater amount of immobilized water. Note that this movement approaches an order of magnitude in brine permeability for a given air permeability. (Redrawn from Juhasz, 1986.)

especially pronounced within the conventional grain-size region for clays, i.e. < 3.9 μm grain diameter. It is exacerbated by the capillary retention exhibited by clay-size particles, which causes a further reduction in the cross-sectional area of pore that is available for free-fluid flow. This latter effect is seen in water-wet reservoir rocks in general, through a disparity between (Klinkenberg-corrected) gas permeability and brine permeability measured on the same core plug (Juhasz, 1986). The disparity increases with increasing clay-mineral content (Fig. 7).

A particular aspect of this problem is that fibrous, pore-bridging clay minerals such as illite can cause a reduction in permeability by a factor of up to 30 relative to a clean sand with otherwise identical pore geometry (Heaviside *et al.*, 1983). The presence of interface-sensitive, pore-bridging clay minerals can be investigated using scanning electron microscope (SEM) photomicrographs of rock pieces that have been subjected to critical-point drying, which avoids the passage of a dragging interface through the pore space (Hall *et al.*, 1978). If these minerals are confirmed to be present, core analysis must proceed only after miscible solvent cleaning of the core plugs, so that the plugs themselves are not dried out, with permeability then being measured to brine. Conventional core preparation involves hot solvent cleaning in a retort followed by oven drying: this procedure would cause the clay mineral strands to collapse against the pore wall, thereby increasing artificially the sample permeability (Pallatt *et al.*, 1984). Since the publication of the definitive literature, this problem has turned out to be far more widespread than originally perceived.

Microporosity

In water-wet rocks, microporosity within clay-mineral overgrowths on quartz grains contributes to the low-resistivity pay sand problem (Hurst & Nadeau, 1995). Rock resistivity is lowered because of the continuous conducting path that the water-filled microporosity presents. Therefore, even though an appropriate shaly sand equation might be used to evaluate water saturation (e.g. equation (3)), the resulting values are high, giving the impression that the rock is water-bearing when it might not be. Indeed, there have been many examples of microporous clay-mineral overgrowths giving rise to low formation resistivities in sands that have produced oil or gas with no water-cut (Worthington, 2000). Electron microscopy and the determination of pore size distribution through mercury injection are vital diagnostic tools.

An interesting example of the effect of overgrowth microporosity upon apparent hydrocarbon saturation has been seen in the Miocene sands of the Attaka Field off East Kalimantan, Indonesia (Partono, 1992). These sands comprise medium-to-coarse, subangular to subrounded, quartz grains. Fragmented clay-alteration products are present and there are traces of microcrystalline pyrite. The clay-mineral content

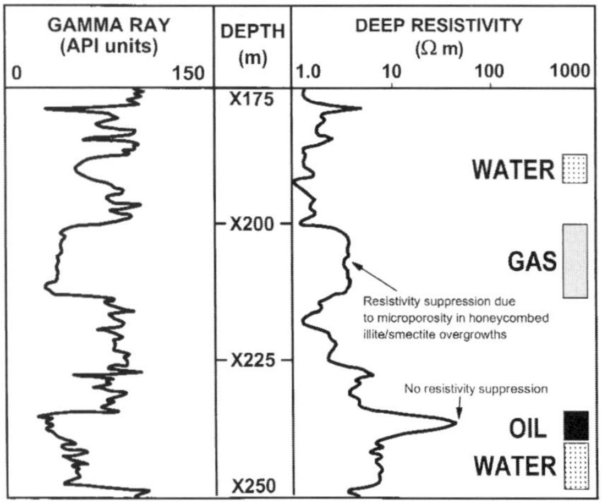

Fig. 8 Example of the effect of clay-mineral overgrowths on the resistivity of a gas-bearing sand in the Attaka field, Indonesia. (After Partono, 1992.)

varies from 18 to 43%, predominantly in the form of interlayered illite–smectite that forms honeycombed overgrowths on quartz grains. The clay-mineral overgrowths have been confirmed through electron microscopy. Where these overgrowths are present, a water zone typically shows a formation resistivity of about 1 Ωm. Where they are absent, water-zone resistivity is at least 10 Ωm. The differences are a result of clay–water interactions within the micropores associated with the clay-mineral overgrowths.

Figure 8 shows a logged interval within the Miocene sands of the Attaka Field. Fluid content has been determined through formation tests. In the lower part of the interval, clay-mineral overgrowths are less abundant: here the water zone has a resistivity close to 10 Ωm. In the upper part of the interval, clay-mineral overgrowths are prevalent: here the water zone has a resistivity close to 1 Ωm. The gas-bearing interval has a suppressed resistivity, which is actually less than that of the lower water zone, and without appropriate models its gas saturation can be computed incorrectly as non-commercial. The resistivity of the oil zone is much greater, even though the measured resistivity of this relatively thin unit might be suppressed by the effect of adjacent beds. This indicates that in these sands clay-mineral morphology is exerting an influence on interpreted water saturation because it takes the form of a connected, water-filled microporous system that causes departures from the implicit assumption of uniform desaturation of a single pore system. Note that the gamma-ray readings are similar over the gas-bearing and oil-bearing intervals. This observation confirms that the influence of microporosity within these clay-mineral overgrowths extends beyond any shaly sand effects in the classic sense.

Capillarity

The presence of structural or dispersed clays increases the capillary retention of a water-wet granular medium with a corresponding increase in irreducible water saturation. Here the word 'clays' is used strictly as an indicator of size. It does not necessarily include clay minerals in the classic sense. Capillary effects transmit through to the saturation-height functions that are used to distribute water saturation (S_w) throughout a gridded reservoir model. For example, a saturation-height algorithm of the J-function kind (Leverett, 1941) is specific to a given rock type, defined in terms of the ratio of permeability to porosity, and this naturally leads to a requirement for different J versus S_w algorithms for different clay(-mineral) contents.

Important though this subject is, comparatively little has been documented. Baptist & White (1957) observed differences in capillary pressure curves that could be attributed to variations in montmorillonite concentration between the fairly clean Tensleep Sandstone ($S_{wirr} \approx 0.15$) and the shaly Frontier Sandstone ($S_{wirr} > 0.3$) in the Rocky Mountain region. Heseldin (1974) accommodated intraformational variations in reservoir quality by relating porosity to water saturation at each of a sequence of given capillary pressures. The regression coefficients and intercepts were themselves related to capillary pressure. Therefore, at a given capillary pressure the values of these 'constant' terms could be inferred and thence water saturation could be calculated for each of a range of porosities. In this way, an average capillary pressure versus height curve could be constructed for each porosity. Johnson (1987) followed a similar procedure using permeability as a reservoir quality indicator in the Argyll Field, North Sea. Here, variations in reservoir quality were attributed to the presence of authigenic clay minerals. In general, however, the effect of clay minerals upon capillary properties remains underresearched.

Laminated shales

Clays and clay minerals present as shale laminations in a sand–shale sequence contribute to the most common form of the low-resistivity pay sand problem (Worthington, 2000). Where the thickness of the laminations is below the spatial resolution of a conventional (as opposed to a micro-imaging) logging tool, the recorded parametric values for the laminations are merely apparent quantities, because they are affected by adjacent beds. Under these conditions, the resistivity of the sand layers is suppressed by the more conductive shales and thence pay intervals can be overlooked. Solutions to this problem have included the deployment of high-resolution logging tools, increased digital sampling of logs to bring out greater detail, and signal enhancement, e.g. through deconvolution and forward modelling. Where no solution is possible through measurement or data processing, a laminated sand–shale model can be used to effect a petrophysical interpretation. The use of such a model requires that realistic assumptions be made about the boundaries and physical properties of the unresolved layers. The boundaries can be inferred from a micro-imaging tool. The layer properties should be based on core analysis.

A recent development that avoids the need for core analysis is the multicomponent induction tool (Kriegshäuser *et al.*, 2000). The use of three orthogonal transmitter–receiver configurations has furnished sufficient data to calculate the resistivities of alternating sand and shale laminae, subject to the assumption of shale and sand homogeneity and provided that the boundaries of the laminae are known. The emphasis on solutions through extractive computation rather than through sharper spatial resolution of logs marks a philosophical departure from conventional approaches to the evaluation of laminated shaly reservoirs (e.g. Allen, 1984).

IMPLICATIONS FOR INTEGRATED FORMATION EVALUATION

The following subsections summarize the influence of clays and clay minerals upon key stages of the reservoir-evaluation process.

Petrophysical partitioning

It is becoming generally accepted that the most effective approach to integrated reservoir description involves some petrophysical partitioning into rock types (e.g. Doveton, 1994; Moss, 1997). The definition of what constitutes a rock type varies from author to author, but for reservoir applications it increasingly hinges around the relationships between porosity, permeability, capillary properties and water saturation. The practice is to separate out petrophysical units based on log responses tied to core description, and then to investigate for each of these units the optimum forms of the (core-calibrated) relationships between reservoir properties. Important log discriminators of petrophysical rock-type are the gamma ray, neutron, density and photoelectric factor (PEF) logs, all of which respond to clay-mineral content. Thus, clay minerals have an impact upon both the identification of petrophysical rock types and the interpretative algorithms used to characterize them. The predictive relationship between porosity and permeability is especially important in this context. An example of how this relationship can be refined by identifying petrophysical rock types or 'electrofacies' using resistivity, gamma-ray and spontaneous-potential (SP) logs has been presented by Elphick & Moore (1999). Here the key indicators of rock type were formation resistivity, shale resistivity, and normalized SP and gamma-ray responses.

Cut-off values

When it is intended to evaluate hydrocarbons in place, the identification of net sand is often based on the gamma-ray log and/or the SP log. The calculation from these logs of a wetted shale fraction V_{sh} has the effect of normalizing log response and thereby permitting multiwell applications. Here the cut-off value of V_{sh} is adopted on the basis of its incremental impact on computed hydrocarbon volume. It is fixed at a level beyond which less stringent cut-offs would not result in significantly higher computed volumes of hydrocarbons in place. Although this cut-off is influenced by the nature of the constituent clay minerals, it usually is presented as a function of reservoir rock-type. Therefore the cut-off can vary both within and between reservoirs. Where the gamma-ray and SP logs have inadequate spatial resolution, e.g. in thinly laminated sand–shale sequences, recourse has been made to a microlog or micro-imaging device, in which case the influence of clay minerals is manifested through conductivity (electrical microdevice) or ultrasonic reflectivity (borehole televiewer). Similar comments apply to cut-offs for net reservoir (V_{sh} plus porosity) and net pay (V_{sh} plus porosity plus water saturation).

Where the aim is to estimate reserves, the cut-off(s) should take account of the ability of the reservoir to sustain economically viable flow rates. Here the traditional emphasis has been on cut-offs of absolute rather than relative permeability. Permeability cut-offs are impacted directly by both clay-mineral content and hydrocarbon fluid type. Note the earlier comments about the need to take account of the permeating fluid.

Relationship between porosity and permeability

Porosity versus permeability relationships are used to predict permeability (K) from porosity (ϕ), a parameter that can be measured downhole. They usually are established through regression analysis of data distributions of log K versus log ϕ or ϕ. These cross-plots generally show a scatter such that a single value of the predictor (porosity) can correspond to more than an order of magnitude of the predictand (permeability). In an effort to improve this precision, the data are often sorted according to some kind of reservoir data partitioning. Ideally, the data partitioning should be guided by geological understanding as expressed through core descriptions. In this way, the resulting facies- or electrofacies-specific porosity versus permeability relationships retain a geological significance.

Porosity versus permeability relationships have been reviewed by Nelson (1994). He went on to query the reasons for different data distributions on a porosity versus permeability cross-plot (Nelson, 2000). In particular, there is little initial dependence of permeability on porosity in newly deposited sands, where grain-packing and sorting effects predominate. An apparent dependence emerges through diagenesis, the effects of which are impacted by compaction. Developing this theme, there have been some interesting studies of the effect of clay-mineral occurrence on the form of the relationship between porosity and permeability. Figure 9 shows a collation by Wilson (1992) in which the data fields on a porosity–permeability cross-plot are mapped by clay-mineral type. In particular, pore-bridging fibrous illite is seen to be a strong permeability reducer.

Relationship between porosity and formation factor

The relationship between porosity ϕ and formation factor F for porous sandstones was expressed by Archie (1942) as follows:

$$F = 1/\phi^m \qquad (6)$$

where the porosity exponent m is considered to be an indicator of pore shape. Equation (6) is known as Archie's first law. Clay(-mineral) effects on electrical conductivity impact directly

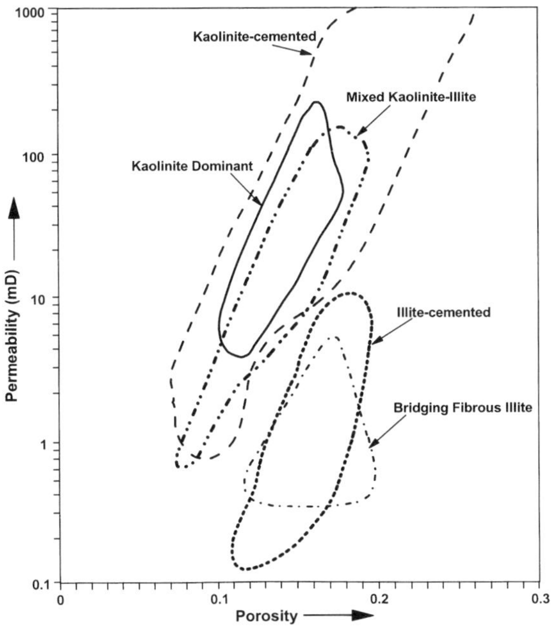

Fig. 9 Porosity versus permeability data fields for Rotliegendes sandstones according to clay-mineral content. (After collation by Wilson, 1992; redrawn from Nelson, 1994.)

on this relationship. Formation factor (F) can be defined as the ratio of the conductivity of a saturating electrolyte (C_w) to the conductivity of a reservoir rock (C_o) that is fully saturated with this electrolyte. Archie found that formation factor was constant for a given rock regardless of the salinity of the saturating electrolyte: hence, F was a factor of the formation rather than of the electrolyte. However, Archie only considered clean (effectively clay-mineral-free) sands saturated with brine. He did not consider sands containing significant concentrations of clay minerals or those saturated with fresher formation waters, which amplify clay(-mineral) effects in the form of an extra conductivity, as per the generic equation (5). This extra conductivity causes C_o to be increased above the level that is compatible with pore geometry alone. Thus, the ratio C_w/C_o is reduced, so that it becomes an apparent formation factor F_a. If this apparent quantity is correlated with porosity, the resulting porosity exponent also will be an apparent quantity m_a, which will be too low,

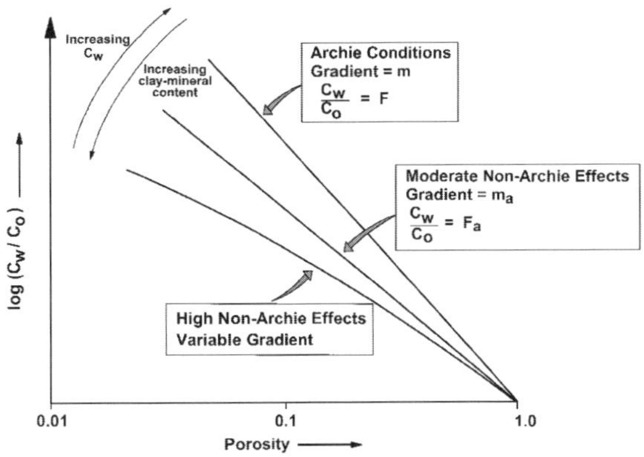

Fig. 10 Schematic effect of clay-mineral content on the relationship between the ratio C_w/C_o and porosity.

and the application of such an equation in well-log analysis will result in incorrect predictions of F from log-derived porosity. These errors transmit through to the evaluation of water saturation (see below). They can be avoided by using a highly saline electrolyte for core measurements of formation factor rather than simulated formation water. However, if this is done, the subsequent evaluation of water saturation must still use a shaly sand equation, such as equation (3) or (4).

Figure 10 illustrates schematically the effect of clay-mineral content and electrolyte conductivity C_w upon the form of the relationship between formation factor and porosity. As one departs from Archie (clean brine-saturated sand) conditions, the trend on a bilogarithmic cross-plot initially remains effectively linear and then becomes distinctly non-linear as shale effects increase further. For this reason, a linear trend on such a cross-plot does not serve to confirm that the Archie conditions are satisfied.

Relationship between resistivity index and water saturation

The relationship between resistivity index I_R and water saturation S_w was expressed by Archie (1942) as follows:

$$S_w = 1/I_R^{1/n} \tag{7}$$

where n is termed the saturation exponent. Equation (7) is known as Archie's second law. Clay(-mineral) effects on electrical conductivity have a direct impact on this relationship, too, but more so than in the case of equation (6). Resistivity index is defined as the ratio of the conductivity of a fully electrolyte-saturated reservoir rock (C_o) to the conductivity that the rock would have if it was partially saturated with the same electrolyte (C_t). This definition can be expressed in the following alternative form:

$$I_R = C_o/C_t = C_w/FC_t \tag{8}$$

Note that formation factor is input to the computation of I_R. Archie found that resistivity index was a constant for a given desaturation level of a particular reservoir rock regardless of the salinity of the saturating electrolyte: hence, I_R was a function of the water saturation rather than of the electrolyte properties. Again, Archie did not consider rocks containing significant amounts of clay(-minerals) and/or fresh waters. The extra conductivity associated with clay(-mineral) effects causes C_t to be increased above the level that is compatible with the partially water-filled pore geometry alone. This increase is greater than that of C_o. Thus, the ratio C_o/C_t is reduced, so that it becomes an apparent resistivity index I_{Ra}. If this apparent quantity is correlated with water saturation, the resulting saturation exponent also will be an apparent quantity n_a, which will be too low, and the application of such a form of equation (7) in well-log analysis could result in incorrect predictions of S_w from log-derived formation resistivity. The problem usually is overcome by correcting values of I_{Ra} to intrinsic values I_R (e.g. Waxman & Thomas, 1974) and by using a shaly sand equation, such as equations (3) and (4).

Figure 11 illustrates schematically the effect of clay-mineral content and electrolyte conductivity C_w upon the form of the relationship between resistivity index and water saturation. As one departs from Archie (clean brine-saturated sand) conditions, the trend on a bilogarithmic cross-plot initially remains effectively linear and then becomes distinctly non-linear as shale effects increase further. Shale effects are greater in the partially water-saturated state than in the fully saturated state. Again, a linear trend on such a cross-plot does not serve to confirm that the Archie conditions are satisfied.

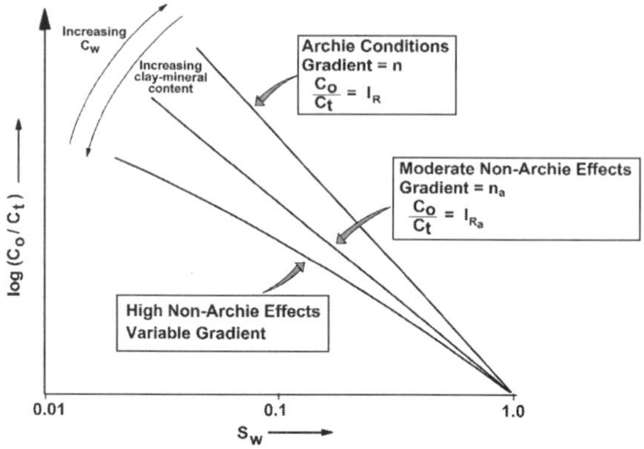

Fig. 11 Schematic effect of clay-mineral content on the relationship between the ratio C_o/C_t and water saturation.

Formation anisotropy

The subject of formation anisotropy has gained attention over the past decade in response to the increase in the drilling of extended-reach and horizontal wells. Most attention has been directed at electrical and acoustic anisotropy, with the petrophysical emphasis on the former. In highly deviated or horizontal wells, resistivity logging tools no longer measure the horizontal formation resistivity but rather a combination of the horizontal and vertical resistivities. The ratio of the vertical to horizontal resistivity is a measure of electrical anisotropy in cases where the reservoir rock is longitudinally isotropic, i.e. there is no directional dependence of resistivity in a direction parallel to the bedding. The square root of this ratio is known as the coefficient of (electrical) anisotropy. This coefficient is a function of scale. The effect of clay minerals on this coefficient is also scale-dependent. Worthington

(1981) examined the variation in the coefficient of anisotropy at the intergranular scale for sandstones that showed a range of electrical shale effects. Essentially, for a given level of water saturation, electrical anisotropy increases with increasing clay-mineral content, where the electrical manifestation of clay minerals is directionally dependent. For the same reason, for a given clay-mineral content, electrical anisotropy increases with increasing hydrocarbon saturation. This same effect can be observed for the limiting case of clean sands, but here the anisotropy is governed primarily by pore shape. At the bedding scale, the longitudinal and transverse resistivities are calculated using resistors in parallel and resistors in series concepts, respectively (Keller, 1966). If the constituent sand and shale beds are themselves internally anisotropic, this microanisotropy can be incorporated with the bedding-scale anisotropy, subject to certain qualifying assumptions.

DISCUSSION

This technical snapshot of the effect of clays and clay minerals upon the physical properties of sandstone reservoirs has adopted the terminology that is in common use in contemporary petrophysical practice, but with the avoidance as far as possible of using the terms 'clay' and 'clay mineral' interchangeably, as much of the petrophysical literature does. The presence of clay minerals within a reservoir rock is handled petrophysically through the adoption of one or other of two interpretative models, the effective porosity model, in which clay-mineral-bound water is included volumetrically within the clay mineral(s), and the total porosity model, in which the same bound water is included with the rest of the interstitial water. Although the procedures used in the two interpretative systems are very different, they both take some account of the compositional effect of clay minerals on measured physical properties. Examples of sources of these clay-mineral effects are the hydrogen content associated with interlayed and structural waters, for the effective porosity system, and the relatively high values of cation exchange capacity associated with interlayered clay minerals, for the total porosity system. Corrections to logs for clay-mineral effects are imperfect, and it is noteworthy that more pronounced departures from petrophysical data trends are seen where these corrections have turned out to be greater. It is also necessary to consider the geometry of clays and clay minerals in order to understand how their presence can have an impact upon reservoir behaviour. At the pore scale, the effects of clays and clay minerals are manifested in permeability characteristics, microporosity and capillary behaviour. At the bedding scale, geometrical effects can prevent a meaningful evaluation of a laminated sand–shale interval where the constituent layers are thinner than the spatial resolution of conventional well-logging tools. Both the compositional and the geometrical effects of clays and clay minerals have an impact upon the *modus operandi* of petrophysical interpretation, especially the use of the key interpretative algorithms such as Archie's laws and the relationship between porosity and permeability. If these effects are not recognized and corrected, serious misevaluations can result.

There is much diverse information available concerning the effects of clays and clay minerals on various physical properties of porous media. A synthesis of the available electrical data has allowed a continuum of behaviour to be identified for clay(-mineral) effects on the resistivity of reservoir rocks (Worthington, 1997). This continuum provides a reference basis for comparing different reservoirs in terms of their electrical characteristics, so that the information from a more fully studied reservoir can be brought to bear beneficially on reservoirs that are at an earlier stage of appraisal. The ongoing drive to contain and manage uncertainty suggests that similar behavioural continua should be investigated for the effects of clay minerals upon other petrophysical properties, especially acoustic data. The benefits of a coordinated exercise of this kind could be considerable.

REFERENCES

Allen, D.F. (1984) Laminated sand analysis. In: *Transactions of the SPWLA 25th Annual Logging Symposium*, Society of Professional Well Log Analysts, Houston, TX, pp. XX1–20.

Archie, G.E. (1942) The electrical resistivity log as an aid in determining some reservoir characteristics. *Trans. Am. Inst. Min. Metall. Eng.*, **146**, 54–62.

Austin, S.P. & Ganley, S.M. (1991) The measurement of the cation exchange capacity of core plugs by a non-destructive 'wet' chemical method. In: *Advances in Core Evaluation II—Reservoir Appraisal* (Ed. by P.F. Worthington and D. Longeron), Gordon and Breach, Reading, pp. 293–308.

Baptist, O.C. & White, E.J. (1957) Clay content and capillary behavior of Wyoming reservoir sands. *Trans. Am. Inst. Min. Metall. Petrol. Eng.*, **210**, 414–416.

Bardon, C. & Pied, B. (1969) Formation water saturation in shaly sands. In: *Transactions of the SPWLA 10th Annual Logging Symposium*, Society of Professional Well Log Analysts, Houston, TX, pp. Z1–19.

Castagna, J.P., Batzle, M.L. & Eastwood, R.L. (1985) Relationships between compressional-wave and shear-wave velocities in clastic silicate rocks. *Geophysics*, **50**, 571–581.

Cook, P.L., Schneeflock, R.D., Bush, J.D. & Marble, J.C. (1990) Trimble Field, Smith County, MS: 100 BCF of by-passed pay at –7000 ft. *Trans. Gulf Coast Assoc. Geol. Soc.*, **40**, 135–145.

Dewan, J.T. (1983) *Essentials of Modern Open-hole Log Interpretation*. PennWell, Tulsa, OK, 361 pp.

Doveton, J.H. (1994) Geologic log analysis using computer methods. *Computer Applications in Geology*, 2, American Association of Petroleum Geologists, Tulsa, OK, 169 pp.

Eberhart-Phillips, D., Han, D.-H. & Zoback, M.D. (1989) Empirical relationships among seismic velocity, effective pressure, porosity, and clay content in sandstone. *Geophysics*, **54**, 82–89.

Ellis, D., Howard, J., Flaum, C., McKeon, D., Scott, H., Serra, O. & Simmons, G. (1988) Mineral logging parameters: nuclear and acoustic. *Techn. Rev.*, **36**(1), 38–52.

Elphick, R.Y. & Moore, W.R. (1999) Permeability calculations from clustered electrofacies: a case study in Lake Maracaibo, Venezuela. In: *Transactions of the SPWLA 40th Annual Logging Symposium*, Society of Professional Well Log Analysts, Houston, TX, pp. AA1–12.

Hall, D.J., Skerrett, E.J. & Thomas, W.D.E. (1978) Critical point drying for scanning electron microscopy: a semi-automatic method of preparing biological specimens. *J. Microbiol.*, **113**(3), 277–290.

Heaviside, J., Langley, G.O. & Pallatt, N. (1983) Permeability characteristics of Magnus reservoir rock. In: *Transactions of the SPWLA 8th European Formation Evaluation Symposium*, Society of Professional Well Log Analysts, Houston, TX, pp. A1–29.

Herron, M.M. & Matteson, A. (1993) Elemental composition and nuclear parameters of some common sedimentary minerals. *Nucl. Geophys.*, **7**, 383–406.

Heseldin, G.M. (1974) A method of averaging capillary pressure curves. In: *Transactions of the SPWLA 15th Annual Logging Symposium*, Society of Professional Well Log Analysts, Houston, TX, pp. E1–8.

Hoyer, W.A. & Spann, M.M. (1975) Comments on obtaining accurate electrical properties of cores. In: *Transactions of the SPWLA 16th Annual Logging Symposium*, Society of Professional Well Log Analysts, Houston, TX, pp. B1–11.

Hurst, A. & Nadeau, P.H. (1995) Clay microporosity in reservoir sandstones: an application of quantitative electron microscopy in petrophysical evaluation. *AAPG Bull.*, **79**, 563–573.

Johnson, A. (1987) Permeability averaged capillary data: a supplement to log analysis in field studies. In: *Transactions of the SPWLA 28th Annual Logging Symposium*, Society of Professional Well Log Analysts, Houston, TX, pp. EE1–18.

Juhasz, I. (1981) Normalized Q_v—the key to shaly sand evaluation using the Waxman–Smits equation in the absence of core data. In: *Transactions of the SPWLA 22nd Annual Logging Symposium*, Society of Professional Well Log Analysts, Houston, TX, pp. Z1–36.

Juhasz, I. (1986) Conversion of routine air-permeability data into stressed brine-permeability data. In: *Transactions of the SPWLA 10th European Formation Evaluation Symposium*, Society of Professional Well Log Analysts, Houston, TX, pp. Y1–15.

Keller, G.V. (1966) Compilation of electrical properties from electrical well logs. *Colorado Sch. Mines Quart.*, **59**(4), 91–110.

Kenyon, W.E. (1997) Petrophysical principles of applications of NMR logging. *Log Anal.*, **38**(2), 21–43.

Kriegshäuser, B., Fanini, O., Forgang, S., *et al.* (2000) Increased oil-in-place in low-resistivity reservoirs from multicomponent induction data. In: *Transactions of the SPWLA 41st Annual Logging Symposium*, Society of Professional Well Log Analysts, Houston, TX, pp. A1–14.

Kuttan, K., Stockbridge, C.P., Crocker, H. & Remfry, J.G. (1980) Log interpretation in the Malay Basin. In: *Transactions of the SPWLA 21st Annual Logging Symposium*, Society of Professional Well Log Analysts, Houston, TX, pp. II1–27.

Leverett, M.C. (1941) Capillary behavior in porous solids. *Trans. Am. Inst. Min. Metall. Eng.*, **142**, 152–169.

Mayer, C. & Sibbit, A. (1980) Global, a new approach to computer-processed log interpretation. *Soc. Petrol. Eng. Pap.*, **9341**, 14 pp.

Minear, J.W. (1982) Clay models and acoustic velocities. *Soc. Petrol. Eng. Pap.*, **11031**, 11 pp.

Moss, B.P. (1997) The partitioning of petrophysical data: a review. In: *Developments in Petrophysics* (Ed. by M.A. Lovell and P.K. Harvey). *Geol. Soc. London Spec. Publ.*, **122**, 181–252.

Nelson, P.H. (1994) Permeability–porosity relationships in sedimentary rocks. *Log Anal.*, **35**(3), 38–62.

Nelson, P.H. (2000) Evolution of permeability–porosity trends in sandstones. In: *Transactions of the SPWLA 41st Annual Logging Symposium*, Society of Professional Well Log Analysts, Houston, TX, pp. MM1–14.

Pallatt, N. & Thornley, D. (1990) The role of bound water and capillary water in the evaluation of porosity in reservoir rocks. In*: Advances in Core Evaluation—Accuracy and Precision in Reserves Estimation*, (Ed. by P.F. Worthington), Gordon and Breach, Reading, pp. 223–237.

Pallatt, N., Wilson, J. & McHardy, B. (1984) The relationship between permeability and the morphology of diagenetic illite in reservoir rocks. *J. Petrol. Technol.*, **36**, 2225–2227.

Partono, Y.J. (1992) Low-resistivity sandstone reservoirs in the Attaka Field. In: *Proceedings of the Indonesian Petroleum Association 21st Annual Convention*, Jakarta, 6–8 October, Vol. 2, pp. 21–34.

Patnode, H.W. & Wyllie, M.R.J. (1950) The presence of conductive solids in reservoir rock as a factor in electric log interpretation. *Trans. Am. Inst. Min. Metall. Petrol. Eng.*, **189**, 47–52.

Prammer, M.G., Drack, E.D., Bouton, J.C. & Gardner, J.S. (1996) Measurements of clay-bound water and total porosity by magnetic resonance logging. *Log Anal.*, **37**(6), 61–69.

Raymer, L.L., Hunt, E.R. & Gardner, J.S. (1980) An improved sonic transit time-to-porosity transform. In: *Transactions of the SPWLA 21st Annual Logging Symposium*, Society of Professional Well Log Analysts, Houston, TX, pp. P1–13.

Theys, P. (1999) *Log Data Acquisition and Quality Control*. Editions Technip, Paris, 453 pp.

Tiab, D. & Donaldson, E.C. (1996) *Petrophysics: Theory and Practice of Measuring Reservoir Rock and Fluid Transport Properties*. Gulf Publishing Co., Houston, TX, 706 pp.

Vernik, L. & Nur, A. (1992) Petrophysical classification of siliciclastics for lithology and porosity prediction from seismic velocities. *AAPG Bull.*, **76**, 1295–1309.

Waxman, M.H. & Smits, L.J.M. (1968) Electrical conductivities in oil-bearing shaly sands. *Soc. Petrol. Eng. J.*, **8**, 107–122.

Waxman, M.H. & Thomas, E.C. (1974) Electrical conductivities in oil-bearing shaly sands: I. The relation between hydrocarbon saturation and resistivity index: II. The temperature coefficient of electrical conductivity. *Soc. Petrol. Eng. J.*, **14**, 213–225.

Wentworth, C.K. (1922) A scale of grade class terms for clastic sediments. *J. Geol.*, **30**, 377–392.

Wilson, M.D. (1992) Inherited grain-rimming clays in sandstones from eolian and shelf environments. In: *Origin, Diagenesis and Petrophysics of Clay Minerals in Sandstones*, Society for Economic Paleontologists and Mineralogists Special Publication 47 (Ed. by D.W. Houseknecht and E.D. Pittman), Tulsa, OK, pp. 213–225.

Winsauer, W.O. & McCardell, W.M. (1953) Ionic double-layer conductivity in reservoir rock. *Trans. Am. Soc. Min. Metall. Petrol. Eng.*, **198**, 129–134.

Worthington, P.F. (1981) The influence of formation anisotropy upon resistivity–porosity relationships. In: *Transactions of the SPWLA 22nd Annual Logging Symposium*, Society of Professional Well Log Analysts, Houston, TX, pp. AA1–25. (Reprinted in *Petrophysics*, **42**(2), 83–92, 2001.)

Worthington, P.F. (1985) The evolution of shaly-sand concepts in reservoir evaluation. *Log Anal.*, **26**(1), 23–40.

Worthington, P.F. (1996) Application of investigative engineering to formation evaluation. 2—Characterisation of reservoir components. *Sci. Drill.*, **5**(6), 249–265.

Worthington, P.F. (1997) A continuum approach to the petrophysical classification and evaluation of reservoir rocks. *Petrol. Geosci.*, **1**, 97–108.

Worthington, P.F. (1998) Conjunctive interpretation of core and log data through association of the effective and total porosity models. In: *Core-Log Integration* (Ed. by P.K. Harvey and M.A. Lovell). *Geol. Soc. London Spec. Publ.*, **136**, 213–223.

Worthington, P.F. (2000) Recognition and evaluation of low-resistivity pay. *Petrol. Geosci.*, **6**, 77–92.

Wyllie, M.R.J., Gregory, A.R. & Gardner, L.W. (1956) Elastic wave velocities in heterogeneous and porous media. *Geophysics*, **21**, 41–70.

Int. Assoc. Sedimentol. Spec. Publ. (2003) **34**, 213–251

Quantitative analysis of clay and other minerals in sandstones by X-ray powder diffraction (XRPD)

S. HILLIER

Macaulay Institute, Craigiebuckler, Aberdeen AB15 8QH, UK, e-mail: S.hillier@macaulay.ac.uk

ABSTRACT

X-ray powder diffraction (XRPD) is an essential tool for the analysis of clay minerals wherever they occur and sandstones are no exception. Quantitative analysis of clay minerals in sandstones is a requirement of many investigations and the potential of XRPD in this respect is unsurpassed. The terms quantitative and semi-quantitative frequently have been used almost interchangeably and without definition. Here definitions and distinctions between these terms are offered in the context of analysis by XRPD. Application of such definitions would better enable those who use the data to identify its limitations and determine whether it is fit for the purpose, or not.

The theory that underlies a large set of the published methods of quantitative analysis by XRPD is outlined. As far as possible the practical aspects of the theory have been emphasized so that the value of understanding it is apparent. Two common ways of preparing samples for quantitative analysis are described in detail, namely the analysis of orientated clay fractions and the analysis of whole-rock samples as random powders. Issues of sample preparation, measurement of peak intensities, validation and uncertainty, and lower limits of detection are discussed for each method and illustrated by examples. Additionally, the identification of the clay minerals encountered most commonly in sandstones is outlined.

Studies of both orientated clay fractions and whole-rock random powders provide complementary information. For well-characterized clay minerals both methods are capable of accurate quantitative results. Nonetheless, if unknown or uncontrolled, many sample and procedural factors may intervene to place results in the semi-quantitative category.

INTRODUCTION

Whenever an analysis of clay minerals is required, it is almost certain that X-ray powder diffraction (XRPD) will be one of the principal tools used. Indeed, many aspects of clay mineral identification, such as interstratification of two or more types, are not readily accessible by any other technique. Clay minerals in sandstones are no exception, and although textural information obtained by electron microscopy may be equally important, more often than not, it is impossible to make an analysis of the abundance or even of the kinds of clay minerals present without recourse to XRPD.

That is not to say that XRPD is the only technique that should be used; the study of clay minerals is always most fruitful when a range of complementary techniques is utilized (Wilson, 1987). Nonetheless, XRPD is undoubtedly one of the most powerful techniques for the analysis of clay minerals. Indeed, for quantitative analysis it is unsurpassed and its utility is augmented by the fact that it is so widely available.

As with any material containing clay minerals, there are essentially two common approaches to quantitative analysis. The first of these is the analysis of some particular 'clay fraction' physically separated from the rest of the rock. Usually, such fractions are prepared for analysis by making them into an orientated layer of clay supported on a substrate. Quantitative analysis of XRPD patterns obtained from clay minerals prepared in this fashion is reported in many publications. The second approach involves the preparation of a so-called random powder and is used when information is also required about the total amounts of clays as well as any non-clay minerals present. Any fraction may be analysed as a random powder but most often it is applied to the analysis of unfractionated 'bulk' or 'whole-rock' samples.

There are advantages and disadvantages to both the clay fraction and the whole-rock approaches. The main advantage of using a clay fraction is one of increased detail and resolution. This is gained not only from having concentrated the clays but also because purposely orientating them enormously enhances the intensity of diffraction from the basal planes of platy clay minerals. This is a great advantage because it is the detailed analysis of the '*d*-spacings' of these planes that is diagnostic of most of the different types of clay minerals. As far as disadvantages are concerned, the problem of relating the clay fraction to the bulk sample and answering the question 'is it representative?' may be the main drawback (Towe, 1974). In some cases this may not be an issue because one may be concerned only with looking for trends. All that is then required is the comparison of like with like, in other words, the method need only be precise. Advantages and disadvantages with the 'whole-rock' method are essentially the converse of these. Both detailed identification and detection limits for clay minerals suffer in 'whole-rock' samples. This is both because of dilution with other minerals and because in a random (or less well-orientated) powder, the intensity of diffraction from the characteristic basal planes of clay minerals will be very much weaker. If the analytical procedure is reliable, the whole-rock approach is capable of providing accurate data on the amounts of clay minerals present. However, depending upon their abundance it may not be possible to distinguish, or even identify, certain types of clay minerals when a whole-rock is examined. Instead whole groups must sometimes simply be lumped together into categories such as 'total dioctahedral clay'. In essence, the two approaches to the quantitative analysis of clay minerals are complementary and the best approach of all is a combined one. For example, the detail obtained from a clay fraction analysis may allow one to identify the clay minerals as a mixture of ordered mixed-layer illite(70)–smectite together with kaolinite. The absolute amounts of these clays present in the sandstone can then be determined from the analysis of the whole-rock. This may indicate that their proportions in the sandstone as a whole are quite different from those in the clay fraction.

In the literature there are many specific methods described for both the analysis of clay fractions or whole-rock samples. Rather than review these methods one by one, this contribution attempts to establish some important definitions and to outline some of the underlying theory upon which a large set of the published methods are based. As far as possible, an attempt is made to present the theory in its practical context, so that the value of understanding it is apparent. Specific approaches to both clay fraction and whole-rock analyses that have been used successfully by the author are then illustrated by way of examples. For each method there is also some discussion of the paramount importance of sample preparation, but by and large the reader is directed to other texts that treat this topic in the detail it deserves. Examples are used that show how the issues of validation, uncertainty and detection limits can be addressed. Applied judiciously, the specific methods illustrated are useful tools and it is hoped that newcomers to the subject will find enough here to feel confident to attempt quantitative analysis of clay minerals in sandstones. It should be emphasized, however, that the methods described are by no means the only methods that might be used, indeed the literature on quantitative analysis is vast. Hopefully some who are new to the subject will be stirred to explore quantitative analysis of clay minerals further and delve confidently into the more detailed texts by Brindley & Brown (1980), Pevear & Mumpton (1989), Bish & Post (1989); Amonette & Zelazny (1994); Zevin & Kimmel (1995) and Moore & Reynolds (1997).

WHAT IS QUANTITATIVE ANALYSIS?

Before considering the details of any method it is necessary to establish some important definitions. For some time the area of quantitative analysis by XRPD, especially in the earth science literature, has been plagued by the almost interchangeable use of the terms quantitative and semi-quantitative, often with no attempt to define what is actually meant. In some papers, methods with obvious shortcomings are described as quantitative, in others authors more keenly aware of the pitfalls back away from this description and substitute semi-quantitative in its place. The result is an inconsistency that undoubtedly reinforces the perception held by many that XRPD methods are not routinely capable of accurate results. Some discussion of this problem was made by Hughes *et al.* (1994), who suggest that the misleading term semi-quantitative should be replaced with the term 'quantitative representation'. According to them, this description should be applied to all methods that generate a set of numbers that represent the mineral quantities present, but that do not necessarily reflect an accurate measure of those quantities. Thus quantitative representation requires only that methods are precise. Precision is concerned only with the reproducibility of a result. Undoubtedly, this is a useful description for a certain class of methods, but this author would narrow the definition of *quantitative representation* further and suggest that it should be applied only to *methods that use measures or proxies other than quantity to represent quantity*. In other words, only those methods that do not provide results in units of quantity should be classed as examples of quantitative representation. Quantitative representation would then include methods such as those used to portray trends in abundance, for example by measurements of differences in peak areas, or peak heights. As emphasized by Hughes *et al.* (1994) precise quantitative representation may be entirely fit for the purpose in hand, there often being no need to know the actual quantities present. This narrowing of the definition from that proposed by Hughes *et al.* (1994) stems naturally from the definitions for quantitative and semi-quantitative analysis that are offered here. Dictionary definitions of the word 'quantitative' indicate that it is usually defined as '. . . expressed in terms of quantity . . .'. Quantitative analysis therefore is concerned with measuring quantity. However, as with any measurement it is incomplete, and perhaps of little use, unless there is also some statement about how confident one is of the accuracy of the measurement, otherwise known as 'measurement uncertainty' or simply 'uncertainty'. At this point it is important to understand the distinction between uncertainty and error. Error is the difference between a result and a 'true' value. Uncertainty defines a range of values within which a true value may lie, at a given level of confidence, e.g. 95%. Uncertainty should be quoted when a validated method is used, for example, to measure the concentration of kaolinite in a sandstone. Errors can be determined only when true values are known. For XRPD, uncertainty may be determined from an analysis of the errors associated with a method, for example, by its application to carefully prepared mixtures, or by participation in round-robin events. Precision and accuracy in the context of XRPD are discussed by Bish & Chipera (1995).

In the light of these concepts, a *quantitative*

analysis is defined as *one expressed in terms of quantity and including some reasonable statement of measurement uncertainty*, e.g. 30.0 wt% ± 3.3 at 95% confidence level. *Semi-quantitative* may then be defined as *an analysis expressed in terms of quantity but for which the uncertainty is not known or unreasonably large, although precision may be assured.* With both definitions the qualifier 'reasonable' is necessary because an evaluation of uncertainty alone is clearly an insufficient distinction between them. Obviously, determining what constitutes reasonable uncertainty for a quantitative analysis by XRPD must be a subjective decision. It nevertheless should be an informed one. Hillier (2000) discussed a range of well-documented studies where accuracy was assessed and found that they generally endorsed the suggestion by Calvert *et al.* (1989) that results within 3% absolute of the true value were 'highly accurate'. Estimates such as this, however, do not take account of the fact that errors are naturally correlated with concentration (see later). Based on the author's attempts to evaluate relationships between concentration and error it is suggested that only methods for which uncertainty is less than $\pm X^{0.5}$ at 95% confidence level, where X is the concentration in wt%, should be classified as quantitative. This is of course simply a suggestion, and it is hoped that further studies will provide a better one. It does, however, set a marker by which the quality of quantitative XRPD methods may be judged.

SOME THEORY

Basic equations and concept of reference intensity ratios

The following is based largely on more detailed treatments given in Snyder & Bish (1989) and Snyder (1992). The basic equation for all forms of quantitative analysis by XRPD was developed by Klug & Alexander (1974)

$$I_\alpha = \frac{K_\alpha X_\alpha}{\rho_\alpha \mu_m} \tag{1}$$

This equation relates the intensity I of a diffraction line (i.e. a peak in a diffraction pattern) from a component α in a mixture to the weight fraction (X) of that component, its density (ρ), the mass attenuation coefficient (μ) of the mixture (m), and a host of terms, collected in the constant (K), which are fixed for a given experimental set-up and diffraction line. The basic problem in using this equation to make quantitative analyses concerns the term for the mass attenuation coefficient of the mixture. Because its computation requires knowledge of the weight fraction of each component in the mixture, the equation is underdetermined. All methods of quantitative analysis are ways of obtaining extra information about the sample so that equation (1) is soluble. One of the most general methods devised is that known as the internal standard method. For this method the basic working equation is derived by dividing two equations of the form given in (1) to obtain

$$\frac{I_\alpha}{I_\beta} = k\frac{X_\alpha}{X_\beta} \tag{2}$$

where the subscripts α and β denote phase α and phase β respectively. In this equation the constant k is the slope of the straight line resulting from a plot of

$$X_\beta \frac{I_\alpha}{I_\beta} \text{ versus } X_\alpha \tag{3}$$

It also should be clear from equation (2) that if a binary mixture is prepared containing equal weight fractions of phase α and phase β the constant k is obtained simply as the ratio of two intensities. The ratio of two such intensities is the basis of many methods of quantitative analysis. For example, one might determine that a given peak from the clay mineral smectite diffracts with four times the intensity of a given peak from an equal weight of kaolinite so that the constant k would be equal to 4. Measurements of these constants can be found throughout the literature. Often, however, they are given different names and the phase that is arbitrarily set to unity and the basis upon which they are normalized, may vary. In the clay mineral literature the illite 001 peak is often chosen as the basis for normalization. In other cases a quartz peak or the 113 peak from corundum has been chosen. The most important point to understand is that essentially they are all one and the same thing, i.e. measures of the relative diffracting power of two phases. In some cases they have been called 'weighting factors' (Biscaye, 1965), in other cases 'intensity factors' (Shultz, 1964; Cook *et al.*, 1975), in other cases 'ki' factors (Chung, 1974a,b; Wilson, 1987), in other cases 'MIFs' (mineral intensity factors) (Reynolds, 1989), and more generally they are known as 'reference intensity ratios' or RIRs. In its most general form (e.g. see Snyder, 1992; Jenkins & Snyder, 1996) a RIR is given by

$$\mathrm{RIR}_{i,s} = \left(\frac{X_s}{X_i}\right)\left(\frac{I_{(hkl)i}}{I_{(hkl)s}}\right)\left(\frac{I^{\mathrm{rel}}_{(hkl)s}}{I^{\mathrm{rel}}_{(hkl)i}}\right) \quad (4)$$

where X denotes weight fraction, I intensity, I^{rel} relative intensity and the subscripts indicate phase i and the standard phase s, respectively. When the standard phase is corundum and intensity measurements relate to the strongest peak from the phase i, RIR values are known formally as I/I_{cor} or $\mathrm{RIR}_{\mathrm{cor}}$.

Determining reference intensity ratios

The usual way to determine a RIR is to prepare a 50 : 50 mixture of the pure mineral with the standard, run an XRPD pattern of this mixture, and measure the ratio of the intensity of the strongest peak from the mineral to that of the strongest peak from the standard. Another way in which RIRs may be obtained is to run pure mineral samples and the standard under *exactly* the same conditions and multiply the measured peak intensity ratio by the ratio of the mass attenuation coefficients of the two phases

$$\mathrm{RIR}_{i,s} = \left(\frac{I_{(hkl)i}}{I_{(hkl)s}}\right)\left(\frac{\mu_i}{\mu_s}\right)\left(\frac{I^{\mathrm{rel}}_{(hkl)s}}{I^{\mathrm{rel}}_{(hkl)i}}\right) \quad (5)$$

Running the standard phase and the pure mineral consecutively should satisfy the criteria of *exactly* the same conditions, provided it is known that changes in the intensity response of the diffractometer between the two runs will be negligible.

The RIRs also may be determined simultaneously with less effort (Chung, 1975), and are corrected easily for the presence of impurities if the RIRs of the impurities are known. For example, a 1 : 1 mixture of corundum and a sample believed to be pure albite is prepared but it is found subsequently that the 'pure' albite is contaminated with quartz. If an accurate $\mathrm{RIR}_{\mathrm{cor}}$ for quartz is known its concentration in the sample of albite can be calculated by designating the 50% of added corundum as the internal standard in equation (8). A RIR for any selected albite peak is then obtained by

$$\mathrm{RIR}^{\mathrm{corrected}}_{\mathrm{albite}} = \frac{\mathrm{RIR}^{\mathrm{measured}}_{\mathrm{albite}}}{W_{\mathrm{albite}}} \quad (6)$$

where, in this example, W is the weight fraction of albite in the 'pure' albite standard. Because RIRs vary with the experimental set-up, for quantitative work they invariably must be determined in the laboratory in which they are to be used. For some purposes, however, it may suffice to take them from the literature or from sources such as the International Center for Diffraction Data (ICDD) Powder Diffraction File (PDF), which includes numerous values for $\mathrm{RIR}_{\mathrm{cor.}}$ Some tables of $\mathrm{RIR}_{\mathrm{cor}}$ also have been published in the journal *Powder Diffraction* (Davis *et al.*, 1989). However, unless verified otherwise, the results obtained from the use of borrowed RIRs should be considered only semi-quantitative at best. Another approach by which RIRs may be obtained is by calculation of diffraction patterns. For example, for random powder patterns of non-clay minerals, the POWD series of programs (Smith *et al.*, 1982) has been widely used. Furthermore, as well as including experimentally measured RIRs, the 1998 release of the ICDD PDF includes $\mathrm{RIR}_{\mathrm{cor}}$ values for some 6000 minerals for which XRPD patterns have been calculated with the POWD12 program. Many of the values in the PDF, however, should be regarded with extreme caution and their use needs to be considered very carefully indeed. For example, a whole series of patterns that are calculated for quartz have RIRs listed as 0.6. Experimentally determined values for quartz usually range from 3 to 5, so these calculated values are clearly erroneous. The primary advantage of calculating RIRs is that it is not necessary to have pure specimens of the minerals to hand. Indeed, where clay minerals are concerned it is a formidable task to assemble a collection of pure clays, let alone go to the trouble of preparing mixtures. Fortunately, owing to the efforts of Reynolds (1985, 1989) it is a relatively simple matter to calculate RIRs for the 001 clay-mineral peaks using his program NEWMOD.

Another feature to realize is that irrespective of how they are obtained or expressed it is easy to convert a RIR from one basis to another. The procedure is simply to find a common denominator and divide the one RIR by the other. This can be useful in two situations. Firstly one may wish to switch to a different internal standard in order to circumvent problems that may arise with overlapping and interfering peaks. Secondly, one can use programs such as NEWMOD to calculate RIRs for clay minerals that are difficult to obtain in pure form and then convert them to RIRs for use with whole-rock internal standard methods. By way of example, assume that a RIR is calculated for an Fe-rich chlorite 002 peak relative to the illite 001 using NEWMOD. Assume also that the RIR of the corundum 113 peak relative to the illite 001 peak has been measured on a 1 : 1 mixture of a sample of pure illite with corundum (note that this RIR is expressed with illite as denominator, i.e. the reciprocal of a conventional RIR_{cor}). Then the RIR for an Fe-rich chlorite 002 relative to corundum is given by

$$RIR_{ch,c} = \frac{RIR_{ch,i}}{RIR_{c,i}} \tag{7}$$

Caution is necessary, however, because sometimes RIRs are expressed with the standard phase as the numerator and sometimes as the denominator. Furthermore, in the example given one would have to be reasonably confident that the calculated and the real illite were well matched.

Using reference intensity ratios

Regardless of how they are derived there are basically two ways in which RIRs are used to make quantitative analyses. One of these ways is called the added internal standard RIR method and the other is called the normalized RIR method (Snyder, 1992). In terms of RIR_{cor} the concentration of any phase (i) in a sample spiked with a known amount of corundum is obtained by rearranging equation (4)

$$X_i = \left(\frac{X_{cor}}{RIR_{cor}}\right)\left(\frac{I_{(hkl)i}}{I_{(113)cor}}\right) \tag{8}$$

Thus the measured peak intensity of the phase to be quantified is divided by the measured peak intensity of the main corundum 113 peak and multiplied by the added weight percent of corundum divided by the RIR_{cor} for the phase in question. The result is the concentration of the phase in the spiked sample. This must be divided by the fraction that is sample to obtain the concentration in the original, e.g. divide by 0.8 in a mixture spiked with 20% corundum. Although this equation is given in terms of the main corundum 113 peak, in practice this peak might be so severely overlapped by peaks from other phases in the sample that it cannot be measured. This rarely is a problem because the intensity of the corundum 113 peak may be determined from the intensity of any measurable corundum peak simply by dividing by its relative intensity. In fact, regardless of the internal standard that is chosen it is good practice to determine its intensity from appropriately scaled measurements of as many of its peaks as possible. Following this procedure will give more confidence that the intensity measurement of the standard is accurate, because discrepant values, such as those resulting from obscured peak overlap, should become obvious. Furthermore, in a very complex pattern where many overlaps are suspected, one can elect to base the quantification on a measurement of a peak from the standard that gives the lowest intensity value for the main peak after scaling. This assumes that the lowest value signifies a peak that has the least overlap problem and, preferably, is not overlapped at all. This is, of course, only justified if it is reasonable to assume that other factors such as preferred orientation can be ignored.

With the RIR method in its added internal standard form, each mineral phase is quantified independently of the other minerals present. Thus it is not necessary to quantify all of the minerals present in a mixture. Indeed, it is not even necessary to identify all of them. If X-ray amorphous materials, such as organic matter or volcanic glass, are absent from the sample then identification and quantification of all the crystalline phases present by this method should result in an analysis total close to 100 wt%. If amorphous materials are known to be present, or expected, and all crystalline phases have been identified and quantified, the amorphous content itself may be estimated by difference. This form of RIR method, which relies upon the addition of a so-called 'internal standard', is also sometimes known as the 'matrix flushing' method after Chung (1974a), because matrix absorption effects are 'flushed' out of the equation for quantitative analysis. Chung (1974a) was the first to point out that the internal standard method formulated as in equation (8) prescribes the slope of the calibration curve for every component. Hence it is not necessary to go to the trouble of preparing such curves.

Even without the addition of a known amount of an internal standard to a sample it is still possible to compute the ratio of the weight fractions of any two phases in a mixture, provided their RIRs are known. If follows that if RIRs are available for all phases present and it is reasonable to assume, as an additional constraint, that all these phases sum to 100 wt%, then the concentration of any phase in the mixture can be calculated by the normalized RIR method

$$X_i = \left(\frac{\mathrm{RIR}}{I_{(hkl)i}} \sum_{i=1}^{n} \frac{I_{(hkl)i}}{\mathrm{RIR}} \right)^{-1} \qquad (9)$$

where $I_{(hkl)i}$ denotes the intensity of a specific peak from mineral *i*. This method is also sometimes known as the 'adiabatic' method after Chung (1974b) (adiabatic = nothing lost, nothing gained), but the normalized RIR method is the preferred name (Jenkins & Snyder, 1996). It must be stressed, however, that this method is applicable only if all the components in the mixture are identified and analysed. When this is true it is not necessary to go to the trouble of adding a known amount of internal standard. Various manifestations of normalized RIR methods almost certainly account for the most common form of method applied to the analysis of clay fractions prepared as orientated mounts. An example is the often quoted study by Biscaye (1965) who used factors of 1, 4, 2 and 2 for the glycolated smectite 001, the illite 001, and the chlorite 002 and kaolinite 001, respectively, the latter two peaks being proportioned to the intensities of the 'resolved' 3.5 Å kaolinite–chlorite doublet.

The main drawback of any normalized RIR method is the normalization assumption itself. As well as giving a false sense of accuracy because of neat 100% totals, any errors in one phase determination, as well as any systematic errors, are hidden and distributed across all other phases. As far as clay fractions are concerned, only rarely do they contain nothing other than clay minerals. When other phases, either crystalline or amorphous are present, analyses based on a normalized method rapidly deteriorate into nothing more than relative abundance. Thus the accuracy of normalized methods is almost always suspect and, in terms of the definitions presented previously, in most cases they fall quickly into the semi-quantitative category. If accuracy is of prime concern then an internal standard RIR method is a far superior choice. With this method an analysis total that sums to ≈ 100% provides a fair degree of confidence that the analysis is both complete and accurate. Moreover, the analysis total is about the only check there is. The only drawback is that adding the internal standard dilutes the sample.

The success of the internal standard method hinges on accurate measurement of the intensity of the internal standard. From equation (8) it can be seen that error in the measurement of the intensity of the internal standard will produce a constant relative error in the determination of the analytes. Thus in terms of absolute error this factor is most significant for phases present in large concentrations. The more standard added, the more accurate its intensity

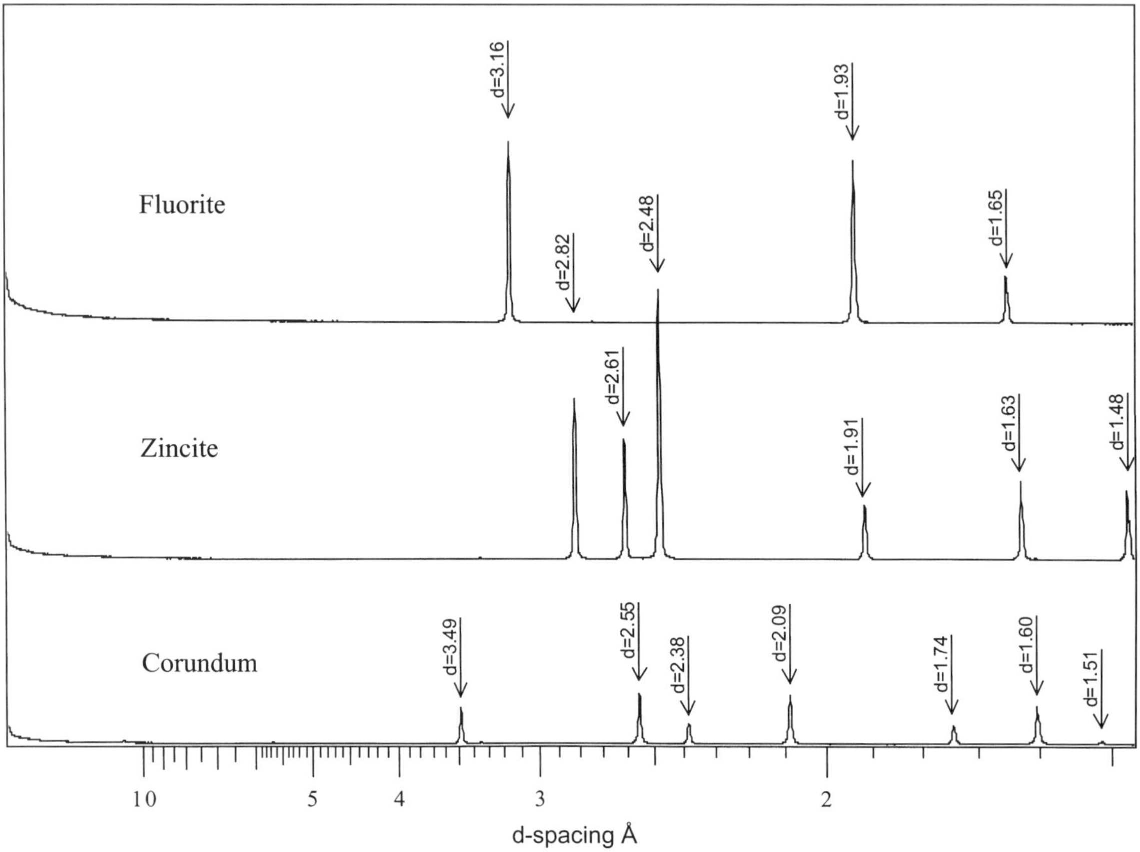

Fig. 1 XRPD patterns (random powders) and peak positions of some possible standards for the internal standard reference intensity ratio (RIR) method. Patterns are offset for clarity and intensity of each pattern is scaled in proportion to RIR_{cor} (see Table 1).

measurement is likely to be. Nevertheless, one must also try to avoid overdilution of the sample with too much of the standard. For most substances an addition of the order of about 20 wt% is an appropriate compromise.

Choice of standard is another decision that must be made. Probably the most widely used are corundum (Al_2O_3), zincite (ZnO) and fluorite (CaF_2). Patterns of these materials are illustrated in Fig. 1, and details of possible sources, indicative values for RIR_{cor}, mass attenuation coefficients for Cu and Co radiation and common peak interference problems are listed in Table 1. In practice several factors will determine the choice of standard. Often the main consideration is peak interference, but similarity of attenuation coefficient to the sample and the diffracting ability of the standard also may be important. If contamination of the sample or its recovery for other purposes is an issue then a salt such as KCl may be used, as it can be washed out of the sample afterwards (D. Moore, personal communication, 2001).

Lower limits of detection

As with any form of quantitative analysis, another value worth knowing is the lower limit of detection (LLD). The LLD is variously defined but one commonly accepted definition is the concentration that will produce a peak equivalent to some defined number of standard

Table 1 Details of some potential random powder standards for the internal standard RIR method.

Standard	μ Cu, μ Co	RIR_{cor} (indicative)	Possible source	Common peak interference problems
Corundum (Al_2O_3)	31, 48	1.0	Aldrich 1344-28-1 (minor gamma alumina contamination 1%)	Major amounts of calcite causes severe overlap problems with several peaks
Zincite (ZnO)	50, 78	5.5	Fischer Scientific Z/1350/53	Halite can overlap one peak and several peaks sit in areas where there may be overlap problems with non-basal clay peaks
Fluorite (CaF_2)	91, 137	3.6	BDH 275944X	Overlaps some important albite peaks

μ = mass attenuation coefficient; RIR = reference intensity ratio.

deviations of the background (Davis, 1987). Thus at two standard deviations (95% probability) the LLD of phase *j* may be evaluated as

$$LLD_j = \frac{2\sigma_b}{I_p/W_j} \tag{10}$$

where (σ_b) is the standard deviation of the number (N) of background counts accumulated, and is given by the $\sqrt{N}$, and the term I_p/W_j expresses peak counts per unit weight of component *j*. Zevin & Kimmels (1995), however, point out that the probability of detecting a signal of two standard deviations above background also needs to be considered and that allowance must be made for the fact that both the total intensity and the background intensity must be measured to evaluate I_p. Taking these factors into account at the same probability level (95%), the LLD is defined as

$$LLD_j = \frac{4\sqrt{2}\,\sigma_b}{I_p/W_j} \tag{11}$$

This is approximately equivalent to six standard deviations above background. A further important point to make is that LLDs are inversely proportion to their respective RIRs. In other words, the best diffractors (largest RIRs) have the lowest detection limits. This relationship reflects the fact that an RIR is simply a relative measure of diffracting power. It is, however, a very useful relationship because when the added internal standard form of the RIR method is used the measured LLD for the standard can be used to estimate the LLD of any other phase in the sample, be it present or not. The only requirements are that the RIR of the phase of interest must be known and the character of the background should not be substantially different from that in the region where it is measured for the standard phase. Thus

$$LLD_a = LLD_s\, RIR_a^{-1} \tag{12}$$

where the subscripts denote phase *a* and the standard phase *s*, respectively. Use of the measured LLD for the added internal standard and equation (12) to estimate the LLD for another phase explicitly accounts for the fact that LLDs depend on factors such as the mass attenuation coefficient of the sample. If the sample has a high mass attenuation coefficient, the sensitivity (c.p.s. per wt%) and hence the detection limit for any phase in the sample will be lower than for the same phase in a sample with a lower mass attenuation coefficient (Chung, 1974b). Furthermore, the detection limit for any counting strategy also may be calculated because the precision depends only on the square root of the counting time. Thus the LLD obtained by counting twice as long is obtained by dividing by the square root of 2, or by the square root of 10 for a tenfold increase in the counting time. In combination with equation (11) or (12) this allows one to very quickly determine the counting strategy that must be used to obtain a given LLD appropriate for the task in hand. It also is important to note that the square root dependence places a practical limit on improving a LLD by simply counting for longer. If LLDs must be pushed as low as possible then other strategies that increase the number of counts in the peak and/or the peak to background ratio may be necessary.

Full-pattern methods

In addition to the use of simple methods that rely on the measurement of a small number of peaks there has been a recent upsurge of interest in so-called full-pattern fitting methods, foremost amongst which is the Rietveld method. Essentially, the Rietveld method involves fitting the whole of the observed diffraction pattern with a synthetic pattern that is a sum of patterns calculated for each phase in the sample (Snyder & Bish, 1989; Bish & Post, 1993; Bish, 1994; Mumme *et al.*, 1996). The difference between the synthetic pattern and the observed pattern is minimized by an interactive refinement/optimization procedure. In effect, the Rietveld method uses calculated standards with the advantage that such standards are tailored to the actual minerals present during the refinement process. Concentrations for the phases present in the sample are obtained from the final values of the refined Rietveld scale factors. Snyder (1992) has shown how the Rietveld method may also be described in RIR notation. Essentially, the Rietveld scale factors play the role of RIRs, however, quantitative analysis may be performed with no need to measure RIRs in advance. Thus without the use of an internal standard the Rietveld method is analogous to a normalized RIR method. Hence it also is subject to the same sort of systematic errors as well as errors in model parameters that are swept into the scale factors, potentially resulting in inaccurate analyses. Additionally, the Rietveld method can be used with an internal standard. It seems clear that that is a practice which should be recommended. The inclusion of a standard may be the only way to detect many potential problems, and thereby gauge the reasonableness of an analysis of an unknown. In experienced hands there is no doubt that the Rietveld method has enormous potential as a tool for accurate phase analysis. Eventually, it will probably become the method of choice. However, it can take a considerable amount of time to set up and in the hands of the inexperienced it can easily turn into a black box producing nonsense with an air of sophistication. Furthermore, phases with one- and two-dimensional diffraction effects, such as most clay minerals, cannot yet be analysed routinely using the Rietveld method. For such phases the much simpler and quicker peak-based methods undoubtedly still have a substantial role to play.

A final point on theory

A reasonable grasp of the theory presented above should represent a firm basis upon which to dig deeper into the more comprehensive and detailed treatments from which most of the above was borrowed. Daunting though some of it may at first appear to the uninitiated, quantitative analysis can be accomplished with little more than proper attention to sample preparation, a few accurately measured RIRs and the programming of equation (8) or (9) into a spreadsheet.

ANALYSIS OF ORIENTED CLAY FRACTIONS

The first steps that are required for the quantitative analysis of clay minerals in orientated preparations are appropriate procedures for sample preparation and identification of the clay minerals present in the sample. Sample preparation is the key to the success of many methods and the quantitative analysis of clay minerals prepared as orientated samples is no exception. Only an outline of the issues related to samples preparation can be given here. For further details the reader is directed to the very clear and comprehensive treatment of all aspects of the subject presented by Moore & Reynolds (1997). Likewise, a reliable and accurate identification of the clay minerals present in a sample is also a crucial, but surprisingly neglected aspect of quantitative analysis. In essence, one cannot purport to make an accurate quantitative analysis of a clay mineral if one does not know enough about the kind of clay that it is (Ottner *et al.*, 2000). As an example, consider that the RIRs of peaks from some common clay minerals can differ by a factor of four depending on differences in chemistry or mixed-layer proportions. Clearly then, schemes that classify all clay minerals into broad categories such as smectite, illite, kaolinite and chlorite, without mention of further details of identification, can never be more than semi-quantitative, and may be poor at that.

Issues of sample preparation

As emphasized by Moore & Reynolds (1997) there is no single 'correct' way to prepare samples for XRPD. Where clay fractions are concerned, this especially is true of the first stages involving the disaggregation of the sample and the dispersion and separation of a clay-sized fraction. Nevertheless, when quantitative analysis is a requirement some properties of the sample presented to the X-ray beam are crucial to its success. It is not the intention here to review the techniques of sample preparation in any detail. They are covered more than adequately elsewhere (Brindley & Brown, 1980; Moore & Reynolds, 1997). Here mention is made only of techniques that are often necessary and/or useful when clays in sandstones are the focus of attention. The reader may follow up these leads in the previously cited references.

With sandstones the first stage of separating a clay fraction may involve nothing more than crushing the sample to liberate the clay. The clay is then dispersed in deionized water or distilled water and a clay fraction of a specified size obtained. Size fractionation is accomplished either by timed sedimentation, or by centrifugation, according to Stokes' law. For fractions of < 2 μm or larger, sedimentation under gravity is convenient. For smaller fractions a centrifuge is essential. Sometimes the clay minerals in sandstones are studied using a conventional < 2 μm Stokes' size fraction, but some workers advocate the use of coarser < 6 μm or < 4 μm fractions. This is because many clay minerals in sandstones are larger than the conventional defined clay size (< 2 μm). This is evident when the dimensions of many forms of pore-filling chlorite or kaolinite are measured by electron microscopy. Fractions larger than < 2 μm therefore may be more 'representative' of the population of clay minerals present in some sandstones.

Unless they are very indurated, the crushing and disaggregation of sandstone to liberate the clays may be carried out conveniently using nothing more than a hammer and a steel plate, with the sample rolled up in a plastic bag. The bag will be partially destroyed in the process, but remains intact enough to pour the crushed sample from it into a beaker ready for dispersion. Whenever possible, rotary grinding, especially dry grinding of the sample in any form of mill, should be avoided because there is always the risk of damaging the clay minerals. Prolonged grinding will also generate new clay-sized material. Even hand grinding may be deleterious, so the best policy is always to do as little as possible to liberate the clays. Furthermore, any form of crushing or grinding may pose a problem if the samples are required, perhaps primarily, for other studies such as isotopic analysis. In this case, the clay fraction of porous sandstones may be liberated using the freeze–thaw action of water using a temperature cycling chamber, although the procedure may require weeks to complete (Zwingmann, 1995).

In some instances, such as with samples from boreholes, it may be necessary to wash out soluble salts, usually sodium and potassium chlorides, the remnants of drilling and formation fluids. The presence of these salts will otherwise prevent dispersion. With consolidated samples, one of the simplest ways to do this is by soaking in deionized water before crushing. Alternatively the samples, including friable samples, may be sealed in dialysis tubing. The tubing is rinsed internally, knotted at one end, filled with a sample, knotted at the other end, then soaked overnight in a large beaker full of deionized water. With samples from boreholes it is often also necessary to remove hydrocarbons. The presence of hydrocarbons tends to stick minerals together into undispersable aggregates, but again the golden rule is to do as little as possible to obtain a dispersible clay fraction. Hydrocarbons may be removed by washing the sample with a solvent such as xylene or dichloromethane, but less expensive alternatives such as washing with a household bleach solution may be just as effective. If solvents are used the sample should be washed with acetone prior to redispersal in water. In all cases one should be acutely aware that chemical treatment of any kind may alter, even destroy, the original clays.

Once the clay fraction is obtained the next stage is to prepare it in a manner suitable for placing into the diffractometer. Invariably, this involves the concentration of the sample from suspension and the formation of some form of flat layer of clay supported on a substrate. Substrates vary but the most common are glass slides or ceramic plates. More rarely flat steel plates, molybdenum plates, or silver membrane filters are used. Metal plates can be used to advantage when heating clays to temperatures in excess of 500°C. The formation of the clay layer on the substrate may be accomplished by numerous techniques. All of them are based around sedimentation, centrifugation, smearing of a clay paste, direct filtration, or transfer from a filter. In each case the goal is to deposit a uniform, flat and smooth layer of clay.

From the point of view of quantitative analysis the most important characteristics of this clay layer are its homogeneity and thickness. Methods which allow particle size segregation to develop as the sample is formed, such as by any form of direct sedimentation, should be avoided (Gibbs, 1965). This is because different clay minerals have different particle size distributions. Some, such as smectites, tend to be finer in size and so are concentrated at the surface of the layer; others, such as kaolinite, tend to be coarser in size and so are concentrated at the base of the layer. As most diffraction is contributed from the surface of the clay layer presented to the X-ray beam, especially at low angles of incidence (θ), the intensities recorded from such an inhomogeneous sample will not be proportional to abundance. According to Gibbs (1965) segregation of minerals during sample preparation may produce differences in the measured peak area intensity ratios of the 17 Å/10 Å peaks from smectite and illite, respectively, of up to 250%. If, for the sake of argument, we assume that the smectite 17 Å peak diffracts with about five times the intensity of an illite 10 Å peak, then in a 50 : 50 mixture of smectite and illite by weight, such differences would result in overestimation of the smectite content by about 20 wt% (40% relative error), and the illite would be underestimated by this amount. Many studies subsequent to Gibbs (1965) have remade this point, as discussed by McManus (1991).

Probably the best method for preparing orientated samples for quantitative purposes is the filter peel method (Drever, 1973). The author has found that one of the keys to success with this method is to use flimsy, and hence easily peeled, filters such as the isopore membrane filter marketed by Millipore (HTTP04700). Just as the last of the filtrate passes through the filter place a glass slide on to the still damp deposit of clay. Turn off the vacuum and press the slide gently into the clay deposit. Then lift two of the opposite edges of the filter and fold them flat over the back of the glass slide gently pressing them down. Using one or both of the remaining filter edges lift the glass slide clear of the filtration apparatus. When completely dry, peel the filter away leaving the clay deposit on the glass slide. The prime advantage of the filter peel method is that, provided sedimentation times are no more than a few minutes, the sample should be homogeneous, furthermore it is the initially deposited clay that forms the surface of the sample. If filtration times are longer than a few minutes, the suspension must be agitated to avoid differential sedimentation. The technique works well with fractions of $< 2\ \mu m$ or finer. Coarser fractions, or particularly uncohesive samples, may prove problematic.

The other property that must be controlled is sample thickness. In essence, the sample must be thick enough to completely absorb and diffract the X-ray beam for the entire angular range (2θ) over which peaks will be measured and compared. At some angle, a sample that is too thin will begin to transmit some of the X-ray beam. The result is that the recorded intensities for peaks at this angle and beyond will be lower than they should be. The amount of sample required to make a layer of clay that is effectively infinitely thick can be easily calculated (Moore & Reynolds, 1997). It is a function, however, of the mass attenuation coefficient (μ) of the sample, which depends on its chemical composition. Thus the thickness required depends on what minerals are present, their proportions and the radiation used. With Cu radiation for a sample dominated by kaolinite ($\mu \approx 32$) the amount of clay required to obtain the correct intensity for the kaolinite 002 peak, at approximately 25° 2θ, is 15 mg cm^{-2}. In contrast, for a sample dominated by an Fe-rich chlorite ($\mu \approx 100$) obtaining the correct intensity for the chlorite 004, also at approximately 25° 2θ, requires only 5 mg cm^{-2}. One advantage of using Co radiation is that the mass attenuation coefficients of most common clay minerals cover a very much narrower range than if Cu radiation is used. Hence thickness considerations are more uniform. In practice it is quite difficult to prepare samples that are thick enough for the requirements of quantitative analysis so that one should always tend towards making samples thicker when the sample allows. Generally, for a clay sample prepared on a glass slide, if the clay layer appears transparent when held up to a light it will definitely not be thick enough for the purposes of quantitative analysis.

Identification of common clay minerals in sandstones

Details of procedures for the identification of clay minerals are treated in depth in texts such as Brindley & Brown (1980) and Moore & Reynolds (1997). As well as drawing upon one's accumulated experience, the precise identification of clay minerals often requires one to pore over texts such as these and the literature they cite. However, the subset of clay minerals most commonly encountered in sandstones is limited enough that it is useful to outline their identification and illustrate the process with examples of real diffraction patterns.

In essence the identification of clay minerals by XRPD is based on recording a series of diffraction patterns before and after various treatments. The positions, intensities and shapes of the various peaks in a diffraction pattern and changes observed in these features in response to the various treatments applied form the basis of all identifications. A common scheme is to record diffraction patterns in an untreated air-dried state, after treatment of the sample with ethylene glycol, and after heating the sample to various temperatures. The response to treatment with 10% HCl at 90°C for 30 min is also a useful test. Table 2 summarizes the basic information that is diagnostic for the identification of the clay minerals illite, glauconite, kaolinite, smectite, chlorite, and the mixed-layer clay minerals illite–smectite and chlorite–smectite.

Illite

Illite is characterized by a series of peaks at about 10, 5 and 3.3 Å, that at 5 Å being about one-third the intensity of the other two. Confusion sometimes arises because of the use of the term illite both for a specific mineral and in a more general sense to refer to the more heterogeneous micaceous component of sedimentary rocks (Bailey *et al.*, 1986). In terms of the precise definition of the mineral species illite, glycolation and heating to temperatures less than 500°C should produce no change at all in peak positions and shapes. In practice, the general term is used more often because there is almost invariably some degree of subtle change. Glauconite often can be distinguished from

Table 2 Effect of various treatments on the basal spacings (Å) of some clay minerals commonly found in sandstones.

<table>
<tr><th></th><th colspan="5">Treatment</th></tr>
<tr><th>Mineral</th><th>Air-dried</th><th>Glycol</th><th>Heated to 375°C</th><th>Heated to 550°C</th><th>HCl</th></tr>
<tr><td>Illite</td><td rowspan="2">10</td><td rowspan="2">10: Small amount of expandability often apparent</td><td rowspan="6">10</td><td rowspan="6">10</td><td>Insoluble</td></tr>
<tr><td>Glauconite</td><td>Soluble</td></tr>
<tr><td>Smectite (Na)</td><td>12*</td><td rowspan="2">17</td><td rowspan="2">Fe-rich species soluble</td></tr>
<tr><td>Smectite (Ca, Mg)</td><td>15*</td></tr>
<tr><td>Illite–smectite (random)</td><td rowspan="2">15–10*</td><td>17
Broad, higher orders non-integral</td><td rowspan="2">Insoluble</td></tr>
<tr><td>Illite–smectite (ordered)</td><td>15–10
Higher orders non-integral unless near 50 : 50</td></tr>
<tr><td>Kaolinite, dickite</td><td>7.15</td><td>7.15</td><td>7.15</td><td>Destroyed</td><td>Insoluble</td></tr>
<tr><td>Chlorite (Fe-rich)</td><td>14.2</td><td>14.2</td><td>14.2</td><td>14.2
Enhanced</td><td rowspan="4">Soluble</td></tr>
<tr><td>Berthierine</td><td>7</td><td>7</td><td>Often sensitive</td><td>Destroyed</td></tr>
<tr><td>Chlorite (Mg-rich)</td><td>14.1</td><td>14.1</td><td>14.1</td><td>14.1
Enhanced</td></tr>
<tr><td>Corrensite</td><td>28</td><td>31
Integral higher orders</td><td>24</td><td>12</td></tr>
<tr><td>Chlorite (dioctahedral)</td><td>14.1</td><td>14.1</td><td>14.1</td><td>14.1
Enhanced</td><td rowspan="2">Insoluble</td></tr>
<tr><td>Tosudite</td><td>28</td><td>31
Integral higher orders</td><td>24</td><td>12</td></tr>
</table>

* Function of relative humidity.

illite by its relatively weak 002 peak, perhaps only one-fifth the intensity of the 001 peak compared with one-third or more for illite (Fig. 2).

Smectite

Smectite is identified by a strong peak at about 17 Å following glycolation, which collapses to about 10 Å after heating. Random mixed-layer illite–smectites show similar behaviour so it is also essential to confirm that the positions and intensities of the other peaks are correct for pure smectite. In particular, the position of the 003 peak should be located at about 5.65 Å and the 001 peak should be sharp (Fig. 3). The primary division of the smectites is into dioctahedral and trioctahedral types. Without recourse to random powder patterns a useful indication of which type is present may be obtained from the response to treatment with HCl. Trioctahedral species are usually soluble in HCl, but caution must be applied because Fe-rich dioctahedral species are also soluble.

Illite–smectite

The precise identification of mixed-layer illite–smectite is a subject in itself (Brindley & Brown, 1980; Środon, 1980; Moore & Reynolds, 1997). Nevertheless, the first steps in the distinction between so-called randomly interstratified and ordered varieties are relatively straightforward. Superficially, most natural random mixed-layer

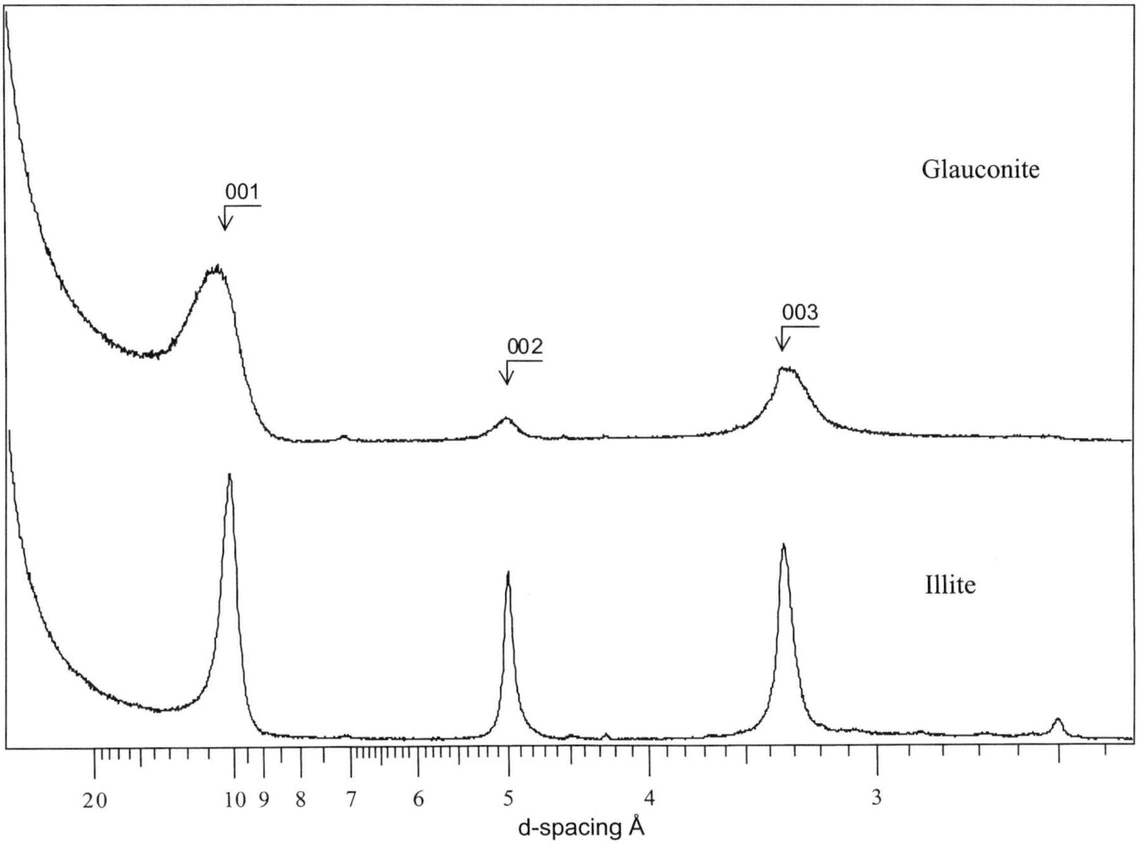

Fig. 2 XRPD patterns (orientated, air-dried) of illite and glauconite showing differences in relative intensities.

illite–smectite resembles smectite. Glycolated patterns are the most diagnostic (Fig. 4). The relative intensity of the 17 Å peak decreases with increasing proportions of illite and the proportions may be crudely estimated from the position of the peak, which migrates from 5.65 Å towards the pure illite peak position at 5 Å. In diagenetically altered sandstones most mixed-layer illite–smectites have compositions between the smectite end-member and a composition of about 60% illite. More illitic compositions tend to show features indicative of ordered interstratification. Glycolated patterns of ordered illite–smectite usually show the presence of a broad peak somewhere between 17 Å and 10 Å and once again the position of the peak migrating towards 5 Å can be used to estimate the proportions of smectite and illite layers (Fig. 5). Although the glycolated XRPD patterns of illite–smectites may appear complex, all of them collapse to approximately 10 Å after heating to yield a simple pattern very similar to that of illite (Figs 4 & 5). Indeed, they often can be quantified more easily following collapse by heating.

Kaolinite

Kaolinite is identified primarily by two peaks, one at about 7.15 Å and the other at about 3.58 Å (Fig. 6). A third peak, the 003, which is much weaker, is also useful because it is not normally intense enough to be observed for trioctahedral species such as chlorite and

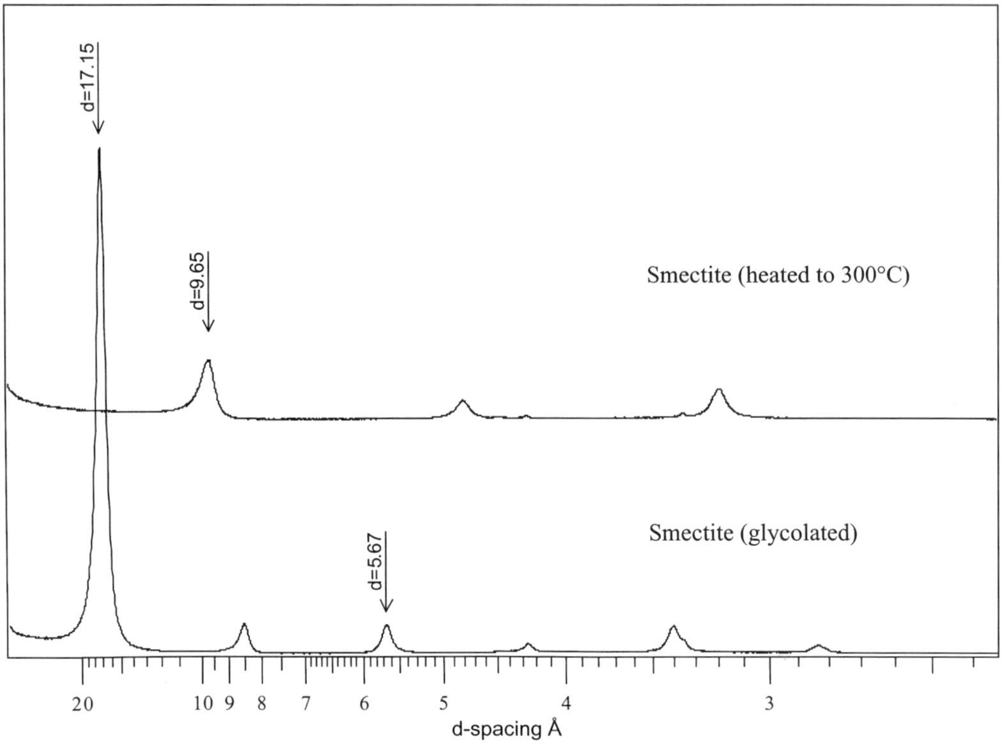

Fig. 3 XRPD patterns (orientated) of pure smectite, note sharpness of the peak at 17 Å in the glycolated trace.

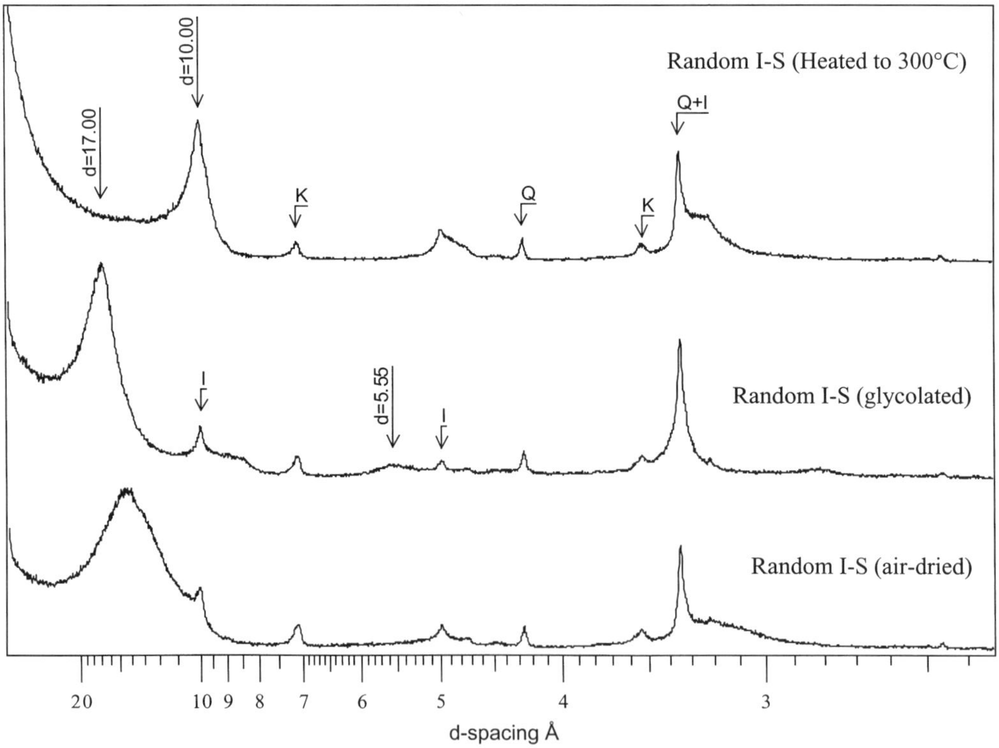

Fig. 4 XRPD patterns (orientated) of a random mixed-layer illite–smectite. Impurities, I = illite, K = kaolinite, Q = quartz.

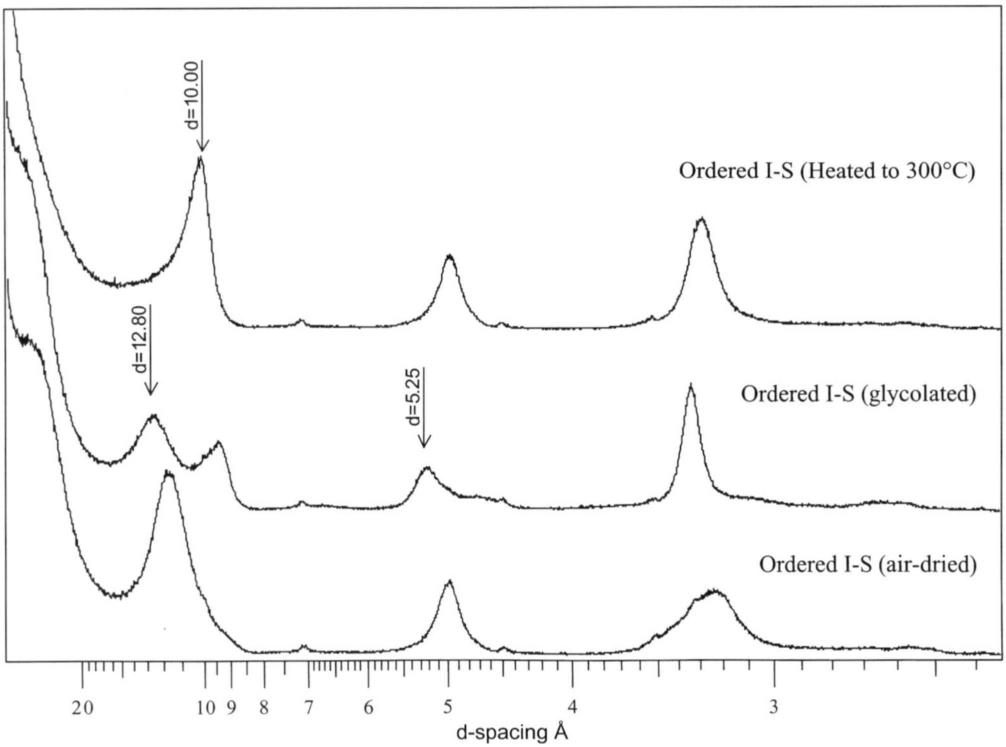

Fig. 5 XRPD patterns (orientated) of an ordered mixed-layer illite–smectite.

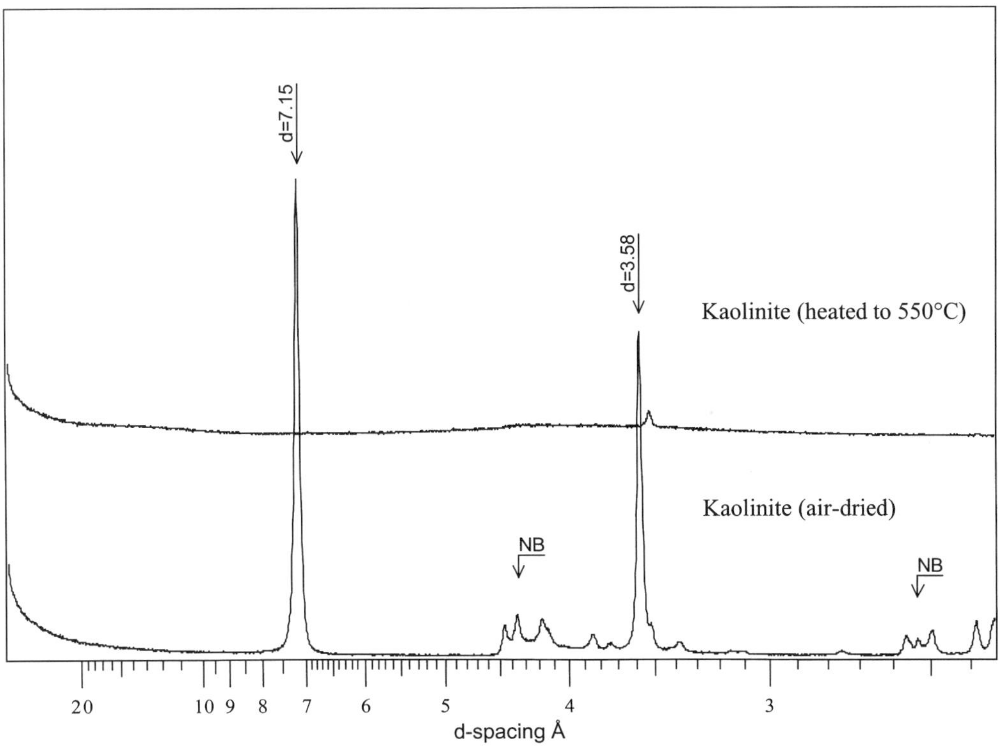

Fig. 6 XRPD patterns (orientated) of kaolinite, before and after heating to 550°C. NB = non-basal (*hkl*) peaks from kaolinite.

berthierine, which otherwise share similarly positioned peaks. Distinguishing kaolinite from chlorites and/or berthierine can pose problems, especially when mixed together in the same sample. Often, the precise position of the kaolinite 002 peak is diagnostic and may be resolved easily from the 004 peak of Fe-rich chlorites. Mg-rich chlorite and some serpentine group species such as berthierine may have spacings that are less distinct from that of kaolinite. In these cases, the relative intensities and the response to other treatments is essential extra information. Furthermore, certain mineral associations are uncommon, for example, it is rare to observe Mg-rich trioctahedral chlorite in the same sample as kaolinite. When chlorite and kaolinite are observed in the same rock, provided they are both authigenic, the chlorite will most likely be a chamosite, because Fe-rich chlorites (chamosites) are always the most aluminous trioctahedral chlorites. In general, heating to 550°C is a useful test because kaolinite is converted to an X-ray amorphous phase, although berthierine is similarly affected. Kaolinite probably is the most common mineral of the kaolin group found in sandstones, but the polytype dickite also is common. These cannot be distinguished from basal reflections alone. Thus unless one goes to the trouble of identifying which is present, such as by a study of the non-basal reflections, the group term kaolin is the correct name to use.

Chlorite

The chlorites are a very diverse group of minerals. In sandstones, trioctahedral species tend to be much more common than their dioctahedral analogues. They range from Fe-rich chamosites to Mg-rich clinochlores. Chamosites have odd-order basal peaks of much weaker intensity than even-order peaks (Fig. 7). In sandstones it is also common for some chamosites, especially those that occur as pore-linings, to be interstratified with a 7 Å mineral, probably berthierine. This interstratification broadens the odd-order peaks relative to the even orders. For clinochores all peaks will be more nearly equal in intensity (Fig. 7), although the range depends on composition. All trioctahedral species are soluble in HCl. Dioctahedral species are often indicated by unusual relative intensities (Fig. 7). In particular, an 003 peak that is greater in intensity than the 002 and 004 peaks is a sure indication that the chlorite is dioctahedral, at least in part. Nearly all chlorites dehydroxylate when heated to temperatures of about 550°C. This results in a marked increase of the intensity of the peak at about 14 Å relative to other peaks, which may disappear from the pattern entirely. In detail there is also a notable contraction of the spacing, for example to ≈ 13.9 Å. Although this may seem trivial it results in a significant displacement of higher order peaks from their positions in the undehydroxylated state.

Chlorite–smectite

The most common forms of mixed-layer chlorite–smectites reported from sandstones are the trioctahedral mineral corrensite and its dioctahedral analogue tosudite. Both minerals can be identified readily in glycolated patterns by the presence of peaks at about 31 Å resulting from the ordered interstratification of smectite (17 Å) and chlorite (14 Å) layers (Fig. 8). Upon heating both collapse to give spacings that are submultiples of 24 Å. As with the chlorites they may be distinguished from each other by their response to HCl. They also can be distinguished by the fact that the 004 peak is always more intense than the 006 peak for corrensite, whereas in tosudite this relationship is reversed (Fig. 8). Such changes in relative intensities are analogous to the differences shown by trioctahedral and dioctahedral chlorites and result from differences in scattering power owing to the occupancy of the octahedral sheets. Both corrensite and tosudite are common in red beds and once again mineral associations such as kaolinite with tosudite (aluminous system) may provide the initial clues as to which variety has been found. Both dioctahedral and trioctahederal forms of chlorites and mixed-layer chlorite–smectites can be distinguished based on the

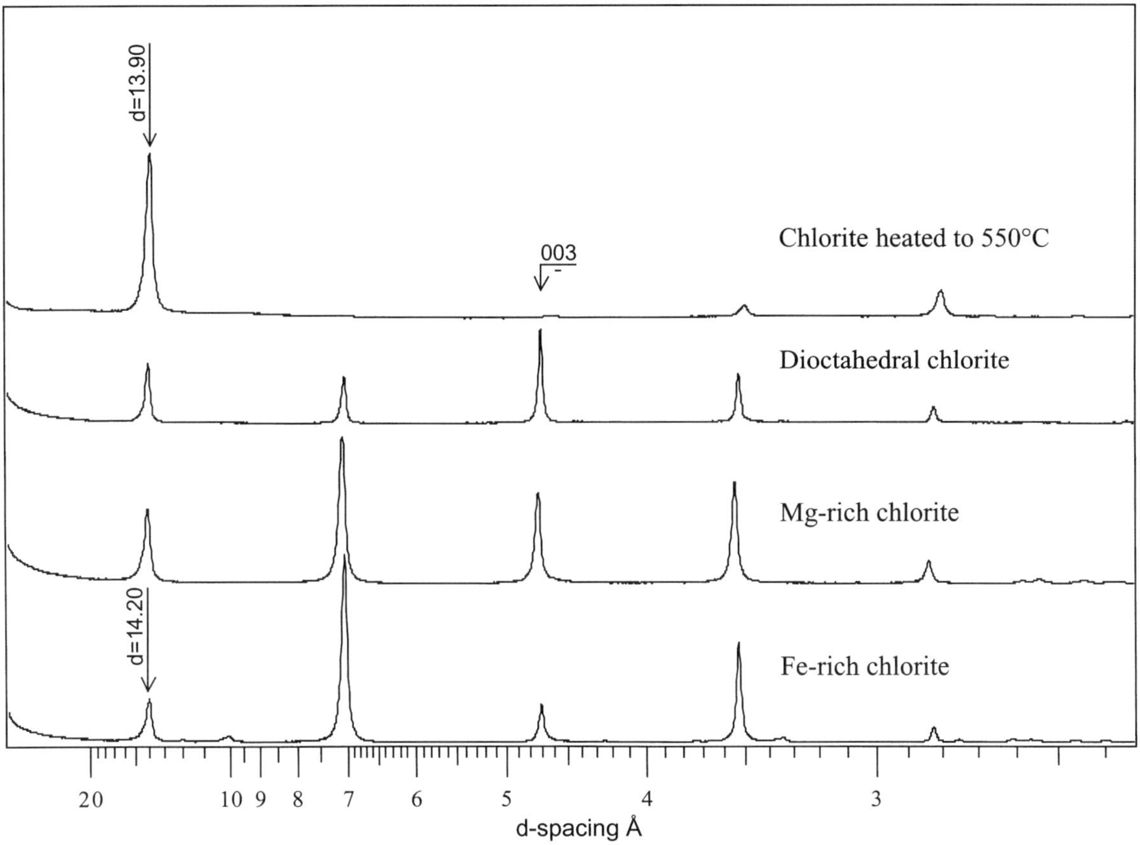

Fig. 7 XRPD patterns (orientated, air-dried) of chamosite (Fe-rich chlorite), clinochlore (Mg-rich chlorite), dioctahedral chlorite (Al-rich chlorite) and chlorite after heating to 550°C.

position of the 060 peaks, but this requires random powder data.

Measurement of peak intensities

All forms of quantitative analysis by XRPD rely on measuring peak intensities. For clay minerals the most important point to make is that intensities must be measured as integrated intensities, that is peak areas. Measures such as peak height are inappropriate because of the enormous natural variation in clay mineral peak shapes and widths. This variability reflects the very nature of clay minerals, which are characterized by wide variations in features such as particle size, stacking faults and other forms of so-called order–disorder. Although these features affect peak shapes and widths, integrated peak intensity is not affected, so that this is the quantity upon which analysis must be based. The principle is illustrated in Fig. 9, which compares the peak heights and peak areas of two kaolinites present at the same concentration. Peaks from some clay minerals, especially mixed-layer varieties, are often very broad. Measuring the intensity of such broad peaks often presents special problems, usually related to defining the background. This especially is true if the peaks are both broad and situated in a region where the background is sloping and/or curved. Additionally, it is not always possible to measure the peaks one would like to. This can come about either because the peaks are of low intensity, or

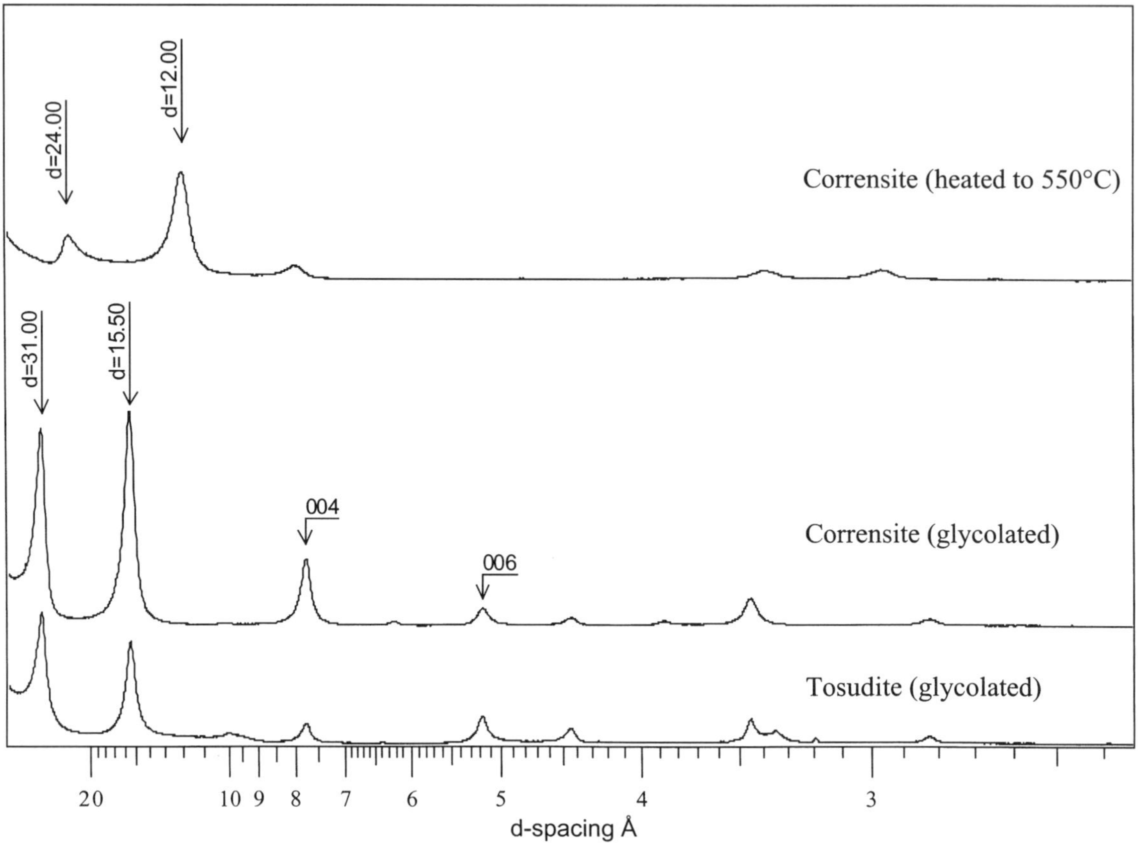

Fig. 8 XRPD patterns (orientated) of glycolated corrensite and glycolated tosudite.

because they are severely overlapped by other peaks. Moore & Reynolds (1997) emphasize that it is best to avoid peaks in the low-angle region and always best to use peaks that are as close together as possible, because this will minimize many angle-dependent systematic errors. Where possible one should always heed this advice, but in practice one is frequently unable to.

Where peak overlap is a problem there are a number of approaches one may try. Depending on the complexity of the overlap, these range from simple procedures, such as measuring the half that is resolved and doubling it, to peak fitting (decomposition) using sophisticated software. In the wrong hands, peak-fitting software can produce ludicrous results and its use must always be carefully constrained.

Quantitative method for orientated samples

The method described for quantitative analysis by XRPD of clay minerals in orientated samples is a normalized RIR method based on equation (9). The RIRs are determined for each and any clay mineral from one-dimensional XRPD patterns calculated using the NEWMOD program (Reynolds, 1985). The RIRs obtained from calculated XRPD patterns are analogous to those that can be determined from prepared mixtures of pure clay minerals. As described by Moore & Reynolds (1997), calculated RIRs do not eliminate any of the sources of error in quantitative analysis, but neither do they add any new ones. They do, however, have two important advantages. Firstly, they eliminate the need to have on hand a collection of pure clay minerals in

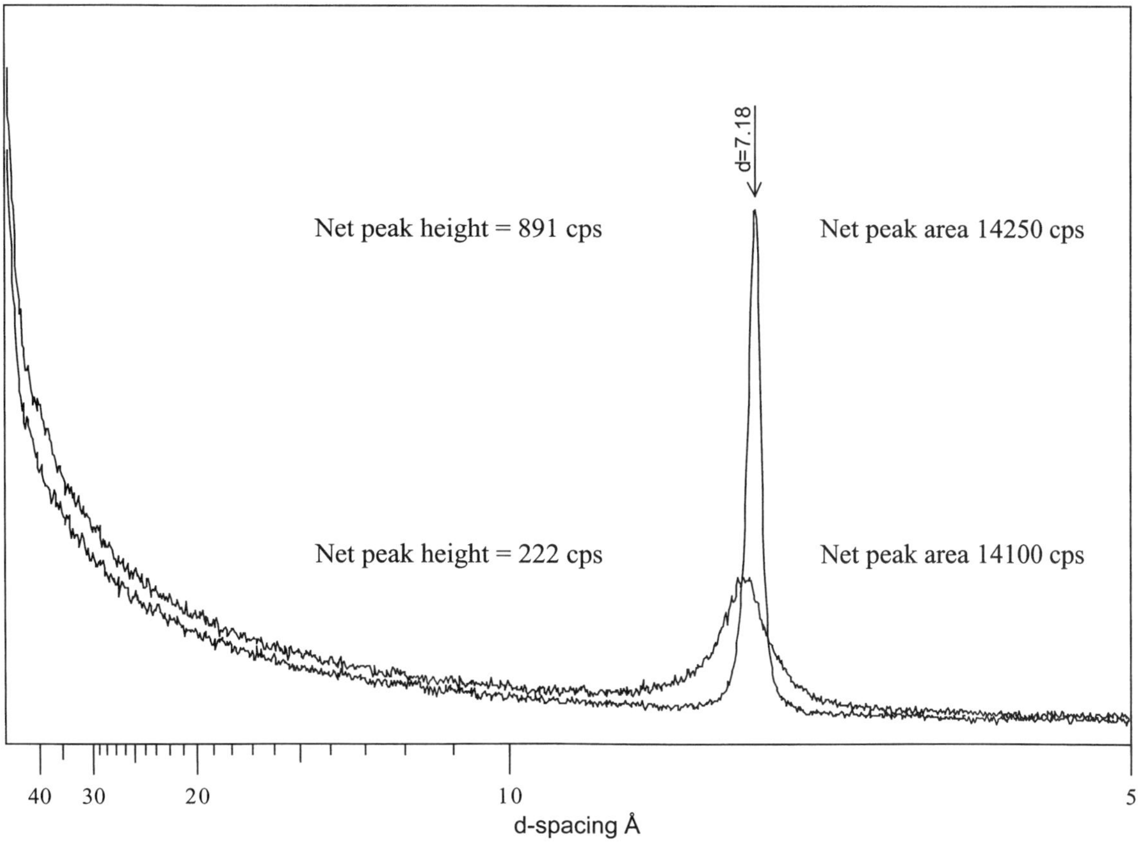

Fig. 9 Comparison of peak heights and peak area measurements for two kaolinites of different 'crystallinity' present at exactly the same concentration. XRPD patterns are random powders.

the formidable quantities necessary to make up mixtures and determine RIRs. Secondly, RIRs can be tailored to the minerals that need to be quantified by adjusting the chemical or structural parameters for the calculated patterns. The first of these advantages is particularly evident when the analysis must include mixed-layer clay minerals. Mixed-layer clays are so ubiquitous that this is a very useful advantage. Furthermore, it is rare that mixed-layer clays are available in pure form so that often the only means of obtaining suitable RIRs is by calculation. The second advantage means that the library of RIRs is easily refined and extended so that it is never necessary to accept an inappropriate standard, such as an RIR determined for an Fe-free illite when it is known that the illite in question is Fe-rich.

The RIRs calculated using the NEWMOD program have been published in several texts (e.g. Reynolds, 1989; Hughes *et al.*, 1994; Moore & Reynolds, 1997), and some values are gathered in Table 3. It is important to note that these values have been calculated for a particular set of experimental conditions. For accurate work the only acceptable approach is to calculate RIRs that are appropriate for the exact experimental conditions used to obtain the diffraction patterns. Table 3 includes a set of RIRs appropriate to the collimation and sample conditions of a Siemens D5000 diffractometer with Co Kα radiation, as used at the Macaulay Institute. These

Table 3 Comparison of some reference intensity ratios (RIRs) calculated with NEWMOD for various basal 00*l* clay mineral peaks.

Mineral	Peak	1	2	3
Montmorillonite (no Fe)	001	5.10		5.0 (0.1 Fe)
	003	0.36	0.59	
	005	0.93	0.77	
Illite (0.1 Fe, 0.75 K)	001	1.00	(1.0)	1.0
	002	0.37	0.36	
	003	0.77	0.75	
Chlorite (3 Fe)	001	0.55		
	002	6.59		6.25 (high Fe)
	003	0.64	0.84 (1 Fe)	
	004	2.65		
Chlorite (3 Mg)	001	0.65		
	002	0.96		0.94 (low Fe)
	003	1.01	0.84 (1 Fe)	
	004	0.97		
Kaolinite	001	2.40		2.32
	002	1.70	1.64	
	003	0.14	0.16	

(1) Macaulay Institute; (2) Moore & Reynolds (1997) normalized here assuming 001/003 of 0.75 for illite; (3) Hughes *et al.* (1994).

are goniometer radius = 21.75 cm, axial divergence Soller 1 = 7.3°, Soller 2 = 2.3°, sample length 3 cm, quartz reference intensity = 15 000 c.p.s., sigma star = 12°, mustar (mass attenuation coefficient) = 50. Some of these parameters, such as the axial divergence, influence the RIRs significantly. It therefore is important to stress that they apply only to the Siemens D5000 collimated as described. Nonetheless, it is clear from a comparison of the values in Table 3 that even RIRs calculated for different conditions are broadly comparable, and if semi-quantitative results are acceptable the use of published values may be justified. Moore & Reynolds (1997) have described the use of the NEWMOD program to calculate RIRs in detail. Essentially, all that is required is to calculate diffraction patterns for a predefined set of experimental conditions and instrumental parameters, measure peak areas from these diffraction patterns, and ratio these areas to the area of the peak chosen as the basis for normalization, such as the illite 001 peak. Provided the experimental conditions are not changed, this set of RIRs need never be determined again.

An example of the data and the calculations required to make an analysis are given in Table 4. Measured peak intensities are divided by their respective RIRs. If more than one peak is used for a mineral, averages are taken. The resulting values are then simply expressed as a percentage of the total.

Validation and uncertainty

The method of quantitative analysis based on RIRs calculated with the NEWMOD program has been validated in detail by its originator R.C. Reynolds Jr. (Reynolds, 1989). Reynolds (1989) showed that calculated absolute intensities were typically within ±5% of measured intensities for numerous samples of kaolinites (12), illites (11) and mixed-layer illite–smectites (22). Therefore, applied as described there is no reason to suppose that this method will not work provided the XRD patterns are calculated for the instrument on which the samples are measured.

Nonetheless, comparison of the results of applying the method to artificial mixtures

Table 4 Example of the calculations required for the normalized reference intensity ratio (RIR) method.

	Peak						
	Smectite 001	Chlorite 001	Chlorite 004	Kaolinite 002	Kaolinite 003	Illite 001	Totals
Intensity	1923	722	1239	421	19	1878	
RIR	5.0	0.6	1.8	1.7	0.1	1.0	
Intensity/RIR	384	1204	700	247	197	1878	
Averages	384		952		222	1878	3438
Wt% found	11.2		27.7		6.5	54.6	100.0

Table 5 Weight per cent found, weight per cent known and absolute and relative errors for four prepared mixtures of illite, kaolinite, smectite and chlorite.

Mix		Smectite	Chlorite	Kaolinite	Illite
1	Per cent found	1.6	8.4	60.8	29.2
	Per cent known	2.6	11.4	56.7	29.3
	Absolute error	−1.0	−3.0	4.1	−0.1
	Relative error	−38.7	−26.2	7.2	−0.2
2	Per cent found	11.2	27.7	6.5	54.6
	Per cent known	10.5	28.5	2.9	58.1
	Absolute error	0.7	−0.8	3.6	−3.5
	Relative error	6.6	−2.9	126.1	−6.0
3	Per cent found	31.6	51.9	13.2	3.3
	Per cent known	30.6	55.6	10.4	3.4
	Absolute error	1.0	−3.7	2.9	−0.2
	Relative error	3.2	−6.6	27.6	−4.9
4	Per cent found	66.0	1.8	22.4	9.7
	Per cent known	56.2	3.6	29.8	10.5
	Absolute error	9.8	−1.7	−7.4	−0.7
	Relative error	17.5	−48.7	−24.8	−6.8

where the answer is known is a useful exercise that should be encouraged, because this enables validation of the method as applied in any particular laboratory, or for any particular study. It also should provide a feel for other potential sources of error, such as those resulting from the inherent characteristics of the sample and the methods of sample preparation. For many clay minerals, validation by this method is not possible because they are practically unattainable in pure form. However, with careful selection and preparation it is not especially difficult to obtain samples of clays such as illite, chlorite, kaolinite and smectite of a purity sufficient for these purposes.

To demonstrate validation, four artificial mixtures were prepared as orientated slides, each mixture containing four different clay minerals in various proportions. The samples used for this purpose were all $< 2\ \mu m$ size fractions determined to be essentially pure by examination of their XRPD patterns. In order to assure validation across a range of abundances the mixtures were prepared so that each of the four clay minerals occurred at a concentration of approximately 3, 10, 30 and 60% by weight. The XRPD patterns obtained from these mixtures after glycolation are compared in Fig. 10.

A comparison of the weight per cent known and the weight per cent found for all four mixtures is presented in Table 5 and graphically in

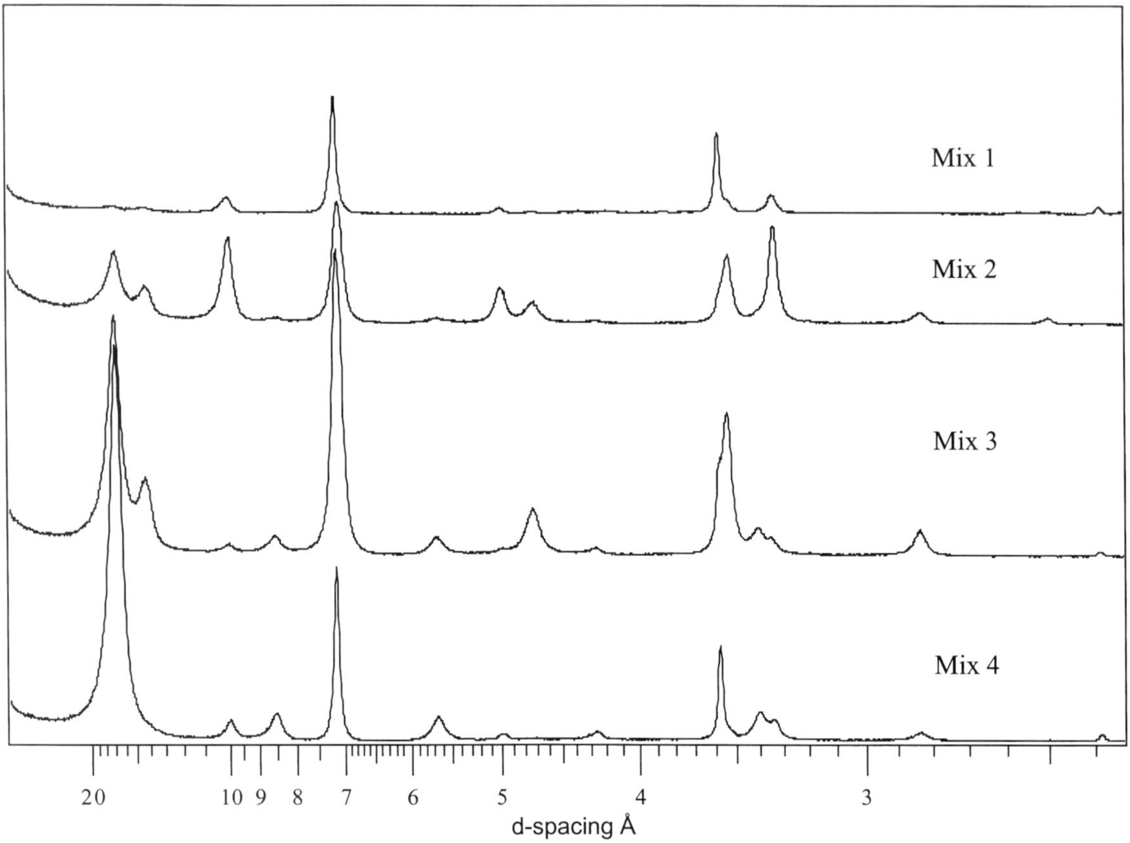

Fig. 10 Glycolated XRPD patterns (orientated) of four artificially prepared mixtures of kaolinite, illite, smectite and chlorite. For details see Table 5.

Fig. 11. In all cases, absolute errors, i.e. the difference between the weight per cent known and the weight per cent found, are typically plus or minus a few weight per cent or less. Results such as these are encouraging because they demonstrate that a method based entirely on calculated RIRs works. It should, however, be emphasized that validation is always an important step. Indeed, if quantitative results are essential then validation on prepared mixtures is about the only practical way to assess the uncertainty associated with the method. In the example given, three out of sixteen (≈ 20%) of the results fall outside the suggested uncertainty envelope for a method to be counted as quantitative. Thus although, on the whole, the results appear reasonably good, we can see that even for this fairly simple system, if the method was applied to real unknowns, we would be wise to classify it as semi-quantitative.

If validation is not attempted, and/or the method is validated on some clay minerals but extended to other kinds, then clearly the results should be classified as semi-quantitative. This is not to say that high-quality truly quantitative results are not obtainable using the method described, rather it is to emphasize that one must take the utmost care with the details of the method if this is the goal. For the normalized RIR method, it also will naturally deteriorate to semi-quantitative if the assumption that all phases are analysed is invalid. As far as clay fractions from sandstones are concerned, substantial amounts of quartz, feldspars and

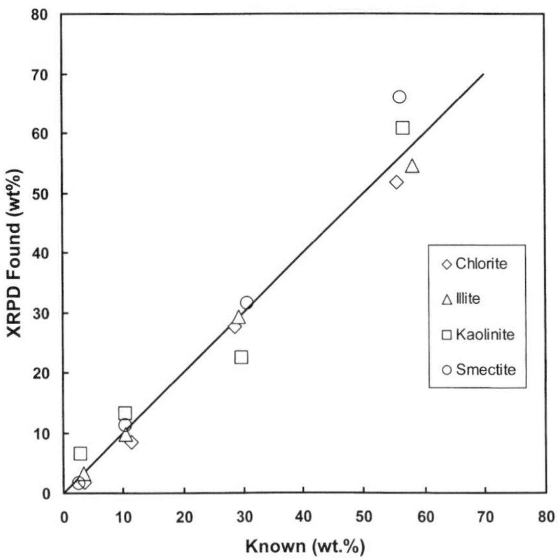

Fig. 11 Comparison of weight per cent known versus weight per cent found for the four clay mineral mixtures. For details see Table 5.

carbonates often are present, thereby invalidating this assumption. Some schemes claim to be able to quantify both clay and non-clay minerals in orientated preparations. The assumption that underlies these claims is that all minerals in the sample are consistently orientated to the same degree. Indeed, unless corrections are applied, this basic assumption underlies all quantitative methods. For orientated preparations this is not unreasonable where analysis of platy clay minerals is concerned. It is, however, unreasonable to suppose that minerals such as quartz, feldspars and carbonates will adopt a consistent orientation. If such minerals must be analysed along with clay minerals then the random powder approach should be used.

Lower limits of detection

For orientated specimens, there are probably few instances where the lower limits of detection (LLD) need to be evaluated specifically, but it is always useful to have an idea of what it is likely to be. For the clay-mineral mixtures described previously, evaluation of equation (11) for the most intense peaks for each mineral indicates LLDs of some fraction of a weight per cent. Notably, these values are considerably lower than what might be described as the generally perceived 'typical' detection limits of XRPD of 'a few weight per cent'. Obviously, the values obtained in practice will depend on the intensity obtained from a particular diffractometer. Nonetheless, the measured XRPD patterns for these clay mineral mixtures were obtained by counting for 1 s per step, on a fairly standard diffractometer, and in that sense they ought to be quite typical. In practice, however, one may encounter various peak interferences, so that for reasonably 'crystalline' clays a nominal value of about 1 wt% probably is realistic. For much broader peaks from mixed-layer clays then a LLD of a few weight per cent might be a more reasonable nominal value. To illustrate the point, Fig. 12 shows an XPRD pattern obtained from an orientated sample prepared and measured using identical methods to those described previously. The sample consists of a mixture of 1 wt% each of smectite, illite and chlorite in a matrix of kaolinite. When displayed on a linear axis at full scale, peaks from the minerals present at the 1 wt% level are not readily apparent, but when the intensity scale is magnified such as in log form, peaks arising from the smectite, illite and chlorite are clearly visible.

ANALYSIS OF RANDOM POWDERS

Useful though it may be, quantitative analysis of clay fractions from sandstones always begs the question as to what, if any, relationship such analyses have to the population of clay minerals in the sandstone as a whole. To attempt to answer such questions, sandstones must be analysed as so-called bulk or whole-rock samples. Additionally, such analyses also have the potential to provide a wealth of information on the nature and abundance of all of the other minerals of which sandstones are composed. Analysis of whole-rock samples generally requires that the sample is prepared as a random powder. It must not be forgotten,

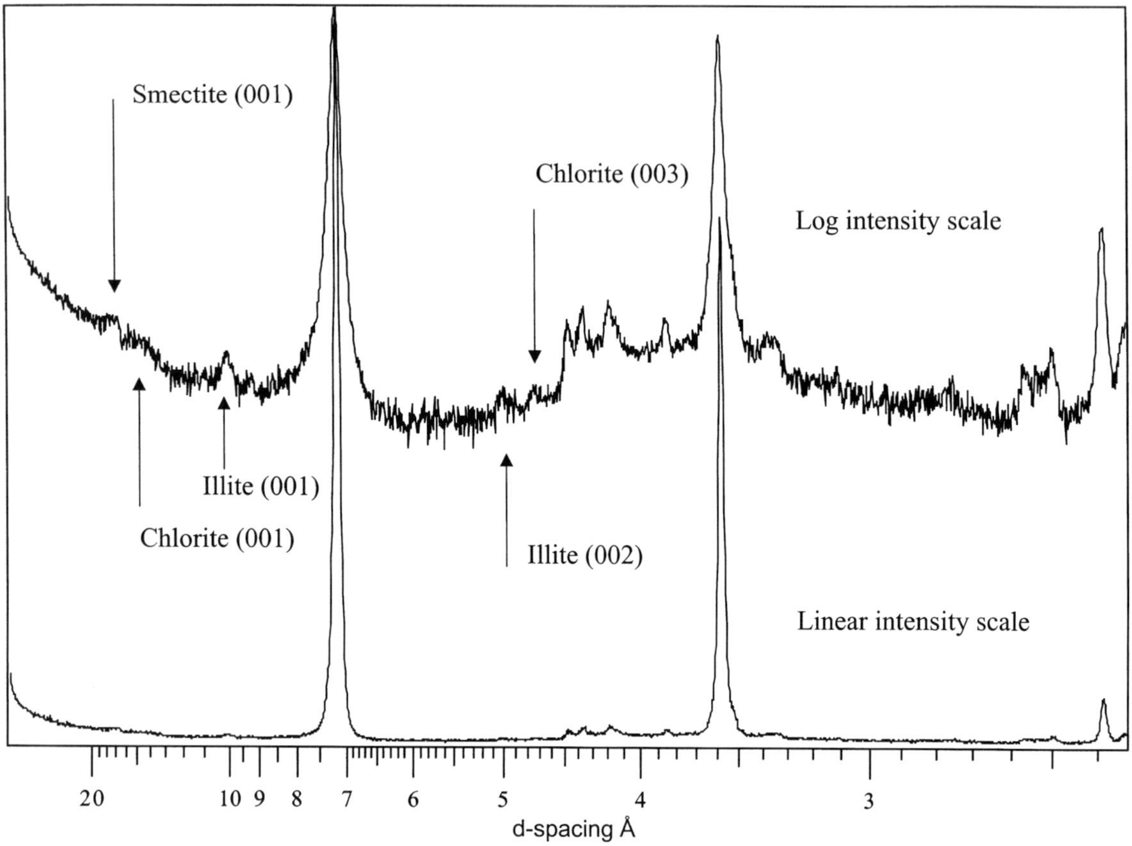

Fig. 12 XRPD patterns (orientated) illustrating detection of 1 wt% smectite, illite and chlorite in a matrix of kaolinite.

however, that the random powder method often can be applied profitably to any fraction of a sample.

Issues of sample preparation

There are many factors that should be considered when preparing random powder samples (Brindley & Brown, 1980; Bish & Chipera, 1988; Bish & Reynolds, 1989; Bish, 1994). However, when quantitative analysis is the aim there essentially are two very important features that must be controlled. The first concerns the size of the powder particles, and the size of the diffracting domains, sometimes referred to as crystallite size. For simplicity we will assume that these are equivalent. The second is texture within the powder sample as it is presented to the X-ray beam. This is otherwise known as preferred orientation. For any method of quantitative analysis by XRPD, reproducibility of diffraction data is paramount. If either or both of particle size and preferred orientation are not controlled, the reproducibility of the diffraction data simply will not be good enough for quantitative analysis.

As far as particle size is concerned, invariably it needs to be reduced to the order of 10 μm or less. There are a variety of reasons why this is important (Bish, 1994) but foremost amongst these is one of particle statistics. If particle size is coarser than 10 μm there will simply be an insufficient number of particles in the right orientation to obtain reproducible intensity measurements, especially for phases present at low concentrations. Hand grinding is not sufficient

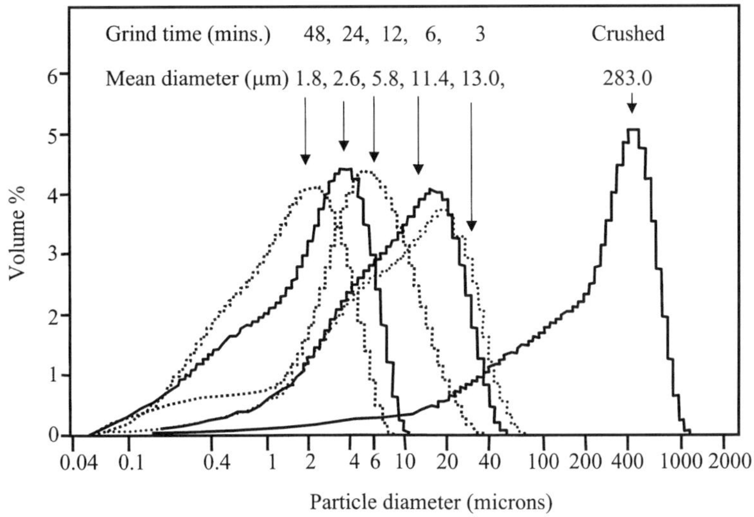

Fig. 13 Particle size distribution of aliquots of quartz ground for increasing lengths of time in a McCrone mill as determined by a laser particle size analyser.

to reduce particle size to meet these requirements and some form of milling is essential. However, one also must be aware that excessive milling may also produce undesirable effects such as amorphitization. To avoid such problems one should experiment with a mill to determine the optimum time for grinding. Wet grinding is also preferable because both mixing and heat dissipation are better. There are a great variety of mills in use, but one of the most popular for XRPD applications is the rod mill sold by McCrone. This mill is very effective at reducing particle size without altering the diffraction characteristics of the sample. Figure 13 shows the particle size distribution as measured by laser diffraction for a pure sample of quartz, hand crushed to sand size then ground as subsamples of 2 g for 3, 6, 12, 24 and 48 min in 6 ml of water. These results indicate that a grinding time of 12 min produces a particle size reduction for which most particles are < 10 μm in size. Comparison of integrated X-ray intensities measured on all five subsamples (r.s.d. 2.6%) indicates no evidence for amorphitization, even after 48 min grinding time. However, the characteristics of individual mills should be checked, as even the same make of mill may behave differently in another laboratory (J. Środon, personal communication, 1999).

One only has to look at the number of publications and variety of methods of preparing a random powder mount to realize that eliminating preferred orientation is a difficult task. Nonetheless, the effort expended to find ways of making a random powder is also an indication of just how important it is. Making a random powder is difficult because most minerals have anisotropic shapes. For example, clay minerals tend to be platy, and with the slightest amount of pressure will adopt a preferred orientation. Diffraction from some planes is then overrepresented, whereas for others it is diminished. As described above, this is in fact used to great advantage to identify different clay minerals by purposely inducing preferred orientation and thereby enhancing the diffraction from the characteristic basal planes.

Preferred orientation is, however, much easier to induce than to avoid. Inevitably, even the most careful packing of a powder sample into a standard cavity holder will result in some degree of preferred orientation. Furthermore, the amount of preferred orientation is unlikely to be reproducible on the same sample let alone from one sample to another. If all minerals were affected in the same way, and to the same degree, preferred orientation would not pose a problem. However, some minerals are more

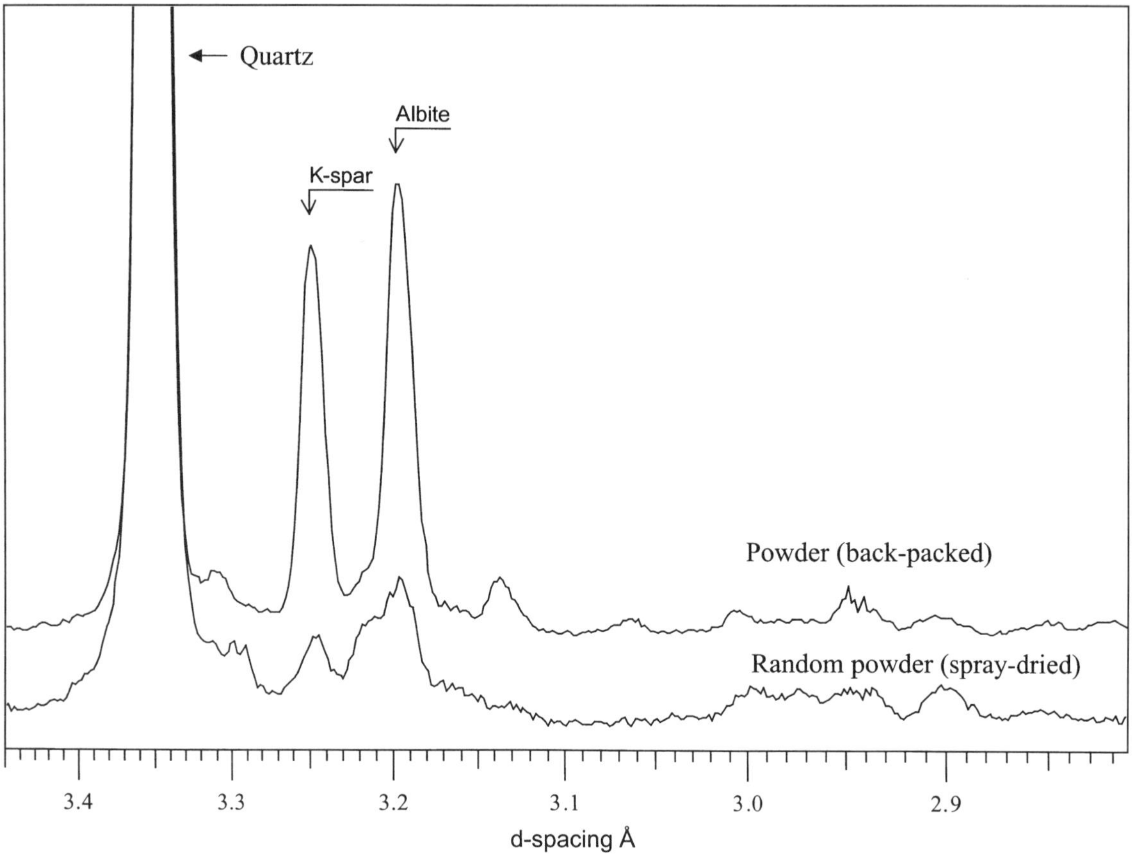

Fig. 14 Comparison of a random powder and a textured powder of the same sample in the region of the most intense K-feldspar and plagioclase feldspar peaks. Patterns offset for clarity.

susceptible than others and not all minerals orientate in the same way. The result is variation in peak intensities that are unrelated to concentration. Clay minerals are particularly prone, but many other minerals, notably carbonates and feldspars, are also very susceptible. Figure 14 illustrates this point by comparing the intensities obtained for feldspars from the same sandstone prepared as a random powder and as a powder in which texture is present. There can be little doubt that the problem posed by preferred orientation has been one of the greatest obstacles to the development of reliable methods of quantitative analysis of random powders (Bish & Chipera, 1988).

Of the wide variety of methods proposed for preparing random powders it probably would be fair to say that most laboratories routinely use some variant of either side or back-loading of a standard powder holder. This is simply because such procedures are relatively straightforward to perform. In many cases, however, it must be doubtful that any of these techniques approach a random orientation, although this may be of secondary importance because reproducibility is the key issue. However, if the quantitative method is to be used by more than one person, perhaps over a long period of time where expertise will come and go, then guarantees of reproducibility are unlikely.

One alternative to back and side-loading is a method known as spray drying. This method consists of spraying a sample as an aqueous suspension into a heated chamber, where it

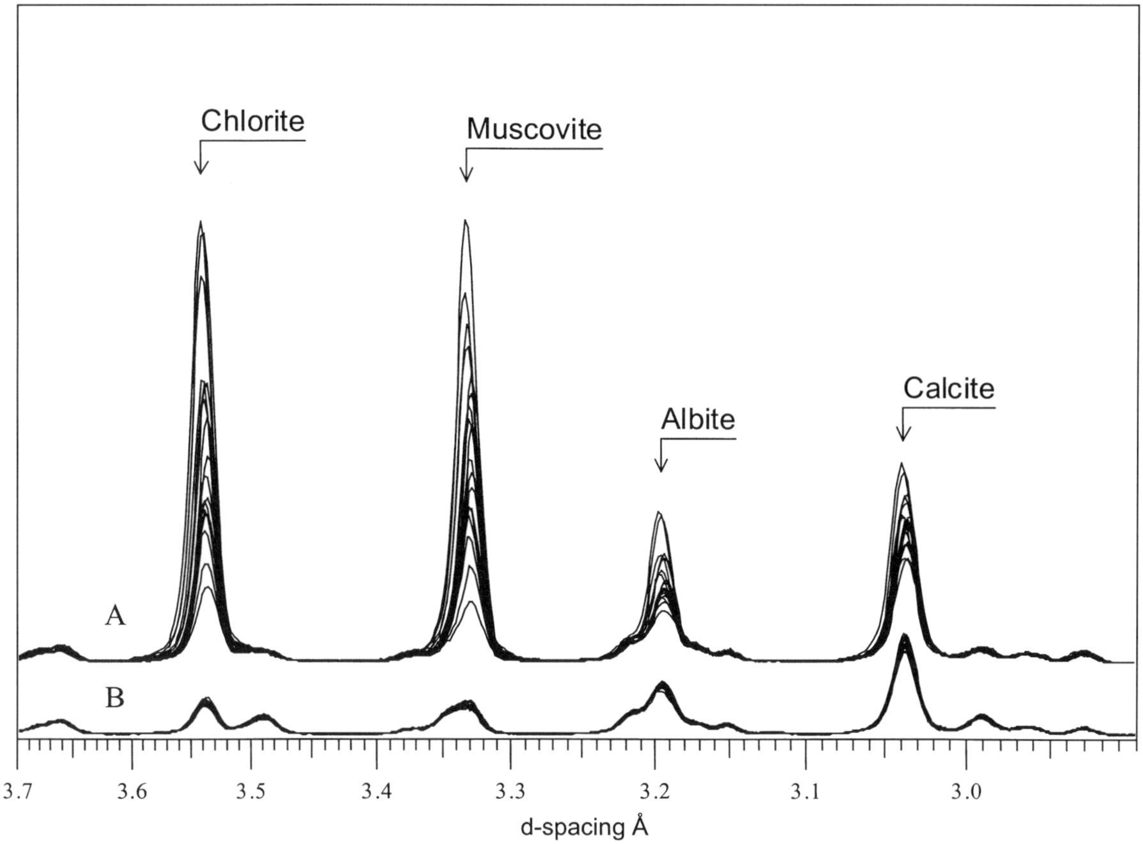

Fig. 15 XRPD patterns of a synthetic mixture of 25% chlorite, 25% muscovite, 25% albite and 25% calcite, all minerals that frequently exhibit preferred orientation. (A) XRPD patterns from 18 separate loadings of a freeze-dried portion of the mixture by three different operators. (B) XRPD patterns from 18 different loadings of a spray-dried portion of the same mixture by three different operators. Sets of patterns offset for clarity.

dries in the form of spherical spray droplets. The resulting dry product consists of thousands of tiny spherical agglomerates of the sample components. Both the arrangement of any component within the spherical agglomerates and of the agglomerates themselves ensure that preferred orientation is eliminated. Spray drying is a well-known industrial process. Although there have been several attempts to use it for XRPD (Flörke & Saalfeld, 1955; Jonas & Kuykendall, 1966; Hughes & Bohor, 1970; Smith *et al.*, 1979) it has not been widely adopted. This appears largely to be the result of problems of sample recovery. However, Hillier (1999) has shown how small samples ≈ 1 g may be spray dried routinely using an airbrush and a sheet of paper to collect the sample. Hopefully this technique will now be adopted much more widely.

One of the main advantages of spray drying is that, as a consequence of eliminating preferred orientation, the XRPD patterns from spray-dried samples are extremely reproducible. By way of example, Fig. 15 compares the XRPD patterns obtained by three different operators who emptied and loaded two portions of the same sample six times each. One portion was spray dried and the other freeze-dried. For the freeze-dried portion (Fig. 15A) no two runs were the same. Additionally, the extent of preferred orientation for each phase, as indicated by enhanced intensity, is inconsistent between phases. In contrast, the diffraction pattern from the spray-dried portion is reproducible by, and between, all three operators. A further practical advantage is that in combination with wet grinding, samples may be spray-dried directly from the mill. Spray-dried powders are also much quicker to load because they are simply poured into the cavity holder in excess, shaken from side to side to obtain good packing, and the surplus removed with a swipe from the edge of a spatula. Tests on quartz have shown that the individual granules are solid and that there is no loss of intensity with respect to that obtained from a 'normal' powder sample, such as might be expected if the granules were hollow or insufficiently packed.

Measurement of peak intensities

As described for the analysis of orientated clay minerals the best measure of intensity is integrated intensity. Differences in 'crystallinity' between examples of the same mineral produce significant changes in peak width and peak height, but in theory integrated intensity will be constant. Overlapping peaks also can be dealt with as described previously. Furthermore, for a random powder, diffraction data are normally obtained over a much larger 2θ range so that the choice of available peaks may be greater. If choice allows, it is often easier and more accurate to measure a smaller peak free from interferences than attempt to decompose complex overlaps. Such peaks also may be used to make overlap corrections in other parts of the pattern. The procedure is simply to scale the intensity of the non-overlapped peak from phase (a) (based on a knowledge of the relative intensities of phase (a)) so that the appropriate contribution of (a) may be subtracted from another region of the pattern where it is overlapping with (b). For minerals that generally are poor diffractors (small RIRs), such as most clay minerals, the intensity will tend to be very weak in a random powder. Therefore, diffraction data are best obtained under conditions that will optimize intensity, perhaps at the expense of resolution. In favourable circumstances it may be possible to measure more than one peak from various minerals. This is a practice to be encouraged because consistency in the calculated weight fractions provides assurance that the intensity measurements are reasonable. Conversely, inconsistency is often the only way to recognize that something is not right.

Quantitative method for random powders

The method illustrated for the quantitative analysis of random powders is the internal standard RIR method. As discussed previously, this is one of the best general methods, because the addition of the standard and the analysis of each mineral with respect to it alone allows the final total to be used as an indicator of success. Here it is illustrated for random powders, but with the use of an orientating internal standard, such as pyrophyllite or talc, it can be applied just as easily to orientated clay fractions (Mossman *et al.*, 1967; Quakernat, 1970; Heath & Pisias, 1979). The calculations necessary are based on equation (8) and are illustrated in Table 6, which contains a set of data and results for one of the artificial mixtures described by Hillier (2000). Weight per cents are obtained simply by dividing measured peak intensities by the intensity for the standard phase and multiplying by the RIR scaled for the amount of added standard. Where more than one peak is measured, averages are taken. The results are the concentrations in the spiked sample. If the example was a real sandstone these results would each be divided by 0.9 (to allow for 10% added corundum standard) to obtain the true concentrations. These calculations are easily programmed into a spreadsheet. The importance of obtaining consistent intensity measurements for the standard is also illustrated and one may easily add measures of consistency such as relative standard deviation.

Assuming appropriate sample preparation, all that is required for this method to work is a set of accurately measured RIRs. Table 7 compares RIRs from a variety of different sources.

Table 6 Example of the calculations required for the internal standard reference intensity ratio (RIR) method.

(a) Corundum standard

Wt% STD	Peak position	Intensity	Relative intensity	Normalized intensity	Average	RSD
10	3.48		0.58			
	2.55		0.88			
	2.38		0.37			
	2.08	52.5	1.00	52.5	53.5	1.8
	1.74	24.5	0.45	54.4		
	1.60		0.94			

(b) Minerals in sample

Mineral	Peak position	RIR	Wt% STD/RIR	Intensity	Wt% found	Wt% found (averaged)	Wt% known
Quartz	3.34	4.04	2.5	701.3	32.5		
	4.26	0.82	12.2	127.2	29.0	32.0	29.4
	1.81	0.47	21.1	87.2	34.5		
Calcite	3.03	2.73	3.7	155.0	10.6	10.6	10.0
Dolomite	2.89	1.94	5.2	106.9	10.3	10.3	10.2
Albite	3.19 (cluster)	2.07	4.8	130.8	11.8	11.2	13.2
	4.03	0.62	16.2	35.1	10.6		
Mica	10.00	0.23	43.5	8.6	7.0	7.0	8.2
Kaolin	3.58	0.53	18.9	28.8	10.2	10.2	10.0
Chlorite	3.52	0.54	18.5	28.7	9.9		
	4.72	0.29	34.1	15.6	9.9	9.9	9.0
Total						91.2	90.0

STD = standard; RSD = relative standard deviation.

As mentioned previously, measured values depend on the sample and experimental conditions used, including the radiation and whether or not a diffracted beam monochromator is fitted (Hubbard & Smith, 1977). Some sources listed in Table 7 do not report the experimental conditions, so it is assumed that they are for copper radiation without a monochromator. This is the convention adopted for values reported in the PDF. Measurements made by the author with cobalt radiation and a diffracted beam monochromator at the Macaulay Institute have been converted to copper with no monochromator following the procedure described by Hubbard & Smith (1977). Measurements referenced to quartz peaks also have been converted to a corundum base, using the Maculay Institute RIR_{cor} for quartz, and equation (7). Clearly, there is significant variation in these values from different sources. Nonetheless, given that the experimental conditions will have been different it is encouraging that at least a certain degree of consistency is apparent. Thus quartz is clearly the strongest diffractor with a value near 4. Along with its abundance, this simple fact explains why quartz, somewhat

Table 7 Comparison of some reference intensity ratios (RIRs) from the literature for some common non-clay minerals in sandstones. For columns 3 and 4 (because the basis for normalization reported in the publications is changed from quartz to corundum), to facilitate comparison, the Macaulay RIRcor for quartz is used. Therefore the value for quartz is forced to correspond exactly with the Macaulay measured value and hence is enclosed in parenthesis.

	Source					
Mineral (peak)	**1**	**2**	**3**	**4**	**5**	**6**
Quartz (3.34)	4.5	4.0	(4.5)	(4.5)	4.1	3.4
K-feldspar (3.24)	1.0	0.7	2.1	1.0	1.3	0.8
Plagioclase (3.19)	2.2	–	2.1	1.6	1.7	2.1
Calcite (3.03)	3.1	2.0	3.8	2.7	3.0	3.2
Dolomite (2.89)	2.7	1.4	3.5	2.9	2.7	2.5
Pyrite (2.72)	1.8	2.0	1.8	1.9	0.7	1.6

(1) Macaulay Institute, converted from cobalt with diffracted beam monochromator to copper without monochromator; (2) Davis & Johnson (1989); (3) Hughes *et al.* (1994), basis changed from quartz to corundum; (4) Cook *et al.* (1975), basis changed from quartz to corundum; (5) Davis *et al.* (1989), means of all values reported as integrated intensity measurements; (6) PDF numbers 46-1045 (*), 09-0466 (*), 71-0957 (C), 72-1651 (C), 84-1208 (C), 42-1340 (*).

unfortunately, dominates the diffraction pattern of sandstones. The values for feldspars are more variable. This may be related to their tendency towards preferred orientation, compositional variability (Chipera & Bish, 1995) and the position of the most intense peaks in clusters close to the main peak for quartz, making intensity measurement difficult. Nonetheless, a value of about 1 appears appropriate for K-feldspar, which clearly is a weaker diffractor than plagioclase with a value of about 2. For the carbonates, a value of about 3 seems most consistent for calcite, whereas for dolomite a slightly lower value around 2.6 is suggested. Lastly, pyrite is quite consistently near 1.7.

It must be emphasized that this comparison is not made to suggest that certain values are correct and/or better than others. It is made simply to indicate the sort of values that one might expect to obtain for various common phases that will be encountered in sandstones. It serves also to reinforce the point that analyses can be semi-quantative only if RIRs are borrowed from the literature. For quantitative analysis they must be measured accurately on the equipment that will be used. Nonetheless, the degree of consistency that does exist perhaps can be taken as a prompt that more effort should be made to determine RIRs in relation to an instrument sensitivity (intensity) standard, as suggested by Schriener (1995). In that way RIRs might become transferable from one laboratory to another.

With single-peak methods one must always be aware of potential peak interference. Furthermore, for some peaks, such as the important non-basal clay peaks, one may have to devise a standardized strategy for measuring their intensity because they are often present as broad two-dimensional diffraction bands tailing off asymmetrically to higher angles. The main point to make here is that it is not necessary to measure all of a particular peak or band. Some consistently measurable part is just as good. Some of the best peaks to use for a variety of common minerals are listed in Table 8 along with notes on interferences and measurement strategies.

Validation and uncertainty

The author's results of applying both the RIR internal standard method and the Rietveld method to a variety of samples containing a wide range of clay and non-clay minerals are shown in Fig. 16. This figure includes data from three studies. The study published by Hillier (2000) accounts for 140 points; an unpublished study presented as a conference poster by Hillier (1998) accounts for a further 57 points;

Table 8 Some possible peaks for random powder reference intensity ratio (RIR) methods including major potential interferences and other comments.

Mineral	Peak(s) (Å)	Interferences	Comments
Albite	4.03		Rare interferences, but low intensity
	3.19	Fluorite	Complex group of peaks, difficult to measure intensity
Barite	3.10		
Calcite	3.03	K-feldspar	Unless very abundant most other peaks have insufficient intensity
Dolomite	2.89		Unless very abundant most other peaks have insufficient intensity
K-feldspar	3.49	Corundum	
	3.24	Quartz	Close to major quartz peak and often difficult to measure
	2.16	Minor corundum, dolomite	Low intensity
Pyrite	2.70	Haematite, aragonite	Rare to find haematite and pyrite together
Quartz	4.26	K-feldspar, clay minerals	Often 'sits' on top of non-basal clay band
	3.34	Illite, illite–smectite	Can be difficult to define peak tails from background
	1.81	Dolomite, KCl	Low intensity but reliable owing to few interferences
Siderite	2.79	Zincite	
Haematite	2.70	Pyrite, aragonite	Rare to find haematite and pyrite together
Chlorite	7.1		Intensity varies considerably with composition
	3.52	Anatase	Useful when kaolinite present
	4.6	All trioctahedral clays, K-feldspar	About 1/3 the intensity of the equivalent dioctahedral clay band. Difficult to measure precisely
Illite	10.0		Very weak if the powder is truly random
	4.45	All dioctahedral clays, barite, quartz in the 'tail'	Most dioctahedral clays diffract with similar intensity in this region unless Fe-rich (see Fig. 19). Peak is asymmetric so only some part can be measured
Illite–smectite	4.45	All dioctahedral clays, barite, quartz in the 'tail'	Most other peaks far too weak, and as per comments for illite
Kaolinite/dickite	7.1	Chlorite	Useful when chlorite is not present for determining kaolin contribution to total dioctahedral clays measured using 4.45 non-basal clay band
	3.58	Siderite	Useful when chlorite is also present
	4.45	All dioctahedral clays, barite, quartz in the 'tail'	As per comments for illite and illite–smectite
Smectite	17.0		A strong peak, but random powder must be glycolated
	4.45	All dioctahedral clays, barite	Most other peaks far too weak unless glycolated 17 Å peak used

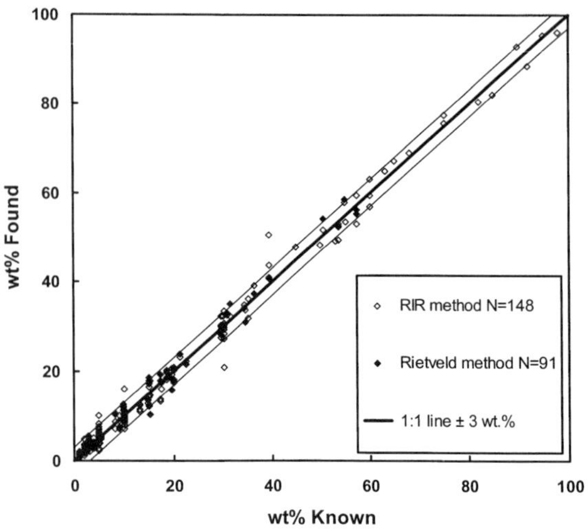

Fig. 16 Comparison of weight per cent found versus weight per cent known for a variety of clay and non-clay minerals in recent studies by the author, see text for details.

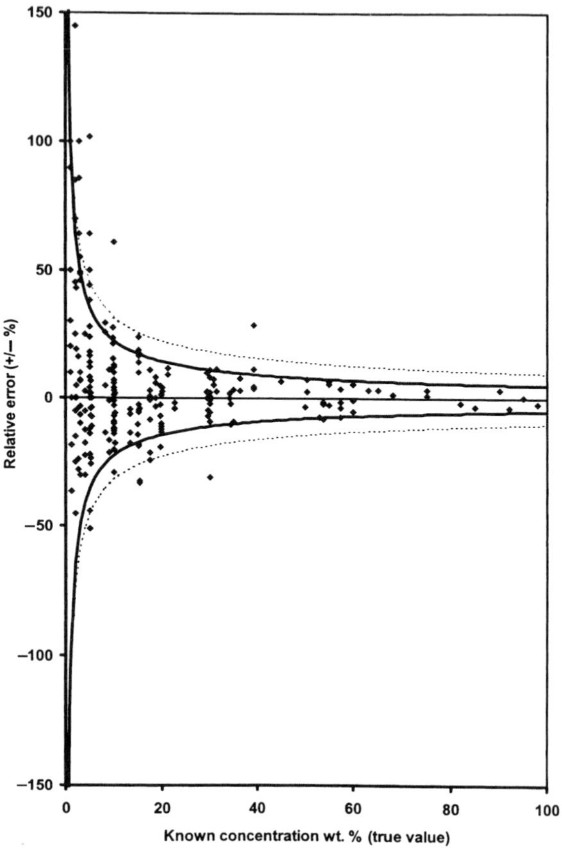

Fig. 17 Relative error versus concentration for the same 239 points illustrated in Fig. 16. Dotted lines = $\pm 100X^{-0.5}$, solid lines = $\pm 100X^{-0.65}$, where X = concentration in wt%.

and submissions by the author (participant no. 172) to the round-robin on 'Quantitative Phase Analysis' organized by the International Union of Crystallography (Madsen, 1999) accounts for another 42 points. Thus Fig. 16 summarizes results for 239 separate data points in terms of the percentage known versus percentage found. The spread of data about the 1 : 1 line gives an indication of the sort of accuracy that can be achieved by the internal standard RIR method. Most data fall within ±3 wt% of the true value. As mentioned previously, however, error tends to be correlated with concentration and statements such as ±3 wt% do not reflect this. Figure 17 illustrates this point, showing the same 239 points plotted as relative error versus concentration.

Data such as these may be used to derive general models of the relationship of uncertainty to concentration. Earlier it was suggested that to qualify as a quantitative XRPD method, uncertainty should be less than $\pm X^{0.5}$ at the 95% confidence level, where X = concentration in weight per cent. The dotted lines in Fig. 17 illustrate this relationship in the appropriate form for relative error ($\pm 100X^{-0.5}$) at the 95% confidence level. Note that any other level of confidence may be obtained simply by scaling the first term by the required number of standard deviations, e.g. $\pm 50X^{-0.5}$ at 67% confidence. Equations of similar form ($y = cx^b$, where c and b are constants) that relate precision and accuracy with concentration for a variety of chemical analysis methods were discussed by Hughes & Hurley (1987). An analysis of the author's data shown in Fig. 17 suggests that equations of this form also may be suitable for describing the relationship of uncertainty to concentration for quantitative XRPD methods. Fitting such a model to data such as those shown in Fig. 17 by statistical analysis of the

Table 9 Calculation of the lower limits of detection for minerals present in 'Mix 2' of Hillier (2000).

	Position	Wt%	Peak counts	Back counts	Counts/wt%	LLD wt% (measured)	LLD wt% (estimated)	RIR
Albite	3.19	13.2	13 397	5041	1014.9	0.40	0.21	2.07
Calcite	3.03	10.0	15 531	3712	1553.1	0.22	0.16	2.73
Dolomite	2.89	10.2	11 775	2634	1154.4	0.25	0.22	1.94
Quartz	4.26	29.4	13 992	2321	475.9	0.57	0.54	0.80
Quartz	3.34	29.4	69 342	5032	2358.6	0.17	0.11	4.04
Quartz	1.81	29.4	9738	2277	331.2	0.81	0.83	0.52
Chlorite	3.52	9.0	3014	1301	334.8	0.61	0.76	0.54
Kaolinite	3.58	10.0	2701	1951	270.1	0.93	0.81	0.53
Muscovite	10.00	8.2	812	3094	99.0	3.18	1.88	0.23
Corundum	2.08	10.0	5561	1803	556.1	0.43	0.43	1.00

LLD = lower limits of detection; RIR = reference intensity ratio.

errors for specific concentration ranges, is one way to provide a generalized estimate of the uncertainty associated with a quantitative method for a range of analytes. The solid lines shown on Fig. 17 ($\pm 100X^{-0.65}$) illustrate a possible fit to these data at a 95% confidence level. Such an estimate is, however, controlled by the phases that have the largest associated errors. For phases that are easy to analyse accurately uncertainty will be overestimated. It is of course possible to determine the specific relationship for a single analyte but this requires considerable effort. One possible approach is to prepare a number of replicate samples for chosen low, moderate and high concentrations such as 3, 15 and 75 wt%. Then, for each concentration the standard deviation of the relative error is determined and plotted against concentration. By fitting an equation of the form given above to these data a relationship may be obtained for uncertainty at any concentration. For well-defined analytes, such procedures incorporate all sources of uncertainty associated with all stages of a particular method, e.g. weighing errors, counting statistics, peak intensity measurement, etc. The procedure thereby involves less effort than attempting to evaluate all these contributions to uncertainty individually. However, for some minerals one must always remember that factors such as solid solution may introduce yet another source of uncertainty, although it is not especially difficult to devise RIRs that take account of this.

LLDs in random powders

Based on equation (11) the calculation of the LLDs for the minerals present in one of the prepared mixtures described by Hillier (2000) are presented in Table 9. The table also includes values calculated using only the measured LLD for the 10% corundum addition, the appropriate RIR for each peak and equation (12). The relationship of these detection limits to the RIR is also illustrated in Fig. 18. These data were obtained from a fairly standard diffractometer and for most phases encountered in sandstones one should expect LLDs of some fraction of a weight per cent. The only exception is mica based on its 10 Å peak. In a random powder the non-basal peaks from mica, and from other illitic clays, are stronger diffractors than the basal peaks (Fig. 19). Furthermore, as mentioned in Table 8, unless they are Fe-rich, most dioctahedral clay minerals diffract with similar intensity at the position of the 02,11 'clay band' (Fig. 19). Thus in practice, the detection and quantification of 1% of illitic clay in a matrix

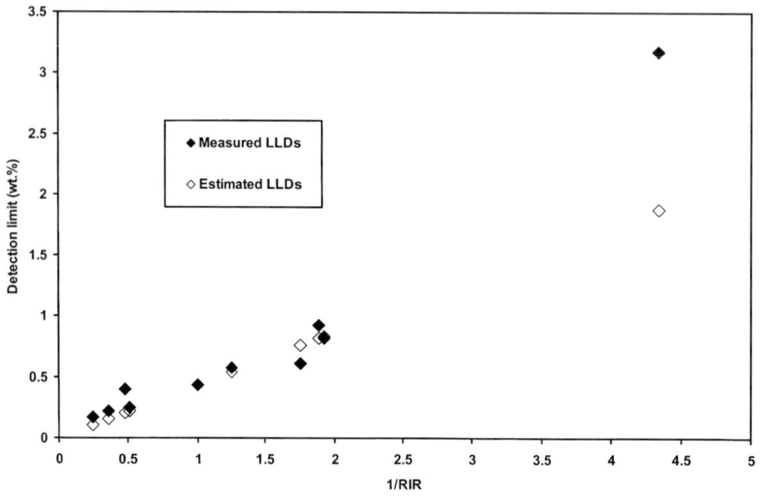

Fig. 18 Experimentally measured lower limits of detection for various minerals versus the reciprocal of their respective reference intensity ratios (RIRs) compared with lower limits of detection calculated from the measured lower limit of detection for the internal standard and the appropriate RIR for each mineral.

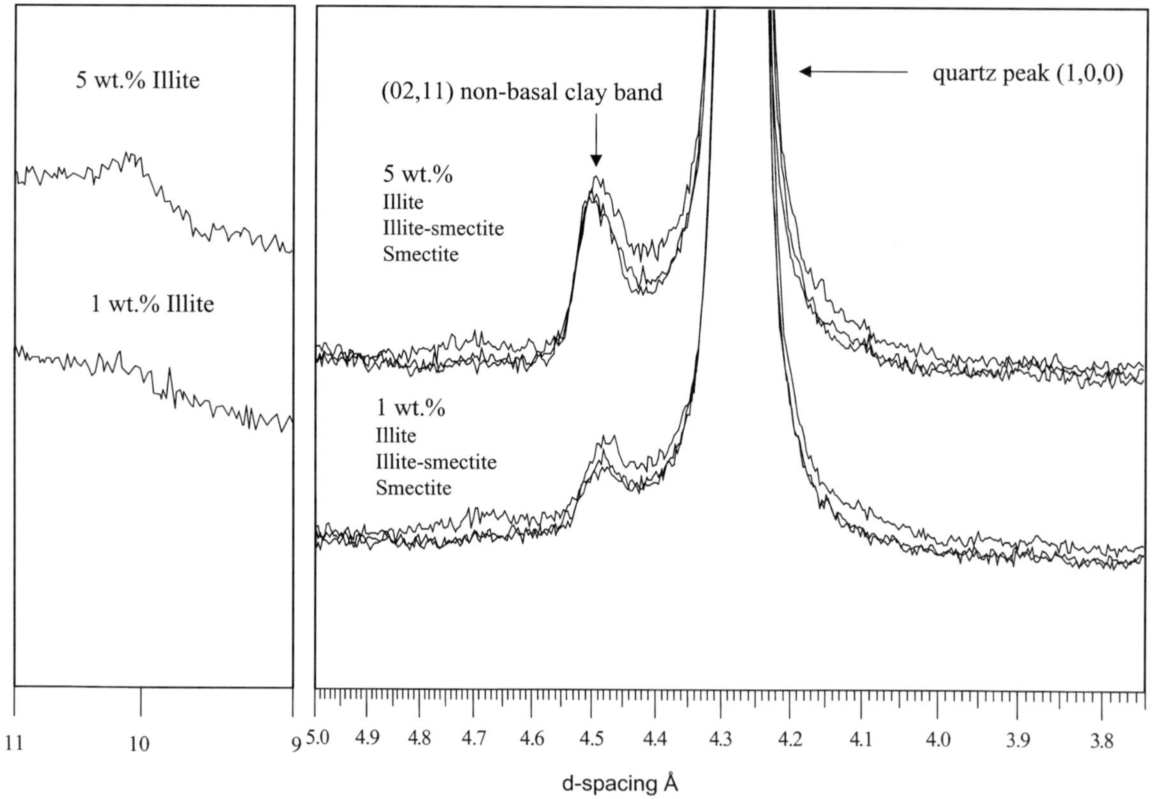

Fig. 19 Diffraction patterns (random powders) obtained from prepared mixtures of 5 wt% and 1 wt% of three dioctahederal clay minerals in a matrix dominated by quartz with 20 wt% added corundum. The clay minerals are an illite, an ordered mixed-layer illite–smectite and a montmorillonite. The region of the 10 Å peak for the illite sample is also shown as an insert (at the same *y* scale), to illustrate the fact that the 02,11 peak is a stronger and thereby more easily detected peak in a random powder. All samples were prepared by spray drying.

dominated by quartz should be achievable routinely.

CONCLUDING REMARKS

Probably all sandstones contain some clay. The techniques described here allow one to determine how much. They are, of course, equally applicable to all kinds of rocks. It should never be forgotten that XRPD is not the only way that quantitative analysis of clay minerals may be made. Nevertheless, properly applied it has the potential to be the most accurate. Studies of both clay fractions and whole-rock samples provide complementary information. The data from orientated clay fractions are necessary to provide the detailed identification of the kinds of clay minerals present. Analysis of whole-rocks as random powders enables the absolute amounts of the different clay minerals to be determined. Random powders also can provide a wealth of information on the other minerals that the sandstone contains. I have tried to show that quantitative analysis of both kinds of sample is not that difficult to make. If proper attention is paid to sample preparation, even simple single peak methods are capable of accurate results. However, the need to validate the methods used should never be forgotten. In review the author was reminded (D. Moore, personal communication) of the many caveats offered by George Brindley in his chapter on quantitative clay mineral analysis in the classic Brindley & Brown monograph (Brindley & Brown, 1980). It is certainly good advice to always have these caveats in mind. Possibly the most important one to reiterate is that 'careful characterisation of the minerals to be measured . . . (is) . . . as important as their quantitative estimation'. It may be possible to analyse poorly characterized clay minerals precisely, but insufficient characterization precludes accurate analysis. In some cases this may be entirely fit for the purpose, but in many others it may mean we are unable to answer the questions that prompted the analysis in the first place.

ACKNOWLEDGEMENTS

The assistance of Caroline Thomson in the laboratory has been invaluable and the manuscript was improved by the comments of Derek Bain and the constructive reviews of Dewey Moore and David Doff. The author acknowledges the financial support of the Scottish Executive Environment and Rural Affairs Department (SEERAD).

REFERENCES

Amonette, J.E. & Zelazny, L.W. (1994) *Quantitative Methods in Soil Mineralogy.* Soil Science Society of America, Madison, WI.

Bailey, S.W., Alietti, A., Formoso, M.L.L., *et al.* (1986) Report of the AIPEA nomenclature committee (illite, glauconite, volkonskoite). *Assoc. Int. Pour Etud. Argiles Newsl. Suppl.* **22** (February), 1–2.

Biscaye, P.E. (1965) Mineralogy and sedimentation of recent deep-sea clay in the Atlantic Ocean and adjacent seas and oceans. *Geol. Soc. Am. Bull.*, **76**, 803–832.

Bish, D.L. (1994) Quantitative X-ray diffraction analysis of soils. In: *Quantitative Methods in Soil Mineralogy* (Ed. by J.E. Amonette and L.W. Zelazny), Soil Science Society of America, Madison, WI, pp. 267–295.

Bish, D.L. & Chipera, S.J. (1988) Problems and solutions in quantitative analysis of complex mixtures by X-ray powder diffraction. *Adv. X-ray Anal.*, **31**, 295–308.

Bish, D.L. & Chipera, S.J. (1995) Accuracy in quantitative X-ray powder diffraction analyses. *Adv. X-ray Anal.*, **38**, 47–57.

Bish, D.L. & Post, J.E. (1989) *Modern Powder Diffraction.* Reviews in Mineralogy, Vol. 20, Mineralogical Society of America Washington, DC.

Bish, D.L. & Post, J.E. (1993) Quantitative mineralogical analysis using the Rietveld full-pattern fitting method. *Am. Mineral.*, **78**, 932–940.

Bish, D.L. & Reynolds, R.C. Jr. (1989) Sample preparation for X-ray diffraction. In: *Modern Powder Diffraction* Reviews in Mineralogy, Vol. 20 (Ed. by D.L. Bish and J.E. Post), Mineralogical Society of America, Washington, DC, pp. 73–99.

Brindley, G.W. & Brown, G. (1980) *Crystal Structures of Clay Minerals and their X-ray Identification.* Mineralogical Society Monograph 5, London.

Calvert, C.S., Palkowsky, D.A. & Pevear, D.R. (1989) A combined X-ray powder diffraction and chemical method for the quantitative mineral analysis of geological samples. In: *Quantitative Mineral Analysis of Clays*, Clay Mineral Society Workshop Lectures Vol. 1 (Ed. by D.R. Pevear & F.A. Mumpton), Clay Minerals Society, Evergreen, CO, pp. 154–166.

Chipera, S.J. & Bish, D.L. (1995) Multireflection RIR and intensity normalizations for quantitative analyses: applications to feldspars and zeolites. *Powder Diffract.*, **10**, 47–55.

Chung, F.H. (1974a) Quantitative interpretation of X-ray diffraction patterns of mixtures. I. Matrix-flushing method for quantitative multicomponent analysis. *J. Appl. Crystall.*, **7**, 519–525.

Chung, F.H. (1974b) Quantitative interpretation of X-ray diffraction patterns of mixtures. II. Adiabatic principle of X-ray diffraction analysis of mixtures. *J. Appl. Crystall.*, **7**, 526–531.

Chung, F.H. (1975) Quantitative interpretation of X-ray diffraction patterns of mixtures. III. Simultaneous determination of a set of reference intensities. *J. Appl. Crystall.*, **8**, 17–19.

Cook, H.E., Johnson, P.D., Matti, J.C. & Zemmels, I. (1975) Methods of sample preparation and X-ray diffraction data analysis, X-ray mineralogy laboratory, Deep Sea Drilling Project, University of California, Riverside. In: *Initial Reports of the Deep Sea Drilling Project 28* (Ed. by D.E. Hayes, L.A. Frakes, *et al.*), US Government Printing Office, Washington, DC, pp. 999–1007.

Davis, B.L. (1987) The estimation of limits of detection in RIM quantitative X-ray diffraction analysis. *Adv. X-ray Anal.*, **31**, 317–323.

Davis, B.L. & Johnson, L.R. (1989) Quantitative mineral analysis by X-ray transmission and X-ray diffraction. In: *Quantitative Mineral Analysis of Clays*, Clay Mineral Society Workshop Lectures Vol. 1 (Ed. by D.R. Pevear and F.A. Mampton), Evergreen, CO, pp. 104–117.

Davis, B.L., Smith, D.K. & Holomany, M.A. (1989) Tables of experimental reference intensity ratios Table No. 2, December 1989. *Powder Diffract.*, **4**, 201–204.

Drever, J.I. (1973) The preparation of oriented clay mineral specimens for X-ray diffraction analysis by a filter-membrane peel technique. *Am. Mineral.*, **58**, 553–554.

Flörke, O.W. & Saalfeld, H. (1955) Ein Verfahren zur Herstellung texturfreier Röntgen-Pulverpräparate. *Z. Kristallogr.*, **106**, 460–466.

Gibbs, R.J. (1965) Error due to segregation in quantitative clay mineral X-ray diffraction mounting techniques. *Am. Mineral.*, **50**, 741–751.

Heath, G.R. & Pisias, N.G. (1979) A method for the quantitative estimation of clay minerals in north Pacific Deep-Sea sediments. *Clay. Clay Mineral.*, **27**, 175–184.

Hillier, S. (1998) Accurate quantitative analysis of the mineral components of 'drilling muds' by X-ray powder diffraction. *Minerals in the Oil Industry Conference*, Aberdeen, 27 October.

Hillier, S. (1999) Use of an air-brush to spray dry samples for X-ray powder diffraction. *Clay Mineral.*, **34**, 127–135.

Hillier, S. (2000) Accurate quantitative analysis of clay and other minerals in sandstones by XRD; comparison of a Rietveld and a reference intensity ratio (RIR) method, and the importance of sample preparation. *Clay Mineral.*, **35**, 295–306.

Hubbard, C.R. & Smith, D.K. (1977) Experimental and calculated standards for quantitative analysis by powder diffraction. *Adv. X-ray Anal.*, **20**, 27–39.

Hughes, H. & Hurley, P.W. (1987) Precision and accuracy of test methods and the concept of K-factors in chemical analysis. *Analyst*, **112**, 1445–1449.

Hughes, R. & Bohor, B. (1970) Random clay powders prepared by spray-drying. *Am. Mineral.*, *55*, 1780–1786.

Hughes, R.E., Moore, D.M. & Glass, H.D. (1994) Qualitative and quantitative analysis of clay minerals in soils. In: *Quantitative Methods in Soil Mineralogy* (Ed. by J.E. Amonette and L.W. Zelazny), Soil Science Society of America, Madison, WI, pp. 330–359.

Jenkins, R. & Snyder, R.L. (1996) *Introduction to X-ray Powder Diffractometry*. Wiley, New York.

Jonas, E.C. & Kuykendall, J.R. (1966) Preparation of montmorillonites for random powder diffraction. *Clay Mineral.*, **6**, 232–235.

Klug, H.P. & Alexander, L.E. (1974) *X-ray Diffraction Procedures for Polycrystalline and Amorphous Materials*. Wiley, New York.

Madsen, I. (1999) *Quantitative Phase Analysis Round Robin*. Newsletter No. 22 (November), Commission on Powder Diffraction, International Union of Crystallography, New York, pp. 3–5.

McMannus, D.A. (1991) Suggestions for authors whose manuscripts include quantitative clay mineral analysis by X-ray diffraction. *Mar. Geol.*, **98**, 1–5.

Moore, D.M. & Reynolds, R.C. Jr. (1997) *X-ray Diffraction and the Identification and Analysis of Clay Minerals*, 2nd edn. Oxford University Press, New York.

Mossman, M.H., Freas, D.H. & Bailey, S.W. (1967) Orienting internal standard method for clay mineral X-ray analyses. *Clay. Clay Mineral.*, **15**, 441–453.

Mumme, W.G., Tsambourakis, G., Madsen, I.C. & Hill, R.J. (1996) Improved petrological modal analyses from X-ray powder diffraction data by use of the Rietveld method I. Selected sedimentary rocks. *J. Sed. Res.*, **66**, 132–138.

Ottner, F., Gier, S., Kuderna, M. & Schwaighofer, B. (2000) Results of an inter-laboratory comparison of methods for quantitative clay analysis. *Appl. Clay Sci.*, **17**, 223–243.

Pevear, D.R. & Mumpton, F.A. (Eds) (1989) *Quantitative Mineral Analysis of Clays*. Clay Mineral Society Workshop Lectures Vol. 1, Clay Minerals Society, Evergreen, CO.

Quakernaat, J. (1970) Direct diffractometric quantitative analysis of synthetic clay mineral mixtures with molybdenite as orientation-indicator. *J. Sed. Petrol.*, **40**, 506–513.

Reynolds, R.C. Jr. (1985) *NEWMOD, a computer program*

for the calculation of one-dimensional diffraction patterns of mixed-layered clays. R.C. Reynolds, Jr, Hannover, NH.

Reynolds, R.C. Jr (1989) Principles and techniques of quantitative analysis of clay minerals by X-ray powder diffraction. In: *Quantitative Mineral Analysis of Clays*, Clay Mineral Society Workshop Lectures Vol. 1. (Ed. by D.R Pevear and F.A. Mumpton), Clay Minerals Society, Evergreen, CO, pp. 4–36.

Schreiner, W.N. (1995) A standard test method for the determination of RIR values by X-ray diffraction. *Powder Diffract.*, **10**, 25–33.

Schultz, L.G. (1964) Quantitative interpretation of mineralogical composition from X-ray and chemical data for the Pierre Shale. *US Geol. Surv. Prof. Pap.*, **391-C**, 1–31.

Smith, D.K., Nickols, M.C. & Zolensky, M.E. (1982) *'POWD10' a FORTRAN IV Program for Calculating X-ray Powder Diffraction Patterns—Version 10.* The Pennsylvanian State University, University Park, PA, 103 pp.

Smith, S.T., Snyder, R.L. & Brownell, W.E. (1979) Minimisation of preferred orientation in powders by spray-drying. *Adv. X-ray Anal.*, **22**, 77–87.

Snyder, R.L. (1992) The use of reference intensity ratios in X-ray quantitative analysis. *Powder Diffract.*, **7**, 186–193.

Snyder, R.L. & Bish, D.L. (1989) Quantitative analysis. In: *Modern Powder Diffraction*, Reviews in Mineralogy Vol. 20 (Ed. by D.L. Bish and J.E. Post), Mineralogical Society of America, Washington, DC, pp. 101–144.

Środon, J. (1980) Precise identification of illite/smectite interstratifications by X-ray powder diffraction. *Clay. Clay Mineral.*, **28**, 401–411.

Towe, K.M. (1974) Quantitative clay petrology: the trees but not the forest. *Clay. Clay Mineral.*, **22**, 375–378.

Wilson, M.J. (1987) *A Handbook of Determinative Methods in Clay Mineralogy.* Blackie & Son, New York.

Zevin, L.S. & Kimmel, G. (1995) *Quantitative X-ray Diffractometry.* Springer-Verlag, New York.

Zwingmann, H. (1995) *Etude des conditions de mise en place des gaz naturels dans les réservoirs gréseux (Rotliegende d'Allemagne). Aspects minéralogiques, morphologiques, géochimiques et isotopiques.* These, L'Universite Louis Pasteur, Strasbourg, 197 pp.

Int. Assoc. Sedimentol. Spec. Publ. (2003) **34**, 253–287

A review of radiometric dating techniques for clay mineral cements in sandstones

P.J. HAMILTON

Centre of Excellence in Mass Spectrometry, Applied Geology, Curtin University and CSIRO Division of Petroleum Resources, PO Box 1130, Technology Park, Bentley WA 6102, Australia, e-mail: Joe.Hamilton@per.dpr.csiro.au

ABSTRACT

The principles underlying the application of radiometric dating methods to determining the timing of clay-mineral cementation in sandstones are reviewed. Most previous work has centred on the use of the Rb–Sr and K–Ar schemes for age determinations on separated glauconites, illites and mixed-layer illite–smectites. Some of the difficulties encountered in attempting to date clay-mineral diagenesis include other mineral contaminants, potentially contaminant elements in cation exchange sites, separation of multiple growths of a particular mineral formed at different times in the same sandstone, isolation of the mineral phase of interest in a pure state and the need for three or more cogenetic phases required for the application of some radiometric schemes. As experience has been gained, the need to carefully characterize the phase being analysed, increased measurement sensitivity and increasingly sophisticated methods of sample preparation have resulted in clay mineral dates that can be related to a diagenetic age with more confidence. Additionally, much has been learnt of the processes of clay-mineral diagenesis.

More recently there has been a shift in emphasis in clay-mineral geochronology to greater use of the $^{40}Ar/^{39}Ar$ method for dating clay-mineral diagenesis in fine-grained sediments. The associated advantage of being more suitable for extremely small samples, on the order of tens of micrograms that can be readily screened for contaminants, is an attractive feature for potential application to sandstones. Although considerable research remains to be done, there is considerable scope with this technique for application to sandstones when coupled with laser microprobe heating of clay cements *in situ* in thin-sections.

INTRODUCTION

The field of isotope geochemistry has a wealth of application to the study of sedimentary basins, ranging from determining sedimentary provenance through isotopic characterization of mineral detritus to constraining the timing and physico-chemical conditions of diagenetic mineral-forming events. This article reviews the radiometric isotope methods available to constrain or determine the age of formation of clay-mineral cements in sandstones. Clauer & Chaudhuri (1995) and Emery & Robinson (1993) have also published recent reviews of this subject matter. The methods used most commonly for dating diagenetic clay-minerals are the rubidium–strontium (Rb–Sr), potassium–argon (K–Ar) and argon 40–39 ($^{40}Ar/^{39}Ar$) geochronometers. The bias of this paper therefore is towards discussion of these methods and towards their application to dating illitic and glauconitic clays, which are the diagenetic clay phases most commonly analysed. Also reviewed are a number of recent technical developments in $^{40}Ar/^{39}Ar$ dating methods, so far mainly applied to fine-grained sediments but with exciting potential for application to clay cements in sandstones.

It is appropriate at the start of this paper to examine briefly why it is that there is any interest in determining the age of clay-mineral cement formation in sandstones. During sediment burial, detrital mineral grains and pore fluids undergo readjustments (diagenesis) to changes in pressure, temperature and pore-fluid chemical environment. Diagenesis that results in the precipitation of new minerals in pore space is termed authigenesis and results in loss of porosity. Additionally such minerals may result in loss of permeability by acting as cements between detrital framework grains, bridging across and even occluding pore spaces and pore throats.

Loss of porosity and permeability through clay diagenesis is deleterious to deliverability of a potential hydrocarbon reservoir (Stalder, 1973; McHardy *et al.*, 1982; Pallat *et al.*, 1984; Macchi, 1987). There is therefore a practical interest in tracking environmental conditions whereby reservoir quality is modified and determining when these modifications occur with respect to the timing of migration and entrapment of oil and gas. A major impetus for wishing to quantify these aspects of diagenesis, including the timing of diagenetic clay cementation, has been to develop predictive models (Meshri & Ortoleva, 1990; Emery & Robinson, 1993; Le Gallo *et al.*, 1998) for porosity and permeability. Such models also may enable prediction of the occurrence and distribution of clay-mineral cements that can give rise to formation damage problems such as swelling clays and migration of fine clay particles, so resulting in loss of reservoir quality.

One approach to determining the timing of clay diagenesis is to refer the timing of the particular cement to the relative timing of the diagenetic sequence deduced from petrographic observations and superimposed on a burial history model (Fig. 1). Quantitative time and temperature information that may be available are superimposed on the model. This can include fluid-inclusion temperatures from quartz and carbonate cements and radiometric dates from non-clay diagenetic phases. The latter can include K-feldspar overgrowths (Girard *et al.*, 1988) and diagenetic minerals hosted by stylolites and fault planes (Thomas *et al.*, 1993; Vrolijk & van der Pluijm, 1999). However, the opportunities to constrain timing of clay-mineral cementation through such radiometric ages would not be frequent. Further, using petrographic observations to deduce the relative timing of a particular cement is a subjective exercise, often confounded by the presence of ambiguous textural relationships between minerals, particularly where there have been multiple generations of one or more of the

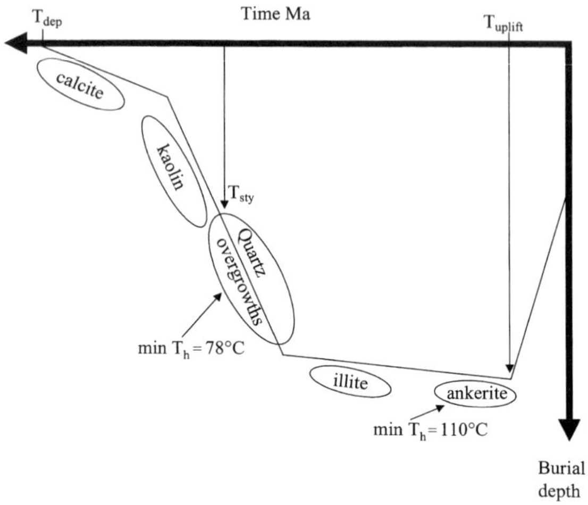

Fig. 1 Burial history model upon which has been superimposed the relative timing of the diagenetic sequence deduced from petrographic observations as well as quantitative temperature and time information available from fluid inclusions in quartz and carbonate cements and a radiometric date determined from K-feldspar overgrowths. The timing of faulting and/or stylolitization (T_{sty}) may be constrained from radiometric ages on synkinematic illite/muscovite. The depositional age (T_{dep}) and timing of uplift (T_{uplift}) would be constrained from the local geological context.

diagenetic phases. The integration of petrographic observations with use of geochronological methods for determining the timing of clay-mineral cementation offers the possibility of removing some of the ambiguity of this approach.

Dating diagenetic clays also has resulted in aspects of their mechanisms of formation being proposed (e.g. Hamilton *et al.*, 1992; Robinson *et al.*, 1993; Clauer & Chaudhuri, 1995; Darby *et al.*, 1997). Further, there is an enormous challenge presented by attempting to analyse particles of extremely small size, as small as tens of nanometres and less, in a highly complex physico-chemical matrix. In this regard it will be apparent in the discussion below that there has been a trend towards more refined mineral characterization coupled with decreasing sample size and increasingly sensitive instrumental capability. Dating of clay minerals in thin-section and extracting age information from analysis of complex clay-mineral mixtures are now real possibilities.

CLAY MINERALOGY

Clays are hydrous aluminosilicates and belong to the phyllosilicate group of minerals (Deer *et al.*, 1967; Bailey, 1984, 1988; Velde, 1985). In addition to aluminium (Al) and silicon (Si) they may also contain other cations, including alkali and alkaline earth metals. They have a sheet-like structure in which the 'building blocks' are either tetrahedra or octahedra. These are linked into planar layers by sharing of oxygen (O) ions between Si or Al ions of the adjacent tetrahedra or octahedra. The tetrahedron results from the close packing of four O ions with the space between them occupied by a Si or Al ion. The octahedron results from the close packing of six anions that are dominantly O but can also include some hydroxyl (OH) ions. The space between them is occupied mainly by Si or Al, but there also can be other cations such as iron (Fe), calcium (Ca) and magnesium (Mg). Tetrahedral and octahedral sheets are bound to each other in layers that extend for tens to thousands of nanometres (nm) in the *a* and *b* crystallographic directions. The layers are stacked on top of each other in the *c* direction. Clay minerals that are mentioned in the context of radiometric dating in this article are described briefly below.

Kaolinite is comprised of one tetrahedral layer linked to one octahedral layer and so is termed a 1 : 1 structure: O–H–O bonds connect the two layers. The chemical formula of kaolinite is $Al_2Si_2O_5(OH)_4$.

Illite is comprised of one octahedral layer sandwiched between two tetrahedral layers and so is termed a 2 : 1 structure. The middle O–OH layers are shared by both the tetrahedral and octahedral layers. Two illite layers are connected by O–K–O bonds. The K^+ is required for charge balance accompanying the substitution of Al^{3+} for some Si^{4+} in the tetrahedra and the substitution of divalent cations for Al^{3+} in the octahedra. The O–K–O bonding is strong and prevents the swelling of illitic clays. Illites can have different structures termed polytypes that reflect different ways in which layers are stacked. The 1M and 1Md polytypes are those prevalent for low-temperature (i.e. diagenetic) illites and also glauconites. At metamorphic temperatures of > 250°C, illites are of the 2M polytype. The general chemical formula for illite is $K_yAl_4(Si_{8-y},Al_y)O_{20}(OH)_4$, where *y* typically is between 1 and 1.5. It is chemically similar to muscovite but with less K and more Si.

Smectite also is a 2 : 1 clay but with some substitution of divalent cations for Al^{3+} in the octahedra. There is less binding of layers together by K^+ than in illite. Cations that do bind layers together are completely exchangeable and so should reflect the chemistry of the aqueous medium with which the smectite was last in contact. These interlayer cations are variably hydrated and this results in the swelling characteristic of smectitic clays. The term smectite actually refers to a group of similar minerals with a general chemical formula of $(1/2Ca, Na)_{0.7}(Al,Mg,Fe)_4(Si,Al)_8O_{20}(OH)_4 \cdot nH_2O$.

Mixed-layer, or interstratified, clay minerals are mixtures of two (or less commonly three) different mineral layers in a single structure

(Środón, 1999a). Most mixed-layer clays contain smectite as a swelling component. Of the mixed-layer clays that occur, *illite–smectite* has been the focus of radiometric dating applications. During burial diagenesis mixed-layer illite–smectite becomes progressively more illite-rich. There has been considerable debate concerning whether the illitization process occurs by solid state transformation or dissolution/crystallization mechanisms (Boles & Franks, 1979; Ahn & Peacor, 1986; Huang *et al.*, 1993; Inoue & Kitagawa, 1994; Elliott & Matisoff, 1996; Altaner & Ylagan, 1997; Środón, 1999a). Geochronological studies, beginning with that of Perry (1974), have made a major contribution to this debate. However, the bulk of these studies have been, and continue to be, directed towards analysis of the illitization of mixed-layer illite–smectite in shales and bentonites (e.g. Elliott & Aronson, 1987; Velde & Renac, 1996; Dong *et al.*, 2000) as opposed to sandstones. Mention is made later of a mixed-layer chlorite–smectite, termed *corrensite*, which has been subjected to radiometric dating.

Chlorite has a 2 : 1 : 1 structure comprised of a negatively charged 2 : 1 structure interlayered with an additional octahedral layer that is positively charged and comprised of cations and hydroxyl ions. A general chemical formula of chlorite is $(Mg,Al,Fe)_{12}[(Si,Al)_8O_{20}](OH)_{16}$.

Glauconite is a term that Odin & Matter (1981) have suggested be restricted to occurrences of dark green, Fe-rich, mica-type clay mineral of marine origin and with $K_2O > 6\%$. The term glaucony is recommended as a facies term to describe the group of Fe-rich marine clays that range in composition from glauconitic smectite to glauconitic mica. It is similar in structure to illite with some mixed layering of an expandable component. The glauconitic mineral chemical formula is $(K,Na,Ca)_{1.2-2.0}(Al,Mg,Fe)_4(Si_{7-7.6}Al_{1-0.4}O_{20})(OH)_4 \cdot nH_2O$.

Fundamental particles

The development of understanding of clay-mineral particle size has been crucial to the understanding of many aspects of clay-mineral diagenesis, including interpretation of X-ray diffraction (XRD) data for mixed-layer clays (Nadeau *et al.*, 1984a, b), illitization of illite–smectite (Nadeau *et al.*, 1985) and $^{40}Ar/^{39}Ar$ and K–Ar radiometric age results (e.g. Clauer *et al.*, 1997; Smith *et al.*, 1998; Dong *et al.*, 2000). The fundamental particle is considered to be the smallest unit that a clay-mineral particle may occupy (Nadeau, 1985, 1999; Nadeau *et al.*, 1985) and its size determines properties such as surface area, cation exchange capacity and degree of open- or closed-system behaviour with respect to radiometric isotope systems.

SAMPLE PREPARATION

Mineral separation

There are two basic approaches to clay-mineral isolation. One is purely physical and the other involves both chemical and physical treatments. The first involves sample disaggregation by a freeze–thaw technique proposed by Liewig *et al.* (1987). This involves a sample being immersed in water in a closed container and subjected to a repeated freeze–thaw cycle leading to disaggregation of the rock into its constituent grains. Size fractionation is then undertaken, first by separating coarser particles (> 50 μm) by sieving, followed by settling in water using Stokes' law to separate silt-sized particles and then using ultra-high-speed centrifugation to size fractionate the finest particles (< 2 μm). Advantages include lack of any heat or chemical treatment and the supposed retention of the integrity of original grain shape and size. Disadvantages include the non-removal of organic matter and potential flocculants. Although potentially more of a problem in finer grained sediments, organic matter may represent a contamination source for Ar (Hamilton *et al.*, 1989), given the strong association between organic matter and noble gases (of which Ar is one). In addition, organic matter is a major source of volatiles and could result in a need for extensive gas purification prior to isotopic analysis of Ar.

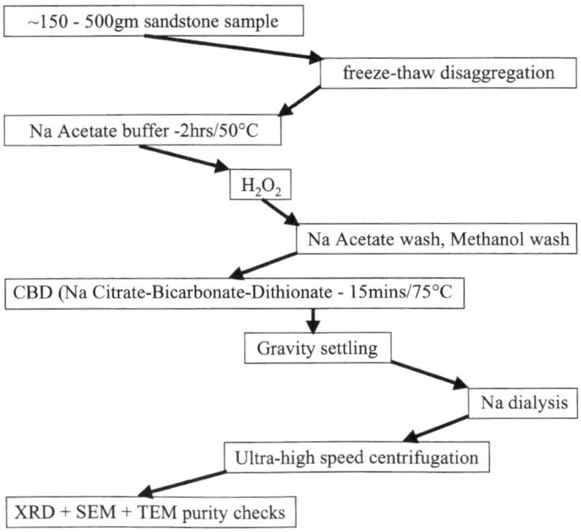

Fig. 2 Physico-chemical procedure for separation of clay minerals from sandstones.

The other set of procedures was developed by soil scientists (Jackson, 1979) and is outlined in Fig. 2. Recently, Menegatti *et al.* (1999) have recommended a disodium peroxodisulphate treatment for removal of organic matter. It appeared to have no impact on the K–Ar or Rb–Sr isotope systematics of the clay minerals investigated. The physico-chemical approach does have the advantage of removing rock constituents that may cause clay flocculation and would otherwise reduce the yield of finer sized clay particles. The use of many of the chemical procedures indicated in Fig. 2 will result in displacement of the exchangeable cations and may obviate the leachate–residue approach sometimes used for Rb–Sr dating of some clay minerals (see later).

Emery & Robinson (1993) raise the possibility that salts residual from pore waters in a sandstone sample may not be entirely removed by the separation procedures and constitute a source of contamination. Dong *et al.* (1997) subjected thin-section samples for laser probe $^{40}Ar/^{39}Ar$ dating to prolonged water soakings in order to remove a suspected water-soluble salt contaminant.

Sample purification

A major difficulty in geochronological analysis of clay minerals is their separation from the host sandstone in a pure state. Not only must the mineral be free of contaminants but also it must be isolated in sufficient quantity for isotopic analysis, possibly in duplicate. In addition, quantities are required for purity assessment and confirmation of compositional identity using XRD and transmission electron microscopy (TEM) analysis. Indeed, sample characterization is essential as an aid to interpretation of radiometric data for clay mineral separates (Clauer *et al.*, 1992a). Diagenetic clay minerals often are present in sandstones only in volumetrically small amounts and separation techniques need to be efficient. For some clay minerals it may be important to retain grain shape. Separation procedures also must be such as to not perturb the isotope system to be investigated.

Mineral contaminants

It is a basic assumption in geochronology that a mineral sample to be dated contains no other minerals that may contribute amounts of the parent and/or daughter element to be analysed. There are two approaches to managing contaminant phases when they occur. Firstly the contribution the contaminant makes to the mineral system being analysed can be quantified by XRD analyses of various mixtures of mineral and contaminant, plotting these against apparent age and extrapolating to end-member ages for contaminant and diagenetic phases (e.g., Mossmann, 1991; Pevear, 1992, 1999). On such a plot the mixing trend is not simply a rectilinear relationship (Środoń, 1999b, 2000; Ylagan *et al.*, 2000). At low levels of contamination (< 5%) mixing trends can appear to be straight lines (see fig. 3 in Hamilton *et al.*, 1989) and there is little error in extrapolation of the trend to obtain the age of the diagenetic end-member. However, where contamination levels are high, an age for the diagenetic end-member derived by straight-line extrapolation can be seriously in error (see also Fig. 8 later).

The other approach, where appropriate, is removal of the contaminant phase by either dissolution or leaching. A wide range of chemicals has been assessed for this purpose, particularly in regard to sample treatments related to dating glauconites. The chemical treatments used have included hydrochloric acid, acetic acid, ammonium acetate, sodium–EDTA, ammonium–EDTA, sodium citrate and ion exchange resin. Various authors (Morton & Long, 1980, 1982; Clauer, 1982; Kralik, 1984; Clauer *et al.*, 1992b, 1993) have assessed the effects of these treatments and they have been reviewed in detail by Clauer & Chaudhuri (1995). Although most treatments appear effective in removing carbonates, the most common soluble phases associated with clay-mineral separates, they are also intended as a way of removing exchangeable cations prior to isotopic analysis. This is discussed in the following section and in later sections dealing with dating applications to specific minerals.

Exchangeable cation contaminants

A characteristic of clay minerals that distinguishes them from other silicates is the presence of unbalanced negative charges on the edges of unit cells. These negative charges derive from isomorphous substitution within the structure (e.g. substitution of Al^{3+} for Si^{4+}), from broken bonds at the edges and surfaces of the clay structure and from the dissociation of accessible hydroxyl groups. These sites of negative charge attract sufficient positively charged ions from the surrounding pore-fluid medium in order to acquire neutrality. Cations adsorbed to the external surfaces and absorbed between the layers of clay structures are exchangeable in that they can exchange position with other cations in the pore-fluid. The quantity of cations that a clay can adsorb is called the cation exchange capacity (CEC) and has units of milliequivalents (mEq) per 100 g or centimoles per kilogram. Cation exchange capacity ranges from low values of *c.* 3–10 mEq per 100 g for kaolinites to *c.* 10–40 mEq per 100 g for illites, glauconites and chlorites to high values of *c.* 60–150 mEq per 100 g for smectites.

As the exchangeable cations can include K^+, Rb^+ and Sr^{2+} this component of clay structure can be a source of contamination. Clay mineral isolation procedures that include the chemical treatments indicated in Fig. 2 are designed to displace the exchangeable cations with other cations such as Na^+ and NH_4^+. This has the distinct disadvantage of destroying any chronological or environmental information that could otherwise be deduced from the isotope systematics of the exchangeable Rb and Sr. Of major interest in the exchangeable component is the extent to which it may reflect the pore-fluid environment at the time of closure of a mineral isotope system or whether cations and associated daughter isotope products continue to equilibrate with the external environment.

Comparable environmental information is not recoverable for the K–Ar isotope systematics of the exchangeable component of a clay. Radiogenic ^{40}Ar atoms produced from K atoms in exchangeable sites are not bound to an exchange site on the clay surface and therefore will diffuse into the pore-fluid system. Thomson & Hower (1973) and Aronson & Douthitt (1986) recommended removal of this easily exchangeable K (termed K3) prior to K–Ar analysis. Thomson & Hower (1973) also suggested that the K located in hydrated layers (termed K2), which is less easily exchanged than K3 in exchangeable sites on clay surfaces, also did not contribute to radiogenic ^{40}Ar that could be accumulated. They suggested that K2 also be removed by acid dissolution prior to K–Ar analysis. Aronson & Douthitt (1986) tested this hypothesis by K–Ar analyses of residues at successive stages of acid dissolution of an illite–smectite. Although the Ar data are rather scattered it seemed that the K2 site had been retentive of its daughter ^{40}Ar. They suggested that for this sample the K2 site was strongly bound to the lattice structure and comprised an intermediate structural stage in the conversion of interlayer sites between smectite layers.

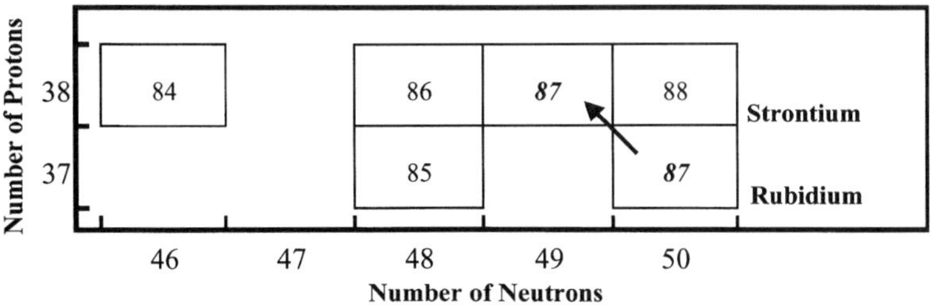

Fig. 3 Portion of the periodic table showing the naturally occurring isotopes of Rb and Sr as well as indicating (by arrow) the nuclear transformation accompanying the decay of ^{87}Rb.

GEOCHRONOLOGY: ISOTOPE SYSTEMATICS

There are several textbooks on isotope geology in which principles, methods and applications are discussed at length (Dalrymple & Lanphere, 1969; Faure, 1986; Clauer & Chaudhuri, 1995; Dickin, 1995; McDougall & Harrison, 1999). Chapter 5 of Emery & Robinson (1993) provides a useful overview of the application of radiogenic isotope systems to the study of sediments. Some description is given below but only in so far as it relates to the subject matter of this paper.

Geochronology is the study of the timing of formation of rocks and minerals and of the timing of the processes that affect them. Naturally occurring radioactive isotopes provide the basis for absolute methods of age determination. Isotopes are atoms of an element that have differing numbers of neutrons in the nucleus (see Fig. 3 & 6). Some isotopes are inherently unstable and decay to a stable isotope. Decay is characterized by loss or gain of atomic particles and emission of thermal energy. It is described statistically by the rate of decay of the number of atoms of a radioactive isotope ($-\mathrm{d}N/\mathrm{d}t$) being proportional to the number (N) of those atoms present (Rutherford & Soddy, 1903), i.e.

$$-\frac{\mathrm{d}N}{\mathrm{d}t} = \lambda N \qquad (1)$$

where λ is the decay constant, or probability of any particular atom decaying per unit time. Integration of equation (1) leads to

$$N = N_\mathrm{i}\mathrm{e}^{-\lambda t} \qquad (2)$$

where N_i is the initial number of radioactive atoms present when $t = 0$, and N is the number remaining after time $t = t$. The half-life, $t_{1/2}$, is the amount of time for a number (N) of radioactive atoms to decay to half that number ($N/2$). The decay constant λ is inversely proportional to the half-life of the radioactivity, as can be seen from the equations below.

When $t = t_{1/2}$

$$N = \frac{N_i}{2} \qquad (3)$$

Substitution of the above into equation (2) and integration gives

$$\frac{\ln 2}{\lambda} = t_{1/2} = \frac{0.693}{\lambda} \qquad (4)$$

The number of daughter nuclei (D^*) formed in time t can be expressed as

$$D^* = N_\mathrm{i} - N \qquad (5)$$

$$D^* = N(\mathrm{e}^{\lambda t} - 1) \qquad (6)$$

This can be rearranged in the form

$$t = \frac{1}{\lambda}\cdot\ln\left(1 + \frac{D^*}{N}\right) \qquad (7)$$

The total number of daughter atoms (D) is equal to the number of non-radiogenic daughter atoms initially present (D_i) plus those produced by decay, i.e.

$$D = D_\mathrm{i} + N(\mathrm{e}^{\lambda t} - 1) \qquad (8)$$

Knowledge of λ and D_i, and measurement of the isotope abundances, D and N, allows the value for t to be calculated and will represent the age of the system.

In practice it is isotope ratios that are measured directly rather than the absolute isotope abundances. Isotope ratios are measured in a mass spectrometer, which produces ions from the element to be analysed. These are accelerated by application of high voltage into a flight tube that passes through a magnetic field. Ions of different mass-to-charge ratio follow different paths through the magnetic field and can be focused into one or more collectors, where the intensity of each ion beam can be measured and related to an isotope ratio.

THE AGE CONCEPT

In order for a decay scheme to retain age information the stable daughter product must be quantitatively retained in the rock or mineral of interest. Total or partial loss of the daughter isotope may occur during chemical or physical alteration processes, such as recrystallization, dissolution and thermally activated diffusion. With respect to thermal diffusive loss, there will be a critical temperature above which the daughter isotope begins to be lost by diffusion. Below this temperature, accumulation occurs quantitatively. It is termed the closure temperature and a radiometric 'age' is the time at which the rock or mineral began to accumulate the daughter isotope below this temperature. Closure temperature is different for different decay schemes and for different minerals. For most diagenetic scenarios and decay schemes in common use diagenetic minerals form below their closure temperatures and the age recorded will be that of mineral formation. However, diagenetic clay minerals may recrystallize during burial, be heated to a temperature above that of closure or undergo structural and/or chemical modifications. All such processes may result in a partial or total 'resetting' of the age. Interpretation then becomes a matter of whether the 'age' has real temporal meaning in the context of geological environment of the samples subjected to geochronological analysis.

Rb–Sr METHOD

Isotope systematics

Of the two naturally occurring rubidium isotopes, ^{87}Rb comprises 27.83% and is unstable. It decays, with a half-life of 48.8×10^9 yr, by electron (β^-) emission to form stable ^{87}Sr (Fig. 3). The value of the decay constant, λ, is 1.42×10^{-11} yr^{-1} (Steiger & Jäger, 1977). Rubidium readily substitutes for potassium (K), another alkali metal, because of the geochemical similarity (i.e. same ionic charge and similar ionic radius) and therefore is particularly enriched in K-bearing minerals. Similarly, strontium (Sr), an alkaline earth metal, substitutes readily for calcium (Ca).

For the Rb–Sr system equation (8) becomes:

$$^{87}\mathrm{Sr} = {}^{87}\mathrm{Sr}_i + {}^{87}\mathrm{Rb}(e^{\lambda t} - 1) \qquad (9)$$

where ^{87}Sr and ^{87}Rb are concentrations in the phase of interest, $^{87}\mathrm{Sr}_i$ is the concentration at the time of closure, λ is the decay constant for ^{87}Rb in yr^{-1} and t is the time elapsed in years since the mineral last became a closed system. The ^{87}Sr and ^{87}Rb concentrations are measured with reference to the stable ^{86}Sr isotope, the abundance of which relative to the other non-radiogenic Sr isotopes is constant. Dividing through by ^{86}Sr concentration, equation (9) becomes:

$$\frac{^{87}\mathrm{Sr}}{^{86}\mathrm{Sr}} = \left(\frac{^{87}\mathrm{Sr}}{^{86}\mathrm{Sr}}\right)_i + \frac{^{87}\mathrm{Rb}}{^{86}\mathrm{Sr}}(e^{\lambda t} - 1) \qquad (10)$$

Analytical methods

Following sample digestion with hydrofluoric acid and dissolution in hydrochloric acid, Rb and Sr are separated by ion exchange chromatography. Prior to acid digestion, a known weight of sample is spiked with known amounts of ^{87}Rb- and ^{84}Sr enriched solutions. The Rb and Sr are analysed by mass spectrometry using the isotope dilution method to yield the concentrations and the $^{87}\mathrm{Sr}/^{86}\mathrm{Sr}$ ratio. Ratios of $^{87}\mathrm{Sr}/^{86}\mathrm{Sr}$ are normalized to $^{86}\mathrm{Sr}/^{88}\mathrm{Sr} = 0.1194$, a procedure which corrects not only

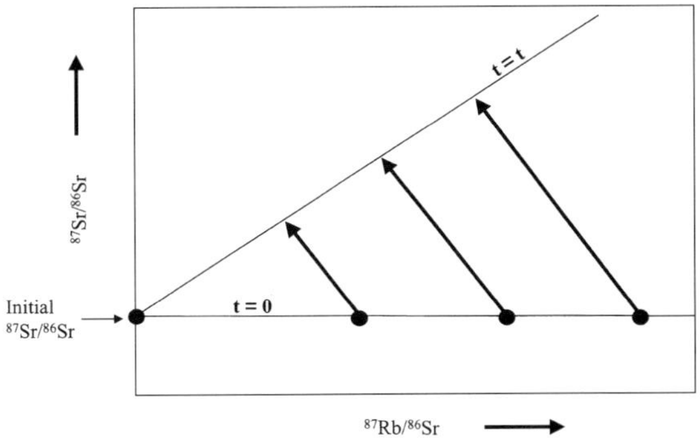

Fig. 4 The Rb–Sr isochron diagram. A suite of cogenetic samples at the time of formation ($t = 0$) differ in $^{87}Rb/^{86}Sr$ ratio but are in isotopic equilibrium with the same $^{87}Sr/^{86}Sr$. With time the amounts of ^{87}Rb and the $^{87}Rb/^{86}Sr$ ratios decrease with proportional increases in the amounts of ^{87}Sr and the $^{87}Sr/^{86}Sr$ ratios. After time $t = t$ the samples are aligned along a straight line with a slope that is proportional to t and which has an intercept on the y-axis equal to the starting composition of $^{87}Sr/^{86}Sr$.

for isotopic fractionation during the course of mass spectrometric analysis but also for any minor fractionations that the sample may have experienced previously in nature or during the preparation procedures.

Application of the Rb–Sr system—the isochron method

Equation (10) is that of a straight line ($y = mx + c$), where $^{87}Sr/^{86}Sr = y$, $^{87}Rb/^{86}Sr = x$, $(e^{\lambda t} - 1) = m$ = slope, and $(^{87}Sr/^{86}Sr)_i = c$ = intercept on the y-axis when $x = 0$. Plotting $^{87}Sr/^{86}Sr$ and $^{87}Rb/^{86}Sr$ data yields a diagram that may be used in two ways for dating purposes—isochrons and model ages. The first method uses what is termed an isochron plot and involves plotting the measured $^{87}Sr/^{86}Sr$ and $^{87}Rb/^{86}Sr$ ratios for a suite of samples believed to be cogenetic (Fig. 4). The regression line through the data points is called an isochron and the calculated slope is used to calculate the age. The intercept on the y-axis when $^{87}Rb/^{86}Sr = 0$ is the initial $^{87}Sr/^{86}Sr$.

There are several conditions that must prevail in order for this approach to be viable:

1 The isotope composition of rubidium is everywhere the same at the present day. Similarly the non-radiogenic isotope ratios of strontium are everywhere the same.
2 The samples analysed formed at the same time and were in isotopic equilibrium, i.e. with the same initial $^{87}Sr/^{86}Sr$ ratio.
3 The samples have remained closed to Rb and Sr gain or loss from time $t = 0$ to the present.
4 There is sufficient spread in $^{87}Rb/^{86}Sr$ for the isochron to be statistically well defined.
5 A minimum of three samples satisfying the above conditions is required.

Some of these assumptions are discussed briefly below.

Constancy of Rb isotopic composition

Apart from variations in isotopic composition that may arise from radioactivity the small mass differences between the isotopes of an element give rise to fractionation during physical, chemical and biological processes. Isotopic fractionation is dependent on the *relative* mass difference between the isotopes and therefore is larger for lighter elements than for heavier ones. Fractionation during mass spectrometric analysis is observable and can be corrected for by monitoring the difference between measured ratios for a non-radiogenic isotope pair and an assumed ratio for that isotopic pair (Dickin, 1995). Where there are only two isotopes, as for Rb, this procedure cannot be used. Instead Rb samples are analysed under similar conditions so as to achieve similar levels of fractionation and it is assumed that $^{85}Rb/^{87}Rb$ is everywhere the same at 2.593 (Catanzaro *et al.*, 1969). McDougall & Harrison (1999) most recently reviewed the data pertaining to the level of variation in potassium (K) isotope composition and concluded that essentially it is constant, with variations in $^{40}K/K$ being less than 1%. Rubidium, at more than double the atomic mass of K, can be expected to show even less variation in natural isotopic composition. It is reasonably assumed that natural variation in the isotopic composition of Rb has no significant impact on the precision of Rb–Sr ages.

Cogenetic sample suites

The Rb–Sr method often has been applied to the dating of igneous rocks (Faure, 1986; Dickin, 1995), where whole-rock sample suites and/or suites of separated minerals may be easily found that satisfy the conditions of having formed at the same time and in isotopic equilibrium. However, in the diagenetic environment, the timing of events where new phases are formed or pre-existing ones are altered is usually sequential rather than contemporaneous. Thus the cogenetic formation of three or more diagenetic phases is unlikely and the isochron approach is rarely applicable to dating diagenesis, although some exceptions are mentioned later. Straight lines on isochron diagrams can arise through mixing of two components with different isotope systematics in different proportions and any age significance should be interpreted with caution (Clauer & Chaudhuri, 1995).

Application of the Rb–Sr system—model ages

The second use of the $^{87}Sr/^{86}Sr$ versus $^{87}Rb/^{86}Sr$ plot is in the calculation of model ages.

The isochron equation (10) can be rewritten as

$$t = \frac{1}{\lambda}\ln\left[\frac{\frac{^{87}Sr}{^{86}Sr} - \left(\frac{^{87}Sr}{^{86}Sr}\right)_i}{\frac{^{87}Rb}{^{86}Sr}} + 1\right] \quad (11)$$

This equation can be used to calculate a model age (t) for a single sample if an initial $^{87}Sr/^{86}Sr$ ratio can be assumed. Graphically (Fig. 5) this equates to extrapolating from the data point of a phase with high $^{87}Rb/^{86}Sr$ ratio to an assumed initial $^{87}Sr/^{86}Sr$ ratio and calculating the age from the slope of the extrapolation line. In practice, however, it is difficult to constrain the $^{87}Sr/^{86}Sr$ ratio that would have prevailed during clay diagenesis. The exception would be the instance of clay-mineral formation known to have occurred at or very near the sea bed and at a time when the seawater $^{87}Sr/^{86}Sr$ ratio was well defined and reasonably could be assumed to be that with which the clay mineral formed.

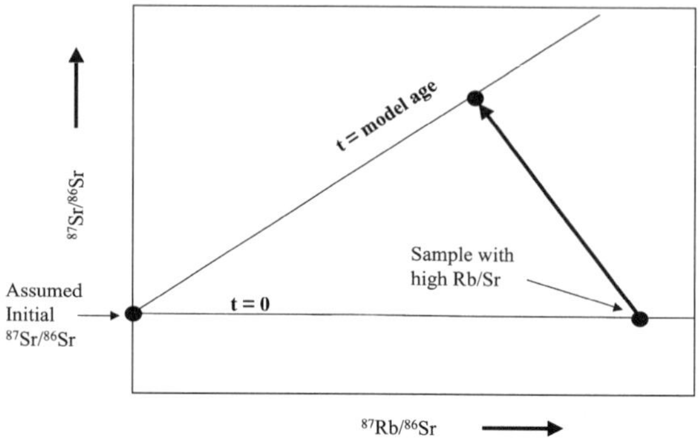

Fig. 5 Illustration of the use of a $^{87}Sr/^{86}Sr$ versus $^{87}Rb/^{86}Sr$ plot to calculate a model age (t) for a sample with a high $^{87}Rb/^{86}Sr$ ratio and an assumed initial $^{87}Sr/^{86}Sr$ ratio.

Table 1 Constants relevant to the ^{40}K–^{40}Ar decay scheme (Beckinsale & Gale, 1969; Garner *et al.*, 1975).

Parameter	Value
λ_e = decay constant – electron capture	$0.581 \times 10^{-10}\ yr^{-1}$
λ_{β^-} = decay constant – electron emission	$4.962 \times 10^{-10}\ yr^{-1}$
λ = decay constant – total	$5.543 \times 10^{-10}\ yr^{-1}$
Branching ratio = $\lambda_e/\lambda_{\beta^-}$	0.117
λ_e/λ = fraction of ^{40}K decay to ^{40}Ar	0.1048
$t_{1/2}$ = half-life	1.250×10^9 yr
^{40}K atomic abundance	0.01167%

K–Ar METHOD

Isotope systematics

Potassium has three naturally occurring isotopes (Fig. 6). Of these only ^{40}K is radioactive, has a half-life of 1.250×10^9 years and is the least abundant, comprising only 0.01167% of total K (Table 1). The decay is branched (Fig. 7) with 89.5% of the ^{40}K decaying to form ^{40}Ca and 10.5% to ^{40}Ar. Decay to ^{40}Ca is by electron (β^-) emission and forms the basis for the K–Ca geochronometer (see later). Most of the ^{40}K decay to $^{40}Ar^*$ (radiogenic ^{40}Ar) occurs through capture of an extranuclear electron by a proton,

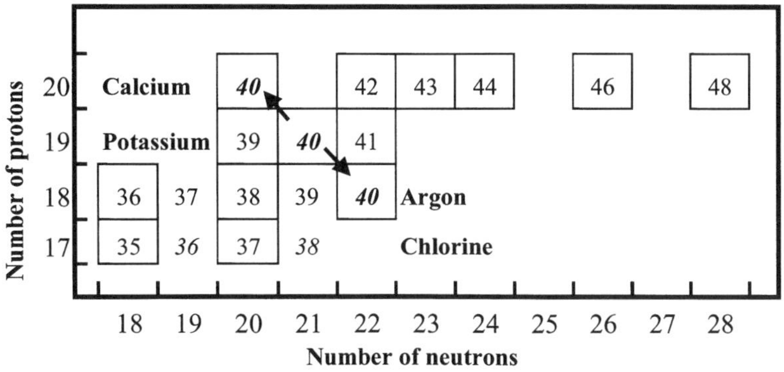

Fig. 6 Portion of the periodic table showing the naturally occurring isotopes of Ca, K, Ar and Cl as well as indicating the nuclear transformations (by arrows) accompanying the decay of ^{40}K.

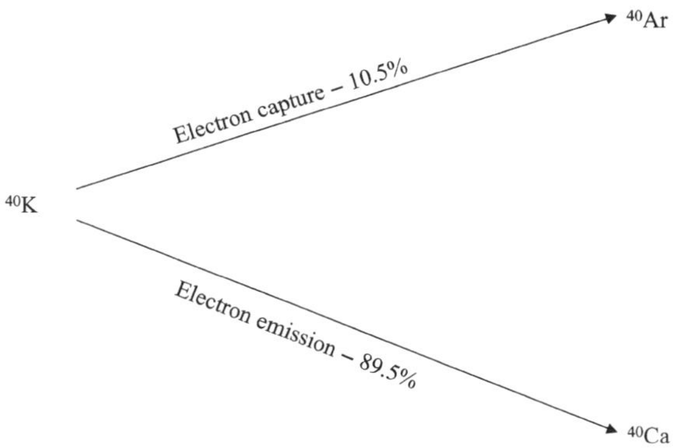

Fig. 7 The branched decay of ^{40}K to ^{40}Ca and ^{40}Ar.

transforming it to a neutron with emission of a neutrino. A very small proportion (0.001%) of decay occurs by emission of a positron (β^+). The various constants associated with ^{40}K radioactivity are listed in Table 1.

The fraction of ^{40}K that decays to Ar (f_{Ar}) is the ratio of the partial decay constant (λ_e) to the total decay constant (λ), i.e.

$$f_{Ar} = \frac{\lambda_e}{\lambda_e + \lambda_{\beta^-}} = \frac{\lambda_e}{\lambda} \tag{12}$$

For the K–Ar system, equation (7) is modified to reflect the branched decay and yields

$$t = \frac{1}{\lambda}\ln\left[\frac{^{40}\text{Ar}^*}{^{40}\text{K}} \cdot \left(\frac{\lambda}{\lambda_e}\right) + 1\right] \tag{13}$$

Thus, measurement of ^{40}K and ^{40}Ar* and knowing the decay constants allows an age (t) to be determined.

Analytical methods

Potassium and ^{40}Ar concentrations are determined on separate sample aliquots and in duplicate if sample availability permits. Potassium concentration may be determined in a number of ways, although flame photometry and atomic absorption emission spectrometry are the most common. The concentration of ^{40}Ar is determined by mass spectrometry using the isotope dilution technique. A sample is preheated *in vacuo* to drive off adsorbed atmospheric gases. There appears to be no concensus on the best preheating procedure for minimizing atmospheric Ar contributions. There are significant differences in published temperatures and times used for the preheating process, ranging, for example, from 80°C for several hours (Zwingmann *et al.*, 1998) to 160°C for 24 h (Robinson *et al.*, 1993). Argon makes up about 1% of Earth's atmosphere and this component may comprise a significant proportion of the total Ar measured. The atmospheric Ar is derived from that adsorbed on the sample and from interior surfaces of the gas extraction line and the mass spectrometer. The sample is then heated until fused in order to release the gases from the lattice structures of the sample. A known amount of a ^{38}Ar spike is added to the released gases prior to the Ar being isolated in a pure state for mass spectrometric analysis.

Two Ar isotope ratios are measured, ^{40}Ar/^{38}Ar and ^{38}Ar/^{36}Ar. From these, and the known amounts of ^{38}Ar spike and sample used, the total amounts of ^{40}Ar and ^{36}Ar can be calculated. The total ^{40}Ar will include contributions of ^{40}Ar other than that produced from ^{40}K decay in the sample, i.e.

$$^{40}Ar_t = {}^{40}Ar_a + {}^{40}Ar_i + {}^{40}Ar^* \tag{14}$$

where $^{40}Ar_t$ is the measured amount of ^{40}Ar following sample fusion, $^{40}Ar_a$ is the ^{40}Ar derived from the atmosphere and released during sample fusion, $^{40}Ar_i$ is the 'excess' ^{40}Ar, i.e. ^{40}Ar trapped in the sample at the time it was formed, and $^{40}Ar^*$ is the ^{40}Ar formed by ^{40}K decay in the sample.

The relationship between total ^{40}Ar measured and ^{40}K is derived from equations (7) and (14):

$$^{40}Ar_t = {}^{40}Ar_a + {}^{40}Ar_i + {}^{40}K\left(\frac{\lambda_e}{\lambda}\right) \cdot (e^{\lambda t} - 1) \tag{15}$$

Dividing through by ^{36}Ar and using the atmospheric $^{40}Ar/^{36}Ar$ ratio of 295.5 (Nier, 1950), equation (14) can be expressed as:

$$\left(\frac{^{40}Ar_t}{^{36}Ar}\right) = \left(\frac{^{40}Ar^*}{^{36}Ar}\right) + 295.5 + \left(\frac{^{40}Ar_i}{^{36}Ar}\right) \tag{16}$$

and equation (15) as

$$\left(\frac{^{40}Ar_t}{^{36}Ar}\right) = 295.5 + \left(\frac{^{40}Ar_i}{^{36}Ar}\right) + \left(\frac{^{40}K}{^{36}Ar}\right) \cdot \left(\frac{\lambda_e}{\lambda}\right) \cdot (e^{\lambda t} - 1) \tag{17}$$

Application of the K–Ar system—the assumptions

There are a number of assumptions that are made in applications of the K–Ar decay scheme.

1 The isotope composition of K is everywhere the same at the present day. The matter of K isotope ratio variability has been alluded to earlier and discussed by Hamilton *et al.* (1989) and McDougall & Harrison (1999), who concluded that the extent of natural $^{40}K/K$ variability is so small as to have negligible effect on measured K–Ar ages.

2 An analysed sample contains no K- or radiogenic Ar-bearing contaminants. This probably is the single most difficult impediment to successful use of the K–Ar decay scheme for dating clay-mineral cements. Many publications (e.g. Liewig *et al.*, 1987; Hamilton *et al.*, 1989; Clauer *et al.*, 1992a; Pevear, 1992, 1999) present and discuss examples of contamination problems that may arise in K–Ar dating of diagenetic illite. Volumetrically small amounts of contamination with old K-bearing phases may have a disproportionate effect because of the high radiogenic ^{40}Ar in such contaminants. For example 1 wt% addition of 1800 Ma K-feldspar to 50 Ma diagenetic illite would result in an apparent age of 87 Ma—more than 50% in error. This volumetrically low level of contamination may be difficult to detect by XRD, but may be revealed more readily by TEM analysis.

Hamilton *et al.* (1992) noted that fine-grained illite supposedly of diagenetic origin separated from Brent Group sandstones with detrital illite often gave older ages than for those from 'clean' sandstones and ascribed this effect to contamination. As a first step it is recommended (Emery & Robinson, 1993) that separation of a specific clay mineral for dating not be attempted for sandstone samples containing potential contaminants such as detrital mica, illite and feldspar. This does not necessarily mean that samples with more than one K-bearing diagenetic clay mineral should be avoided. Use of a high-gradient magnetic separation technique may assist in separating such mixtures of K-bearing authigenic clays. Recently Liewig & Clauer (2000) have physically isolated different morphologies of authigenic illite, which from the K–Ar results appear to have formed at different times.

Even inclusion of other phases that supposedly are neither K- nor Ar-bearing may provide sources of contamination. It is not uncommon for admixed kaolinite to be reported as present in illite separates that have been analysed for K–Ar age. Kaolinite has no essential K in its structure. However, Emery & Robinson (1993) report a Rb content of 83 ppm in a kaolinite leached with 1 N ammonium acetate from a Permian sandstone from the southern North Sea. If this Rb is indeed part of the kaolinite crystal structure then a low level of essential K

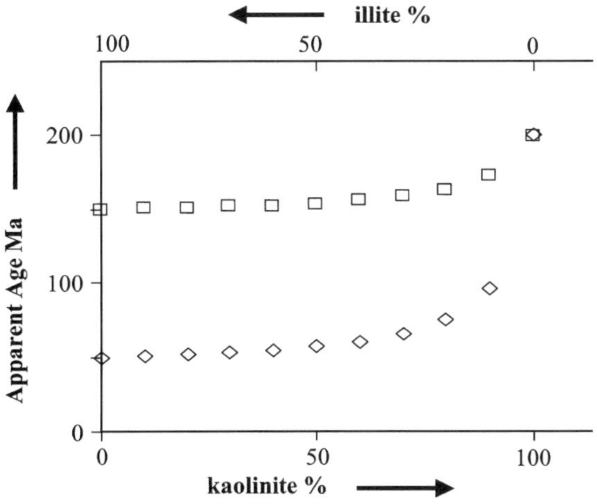

Fig. 8 Illustration of the effect on apparent illite age of mixing 200 Ma kaolinite with 0.25 wt% K with 150 Ma and 50 Ma illites with 5 wt% K.

is likely, given the geochemical similarity with Rb. Figure 8 shows the effect of admixing 200 Ma kaolinite of 0.25 wt% K content with 150 Ma and 50 Ma illites of 5 wt% K. Because the kaolinite is so low in K and radiogenic ^{40}Ar, large amounts (30–50%) of this contaminant may be tolerated without the mixed age being more than a few per cent in error from the true illite age. Contamination with kaolinite at these levels could be responsible for some of the 'noise' observed for regional data sets of illite ages.

3 An analysed sample consists of a single, homogeneous K-bearing mineral. This refers to use of ^{40}K–^{40}Ar dating where the K and Ar have to be measured on separate aliquots requiring sample homogeneity. Where sample size permits, analyses are conducted in duplicate to ensure sample heterogeneity.

4 There has been no gain or loss of K or $^{40}Ar^{*}$, other than by decay of ^{40}K, since the last time the mineral became a closed system. Velde & Renac (1996) have discussed this issue in the context of the meaning of K–Ar ages for samples that have experienced illitization of mixed-layer illite–smectite to varying degrees. As this is a process that involves the reorganization of crystal lattice structures that have previously hosted sites of ^{40}K decay, there will be opportunities for open-system loss of daughter Ar isotope product as well as redistribution of K within the new lattice structures.

Lattice reorganization associated with the polytypic transition of 1Md to 2M illite does promote loss of Ar (Hunziker *et al.*, 1986), such that once the transition has been fully accomplished the K–Ar age is entirely reset. Part of the age resetting appears to be initiated by thermal diffusive loss of Ar starting at about 180°C, before there is any evidence of the polytypic transition having been started (Hunziker *et al.*, 1986). The question of whether K–Ar ages of diagenetic clays can be lowered or even reset by thermal diffusive loss of Ar has been addressed by many authors (e.g. Evernden *et al.*, 1961; Odin & Bonhomme, 1982; Zimmermann & Odin, 1982; Hunziker, 1986; Hamilton *et al.*, 1989; Emery & Robinson, 1993; Ziegler *et al.*, 1994; Clauer *et al.*, 1995, 1996; Spötl *et al.*, 1996). The results of modelling thermal diffusive loss of Ar (Hamilton *et al.*, 1989) indicated that at the temperatures of generally less than about 180°C that prevail in most hydrocarbon reservoir sandstones there will be little effect on illite K–Ar ages through thermal diffusive loss of Ar. Clauer *et al.* (1995; and see also resultant discussions by Clauer *et al.* (1996) and Spötl *et al.* (1996)) correctly note that a meaningful K–Ar age obtains for an authigenic clay only when there is no diffusive loss of radiogenic daughter product. Clauer *et al.* (1995) further state that this can prevail only if the illite crystallized at a temperature greater than that reached at maximum burial. The inference of this statement is that for an example of illite authigenesis begun at 80°C, terminating at 120°C and then experiencing a subsequent burial temperature of say 150°C, there would be diffusive loss of Ar during the burial temperature interval of 120–150°C—in contradiction to the results of the diffusion modelling of Hamilton *et al.* (1989). The possibility of post-crystallization thermal diffusive loss of Ar from authigenic clays has been raised by some recent K–Ar illite age studies (e.g. Ziegler *et al.*, 1994; Liewig & Clauer, 2000). The strongest supporting evidence comes from the former study, wherein oxygen isotope compositions of authigenic illites from the Rotliegend of the southern North Sea appear to have last equilibrated with present-day formation water. Coeval ankerite retains oxygen isotope evidence of precipitation from an earlier more ^{18}O-depleted formation water. In order for the illites to have re-equilibrated their oxygen isotope composition, it would require breaking of the Si–O tetrahedrally coordinated bonds, a process that is likely to have promoted diffusive loss of Ar during such lattice rearrangement.

5 Any Ar incorporated into the mineral at the time of closure has the $^{40}Ar/^{36}Ar$ ratio of present-day atmospheric Ar. That is, there is no ^{40}Ar in excess to that accounted for by ^{40}K decay since closure and by atmospheric contamination.

Application of the K–Ar system—the isochron method

For the K–Ar system, equation (17) is the isochron equation

$$\left(\frac{^{40}Ar_t}{^{36}Ar}\right) = 295.5 + \left(\frac{^{40}Ar_i}{^{36}Ar}\right) + \left(\frac{^{40}K}{^{36}Ar}\right) \cdot \left(\frac{\lambda_e}{\lambda}\right) \cdot (e^{\lambda t} - 1) \quad (17)$$

As noted previously for the isochron equation, it has the form of a straight line; $y = 295.5 + c + xm$. When the K–Ar data for a suite of cogenetic samples are plotted on the isochron diagram (Fig. 9) they will define a straight line with a slope, m, that yields the age, t, and an intercept with a value of $(295.5 + c)$. The value of c is $^{40}Ar_i/^{36}Ar$ and represents the excess ^{40}Ar trapped in the samples at the time of formation. The intercept of the extrapolated regression line on the isochron diagram does, however, usually have a large error, and recognition of excess ^{40}Ar often will be highly uncertain.

As in discussion of the isochron method for the Rb–Sr system there are certain conditions to be met in order for a set of K–Ar data to conform to an isochron. These include:

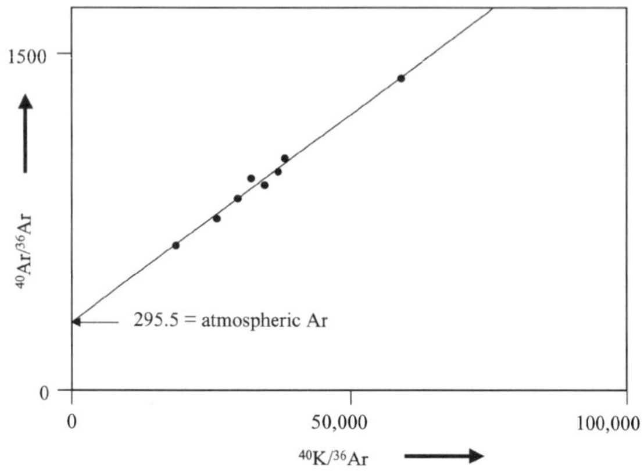

Fig. 9 K–Ar isochron diagram plotted as $^{40}Ar/^{36}Ar$ versus $^{40}K/^{36}Ar$ for a suite of supposedly cogenetic samples.

1 the samples analysed formed at the same time and were in isotopic equilibrium;
2 there is sufficient spread in $^{40}K/^{36}Ar$ for the isochron to be statistically well defined;
3 a minimum of three samples satisfying the above conditions is required.

There are two major obstacles to the existence and recognition of an isochronous relationship for the K–Ar systems of a suite of diagenetic minerals.

First, and as discussed for the Rb–Sr system, in the diagenetic environment, cogenetic formation of three or more suitable diagenetic phases is unlikely and the isochron approach is likely to be rarely applicable. Second, many diagenetic phases form over a significant period of time. Different size fractions of diagenetic illite, for example, may show increasing K–Ar age with increase in grain size. If not resulting from the presence of contaminants in the coarser fractions, then the inference is that the illite has grown over a significant period of time, typically of at least 5–10 Ma (Lee *et al.*, 1989; Hamilton *et al.*, 1992; Hogg *et al.*, 1993).

The isochron approach applied to a suite of minerals, if they are indeed cogenetic, can be used to assess whether any excess ^{40}Ar exists in the minerals analysed. If none of the Ar present in the diagenetic environment is incorporated into a suite of cogenetic diagenetic mineral structures then the term $^{40}Ar_i/^{36}Ar$ disappears from the isochron equation (17), all of the ^{36}Ar is of atmospheric origin and the *y*-intercept will have the value of atmospheric $^{40}Ar/^{36}Ar$ at 295.5 (Fig. 9). The data array on the isochron diagram is then a result of mixing between the samples and atmospheric Ar.

An intercept on the *y*-axis of the K–Ar isochron diagram at greater than the atmospheric $^{40}Ar/^{36}Ar$ of 295.5 would be indicative of the incorporation of an excess ^{40}Ar component in the sample suite at the time of formation. This requires all the samples to have formed in isotopic equilibrium and therefore have the same $^{40}Ar/^{36}Ar$ at the time of closure.

One method of assessing the presence of excess ^{40}Ar for a suite of samples is to use a plot of total ^{40}Ar against ^{40}K (Shafiqullah & Damon, 1974). A positive intercept on the *y*-axis at zero ^{40}K is indicative of the presence of excess ^{40}Ar (Fig. 10). Robinson *et al.* (1993) used this type of plot for K–Ar data for illites from Permian sandstones of the southern North Sea (Fig. 10). The positive intercept on the *y*-axis has an associated error that includes the origin and therefore is equivocal as to whether the presence of excess ^{40}Ar is indicated. Use of such plots assumes the sample suite is cogenetic. The scatter on Fig. 10 suggests that this is not the case and given that the samples used are from different depths, from different wells and of differing grain size it is unlikely that they were all formed at the same time. Therefore the assumptions underlying use of isochron plots are not met and the data set is unsuitable for assessing the presence of excess ^{40}Ar. Incorporation of excess ^{40}Ar into diagenetic clay minerals has not yet been documented.

Application of the K–Ar system—model ages

Model ages for a single K-bearing mineral sample may be calculated using

$$t = \frac{1}{\lambda}\ln\left[\frac{^{40}Ar^*}{^{36}Ar}\cdot\left(\frac{\lambda}{\lambda_e}\right)+1\right] \qquad (13)$$

Single sample K–Ar ages for illitic mineral separates has been the most commonly applied geochronological method for dating diagenetic events in sandstones from hydrocarbon reservoirs. The assumptions underlying its use have been outlined and discussed in preceding sections. Despite the various limitations it has proven successful in revealing patterns of regional fluid flow in basins, the timing of hydrocarbon accumulations and improving the understanding of diagenetic processes (e.g. Lee *et al.*, 1985, 1989; Jourdan *et al.*, 1987; Burley & Flisch, 1989; Girard *et al.*, 1989; Glasman *et al.*, 1989a, b; Hamilton *et al.*, 1987, 1989, 1992; Walgenwitz *et al.*, 1990; Burley & MacQuaker, 1992; Clauer *et al.*, 1992a; Mossmann *et al.*, 1992; Hogg *et al.*, 1993; Ziegler *et al.*, 1994; Darby *et al.*, 1997; Zwingmann *et al.*, 1998; Liewig & Clauer, 2000).

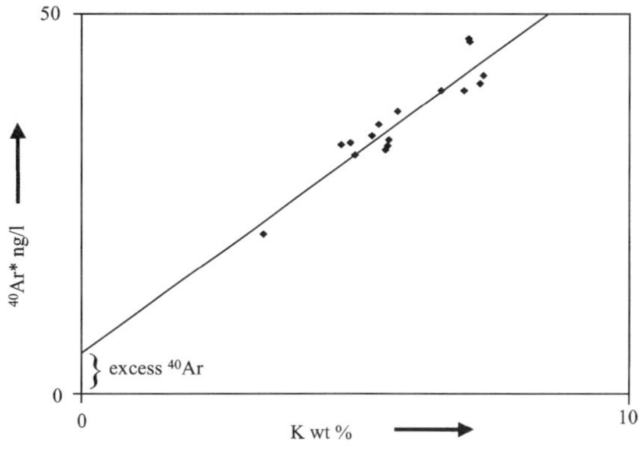

Fig. 10 Plot of total ^{40}Ar versus ^{40}K. The positive intercept on the *y*-axis indicates the existence and amount of excess ^{40}Ar. The error on this intercept associated with the regression line includes the origin so the evidence for excess ^{40}Ar is equivocal (after Robinson *et al.*, 1993).

$^{40}Ar/^{39}Ar$ METHOD

There has been extensive application of this method to the study of a wide range of planetary materials, ranging from meteorites and lunar rocks to terrestrial igneous and metamorphic rocks. It has yet to be applied extensively to the study of the timing of clay diagenesis, particularly in sandstones, although in recent years there have been some promising developments. These include the possibility of *in situ* dating of clays in thin-sections using a laser microprobe.

The $^{40}Ar/^{39}Ar$ method has the potential to overcome some of the drawbacks and limitations of the K–Ar system for dating diagenetic clays. One of the major drawbacks with application of the K–Ar method to dating diagenetic clays is the need for separate measurement of the K and Ar, requiring separate sample aliquots. Further, these analyses should be undertaken in duplicate, if sample size permits, in order to ensure homogeneity of K and Ar distribution in the separated mineral. This in turn often necessitates relatively large sample sizes of tens if not hundreds of milligrams and the associated time-consuming processes of mineral separation, grain-size fractionation and monitoring sample purity. In marked contrast is the potential with $^{40}Ar/^{39}Ar$ dating of obtaining ages for individual glaucony grains (Smith *et al.*, 1993), ages for clay-mineral sample sizes of 1–100 μg (e.g. Dong *et al.*, 1995; Onstott *et al.*, 1997) and ages for clay-minerals *in situ* on thin-sections (Dong *et al.*, 1997).

Isotope systematics and analytical methods

Several texts present aspects of the method, the most detailed and recent being that of McDougall & Harrison (1999). Some specific applications to diagenetic clays are usefully summarized by Emery & Robinson (1993). Merrihue & Turner (1966) were the first to document the principles of $^{40}Ar/^{39}Ar$ dating following the observation of the production of ^{39}Ar from ^{39}K by neutron irradiation of meteorite samples in a nuclear reactor. Potassium-39 is a stable isotope of K and is the most abundant at 93.26% (Garner *et al.*, 1975). This means that the ^{39}Ar can act as a proxy for ^{39}K. Assuming that $^{40}K/^{39}K$ is constant, then both the ^{40}K and ^{40}Ar in a sample can be measured by an Ar isotope ratio alone. As isotope ratios can be measured much more precisely than elemental abundances, the $^{40}Ar/^{39}Ar$ method offers the possibility of obtaining more precise ages and using much smaller sample sizes than possible for the K–Ar method. It should be noted that a $^{40}Ar/^{39}Ar$ age does not offer additional information compared with a K–Ar age and is still subject to the same uncertainties and assumptions. Other nuclear reactions occur during irradiation that produce Ar isotopes that are interferences and need to be corrected for. This particularly is so for minerals with a low K/Ca ratio because of the ^{36}Ar and ^{39}Ar isotopes produced by neutron irradiation of ^{40}Ca.

A $^{40}Ar/^{39}Ar$ age is calculated from the equation

$$t = \frac{1}{\lambda}\left(\frac{^{40}\mathrm{Ar}^*}{^{36}\mathrm{Ar}} \cdot J + 1\right) \qquad (18)$$

where J is a dimensionless parameter that depends on the $^{40}K/^{39}K$ ratio (assumed constant), ^{40}K decay constants, time of irradiation, neutron flux at the sample site in the reactor and neutron capture cross-section for the conversion of ^{39}K to ^{39}Ar. Because severe flux gradients can exist in the reactor, it is difficult to calculate the exact neutron dose that a sample has received. Following Merrihue & Turner (1966) and Mitchell (1968) it is normal practice to include standards of known K/Ar age with the samples to be irradiated. These standards are effectively flux monitors. For each standard, using the known age and measured $^{40}Ar^*/^{39}Ar$ ratio, equation (18) yields a J value appropriate to the position of that standard in the neutron flux. Values of J can then be interpolated for each sample from the known spatial relationships between standards and samples.

Step heating

Rather than releasing all the Ar for analysis by sample fusion, it is possible to release the gas for analysis in discrete volumes by progressive heating to higher temperatures. A $^{40}Ar^*/^{39}Ar$ age can be calculated for each gas volume. The method proposed by Merrihue & Turner (1966) is called step heating and previously had been used to study Xe gas release from meteorite samples that had been irradiated. A $^{40}Ar/^{39}Ar$

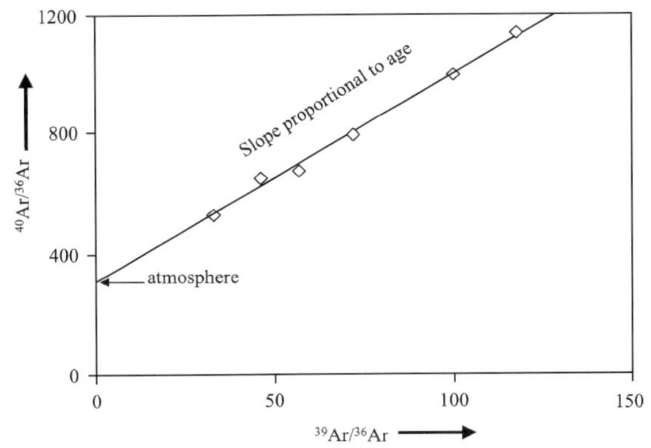

Fig. 11 $^{40}Ar/^{36}Ar$ versus $^{39}Ar/^{36}Ar$ isochron diagram for a step-heated sample.

age based on analysis of all the Ar released by sample fusion is called a 'total gas age' and should be identical to the K–Ar age. Unlike a single K–Ar age or total gas age, the step-heating data can be plotted on an isochron diagram comprising a plot of $^{40}Ar/^{36}Ar$ versus $^{39}Ar/^{36}Ar$ (Fig. 11). An intercept on the *y*-axis at greater than the atmospheric $^{40}Ar/^{36}Ar$ ratio of 295.5 would be indicative of excess Ar. A disadvantage of this plot is the use of ^{36}Ar as common to both axes and yet this isotope is generally measured with the poorest precision compared with the other Ar isotopes. An isochron plot of $^{36}Ar/^{40}Ar$ versus $^{39}Ar/^{40}Ar$ (Fig. 12) is preferable (Roddick *et al.*, 1980) as ^{40}Ar, common to both axes, is measured with much better precision. The age is derived from the value of $^{39}Ar/^{40}Ar$ at the intercept of the regression line with the *x*-axis. The trapped or initial $^{36}Ar/^{40}Ar$ is indicated by the *y*-intercept.

The other way of displaying step-heating data is as an age spectrum or $^{40}Ar/^{39}Ar$ release pattern comprising a plot of $^{40}Ar/^{39}Ar$ age for each gas aliquot against the cumulative amount of ^{39}Ar released (Fig. 13). Radiogenic ^{40}Ar and ^{39}Ar are both derived from K and will occupy similar sites within mineral crystals and have similar diffusion rates during gas extraction. Thus for a mineral system that has been closed to diffusive loss of Ar, the step-heating process should yield the two Ar isotopes in the same ratio for each gas aliquot, and therefore all

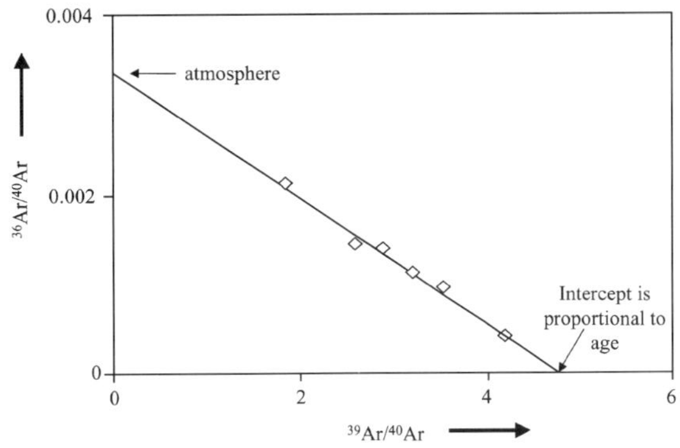

Fig. 12 $^{36}Ar/^{40}Ar$ versus $^{39}Ar/^{40}Ar$ isochron diagram for a step-heated sample.

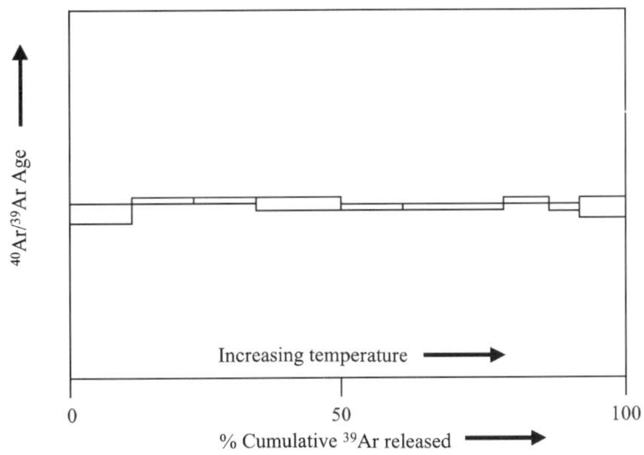

Fig. 13 $^{40}Ar/^{39}Ar$ age for each gas aliquot against cumulative amount of ^{39}Ar released during a step-heating experiment.

will also have the same age. The resultant age spectrum has a flat plateau indicative of closed-system behaviour since mineral crystallization (Fig. 13).

The calculation of the age for each gas aliquot is dependent on the assumption that trapped Ar has atmospheric composition. If this is not the case then the Ar age spectrum will be much more complex (Fig. 14) and the presence of excess Ar will be indicated by *y*-intercepts on the isochron plots of higher non-radiogenic ^{40}Ar contents than for the atmosphere (Figs 11 & 12). The straight lines on these plots are mixing lines between the radiogenic Ar component and the trapped component. The age is

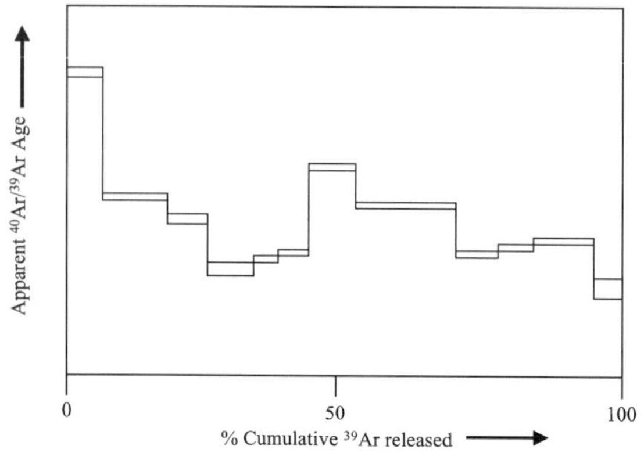

Fig. 14 $^{40}Ar/^{39}Ar$ age spectrum plot where no plateau age is apparent and individual step ages have no meaning. Such profiles are indicative of sample disturbance since closure and the presence of excess ^{40}Ar.

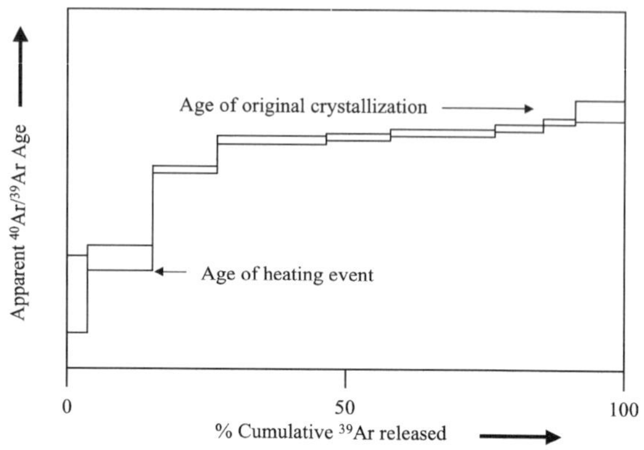

Fig. 15 $^{40}Ar/^{39}Ar$ age spectrum plot for a sample that has experienced some thermal diffusive loss, with the age of the heating event being reflected in the low-temperature step-heating results, whereas the original age of crystallization is recorded in the higher temperature steps.

obtained from the slope of the regression line on the $^{40}Ar/^{36}Ar$ versus $^{39}Ar/^{36}Ar$ plot and from the intercept on the *x*-axis from the $^{36}Ar/^{40}Ar$ versus $^{39}Ar/^{40}Ar$ plot.

A significant advantage of the step-heating technique is that it allows recognition of portions of minerals that differ in their susceptibility to thermal diffusive loss of Ar. The edges of grains and crystal defects will yield their Ar more rapidly and at lower temperatures than the interior of grains. If a sample has been heated sufficiently after crystallization to cause thermal diffusive loss of Ar, then less Ar-retentive sites will have lower $^{40}Ar^*/^{40}K$ ratios than the more retentive sites. As the diffusion rate for K is much lower than for Ar, the different sites are, however, likely to retain all their K. Thus after irradiation all sites will have the full complement of ^{39}Ar derived from K, but will vary in radiogenic ^{40}Ar depending on site retentivity during heating and thermal diffusion. This will be revealed by successively increasing $^{40}Ar^*/^{39}Ar$ ratios during step heating, resulting in an age spectrum that has a staircase shape (Fig. 15). The lower temperature gas release will be mainly from the sites that are more readily degassed. The $^{40}Ar/^{39}Ar$ age corresponding to the low-temperature gas release may be indicative of the age of the heating event, whereas the ages corresponding to the highest temperature releases may constrain the original age of crystallization. Such interpretations of non-uniform age spectra depend on the applicability of a model to the diffusion behaviour of Ar in the mineral phase of interest. Diffusion mechanisms and domains during step heating *in vacuo* need to mimic the behaviour in nature if thermochronological information is to be reconstructed from such age spectra. Hassanipak & Wampler (1996) found that two activation energies were required to model the stepwise release of Ar from glauconite and illite samples and suggested that this related to the mineralogical heterogeneity common to these minerals. Release of Ar from many clay minerals is likely to be related variously to the degree of crystallinity, or density of crystal defects, grain size, composition and temperatures of dehydroxylation. Quantitative interpretation of such age spectra for diagenetic clays is likely to be difficult. However, if diffusive Ar loss from finest size illite-rich and glauconite-rich clay separates from sandstones is insignificant at temperatures of less than about 180°C, then such age spectra are unlikely.

Laser microprobe heating

Kelley (1995) and McDougall & Harrison (1999) provide useful summaries of the development of laser microprobe heating applied to $^{40}Ar/^{39}Ar$ dating. One approach to laser-use is the creation of ablation pits by pulsing laser beam incidence on the surface of the material to be analysed. This is followed by on-line mass spectrometric analysis to yield a total fusion $^{40}Ar/^{39}Ar$ age from minute (< 50 μm diameter) preselected and petrographically well characterized portions of rocks and minerals. York *et al.* (1981) were the first to demonstrate the use of a continuous wave laser for progressive heating of a sample surface to reproduce the step-heating method used for conventional $^{40}Ar/^{39}Ar$ dating. Most applications thus far have been applied to geochronological studies of meteorites, lunar rocks and terrestrial igneous and metamorphic rocks. Increased sensitivity of mass spectrometer analysis have parallelled the advances in laser technology such that sample sizes for laser microprobe dating need be in the region of only a few micrograms to obtain a total $^{40}Ar/^{39}Ar$ fusion age (Onstott *et al.*, 1997). Using these and other technical advances, there have been only a very few applications to dating clay diagenesis and these have used fine-grained sediments rather than sandstone. The results of these few applications, which are discussed later, are encouraging with respect to the possibility of obtaining true ages for diagenetic illitic clays in sandstones.

Irradiation in vacuo

Some of the potential problems with neutron irradiation of fine-grained samples involve loss of Ar isotopes from the mineral lattices (see below). Encapsulation in evacuated quartz vials prior to irradiation allows such Ar to be retained (Hess & Lippolt, 1986; Foland *et al.*, 1992; Smith *et al.*, 1993). At a minimum the encapsulation procedure allows a $^{40}Ar/^{39}Ar$ total gas age to be obtained from analysis of gas released from both the vial and the mineral sample during heating.

$^{40}Ar/^{39}Ar$ dating of clays—some important issues

There are some aspects of neutron irradiation and step heating for $^{40}Ar/^{39}Ar$ dating of clays that are important to consider. Of relevance is the discussion provided in McDougall & Harrison (1999) of the need for the irradiation parameters to be optimized.

Temperature effects during irradiation

Mitchell (1968) documented sample temperatures of up to *c.* 200°C during neutron irradiation and for most minerals this would not be considered sufficient to result in diffusional loss of Ar. However, because of the fine-grained nature of clay particles diffusional transport distances can be orders of magnitude smaller than in most other minerals subjected to irradiation. There are several examples where diffusive loss of Ar from clay particles during irradiation can be shown to have occurred. Halliday (1978) noted that the amount of loss of radiogenic ^{40}Ar of 25–35% from irradiated clay concentrates correlated with the J value, and attributed the losses to diffusion resulting from heating during irradiation. Hess & Lippolt (1986) also observed ^{40}Ar losses during irradiation and these similarly also have been ascribed to thermal diffusive losses during irradiation.

Mechanical compaction of sample clays into coarse aggregates could artificially increase the diffusional transport distance and lessen the potential for diffusive Ar loss. It has been argued that with both encapsulation and use of compacted clay separates some of the step-heating age-spectra data obtained may have geological meaning. These procedures also have been adopted in order to overcome the problems of ^{39}Ar recoil from fine-grained material (see later).

Phase transitions during step heating in vacuo

Interpretation of data from the step-heating technique is based on the assumption that release patterns of Ar can be interpreted as being a consequence of diffusion that has been thermally activated. Progressive Ar release with increasing temperature reflects the increasingly retentive domains within a mineral structure. Recognition of different domains of Ar retentivity through step heating of multimineralic assemblages may enable the distinction of and correction for contaminants in diagenetic clay separates (Onstott *et al.*, 1997). However, an assemblage of domains in a clay-mineral structure subjected to rapid heating and high temperatures in a vacuum may lose Ar not only by diffusion but also through lattice rearrangements that would accompany dehydroxylation and possibly polytypic transitions. All clay minerals contain structural hydroxyl groups and experience one or more dehydroxylation reactions with increasing temperature (Michael & McWhinnie, 1989; Drits *et al.*, 1996; Murad & Wagner, 1996). Evernden *et al.* (1960) and Foland *et al.* (1984) indicate that the pattern of Ar release during step heating of glauconies reflects mineral decomposition and that there is a close relationship between Ar loss and dehydroxylation in glaucony samples. Stepwise heating release of Ar is strongly related to mineral decomposition, including hydroxyl loss in biotites (Lo *et al.*, 2000) and muscovites (Sletten & Onstott, 1998) and their $^{40}Ar/^{39}Ar$ age spectra are often complex. In hornblendes Ar release and dehydroxylation are not coupled, with most of the Ar not being released until the hornblende is almost completely dehydrogenated. The $^{40}Ar/^{39}Ar$ age spectra for hornblendes appear to be meaningful. In contrast the view has been expressed in some recent publications (see later) that the step-heating profiles from mixtures of clay of differing ages do have structures from which it may be possible to extract $^{40}Ar/^{39}Ar$ ages of end-member components. It would seem advantageous for any step-heating Ar release patterns of mixtures of clay minerals to be examined in the context of their dehydroxlation reactions and other phase changes and grain size distributions (Sletten & Onstott, 1998).

^{39}*Ar recoil*

In order to avoid irradiation damage to a mineral lattice, that could result in changes to Ar retentivity and render difficult the interpretation of $^{40}Ar/^{39}Ar$ age spectra, a neutron flux of $< 10^{18}$ n cm^{-2} is used (McDougall & Harrison, 1999). However, this does not circumvent the problem that is termed recoil and relates to the ejection of Ar isotopes from grain surfaces. The neutron interactions that produce daughter Ar isotopes impart sufficient energy for them to be displaced from the lattice site of the parent atoms and if near the surface of a grain may be ejected into the surroundings. Depending on the neutron energy (which varies quite widely) the recoil distance for ^{39}Ar can range quite widely, but has an average value of about 0.1 μm (Turner & Cadogan, 1974, Onstott *et al.*, 1995).

The deficit in ^{39}Ar owing to recoil-loss is the difference between the measured ^{39}Ar in the sample and that expected for its K content. Other Ar isotopes are produced by neutron irradiation and these also are subject to recoil loss. As clay particles have dimensions of the same order as ^{39}Ar recoil distance it is readily appreciated that recoil has been a major impediment to many attempts to apply the $^{40}Ar/^{39}Ar$ method to dating clay minerals. Brereton *et al.* (1976) suggested that $^{40}Ar/^{39}Ar$ ages greater than K–Ar ages for glauconites were a result of ^{39}Ar recoil loss during irradiation. Foland *et al.* (1984) confirmed this feature of $^{40}Ar/^{39}Ar$ ages of glauconites with ^{39}Ar losses of between 17% and 29%. By using relatively low neutron fluxes and demonstrating retention of all ^{40}Ar they confirmed that the ^{39}Ar losses were the result of recoil rather than loss owing to radiation damage during irradiation. The ^{39}Ar that is lost from a lattice site may be implanted into a neighbouring crystal lattice site, but some will terminate on or between particle surfaces or in the sample container. Thus as well as the loss of ^{39}Ar from tightly bound lattice sites the distribution of ^{39}Ar in the sample after irradiation does not reflect the distribution of the parental ^{40}K. These authors further suggest that the $^{40}Ar/$

^{39}Ar age spectra obtained from step heating are meaningless, including those that appear to have reasonable plateaux segments, and reflect the effects both of recoil and mineral decomposition during the vacuum heating procedure.

Irradiation in vacuo

This is a procedure now commonly adopted that overcomes some of the problems of ^{39}Ar recoil loss (Hess & Lippolt, 1986; Foland *et al.*, 1992; Smith *et al.*, 1993). The sample is placed in a quartz vial that is then evacuated and sealed prior to irradiation. The purpose of the vial is to retain any ^{39}Ar ejected from the sample during irradiation. Total ^{40}Ar and ^{39}Ar degassed from both the vial and the sample are measured by mass spectrometry after irradiation. In this way $^{40}Ar/^{39}Ar$ total fusion ages are obtained that should agree with K–Ar ages but with the advantage of using very much smaller sample aliquots.

Kapusta *et al.* (1997) modified the vacuum-encapsulated technique for a glauconite standard and other clay-sized fractions of mixed mineralogies by hydraulic pressing of samples at 2.5 kg m^{-2} under low vacuum prior to encapsulation and irradiation for $^{40}Ar/^{39}Ar$ step-heating analysis. They demonstrated that the ^{39}Ar deficit in the analysed samples was reduced by reducing the adsorbed water content by one, or a combination, of preheating the sample at 110°C, compressing the samples and irradiation *in vacuo* as apposed to in air.

OTHER GEOCHRONOMETERS

There are many other geochronometers that are based on decay schemes with half-lives appropriate to the dating of diagenetic clays in sandstones. These include uranium–thorium–lead, rhenium–osmium, lutetium–hafnium, samarium–neodymium and potassium–calcium. For a variety of reasons none of these have ever been applied for this purpose. As with the Rb–Sr and K–Ar systems, lack of cogenetic samples of appropriate chemistries is a normal characteristic of diagenetic mineral assemblages and does not encourage the application of other geochronometers. Lack of constraint on initial ratios for model age calculation using these other geochronometers is another disincentive. Some of the procedures for separation and isotopic analyses of the parent and daughter elements are more technically challenging than for the more frequently applied Rb–Sr and K–Ar systems.

The K–Ca system

For the ^{40}K branched decay to ^{40}Ca (Fig. 7) the radiogenic daughter product is also the most common of the naturally occurring Ca isotopes. Application of the K–Ca geochronometer therefore is best suited to minerals with high K/Ca ratios (Marshall & DePaolo, 1982) and K-bearing clay minerals might appear to be amenable to its use. However, K–Ca dating of diagenetic clays has not yet been attempted. Few laboratories have established the technique so that knowledge of $^{40}Ca/^{42}Ca$ in the crust is sparse but would need to be constrained for calculation of model K–Ca ages. Calcium isotope ratio measurements are also more difficult than for many other elements, as the correction procedure for natural and instrumental fractionations is more complex (Dickin, 1995).

The Sm–Nd system

This geochronometer is based on the alpha-decay of ^{147}Sm to ^{143}Nd, with a relatively long half-life of 106 billion years. There is a general lack of dispersion of Sm/Nd ratios in sediments because both elements are of the same rare earth element (REE) group and therefore geochemically similar. This, together with the long half-life and the lack of constraint on initial $^{143}Nd/^{144}Nd$ for model age calculation, has militated against application of this geochronometer to dating diagenetic clays in sandstones. There has been some limited application to dating Precambrian and Palaeozoic sediments (Stille & Clauer, 1986; Bros *et al.*, 1992; Ohr *et al.*, 1994). The method does not yield precise ages and so far has been applied only to fine-grained sediments. It is evident that large fractionations of Sm/Nd ratio can accompany diagenetic reactions involving REE-enriched phases with high Sm/Nd ratios, such as apatite (Ohr *et al.*, 1994) or bitumen (Bros *et al.*, 1992). If a clay mineral were to form diagenetically at the same time, there is then the basis for an isochron approach to dating the diagenetic event. This approach has yet to be investigated for sandstones but will have only limited application as the event to be dated will need to be old (> 400 Ma; Ohr *et al.*, 1994) because of the long half-life of ^{147}Sm.

RADIOMETRIC DATING OF SPECIFIC DIAGENETIC CLAYS

Chlorite

Only one attempt to date diagenetic chlorite has been published (Emery & Robinson, 1993). Chlorite separates from five Triassic sandstones of the Central North Sea region were subjected to ^{40}Ar/^{39}Ar step heating. Neither the data nor isochron plots are provided in this reference but age spectra are shown. All five samples yielded 'saddle' shaped age spectra, with no well-defined plateau ages and with many of both the high- and low-temperature steps yielding ages older than that of deposition. None of the computed total fusion ^{40}Ar/^{39}Ar ages are geologically meaningful. For many of the intermediate temperature steps comprising between *c.* 20 and 50% of ^{39}Ar release, minimum ages of 41–65 Ma were realized. The authors note the compatibility of these 'ages' with an age of 74 ± 3 Ma for a diagenetic illite–smectite that petrographically appears to pre-date the chlorite. It seems likely though that the data are geologically meaningless, with the possibility that the data reflect the presence of excess ^{40}Ar and/or ^{39}Ar recoil. Further, even if a ^{40}Ar/^{39}Ar data set were obtained with a good, 'geologically reasonable' plateau age, without knowledge of where the K is precisely located in the chlorite structure it would be difficult to assign meaning to the ^{40}Ar/^{39}Ar data. It would be worthwhile to repeat this experiment using irradiation of vacuum encapsulated samples if coupled with detailed analytical documentation of the structural location of the K. Many analyses of diagenetic chlorites indicate low but significant and measurable K contents (Curtis *et al.*, 1985; Whittle, 1986; Hillier & Velde, 1991). Although such K may represent contamination such as intergrown illite it is acknowledged that it may be residual from a precursor smectitic phase.

Mixed-layer chlorite–smectite (corrensite)

Janks *et al.* (1992) obtained three clay separates (< 1.0 μm) with high contents of corrensite from Permian sandstones. The separates were left in an aggregated state prior to irradiation in evacuated vials in an attempt to reduce ^{39}Ar recoil losses. Ar was released by step heating with a laser microprobe for isotopic analysis. The age spectra are highly irregular, reflecting the presence of other diagenetic and detrital phases with different Ar release profiles. Early gas releases seemed to occur at low temperatures of < 300°C, much lower than expected for illite (Bray *et al.*, 1987) and with ages of *c.* 250–270 Ma. Although difficult to assign confident interpretation to these age spectra, they did indicate that age spectra for mineral mixtures might have sufficient structure to extract age information for one or more components of the mixture.

Illite and mixed-layer illite–smectite

Rb–Sr and K–Ar

Illitic diagenetic clays in sandstones have been subjected to radiometric age determinations more than any other diagenetic clay. This reflects their ubiquitous occurrence and, being characterized by essential K and a high Rb/Sr ratio, their suitability for dating based on the ^{40}K–^{40}Ar and ^{87}Rb–^{87}Sr decay schemes. In addition, illite often appears to be the one of the latest if not the last mineral to form prior to hydrocarbon charge to a reservoir (Lee *et al.*,

1985, 1989). It has been argued that if illite diagenesis ceases at the time pore-water is displaced by hydrocarbon then age determinations on last-formed illite (= finest grained?) provides constraints on the timing and nature of hydrocarbon charge.

A very considerable number of publications (e.g. Lee *et al.*, 1985, 1989; Thomas, 1986; Hamilton *et al.*, 1987, 1989, 1992; Jourdan *et al.*, 1987; Liewig *et al.*, 1987; Burley & Flisch, 1989; Ehrenberg & Nadeau, 1989; Glasman *et al.*, 1989a, b; Scotchman *et al.*, 1989; Barnes *et al.*, 1992; Burley & MacQuaker, 1992; Mossmann *et al.*, 1992; Hogg *et al.*, 1993; Robinson *et al.*, 1993, Ziegler *et al.*, 1994; Clauer *et al.*, 1995, 1997; Darby *et al.*, 1997; Liewig & Clauer, 2000) present K–Ar data for diagenetic illitic clay from sandstones and in particular for those from North Sea hydrocarbon fields. In addition to attempting to constrain the timing of hydrocarbon charge to reservoirs, illite ages were also used to try to address the time span of the illite diagenesis by comparing ages for different size fractions. Low levels of contamination in coarser size fractions present a major difficulty with routine use of this approach. A consensus has emerged that illite ages, even for finest grained illites, are average ages and that contamination with older K-bearing detritus will often present difficulties of data interpretation. Physical separation of illitic clays of different generations is often a difficult, if not impossible, task. Trends of decreasing K–Ar illite ages with increasing depth often reflect the increasing relative abundance of more recently formed authigenic illite and/or the progressive illitization of mixed layer illite–smectite. Illitic clays from Jurassic sandstones in the North Sea at depths less than *c.* 3 km are often dominated by illite–smectites that are detrital and a product of Jurassic weathering (Burley & MacQuaker, 1992). Consequently they have K–Ar ages much older than those of illite diagenesis and are a source of contamination. Whatever the mechanism of illitization of illite–smectite mixed-layer clay it is likely that until the pure illite end-product is reached, K–Ar ages are likely to inherit some older age from earlier stages of the transformation process (Velde & Renac, 1996).

Nevertheless, Walgenwitz *et al.* (1990) were able to document K–Ar age distinction between different clay mineralogies from Cretaceous sandstones from offshore Angola. Three illite-rich samples (with small amounts of admixed chlorite) from samples over a 5.6 m depth interval from one well gave K–Ar ages of 17–20 Ma. Four other illite-rich samples from samples (with variable amounts of admixed chlorite) over a 15 m depth interval from the same well gave K–Ar ages of 48–55 Ma. The latter age range is similar to a range of 52–60 Ma for mixed-layer illite–smectite separates (with variable amounts of admixed chlorite) obtained from samples over a 1.5 m depth interval from a different well. Girard *et al.* (1989) had earlier obtained similar results from the same region.

K–Ar ages for finest-size illite from Brent Group sandstones that had minimal detrital clay content appeared to correlate with the modelled time of hydrocarbon charge (Hamilton *et al.*, 1992). K–Ar illite ages when examined on a field-wide or regional basis often appear to relate to structural control of fluid flow events that seem to promote illite diagenesis (Robinson *et al.*, 1993; Darby *et al.*, 1997; Zwingmann *et al.*, 1998, 1999). Thus, despite the difficulties associated with K–Ar dating of illite diagenesis the data acquired have been of some use in addressing issues of timing of hydrocarbon charge and fluid flow regimes during basin evolution. More importantly the experience gained will be of great benefit in better defining clay diagenesis in sandstones when coupled with recent advances in $^{40}Ar/^{39}Ar$ analyses of clay minerals from fine-grained rocks (see later).

Clauer *et al.* (1992a) used a review of K–Ar dating of diagenetic illites to recommend appropriate procedures to be followed in order to obtain the most useful data. In many cases, use of the freeze–thaw technique (Liewig *et al.*, 1987) for disaggregation reduces the generation of fine particles from detrital grains and hence the possibility of contamination of fine-size clay separates. Critical point drying of core samples preserves clay morphologies and use of scanning electron microscopy (SEM) enables them to be observed. Scanning electron microscopy also allows potential contaminant phases to be noted, textural relationships to other phases to be documented and timing of illite diagenesis relative to other diagenetic phases to be deduced. Monitoring for contaminant phases by use of XRD and TEM is essential and their quantification can be used to correct data from contaminated clay separates in order to obtain ages for pure end-member diagenetic clays (Liewig *et al.*, 1987; Clauer *et al.*, 1992a; Pevear, 1992, 1999). Analysis by XRD also provides the means for documenting the nature and composition of mixed-layer clays and recognizing illite polytypes.

Clauer *et al.* (1993) conducted a number of leaching treatments in order to assess their effects on the Rb–Sr and K–Ar isotope systematics of diagenetic illite. They noted that none of the reagents used appeared to preferentially remove radiogenic ^{40}Ar. The initial impetus for leaching treatments of clays was removal of contaminant carbonate phases (e.g. Bofinger *et al.*, 1968). For both K–Ar and Rb–Sr analyses Clauer *et al.* (1993) recommended 1 N HCl as the most efficacious treatment. It was demonstrated in this and other studies that the Rb–Sr data of untreated samples, leached samples and residues from leaching are often colinear on an isochron plot and can reflect the age of crystallization. The inference is that any exchangeable Sr present has been present since crystallization and represents the ^{87}Sr/^{86}Sr in the pore-fluid environment at the time of crystallization.

^{40}Ar/^{39}Ar

There are few published studies of ^{40}Ar/^{39}Ar dating of diagenetic illite from sandstones. Emery & Robinson (1993) present a ^{40}Ar/^{39}Ar step-heating age spectrum for diagenetic illite separated from a Permian Rotliegend Sandstone from the southern North Sea. The K–Ar age for the sample is 155 ± 3 Ma but both the reasonably well defined plateau ^{40}Ar/^{39}Ar age and the total fusion age are older at 240 Ma and 217 Ma, respectively. These anomalously old ages are attributed to ^{39}Ar recoil loss. The existence of a plateau age on the age spectrum plot suggests that the recoil loss had been fairly uniform through the fine-sized illite particles. A repeat of the analyses using vacuum encapsulation would be worthwhile.

Thomas *et al.* (1993) have undertaken ^{40}Ar/^{39}Ar dating of authigenic muscovites formed in and close to stylolites in Jurassic Norphlet Formation sandstones from offshore Alabama wells. These were total gas ages obtained from analysis of gas released by laser microprobe fusion of predetermined areas of 50 μm or greater diameter occupied by authigenic muscovite in thin-section wafers that previously had been irradiated. By avoiding fusion of the edges of such areas and disaggregation of the rock the authors believe that the problem of ^{39}Ar recoil is minimized. A recoiling ^{39}Ar atom that leaves one muscovite particle is able to enter another without loss to the non-solid environment. Similarly, Bray *et al.* (1987) noted that by avoiding dispersion of aggregated illites they minimized the ^{39}Ar recoil effect and obtained laser probe ^{40}Ar/^{39}Ar ages in agreement with K–Ar ages.

Glauconitic clays

Rb–Sr and K–Ar systems

Radiometric ages for glauconies are numerous and represent *c.* 40% of the absolute age data base for the geological time-scale of the past 250 Ma. This is as a consequence of the belief that glaucony forms at or very close to the sediment–water interface in shallow-marine environments (Odin & Morton, 1988) and so may provide material suitable for determining stratigraphical ages. Thus, with their abundant K and Rb and high Rb/Sr ratios, glauconitic minerals are potentially suitable for constraining ages of sedimentation. Glauconies frequently occur as ovoid pellets that can exhibit a wide range in colour, texture and chemical composition, even in a single sample. The pellets are often easy to separate by electromagnetic means and hand picking. However, complications for mineral separation and data interpretation can arise with the presence of reworked glaucony pellets and intergranular glauconite cement. The process of glauconitization (Odin & Matter, 1981; Clauer *et al.*, 1992b; Clauer & Chaudhuri, 1995) appears to involve a gradual transition beginning with detritus, including clay minerals, that is converted to syndepositional Fe-rich smectite through a closed-system dissolution–precipitation process. This is followed by a process whereby K content increases as the glaucony structure transforms to a more illitic glauconite in an environment apparently more open to exchange with penecontemporaneous sea water. Odin & Dodson (1982) estimate the time for this process to range from 10^3 to 10^6 yr. The process often does not proceed to completion in all glaucony grains. Where the process has halted at an early stage with low to intermediate K (less than about 5.5% K_2O) there may still be some inheritance of a K-bearing detrital precursor and a high expandable component, which renders such a sample unsuitable for dating. In a single sediment sample there therefore can be a considerable range in appearance and chemistry of glaucony grains.

Morton & Long (1982), Odin (1982) and Odin & Rex (1982) have suggested various criteria for recognizing glaucony samples likely to have recorded meaningful Rb–Sr and K–Ar ages. These criteria include sample purity, low abundance of expandable layers, high K content and enclosure in an unweathered matrix of low permeability. Remnants of K- and Rb-bearing detritus are the likely cause for K–Ar and Rb–Sr ages for some glauconites being older than ages of deposition of host sediment (Keppens & Pasteels, 1982; Smalley *et al.*, 1986). High K content is important as at low to intermediate K content there is more of the smectitic end-member, which is expandable, has high cation exchange capacity and is more susceptible to open-system behaviour. However, even samples with high K content can contain a significant expandable component of 2–10% (Hower, 1961) and be susceptible to cation exchange.

Morton & Long (1982) consider the Rb–Sr system of glaucony pellets in terms of four domains:

1 expandable clay layers;
2 non-expandable clay layers;
3 individual mineral particles of small size (< 2 μm) and composed of 1 and 2 above;
4 the entire pellet composed of 3 above but often other phases as well, which may be diagenetic (e.g. carbonates, iron oxides, apatite) or detrital.

A major issue for dating glauconites is the contamination that might be associated with non-glauconitic phases and with cation exchangeable sites. With respect to the Rb–Sr system, some contaminants (early diagenetic carbonate and apatite) are often assessed as to whether they might have been in isotopic equilibrium with the glaucony at the time of isotopic closure.

Morton & Long (1982) recommended treatment of glaucony pellets and size fractions derived therefrom with 1 M NH_4Oac for 24 h to remove exchangeable cations prior to Rb–Sr isotopic analysis. They claimed that the resultant Rb–Sr isochrons for various glaucony populations from the Llano Uplift of central Texas reflected recrystallization events as a result of meteoric water flushing consequent to regional emergence above sea-level.

Clauer & Chaudhuri (1995) note the care that needs to be taken in preserving sample weights between leaching or ion exchange treatments because of the ease with which small sample weights may be lost from very fine-grained materials. There also is the issue of hysteresis effects with ion exchange reactions with clays (Verburg & Baveye, 1994). As noted by Grant *et al.* (1984), not all supposedly exchangeable cations may be purged with the use of cation exchange reagents.

Smalley *et al.* (1986), Rundberg & Smalley (1989) and Smalley & Rundberg (1990) present several isochron examples where hand-picked glaucony populations of different colours together with coexisting carbonate bioclasts indicate Cenozoic ages of sedimentation. The ages were in agreement with sedimentation ages deduced from Sr isotope stratigraphy. Their glaucony samples had been subjected to brief washing with dilute HCl followed by 24 h exposure to 1 M NH4Oac. Clauer & Chaudhuri (1995) suggest treatment of glaucony samples with dilute HCl as most efficacious for contaminant removal. Further, they note that straight lines joining the Rb–Sr isotopic analyses of both the acid leachate and residue plotted on the isochron diagram often have slopes that can be interpreted as recording the age of crystallization. As a means of confirming the age significance of such plots they suggest also plotting the untreated sample, which should be collinear. This will be so, provided there are no weight losses during the analytical procedure and both the exchangeable cation component and any non-glauconite phase (e.g. coexisting carbonate) were in isotopic equilibrium at the time of closure. Odin & Hunziker (1982) and Harris & Fullagar (1989) demonstrate that with judicious selection of glaucony samples, both K–Ar and Rb–Sr ages can be obtained that are in agreement and closely coincide with the age of sedimentation. However, it should be noted that the Rb–Sr systems of glauconites can be reset at some time after deposition during emergence above sea-level (Morton & Long, 1982). Given that the glauconitization process is confined to shallow-marine environments, glauconites will be prone to this type of resetting mechanism for a period of possibly millions of years after initial sedimentation. Secondly, glauconite grain populations are heterogeneous so that isotopic data for a hand-picked population will be averaged over a large number of grains that could have achieved closure at slightly different times and strictly were not in isotopic equilibrium with each other. Some of the $^{40}Ar/^{39}Ar$ data discussed below are relevant to these observations.

$^{40}Ar/^{39}Ar$ system

The obvious grain-to-grain heterogeneity of glaucony populations and the limit this places on the precision of an isotopic age from such a population led Smith *et al.* (1993, 1998) to undertake $^{40}Ar/^{39}Ar$ dating of individual glaucony grains. Individual grains were vacuum encapsulated in quartz vials prior to irradiation. Ar released from the crushed vial was measured mass spectrometrically prior to laser probe step heating of the glaucony grain. Smith *et al.* (1993) noted total Ar release during step heating between 200 and 600°C. In this earlier study, integrated $^{40}Ar/^{39}Ar$ ages (based on gas release from both vial and sample grain) agreed well with published K–Ar ages for bulk samples. Two further observations led to a model of recoil behaviour in glauconite grains. First they documented a substantial reduction in ^{39}Ar recoil losses from a range of 27–61% for irradiation in air to 11–21% for *in vacuo* irradiation. Secondly, they noted, as had Hess & Lippolt (1986), that the $^{37}Ar/^{39}Ar$ ratios were the same in the vial gas as in the residual grain, even though the ^{37}Ar recoil distance is about four times that of ^{39}Ar. They suggested a 'Velcro' model in which Ar recoil loss depends on clay lamina thickness. It envisages that recoiling Ar atoms once ejected from a clay lamina of say *c.* 10 nm thickness can reinject an adjacent lamina if it has sufficient energy. In this way a proportion of recoiling atoms will become 'stuck' in a clay lamina even though they may have been previously ejected from laminae several times. The model provides an explanation for the lower recoil ^{39}Ar loss for irradiation *in vacuo* compared with in air and for the equivalence of $^{37}Ar/^{39}Ar$ ratios in vial gas and residual sample grains.

In a second, similar, study Smith *et al.* (1998) investigated the intergrain $^{40}Ar/^{39}Ar$ age variability for three glaucony populations that had been used for time-scale calibration of stage boundaries. Individual grains of each population showed differences in colour and amounts of ^{39}Ar recoil loss but would all be classified as suitably evolved to yield a meaningful age date by conventional means. Each population showed a spread of 5–8 Ma, with the oldest of a population providing an acceptable estimate of the stage boundary. The implied continued evolution of the glaucony grains for another 5–8 Ma suggests that continuation of the glauconitization process may be related to sea-level changes within this period.

Use of a sample-specific *J* value

Kapusta *et al.* (1997) suggested a method for obtaining more meaningful total fusion $^{40}Ar/^{39}Ar$ ages from step heating data for glauconite and other clay fractions. Normally the parameter *J* is calculated from a monitor mineral of known K–Ar age. A modified *J* value specific to the mineral being analysed can be obtained by using another aliquot of the same sample as a monitor. The K–Ar age required for the calculation of *J* from the sample monitor is assumed to equate to the total fusion $^{40}Ar/^{39}Ar$ age. The method was demonstrated successfully for the GLO glauconite standard.

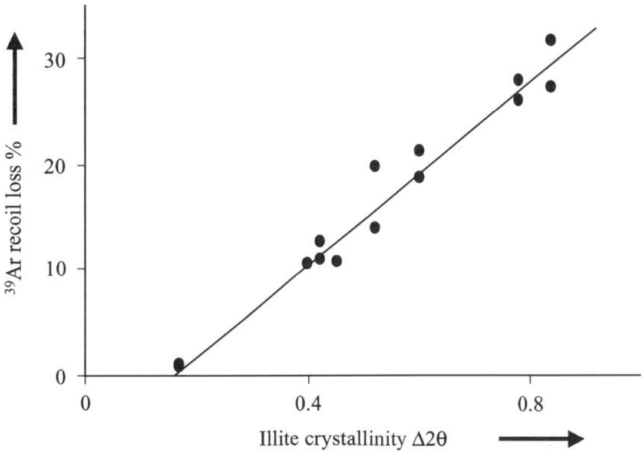

Fig. 16 Relationship between illite crystallinity and amount of ^{39}Ar recoil loss for two illite-rich bentonites and for illite-rich size fractions from a shale and a siltstone of varying diagenetic to metamorphic grade (data from Dong *et al.*, 1995). Data for muscovite/illite separates from hydrothermal alteration associated with a mercury deposit and illite–smectite separates from Gulf Coast K-bentonites also conform to this trend (Hall *et al.*, 1997; Dong *et al.*, 2000).

RECENT DATING OF CLAY DIAGENESIS IN SHALES: RELEVANCE TO SANDSTONES

There have been some significant recent advances in the $^{40}Ar/^{39}Ar$ dating of illitic clays. The applications have been directed mainly at fine-grained sediments such as bentonites and shales, which have implications for such application to illites in sandstones.

The retention age concept

Several studies have demonstrated that the structure of a clay sample exerts considerable control on the amount of ^{39}Ar recoil lost during irradiation. The more densely packed the layers of illite lattice structure are, the less the ^{39}Ar recoil loss. Thus reducing adsorbed water content, compressing samples and using samples of low expandable clay content all result in lower ^{39}Ar recoil losses. This suggests that recoil is related to the ratio of surface area to volume. Illite crystallinity has been found to be correlated with degree of ^{39}Ar recoil loss (Fig. 16; Dong *et al.*, 1995; Hall *et al.*, 1997), and also is variously related to crystal size, density of defects, degree of order and expandable clay proportion. It therefore is a measure of the exposure of surface of the 2 : 1 layer structure to pathways of recoil escape from the clay lattice. A ^{39}Ar atom formed from a parent K atom on the exterior surface of a clay crystallite is likely to be lost to the exterior environment. Conversely a recoiling ^{39}Ar formed at an interlayer site within the crystallite, which might comprise tens of 2 : 1 layers, is more likely to be retained by being implanted to another 2 : 1 layer. This also mirrors, in a simplistic manner, the different retentivities for radiogenic ^{40}Ar. Atoms of ^{40}K in exchangeable sites will have daughter ^{40}Ar atoms that are readily lost by diffusion into the external environment. In contrast, that formed from ^{40}K that is in more strongly bonded lattice sites is likely to be retained. For this reason the more exchangeable K should always be removed by leaching prior to a K–Ar analysis.

Based on step-heating $^{40}Ar/^{39}Ar$ data from illitic clay separates that had been vacuum encapsulated prior to irradiation, Dong *et al.* (1995) proposed that meaningful interpretation could be obtained from 'retention ages'. A retention age is computed from the total ^{40}Ar and ^{39}Ar retained in the sample after irradiation and does not include ^{39}Ar lost by recoil. The clays analysed were crystalline illitic separates from fine-grained sediments comprising bentonites, a shale and a siltstone. The samples had experienced either a high grade of diagenesis or low-grade metamorphism and did not contain a significant smectite component. The retention ages are claimed to be in agreement with other geochronological data and are interpreted as reflecting the time of sedimentation or diagenesis.

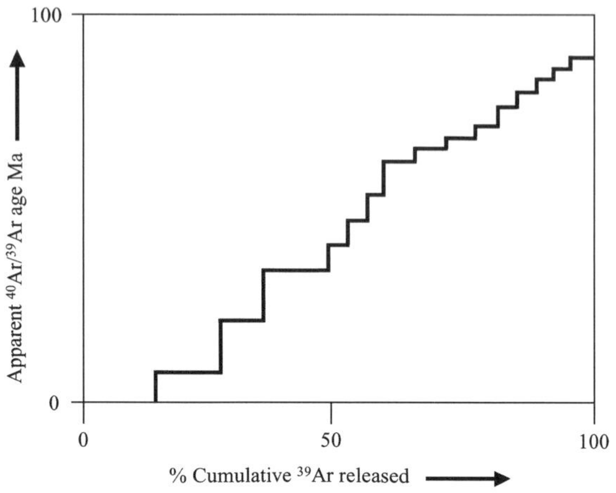

Fig. 17 Generalized shape of $^{40}Ar/^{39}Ar$ age spectrum plots for whole-rock bentonite samples and their illite-rich mixed-layer illite–smectite clay separates (after Dong *et al.*, 2000).

Unravelling end-member $^{40}Ar/^{39}Ar$ ages

The advances of the group based at the University of Michigan (Dong *et al.*, 1995; Hall *et al.*, 1997) in obtaining $^{40}Ar/^{39}Ar$ ages for illitic clays initially were based on application to fine-grained rocks affected by tectonically driven alteration and by hydrothermal alteration. Fine-grained samples affected by burial diagenesis alone were next investigated by this same group (Dong *et al.*, 2000). These present a different set of challenges in that, with lower illite crystallinity, generally smaller crystallite size and greater smectite proportions, more open-system behaviour and greater ^{39}Ar recoil loss can be expected. In addition, there is the challenge of such fine-grained samples often containing several K-bearing diagenetic and detrital phases that are difficult to separate by conventional means. Dong *et al.* (2000) undertook $^{40}Ar/^{39}Ar$ step heating analyses of whole-rock and size separates from bentonites and of $< 0.2\ \mu m$ separates from Texas Gulf Coast shales. The bentonites are supposedly free of detrital components, comprise illite-rich diagenetic illite–smectite (and provide a basis for comparison with the shale data). The amount of smectite in the bentonite illite–smectite size-separates ranged from 20 to 40%. The shale separates comprise illite-rich diagenetic illite–smectite (15% smectite) and detrital illite estimated at 2–3% abundance.

The staircase-shaped age spectra (Fig. 17) for the bentonites and associated size separates are all similar and interpreted as mixing of two different Ar release patterns from smectitic and illitic components of the mixed-layer illite–smectites. The shale separates have total gas and retention ages mostly greater than the depositional ages and therefore represent mixtures of the diagenetic and detrital components. The age spectra (Fig. 18) appeared to have two distinct segments—a staircase profile at intermediate temperatures and a peak at higher temperatures, and probably represent the overlap of two different release profiles of the two end-member components. Based on the similarity of the intermediate temperature segment to the staircase shape of the bentonite data, this was attributed to a dominance of Ar release from the diagenetic component of the shales. The peak sections are ascribed to the generally higher temperature release that characterizes better crystallized illite (e.g. Hunziker *et al.*, 1986). Through a process of normalizing the $^{40}Ar/^{39}Ar$ and ^{39}Ar release data for each temperature step to the totals for each sample and comparing the resultant normalized age profiles to those of well-crystallized illite data from Dong *et al.* (1995) and the diagenetic bentonite profiles, Dong *et al.* (2000) demonstrated how a diagenetic illite–smectite age could be resolved

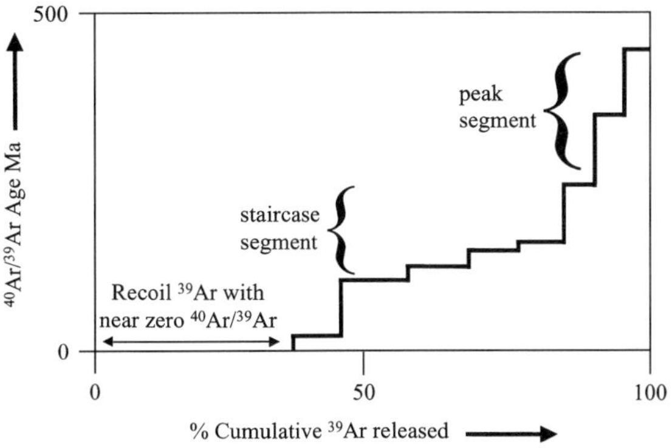

Fig. 18 Generalized shape of $^{40}Ar/^{39}Ar$ age spectrum plots for illitic separates from Texas Gulf Coast shales (after Dong *et al.*, 2000). (Note the difference in vertical scale compared with that of Fig. 17.)

for the shale data. With further research, the method offers the possibility of age distinction of diagenetic and detrital end-member minerals because of the different profiles of Ar release with increasing temperature.

Onstott *et al.* (1997) analysed illite-rich separates (< 0.02 μm and 0.2–2 μm) and powdered muscovites at sample sizes of 1–100 μg, a factor of 10^3 less than required for K–Ar analysis. Samples were vacuum encapsulated for irradiation and a laser microprobe was used for step-heating release of Ar. The integrated $^{40}Ar/^{39}Ar$ age, equivalent to the total fusion age, was within error of the K–Ar age for each sample despite the disparity in sample size. These authors were able to successfully reproduce the age spectrum of a muscovite–illite mixture based on a model of the recoil process and knowledge of the grain-size distribution and composition of the mixture model. Argon release is related to processes of dehydroxylation, and the temperatures at which this occurs depend on grain size, number and type of interlayer sites and lattice structure. They are different between mixed-layer illite–smectite, illite and muscovite (Kodama & Brydon, 1968; Drits *et al.*, 1995; Onstott *et al.*, 1997; Sletten & Onstott, 1998). It would seem that at least for two-component detrital–diagenetic mineral mixtures, age resolution of step-heating spectra may be possible, although calibrations against a range of compositions and grain sizes are required (Onstott *et al.*, 1997; Sletten & Onstott, 1998).

Thin-section dating

The realization of meaningful $^{40}Ar/^{39}Ar$ ages from extremely small illitic mineral separates led Dong *et al.* (1997) to attempt direct dating of such clays in thin-section. Suitable areas for analysis, rich in diagenetic illite and devoid of detrital muscovite, were identified from SEM observations on thin-sections. Targeted fragments were cut from the sections and vacuum encapsulated for irradiation. Argon released by laser fusion of 50–60 μm spots on these fragments yielded retention ages in agreement with previous ages as well as with those from very small sample aliquots (14–77 μg) also analysed in this study.

Clay dating of fault movements

An indirect method for constraining the timing of clay diagenesis in sandstones is to determine the age at which a sandstone has been affected by stylolitization and/or faulting. This requires the presence of clays that have formed or recrystallized as a consequence of deformation and are closely associated with the deformation feature. One example of dating diagenetic muscovites within and associated with zones of stylolitization has been discussed earlier (Thomas *et al.*, 1993). Such an age can then be related in a relative sense to the timing of clay diagenesis deduced from petrographic observation of the sandstone hosting the deformation feature.

Vrolijk & van der Pluijm (1999) showed that deformation associated with faulting appeared to lower the kinetic barrier to illitization of mixed-layer illite–smectite from the observation that the process was more advanced in clay-rich fault gouge than in the host rocks. Combined XRD and K–Ar age data were used to support the hypothesis of deformation-related illitization. Such samples presumably would be amenable to $^{40}Ar/^{39}Ar$ *in situ* laser probe dating, as has been undertaken for higher temperature fault-related formation of other K-bearing minerals (e.g. Cheilletz *et al.*, 1999).

SUMMARY

The above review highlights that the Rb–Sr, K–Ar and $^{40}Ar/^{39}Ar$ radiometric dating methods have been used with considerable success for constraining the timing of clay cementation in sandstones. Of paramount importance in attempting to date any diagenetic clay by any of the methods available is the fullest possible mineralogical, compositional and structural characterization of the clay sample. This is a prerequisite for both appropriate sample treatment prior to isotopic analysis and for data interpretation.

In particular, enormous progress has been made in the application of the $^{40}Ar/^{39}Ar$ method to the study of clay diagenesis and metamorphism in fine-grained sediments. Such application has yet to be made for sandstones. With ultrasmall sample-size requirements coupled with the possibility of laser spot fusion ages on thin sections there will be many opportunities for generating better constraints on the temporal aspects of clay-mineral cement formation in sandstones. Clay-mineral cements such as kaolinite and chlorite should become more amenable to $^{40}Ar/^{39}Ar$ dating, even with their very low K contents. Obtaining ages from such material is becoming easier in a technical sense but there will be an increasing challenge in trying to interpret such data. Understanding the lattice location of K in low-K minerals, for example, will be crucial to understanding $^{40}Ar/^{39}Ar$ ages obtained from them.

The retention age concept of Dong *et al.* (1995), whereby $^{40}Ar/^{39}Ar$ recoil loss can be ignored appears to yield meaningful ages for fine-grained sediments. A large number of such retention ages could be generated by analysis of ultrasmall aliquots from a single fine-grained size fraction of a diagenetic illite that previously would have been used to generate a single K–Ar age. This would reduce the 'averaging' effect and, from the likely spread of ages obtained, offer better constraints on the timing of cessation of illite diagenesis. It may be possible also to address the question of the continuation of illite diagenesis in water-wet portions of hydrocarbon sandstone reservoirs. Through the recent advances in $^{40}Ar/^{39}Ar$ geochronology, long-standing issues such as the timing of illitization of mixed-layer illite–smectite, the meaning of fine-size illite ages and the processes of glauconitization may be better addressed at an increasingly and appropriately smaller scale.

Studies of fine-grained sediments, including those of metamorphic grade, have demonstrated a convergence of scales at which sampling, sample observations and characterizations and isotopic measurements can be made. Although it is unlikely that these scales of measurement and observation will be coincident, they have converged by orders of magnitude over the last two decades and auger well for future applications to clay diagenesis in sandstones. This particularly is so for the $^{40}Ar/^{39}Ar$ method, which has the smallest sample size requirement and is likely to be the focus for clay-mineral dating in the next decade.

REFERENCES

Ahn, J.H. & Peacor, D.R. (1986) Transmission and electron microscopy of the smectite-to-illite transition. *Clay. Clay Mineral.*, **34**, 165–179.

Altaner, S.P. & Ylagan, R.F. (1997) Comparison of structural models of mixed-layer illite/smectite and reaction mechanisms of smectite illitization. *Clay. Clay Mineral.*, **45**, 517–533.

Aronson, J.L. & Douthitt, C.B. (1986) K–Ar systematics of an acid-treated illite–smectite: implication for evaluating age and crystal structure. *Clay. Clay Mineral.*, **34**, 473–482.

Bailey, S.W. (Ed.) (1984) *Micas*. Reviews in Mineralogy 13, Mineralogical Society of America, Washington, DC, 584 pp.

Bailey, S.W. (Ed.) (1988) *Hydrous Phyllosilicates (Exclusive of Micas)*. Reviews in Mineralogy 19, Mineralogical Society of America, Washington, DC, 725 pp.

Barnes, D.A., Girard, J.-P. & Aronson, J.L. (1992) K–Ar dating of illite diagenesis in the middle Ordovician St. Peter Sandstone, central Michigan Basin, USA: implications for thermal history. In: *Origin, Diagenesis and*

Petrography of Clay Minerals in Sandstones, Society for Economic Paleontologists and Sedimentologists Special Publication 47 (Ed. by D.W. Houseknecht and E.D. Pittman), Tulsa, OK, pp. 35–48.

Beckinsale, R.D. & Gale, N.H. (1969) A reappraisal of the decay constants and branching ratio of ^{40}K. *Earth Planet. Sci. Lett.*, **6**, 289–294.

Bofinger, V.M., Compston, W. & Vernon, M.J. (1968) The application of acid leaching to the Rb–Sr dating of a middle Ordovician shale. *Geochim. Cosmochim. Acta*, **32**, 823–833.

Boles, J.R. & Franks, S.G. (1979) Clay diagenesis in Wilcox sandstones of southwest Texas: implications of smectite diagenesis on sandstone cementation. *J. Sed. Petrol.*, **49**, 55–70.

Bray, C.J., Spooner, E.T.C., Hall, C.M., *et al.* (1987) Laser probe $^{40}Ar/^{39}Ar$ and conventional K–Ar dating of illites associated with the McLean unconformity-related Uranium deposits, North Saskatchewan, Canada. *Can. J. Earth Sci.*, **24**, 10–23.

Brereton, N.R., Hooker, P.J. & Miller, J.A. (1976) Some conventional potassium–argon and ^{40}Ar–^{39}Ar age studies of glauconite. *Geol. Mag.*, **113**, 329–340.

Bros, R., Stille, P., Gauthier-Lafaye, F., Weber, F. & Clauer, N. (1992) Sm–Nd dating of Proterozoic clay materials: an example from the Francevillian sedimentary series (Gabon). *Earth Planet. Sci. Lett.*, **113**, 207–218.

Burley, S.D. & Flisch, M. (1989) K/Ar geochronology and the timing of detrital I/S clay illitization and authigenic illite precipitation in the Piper and Tartan fields, outer Moray Firth, UK North Sea. *Clay Mineral.*, **24**, 285–315.

Burley, S.D. & MacQuaker, J.H.S. (1992) Authigenic clays, diagenetic sequences and conceptual diagenetic models in contrasting basin-margin and basin-centre North Sea Jurassic sandstones and mudstones. In: *Origin, Diagenesis and Petrography of Clay Minerals in Sandstones*, Society for Economic Paleontologists and Mineralogists Special Publication 47 (Ed. by D.W. Houseknecht and E.D. Pittman), Tulsa, OK, pp. 81–110.

Catanzaro, E.J., Murphy, T.J., Garner, E.L. & Shields, W.R. (1969) Absolute isotopic abundance ratio and atomic weight of terrestrial rubidium. *J. Res. Nat. Bur. Stand.*, **73A**, 511–516.

Cheilletz, A., Ruffet, G., Marignac, C., *et al.* (1999) $^{40}Ar/^{39}Ar$ dating of shear zones in the Variscan basement of Greater Kabylia (Algeria). Evidence of an Eo-Alpine event at 128 Ma (Hauterivian–Barremian boundary): geodynamic consequences. *Tectonophysics*, **306**, 97–116,

Clauer, N. (1982) The rubidium–strontium method applied to sediments: certitudes and uncertainties. In: *Numerical Dating in Stratigraphy* (Ed. by G.S. Odin), Wiley, Chichester, pp. 245–276.

Clauer, N. & Chaudhuri, S. (1995) *Clays in Crustal Environments: Isotope Dating and Tracing.* Springer-Verlag, Bertin, 359 pp.

Clauer, N., Cocker, J.D. & Chaudhuri, S. (1992a) Isotopic dating of diagenetic illites in reservoir sandstones: influence of the investigator effect. In: *Origin, Diagenesis and Petrography of Clay Minerals in Sandstones*, Society for Economic Paleontologists and Mineralogists Special Publication 47 (Ed. by D.W. Houseknecht and E.D. Pittman), Tulsa, OK, pp. 5–12.

Clauer, N., Keppens, E. & Stille, P. (1992b) Sr isotopic constraints on the process of glauconitization. *Geology*, **20**, 133–136.

Clauer, N., Chaudhuri, S., Kralik, M. & Bonnot-Courtois, C. (1993) Effects of experimental leaching on Rb–Sr and K–Ar isotopic systems and REE contents of diagenetic illite. *Chem. Geol.*, **103**, 1–16.

Clauer, N., O'Neil, J.R. & Furlan, S. (1995) Clay minerals as records of temperature conditions and duration of thermal anomalies in the Paris Basin, France. *Clay Mineral.*, **30**, 1–13.

Clauer, N., O'Neil, J.R., Furlan, S. & Mossman, J.R. (1996) Clay minerals as records of temperature conditions and duration of thermal anomalies in the Paris Basin, France: a reply to the discussion by Spötl *et al. Clay Mineral.*, **31**, 209–215.

Clauer, N., Srodon, J., Francu, J. & Sucha, V. (1997) K–Ar dating of illite fundamental particles separated from illite/smectite. *Clay Mineral.*, **32**, 181–2196.

Curtis, C.D., Hughes, C.R., Whiteman, J.A. & Whittle, C.K. (1985) compositional variation within some sedimentary chlorites and some comments on their origin. *Mineral. Mag.*, **49**, 375–386.

Dalrymple, G.B. & Lanphere, M.A. (1969) *Potassium–Argon Dating.* Freeman, San Fransisco, 258 pp.

Darby, D., Wilkinson, M., Fallick, A.E. & Haszeldine, R.S. (1997) Illite dates record deep fluid movements in petroleum basins. *Petrol. Geosci.*, **3**, 133–140.

Deer, W.A., Howie, R.A. & Zussmann, J. (1967) *Rock-forming Minerals*, 3, *Sheet Silicates.* Longmans, Green & Co., London, 270 pp.

Dickin, A.P. (1995) *Radiogenic Isotope Geology.* Cambridge University Press, Cambridge, 490 pp.

Dong, H., Hall, C.M., Peacor, D.R., & Halliday, A.N. (1995) Mechanism of Argon retention in clays revealed by laser ^{40}Ar–^{39}Ar dating. *Science*, **267**, 355–359.

Dong, H., Hall, C.M., Halliday, A.N. & Peacor, D.R. (1997) Laser ^{40}Ar–^{39}Ar dating of microgram-size illite samples and implications for thin section dating. *Geochim. Cosmochim. Acta*, **61**, 3803–3808.

Dong, H., Hall, C.M., Peacor, D.R., Halliday, A.N. & Pevear, D.R. (2000) Thermal $^{40}Ar/^{39}Ar$ separation of diagenetic from detrital illitic clays in Gulf Coast shales. *Earth Planet. Sci. Lett.*, **175**, 309–325.

Drits, V.A., Besson, G. & Muller, F. (1995) An improved model for structural transformations of heat-treated aluminous dioctahedral 2 : 1 layer silicates. *Clay. Clay Mineral.*, **43**, 718–731.

Drits, V.A., Besson, G. & Muller, F. (1996) X-ray diffraction study of structural transformations of aluminous 2 : 1 layer silicates during dehydroxylation. *J. Phys.*, **6**, 91–102.

Ehrenberg, S.N. & Nadeau, P.H. (1989) Formation of diagenetic illite in sandstones of the Garn Formation, Haltenbanken area, mid-Norwegian continental shelf. *Clay Mineral.*, **24**, 233–253.

Elliott, W.C. & Aronson, J.L. (1987) Alleghanian episode of K-bentonite illitization in the southern Appalachian Basin. *Geology*, **15**, 735–739.

Elliott, W.C. & Matisoff, G. (1996) Evaluation of kinetic models for the smectite to illite transformation. *Clay. Clay Mineral.*, **44**, 77–87.

Emery, D. & Robinson, A. (1993) *Inorganic Geochemistry: Applications to Petroleum Geology.* Blackwell, Oxford, 254 pp.

Evernden, J.F., Curtis, G.H., Kistler, R.W. & Obradovitch, J. (1960) Argon diffusion in glauconite, microcline, sanidine, leucite and phlogopite. *Am. J. Sci.*, **38**, 583–604.

Evernden, J.F., Curtis, G.H., Obradovitch, J. & Kistler, R.W. (1961) On the evaluation of glauconite and illite for dating sedimentary rocks by the potassium–argon method. *Geochim. Cosmochim. Acta*, **23**, 78–99.

Faure, G. (1986) *Principles of Isotope Geology*, 2nd edn. Wiley, New York, 589 pp.

Foland, K.A., Linder, J.S., Laskowski, T.E. & Grant, N.K. (1984) ^{40}Ar–^{39}Ar dating of glauconies: measured ^{39}Ar recoil loss from well crystallized specimens. *Isotope Geosci.*, **2**, 241–264.

Foland, K.A., Hubacher, F.A. & Aregart, G.B. (1992) ^{40}Ar–^{39}Ar dating of very fine-grained samples: an encapsulated vial procedure to overcome the problem of ^{39}Ar recoil loss. *Chem. Geol.* (*Isotope Geosci. Sect.*), **102**, 269–276.

Garner, E.L., Machlan, L.A. & Barnes, I.L. (1975) The isotopic composition of lithium, potassium and rubidium in some Apollo 11, 12, 13, 14, 15 and 16 samples. In: *Proceedings of the 6th Lunar Science Conference*, Pergamon, Oxford, pp. 1845–1855.

Girard, J.-P., Aronson, J.L. & Savin, S.M. (1988) Separation, K–Ar dating and ^{18}O/^{16}O ratio measurements of diagenetic K feldspar overgrowths: an example from the Lower Cretaceous arkoses of the Angola margin. *Geochim. Cosmochim. Acta*, **52**, 2207–2214.

Girard, J.-P., Aronson, J.L. & Savin, S.M. (1989) Diagenesis of the Lower Cretaceous arkoses of the Angola margin: K–Ar dating and ^{18}O/^{16}O evidence. *J. Sed. Petrol.*, **59**, 519–538.

Glasman, J.R., Clark, R.A., Larter, S.R., Briedis, N.A. & Lundegard, P.D. (1989a) Diagenesis and hydrocarbon accumulation, Brent Sandstone (Jurassic), Bergen High area, North Sea. *AAPG Bull.*, **73**, 1341–1360.

Glasman, J.R., Lundegard, P.D., Clark, R.A., Penny, B.K. & Collins, I.D. (1989b) Geochemical evidence for the history of diagenesis and fluid migration: Brent Sandstone, Heather field, North Sea. *Clay Mineral.*, **24**, 255–284.

Grant, N.K., Laskowski, T.E. & Foland, K.A. (1984) Rb–Sr and K–Ar of Palaeozoic glauconites from Ohio and Missouri, USA. *Isotope Geosci.*, **2**, 217–239.

Hall, C.M., Higueras, P.L., Kesler, S.E., *et al.* (1997) Dating of alteration episodes related to mercury mineralization in the Almaden district, Spain. *Earth Planet. Sci. Lett.*, **148**, 287–298.

Halliday, A.N. (1978) ^{40}Ar–^{39}Ar step heating studies of clay concentrates from Irish orebodies. *Geochim. Cosmochim. Acta*, **42**, 1851–1858.

Hamilton, P.J., Fallick, A.E., MacIntyre, R.M. & Elliot, S. (1987) Isotopic tracing of the provenance and diagenesis of Lower Brent Group sands, North Sea. In: *Petroleum Geology of Northwest Europe* (Ed. by J. Brooks and K. Glennie), Graham & Trotman, London, pp. 939–949.

Hamilton, P.J., Kelley, S. & Fallick, A.E. (1989) K–Ar dating of illite in hydrocarbon reservoirs. *Clay Mineral.*, **24**, 215–231.

Hamilton, P.J., Giles, M. & Ainsworth, P. (1992) K–Ar dating of illites in Brent Group reservoirs: a regional perspective. In: *Geology of the Brent Group* (Ed. by A.C. Morton, R.S. Haszeldine, M.R. Giles and S. Brown). *Geol. Soc. London Spec. Publ.*, **61**, 388–400.

Harris, W.B. & Fullagar, P.D. (1989) Comparison of Rb–Sr and K–Ar dates of Middle Eocene bentonite and glauconite, southeastern Atlantic coastal plain. *Geol. Soc. Am. Bull.*, **101**, 573–577.

Hassanipak, A.A. & Wampler, J.M. (1996) Radiogenic argon released by stepwise heating of glauconite and illite: the influence of composition and particle size. *Clay. Clay Mineral.*, **44**, 717–726.

Hess, J.C. & Lippolt, H.J. (1986) Kinetics of argon isotopes during neutron irradiation: ^{39}Ar loss from minerals as a source of error in ^{40}Ar/^{39}Ar dating. *Chem. Geol. (Isotope Geosci. Sect.)*, **59**, 223–236.

Hillier, S. & Velde, B. (1991) Octahedral occupancy and the chemical composition of diagenetic (low temperature) chlorite. *Clay Mineral.*, **26**, 149–168.

Hogg, A.J.C., Hamilton, P.J. & MacIntyre, R.M. (1993) Mapping diagenetic fluid flow within a reservoir: K–Ar dating in the Alwyn area (UK North Sea). *Mar. Petrol. Geol.*, **10**, 279–294.

Hower, J. (1961) Some factors affecting the nature and origin of glauconite. *Am. Mineral.*, **46**, 313–334.

Huang, W.-L., Longo, J.M. & Pevear, D.R. (1993) An experimentally derived kinetic model for smectite-to-illite conversion and its use as a geothermometer. *Clay. Clay Mineral.*, **41**, 162–177.

Hunziker, J.C., Frey, M., Clauer, N., *et al.* (1986) The evolution of illite to muscovite: mineralogical and isotopic data from Glarus Alps, Switzerland. *Contrib. Mineral. Petrol.*, **92**, 157–180.

Inoue, A. & Kitagawa, R. (1994) Morphological characteristics of illitic clay mineral from a hydrothermal system. *Am. Mineral.*, **79**, 700–711.

Jackson, M.L. (1979) *Soil Chemical Analysis—Advanced Course*, 2nd edn. Published by the author, Madison, WI, 895 pp.

Janks, J.S., Yusas, M.R. & Hall, C.M. (1992) Clay mineralogy of an interbedded sandstone, dolomite and anhydrite: the Permian Yates Formation, Winkler County, Texas. In: *Origin, Diagenesis and Petrography of Clay Minerals in Sandstones*, Society for Economic Paleontologists and Mineralogists Special Publication 47, (Ed. by D.W. Houseknecht and E.D. Pittman), Tulsa, OK, pp. 145–157.

Jourdan, A., Thomas, M., Brevart, O., *et al.* (1987) Diagenesis as the control of the Brent Sandstone reservoir properties in the Greater Alwyn area (east Shetland Basin). In: *Petroleum Geology of Northwest Europe* (Ed. by J. Brooks and K. Glennie), Graham & Trotman, London, pp. 951–961.

Kapusta, Y., Steinitz, G., Akkerman, A., *et al.* (1997) Monitoring the deficit of ^{39}Ar in irradiated clay fractions and glauconites: modeling and analytical procedure. *Geochim. Cosmochim. Acta*, **61**, 4671–4678.

Kelley, S. (1995) Ar–Ar dating by laser microprobe. In: *Microprobe Techniques in the Earth Sciences*, Mineralogical Society Series 6 (Ed. by P.J. Potts, J.F.W. Bowles, S.J.B. Reed and M.R. Cave), Chapman & Hall, London, pp. 327–358.

Keppens, E. & Pasteels, P. (1982) A comparison of rubidium–strontium and potassium–argon ages on glauconies. In: *Numerical Dating in Stratigraphy* (Ed. by G.S. Odin), Wiley, Chichester, pp. 225–239.

Kodama, H. & Brydon, J.E. (1968) Dehydroxylation of microcrystalline muscovite. *Trans. Faraday Soc.*, **64**, 3112–3119.

Kralik, M. (1984) Effects of cation-exchange treatment and acid leaching on the Rb–Sr system of illite from Fithian, Illinois. *Geochim. Cosmochim. Acta*, **48**, 527–533.

Le Gallo, Y., Bildstein, O. & Brosse, E. (1998) Coupled reaction flow modelling of diagenetic changes in reservoir permeability, porosity and mineral compositions. *J. Hydrol.*, **209**, 366–388.

Lee, M., Aronson, J.L. & Savin, S.M. (1985) K–Ar dating of Rotliegendes Sandstone, Netherlands. *AAPG Bull.*, **68**, 1381–1385.

Lee, M., Aronson, J.L. & Savin, S.M. (1989) Timing and conditions of Permian Rotliegende Sandstone diagenesis, southern North Sea: K–Ar and oxygen isotope data. *AAPG Bull.*, **73**, 195–215.

Liewig, N. & Clauer, N. (2000) K–Ar dating of varied microtextural illite in Permian gas reservoirs, northern Germany. *Clay Mineral.*, **35**, 271–281.

Liewig, N., Clauer, N. & Sommer, F. (1987) Rb–Sr and K–Ar dating of clay diagenesis in Jurassic sandstone oil reservoir, North Sea. *AAPG Bull.*, **71**, 1467–1474.

Lo, C.-H., Lee, J.K.W. & Onstott, T.C. (2000) Argon release mechanisms of biotite in vacuo and the role of short-circuit diffusion and recoil. *Earth Planet. Sci. Lett.*, **165**, 135–166.

Macchi, L. (1987) A review of sandstone illite cements and aspects of their significance to hydrocarbon exploration and development. *Geol. J.*, **22**, 333–345.

Marshall, B.D. & DePaolo, D.J. (1982) Precise age determination and petrogenetic studies using the K–Ca method. *Geochim. Cosmochim. Acta*, **46**, 2537–2545.

McDougall, I. & Harrison, T.M. (1999) *Geochronology and Thermochronology by the $^{40}Ar/^{39}Ar$ Method*, 2nd edn. Oxford University Press, Oxford, 269 pp.

McHardy, W.J., Wilson, M.J. & Tait, J.M. (1982) Electron microscope and X-ray diffraction studies of filamentous illitic clay from sandstones of the Magnus field. *Clay Mineral.*, **17**, 23–29.

Menegatti, A.P., Fruh-Green, G.L. & Stille, P. (1999) Removal of organic matter by disodium peroxodisulphate: effects on mineral structure, chemical composition and physico-chemical properties of some clay minerals. *Clay Mineral.*, **34**, 247–257.

Merrihue, C. & Turner, G. (1966) Potassium–argon dating by activation with fast neutrons. *J. Geophys. Res.*, **71**, 2852–2857.

Meshri, I. & Ortoleva, P.J. (Eds) (1990) Prediction of reservoir quality through chemical modelling. *AAPG Mem.*, **49**, 175 pp.

Michael, P.J. & Mcwhinnie, W.R. (1989) Mossbauer and ESR studies of the thermochemistry of illite and montmorillonite. *Polyhedron*, **8**, 2709–2718.

Mitchell, J.G. (1968) The argon-40/argon-39 method for potassium–argon age determination. *Geochim. Cosmochim. Acta*, **32**, 781–790.

Morton, J.P. & Long, L.E. (1980) Rb–Sr dating of Palaeozoic glauconite from the Llano region, central Texas. *Geochim. Cosmochim. Acta*, **44**, 663–672.

Morton, J.P. & Long, L.E. (1982) Rb–Sr ages of glauconite recrystallisation: dating times of regional emergence above sea level. *J. Sed. Petrol.*, **54**, 495–506.

Mossmann, J.R. (1991) K–Ar dating of authigenic illite–smectite clay material: application to complex mixtures of mixed layer assemblages. *Clay Mineral.*, **26**, 189–198.

Mossmann, J.R., Clauer, N. & Liewig, N. (1992) Dating thermal anomalies in sedimentary basins: the diagenetic history of clay minerals in the Triassic sandstones of the Paris Basin, France. *Clay Mineral.*, **27**, 211–226.

Murad, E. & Wagner, U. (1996) The thermal behaviour of an iron rich illite. *Clay Mineral.*, **31**, 45–52.

Nadeau, P.H. (1985) The physical dimensions of fundamental clay particles. *Clay Mineral.*, **20**, 499–514.

Nadeau, P.H. (1999) Fundamental particles; an informal history. *Clay Mineral.*, **34**, 185–191.

Nadeau, P.H., Tait, J.M., McHardy, W.J. & Wilson, M.J. (1984a) Interstratified XRD characteristics of physical mixtures of elementary clay particles. *Clay Mineral.*, **19**, 67–76.

Nadeau, P.H., Wilson, M.J., McHardy, W.J. & Tait, J.M. (1984b) Interstratified clays as fundamental particles. *Science*, **225**, 923–925.

Nadeau, P.H., Wilson, M.J., McHardy, W.J. & Tait, J.M. (1985) The nature of some illitic clays from bentonites and sandstones: implications for the conversion of smectite to illite during diagenesis. *Mineral. Mag.*, **49**, 393–400.

Nier, A.O. (1950) A redetermination of the relative abundances of the isotopes of carbon, nitrogen, oxygen and potassium. *Phys. Rev.*, **77**, 789–793.

Odin, G.S. (1982) How to measure glaucony ages. In: *Numerical Dating in Stratigraphy* (Ed. by G.S. Odin), Wiley, Chichester, pp. 387–403.

Odin, G.S. & Bonhomme, M.G. (1982) Argon behaviour in clays and glauconies during preheating experiments. In: *Numerical Dating in Stratigraphy* (Ed. by G.S. Odin), Wiley, Chichester, pp. 333–343.

Odin, G.S. & Dodson, M.H. (1982) Zero isotopic age of glauconies. In: *Numerical Dating in Stratigraphy* (Ed. by G.S. Odin), Wiley, Chichester, pp. 277–305.

Odin G.S. & Hunziker, J.C. (1982) Radiometric dating of the Albian-Cenomanian boundary. In: *Numerical Dating in Stratigraphy* (Ed. by G.S. Odin), Wiley, Chichester, pp. 537–556.

Odin, G.S. & Matter, A. (1981) De glauconiarum origine. *Sedimentology*, **28**, 611–641.

Odin, G.S. & Morton, A.C. (1988) Authigenic green particles from marine environments. In: *Diagenesis, II*, Developments in Sedimentology 43 (Ed. by G.V. Chilingarian and K.H. Wolf), Elsevier, Amsterdam, pp. 213–264.

Odin, G.S. & Rex, C. (1982) Potassium–argon dating of washed, leached, weathered and reworked glauconies. In: *Numerical Dating in Stratigraphy* (Ed. by G.S. Odin), Wiley, Chichester, pp. 363–385.

Ohr, M., Halliday, A.N. & Peacor, D.R. (1994) Sr and Nd isotopic evidence for punctuated clay diagenesis, Texas Gulf Coast. *Geochim. Cosmochim. Acta*, **58**, 289–312.

Onstott, T.C., Miller, M.L., Ewing, R.C. & Walsh, D. (1995) Recoil refinements: implications for the $^{40}Ar/^{39}Ar$ dating technique. *Geochim. Cosmochim. Acta*, **59**, 1821–1834.

Onstott, T.C., Mueller, C., Vrolik, P.J. & Pevear, D.R. (1997) Laser ^{40}Ar–^{39}Ar microprobe analyses of fine grained illite. *Geochim. Cosmochim. Acta*, **61**, 3851–3861.

Pallat, N., Wilson, J. & McHardy, B. (1984) The relationship between permeability and the morphology of diagenetic illite in reservoir rocks. *J. Petrol. Technol.*, **36**, 2225–2227.

Perry, E.A. (1974) Diagenesis and the K–Ar dating of shales and clay minerals. *Geol. Soc. Am. Bull.*, **85**, 827–830.

Pevear, D.R. (1992) Illite age analysis, a new tool for basin thermal history analysis. In: *Proceedings of the 7th International Symposium on Water–Rock Interaction* (Ed. by Y.K. Kharaka and A.S. Maest), Balkema, Rotterdam, pp. 1251–1254.

Pevear, D.R. (1999) Illite and hydrocarbon exploration. *Proc. Nat. Acad. Sci. USA*, **96**, 3440–3446.

Robinson, A.G., Coleman, M.L. & Gluyas, J.G. (1993) The age of illite cement growth, Village Fields area, southern North Sea: evidence from K–Ar ages and $^{18}O/^{16}O$ ratios. *AAPG Bull.*, **77**, 68–80.

Roddick, J.C., Cliff, R.A. & Rex, D.C. (1980) The evolution of excess argon in Alpine biotites—a ^{40}Ar–^{39}Ar analysis. *Earth Planet. Sci. Lett.*, **48**, 185–208.

Rundberg, Y. & Smalley, P.C. (1989) High resolution dating of Cenozoic sediments from northern North Sea using $^{87}Sr/^{86}Sr$ stratigraphy. *AAPG Bull.*, **74**, 298–308.

Rutherford, E. & Soddy, F. (1903) Radioactive change. *Philos. Mag. Ser. 6*, **5**, 576–591.

Scotchman, I.C., Johnes, L.H. & Miller, R.S. (1989) Clay diagenesis and oil migration in Brent Group sandstones of the NW Hutton field, UK North Sea. *Clay Mineral.*, **24**, 339–374.

Shafiqullah, M. & Damon, P.E. (1974). Evaluation of K–Ar methods. *Geochim. Cosmochim. Acta*, **38**, 1341–1358.

Sletten, V.W. & Onstott, T.C. (1998) The effect of instability of muscovite during *in vacuo* heating on $^{40}Ar/^{39}Ar$ step heating spectra. *Geochim. Cosmochim. Acta*, **62**, 123–141.

Smalley, P.C. & Rundberg, Y. (1990) High resolution dating of Cenozoic sediments from northern North Sea using $^{87}Sr/^{86}Sr$ stratigraphy: reply. *AAPG Bull.*, **74**, 1287–1290.

Smalley, P.C., Nordaa, A. & Raheim, A. (1986) Geochronology and palaeothemometry of Neogene sediments from the Voring Plateau using Sr, C and O isotopes. *Earth Planet. Sci. Lett.*, **78**, 368–378.

Smith, P.E., Evensen, N.M. & York, D. (1993) First successful ^{40}Ar–^{39}Ar dating of glauconies: argon recoil in single grains of cryptocrystalline material. *Geology*, **21**, 41–44.

Smith, P.E., Evensen, N.M., York, D. & Odin, G.S. (1998) Single grain ^{40}Ar–^{39}Ar ages of glauconies: implications for the geologic time scale and global sea level variations. *Science*, **279**, 1517–1519.

Spötl, C., Worden, R.H. & Walgenwitz, F. (1996) Clay minerals as records of temperature conditions and duration of thermal anomalies in the Paris Basin, France: Discussion. *Clay Mineral.*, **31**, 203–208.

Środón, J. (1999a) Nature of mixed-layer clays and mechanisms of their formation and alteration. *Ann. Rev. Earth Planet. Sci.*, **27**, 19–53.

Środón, J. (1999b) Extracting K–Ar ages from shales: a theoretical test. *Clay Mineral.*, **33**, 375–378.

Środón, J. (2000) Reply to discussion of 'Extracting K–Ar ages from shales: a theoretical test'. *Clay Mineral.*, **35**, 605–608.

Stalder, P.J. (1973) Influence of crystallographic habit and aggregate structure of authigenic clay minerals on sandstone permeability. *Geol. Mijnbouw*, **52**, 217–220.

Steiger, R.H. & Jäger, E. (1977) Subcommission on geochronology: convention on the use of decay constants in geo- and cosmochronology. *Earth Planet. Sci. Lett.*, **36**, 359–362.

Stille, P. & Clauer, N. (1986) Sm–Nd isochron age and provenance of the argillites of the Gunflint Iron Formation in Ontario, Canada. *Geochim. Cosmochim. Acta*, **50**, 1141–1146.

Thomas, A.R., Dahl, W.M., Hall, C.M. & York, D. (1993) ^{40}Ar–^{39}Ar analyses of authigenic muscovite, timing of stylolitisation and implication for pressure solution mechanisms: Jurassic Norphlet Formation, offshore Alabama. *Clay. Clay Mineral.*, **41**, 269–279.

Thomas, M. (1986) Diagenetic sequences and K–Ar dating in Jurassic sandstones, Central Viking Graben: effects on reservoir properties. *Clay Mineral.*, **21**, 695–710.

Thompson, G.R. & Hower, J. (1973) An explanation for low radiometric ages from glauconite. *Geochim. Cosmochim. Acta*, **37**, 1473–1491.

Turner, G. & Cadogan, P.H. (1974) Possible effects of ^{39}Ar recoil in $^{40}Ar/^{39}Ar$ dating. *Proc. Lunar Planet. Sci. Conf.*, **5**, 1601–1615.

Velde, B. (1985) *Clay Minerals. A Physico-chemical Explanation of their Occurrence.* Developments in Sedimentology 40, Elsevier, Amsterdam, 427 pp.

Velde, B. & Renac, C. (1996) Smectite to illite conversion and K–Ar ages. *Clay Mineral.*, **31**, 25–32.

Verburg, K. & Baveye, P. (1994) Hysteresis in the binary exchange of cations on 2 : 1 clay minerals: a critical review. *Clay. Clay Mineral.*, **42**, 207–220.

Vrolijk, P. & Pluijm, van der B.A. (1999) Clay gouge. *J. Struct. Geol.*, **21**, 1039–1048.

Walgenwitz, F., Pagel, M., Meyer, A., Maluski, H. & Monie, P. (1990) Thermochronological approach to reservoir diagenesis in the offshore Angola Basin, $^{40}Ar/^{39}Ar$ and K–Ar investigation. *Am. Assoc. Petrol. Geol. Bull.*, **74**, 547–563.

Whittle, C.K. (1986) Comparison of sedimentary chlorite compositions by X-ray diffraction and analytical TEM. *Clay Mineral.*, **21**, 937–947.

Ylagan, R.F., Pevear, D.R. & Vrolijk, P.J. (2000). Discussion of 'Extracting K–Ar ages from shales: a theoretical test'. *Clay Mineral.*, **35**, 599–604.

York, D., Hall, C.M., Yanase, Y., Hanes, J.A. & Kenyon, W.J. (1981) ^{40}Ar–^{39}Ar dating of terrestrial minerals with a continuous laser. *Geophys. Res. Lett.*, **8**, 1136–1138.

Ziegler, K., Sellwood, B.W. & Fallick, A.E. (1994) Radiogenic and stable isotope evidence for the age and origin of authigenic illites in the Rotliegend, southern North Sea. *Clay Mineral.*, **29**, 555–565.

Zimmermann, J.L. & Odin, G.S. (1982) Kinetics of the release of Argon and fluids from glauconies. In: *Numerical Dating in Stratigraphy* (Ed. by G.S. Odin), Wiley, Chichester, pp. 345–362.

Zwingmann, H. Clauer, N. & Gaupp, R. (1998) Timing of fluid flow in a sandstone reservoir of the North German Rotliegend (Permian) by K–Ar dating of related hydrothermal illite. In: *Dating and Duration of Fluid Flow and Fluid–Rock Interaction* (Ed. by J. Parnell). *Geol. Soc. London Spec. Publ.*, **44**, 91–106.

Zwingmann, H., Clauer, N. & Gaupp, R. (1999) Structure related geochemical (REE) and isotopic (K–Ar, Rb–Sr and $\delta^{18}O$) characteristics of clay minerals from Rotliegend sandstone reservoirs (Permian, northern Germany). *Geochim. Cosmochim. Acta*, **63**, 2805–2823.

Chlorite case study

Int. Assoc. Sedimentol. Spec. Publ. (2003) **34**, 291–316

Chlorite authigenesis and porosity preservation in the Upper Cretaceous marine sandstones of the Santos Basin, offshore eastern Brazil

S.M.C. ANJOS[1], L.F. DE ROS[2] *and* C.M.A. SILVA[1]
[1]*PETROBRAS Research Center, Ilha do Fundão, Cidade Universitária, Q. 7, 21949–900, Rio de Janeiro, RJ, Brazil, e-mail: anjos@cenpes.petrobras.com.br*
[2]*Universidade Federal do Rio Grande do Sul, Instituto de Geociências, Av. Bento Gonçalves, 9500, 91501–970, Porto Alegre, RS, Brazil, e-mail: lfderos@inf.ufrgs.br*

ABSTRACT

The main reservoirs in Santos Basin, eastern Brazilian margin, are Upper Cretaceous, marine turbidite and shelf sandstones that show porosity greater than 20% below 4000 m depth. This anomalous porosity is ascribed mostly to the inhibitory effect of chlorite rims on quartz cementation, and pressure dissolution. The sandstones are mostly fine-grained lithic arkoses with up to 24% basic and acidic volcanic rock fragments (VRF) derived from the uplift and erosion of the Lower Cretaceous Serra Geral Formation of the Paraná Basin. The chlorite pore-linings were formed by the mesogenetic replacement of precursor smectitic coatings derived from the eogenetic alteration of VRF, and by the direct neoformation of chlorite rims. Ions required for these reactions were derived from detrital smectites (associated mudrocks, matrix, inherited coatings), VRF and derived eogenetic smectites, and Aptian evaporites. The best reservoirs are the Ilhabela Member turbidites, where moderate eogenetic smectite formation was followed by limited compaction, chloritization and chlorite neoformation. Sandstones with abundant precursor smectite were extensively obstructed by abundant chlorite. Sandstones with scarce chlorite often suffered intense quartz cementation and pressure dissolution during deep burial. Turbidites show better development of chlorite rims and porosity than the Juréia Formation shelf deposits, apparently owing to smectite generation from VRF alteration within a more closed eogenetic system and, potentially, to diffusion from interlayered mudrocks.

INTRODUCTION

Some deep sandstone reservoirs contain abnormally high porosity values, which in some cases are interpreted to be a result of the inhibition of quartz cementation and/or pressure dissolution by the presence of chlorite pore-linings (e.g. Pittman & Lumsden, 1968; Heald & Larese, 1974; Dutton, 1977; Tillman & Almon, 1979; Pittman *et al.*, 1992; Ehrenberg, 1993; Spötl *et al.*, 1994). Porosities as high as 25% have been reported at depths greater than 4000 m, and as deep as 7000 m in sandstones such as the Tuscaloosa in Louisiana (Thomson, 1979; Ryan & Reynolds, 1996; Weedman *et al.*, 1996), and the Norphlet in Mississippi and Alabama

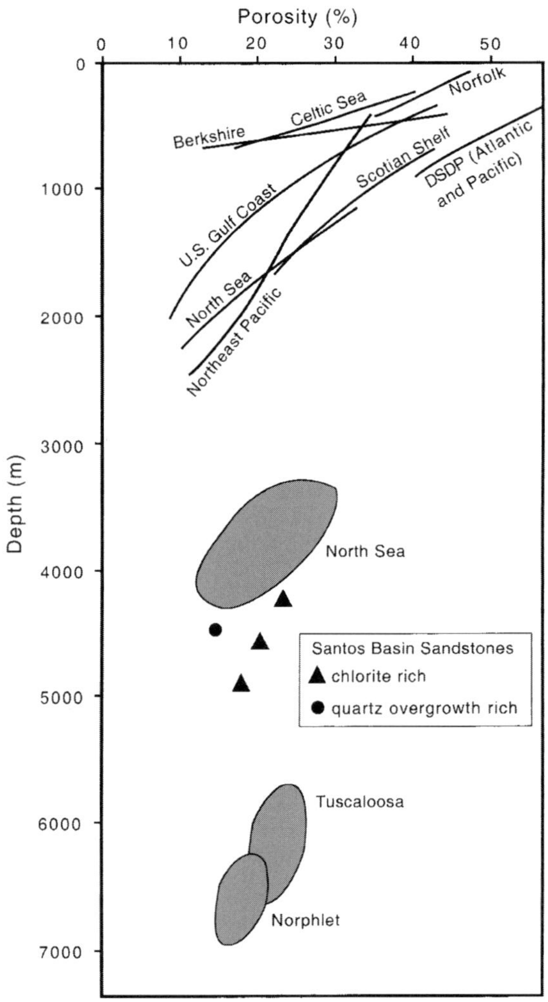

Fig. 1 Steep porosity decline trends with depth in various basins, as compared with occurrences of deep, anomalously porous chlorite-rich sandstones from the Santos Basin (this study), the North Sea (Ehrenberg, 1993), the Tuscaloosa (Thomson, 1979; Weedman *et al.*, 1996), & the Norphlet formations (Dixon *et al.*, 1989; Thomson & Stancliffe, 1990). The points of Santos Basin sandstones are average values of four different boreholes.

(Dixon *et al.*, 1989; Thomson & Stancliffe, 1990), contrasting with the common trends of porosity decrease with depth of clastic reservoirs (Fig. 1). Characterization of genetic, petrographic and petrophysical aspects of these abnormally porous, chlorite-rich sandstones, and the prediction of their spatial distribution is of great importance for the exploration of many basins worldwide.

Santos Basin is the largest among the eastern offshore Brazil, passive margin basins (Fig. 2). The main targets for petroleum exploration in the basin are Upper Cretaceous, marine turbidite and shelf sandstones that form reservoirs up to 60 m thick. They are characterized by anomalously high porosity (greater than 25%), considering their present depths between 4000 and 5000 m. Chlorite is the most abundant diagenetic constituent of these sandstones, and is interpreted to be the main factor in porosity preservation. The sandstones are rich in volcanic rock fragments (VRF), alteration and dissolution of which is interpreted to be the main source of Fe, Mg, Si and Al ions for the observed chlorite authigenesis (Sombra *et al.*, 1990; Anjos *et al.*, 1998).

The objective of this paper is to characterize the origin and diagenetic evolution of authigenic chlorite in the Upper Cretaceous sandstones of the Santos Basin, Brazil.

GEOLOGICAL SETTING

The Santos Basin is located on the southeastern Brazilian continental margin, limited to the northeast by the Cabo Frio Arch and to the southwest by the Florianópolis Platform (Fig. 2). The total area of the basin at water depths < 2000 m is approximately 275 000 km^2 (Pereira & Macedo, 1990; Pereira & Feijó, 1994).

The basin originated during the Lower Cretaceous South Atlantic rifting. Rift sequences and structural evolution are poorly constrained, but allegedly were influenced by transtensional conditions related to the Rio de Janeiro, Curitiba and Florianópolis transfer zones. At the end of the rift stage, during the Aptian, an evaporitic sequence, approximately 1000 m thick, was deposited under transitional conditions in a narrow proto-oceanic sea. The salts were covered by a wide marine carbonate platform of Albian age, followed by clastic marine sedimentation.

The main clastic reservoirs of the basin are Santonian to Maastrichtian turbidite sandstones

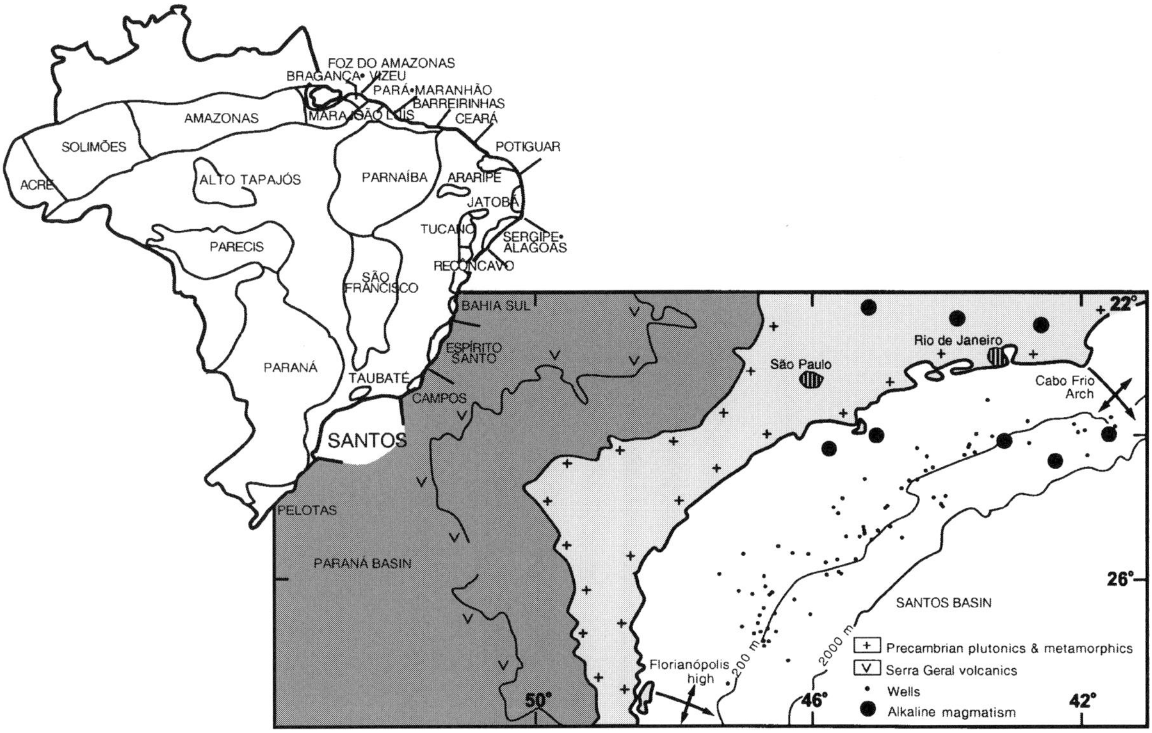

Fig. 2 Location of Santos Basin in offshore southeastern Brazil, showing studied wells, main occurrences of the Upper Cretaceous alkaline volcanism, and outcropping areas of Precambrian crystalline rocks and of the Paraná Basin Serra Geral Lower Cretaceous volcanics.

of the Ilhabela Member of the Itajaí-Açu Formation, and shallow-marine sandstones of the Juréia Formation (Fig. 3). Until recently, the most important discovery of the basin was the Merluza Field, which has produced gas and condensate from Santonian Ilhabela turbidites since 1993.

The basement and immediate source terrains for the sandstones are granitic–gneissic, Precambrian rocks that crop out along the northern and western margins of the basin (Fig. 2). Further west are sedimentary rocks of the adjacent intracratonic Paraná Basin (Fig. 2), covered for most of its extent by the approximately 1 km thick succession of Lower Cretaceous tholeiitic volcanics of the Serra Geral Formation (Mantovani *et al.*, 1985; Piccirilo *et al.*, 1988; Mizusaki *et al.*, 1992).

Basic alkaline magmatism occurred along the northern part of the basin and the adjacent margin, mostly along the Cabo Frio Arch (Fig. 2) during the Late Cretaceous (dated at 92 ± 9 to 87 ± 12 Ma; Mizusaki *et al.*, 1992).

Intense deformation, related to the halokinesis of the Aptian evaporites, started early in the Albian. The increasing halokinetic movements during the Late Cretaceous created large structures in the deep part of the basin and influenced the sedimentation patterns. The Merluza turbidite reservoirs were deposited within an elongated trough generated by halokinesis during deposition of the Albian carbonates (Enciso & Tisi, 1998). Hydrocarbon source rocks are poorly constrained, but are considered to be lacustrine rift shales of the Guaratiba Formation. The geochemical characteristics of the light oils of the basin suggest that these shales were deposited in a brackish, transitional setting (Mello *et al.*, 1991).

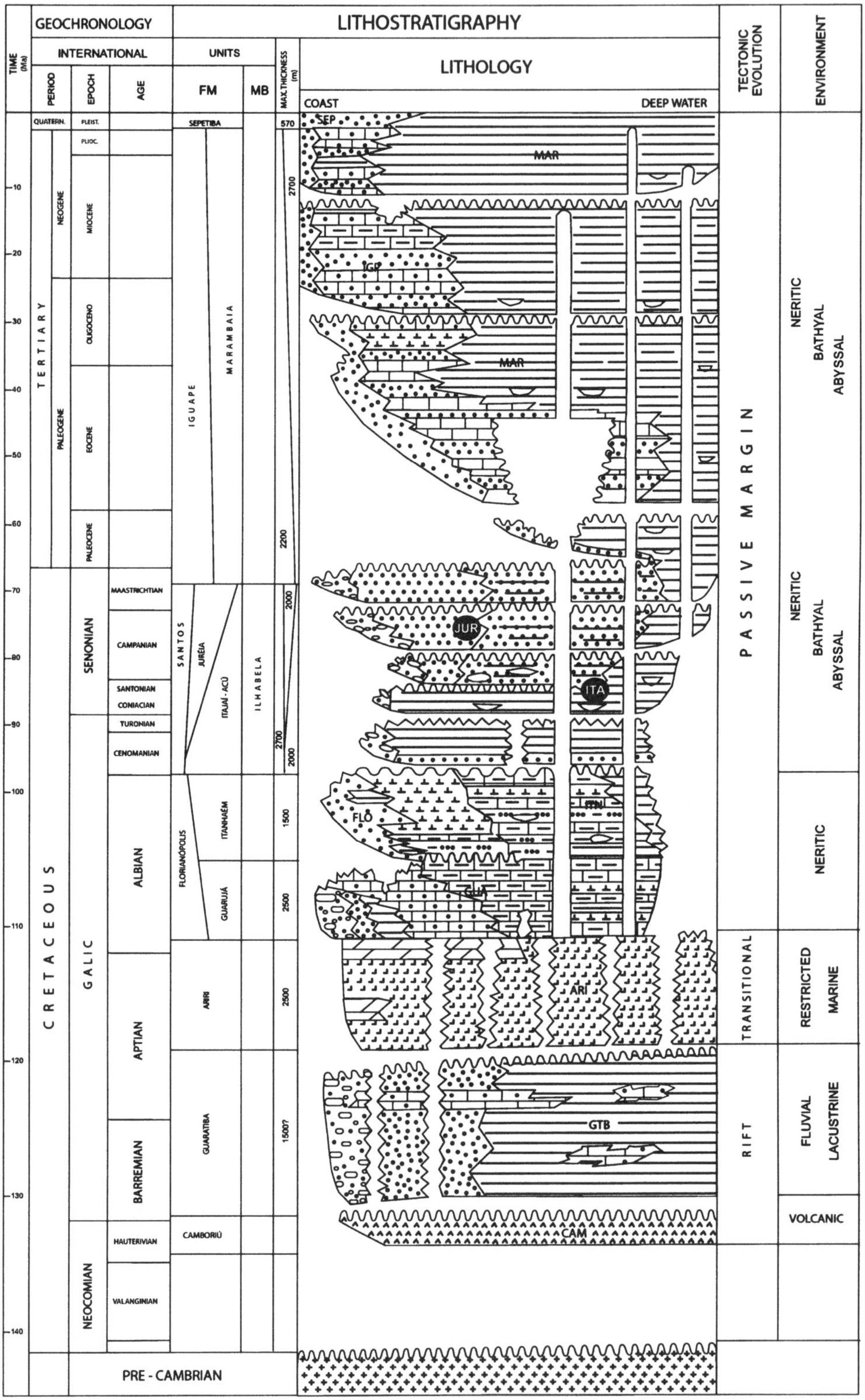

Fig. 3 Stratigraphical column of the Santos Basin, showing the position of the studied sandstones of the Juréia Formation (shelf) and the Ilhabela Member of the Itajaí-Açu Formation (turbidites); JUR and ITA respectively. Modified from Pereira & Feijó (1994).

METHODS

More than 300 m of cores from 25 wells distributed throughout the basin (Fig. 2) were analysed with the aim of characterizing the depositional environments of the sandstones. Detailed petrographic description was performed on 230 thin-sections prepared from representative, blue epoxy-impregnated samples. Seventy-six thin-sections were selected for compositional modal analyses by point counting (300 per thin-section).

X-ray diffraction (XRD) analyses were performed in a Siemens D5000 diffractometer on 246 shale samples and on the fine fraction (< 2 μm) of 104 sandstone samples, in order to determine the clay mineralogy. A speed of 6° min^{-1}, $K\alpha$ copper radiation, 40 kV and 30 mA of filament current were used. Chlorite polytypes were determined from the < 2 μm fraction from a few selected samples by the slow scanning of 30 to 57°2θ of randomly orientated samples.

Scanning electron microscopy (SEM) was executed in a JEOL JXA 840A electron microscope under secondary and backscattered electron (BSE) modes in 40 sandstone samples, in order to characterize the habits of authigenic chlorite and its paragenetic relationships with other diagenetic constituents. Semi-quantitative, energy-dispersive spectrometry (EDS) analyses of the clay minerals, carbonates and feldspars were performed in coupled Voyager/NORAN equipment.

Quantitative analyses of chlorite elemental composition on seven representative thin-sections were obtained in a CAMECA Camebax SX 50 wavelength spectrometry microprobe (WDS) equipped with four spectrometers and one back-scattered electron detector (BSE). Thin-sections for WDS and BSE/EDS analyses were polished and covered with a thin carbon coat. Operating conditions were: 20 kV acceleration voltage, 8 nA measured beam current and a 1–10 μm beam diameter (depending on the extent of homogeneous areas). Precision of analyses was better than 0.1 oxide mol.%.

Carbon and oxygen stable isotope ratios were analysed in authigenic carbonates associated with chlorite in 14 samples with calcite and two samples with dolomite. The samples were reacted with 100% phosphoric acid for 1 h at 25°C. The released CO_2 was analysed using a Delta E Finnigan MAT mass spectrometer. The phosphoric acid fractionation factor used was 1.01025 (Friedman & O'Neil, 1977). Stable isotope data are presented in the normal δ notation relative to PDB (Pee Dee Belemnite) (Craig, 1957). Precision (1 σ) was better than ±0.05‰ for both $\delta^{13}C$ and $\delta^{18}O$. The equations of Irwin *et al.* (1977) were used to calculate the precipitation temperature of calcite and dolomite.

Fluid inclusion analyses were performed on calcite cement covering chlorite rims in two representative samples using a Chaixmeca heating and freezing stage, calibrated at regular intervals using pure standards. The accuracy was ±1°C for temperatures between +20°C and +150°C.

Stable oxygen and hydrogen isotope ratios were determined in chlorites from two samples of sandstones from different wells to provide information on the temperature and composition of precipitating fluids. An end-member ferroan chlorite from a turbidite reservoir and a magnesian chlorite from a shelf sandstone were selected to represent the variability of chlorite compositions and settings. The compositions of chlorite separates in two samples were verified by X-ray diffraction and scanning electron microscopy prior to isotopic analysis. Oxygen was extracted from chlorite using the BrF_5 method of Clayton & Mayeda (1963), and quantitatively converted to CO_2 over red-hot graphite. An oxygen isotope, CO_2–H_2O fractionation factor of 1.0412 was used to calibrate the reference gas introduced into the VG-Micromass 602C mass spectrometer. A $\delta^{18}O_{SMOW}$ value of +9.7‰ was obtained for the silica standard NBS-28 during these experiments. Reproducibility on duplicate analyses was never larger than ±0.2‰. The temperatures of chlorite formation were calculated applying the Wenner & Taylor (1971) equation and assuming a $\delta^{18}O_{water}$ of +2.0‰ SMOW (Standard Mean Ocean Water).

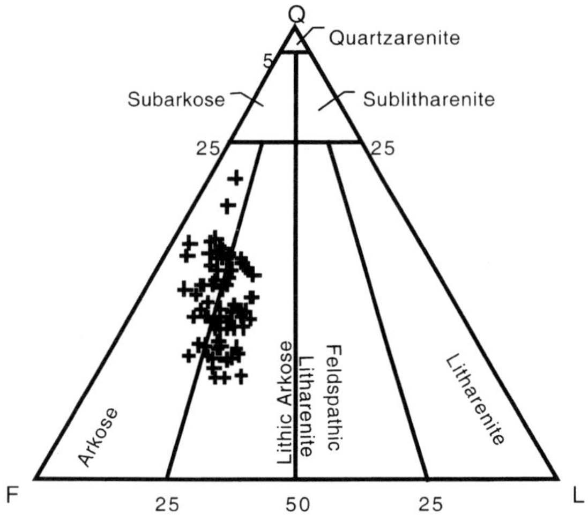

Fig. 4 Major detrital composition of the Santos Basin Upper Cretaceous sandstones, plotted on a Folk (1968) diagram. Rock fragments are essentially volcanic.

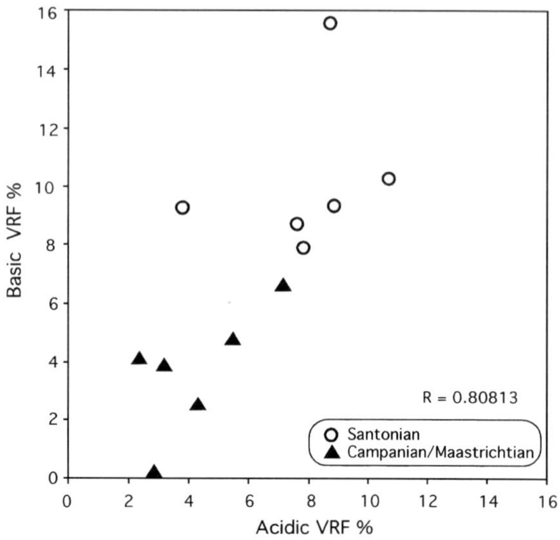

Fig. 5 Average contents of basic and acidic volcanic rock fragments in Santos sandstones. The fairly good correlation indicates a common source, with decreasing supply with time.

TEXTURE, DETRITAL COMPOSITION AND PROVENANCE

Juréia shelf sandstones are predominantly cross-bedded, moderately sorted, fine to coarse grained, with a coarsening upward pattern. Locally, conglomeratic, mud-intraclastic, coarse-grained bioclastic or fine-grained bioturbated deposits are interbedded with the dominant facies. This facies association is interpreted to represent barrier bar, tidal channel, shoreface and beach deposits (Sombra *et al.*, 1990).

The Ilhabela turbidite sandstones show diverse textural characteristics. The medium- to coarse-grained and moderately to poorly sorted turbidites, commonly with coarse-tail grading and amalgamated cycles, are interpreted as high-density turbidity current deposits accumulated in chanellized lobes. The fine-grained and well-sorted turbidites are interpreted as distal, low-density deposits from non-chanelized lobes (Sombra *et al.*, 1990).

Both the turbidite and shelf sandstones are arkoses and lithic arkoses (Fig. 4), which contain as much as 24% volcanic rock fragments (VRF). Monocrystalline quartz and feldspar grains, as well as coarse-crystalline plutonic rock fragments, micas, and the detrital heavy mineral assemblage of zircon, garnet, titanite, opaques, tourmaline and hornblende are interpreted to be derived from the granitic–gneissic Precambrian source areas located along the northern and western margins of the basin (Fig. 2).

Intrabasinal grains include clay intraclasts that generally are strongly compacted to pseudomatrix and chloritized, and are more common in the turbidites. Carbonate bioclasts of molluscs, forams and ostracods, and microcrystalline intraclasts are recrystallized and dissolved to variable degrees, and are more common in the shallow-marine facies.

Identified as the main source of the abundant chlorite authigenesis in the sandstones, the volcanic rock fragments were subjected to detailed petrographic characterization.

Basic VRF (maximum 20%) and acidic VRF (maximum 12%) are widely distributed in samples from every facies. The average content of basic and acidic VRF presents a good correlation in all samples analysed (Fig. 5), suggesting that the two types of fragments have the same provenance.

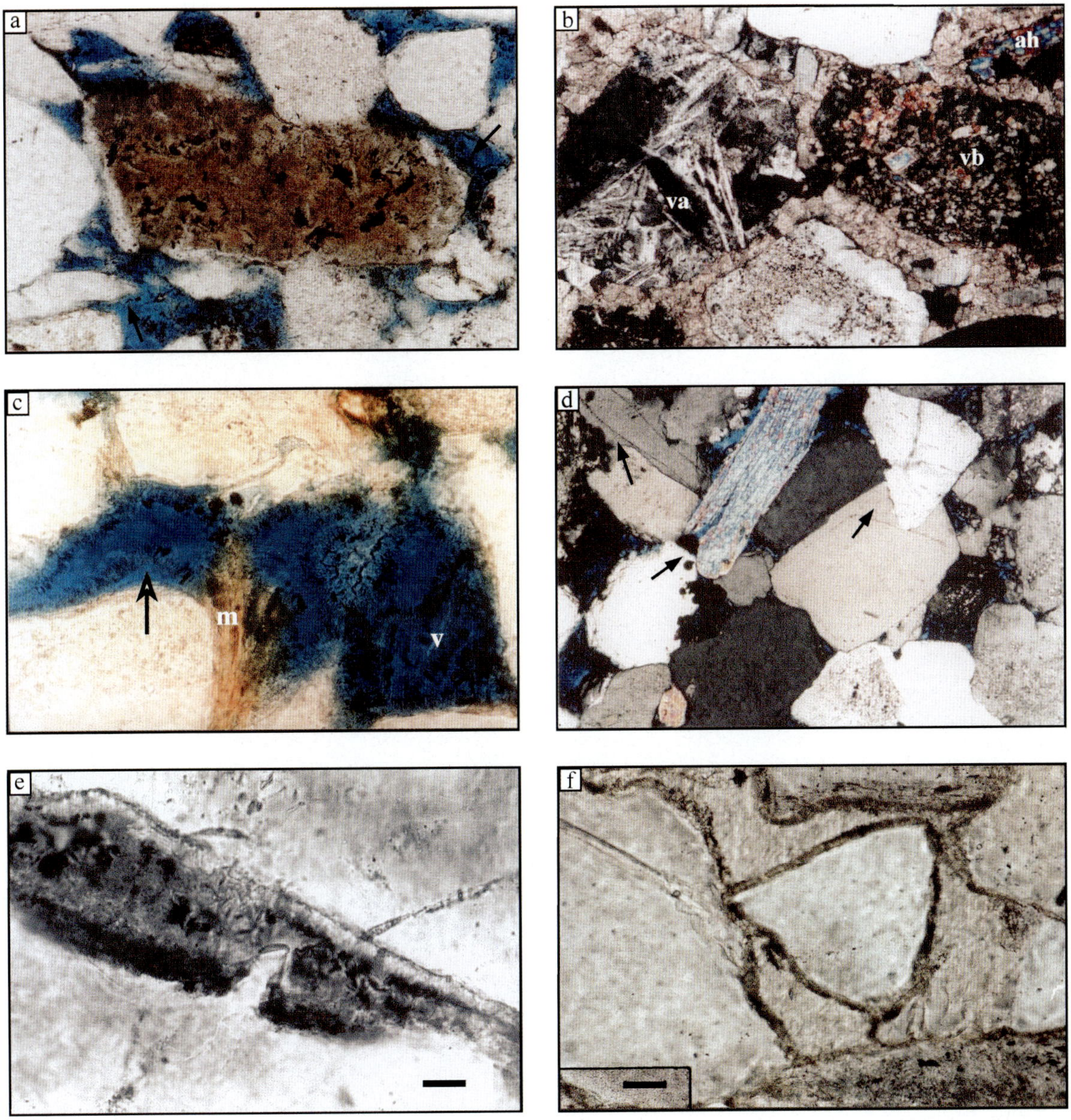

Plate 1 Optical photomicrographs of Santos Basin sandstones. (a) Hemicrystalline basic volcanic rock fragment (VRF) intensely replaced by chlorite and titanium oxide; chlorite pore-linings (arrows); uncrossed polarizers (//P); photograph field is 0.45×0.67 mm. (b) Acidic VRF with granophyric texture (va) and basic holocrystalline VRF (vb), cemented by pre-compactional pore-filling dolomite and anhydrite (ah); crossed polarizers (XP); photograph field is 0.9×1.35 mm. (c) Chlorite rims in a porous Ilhabela sandstone lining quartz grains, remnants of a dissolved volcanic fragment (v), and chloritized mica (m); rims are locally detached from grains (arrow); //P; photograph field is 0.22×0.34 mm. (d) Juréia sandstone devoid of chlorite rims, intensely cemented by quartz overgrowths (arrows); XP; photograph field is 0.45×0.67 mm. (e) Small prismatic quartz outgrowth developed in an interruption of the chlorite coating; //P; scale = 12.5 μm. (f) Poikilotopic calcite cement covering continuous, pre-compactional chlorite coatings; diffraction analyses indicate a composition of corrensite and chlorite; //P; scale bar = 50 μm.

Basic VRF textures are of holohyaline (originally composed of basaltic glass, except for tiny crystallites), hemicrystalline (formed by plagioclase crystals immersed in an originally vitreous mesostasis) and holocrystalline types (originally formed by plagioclase and pyroxene). Holohyaline and hemicrystalline VRF were more intensely dissolved or largely replaced by chlorite, Ti oxides and/or Fe oxides than holocrystalline VRF (see Plate 1a, facing p. 296). However, degree and types of alteration seem to vary largely in samples from the same core, apparently in response to differential composition and texture of the VRF, and original permeability of the sands.

The acidic VRF are all holocrystalline, and were substantially less affected by diagenetic alteration processes than the basic VRF. They commonly show granophyric textures with a micrographic intergrowth of K-feldspar and quartz and sanidine or quartz phenocrysts (see Plate 1b, facing p. 296), and granular aphyric textures.

The Upper Cretaceous olivine basalts and diabases of the basic alkaline, intrabasinal magmatism initially were considered the main source for the VRF in Santos sandstones. Other possible sources considered were the Lower Cretaceous volcanics of the Serra Geral Formation, and the equivalent intrabasinal rift-phase, Lower Cretaceous tholeiitic basalts of the Camboriú Formation (Fig. 3).

The granophyric textures of the acidic VRF are very similar to the textures of the acidic volcanic rocks that cover the basalts in the Serra Geral Formation (see Plate 1b, facing p. 296). These acidic granophyric rocks show no resemblance in texture and composition to any product of the intrabasinal alkaline volcanism. The origin of the micrographic intergrowth is the crystallization of a K-feldspar/quartz eutectic from an acidic melt, which is not possible from alkaline magmas, which are silica-deficient.

Palinspastic reconstructions indicate that the areal extension of the Serra Geral volcanics and the whole Paraná Basin were substantially larger than today (Almeida & Carneiro, 1998). The epeirogenic uplift of the coastal range ('Serra do Mar') along eastern Brazil during the Late Cretaceous and Paleogene produced a generalized erosion and westward retreat of the Serra Geral Formation scarp and of the eastern boundary of the Paraná Basin (Almeida & Carneiro, 1998).

DIAGENESIS

Diagenetic composition and paragenesis

The porosity and permeability of Santos clastic reservoirs are strongly controlled by diagenesis. The main diagenetic processes identified were compaction, alteration and dissolution of VRF, and precipitation of smectite, chlorite, calcite and quartz. The definition of precise paragenetic relationships was not possible for all the diagenetic phases observed, owing to the scarcity of geothermometric and geochronological data, and to the complex and recurrent nature of several textural relationships. Based on optical and electron microscopy, and on elemental and isotopic composition, a simplified representation of the paragenetic relationships among the diagenetic phases in shelf and turbidite sandstones is depicted in Fig. 6.

Chlorite, mostly as coatings and rims, is the main diagenetic constituent in the deep turbidite reservoirs of Santos Basin, with contents up to a maximum of 19.7 vol.%. Sandstones affected by intense bioturbation, or rich in pseudomatrix from the compaction of mud intraclasts may show up to 42 vol.% of chloritized matrix. Chlorite prevails in the Ilhabela turbidites, whereas quartz overgrowths (maximum 8.6%) are more abundant in the Juréia shelf sandstones. Sandstones with continuous chlorite pore-lining display limited compaction and have only trace amounts of quartz cementation (see Plate 1c, facing p. 296), whereas sandstones devoid of chlorite coatings are pervasively cemented by quartz overgrowths (see Plate 1d, facing p. 296). Sandstones with thin and/or discontinuous chlorite coatings present small quartz outgrowths (see Plate 1e, facing p. 296) rather than well-developed overgrowths.

Fig. 6 Paragenetic diagram for the main diagenetic processes occurring in the shelf (Juréia Formation) and turbidite (Ilhabela Member, Itajaí-Açu Formation) sandstones, as interpreted from petrographic relationships.

Table 1 Isotopic ratios and calculated precipitation temperatures for diagenetic carbonates in Santos sandstones.

Carbonate	Facies	δ^{13}C (PDB)	δ^{18}O (PDB)	T°C ($\delta_{water} = -1.2$)	T°C ($\delta_{water} = +2$)	Interpretation
Dolomite	Shelf	0.06	−1.96	36.2		Marine (bioclast dissolution)
Dolomite	Shelf	−0.21	−3.31	44.4		Marine (bioclast dissolution)
Calcite	Shelf	−0.98	−5.96	40.1		Marine (recrystallization)
Calcite	Shelf	−1.33	−9.24	59.8		Marine (recrystallization)
Calcite	Shelf	0.11	−7.79	50.7		Fermentation (mixed?)
Calcite	Turbidite	−12.44	−4.71	33.4		Sulphate reduction
Calcite	Turbidite	−5.57	−7.58	49.5	70.1	Decarboxylation (mixed?)
Calcite	Turbidite	−4.42	−8.8	57.0	78.7	Decarboxylation (mixed?)
Calcite	Shelf	−5.89	−8.64	56.0	77.5	Decarboxylation (mixed?)
Fe-calcite	Shelf	−19.13	−8.85	55.4	76.9	Decarboxylation (mixed?)
Calcite	Shelf	−6.53	−9.29	59.5	81.6	Decarboxylation
Calcite	Shelf	−6.27	−9.35	60.5	82.7	Decarboxylation
Calcite	Turbidite	−9.26	−9.17	59.3	81.4	Decarboxylation
Calcite	Turbidite	−6.1	−9.82	63.6	86.2	Decarboxylation
Calcite	Turbidite	−5.89	−9.85	63.8	86.4	Decarboxylation
Calcite	Shelf	−11.32	−11.61	75.9	100.1	Decarboxylation

Carbonates are the third diagenetic constituent in abundance, predominating in the shelf deposits. Calcite is dominant, occurring as coarse, pore-filling cement on top of chlorite coatings and rims (see Plate 1f, facing p. 296), and replacing, to a limited extent, feldspar and volcanic grains. Isotopic data from the diagenetic calcite presented in Table 1 indicate a greater

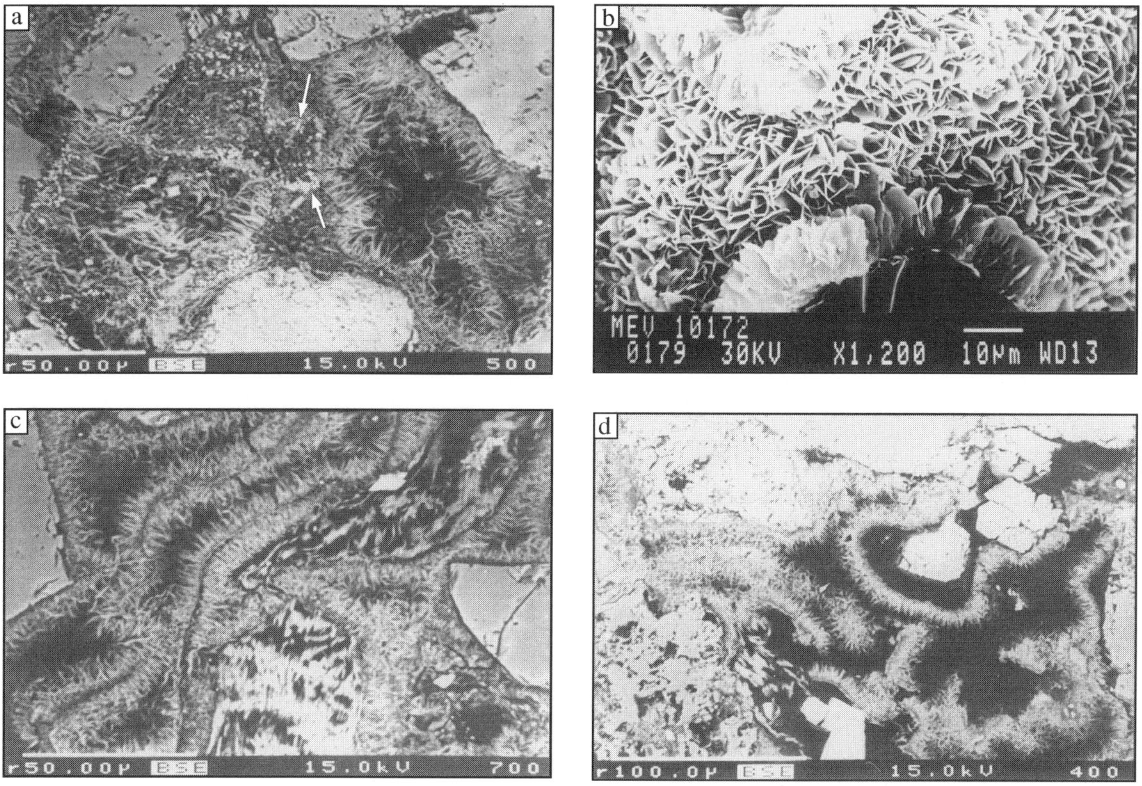

Fig. 7 Backscattered electron (BSE) and secondary scanning electron microscopy (SEM) photomicrographs of the authigenic chlorites: (a) thick chlorite rims lining the pores, and chaotic mesh of chlorite and anatase (arrows) replacing altered volcanic rock fragments (BSE); (b) detail of a chlorite rim made of thin platelets arranged edge-to-face and perpendicularly to grains surfaces (SEM); (c) isopachous double chlorite rims grown on both sides of chloritized coatings, which were covering heavy mineral and volcanic grains that are now partially dissolved (BSE); (d) isopachous chlorite rims externally surrounding chloritized coatings, which were covering heavy mineral and feldspar grains that are now partially dissolved—radial, rosette chlorite aggregate partially filling the pore at the centre (BSE).

homogeneity of processes and sources for the cements in the turbidite sandstones than for those in the shelf sandstones. The latter display a stronger influence of marine and bacterial carbonate in relation to a thermal decarboxylation source, which predominates in the turbidites. Fluid inclusions in pore-filling calcite covering chlorite coatings in two samples yielded homogenization temperatures of approxmately 70°C.

Dolomite is limited to shallow-marine sandstones, where it occurs as pre-compactional pore-lining and pore-filling cement, in places associated with coarsely crystalline anhydrite (see Plate 1b, facing p. 296). Shallow-marine, hybrid bioclastic arenites of the Juréia Formation are strongly cemented by microcrystalline calcite, or by dolomite plus anhydrite. Isotopic data from the dolomite cement of hybrid shelf arenites presented in Table 1 indicate a marine eogenetic source, most probably from the early dissolution of unstable bioclasts. Precompactional, pore-lining dolomite with marine isotopic values is probably derived from the replacement of a Mg-calcite rim cement (Morad, 1998).

Finely crystalline anatase very commonly is associated with chlorite, mostly along the margins of dissolved and chloritized VRF (Fig. 7a), and within chloritized mud intraclasts and derived pseudomatrix. The occurrence of microcrystalline TiO_2 is commonly reported in association with chlorite pore-linings in deeply buried sandstones, being apparently related to the incompatibility between the highly charged Ti^{4+} and the chlorite structure (Ryan & Reynolds, 1997).

Authigenic albite commonly partially to totally replaces feldspar grains and the plagioclase crystals within basic VRF. Such chemically pure diagenetic albite occurs both in shelf and in turbidite sandstones. Microcrystalline halite locally is present within chlorite rims and between grains and rims. The possible presence of coarse halite cement filling intergranular pores in some sandstones could not be verified, as no special coring and sampling procedures were followed to preserve such cement types (cf. McBride *et al.*, 1987; Studlick *et al.*, 1990).

Compaction is remarkably limited for such deeply buried sandstones. That is demonstrated by their average intergranular volume (IGV) of 27%, but some chlorite-poor shelf sandstones show stronger compaction and IGV values as low as 10%.

Chlorite

As the main diagenetic constituent and control on reservoir quality, chlorite is described in detail in this section. Chlorite occurs replacing framework grains, as pore-filling cement and predominantly as pore-lining cement. Among the variable habits of occurrence of chlorite in the Santos sandstones, the most important are:

1 Continuous, isopachous rims on framework grains (see Plate 1c, facing p. 296). This is the most common chlorite habit, particularly in the Ilhabela turbidites. The rims are 5–30 μm thick, with an average of 10 μm, and are made of thin platelets arranged perpendicularly to the surface of the grains, and commonly are arranged edge-to-face (Fig. 7b).

2 More or less continuous, thin (< 10 μm) coatings. The coatings are covered or replaced by the rims (see Plate 1e, facing p. 296; Fig. 7c & d), and locally preserved along intergranular contact areas and beneath pore-filling calcite cement (see Plate 1f, facing p. 296). In some of the preserved coatings, diffraction and microprobe analyses indicate the presence of a corrensite (regular chlorite–smectite mixed-layer) component.

3 Heterogeneously distributed intergranular 'matrix'. This is generated by the chloritization of clay matrix introduced by bioturbation or fluidization (usually in the shelf sandstones), or of pseudomatrix derived from the compaction of clay intraclasts (more commonly in the turbidites), which are replaced by a mesh of chaotically interlocked chlorite platelets.

4 Discontinuous, radiated, rosette or globular aggregates. These occur dominantly in the turbidite sandstones, either in shallow sandstones, discretely and directly on grain surfaces, or associated with rims in deeply buried sandstones (Fig. 7d). Globular aggregates in shallow sandstones mostly are made of corrensite.

5 Remnants of coatings preserved in the concave portions of grain surfaces and absent from the convex areas. These are recognized as clays inherited from the reworking and abrasion of grains previously covered with eogenetic clay coatings from continental, transitional or shallow-marine settings (Wilson, 1992).

6 Replacement of grains, including basic VRF, clay intraclasts, biotite and ferromagnesian heavy minerals, by chlorite in very finely crystalline or chaotic mesh aggregates, commonly associated with finely crystalline anatase (see Plate 1a & c, facing p. 296; Fig. 7a & c).

A peculiar habit of chlorite in some Santos sandstones are double rims formed both in the internal and external sides of coatings surrounding mouldic grain dissolution pores. This indicates that, at these sites, chlorite precipitation at least partly post-dated grain dissolution (Fig. 7c). Conversely, in some places chlorite rims cover only the external side of such coatings around grain dissolution pores (Fig. 7d), suggesting that precipitation took place prior to the dissolution of these grains. In places, rims covering thin coatings are detached from grain surfaces and/or are fragmented owing to

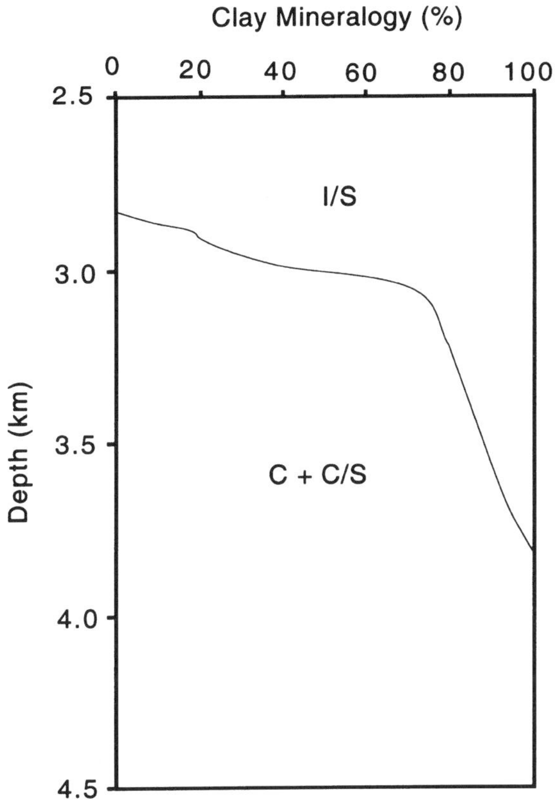

Fig. 8 Relative proportion of clay minerals with depth in Santos Basin sandstones: I/S, illite–smectite; C, chlorite; C/S, chlorite–smectite.

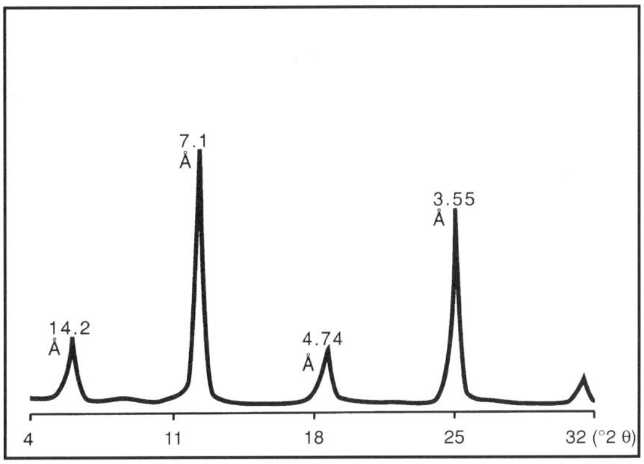

Fig. 9 Typical X-ray diffraction pattern of an orientated, air-dried, fine (< 5 μm) fraction of Santos Basin deep reservoirs, depicting the characteristic peaks of a ferroan chlorite.

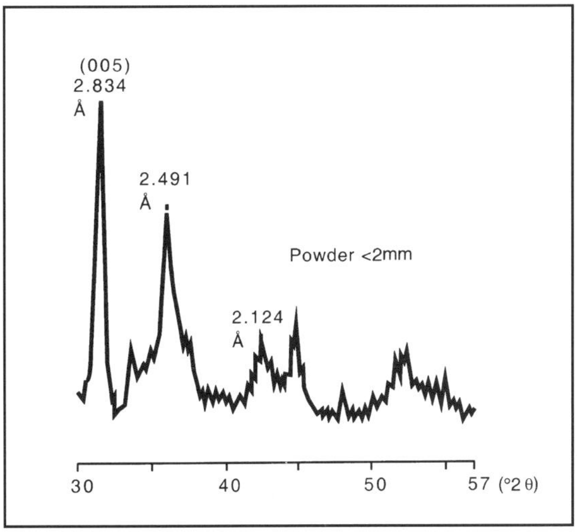

Fig. 10 X-ray diffraction pattern of the < 2 μm, randomly orientated fraction of an Ilhabela Member sandstone, indicating a Ib (β = 90°) chlorite polytype.

compaction and dissolution of the framework (see Plate 1c, facing p. 296; Fig. 7d).

X-ray diffraction analyses of sandstone clay fractions show that chlorite relative content increases with depth (Fig. 8) and that the chlorites are dominantly ferroan (chamosite; Fig. 9).

Representative microprobe analyses of the chlorite rims are presented in Table 2, which shows an FeO content varying from 13 to 27 mol.% and MgO varying from 11 to 22 mol.%. The results show a variation from ferroan chlorites with an FeO/MgO ratio of approximately 20 : 10, to magnesian chlorites with FeO/MgO of approximately 15 : 20. Such magnesian chlorites occur in thick fringes (20 μm average thickness) in sandstones close to domes of the Aptian salts. The paragenetic relationships of the two compositional varieties are the same.

Polytype determination by slow diffractometry from 30 to 57°2θ shows that the ferroan chlorites are of the Ib (β = 90°) polytype (Fig. 10), and the magnesian chlorites of the IIb (β = 97°) polytype. Such relationship between polytype and composition is common in diagenetic chlorites (Hillier, 1994).

Microprobe analyses of the coatings preserved along intergranular contacts and below

Table 2 Representative microprobe analyses and calculated formulae (base on 14 oxygens per formula unit) of authigenic chlorites in Santos Basin sandstones.

Analysis	20C1P2	20C1P3	20C210	20C4P3	54C1P2	54C1P3	54C1P5	54C1P7	54C2P5	54C2P6	54C2P7	54C3P1	54C3P3	54C3P5
SiO_2	28.5	24.4	23.2	23.0	31.1	32.2	28.2	32.1	31.0	32.6	28.7	31.4	28.3	30.4
Al_2O_3	15.4	16.8	18.3	18.6	15.3	14.7	16.2	15.5	15.5	14.8	14.4	15.6	15.7	15.2
FeO	27.5	21.4	22.0	21.7	15.4	14.3	14.2	14.1	14.6	14.0	13.0	14.4	13.8	15.4
MgO	12.7	11.5	11.5	11.5	20.6	22.1	18.9	21.5	20.9	21.7	21.0	22.4	20.4	20.6
TiO_2	0.8	0.0	0.0	0.1	0.0	0.1	0.0	0.0	0.1	0.0	0.0	0.0	0.1	0.1
MnO	0.3	0.0	0.0	0.1	0.0	0.0	0.0	0.0	0.0	0.0	0.0	0.0	0.0	0.0
CaO	0.1	0.2	0.1	0.2	0.1	0.1	0.1	0.1	0.2	0.2	0.2	0.2	0.2	0.2
Na_2O	0.0	0.0	0.0	0.0	0.0	0.0	0.0	0.0	0.1	0.0	0.0	0.0	0.0	0.0
K_2O	0.5	0.0	0.1	0.1	0.0	0.0	0.0	0.0	0.0	0.0	0.0	0.0	0.0	0.0
Total	85.9	74.4	75.4	75.2	82.6	83.6	77.7	83.3	82.3	83.4	77.3	84.1	78.6	81.9
Si	3.11	3.00	2.83	2.81	3.28	3.32	3.15	3.32	3.26	3.36	3.21	3.22	3.13	3.23
Al	1.98	2.43	2.64	2.67	1.89	1.79	2.14	1.89	1.92	1.80	1.90	1.90	2.04	1.90
Fe^{2+}	2.52	2.20	2.25	2.22	1.36	1.23	1.33	1.22	1.28	1.21	1.21	1.24	1.28	1.37
Mg	2.07	2.11	2.09	2.10	3.23	3.41	3.15	3.30	3.27	3.34	3.49	3.44	3.37	3.27
Ti	0.07	0.00	0.00	0.01	0.00	0.00	0.00	0.00	0.01	0.00	0.00	0.00	0.01	0.01
Mn	0.03	0.00	0.00	0.01	0.00	0.00	0.00	0.00	0.00	0.00	0.00	0.00	0.00	0.00
Ca	0.02	0.03	0.02	0.03	0.01	0.01	0.01	0.01	0.02	0.02	0.02	0.03	0.03	0.02
Na	0.00	0.00	0.00	0.00	0.00	0.00	0.00	0.00	0.01	0.00	0.00	0.01	0.00	0.00
K	0.07	0.01	0.02	0.01	0.01	0.01	0.00	0.00	0.00	0.00	0.00	0.00	0.00	0.01
Al^{IV}	0.9	1.0	1.2	1.2	0.7	0.7	0.9	0.7	0.7	0.6	0.8	0.8	0.9	0.8
Al^{VI}	1.1	1.4	1.5	1.5	1.2	1.1	1.3	1.2	1.2	1.2	1.1	1.1	1.2	1.1

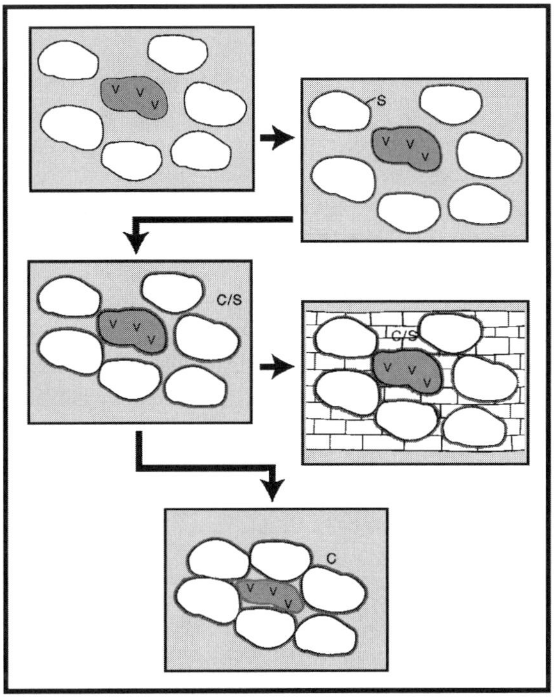

Fig. 11 Schematic representation of the evolution of chlorite authigenesis in Santos sandstones. Eogenetic smectite (S) pore-linings are chloritized through chlorite–smectite (C/S) mixed layers, which are preserved within areas tightly cemented by mesogenetic calcite, with evolution of porous areas leading to the growth of chlorite rims. Shaded particles denoted 'V' are volcanic rock fragments.

occurrence of ferroan chlorites derived from the transformation of smectitic clays in Santos sandstones, as indicated by diffractometric and microprobe analyses, suggests that the FeO/MgO ratio cannot be used directly to determine the precursor clays.

In addition to the replacement of authigenic pore-lining smectites, in some sandstones, abundant chlorite was formed owing to the extensive replacement of smectitic pseudomatrix from the compaction of mud intraclasts, or of smectitic bioturbation matrix.

If the Santos chlorites were derived from the replacement of green ferroan clays such as berthierine, 'verdine', 'odinite', or 'phyllite V' (Odin, 1985, 1988), then remnants of these 7 Å clay minerals, or mixed-layers with serpentine–chlorite structure, would be detected in the shallow sandstones, such as in the Tuscaloosa sandstones of the Gulf Coast (Ryan & Reynolds, 1996), and in some Jurassic sandstones of the North Sea (Hillier, 1994). In that case, the preferential concentration of chlorite should occur in estuarine, deltaic or inner shelf facies, as syngenetic and early eogenetic ooids, precompactional coatings and 'matrix'. Against this argument, the larger chlorite concentrations in Santos sandstones occur as rims in turbidite facies. The absence of ooids of chamosite, considered as important evidence for derivation of chlorites from such 7 Å clay minerals in other units (Stonecipher & May, 1990; Ehrenberg, 1993) also attest against such types of chlorite precursors for the occurrence studied.

Likewise, a derivation of Santos chlorites from the mesogenetic destruction of kaolinite seems unlikely, considering that kaolinite practically is absent from the shallow-buried sandstones, as also is eogenetic siderite or other ferroan carbonates with which interaction with kaolinite might promote such reaction (cf. Hutcheon *et al.*, 1980; Morad *et al.*, 1998).

On the other hand, direct mesogenetic neoformation is interpreted as an important process of chlorite authigenesis in Santos sandstones, as indicated by the high crystallinity of the thick rims (Fig. 7a), as well as by the local occurrence of chlorite rims covering late albite and quartz outgrowths, and by the increasing amount of chlorite with depth. It is, however, impossible to define precisely the relative amounts of chlorite that were directly neoformed and of chlorite derived from the transformation and replacement of smectitic precursors, because neoformed chlorite was commonly precipitated on top of chloritized smectite coatings and rims (Fig. 7c & d). Therefore, the main processes of authigenesis of chlorite within Santos Basin sandstones were the transformation of eogenetic smectite pore-linings and the neoformation directly from pore-fluids during increasing burial.

Sources and sites of authigenesis

The ionic sources for chlorite authigenesis include the alteration of VRF (and other ferromagnesian grains), smectitic clay minerals derived from such alteration, detrital smectitic clays in inherited coatings, syndepositional and bioturbation matrix and from associated mudrocks and Aptian evaporites (Fig. 3). The relative importance of these sources for the authigenesis of chlorite in Santos Basin sandstones is impossible to calculate based on the presently known petrologic, sedimentological, stratigraphical and structural parameters. However, some recognized patterns deserve a specific discussion.

The alteration of VRF was clearly a process of major importance, although there is not a simple, direct relationship between the amount of VRF and chlorite. Volcanic rock fragments are, as discussed previously, distributed without significant areal variation throughout the basin. Such random distribution plus the textural and compositional characteristics of VRF (mostly, the granophyric character of acidic VRF) indicate that they are derived from the erosion of the Early Cretaceous Serra Geral Formation, owing to the uplift of the eastern margin of the Paraná Basin, and not from the Late Cretaceous alkaline magmatism (Fig. 2).

Chlorite distribution varies mostly in relation to present depth (suggesting an accretional authigenesis with increasing burial), and with facies, which will be discussed further. Therefore, the amount of VRF although important, has not directly controlled the intensity of chlorite authigenesis. This occurs because other processes and parameters are superimposed on the bulk chlorite generation from VRF alteration. One possible origin of such interference would be the variation of VRF composition, but the occurrence of basic and acidic VRF shows a very good correlation within the basin (thereby indicating a common Serra Geral source; Fig. 5). The types (e.g. dissolution versus replacement) and the intensity of alteration would provide another influence on chlorite distribution, but no clear relationship was detected in the petrography.

Although the Upper Cretaceous alkaline magmatism (Fig. 2; Mizusaki *et al.*, 1992) was not a significant source for the VRF, it may have had some influence on diagenesis at a local scale. Such an influence could be related to the thermal convection of pore-fluids promoted by thermal anomalies connected to large intrusions and volcanic centres, and to the input of juvenile, late magmatic fluids to the elemental and isotopic composition of mesogenetic pore-fluids.

As there is no detected variation in the amounts of volcanic rock fragments of Juréia shelf and Ilhabela turbidite sandstones, the observed paucity of chlorite and consequent abundance of quartz cementation in the Juréia sandstones must be related to other factors. A probable explanation is that the geochemical system of the shelf sandstones was more open than that of the turbidites, presumably owing to the thicker and laterally better connected bodies of the shelf sandstones. As a result of a more open system, the products of dissolution of unstable volcanic material (basic glass and mafic minerals) could be removed from the sandstones as alteration occurred. Conversely, in the turbidites, Fe, Mg, Si and Al were retained at the local scale within chlorite and/or its smectitic precursors. As a result, the turbidites experienced a stronger eogenetic alteration of VRF to smectites, and consequently more chlorite formation during mesodiagenesis. The occurrence of magnesian chlorites of polytype IIb ($\beta = 97°$) in some samples suggests that evaporites may have acted as a source and/or are responsible for an enhanced thermal regime at these sites.

Therefore, the alteration of VRF from the Lower Cretaceous Serra Geral Formation, although recognized as the primary source for chlorite authigenesis, was not the direct control on the amount of chlorite within each area and sandstone body. Depositional facies (and consequent pore-fluid flow patterns), timing of alteration and associated shales and evaporites may have played a major role on the volumetric distribution of chlorite in the sandstones.

Temperature and timing

Chlorite rims in the Itajaí-Açu sandstones previously were considered as an eogenetic product of VRF alteration (Sombra *et al.*, 1990).

However, the spatial and temporal distribution, variable habits and isotopic composition of chlorite indicate a far more complex genetic scenario. The systematic increase in chlorite amounts with depth, both relative to the clay fraction (Fig. 10), and in absolute volumes, indicates an incremental authigenesis during mesodiagenesis, with increasing depth and temperature.

On the other hand, the paragenetic relationships of authigenic chlorite are complex. In some samples thin chlorite coatings and rims occur along intergranular contacts and are covered by pore-filling calcite with isotopic and fluid-inclusion temperatures indicative of moderate burial. In other samples, thick chlorite rims cover paragenetically late quartz and albite outgrowths. This discordant paragenetic relationship results from the occurrence both of chloritized eogenetic, pre-compactional smectitic pore-linings, and neoformed chlorites in the studied sandstones.

The geothermometric evaluation from the $\delta^{18}O$ values of authigenic chlorites relies upon knowledge of the approximate $\delta^{18}O$ of the precipitating pore-fluids (e.g. Land & Dutton, 1978; Longstaffe, 1984, 1986, 1994; Dutton & Land, 1985; Longstaffe *et al.*, 1992), and is strongly influenced by compositional microheterogeneities related to remnants of eogenetic clay mineral precursors, and to compositional zonations. Such zonations indicate that the chlorites grew during diagenesis by a sort of Ostwald-ripening mechanism, with an incomplete compositional and isotopic re-equilibration with pore-fluids (Jahren, 1991; Jahren & Aagaard, 1993). Chlorite isotopic values therefore represent average values.

Considering those limitations, the precipitation temperature of a ferroan chlorite Ib polytype with a $\delta^{18}O$ value of +11.8‰ in an Ilhabela turbidite reservoir was calculated to be 55°C applying the Wenner & Taylor (1971) equation and assuming a $\delta^{18}O_{water}$ of +2.0‰ SMOW (for a mesodiagenetic fluid moderately evolved from sea water; Land & Fisher, 1987). The precipitation temperature of a magnesian chlorite with a $\delta^{18}O$ value of +9.6‰ in a Juréia sandstone was calculated to be 83°C using the same equation and $\delta^{18}O_{water}$ value. However, as discussed previously, these isotopic and temperature values are influenced by remnants of isotopically heavy, eogenetic precursor smectites, and by the gradual and progressive chlorite authigenesis with increasing depth. This suggests that the authigenesis of chlorite may have started at temperatures as low as 50°C, but apparently has continued down to present depths and temperatures. Unfortunately, there are no available formation water analyses to evaluate whether the chlorites are in equilibrium with the present fluid compositions.

Diagenetic chlorites occur in sandstones as four structural polytypes, that apparently correspond to an order of increasing formation temperature: Ib_d (disordered), Ib ($\beta = 97°$), Ib ($\beta = 90°$) and IIb ($\beta = 97°$) (Hayes, 1970; Hillier, 1994). However, chlorite geothermometry derived directly from polytype determination is still largely qualitative and poorly constrained (Caritat *et al.*, 1993). The transition temperature from the Ib to the IIb polytype may range from 50°C (Walker, 1989) to 200°C (Hayes, 1970), and both polytypes may coexist at advanced burial conditions (Spötl *et al.*, 1994), which is critical for the geothermometric use of polytypes within the narrow and sensitive temperature range of diagenesis.

Empirical geothermometers derived from major compositional variations (e.g. the proportion of tetrahedral Al; Cathelineau, 1988) may yield satisfactory results only if applied to settings similar to those of the original data (e.g. bulk-rock composition, temperature range). They also are very sensitive to the coexisting mineral assemblage, to kinetics and to analytical limitations, especially to small-scale heterogeneities as a result of precursor clay remnants and zonation (Caritat *et al.*, 1993; Spötl *et al.*, 1994). Calculations of chlorite temperature in Santos sandstones using the Cathelineau (1988) geothermometric approach yielded an unrealistic temperature range of 177°C to 377°C. This probably is related to remnants of smectitic precursors, as observed in other diagenetic chlorites (e.g. Spötl *et al.*, 1994).

More evidence for the geothermometry of chlorite authigenesis is supplied by the coexisting calcite cement. The range of precipitation temperatures calculated from the $\delta^{18}O$ values of calcite cements that cover chlorite rims (Table 1) vary from approximately 50 to 75°C, assuming a $\delta^{18}O_{water}$ of −1.2‰ (for marine Cretaceous water; Shackleton & Kennett, 1975), or between 70 and 100°C, assuming more realistically a $\delta^{18}O_{water}$ of +2‰ (for a mesodiagenetic fluid moderately evolved from sea water; Land & Fisher, 1987). The $\delta^{13}C$ values for such pore-filling calcite cement in the Ilhabela turbidites point to a source from thermal decarboxylation or a mixing between decarboxylation and organic matter alteration by bacterial fermentation (Table 1). Pre-compactional calcite cements in the Juréia shelf sandstones present $\delta^{13}C$ values indicative of marine, bacterial sulphate reduction and fermentation processes (Table 1). The common occurrence of eogenetic carbonate cements with marine $\delta^{13}C$ signatures in the Juréia sandstones most probably is related to a larger availability of bioclastic and intraclastic carbonate in the shelf settings.

The fact that these calcite cements are coexisting, but not co-precipitated with the chlorite, prevent the application of a thermodynamic approach to chlorite geothermometry (Hutcheon, 1990). Besides, these theoretical thermodynamic geothermometers may yield an error of as much as ±100°C in the calculated temperatures, which is unacceptable within the narrow range of diagenetic chlorites (Caritat *et al.*, 1993).

Therefore, none of the geothermometers discussed provides a solid evaluation for the temperature of chlorite authigenesis in Santos sandstones. The temperatures derived from isotopic values are probably average or minimum temperatures of formation, considering the mixture with unreplaced precursor clays, and the evidence that chlorite authigenesis was a gradual process.

Reservoir implications

The occurrence of abnormally high porosity values in several deep sandstone reservoirs containing authigenic chlorite has received substantial attention in the literature. This is a result of the interpretation that chlorite pore-linings help to preserve porosity by the inhibition of quartz cementation and/or pressure dissolution (e.g. Pittman & Lumsden, 1968; Heald & Larese, 1974; Dutton, 1977; Thomson, 1979; Tillman & Almon, 1979; Dutton & Land, 1985; Dixon *et al.*, 1989; Pittman *et al.*, 1992; Ehrenberg, 1993; Hillier, 1994; Spötl *et al.*, 1994; Ryan & Reynolds, 1996). None of these studies, however, has provided a convincing explanation of the mechanisms by which chlorite coatings or rims effectively inhibit quartz precipitation on quartz grain surfaces, or intergranular pressure-dissolution. Illite pore linings are not as effective as chlorite pore linings in inhibiting quartz overgrowth precipitation and commonly are engulfed within coeval or later quartz overgrowths (Dutton & Land, 1988; Burley & MacQuaker, 1992; Morad *et al.*, 1994; De Ros *et al.*, 2000). In relation to chemical compaction through pressure-dissolution, illite pore-linings seem to enhance intergranular pressure dissolution (Thomson, 1959; Weyl, 1959; Chilingarian, 1983; Tada & Siever, 1989; Bjørkum, 1996). Therefore, the inhibitory action of chlorite upon quartz diagenesis is not purely a matter of insulation of the grains from the pore-fluids, but probably is related to specific and related microenvironmental geochemical conditions yet to be revealed.

Chlorite pore-linings are beneficial for the quality of deep reservoirs only where they are relatively thin. For larger amounts of chlorite and thicker rims, the permeability decreases sharply and the irreducible water saturation contained in chlorite microporosity increases, thereby affecting oil-saturation evaluation through resistivity logs. The authigenesis of abundant chlorite in many lithic volcaniclastic sandstones commonly results in very poor reservoir characteristics (Galloway, 1979; Surdam & Boles, 1979; Seeman & Scherer, 1984; Mathisen & McPherson, 1991). Sublithic sandstones with a limited amount of VRF are the ideal compositional petrofacies for the development of continuous, but not thick, chlorite rims. Therefore, there is an 'optimum thickness' for chlorite rims to effectively inhibit quartz cementation and pressure dissolution, but without radically decreasing permeability and increasing the proportion of microporosity in reservoirs. A similar interpretation, but in relation to an 'optimum volume' of chlorite coatings, was observed for the Tuscaloosa reservoirs by Pittman *et al.* (1992). This 'optimum thickness' comprehensively varies with grain-size (and consequently pore and pore-throat size), and its quantitative determination would demand detailed and systematic microscopic measurements, preferably within a comparative study of several chlorite-rich reservoirs. Within Santos Basin sandstones, the optimum thickness of chlorite rims seems to be between 10 and 5 μm, similar to the Tuscaloosa Sand.

Models for the prediction of quality of chlorite-rich sandstone reservoirs must consider the complex array of chlorite-forming processes. Models based on a single genetic process (e.g. deltaic–estuarine facies for chlorites from berthierine–odinite–verdine precursors) may work for some specific units and settings, but cannot be applied to others, and clearly are not operational for a complex diagenetic scenario such as that of the Ilhabela and Juréia sandstones.

As VRF distribution does not show significant variation and correlation with chlorite amounts in Santos sandstones, the variables dominantly controlling chlorite authigenesis and reservoir quality are depositional facies, burial history and associated lithology. Juréia shelf sandstones normally show less chlorite, and consequently more quartz cementation and pressure-dissolution than turbidites.

Ilhabela turbidite sandstones show continuous and regular chlorite pore-linings and consequently better reservoir quality potential, apparently because of an eogenetic geochemical environment favourable to the formation of precursor smectitic clay coatings, and potentially also as a result of the diffusive ionic supply from interlayered shales.

Finally, the presence of authigenic chlorite presents other potential reservoir problems. Electrical log low-resistivity anomalies are generated in several reservoirs by the irreducible water contained in the microporosity along chlorite pore-linings. Additionally, chlorite is acid-sensitive, and therefore well stimulation and completion fluids must be designed to prevent the precipitation of iron hydroxide subsequent to its dissolution, usually by the addition of suitable Fe-chelating agents.

CONCLUSIONS

This study has shown that the Upper Cretaceous, anomalously porous, deep clastic reservoirs of the Santos Basin (more than 20% porosity at more than 4000 m) owe their porosity preservation mostly to the inhibitory effect of chlorite rims on quartz cementation and pressure-dissolution.

The sandstones are mostly fine-grained lithic arkoses rich in basic and acidic volcanic rock fragments derived from the uplift and erosion of the Lower Cretaceous Serra Geral Formation of the Paraná Basin. The amounts of basic and acidic volcanic fragments show correlative and similar amounts throughout the basin.

Turbidite sandstones of the Ilhabela Member show greater chlorite content, and consequently higher values of porosity, than shelf sandstones of the Juréia Formation, which commonly are cemented by quartz, calcite or dolomite. Other subordinate diagenetic constituents include albite, TiO_2, pyrite, anhydrite and halite.

Chlorite pore-linings were formed under mesogenetic conditions ($T > 50°C$) from the replacement of precursor smectitic coatings derived from the eogenetic alteration of volcanic rock fragments, followed by the neoformation of chlorite rims on top of the chloritized coatings. The smectitic precursor is indicated by remnants of chlorite–smectite mixed layers preserved below pervasive calcite cements and along tight intergranular contacts. Basic volcanic fragments and other ferromagnesian grains also were chloritized extensively.

Sources of ions for chlorite authigenesis

include detrital smectites (within associated mudrocks, syndepositional and bioturbation matrix and inherited coatings on reworked grains), volcanic rock fragments, other ferromagnesian grains and eogenetic smectites derived from their alteration, and Aptian evaporites. Berthierine–odinite (7 Å ferroan clay minerals) were not involved in chlorite evolution.

The ideal reservoir quality conditions are present in Ilhabela turbidites. In these sandstones, moderate eogenetic smectite formation was followed by limited compaction, chloritization and chlorite neoformation. Sandstones with intense formation of precursor smectite are prone to extensive mesodiagenetic porosity obstruction by abundant chlorite. Sandstones with scarce chlorite are prone to intense quartz cementation and pressure-dissolution during deep burial. Some Juréia shelf sandstones may show moderate permeabilities where a limited eogenetic formation of smectite evolved into discontinuous chlorite authigenesis, and include limited reduction of porosity by quartz cementation and pressure-dissolution.

The reasons for the greater potential for chlorite rim development in turbidite sandstones compared with shelf sandstones, and their better reservoir quality are poorly established. We can only speculate about the conditions of these two diagenetic environments that led to such contrast. Apparently, the products of eogenetic alteration of VRF were retained at the local scale as precursor smectites in the more closed geochemical systems of the turbidites, whereas in the shelf sandstones they were removed. Additionally, the turbidites experienced larger amounts of chlorite authigenesis during burial, owing to diffusion from interlayered mudrocks.

This study addresses the complex factors controlling chlorite authigenesis and reservoir quality in the Upper Cretaceous marine sandstones of the Santos Basin. However, the development of operational reservoir quality prediction models will demand more detailed information on the history of burial, hydrocarbon generation and migration, and on the evolution of major faults and halokinetic structures.

ACKNOWLEDGEMENTS

The authors thank PETROBRAS for access to samples, data, information, and for the licence to publish this work. Special thanks to A. Roisenberg and A.M.P. Mizusaki for information on Cretaceous magmatism, and to A.S. Mexias for help with chlorite formula calculation. L.F. De Ros acknowledges the support of the Brazil National Research Council (CNPq), and Rio Grande do Sul State Research Foundation (FAPERGS). The original manuscript was improved by comments of Jeffry D. Grigsby, and an anonymous reviewer. We all gratefully thank editors Sadoon Morad and Richard Worden for initial encouragement and final patience.

REFERENCES

Almeida, F.F.M. & Carneiro, C.D.R. (1998) Origem e evolução da Serra do Mar. *Rev. Brasil. Geociênc.*, **28**(2), 135–150.

Anjos, S.M.C., Silva, C.M.A. & Allmeida, M.S. (1998) Chlorite prediction in deep reservoirs of Santos Basin, Brazil. In: *1998 AAPG International Conference and Exhibition, Abstracts Volume*, The American Association of Petroleum Geologists, Rio de Janeiro, pp. 158–159.

Bjørkum, P.A. (1996) How important is pressure in causing dissolution of quartz in sandstones? *J. Sed. Res.*, **A66**(1), 147–154.

Boles, J.R. & Franks, S.G. (1979) Clay diagenesis in Wilcox sandstones of southwest Texas: implications of smectite diagenesis on sandstone cementation. *J. Sed. Petrol.*, **49**(1), 55–70.

Borer, J.M. & Harris, P.M. (1991) Depositional facies and model for mixed siliciclastics and carbonates of the Yates Formation, Permian Basin. In: *Mixed Carbonate–Siliciclastic Sequences*, Society of Economic Paleontologists and Mineralogists Core Workshop 15 (Ed. by A.J. Lomando and P.M. Harris), Dallas, TX, pp. 1–133.

Burley, S.T. & MacQuaker, J.H.S. (1992) Authigenic clays, diagenetic sequences and conceptual diagenetic models in contrasting basin-margin and basin-center North Sea Jurassic sandstones and mudstones. In: *Origin, Diagenesis, and Petrophysics of Clay Minerals in Sandstones*, Society of Economic Paleontologists and Mineralogists Special Publication 47 (Ed. by D.W. Houseknecht and E.D. Pittman), Tulsa, OK, pp. 81–110.

Burns, L.K. & Ethridge, F.G. (1979) Petrology and diagenetic effects of lithic sandstones: Paleocene and Eocene Umpqua Formation, southwest Oregon. In: *Aspects of Diagenesis*, Society of Economic Paleontologists and Mineralogists Special Publication 26 (Ed. by P.A. Scholle and P.R. Schluger), Tulsa, OK, pp. 307–317.

Burton, J.H., Krinsley, D.H. & Pye, K. (1987) Authigenesis of kaolinite and chlorite in Texas Gulf Coast sediments. *Clay. Clay Mineral.*, **35**(4), 291–296.

Caritat, P.D., Hutcheon, I. & Walshe, J.L. (1993) Chlorite geothermometry: a review. *Clay. Clay Mineral.*, **41**, 219–239.

Cathelineau, M. (1988) Cation site occupancy in chlorites and illites as a function of temperature. *Clay Mineral.*, **23**, 471–485.

Chan, M.A. (1985) Correlations of diagenesis with sedimentary facies in Eocene sandstones, western Oregon. *J. Sed. Petrol.*, **55**, 322–333.

Chang, H.K., Mackenzie, F.T. & Schoonmaker, J. (1986) Comparisons between the diagenesis of dioctahedral and trioctahedral smectite, Brazilian offshore basins. *Clay. Clay Mineral.*, **34**(4), 407–423.

Chilingarian, G.V. (1983) Compactional diagenesis. In: *Sediment Diagenesis*, NATO ASI Series C: Mathematical and Physical Sciences (Ed. by A. Parker and B.W. Sellwood), Reidel, Dordrecht, pp. 57–168.

Clayton, R.N. & Mayeda, T.K. (1963) The use of bromine pentafluoride in the extraction of oxygen from oxides and silicates for isotopic analysis. *Geochim. Cosmochim. Acta*, **27**, 43–52.

Craig, H. (1957) Isotopic standards for carbon and oxygen correction factors for mass spectrometric analysis of carbon dioxide. *Geochim. Cosmochim. Acta*, **12**, 133–149.

Curtis, C.D. (1990) Aspects of climatic influence on the clay mineralogy and geochemistry of soils, palaeosols and clastic sedimentary rocks. *J. Geol. Soc. London*, **147**, 351–357.

Davies, D.K., Almon, W.R., Bonis, S.B. & Hunter, B.E. (1979) Deposition and diagenesis of Tertiary–Holocene volcaniclastics, Guatemala. In: *Aspects of Diagenesis*, Society of Economic Paleontologists and Mineralogists Special Publication 26 (Ed. by P.A. Scholle and P.R. Schluger), Tulsa, OK, pp. 281–306.

De Ros, L.F., Anjos, S.M.C. & Morad, S. (1994) Authigenesis of amphibole and its relationship to the diagenetic evolution of Lower Cretaceous sandstones of the Potiguar rift basin, northeastern Brazil. *Sed. Geol.*, **88**, 253–266.

De Ros, L.F., Morad, S. & Al-Aasm, I.S. (1997) Diagenesis of siliciclastic and volcaniclastic sediments in the Cretaceous and Miocene sequences of NW African margin (DSDP Leg 47A, Site 397). *Sed. Geol.*, **112**, 137–156.

De Ros, L.F., Morad, S., Broman, C., Césero, P. & Gomez-Gras, D. (2000) Influence of uplift and magmatism on distribution of quartz and illite cementation: evidence from Siluro-Devonian sandstones of the Paraná Basin, Brazil. In: *Quartz Cementation in Sandstones*, International Association of Sedimentologists Special Publication 29 (Ed. by R.H. Worden and S. Morad), Blackwell Science, Oxford, pp. 231–252.

Dixon, S.A., Summers, D.M. & Surdam, R.C. (1989) Diagenesis and preservation of porosity in Norphlet Formation (Upper Jurassic), southern Alabama. *AAPG Bull.*, **73**(6), 707–728.

Dutton, S.P. (1977) Diagenesis and porosity distribution in deltaic sandstones, Strawn Series (Pennsylvanian), North-central Texas. *Trans. Gulf Coast Assoc. Geol. Soc.*, **27**, 272–277.

Dutton, S.P. & Land, L.S. (1985) Meteoric burial diagenesis of Pennsylvanian arkosic sandstones, southwestern Anadarko Basin, Texas. *AAPG Bull.*, **69**(1), 22–38.

Dutton, S.P. & Land, L.S. (1988) Cementation and burial history of a low-permeability quartzarenite, Lower Cretaceous Travis Peak Formation, East Texas. *Geol. Soc. Am. Bull.*, **100**, 1271–1282.

Ehrenberg, S.N. (1993) Preservation of anomalously high porosity in deeply buried sandstones by grain-coating chlorite: examples from the Norwegian continental shelf. *AAPG Bull.*, **77**(7), 1260–1286.

Enciso, G. & Tisi, A.L. (1998) Santos Basin—Merluza Field turbidite model revisited. In: *1998 AAPG International Conference and Exhibition, Abstracts Volume*, The American Association of Petroleum Geologists, Rio de Janeiro, pp. 604–605.

Folk, R.L. (1968) *Petrology of Sedimentary Rocks*. Hemphill's, Austin, TX, 107 pp.

Friedman, I. & O'Neil, J.R. (1977) Compilation of stable isotopic fractionation factors of geochemical interest. In: *Data of Geochemistry*, 6th edn (Ed. by M. Fleischer). *US Geol. Surv. Prof. Pap.*, **440-KK**, 12 pp.

Galloway, W.E. (1979) Diagenetic control of reservoir quality in arc-derived sandstones: implications for petroleum exploration. In: *Aspects of Diagenesis*, Society of Economic Paleontologists and Mineralogists Special Publication 26 (Ed. by P.A. Scholle and P.R. Schluger), Tulsa, OK, pp. 251–262.

Gaupp, R., Matter, A., Platt, J., Ramseyer, K. & Walzebuck, J. (1993) Diagenesis and fluid evolution of deeply buried Permian (Rotliegende) gas reservoirs, northwest Germany. *AAPG Bull.*, **77**(7), 1111–1128.

Goodchild, M.W. & Whitaker, J.H.M. (1986) A petrographic study of the Rotliegendes sandstone reservoir (lower Permian) in the Rough gas field. *Clay Mineral.*, **21**, 459–477.

Hay, R.L. & Wiggins, B. (1980) Pellets, ooids, sepiolite and silica in three calcretes of the southwestern United States. *Sedimentology*, **27**, 559–576.

Hayes, J.B. (1970) Polytypism of chlorite in sedimentary rocks. *Clay. Clay Mineral.*, **18**, 285–306.

Heald, M.T. & Larese, R.E. (1974) Influence of coatings on quartz cementation. *J. Sed. Petrol.*, **44**, 1269–1274.

Hillier, S. (1994) Pore-lining chlorites in siliciclastic reservoir sandstones: electron microprobe SEM and XRD data, and implications for their origin. *Clay Mineral.*, **29**(4), 665–679.

Humphreys, B., Kemp, S.J., Lott, G.K., Bermanto, D.A., Dharmayanti, D.A. & Samsori, I. (1994) Origin of grain-coating chlorite by smectite transformation: an example from Miocene sandstones, North Sumatra back-arc basin, Indonesia. *Clay Mineral.*, **29**(4), 681–692.

Hurst, V.J. & Kunkle, A.C. (1985) Dehydroxylation, rehydroxylation and stablility of kaolinite. *Clay. Clay Mineral.*, **33**(1), 1–14.

Hutcheon, I. (1990) Clay-carbonate reactions in the Venture area, Nova Scotia, Canada. In: *Fluid–Mineral Interactions: A Tribute to H.P. Eugster*, The Geochemical Society of Canada Special Publication 2 (Ed. by R.J. Spencer and I.-M. Chou), Ottawa, Ontario, pp. 199–212.

Hutcheon, I., Oldershaw, A. & Ghent, E.D. (1980) Diagenesis of Cretaceous sandstones of the Kootenay Formation at Elk Valley (southeastern British Columbia) and Mt. Allan (southwestern Alberta). *Geochim. Cosmochim. Acta*, **44**, 1425–1435.

Irwin, H., Curtis, C. & Coleman, M.L. (1977) Isotopic evidence for source of diagenetic carbonates formed during burial of organic-rich sediments. *Nature*, **269**, 209–213.

Jahren, J.S. (1991) Evidence of Ostwald ripening related recrystallization of diagenetic chlorites from reservoir rocks offshore Norway. *Clay Mineral.*, **26**, 169–178.

Jahren, J.S. & Aagaard, P. (1993) Diagenetic illite–chlorite assemblages in arenites. I. Chemical evolution. *Clay. Clay Mineral.*, **40**, 540–546.

Janks, J.S., Yusas, M.R. & Hall, C.M. (1992) Clay mineralogy of an interbedded sandstone, dolomite and anhydrite: the Permian Yates Formation, Winkler County, Texas. In: *Origin, Diagenesis, and Petrophysics of Clay Minerals in Sandstones*, Society of Economic Paleontologists and Mineralogists Special Publication 47 (Ed. by D.W. Houseknecht and E.D. Pittman), Tulsa, OK, pp. 145–157.

Kessler II, L.G. (1978) Diagenetic sequence in ancient sandstones deposited under desert climatic conditions. *J. Geol. Soc. London*, **135**, 41–49.

Land, L.S. & Dutton, S.P. (1978) Cementation of a Pennsylvanian deltaic sandstone: Isotopic data. *J. Sed. Petrol.*, **48**(4), 1167–1176.

Land, L.S. & Fisher, R.S. (1987) Wilcox sandstone diagenesis, Texas Gulf Coast: a regional isotopic comparison with the Frio Formation. In: *Diagenesis of Sedimentary Sequences* (Ed. by J.D. Marshall). *Geol. Soc. London Spec. Publ.*, **36**, 219–235.

Longstaffe, F.J. (1984) The role of meteoric water diagenesis of shallow sandstones: stable isotope studies of the Milk River Aquifer and Gas Pool, southeastern Alberta. In: *Clastic Diagenesis* (Ed. by R. Surdam and D.A. McDonald). *AAPG Mem.*, **37**, 81–98.

Longstaffe, F.J. (1986) Oxygen isotope studies of diagenesis in the Basal Belly River Sandstone, Pembina I-Pool, Alberta. *J. Sed. Petrol.*, **56**, 77–88.

Longstaffe, F.J. (1993) Meteoric water and sandstone diagenesis in the western Canada sedimentary basin. In: *Diagenesis and Basin Development*, American Association of Petroleum Geologists Studies in Geology 36 (Ed. by A.D. Hornbury and A.G. Robinson), Tulsa, OK, pp. 49–68.

Longstaffe, F.J. (1994) Stable isotopic constraints on sandstone diagenesis in the western Canada sedimentary basin. In: *Quantitative Diagenesis: Recent Developments and Applications to Reservoir Geology*, NATO ASI Series C: Mathematical and Physical Sciences (Ed. by A. Parker and B.W. Sellwood), Kluwer Academic Publishers, Dordrecht, pp. 223–274.

Longstaffe, F.J., Tilley, B.J., Ayalon, A. & Connolly, C.A. (1992) Controls on porewater evolution during sandstone diagenesis, Western Canada sedimentary basin: an oxygen isotope perspective. In: *Origin, Diagenesis, and Petrophysics of Clay Minerals in Sandstones*, Society of Economic Paleontologists and Mineralogists Special Publication 47 (Ed. by D.W. Houseknecht and E.D. Pittman), Tulsa, OK, pp. 13–34.

Mantovani, M.S.M., Cordani, U.G. & Roisenberg, A. (1985) Geoquímica isotópica em vulcânicas ácidas da Bacia do Paraná e implicações genéticas associadas. *Rev. Brasil. Geociênc.*, **15**(1), 61–55.

Mathisen, M.E. (1984) Diagenesis of Plio-Pleistocene non-marine sandstones, Cagayan Basin, Philipines: early development of secondary porosity in volcanic sandstones. In: *Clastic Diagenesis*, American Association of Petroleum Geologists Memoir 37 (Ed. by R. Surdam and D.A. McDonald), Tulsa, OK, pp. 177–193.

Mathisen, M.E. & McPherson, J.G. (1991) Volcaniclastic deposits: implications for hydrocarbon exploration. In: *Sedimentation in Volcanic Settings*, Society of Economic Paleontologists and Mineralogists Special Publication 45 (Ed. by R.V. Fisher and G.A. Smith), Tulsa, OK, pp. 27–36.

McBride, E.F., Land, L.S. & Mack, L.E. (1987) Diagenesis of eolian and fluvial feldspathic sandstones, Norphlet Formation (Upper Jurassic), Rankin County, Mississippi, and Mobile County, Alabama. *AAPG Bull.*, **71**(9), 1019–1034.

Mello, M.R., Mohriak, W.U., Koutsoukos, E.A.M. & Figueira, J.C.A. (1991) Brazilian and West African oils: generation, migration, accumulation and correlation. In: *Proceedings of Thirteenth World Petroleum Congress, Buenos Aires*, Argentina, Wiley, Chichester, pp. 153–164.

Mizusaki, A.M.P., Petrini, R., Bellieni, G., Comon-Chiaramonti, P., Dias, J.L., De Min, A. & Piccirilo, E.M.

(1992) Basalt magmatism along the passive continental margin of SE Brazil (Campos Basin). *Contrib. Mineral. Petrol.*, **111**, 143–160.

Moncure, G.K., Lahann, R.W. & Siebert, R.M. (1984) Origin of secondary porosity and cement distribution in a sandstone/shale sequence from the Frio Formation (Oligocene). In: *Clastic Diagenesis*, American Association of Petroleum Geologists Memoir 37 (Ed. by D.A. McDonald and R.C. Surdam), Tulsa, OK, pp. 151–161.

Morad, S. (1990) Mica alteration reactions in Jurassic reservoir sandstones from the Haltenbanken area, offshore Norway. *Clay. Clay Mineral.*, **38**(6), 584–590.

Morad, S. (1998) Carbonate cementation in sandstones: distribution patterns and geochemical evolution. In: *Carbonate Cementation in Sandstones*, International Association of Sedimentologists Special Publication 26 (Ed. by S. Morad), Blackwell Scientific Publications, Oxford, pp. 1–26.

Morad, S. & Aldahan, A.A. (1987) Diagenetic chloritization of feldspars in sandstones. *Sed. Geol.*, **51**, 155–164.

Morad, S., Ben Ismail, H., De Ros, L.F., Al-Aasm, I.S. & Serrhini, N.-E. (1994) Diagenesis and formation water chemistry of Triassic reservoir sandstones from southern Tunisia. *Sedimentology*, **41**, 1253–1272.

Morad, S., De Ros, L.F., Nystuen, J.P. & Bergan, M. (1998) Carbonate diagenesis and porosity evolution in sheet-flood sandstones: evidence from the Middle and Lower Lunde Members (Triassic) in the Snorre Field, Norwegian North Sea. In: *Carbonate Cementation in Sandstones*, International Association of Sedimentologists Special Publication 26 (Ed. by S. Morad), Blackwell Scientific Publications, Oxford, pp. 53–85.

Morad, S., Ketzer, J.M. & De Ros, L.F. (2000) Spatial and temporal distribution of diagenetic alterations in siliciclastic rocks: implications for mass transfer in sedimentary basins. *Sedimentology*, **47** (Millenium Reviews), 95–120.

Moraes, M.A.S. & De Ros, L.F. (1992) Depositional, infiltrated and authigenic clays in fluvial sandstones of the Jurassic Sergi Formation, Recôncavo Basin, northeastern Brazil. In: *Origin, Diagenesis and Petrophysics of Clay Minerals in Sandstones*, Society of Economic Paleontologists and Mineralogists Special Publication 47 (Ed. by D.W. Houseknecht and E.W. Pittman), Tulsa, OK, pp. 197–208.

Odin, G.S. (1985) Significance of green particles (glaucony, berthierine, chlorite) in arenites. In: *Provenance of Arenites*, NATO-ASI Series C: Mathematical and Physical Sciences (Ed. by G.G. Zuffa), D. Reidel, Dordrecht, pp. 279–307.

Odin, G.S. (1988) *Green Marine Clays*. Developments in Sedimentology 45, Elsevier, Amsterdam, 445 pp.

Pereira, M.J. & Feijó, F.J. (1994) 21—Bacia de Santos. *Bol. Geociênc. PETROBRAS*, **8**(1), 219–234.

Pereira, M.J. & Macedo, J.M. (1990) A Bacia de Santos: perspectivas de uma nova província petrolífera na plataforma continental sudeste brasileira. *Bol. Geociênc. PETROBRAS*, **4**(1), 3–11.

Piccirilo, E.M., Melfi, A.J., Comin-Chiaramonti, P., *et al.* (1988) Continental flood volcanism from the Paraná Basin (Brazil). In: *Continental Flood Basalts* (Ed. by J.D. Macdougall), Kluwer Academic Publishers, Dordrecht, pp. 195–238.

Pirrie, D., Ditchfield, P.W. & Marshall, J.D. (1994) Burial diagenesis and pore-fluid evolution in a Mesozoic back-arc basin: the Marambio Group, Vega Island, Antarctica. *J. Sed. Res.*, **A64**(3), 541–552.

Pittman, E.D. & Lumsden, D.N. (1968) Relationships between chlorite coatings on quartz grains and porosity: Spiro Sand, Oklahoma. *J. Sed. Petrol.*, **38**(2), 668–670.

Pittman, E.D., Larese, R.E. & Heald, M.T. (1992) Clay coats: occurrence and relevance to preservation of porosity in sandstones. In: *Origin, Diagenesis, and Petrophysics of Clay Minerals in Sandstones*, Society of Economic Paleontologists and Mineralogists Special Publication 47 (Ed. by D.W. Houseknecht and E.D. Pittman), Tulsa, OK, pp. 241–264.

Purvis, K. (1992) Lower Permian Rotliegend sandstones, southern North Sea: a case study of sandstone diagenesis in evaporite-associated sequences. *Sed. Geol.*, **77**, 155–171.

Remy, R.R. (1994) Porosity reduction and major controls on diagenesis of Cretaceous–Paleocene volcaniclastic and arkosic sandstone, Middle Park Basin, Colorado. *J. Sed. Res.*, **A64**(4), 797–806.

Rossel, N.C. (1982) Clay mineral diagenesis in Rotliegend aeolian sandstones of the southern North Sea. *Clay Mineral.*, **17**, 69–77.

Ryan, P.C. & Reynolds, R.C., Jr. (1996) The origin and diagenesis of grain-coating serpentine–chlorite in Tuscaloosa Formation sandstone, U.S. Gulf Coast. *Am. Mineral.*, **81**, 213–225.

Ryan, P.C. & Reynolds, R.C., Jr. (1997) The chemical composition of serpentine/chlorite in the Tuscaloosa Formation, United States Gulf Coast: EDX vs. XRD determinations, implications for mineralogic reactions and the origin of anatase. *Clay. Clay Mineral.*, **45**(3), 339–352.

Searl, A. (1994) Diagenetic destruction of reservoir potential in shallow marine sandstones of the Broadford Beds (Lower Jurassic), north-west Scotland: depositional versus burial and thermal history controls on porosity destruction. *Mar. Petrol. Geol.*, **11**(2), 131–147.

Seeman, U. (1982) Depositional facies, diagenetic minerals and reservoir quality of Rotliegend sediments in the southern Permian Basin (North Sea): a review. *Clay Mineral.*, **17**, 55–67.

Seeman, U. & Scherer, M. (1984) Volcaniclastics as potential hydrocarbon reservoirs. *Clay Mineral.*, **9**, 457–470.

Shackleton, N.J. & Kennett, J.P. (1975) Paleotemperature history of the Cenozoic and the initiation of Antarctic glaciation; oxygen and carbon isotope analyses in DSDP sites 277, 279, and 281. *Init. Rep. Deep Sea Drill. Proj.*, **29**, 743–755.

Silva, C.M.A. & Anjos, S.M.C. (1996) Diagênese dos arenitos do Membro Mucuri, Cretáceo Inferior das Bacias do Espírito Santo e de Mucuri. *Bol. Geociênc. PETROBRAS*, **10**(1/4), 61–80.

Sombra, C.L., Arienti, L.M., Pereira, M.J. & Macedo, J.M. (1990) Parâmetros controladores da porosidade e da permeabilidade nos reservatórios profundos do Campo de Merluza, Bacia de Santos, Brasil. *Bol. Geociênc. PETROBRAS*, **4**(4), 451–466.

Souza, R.S., De Ros, L.F. & Morad, S. (1995) Dolomite diagenesis and porosity preservation in lithic reservoirs: Carmópolis Member, Sergipe-Alagoas Basin, northeastern Brazil. *AAPG Bull.*, **79**(5), 725–748.

Spötl, C., Houseknecht, D.W. & Longstaffe, F.J. (1994) Authigenic chlorites in sandstones as indicators of high-temperature diagenesis, Arkoma foreland basin, USA. *J. Sed. Res.*, **A64**(3), 553–566.

Stonecipher, S.A. & May, J.A. (1990) Facies controls on early diagenesis: Wilcox Group, Texas Gulf Coast. In: *Prediction of Reservoir Quality Through Chemical Modeling* (Ed. by I.D. Meshri and P.J. Ortoleva). *AAPG Mem.*, **49**, 25–44.

Studlick, J.R.J., Shew, R.D., Basye, G.L. & Ray, J.R. (1990) A giant carbon dioxide accumulation in the Norphlet Formation, Pisgah Anticline, Mississippi. In: *Sandstone Petroleum Reservoirs*, Casebooks in Earth Science (Ed. by J.H. Barwis, J.G. McPherson and J.R.J. Studlick), Springer-Verlag, New York, pp. 181–203.

Sturesson, U. (1992) Volcanic ash: the source material for Ordovician chamosite ooids in Sweden. *J. Sed. Petrol.*, **62**, 1084–1094.

Surdam, R.C. & Boles, J.R. (1979) Diagenesis of volcanic sandstones. In: *Aspects of Diagenesis*, Society of Economic Paleontologists and Mineralogists Special Publication 26 (Ed. by P.A. Scholle and P.R. Schluger), Tulsa, OK, pp. 227–242.

Surdam, R.C., Dunn, T.L., Heasler, H.P. & MacGowan, D.B. (1989) Porosity evolution in sandstone/shale systems. In: *Short Course on Burial Diagenesis* (Ed. by I.E. Hutcheon), Mineralogical Association of Canada, Montreal, pp. 61–133.

Tada, R. & Siever, R. (1989) Pressure solution during diagenesis. *Ann. Rev. Earth Planet. Sci.*, **17**, 89–118.

Tang, Z., Parnell, J. & Ruffell, A.H. (1994) Deposition and diagenesis of the lacustrine-fluvial Cangfanggou Group (uppermost Permian to Lower Triassic), southern Junggar Basin, NW China: a contribution from sequence stratigraphy. *J. Paleolimnol.*, **11**, 67–90.

Thomson, A. (1959) Pressure solution and porosity. In: *Silica in Sediments*, Society of Economic Paleontologists and Mineralogists Special Publication 7 (Ed. by H.A. Ireland), Tulsa, OK, pp. 92–110.

Thomson, A. (1979) Preservation of porosity in the deep Woodbine/Tuscaloosa trend, Louisiana. *Trans. Gulf Coast Assoc. Geol. Soc.*, **29**, 396–403.

Thomson, A. & Stancliffe, R.J. (1990) Diagenetic controls on reservoir quality, eolian Norphlet Formation, South State Line Field, Mississippi. In: *Sandstone Petroleum Reservoirs* (Ed. by J.H. Barwis, J.G. McPherson and J.R.J. Studlick), Springer-Verlag, New York, pp. 205–224.

Tillman, R.W. & Almon, W.R. (1979) Diagenesis of Frontier Formation offshore bar sandstones, Spearhead Ranch Field, Wyoming. In: *Aspects of Diagenesis*, Society of Economic Paleontologists and Mineralogists Special Publication 26 (Ed. by P.A. Scholle and V.P. Schluger), Tulsa, OK, pp. 337–378.

Walker, J.R. (1989) Polytypism of chlorite in very low grade metamorphic rocks. *Am. Mineral.*, **74**, 738–743.

Walker, T.R. (1976) Diagenetic origin of continental red beds. In: *The Continental Permian in Central, West, and South Europe* (Ed. by H. Falke), Reidel, Dordrecht, pp. 240–282.

Weedman, S.D., Brantley, S.L., Shiraki, R. & Poulson, S.R. (1996) Diagenesis, compaction, and fluid chemistry modeling of a sandstone near a pressure seal: Lower Tuscaloosa Formation, Gulf Coast. *AAPG Bull.*, **80**(7), 1045–1064.

Wenner, D.B. & Taylor, H.P. (1971) Temperatures of serpentinization of ultramafic rocks based upon O18/O16 fractionation between coexisting serpentine and magnetite. *Contrib. Mineral. Petrol.*, **32**, 165–185.

Weyl, P.K. (1959) Pressure solution and the force of crystallization—a phenomenological theory. *J. Geophys. Res.*, **64**, 2001–2025.

Wilson, M.D. (1992) Inherited grain-rimming clays in sandstones from eolian and shelf environments: their origin and control on reservoir properties. In: *Origin, Diagenesis, and Petrophysics of Clay Minerals in Sandstones*, Society of Economic Paleontologists and Mineralogists Special Publication 47 (Ed. by D.W. Houseknecht and E.D. Pittman), Tulsa, OK, pp. 209–225.

Kaolinite case studies

Int. Assoc. Sedimentol. Spec. Publ. (2003) **34**, 319–342

Origin and diagenetic evolution of kaolin in reservoir sandstones and associated shales of the Jurassic and Cretaceous, Salam Field, Western Desert (Egypt)

R. MARFIL[1], A. DELGADO[2], C. ROSSI[1], A. LA IGLESIA[3] *and* K. RAMSEYER[4]

[1] *Departamento de Petrología y Geoquímica, Universidad Complutense, 28040 Madrid, Spain, e-mail: marfil@geo.ucm.es*

[2] *Laboratorio de Isótopos estables, Estación Experimental del Zaidín (CSIC), 18008 Granada, Spain*

[3] *Instituto de Geología Económica (CSIC), Universidad Complutense, 28040 Madrid, Spain*

[4] *Geologisches Institut, Universität Bern, CH 3012 Bern, Switzerland*

ABSTRACT

Mineralogical, petrological and stable isotopic data have been used to interpret conditions during the formation of authigenic kaolin within Middle Jurassic (Khatatba Formation) and Lower Cretaceous (Alam El Bueib Formation) reservoir sandstones and source-rock shales of the Salam Oil Field, Western Desert, Egypt. The samples studied were selected from cores of five wells, ranging in depth from 2430 to 3600 m and their present temperature varies from 104 to 130°C respectively. The reservoir sandstones are mostly quartzarenites with abundant kaolin, quartz overgrowths and subordinate illite, deposited in fluvial to shallow-marine environments.

Kaolin abundance increases with depth, being almost the only clay mineral in the fluvial sandstones and coaly shales of the Khatatba Formation. Vermiform and blocky kaolin morphologies occur as *pore-filling* cement and replacement of detrital feldspar and mica, whereas *soil-related matrix* consists of microcrystalline kaolinite. Authigenic illite intergrows with blocky kaolin and replaces vermicular kaolin.

Both kaolin polymorphs, kaolinite and dickite are present. Crystals with vermiform morphology are kaolinite, whereas those with blocky morphology are kaolinite and/or dickite. In the clean sandstones, a general depth-related trend of increasing crystal size and dickite abundance is observed. However, in mudrocks or in low permeability sandstones rich in cement or matrix, kaolin minerals do not show any progressive change in morphology and polymorphism, suggesting that the kaolinite-to-dickite transformation possibly occurred in an open diagenetic system.

In kaolin-separates the isotopic values of hydrogen ($\delta D_{SMOW} = -123‰$ to $-61‰$) and oxygen ($\delta^{18}O_{SMOW} = +13.3‰$ to $+18.3‰$) indicate that *soil-related* and early diagenetic *grain-replacive* kaolinite, which precipitated during a period of uplift, still retain the isotopic signature of meteoric-derived water. Later diagenetic well-ordered kaolinite and dickite are characterized by less negative δD ($-54‰$ to $-43‰$) but similar $\delta^{18}O$, most likely owing to equilibrium with present-day porewater $\delta^{18}O$ at elevated temperatures of about 70°C–90°C.

INTRODUCTION

Authigenic kaolin (kaolinite and/or dickite) are, together with illite and chlorite, the most common clay minerals encountered in sandstone reservoirs. However, the origin and diagenetic behaviour of kaolin minerals are poorly understood and still debated (Morad *et al.*, 1994; Osborne *et al.*, 1994; Beaufort *et al.*, 1998; Hassouta *et al.*, 1999). Petrographic and crystallographic studies (e.g. Kantorowitcz, 1984; Glasmann *et al.*, 1989; Giles *et al.*, 1992; Haszeldine *et al.*, 1992) indicate that during burial diagenesis and organic matter maturation kaolinite increases the degree of stacking order and may eventually transform into dickite.

Two morphologies of kaolinite occur in the Jurassic reservoir sandstones of the North Sea (Osborne *et al.*, 1994): (i) early diagenetic vermiform kaolinite, which is often associated with expanded detrital micas, and (ii) later-formed burial diagenetic blocky kaolinite. Ehrenberg *et al.* (1993), McAulay *et al.* (1994), Morad *et al.* (1994), Lanson *et al.* (1996) and Beaufort *et al.* (1998), pointed out that the blocky morphology of kaolinite is commonly dickite, which formed during burial diagenesis and/or in the vicinity of hydrocarbon source rocks. Morad *et al.* (1994) justify the transformation of kaolinite into dickite based on the severe etching of the pseudohexagonal kaolinite remnants. Beaufort *et al.* (1998) showed that the kaolinite to dickite reaction proceeds by a gradual structural change along two distinct paths: accretion of new material from either dissolution of smaller unstable kaolinite crystals and/or detrital minerals (mainly feldspars), and formation of ordered dickite by a dissolution–crystallization process. Ruiz Cruz & Moreno (1993) and Lanson *et al.* (1996) found that the transformation of kaolinite into dickite proceeds via intermediate structures (mixed layers or intergrowths), and the transformation is more complete in the coarser, better sorted permeable sandstones. A general decrease in kaolinite abundance is often observed below 3300 m burial depth, where illite replaces kaolinite. Morad *et al.* (1994) demonstrated that it is the non-transformed kaolinite that is illitized during deep burial.

In the Jurassic and Cretaceous reservoir sandstones of the Salam Field, Egypt, authigenic clay minerals, mainly kaolin, and quartz cement exert a profound control on reservoir quality (Marfil *et al.*, 2000; Rossi *et al.*, 2001). The objective of this paper is to determine the origin and timing of kaolin formation as well as their transformation during burial, and to relate the occurrence of kaolin to the geochemical evolution of the pore-water.

GEOLOGICAL SETTING

The northern part of the Egyptian Western Desert is characterized by east–west and northeast–southwest orientated graben systems and a northward thickening wedge of Jurassic to Miocene sediments. These sediments consist of continental sandstones in the south and marine deposits to the north, where carbonates are predominant. The north–south orientated graben (Matruh Basin) extends southward to the Khalda and Melehia areas (Fig. 1). The Khalda and Salam Fields lie where the Matruh Graben intersects with another very large graben, the Shushan Basin, in a relative structural high bounded by deep-seated normal faults. Both grabens are filled with up to 6 km of Paleogene, Cretaceous and Jurassic sediments. Part of a pronounced east–west orientated palaeo-ridge (Qattara Ridge) exists in the southernmost part of the Khalda area (Lotfy, 1994). This area is part of the Egyptian unstable shelf, which has been affected by various tectonic episodes, such as Late Jurassic to Early Cretaceous rifting, mid-Cretaceous (Aptian) uplift and erosion, and Late Cretaceous to early Tertiary shear and compression. These tectonic events influenced the depositional pattern, source-rock maturation and hydrocarbon entrapment. It is believed that Shushan and Matruh started as rift basins and later turned into pull-apart basins as a result of strike-slip faulting (EGPC, 1992).

The stratigraphical column in the northern

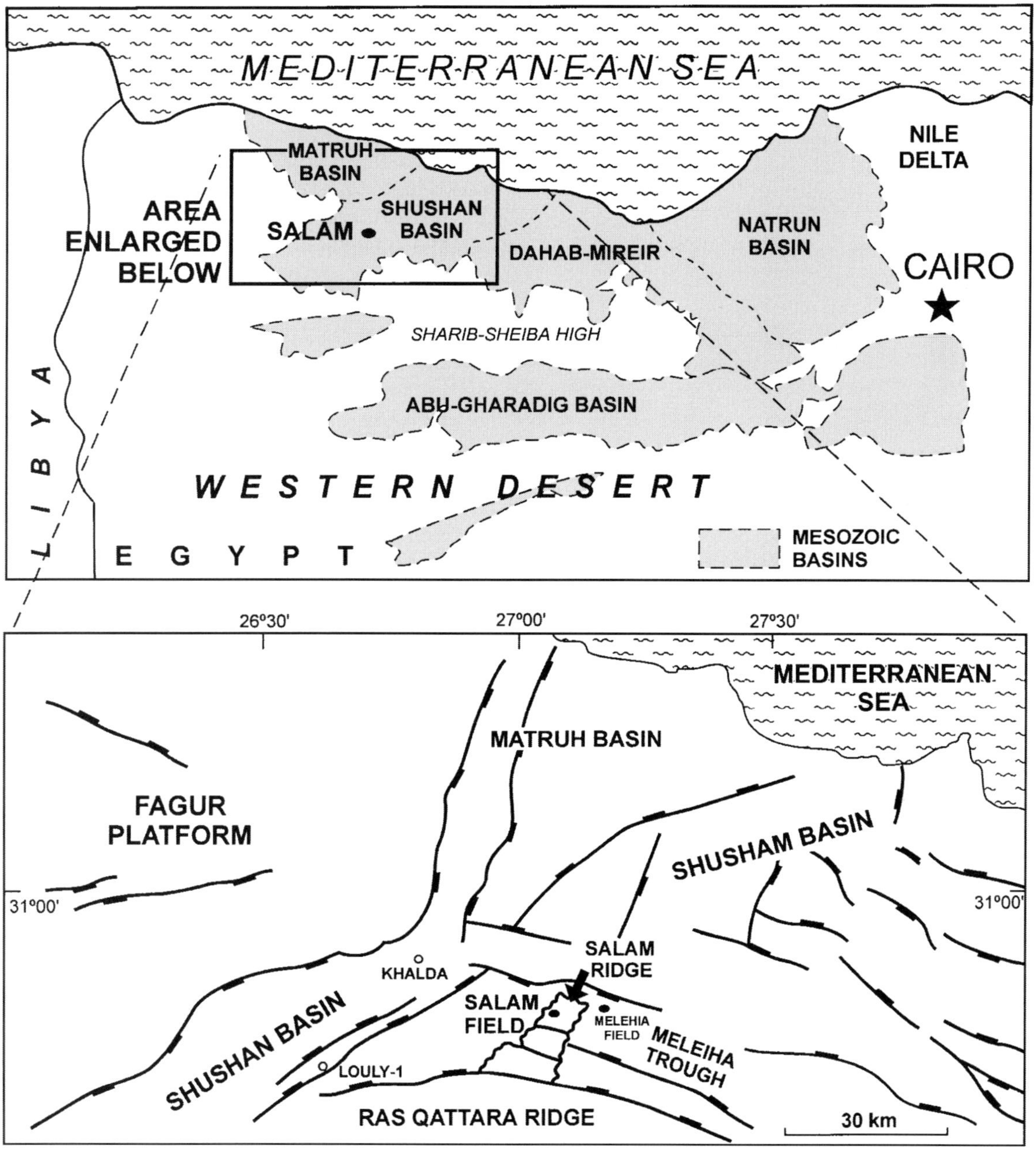

Fig. 1 Location map of the Salam Oil Field in the Western Desert, Egypt, showing the main Mesozoic basins, and main fault systems.

Western Desert includes most of the sedimentary succession from Precambrian basement complex to Recent (Fig. 2). The post-Palaeozoic succession in the Khalda area comprises four major tectonosedimentary cycles separated by unconformities, spanning the Middle Jurassic, Lower Cretaceous, Upper Cretaceous and Eocene to Miocene. Each cycle begins with

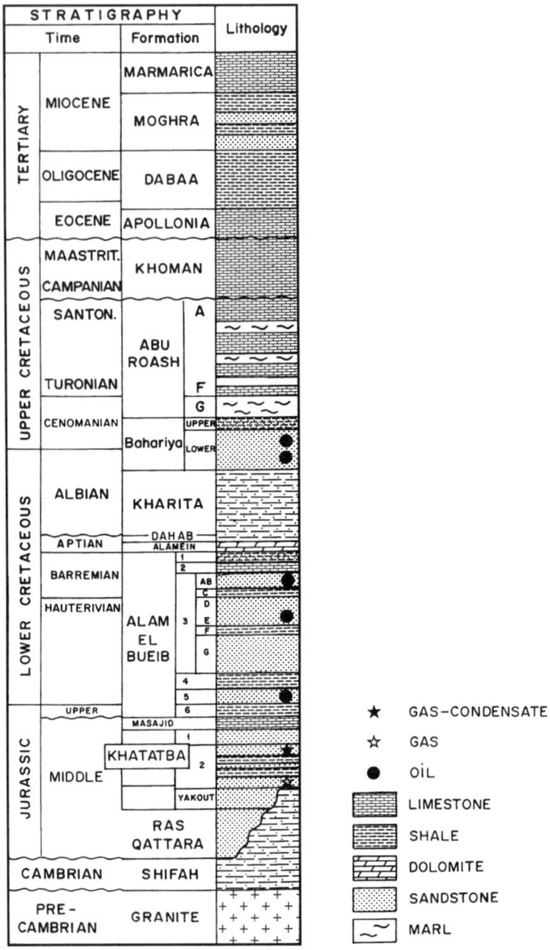

Fig. 2 Generalized stratigraphical column of the Salam Field area. Arbitrary vertical scale.

fluvio-deltaic siliciclastics and terminates with marine carbonates (May, 1991).

The Salam Oil Field lies approximately 460 km west of Cairo and 70 km south of the Mediterranean Coast (Fig. 1). The main reservoirs of the Western Desert, including the Salam Field, are the shallow-marine glauconitic sandstones of the Bahariya (Upper Cretaceous), Alam El Bueib (AEB) (Lower Cretaceous) and Khatatba (Middle Jurassic) Formations. The producing horizons are composed of several layers of sandstone interbedded with mudrocks.

The lower to upper Callovian Khatatba Formation is composed of fluvio-deltaic sandstones and shales, grading upwards and northwestwards into marine shales and limestones of the Upper Jurassic Masajid Formation (El Shazly, 1977; May, 1991). The Masajid Formation is capped by the Cimmerian Unconformity. The Cimmerian orogeny caused the northwards tilting and erosion of the Jurassic succession, and areas such as the Shusan Basin were uplifted (EGPC, 1992). To the south and west carbonaceous shale, coal, marine shale, siltstone, fine-grained sandstone and red shale (Yakout bed) occur at the base of the Khatatba Formation. This formation has good reservoir quality, and contains fair to excellent oil-prone source rocks (type II kerogen) with gas-prone humic organic matter (type III). The source rocks generally are rich in organic matter and mature, so that significant amounts of oil and gas were generated. In the Salam Field, the Khatatba Formation occurs at depths of 3300–3600 m.

The Lower Cretaceous (Neocomian–Barremian) AEB Formation represents rift reactivation during the Early Cretaceous. It was deposited in fluvial to shallow-marine environments. These sediments are capped by transgressive, lower Aptian carbonates of the Alamein Formation (Fig. 2). The unit is composed predominantly of sandstones with siltstone and subordinate shales. Thin carbonate beds occur in most areas. Lignite and carbonaceous shales are common in places, and the coals are mainly vitrinitic and gas prone. The AEB Formation rests unconformably on the Masajid Limestone. In the Salam Field the AEB Formation occurs at depths of 2430–3300 m.

Burial and thermal history

The burial history curve for a representative Salam Field well in the Khalda is shown in Fig. 3 (Marfil *et al.*, 2000). Widespread rapid subsidence occurred during the Cretaceous rifting stage, interrupted in the Paleocene by the Alpine Orogeny, and a phase of regional subsidence during the Tertiary (Taher *et al.*, 1988; Lofty, 1994; Nashaat *et al.*, 1994). The unconformity between Late Cretaceous and

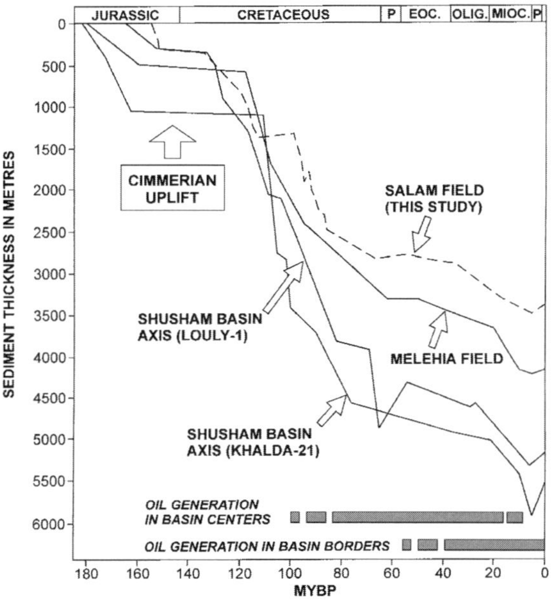

Fig. 3 Representative burial histories for the Khatatba Formation in the Shusan Basin. The burial curve for Salam Field is deduced from the S-4 well. The data for Louly-1, Khalda-21, and Melehia Field are taken from Nashaat *et al.* (1994), Lotfy (1994) and Taher *et al.* (1988), respectively. Khalda-21 and Louly-1 wells are located along the axis of the Shusan Basin, 20–40 km to the southwest and northwest of Salam, respectively. Melehia Field lies 15 km to the northeast of Salam. See Fig. 1 for location of the wells and fields.

Early Tertiary is represented by an erosional hiatus (Said, 1990). The maximum burial depth (*c.* 3500–4000 m) for the AEB and Khatatba Formations is considered to have occurred in the Middle Miocene (Nashaat *et al.*, 1994). The post-Middle Miocene erosion is assumed to represent at least 60 m of uplift in the Melehia area and up to 120–500 m along the Shushan Basin axis (Nashaat *et al.*, 1994).

In the Shusan Basin centre, the Khatatba Formation probably entered the oil window during the Cenomanian–Turonian, passing into the gas window during the Neogene (Lotfy, 1994; Nashaat *et al.*, 1994). In structural highs, such as the Salam ridge, the Khatatba reached the oil window later, during the Eocene (Lotfy, 1994). In the Salam Field, the top of the oil window ($R_0 = 0.6$) is encountered at around 2450 m, and the Khatatba source rocks are still in the oil-generation phase. Oil migration from the Shusan Basin into the Salam structure presumably has occurred since the Palaeocene (Nashaat *et al.*, 1994). Additionally, vertical migration from the local Khatatba source in Salam Field has taken place since the Eocene. As the Salam Trap formed as a result of Palaeocene–Eocene tectonics (El Ayouty, 1990), the deeper reservoirs in Salam (Khatatba) probably have been filled since the Eocene at the earliest.

The present-day geothermal gradient increases on a regional scale from 29.1°C km^{-1} in the Shusan Basin axis to more than 34.6°C km^{-1} over the Qattara ridge, the Fagur platform and other structural highs (Taher *et al.*, 1988; Lotfy, 1994; Nashaat *et al.*, 1994). In the Salam Field, the present-day reservoir temperatures range from 104 to 130°C at depths from 2430 to 3600 m, respectively.

SAMPLES AND ANALYTICAL TECHNIQUES

This study is based on cores from five wells from the Salam oilfield, from which 140 samples have been analysed. The samples are sandstones, siltstones and shales representing the main facies of the AEB and Khatatba Formations from present-day depths of 2400 to 3536 m.

Clay minerals were separated after disaggregation of the samples by gentle crushing and mild ultrasonic treatment. Hydrocarbon extraction was accomplished by treating the disaggregated samples several times with dichloromethane stabilized with amylene. After extraction, the samples were dried at room temperature, and then dispersed in distilled water. Clay minerals were separated by settling according to Stokes' law. The 20–4 and < 4 μm size fractions were selected to obtain the purest kaolin separates.

Thin-sections were prepared after impregnating the samples with blue-dyed epoxy resin. Selected thin-sections were polished and prepared for microprobe analyses and backscatter electron imaging (BSE). Cathodoluminescence petrography was performed using a hot cathode luminoscope (Ramseyer *et al.*, 1989). Scanning electron microscope (SEM) petrography was carried out using a JEOL JSM 6400, equipped with a LINK System energy dispersive X-ray microanalyser (EDX). The samples were examined in secondary electron (SE) and backscatter electron (BSE) modes of imaging.

For transmission electron microscopy (TEM), the clays were dispersed in double-distilled water, a drop of suspension being dried on to a carbon-coated copper grid, and then examined using a JEOL 2000 FX with 200 kV acceleration voltage. The microscope had incorporated a LINK model AN 10 000 energy dispersive X-ray microanalyser.

X-ray diffraction (XRD) analyses were performed on whole-rock samples, the 20–4 μm and the < 4 μm fractions, using a Philips 1710 X-ray powder diffractometer. The samples selected for dickite identification were placed in a cavity mount, levelled off by knife edge without pressing to minimize preferred orientation (Bish, 1990). Semi-quantitative estimates of clay-mineral proportions were determined using methods modified from Schultz (1964). The characterization of kaolin minerals was realized in the 19–40° 2θ Cu Kα range. The diagnostic peaks corresponding to 11-2, 022, 112, 13-2 and (132, 20-4) reflections for dickite (Bailey, 1980) were used. The grade of structural order in kaolinite was determined by using the Hinckley (1963) and Stoch (1974) indexes. The first is based on height measurements of the 020, 1-10 and 11-1 peaks. Normal values range from < 0.5 (disordered) to 1.5 (ordered). The Stoch (1974) index is the ratio between the 020 and 1-10 peak heights. Normal values range from > 1.0 (disordered) to < 0.7 (ordered). According to Galán *et al.* (1994) and Aparicio & Galán (1999) this method is not affected by quartz, feldspar, or iron hydroxide, and therefore its reproducibility is good.

Oxygen and hydrogen stable-isotope analyses were performed on 12 kaolin-rich samples. Kaolin and kaolin-rich fractions (10–15 mg) were reacted with a stoichiometric excess of ClF_3 at 650°C for 12 h (Borthwick & Harmon, 1982; Venneman & Smith, 1990). The released oxygen was converted to CO_2 by reaction with a hot platinized graphite rod (Clayton & Mayeda, 1963). For δD, a uranium line was used (Godfrey, 1962). Samples were degassed by overnight heating at 200°C under high vacuum. The platinum crucible was then heated to approximately 1200°C. The released water was converted to hydrogen by passing over uranium metal at 800°C. The isotope ratios were measured in a Finnigan MAT 251 mass spectrometer. Commercial CO_2 was used as the internal standard for the oxygen analyses of silicates, contrasted with the V-SMOW (Vienna Standard Mean Ocean Water), SLAP (Standard Light Antarctic Precipitation) and GISP (Greenland Ice Sheet Precipitation) water standards, thus giving a value of $\delta^{18}O = +5.1 \pm 0.2‰$ (V-SMOW) for the international NBS-30 (biotite) standard, and a value of $+ 9.6 \pm 0.1‰$ (V-SMOW) for the international NBS-28 (quartz) standard. For δD analyses water was used that previously had been checked against the above-mentioned international water standards.

RESULTS

Bulk composition of sandstones

The sandstones of the Khatatba and AEB Formations are fine- to medium-grained, texturally and compositionally mature quartzarenites, with subordinate subarkoses and quartzwackes (according to the classification of Pettijohn *et al.*, 1973). The most important detrital components in the sandstones are monocrystalline quartz and minor amounts of polycrystalline quartz. Feldspars and micas are scarce except in the upper part (units 1 to 3D) of the AEB Formation, and in some laminated floodplain sandstones and siltstones (unit 2C) of the Khatatba Formation, where feldspars do not usually exceed 5%. Vertical variations in feldspar ranging from 20 to < 2% are observed. Completely and/or partially kaolinized or

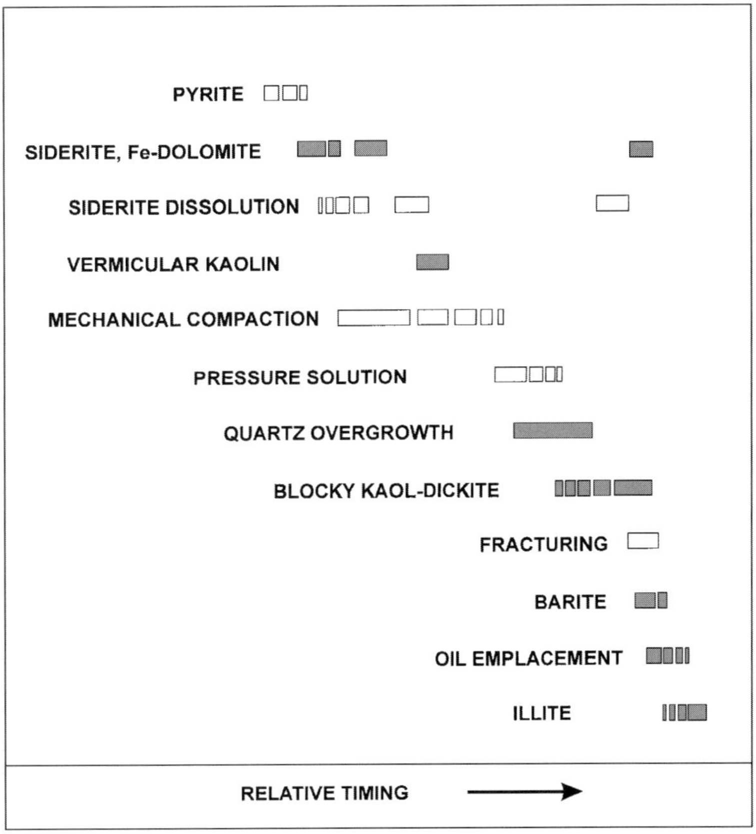

Fig. 4 Diagenetic sequence of the Khatatba and AEB sandstones in the Salam Field. Modified from Marfil *et al.* (2000). Shaded symbols are the principle cements. Non-shaded symbols are other processes that influence the reservoir quality of the sandstones.

dissolved K-feldspar and mica occur in most sandstones. Detrital heavy minerals are zircon, tourmaline, rutile and Fe–Ti oxides. Both kaolin and mixed-layer illite–smectite (I/S) intraclasts are common. Rare glauconite grains are restricted to the shallow-marine facies. Quartz overgrowths in the sandstones range from 2 to 14%. Siderite and dolomite cements vary from 0 to 40% and from 0 to 30%, respectively. Barite and pyrite occur in quantities < 5%. Pyrrhotite cement occurs only in AEB sandstones. Authigenic kaolin abundance, estimated from point counts, ranges from < 1 to 12% (Marfil *et al.*, 2000; Rossi *et al.*, 2001).

Paragenetic sequence

The following paragenetic sequence according to an earlier study by Marfil *et al.* (2000) (Fig. 4) shows: (i) early diagenetic pyrite framboids, siderite and Fe-dolomite cements; (ii) shallow-burial dissolution of siderite and Fe-dolomite cements and feldspars with precipitation of vermicular kaolinite; (iii) the beginning of mechanical compaction; (iv) the beginning of chemical compaction and major quartz precipitation; (v) precipitation of poikilotopic siderite and Fe-dolomite; (vi) dissolution of carbonates and generation of secondary porosity; (vii) a major phase of blocky kaolin precipitation; (viii) fracturing and barite precipitation; (ix) major phase of oil emplacement; (x) pore-bridging illite replaces dickite and kaolinite; (xi) and that precipitation of pyrrhotite–pyrite cements in the AEB Formation has possibly occurred concomitant with the emplacement of oil in the reservoir (Marfil *et al.*, 2000).

The diagenetic processes in both formations are quite similar, except for the higher abundance of late dolomite, and the precipitation of

Table 1 Semi-quantitative analyses obtained by X-ray diffraction of the selected samples.

Sample	Depth (m)	Formation	Rock type	TOC	Size (μm)	% Kaol	% I + I/S	% Q	% Feld	% Py	% Ba
B-119	2512.16	AEB	Sandstone	1.89	20 to 4	27.8	47.9	11.8	1.8	7.9	2.7
B-119	2512.16	AEB	Sandstone		< 4	38.8	47.2	7.2	1.6	5.1	–
BO-73	2621.89	AEB-3C	Siltstone	1.41	20 to 4	41.3	27.8	29.5	1.4	–	–
BO-73	2621.89	AEB-3C	Siltstone		< 4	44.3	40.2	13.2	2.3	–	–
BO-74	2627.68	AEB-3D	Sandstone	1.04	20 to 4	50.8	13.3	32.5	3.5	–	–
BO-74	2627.68	AEB-3D	Sandstone		< 4	51.9	24.9	20.2	3.0	–	–
BO-77	2642.92	AEB-3B	Shale	3.46	< 4	62.4	26.4	10.6	0.5	–	–
BO-81	2991	AEB-3G	Siltstone	1.30	20 to 4	47.0	22.7	28.2	2.1	–	–
BO-81	2991	AEB-3G	Siltstone		< 4	58.7	31.8	9.4	–	–	–
BO-84	3092.2	AEB-5A	Siltstone	0.81	20 to 4	47.0	37.6	6.9	1.2	7.3	–
BO-84	3092.2	AEB-5A	Siltstone		< 4	57.2	36.1	3.7	0.2	2.7	–
BO-85	3373.83	Khatatba-1B	Sandstone	0.33	20 to 4	49.6	–	41.8	0.5	–	8.1
BO-85	3373.83	Khatatba-1B	Sandstone		< 4	42.8	11.8	25.3	–	9.0	11.0
BO-89	3442.11	Khatatba-2C	Sandstone	1.68	20 to 4	49.1	23.5	19.4	–	8.0	–
BO-89	3442.11	Khatatba-2C	Sandstone		< 4	62.3	25.4	11.7	0.3	–	0.3
BO-95	3523.18	Khatatba-2E	Sandstone	0.34	20 to 4	71.3	6.1	9.2	–	–	13.5
BO-99	3534.61	Khatatba-2E	Sandstone	0.44	20 to 4	72.3	–	14.5	0.4	6.5	6.3
BO-100	3536.29	Khatatba-2E	Sandstone	0.39	20 to 4	88.3	–	8.8	1.2	–	1.7
BO-100	3536.29	Khatatba-2E	Sandstone		< 4	91.9	–	3.5	–	4.5	–
BO-85	3373.83	Khatatba-2C	Quartz-s					99.0		1.0	
BO-93	3519.68	Khatatba-2E	Quartz-s					99.0		1.0	

TOC, total organic carbon; I + I/S, illite plus mixed-layer I/S; Q, quartz; Feld, feldspar; Py, pyrite; Ba, barite; quartz-s, quartz separates.

pyrrhotite–pyrite cement in the AEB Formation. The depositional facies (fluvio-deltaic versus shallow marine) exerted a major influence on the composition of the early carbonate cements (siderite versus dolomite) (Rossi *et al.*, 2001).

Distribution, petrography and mineralogy of clay cements in sandstones

Kaolin of widely varying crystal sizes and morphologies is the major clay mineral in sandstones of both formations (Table 1). Kaolin occurs primarily as: (i) *pore-filling* cement, in both primary or secondary porosity (Fig. 5A); (ii) pseudomorphic *replacement* of feldspars and expanded mica, as well as of syndepositional intraclasts of kaolinite and I/S (Fig. 5B); (iii) *soil-related matrix* in sandstones, which show features of infiltrated clay coatings and is interpreted as a product of *in situ* weathering of mica and feldspars (Fig. 5C & D); and (iv) *detrital* kaolinite, mixed with detrital illite in shales and siltstones, showing irregular laminar morphology. Under cathodoluminescence (CL), detrital kaolinite shows a dull and hazy blue luminescence colour whereas authigenic kaolin in the sandstones shows a dark blue luminescence.

Kaolin in both formations is developed mostly as blocky pore-filling cement comprising euhedral, pseudohexagonal plates 5–10 μm wide and 10–20 μm long, that occur in both primary and secondary pores and which were created mainly by dissolution of carbonate cements (Fig. 5E & F). Kaolin also occurs in mouldic pores after feldspar (Fig. 5B). The SEM images (Figs 5A & 6A) show that blocky kaolin is enclosed by, or occurs on the surfaces of, quartz overgrowths. In some places, blocky kaolin is partially or completely impregnated with solid bitumen, which blocks pore throats. In the Khatatba Formation, pore-filling kaolin crystals show a prismatic irregular morphology (Fig. 6A & B).

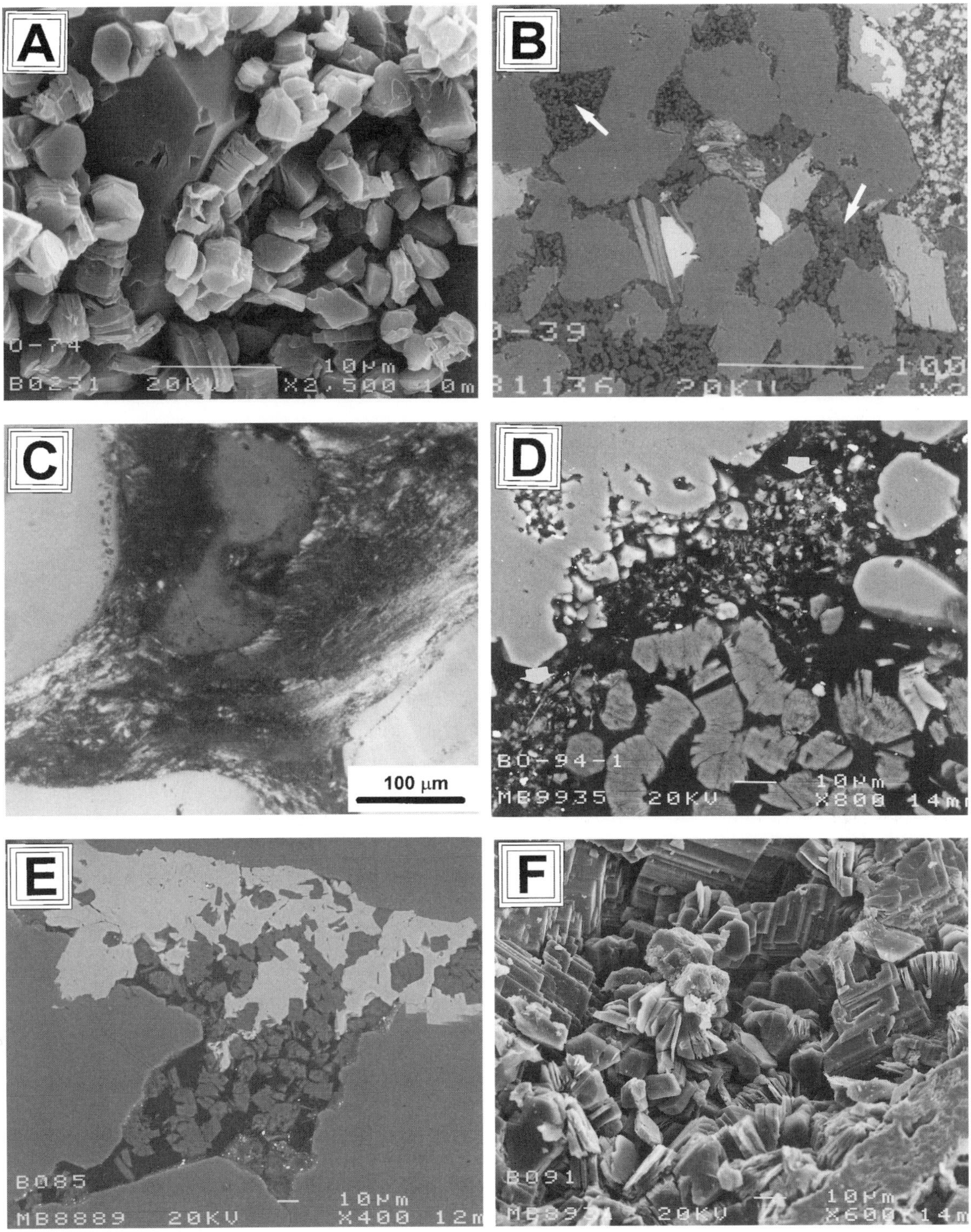

Fig. 5 (A) Scanning electron microscopy (SEM) image of pore-filling blocky kaolinite intercalating with thin flat kaolinite plates. Both kaolinite morphologies post-date quartz overgrowth. BO-74 AEB Formation. (B) SEM image of vermicular kaolin filling feldspar moulds (arrows). Unit 3C, AEB Formation. Scale bar is 5 µm. (C) Photomicrograph of kaolin and illite in soil-related matrix from the BO-94 Khatatba sandstone. (D) Backscatter electron (BSE) micrograph of kaolin in soil-related matrix (arrow) replaced by vermiform kaolin in the BO-94 Khatatba sandstone. (E) BSE image of vermicular kaolin engulfed in early siderite cement, which preserves the vermicular morphology. BO-85 Khatatba sandstone. (F) SEM image of siderite cement engulfing and preserving the vermicular kaolin. BO-91 Khatatba Formation.

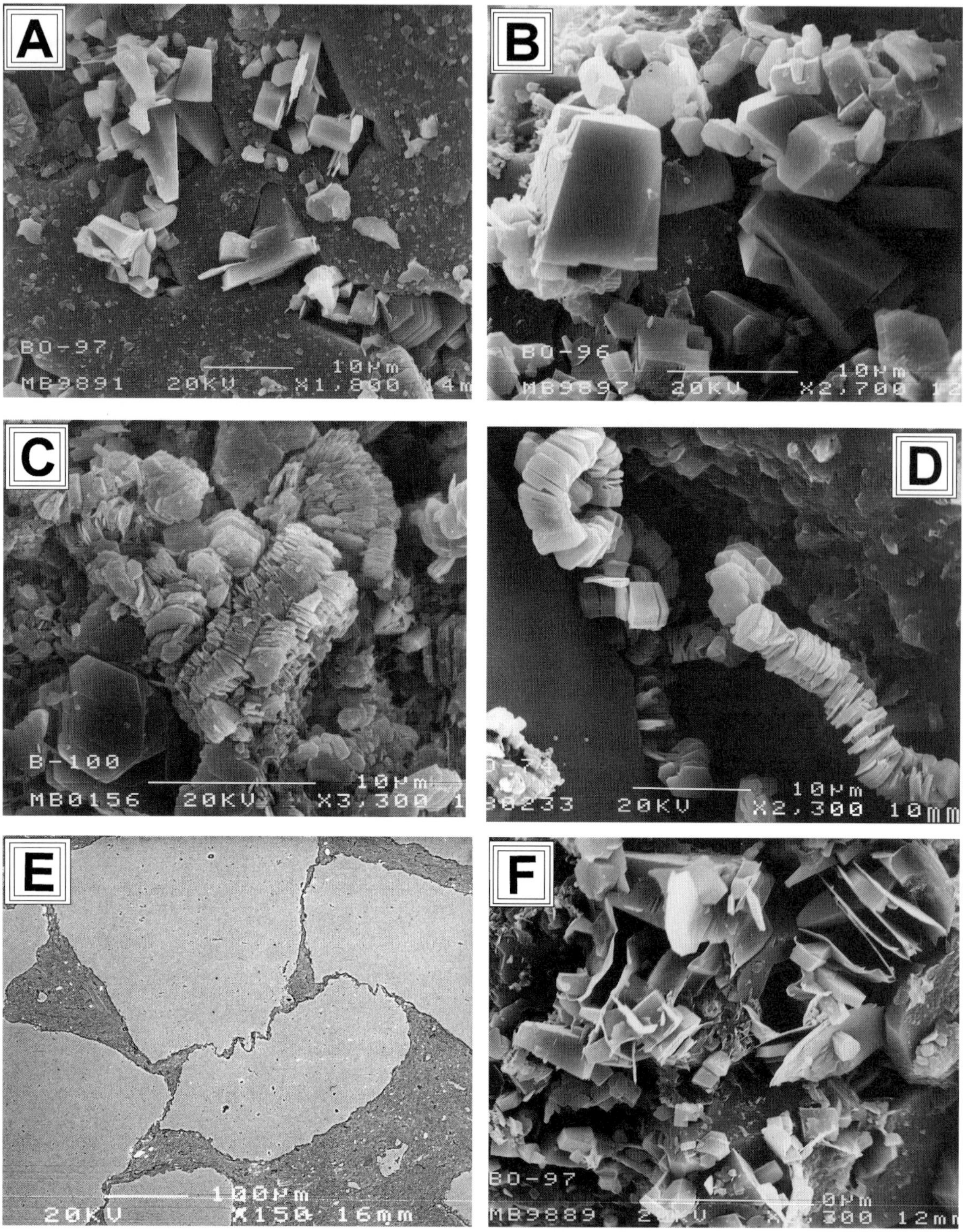

Fig. 6 Scanning electron microscopy (SEM) images showing: (A) and (B) pore-filling kaolin composed of crystals of dickite, intergrowing with quartz cement, Khatatba sandstone, BO-97 and BO-96, respectively. (C) Early vermiform kaolinite, with characteristic curved morphology partly transformed to blocky dickite, Khatatba Formation, sample B-100. (D) Detail of early vermiform kaolinite illustrating bimodal crystal size distribution. (E) BSE image of infiltrated soil-related matrix producing intense corrosion on detrital quartz grains, Khatatba Formation, sample BO-99. (F) Clay aggregate replaced by a variety of crystal sizes of kaolin representing a compound of kaolinite–dickite morphology. Some laminae are curved indicating the presence of illite.

Table 2 Kaolin minerals texture, morphology, polymorphs and kaolinite indexes.

						Kaolin morphology			Kaolinite index	
Sample	Depth (m)	Formation	Rock type	Size (μm)	Kaolin texture	Vermiform	Blocky	Dickite	Hinckley (1963)	Stoch (1974)
B-119	2512.16	AEB	Sandstone	20 to 4	Pore-filling	+	+++	–	–	1.39
B-119	2512.16	AEB	Sandstone	< 4	Replacive			–	0.62	1.43
BO-73	2621.89	AEB-3C	Siltstone	20 to 4	Detrital			–	–	1.41
BO-73	2621.89	AEB-3C	Siltstone	< 4	Detrital			–	–	1.39
BO-74	2627.68	AEB-3D	Sandstone	20 to 4	Pore-filling	–	+++	–	–	1.46
BO-74	2627.68	AEB-3D	Sandstone	< 4	Pore-filling			–	–	1.32
BO-77	2642.92	AEB-3B	Shale	< 4	Detrital			–	–	1.43
BO-81	2991	AEB-3G	Siltstone	20 to 4	Detrital			–	–	1.38
BO-81	2991	AEB-3G	Siltstone	< 4	Detrital			+	0.83	1.44
BO-84	3092.2	AEB-5A	Sandstone	20 to 4	Pore-filling	–	+++	++	0.77	1.2
BO-84	3092.2	AEB-5A	Sandstone	< 4	Pore-filling			–	0.68	1.32
BO-85	3373.83	Khatatba-1B	Sandstone	20 to 4	Pore-filling	–	+++	+++	–	0.81
BO-85	3373.83	Khatatba-1B	Sandstone	< 4	Pore-filling			–	–	1.02
BO-89	3442.11	Khatatba-2C	Sandstone	20 to 4	Pore-filling	–	+++	–	–	1.09
BO-89	3442.11	Khatatba-2C	Sandstone	< 4	Replacive			–	0.76	1.04
BO-95	3523.18	Khatatba-2E	Sandstone	20 to 4	Soil-related	+++	–	–	0.7	0.98
BO-99	3534.61	Khatatba-2E	Sandstone	20 to 4	Soil-related	+++	–	+	–	0.97
BO-100	3536.29	Khatatba-2E	Sandstone	20 to 4	Pore-filling	+	+++	+++	1.01	0.72
BO-100	3536.29	Khatatba-2E	Sandstone	< 4	Replacive			++	1.1	0.69

+++, > 50%; ++, moderate; +, low; –, not detected.

The vermiform kaolin that replaces feldspar consists of large platy crystals stacked vertically, roughly parallel to the (001) face. This results in large composite aggregates of crystals, which usually are curvilinear in the vermiform kaolinite (Fig. 6C). In clean sandstones, vermiform kaolin is composed of bimodal crystals (Fig. 6D). Kaolinized detrital mica displays expanded edges and masses of tightly packed kaolinite sheets, plates and/or laminae.

The soil-related matrix is composed of extremely fine-grained kaolinite and scarce illite less than 1 μm in length, containing corroded silt-sized quartz, carbonaceous material, and abundant Ti-oxides (Fig. 6E). The textures observed include bridges, cutans and massive aggregates filling intergranular pores. Some of the clays are concentrated along thin laminations, and have enhanced intense intergranular pressure dissolution of the quartz grains (Fig. 6E). In some crevasse sandstone facies of the Khatatba Formation, masses of vermicular kaolin fill irregular vertical veins and septarian cracks developed within lumps of carbonaceous organic matter. The Jurassic climate in Egypt was humid and tropical (Keeley & Wallis, 1991), and the abundance of microcrystalline kaolinite as detrital clay in the sandstones probably represents the record of intense tropical weathering (Rossi *et al.*, 2001).

Based on SEM observation (Table 2), kaolin in the sandstones does not show progressive changes in morphology with increasing burial depth. However, in some of the shallower samples, at the top of the AEB Formation, there is pore-filling, blocky kaolin (Fig. 5A), whereas at the base of the Khatatba Formation, mixed vermicular and blocky kaolin with variation of crystal sizes is dominant (Fig. 6C & F). Pore-filling kaolin shows an exponential increase of the crystal thickness with burial depth (Fig. 7). However, in samples with pervasive

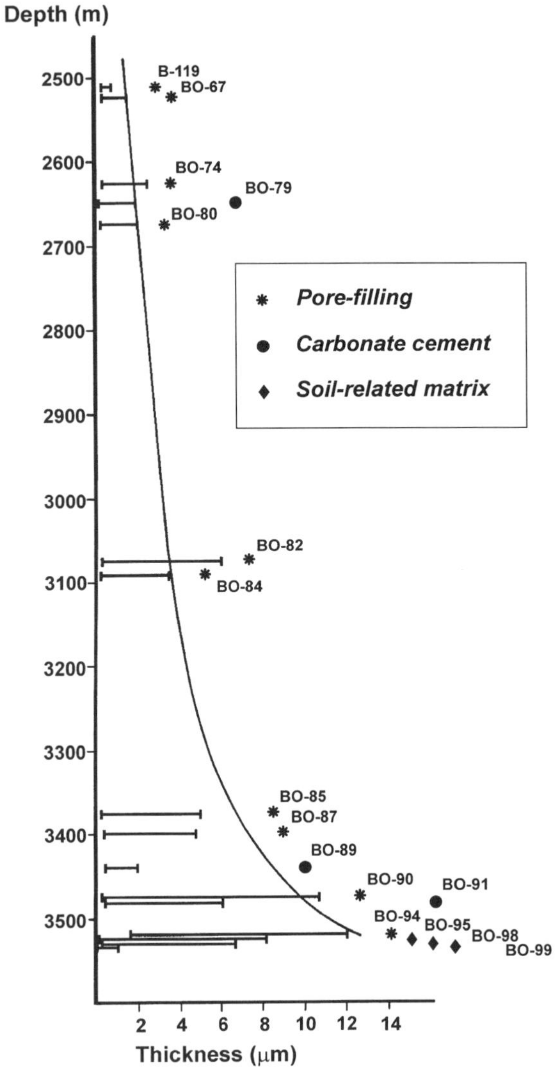

Fig. 7 Increase in the range of the crystal thickness of kaolin with burial depth in the 4–20 μm fraction of samples with: pore-filling blocky kaolin (*); pore-filling, vermicular kaolin engulfed in carbonate cement (•); vermicular kaolin in samples with soil-related matrix (♦). The crystal thickness is represented by horizontal bars. It is remarkable that the crystals from samples with soil-related matrix or carbonate cement are less thick than in the samples with kaolin pore-filling. The measurements were performed in a population of > 50 crystals per sample.

poikilotopic carbonate cement and soil-related matrix the increase of crystal thickness is less pervasive (Figs 5F & 6C).

Authigenic illite occurs as fibrous and lath-like crystals up to 10 μm in length (Fig. 8A). Illite intergrows with quartz and blocky kaolin in the Khatatba sandstones (Fig. 8B), and replaces feldspars in the AEB sandstones (Fig. 8C). In addition, illite develops a pore-occluding honeycomb microfabric when it replaces I/S clay aggregates (Fig. 8D). In some cases, illite laths grew along mica cleavage planes. The TEM observations show filamentous illites with curled edges. Authigenic chlorite and corrensite displaying honeycomb or cardhouse morphologies are minor constituents in marine sandstones of the AEB Formation, and they also replace clay intraclasts.

Composition and microfabric of intercalated shale and siltstone

The interbedded shales, coaly shales and siltstones show similar bulk mineralogical composition in both formations, however, they have a variable clay composition (Table 1). In the Khatatba Formation, the predominant clay mineral is kaolin, with subordinate illite and I/S, whereas in the AEB Formation, the clay minerals are kaolin, illite, I/S, glauconite and chlorite (Marfil *et al.*, 1997). Other components are detrital quartz, mica partially replaced by kaolinite (Fig. 5B), and scarce feldspars. Framboidal pyrite and siderite are abundant in the shales and silstones of both formations.

The microfabric of the shales is diverse, owing to temporal and spatial variations in terrestrial and marine organic matter. The fabric of the marine shales shows moderately preferred particle orientation, and the clay flakes wrap around silt-sized quartz and pyrite. Compressed palynomorphs are generally replaced by pyrite, whereas several framboid clusters have crystallized within the cavities. In the coaly shales no preferred particle orientation is observed, and the humic fragments offer evidence of corrosion by bacteria. The microanalyses reveal high contents in S, Fe, Ti and U.

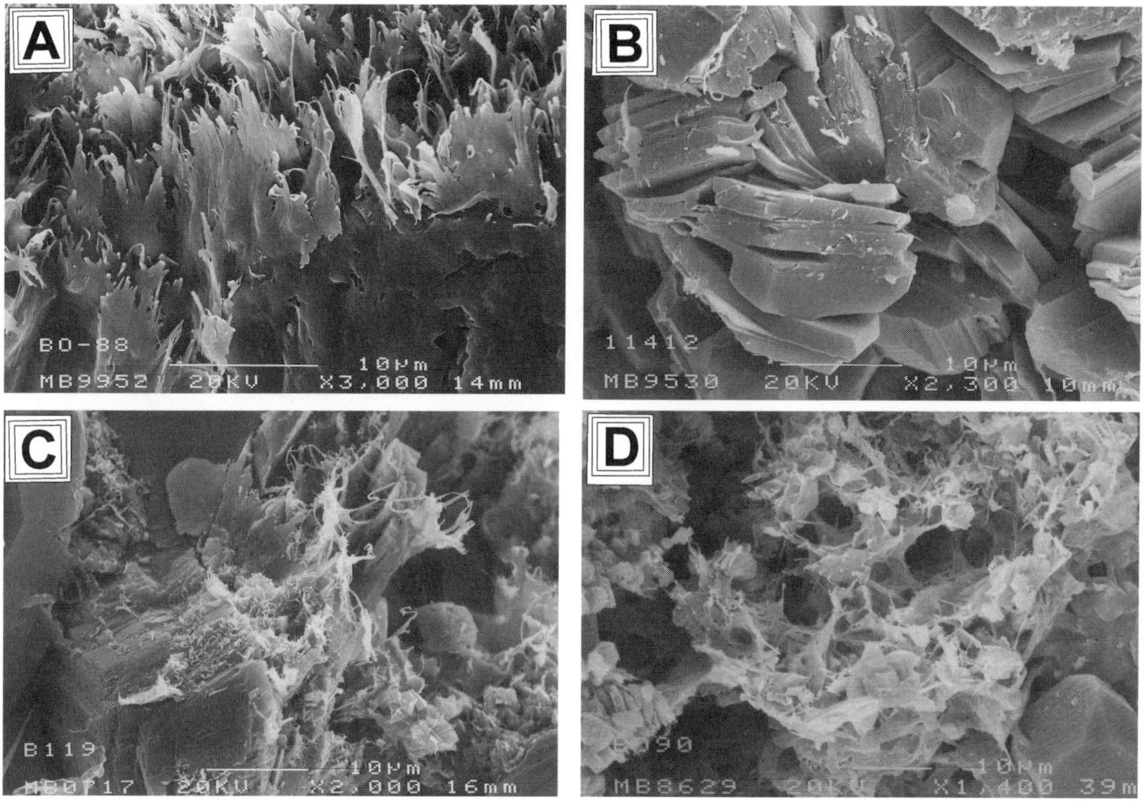

Fig. 8 Scanning electron microscopy (SEM) images of (A) Lath-shaped particles of illite replacing detrital mica. (B) Pore-filling euhedral crystals of dickite displaying blocky morphology and replacing illite fibres, Khatatba Formation, sample BO-91. (C) Feldspar grain replaced by fibrous illite, AEB Formation. (D) Mixed-layer I/S agreggate replaced by fibrous illite.

The random, open microfabric of the siltstones indicates that detrital and organic material were deposited as aggregates or floccules formed by electrochemical attraction (O'Brien & Slatt, 1990). Abundant micropores are still apparent, suggesting that this porous network could have been important in facilitating hydrocarbon migration.

Clay mineralogy

Bulk-rock XRD analyses show kaolin enrichment towards the base of the Khatatba Formation and a lower feldspar content than in the AEB Formation. The mineralogical composition from the 4–20 µm and < 4 µm fractions are given in Table 1. The dominant phases are kaolin minerals, illite and quartz. Dickite was detected in the 4–20 µm fraction, but is scarce in the < 4 µm fraction (Table 2). Crystals with vermiform morphology are kaolinite, whereas crystals with blocky morphology are kaolinite and/or dickite (Table 2). Illite and I/S are more common in the kaolinite-rich samples.

In samples with < 10% quartz, kaolinite crystallinity (Table 2) ranges between 0.7 and 1.10 (Hinckley index), indicating a medium to well-ordered variety (triclinic–pseudomonoclinic). Similar results were obtained using the Stoch index (Table 2). Dickite abundance in the deeper samples is well correlated with the highest crystallinity of kaolinite (Table 2).

All diagnostic reflections for dickite (Bailey, 1980) were identified in the 4 to 20 µm fraction

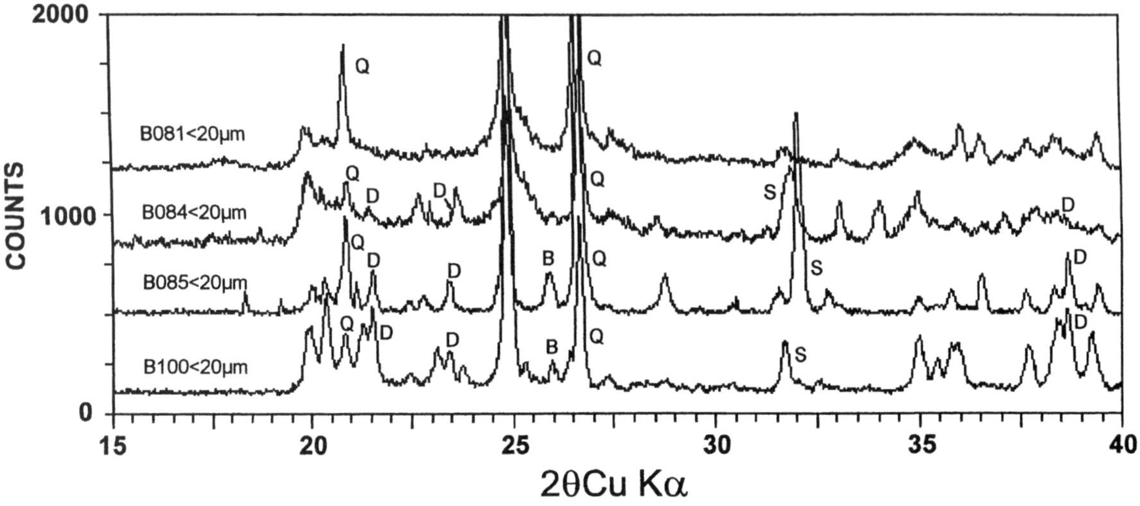

Fig. 9 X-ray diffractograms showing the evolution of the kaolin minerals in the pore-filling cement (4–20 μm fraction) as a function of increasing burial depth: D, dickite; Q, quartz; B, barite; S, siderite.

of the majority of the pore-filling phase of the sandstones in the Khatatba Formation (Fig. 9). These reflections are not present on the diffractograms of the samples with grain-replacive kaolin, soil-related matrix, and detrital kaolinite in the shales. The blocky morphology is common in all analysed pore-filling kaolin and partially in the grain-replacive kaolin in both formations. Vermicular kaolin is engulfed by quartz overgrowths and poikilotopic siderite–dolomite cements. In low-permeability samples such as shales and sandstones rich in soil-related matrix, vermiform kaolinite is preserved from replacement by dickite at all depths.

The progressive evolution of kaolinite to dickite with increasing maximum burial depth is shown in Fig. 9. Pore-filling kaolin of the shallowest sandstones (BO-81, at 2991 m depth) in the AEB Formation is mainly kaolinite, whereas, at the base of the Khatatba Formation (BO-100, at 3536 m depth), dickite and well-ordered kaolinite (Hinckley index = 1.10) dominate. In samples with detrital kaolinite no dickite was detected. In samples with soil-related matrix, only the BO-99 sample (3534 m depth) contains dickite reflections. None of the kaolin-rich samples analysed have shown reflections corresponding to interstratified mixed-layer kaolinite–dickite.

Illitic minerals mostly have a hair-like or fibrous habit and a subordinate lath-shaped morphology, indicating that randomly interstratified I/S is more abundant than ordered I/S (Lanson *et al.*, 1996). Platy particles of illite are scarce in the sandstone and difficult to differentiate from pseudohexagonal kaolin crystals. The platy illite along contacts between quartz grains may represent recrystallized smectitic clay cutans introduced by mechanical infiltration (McBride *et al.*, 1987). This origin also could be proposed for some of the authigenic illite in the Khatatba continental sandstones. In marine sandstones of both formations, illite occurs as thread-like crystals coating quartz overgrowths and showing meniscus fabrics.

The XRD results show that 10 Å clay minerals in both formations are apparently mixtures of authigenic illite and detrital illite–mica or I/S. In the shales, illite and I/S cannot be discriminated by SEM observation, because of the small crystal size and the deformation by mechanical compaction. In the soil-related matrix, as well as in the shales and siltstones, there are lesser amounts of authigenic illite

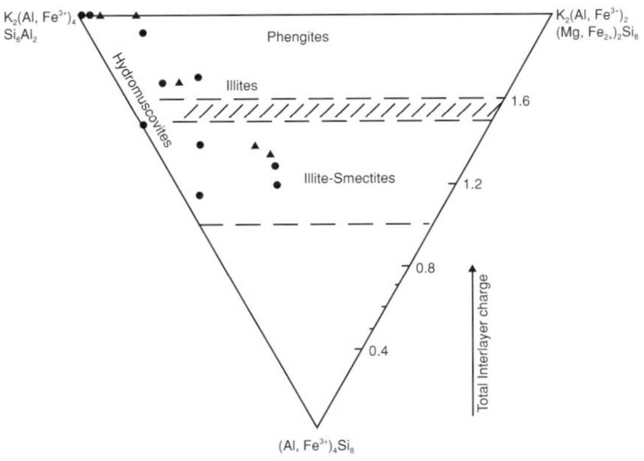

Fig. 10 Chemical composition of the illitic minerals (after Newman & Brown, 1987): ▲, samples from Khatatba Formation; •, samples from the AEB Formation. Hatched area indicates a range of charge between illite '*sensu stricto*' and illite–smectites.

than in the pore-filling kaolin of the sandstones. Illite decreases progressively in abundance from the top of the AEB Formation, to the base of the Khatatba Formation, where it forms < 5%. Diffraction patterns of the oriented < 4 μm fraction show that poorly crystallized illite (with large 001 peak breadth) is dominant in both formations.

In selected samples of both formations, chemical analyses on individual particles of illite have been performed by TEM. These compositions are projected on two well-defined fields: (i) a field of hydromuscovites, and (ii) a field of mixed-layer illite–smectite (Fig. 10).

Stable isotopic data

The measured oxygen and hydrogen isotopic ratios in the fractions richest in kaolin minerals gave δD values ranging from −123‰ to −43‰ and $\delta^{18}O$ values of 13.3‰ to 18.3‰ (Table 3).

Pore-filling kaolin has a $\delta^{18}O$ range from 13.3‰ to 16.6‰ and a δD range of −54‰ to −43‰. Thus this type of kaolin is characterized by a relatively heavy δD. In contrast, grain-replacive, detrital and soil-related kaolins are characterized by a relatively low δD ranging from −123‰ to −61‰. $\delta^{18}O$ values of detrital quartz range from 11.3‰ to 13.3‰. The present-day isotopic composition of the formation water is $\delta^{18}O = -0.59‰$; $\delta D = -13.5‰$.

Contaminants that could alter the $\delta^{18}O$ and δD ratios in these kaolin samples are various amounts of illite and quartz (Table 1). As discussed previously (Marfil *et al.*, 2000), illite originates from deep burial diagenesis at temperatures ranging from 117°C to 140°C. For

Table 3 Oxygen and hydrogen stable isotope ratios.

				Textures		‰ vs SMOW	
Sample	Fraction (μm)	Depth (m)	Formation	Soil/replacive	Pore-filling	$\delta^{18}O$	δD
B-119	< 4	2512.16	AEB	Replacive		+17.8	−93
BO-73	< 4	2621.89	AEB-3C	Detrital		+13.4	−79
BO-74	20 to 4	2627.68	AEB-3D		Pore-filling	+14.3	−54
BO-77	< 4	2642.92	AEB-3G	Detrital		+15.4	−61
BO-81	< 4	2991.00	AEB-5A	Detrital		+16.5	−110
BO-84	< 4	3092.20	Khatatba-1B	Replacive		+18.3	−81
BO-85	20 to 4	3373.83	Khatatba-2C		Pore-filling	+13.3	−43
BO-89	< 4	3442.11	Khatatba-2C	Replacive		+15.3	−101
BO-89	20 to 4	3442.11	Khatatba-2E		Pore-filling	+14.8	−48
BO-95	20 to 4	3523.18	Khatatba-2E	Soil-related		+13.5	−123
BO-99	20 to 4	3534.61	Khatatba-2E	Soil-related		+14.6	−75
BO-100	20 to 4	3536.29	Khatatba-2E		Pore-filling	+16.6	−49
BO-85	Quartz	3373.83	Khatatba-2C			+13.3	–
BO-93	Quartz	3519.68	Khatatba-2E			+11.3	–

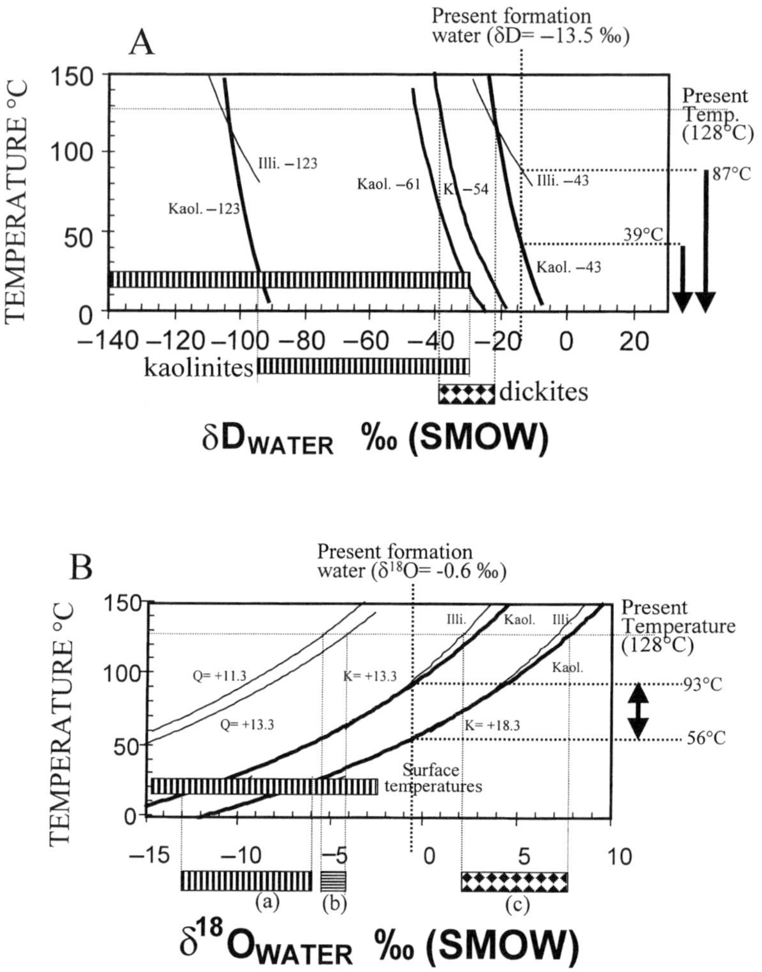

Fig. 11 (A) Hydrogen and (B) oxygen isotope fractionation curves of the maximum and minimum measured kaolinite–illite values. For hydrogen we have used the equation of Sheppard & Gilg (1996) and Capuano (1992) for the kaolinite–water and illite–water systems, respectively. For oxygen we have used the equation of Sheppard & Gilg (1996) for the kaolinite–water and illite–water systems and the equation of Clayton *et al.* (1972) for the quartz–water system. Present-day temperature (128°C) and isotopic composition of the pore-water ($\delta^{18}O = -0.6‰$; $\delta D = -13.5‰$) are considered for the calculation. The areas (a), (b) and (c) indicate the calculated $\delta^{18}O_{pore\text{-}water}$ considering a surface temperature of 15°C (system kaolinite–water (a)) and diagenetic temperatures (systems quartz–water (b) and kaolinite–water (c)). Notice that the presence of neoformed illite has a minor effect in the calculation (plots A and B) and detrital quartz and kaolinite are not in isotopic equilibrium at any temperature, which is consistent with the detrital origin of quartz.

this temperature range the fractionation curves for the systems kaolinite–water and illite–water for both hydrogen and oxygen are very similar (Fig. 11A). In addition, the concentration of hydrogen in illite is only about one-third of that of kaolinite. Consequently, the effect of illite on the δD and $\delta^{18}O$ value is expected to be small. Contamination by quartz is a more severe problem for the $\delta^{18}O$ than for δD values of kaolin. Nevertheless, even considering an extreme case, with 20% detrital quartz ($\delta^{18}O$ = 11.3‰), and the most positive kaolin sample with a $\delta^{18}O$ of 18.3‰ (Table 3), calculations show that the contaminant effect should not be larger than 1.5‰.

DISCUSSION

Kaolin morphologies and kaolinite to dickite transformation

The Khatatba Formation in the Salam Field has a present-day, bottom-hole formation temperature of 128°C at 3525 m depth (Marfil *et al.*, 2000), and mostly dickite and well ordered kaolinite are found. Temperatures in the AEB Formation are lower (116°C at 2654 m depth), and kaolinite is the dominant polymorph in this formation. Thus, dickite occurs at the upper temperature limit recorded in reservoir sandstones from the North Sea (80–130°C;

Ehrenberg *et al.*, 1993; Macaulay *et al.*, 1993; McAulay *et al.*, 1994).

The stability of kaolin polymorphs is a subject of considerable debate. Anovitz *et al.* (1991) concluded that kaolinite is the stable polytype under all geological pressure and temperature conditions, and that the formation of dickite is possible only through non-equilibrium processes. In contrast, Ehrenberg *et al.* (1993) doubt that massive widespread transformation of kaolinite to dickite has taken place outside the stability field of dickite, and they use this transformation as a geothermometer. However, the thermodynamic values from Robie & Hemingway (1995) do not support the temperatures of kaolinite–dickite transformation calculated by Ehrenberg *et al.* (1993).

The kaolinite to dickite transformation is inferred to be a dissolution–reprecipitation mechanism (McAulay *et al.*, 1994; Morad *et al.*, 1994). Morad *et al.* (1994) justify this argument based on the growth of blocky, monoclinic dickite between severely etched pseudohexagonal kaolinite remnants. Similarly, Lanson *et al.* (1996) suggested that the transformation could be a two-step process, consisting of a solid-state transformation without any morphological changes, followed by a dissolution–precipitation reaction involving replacement of vermiform kaolin by blocky crystals.

Distribution of the polymorphs (Table 2) and an increase in the range of crystal thickness with burial depth (Fig. 7) of the sandstones suggest that the kaolinite to dickite transformation occurs over a prolonged time period, similar to the smectite to illite conversion. In addition to time, temperature and pressure, the size of the particles and permeability loss owing to the presence of poikilotopic cements and hydrocarbons also influence the reaction time. Likewise, McAulay *et al.* (1994) observed that the presence of oil coatings may inhibit the kaolinite–dickite transformation. Similarly, Zimmerle & Rosch (1991), Ruiz Cruz & Moreno (1993), and Lanson *et al.* (1996) suggested that dickite is restricted to sedimentary rocks of high porosity and permeability. However, Ehrenberg *et al.* (1993) and Morad *et al.* (1994) inferred that the kinetics of the kaolinite–dickite transformation depend on the nature of the starting material, with microporous clusters of booklets and vermicules in kaolinized feldspars being more reactive than kaolinized mica. This may explain the presence of kaolinite in soil-related matrix and where pore occluding siderite–Fe-dolomite cements are present to suppress or slow down the kaolinite to dickite transformation (Fig. 7). Similarly, the transformation of kaolinized mica to dickite in the AEB siltstones may have been impeded owing to the tight and dense microfabric of the vermicular kaolinite. Nevertheless, at a more advanced stage of diagenesis, kaolinitized mica reacted more readily to form illite rather than dickite, as observed by Ehrenberg *et al.* (1993) and Morad *et al.* (1994).

In the shale samples, the increase in the kaolinite crystallinity with depth occurs, as shown by the increase of the Hinckley and Stoch indexes (Table 2), but the transformation to dickite is limited, possibly owing to the low porosity and permeability.

Origin of soil-related and grain-replacive kaolinite

The relationship between feldspar dissolution and kaolinite abundance could not be determined by XRD owing to the scarce and irregular presence of feldspar. However, there is a general decrease in feldspar content with increasing depth, and feldspars disappear at the base of the Khatatba Formation (Table 1).

The alteration of feldspar to kaolinite in the sandstones may occur under a wide range of conditions: (i) during subaerial exposure of the sandstones; (ii) at shallow burial depth (generally < 1 km) by influx of meteoric water; and (iii) at greater depths (2–4 km) by interaction with acidic water from thermal maturation of organic matter (Curtis, 1983; Surdam *et al.*, 1989). Taking these possibilities into consideration, the *replacement* of framework grains by vermicular kaolinite was probably related to meteoric water infiltration during the Cimmerian and Syrian-arc unconformities, which affected the Khatatba and the AEB Formations, respectively (Keeley & Wallis, 1991; Keeley & Massoud, 1998). The vermicular kaolin post-dates at least one major dissolution event affecting siderite, but pre-dates chemical compaction and probably the precipitation of most quartz overgrowths (Fig. 5B). During the Cimmerian orogeny, the Khatatba sandstones were uplifted and exposed to meteoric water infiltration (Keeley *et al.*, 1990; EGPC, 1992) and the widespread karstification of the Masajid carbonates is indeed a consequence of groundwater flow below the Cimmerian unconformity (Keeley *et al.*, 1990; EGPC, 1992). The Syrian-arc orogeny occurred in the area between Late Cretaceous and middle Eocene times, leading to basin inversion, uplift and erosion (Lotfy, 1994; Nashat *et al.*, 1994). Thus, meteoric water influx seems to be the most effective mechanism for dissolution of carbonate cements and kaolinite precipitation at the expense of feldspars in the AEB Formation.

Origin of pore-filling kaolin

Pore-filling kaolin is located at the surface of quartz overgrowths (Fig. 5A), in post-compactional intergranular pores, and is partially covered and intergrown with illite laths (Fig. 8B) and/or bitumen. This kaolin most likely crystallized during later stages of diagenesis, probably related to the influx of organic acids or CO_2 from adjacent oil and coaly shales (Hayes & Boles, 1992; Macaulay *et al.*, 1993; Lanson *et al.*, 1996). In general, the degree of stacking order of pore-filling kaolin tends to increase with depth. Kaolinite transforms to dickite between 3000 m to 3500 m present depth. The homogenization temperatures of fluid inclusions in quartz overgrowths indicate that the rocks at these depths were subjected to at least 125–140°C (Marfil *et al.*, 2000).

Bjørlykke (1998) suggested that if leaching is related to the generation of CO_2 or organic acids (Surdam *et al.*, 1984) then the sandstones that are close to the source rock would show the most intense leaching. This relationship occurs in the Salam Field, where kaolin tends to increase near the principal source rock (Fig. 2 & Tables 1 & 2), and feldspar is totally dissolved.

Origin and temperature of illite authigenesis

Authigenic illite is intergrown with both pore-filling and grain-replacive kaolin. Illite is formed by the replacement of kaolin (in the Khatatba Formation) and of I/S (Fig. 8C & D) during deep burial diagenesis. Variable chemical composition of the illites in both formations (Fig. 10) indicates mixing of detrital and authigenic illites, in addition to I/S. The sandstones with soil-related matrix have a lower illite content than those with pore-filling clays (XRD data, and Tables 1 & 2). The SEM observations reveal that much of the illite has grown at the expense of I/S and, to a lesser extent, of kaolinite (Fig. 8B–D). In reviewing the temperatures observed for illite development, Giles *et al.* (1992) stated that the kaolinite to illite reaction is controlled by total-time–temperature exposure and not just by temperature alone. The precipitation of authigenic illite in the Jurassic and Cretaceous sandstones of the Salam Field post-dates quartz, barite and bitumen. The homogenization temperatures of the fluid inclusions in quartz overgrowths are 117 to 140°C (Marfil *et al.*, 2000). However, there is a general increase of kaolin and a decrease in illite below 3300 m, which is opposite to the observed tendency in the Jurassic sandstones of the North Sea (Giles *et al.*, 1992; Osborne *et al.*, 1994). This trend probably is related to the low content in feldspar and mica in the sandstones below 3300 m, which were totally kaolinitized during early diagenesis. This is indicated by the presence of vermicular kaolin filling mouldic pores after feldspar in the Khatatba Formation (Rossi *et al.*, 2001).

Stable isotopes

Oxygen and hydrogen stable isotope ratios of clay minerals in sedimentary rocks provide information about the fluids and/or temperatures from which the minerals were formed and/or transformed (Savin & Epstein, 1970a,b; Lawrence & Taylor, 1971; Lee *et al.*, 1989; Kotzer & Kyser, 1995; Delgado & Reyes, 1996; Jeans *et al.*, 1997). In the Khatatba and AEB Formations, isotopic data indicate that kaolin could have formed either at low temperatures (soil-related or early diagenesis) or high temperatures (deep-burial diagenesis). Throughout its geological history, the area was exposed to both marine and meteoric environments, so waters of different isotopic composition may have been involved in the genesis of these minerals. Additionally, the isotopic composition of the pore-water may have been modified during burial diagenesis.

From the equilibrium equation given by Sheppard & Gilg (1996) for hydrogen isotopes ($1000\ln\alpha^{hy}{}_{\text{kaol-water}} = -2.2 \times 10^6 \times T^{-2} - 7.7$), the δD values for kaolinite (soil-related/replacive textures) and a surface temperature of 15 to 25°C, the δD of the pore-water in equilibrium ranges between −94‰ and −28‰ (Fig. 11A). Using the equilibrium equation of Sheppard & Gilg (1996) for oxygen isotopes ($1000\ln\alpha^{oxy}{}_{\text{kaol-water}} = 2.76 \times 10^6 \times T^{-2} - 6.75$), the $\delta^{18}O$ values of the kaolin, and the surface temperature, the $\delta^{18}O$ of the waters ranges from −13‰ to −6‰ (Fig. 11B). If we consider that $\delta^{18}O$ and δD of meteoric water are related by means of the Meteoric Water Line (MWL) ($\delta D = 8 * \delta^{18}O + 10$; Craig, 1961), the $\delta^{18}O$ values of the waters in equilibrium with soil-related/replacement kaolinites would range between −13‰ and −4‰. This range is in good agreement with the calculated $\delta^{18}O$ of the pore-water deduced from the measured $\delta^{18}O$ of the samples and a 15°C precipitation temperature. In addition, the $\delta^{18}O$ and δD values for five kaolinite samples (soil-related matrix/replacive kaolin) cluster along the kaolinite line (Sheppard *et al.*, 1969; Lawrence & Taylor, 1971) (Fig. 12) and hence supporting a meteoric water and low temperature origin for these kaolinites.

Regarding oxygen isotopic signatures of the meteoric water, it is important to stress that during Mesozoic times this area was at low latitude (Ziegler *et al.*, 1983; Keeley & Wallis, 1991). At present, the oxygen isotopic composition of rain water at latitudes between 0° and 20° is typified by values ranging from +1.5‰ and −7‰ (Rozanski *et al.*, 1993). The $\delta^{18}O$ values of the sea water during Mesozoic times were between 2.3‰ and 1.2‰ more negative than present sea water (Shackleton & Kennet, 1974; Scherrer, 1977). Therefore, meteoric water that was derived from the evaporation of the Mesozoic marine waters should have $\delta^{18}O$ compositions between −0.8‰ and −9.3‰. Taking the effect of quartz and illite contamination into account (1.5‰), the calculated range of the isotopic water composition (−13‰ to −6‰ for the soil-related kaolinites) is too negative. Other authors (e.g. Dutta & Suttner, 1986) who have studied kaolinites from this period, also reported anomalously low isotopic values compared with kaolinites that originated in recent weathering profiles at low latitudes. One possible explanation could be the altitude and the continental effect, which lead to more negative rain waters (Fontes, 1980; Rozanski *et al.*, 1993). The palaeogeographical data indicate that the area was not very far from the coast (Keeley & Wallis, 1991), therefore, the possible cause could be humid conditions in mountainous areas. Similarly, Mutti & Weissert (1995) attributed the strongly depleted $\delta^{18}O$ values of meteoric water deduced from early diagenetic carbonates in Triassic sediments from the Southern Alps to be the result of a monsoonal climate in the megacontinent Pangaea.

Some samples with soil-related/replacive textures present δD values near the typical pore-filling textures. These values can be explained easily with reference to the existence of mixtures of diagenetic kaolin–illite and soil-related/replacive kaolin.

Pore-filling kaolin shows δD values between −54‰ and −43‰, indicating higher temperature formation from more positive waters. If we use the present-day isotopic composition of the formation water ($\delta^{18}O = -0.59$‰; $\delta D = -13.5$‰) and the isotopic values of the kaolin, then the precipitation temperature is relatively low ($T <$ 39°C; see Fig. 11A). Still, this low temperature could be explained by the presence of kaolinite that originated in equilibrium with meteoric water at low temperature, i.e. by contamination

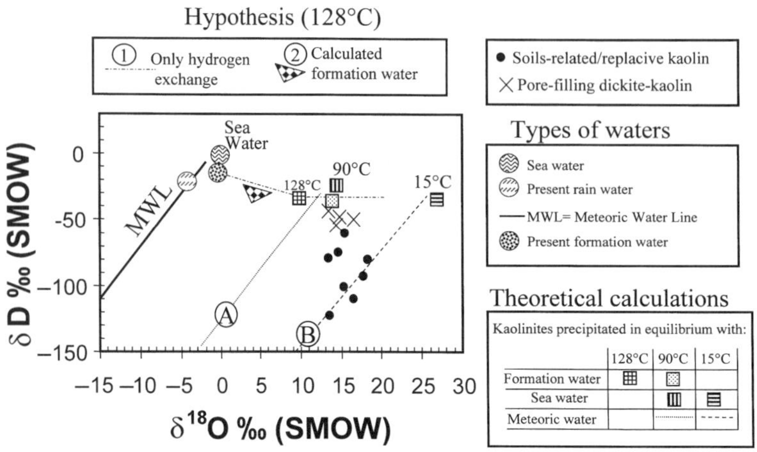

Fig. 12 Plot showing the isotopic composition of samples and the calculated composition of kaolinites that would precipitate at diagenetic temperatures (90°C; 128°C) and surface temperatures from different types of water. Meteoric Water Line (MWL) (Craig, 1961), present rain water (Rozanski *et al.*, 1993) and sea water are shown for reference. Lines represent the theoretical kaolinites in equilibrium with meteoric water at 90°C (line A) and 15°C (line B). Line B is the Meteoric Kaolinite Line named after Savin (1967) and Sheppard *et al.* (1969).

of kaolinite formed during the Kimmerian uplift. However, if this calculation is performed with the $\delta^{18}O$ values of the kaolins (13.3‰ to 16.6‰), then the temperature ranges between 67°C and 93°C. Although this temperature range is lower than the present-day temperature of 128°C, it is consistent with the timing of dickite and ordered kaolinite formation.

If we use the present-day formation temperatures of 128°C and the δD values of the kaolins (−54‰ to −43‰), the calculated δD for the pore-waters would range between −34‰ and −22‰ and for $\delta^{18}O$ between +2.8‰ and +6‰. Thus, either the presence of meteoric water (e.g. from δD) or the presence of formation water enriched in $\delta^{18}O$ is necessary or, alternatively, we should assume that these clay minerals have preserved the $\delta^{18}O$ ratios but exchanged the hydrogen. Relatively narrow ranges in $\delta^{18}O$ and wide ranges of δD have been interpreted by different authors as indicative of a partial isotopic exchange of hydrogen but not of oxygen (Wilson *et al.*, 1987; Greenwood *et al.*, 1994). As a consequence most of the clay minerals would have precipitated under surface conditions, which is in disagreement with the presence of late diagenetic illite and dickite.

The isotopic composition of the pore-filling kaolin can be explained by three different hypotheses.

1 Early diagenetic origin in equilibrium with meteoric water at surface temperature and exchange of hydrogen at high temperature (hypothesis 1 in Fig. 12). This hypothesis is not supported by the petrographic data and is inconsistent with the preservation of δD in soil-related kaolinte.
2 The well-ordered kaolinite and dickite precipitated at present-day temperature conditions (e.g. 128°C). This hypothesis is unlikely as the present-day pore-water is slightly depleted and not strongly enriched in ^{18}O (hypothesis 2 in Fig. 12).
3 Late diagenetic origin in equilibrium with present-day pore-water $\delta^{18}O$ at about 70–90°C. This hypothesis is in good agreement with the isotopic composition of the present-day formation water, the paragenetic sequence and temperature modelling for the Khatatba Formation. Therefore, pore-filling kaolin most probably formed in equilibrium with present-day pore-water $\delta^{18}O$ at elevated temperatures of about 70–90°C during the Syrian-arc orogeny.

CONCLUSIONS

Kaolin and subordinate illite are significant pore-occluding cements in reservoir sandstones of the Jurassic Khatatba and Cretaceous Alam El

Bueib formations of the Salam Oil Field. At a depth interval of 2430 to 3600 m, a mixture of widely varying textures and morphologies of kaolin minerals occur, including (i) pore-filling blocky kaolin that precipitated in reduced primary or secondary porosity of the sandstones, and (ii) vermiform kaolin, which occurs as grain replacive and, less commonly, as soil-related matrix of the sandstones.

Vermiform kaolin was an early precipitate, being enclosed in siderite and Fe-dolomite cements but pre-dating quartz cementation. Conversely, blocky kaolin was formed synchronously with, and subsequent to, precipitation of quartz overgrowths. Kaolinite has a vermiform morphology, whereas ordered kaolinite and/or dickite have a blocky morphology. During deep burial diagenesis kaolinite rather than dickite was replaced by illite. In the Khatatba Formation, below 3300 m, most of the kaolin survived illitization because of a local deficiency in available potassium.

A general depth-related trend of increasing crystal size and dickite abundance in the clean sandstones is remarkable. However, kaolin minerals do not show any progressive change in morphology in mudrocks and in low-permeability sandstones rich in cement or matrix, suggesting that the transformation occurred in an open diagenetic system.

The negative δD and $\delta^{18}O$ values in soil-related/replacive kaolinite indicate the presence of light meteoric waters ($\delta^{18}O_{SMOW} = -13$ to −6‰) during kaolinite precipitation. A humid climate in the hinterland together with a continental and altitude effect may explain the anomalously light δD and $\delta^{18}O$ values of the meteoric water at the low palaeolatitude position during Mesozoic times.

Deep-burial well-ordered pore-filling kaolinite and dickite are characterized by less negative δD but similar $\delta^{18}O$ values relative to soil-related/grain-replacive kaolinite. This kaolin was most probably formed at elevated temperatures of between 70 and 90°C, from pore-waters with a similar stable isotopic composition to present-day formation water.

ACKNOWLEDGEMENTS

This study was funded by the Dirección General de Ciencia y Tecnologia, PB-96-1236-C02-01 and PB-96-1236-C02-02 Projects. The authors wish to thank Khalda Petroleum Co. and Repsol Exploración Egipto for facilitating sampling and providing geological and well data, specially to P. Cámara, J. García-Mallo, F. Martínez-Fresneda, R. Ferrando, J. Suárez and Paloma Bellas. Thanks to Albert Permanyer for his interest and imput at various stages of this research. We are grateful to J.L. Baldonado and A. Rodríguez from the Electron Microscopy Center (Universidad Complutense, Madrid) for assistance with the TEM and SEM studies. The final version of the manuscript benefited from the constructive suggestions and the generous help and patience of Sadoon Morad. The manuscript was greatly improved by the constructive reviews by Shirley P. Dutton and Kitty Milliken.

REFERENCES

Anovitz, L.M., Perkins, D. & Essene, J.E. (1991) Metastability in near-surface rocks of minerals in the system Al_2O_3–SiO_2–H_2O. *Clay. Clay Mineral.*, **39**, 225–233.

Aparicio, P. & Galán, E. (1999) Mineralogical interference on kaolinite crystallinity index measurements. *Clay. Clay Mineral.*, **47**, 12–27.

Bailey, S.W. (1980) Structures of layer silicates. In: *Crystal Structures of Clay Minerals and their X-ray Identification* (Ed. by G.W. Brindley and G. Brown), Mineralogical Society, London, pp. 1–123.

Beaufort, D., Cassagnabere, A., Petit, S., *et al.* (1998) Kaolinite-to-dickite reaction in sandstones reservoirs. *Clay Mineral.*, **33**, 297–316.

Bish, D. (1990) Long-term thermal stability of clinoptilolite: the development of the 'B' phase. *Eur. J. Mineral.*, **2**, 771–777.

Bjørlykke, K. (1998) Clay mineral diagenesis in sedimentary basins: a key to the prediction of rock properties. Examples from the North Sea Basin. *Clay Mineral.*, **33**, 15–34.

Borthwick, J. & Harmon, R. (1982) A note regarding ClF_3 as an alternative to BrF_5 for oxygen isotope analysis. *Geochim. Cosmochim. Acta*, **46**, 1665–1668.

Capuano, R.M. (1992) The temperature dependence of hydrogen isotope fractionation between clay minerals and water: evidence from a geopressured system. *Geochim. Cosmochim. Acta*, **56**, 2547–2554.

Clayton, R.N. & Mayeda, T.K. (1963) The use of bromine pentafluoride in the extraction of oxygen from oxides and silicates for isotopic analisis. *Geochim. Cosmochim. Acta*, **27**, 43–52.

Clayton, R.N., O'Neil, J.R. & Mayeda, T.K. (1972) Oxygen isotope exchange between quartz and water. *J. Geophys. Res.*, **77**, 3057–3067.

Craig, H. (1961) Standards for reporting concentrations of deuterium and oxygen-18 in natural waters. *Science*, **133**, 1833–1834.

Curtis, C.D. (1983) Geochemistry of porosity reduction and enhancement in clastic sediments. In: *Petroleum Geochemistry and Exploration of Europe* (Ed. by J. Brooks), Blackwell, Oxford, pp. 113–125.

Delgado, A. & Reyes, E. (1996) Oxygen and hydrogen isotope compositions in clay minerals: A potential single-mineral geothermometer. *Geochim. Cosmochim. Acta*, **60**, 4285–4289.

Dutta, P.K. & Suttner, L.J. (1986) Alluvial sandstone composition and paleoclimate, II. Authigenic mineralogy. *J. Sed. Petrol.*, **56**, 346–358.

EGPC (1992) *Western Desert, Oil and Gas Fields (Comprehensive Overview)*. Egyptian General Petroleum Corporation, Cairo, 431 pp.

Ehrenberg, S.N., Aagaard, P., Willson, M.J., Fraser, A.R. & Duthie, D.M.L. (1993) Depth-dependent transformation of kaolinite to dickite in sandstones of the Norwegian continental shelf. *Clay Mineral.*, **28**, 325–352.

El Ayouty, M.K. (1990) Petroleum geology. In: *The Geology of Egypt* (Ed. by R. Said), A.A. Balkema, Rotterdam, pp. 567–599.

El Shazly, E.M. (1977) The geology of the Egyptian region. In: *The Ocean Basins and Margins* (Ed. by A.E.M. Nairn, W.H. Kanes and F.G. Stehli), Plenum, New York, pp. 379–444.

Fontes, J.Ch. (1980) Environmental isotopes in groundwater hydrology. In: *Handbook of Environmental Isotope Geochemistry. The Terrestrial Environment* (Ed. by P. Fritz and J. Ch. Fontes), Elsevier, Amsterdam, pp. 75–140.

Galán, E., Aparicio, P., González, I. & La Iglesia, A. (1994) Influence of associated components of kaolin on the degree of disorder of kaolinite as determined by XRD. *Geol. Carpath. Ser. Clays*, **45**, 59–75.

Giles, M.R., Stevenson, S., Martin, S., *et al.* (1992) The reservoir properties and diagenesis of the Brent Group: a regional perspective In: *Geology of the Brent Group* (Ed. by A.C. Morten, R.S. Haszeldine, M.R. Giles and S. Brown). *Geol. Soc. London Spec. Publ.*, **61**, 289–327.

Glassman, J.R., Lundegard, P.D., Clark, R.A., Penny, B.K. & Collins, I.D. (1989) Geochemical evidence for the history of diagenesis and fluid migration: Brent sandstone, HeatherField, North Sea. *Clay Mineral.*, **24**, 255–284.

Godfrey, J.D. (1962) The deuterium content of hydrous minerals from the East-Central Sierra Nevada and Yosemite National Park. *Geochim. Cosmochim. Acta*, **26**, 1215–1245.

Greenwood, P.J., Shaw, H.F. & Fallick, A.E. (1994) Petrographic and isotopic evidence for diagenetic processes in Middle Jurassic sandstones and mudrocks from the Brae area, North Sea. *Clay Mineral.*, **29**, 637–650.

Hassouta, L., Buatier, M.D., Potdevin, J.L. & Liewig, N. (1999) Clay diagenesis in the sandstone reservoir of the Ellon Field (Alwyn, North Sea). *Clay. Clay Mineral.*, **47**, 269–285.

Haszeldine, R.S., Brint, J.F., Fallick, A.E., Hamilton, P.J. & Brown, S. (1992) Open and restricted hydrologies in Brent Group diagenesis: North Sea. In: *Geology of the Brent Group* (Ed. by A.C. Morton, R.S. Haszeldine, M.R. Giles and S. Brown). *Geol. Soc. London Spec. Publ.*, **61**, 401–419.

Hayes, M.J. & Boles, J.R. (1992) Volumetric relations between disolved plagioclase and kaolinite in sandstones: implications for mass transfer in the San Joaquin basin, California. In: *Origin, Diagenesis and Petrophysics of Clay Minerals in Sandstones,* Society of Economic Paleontologists and Mineralogists Special Publication 47 (Ed. by D.W. Houseknecht and E.D. Pittman), Tulsa, OK, pp. 111–123.

Hinckley, A. (1963) Variability in 'crystallinity' values among the kaolin deposits of the Coastal Plain of Georgia and South Carolina. *Clay. Clay Mineral.*, **11**, 229–235.

Jeans, C.V., Fallick, A.E., Fisher, M.J., *et al.* (1997) Clay- and zeolite-bearing Triassic sediments at Kaka Point, New Zealand: evidence of microbially influenced mineral formation from earliest diagenesis into the lowest grade of metamorphism. *Clay Mineral.*, **32**, 373–423.

Kantorowitcz, J. (1994) The nature, origin and distribution of authigenic clay minerals from the Middle Jurassic Ravenscar and Brent Group sandstones. *Clay Mineral.*, **19**, 359–375.

Keeley, M.L. & Massoud, M.S. (1998) Tectonic controls on the petroleum geology of NE Africa. In: *Petroleum Geology of North Africa* (Ed. by D.S. Macgregor, R.T.J. Moody and D.D. Clark-Lowes). *Geol. Soc. London Spec. Publ.*, **132**, 265–281.

Keeley, M.L. & Wallis, R.J. (1991) The Jurassic System in northern Egypt: II. Depositional and tectonic regimes. *J. Petrol. Geol.*, **14**, 49–64.

Keeley, M.L., Dungworth, G., Floyd, C.S., *et al.* (1990). The Jurassic System in northern Egypt. Regional stratigraphy and implication for hydrocarbon prospectivity. *J. Petrol. Geol.*, **13**, 397–420.

Kotzer, T.G. & Kyser, T.K. (1995) Petrogenesis of the Proterozoic Athabasca Basin, northern Saskatchewan, Canada, and its relation to diagenesis, hydrothermal uranium mineralization and paleohydrogeology. *Chem. Geol.*, **120**, 45–89.

Lanson, B., Beaufort, D., Berger, G., Baradat, J. & Lacharpagne, J.C. (1996) Illitization of diagenetic kaolinite to dickite conversion series: late-stage diagenesis of the lower Permian Rotliegend sandstone reservoir. Offshore of the Netherlands. *J. Sed. Res.*, **66**, 501–518.

Lawrence, J.R. & Taylor, H.P. Jr. (1971) Deuterium and oxygen-18 correlation: clay minerals and hydroxides in quaternary soils compared to meteoric waters. *Geochim. Cosmochim. Acta*, **35**, 993–1003.

Lee, M., Aronson, J.L. & Savin, S.M. (1989) Timing and conditions of Permian Rotliegende sandstone diagenesis, southern North Sea: K/Ar and oxygen isotopic data. *AAPG Bull.*, **73**, 195–215.

Lotfy, A. (1994) Jurassic source rock maturity and thermal history modeling of the Khalda West area. North Western Desert, Egypt. In: *12th Exploration and Production Seminar*, Egyptian General Petroleum Corporation, Cairo, pp. 217–233.

Macaulay, C.I., Fallick, A.E. & Haszeldine, R.S. (1993) Textural and isotopic variations in diagenetic kaolinite from the Magnus Oilfield sandstones. *Clay Mineral.*, **28**, 625–639.

Marfil, R., Dorronsoro, C., Rossi, C. & Permanyer, A. (1997) Controles composicionales, diagenéticos y de microfábrica sobre las características de la roca madre jurásica del petróleo del campo de Salam (Western Desert, Egipto). *Geogaceta*, **21**, 159–162.

Marfil, R., Rossi C., Lozano, R.P., Permanyer, A. & Ramseyer, K. (2000) Quartz cementation in Cretaceous and Jurassic reservoir sandstones from Salam Oil Field, Western Desert, Egypt. Constraints on temperature and timing of formation from fluid inclusions. In: *Quartz Cementation in Sandstones*, Intenational Association of Sedimentologists Special Publication 29 (Ed. by R.H. Worden and S. Morad), Blackwell Science, Oxford, pp. 163–182.

May, R.M. (1991) The Eastern Mediterranean Mesozoic Basin: evolution and oil habitat. *AAPG Bull.*, **75**, 1215–1223.

McAulay, G.E., Burley, S.D., Fallick, A.E. & Kusznir, N.J. (1994) Palaeohydrodynamic fluid flow regimes during diagenesis of the Brent Group in the Hutton–NW Hutton reservoirs, constraints from oxygen isotope studies of authigenic kaolin and reverse flexural modelling. *Clay Mineral.*, **29**, 609–629.

McBride, E.F., Land, L.S. & Mack, L.A. (1987) Diagenesis of eolian and fluvial feldspathic sandstones, Norphlet Formation (Upper Jurassic) Rankin County, Mississippi, and Mobile County, Alabama. *AAPG Bull.*, **71**, 1019–1034.

Morad, S., Ben Ismail, H., De Ros, L.F., Al-Aasm, I.S. & Serrihini, N.E. (1994) Diagenesis and formation waters chemistry of Triassic reservoir sandstones from southern-Tunisia. *Sedimentology*, **41**, 1253–1272.

Mutti, M. & Weissert, H. (1995) Triassic monsoonal climate and its signature in Ladinian Carnian carbonate platforms (Southern Alps, Italy). *J. Sed. Res.*, **65**, 357–367.

Nashaat, M., Gamal el Din, A. & Gad, M. (1994) Geochemical material balance and entrapment mechanics in Shusan-Matruh basins, Western Desert, Egypt. In: *12th Exploration and Production Seminar*, Egyptian General Petroleum Corporation, Cairo, pp. 198–216.

Newman, A.C.D. & Brown, G. (1987) The chemical constitution of clays. In: *Chemistry of Clays and Clay Minerals*, Mineralogical Society Monograph 6 (Ed. by A.C.D. Newman), Longman Scientific and Technical, Harlow, pp. 2–129.

O'Brien, N.R. & Slatt, R.M. (1990) *Argillaceous Rock Atlas*. Springer-Verlag, New York, 141 pp.

Osborne, M., Haszeldine, R.S. & Fallick, A.E. (1994) Variation in kaolinite morphology with growth temperature in isotopically mixed pore-fluids, Brent group, UK North Sea. *Clay Mineral.*, **29**, 591–608.

Pettijohn, F.J., Potter, P.E. & Siever, R. (1973) *Sand and Sandstones*. Springer-Verlag, New York, 618 pp.

Ramseyer, K., Fisher, J., Matter, A., Eberhardt, P. & Geiss, J. (1989) A cathodoluminiscence microscope for low intensity luminiscence. *J. Sed. Petrol.*, **59**, 619–622.

Robie, R.A. & Hemingway, B.S. (1995) Thermodynamic properties of minerals and related substances at 298.15 K and 1 bar (10^5 pascals) pressure and at higher temperatures. *US Geol. Surv. Bull.*, **2131**, 461 pp.

Rossi, C., Marfil, R., Ramseyer, K. & Permanyer, A. (2001) Facies related diagenetic control on the porosity of reservoir sandstones, Khatatba Formation, Egypt's Western Desert. *J. Sed. Res.*, **71**, 459–472.

Rozanski, K., Araguás, L. & Gonfiantini, R. (1993) Isotopic patterns in modern global precipitation. In: *Climatic Change in Continental Isotopic Records*, Geophysical Monograph 78 (Ed. by P.K. Swart, *et al.*), American Geophysical Union, pp. 1–36.

Ruiz Cruz, M.D. & Moreno, L. (1993) Diagenetic kaolinite/dickite (Betic Cordilleras, Spain). *Clay. Clay Mineral.*, **41**, 570–579.

Said, R. (1990) *The Geology of Egypt*. Egyptian General Petroleum Corporation, Conoco Hurghada Inc. and Repsol Exploracion, S.A. A.A. Balkema, Rotterdam, 734 pp.

Savin, S.M. (1967) *Oxygen and hydrogen isotope ratios in sedimentary rocks and minerals*. PhD disertation, California Institute Technology, Pasadena, CA, 228 pp.

Savin, S.M. & Epstein, S. (1970a) The oxygen and hydrogen isotope geochemistry of clay minerals. *Geochim. Cosmochim. Acta*, **34**, 25–42.

Savin, S.M. & Epstein, S. (1970b) The oxygen and hydrogen isotope geochemistry of ocean sediments and shales. *Geochim. Cosmochim. Acta*, **34**, 43–63.

Scherrer, M. (1977) Preservation, alteration and multiple cementation of aragonite skeletons from the Cassian beds (Southern Alps), petrographical and geochemical evidence. *Neues Jb. Geol. Palaontol.*, **154**, 3213–3262.

Schultz, L.G. (1964) Quantitative interpretation of mineralogical composition from X-ray and chemical data. *US Geol. Surv. Prof. Pap.*, **391-C**, 31 pp.

Shackleton, N.J. & Kennet, J. (1975) Paleotemperature history of the Cenozoic and the initiation of the Antarctic glaciation: oxygen and carbon isotope analyses in DSDP Sites 227, 279 and 281. *Init. Rep. Deep Sea Drill. Proj.*, **29**, 743–755.

Sheppard, S.M.F. & Gilg, H.A. (1996) Stable isotope geochemistry of clay minerals. *Clay Mineral.*, **31**, 1–24.

Sheppard, S.M.F., Nielsen, R.L. & Taylor, H.P. Jr. (1969) Hydrogen and oxygen isotope ratios of clay minerals from porphyry copper deposits. *Econ. Geol.*, **66**, 515–542.

Stoch, L. (1974) *Clay Minerals.* Geological Publishers, Warsaw, pp. 186–193.

Surdam, R.C., Boese, S.W. & Crossey, L.J. (1984) The chemistry of secondary porosity. In: *Clastic Diagenesis* (Ed. by D.A. McDonald & R.C. Surdam). *AAPG Mem.*, **37**, 127–149.

Surdam, R.C., Crossey, L.J., Hagen, E.S. & Heasler, H.P. (1989) Organic–inorganic interactions and sandstone diagenesis. *AAPG Bull.*, **73**, 1–23.

Taher, M., Said, M. & El-Azhary, T. (1988) Organic geochemical study in Melehia area, Western Desert, Egypt. In: *9th Exploration and Production Seminar*, Egyptian General Petroleum Corporation, Cairo, pp. 1–28.

Venneman, T. & Smith, H. (1990) The rate and temperature of reaction of ClF_3 with silicate minerals, and their relevance to oxygen isotope analysis. *Chem. Geol.*, **86**, 83–88.

Wilson, M.R., Kyser, T.K., Mehnert, H.H. & Hoeve, J. (1987) Changes in the isotopic composition of clays during retrograde alteration. *Geochim. Cosmochim. Acta*, **51**, 869–878.

Ziegler, A.M., Scotese, C.R. & Barret, S.F. (1983) Mesozoic and Cenozoic paleogeographic maps. In: *Tidal Friction and the Earth's rotation*, II (Ed. by P. Broesche and J. Surdermann), Springer-Verlag, Berlin, pp. 240–252.

Zimmerle, W. & Rösch, H. (1991) Petrogenetic significance of dickite in European sedimentary rocks. *Zbl. Geol. Paläontol. Teil II: Paläontol.*, **8**, 1175–1196.

Int. Assoc. Sedimentol. Spec. Publ. (2003) **34**, 343–360

Microscale distribution of kaolinite in Breathitt Formation sandstones (middle Pennsylvanian): implications for mass balance

KITTY L. MILLIKEN

Department of Geological Sciences, University of Texas at Austin, Austin, Texas 78712, USA, e-mail: kittym@mail.utexas.edu

ABSTRACT

Highly discontinuous, patchy distribution at the thin-section scale is an unexplained, but widely observed, characteristic of authigenic kaolinite in sandstones. Detailed petrographic observations of kaolinite (including possible dickite) in Breathitt Formation sandstones (middle Pennsylvanian) reveal information on the timing of chemical reactions and constitute important constraints for calculation of Al mass balance. Kaolinite occurs in both primary and secondary pores. Kaolinite precipitation largely postdates quartz cementation and precedes most grain dissolution. Based on this timing sequence, Al import (into a volume at least as large as a thin-section) was required at one period in the diagenetic history, whereas Al export from the sandstone occurred at another, later period. Distribution of kaolinite into available pore space appears to be random, as the ratio of kaolinite in primary versus secondary pores approximates the ratio of these pore types existing at the time of kaolinite precipitation. The observed Al imbalances together with the lack of a halo distribution around existing and former feldspars supports Al transport that was dominated by advection as opposed to diffusion. Controls on the patchy localization of kaolinite remain unidentified, but possible biogenic controls warrant investigation.

INTRODUCTION: THE AL-MOBILITY ISSUE IN SANDSTONE DIAGENESIS

Controversy over elemental mobility in basinal siliciclastic water–rock interaction has persisted since publication of the seminal paper by Land & Dutton (1978) and the ensuing discussions (Bjørlykke, 1979; Boles & Franks, 1979; Land & Dutton, 1979). Although its volumetric content in sandstones is generally less than that of authigenic quartz or carbonate, authigenic kaolinite and its polymorph dickite ($Al_2Si_2O_5(OH)_4$) are of particular interest in the debate over elemental mobility because precipitation of these phases requires an element (Al) that occurs in extremely low concentrations in subsurface pore fluids (e.g. Giles, 1987; Stoessell, 1987), leading to a substantial conundrum with respect to petrographic evidence for open-system behaviour in diagenetic systems.

Clearly, the precipitation of Al-bearing authigenic phases filling primary pore space in sandstones (e.g. albite, adularia, titanite, zeolites and clays) demands that Al be mobile on some scale. Nonetheless, numerous authors have preferred a model for interrelated detrital feldspar dissolution and kaolinite precipitation that allows for Al mobility on a scale no greater than that of a single thin-section (e.g. Giles, 1987; Stoessell, 1987; Giles & De Boer, 1990; Stoessell & Pittman, 1990; Hayes & Boles, 1992). Data on the microscale distribution of kaolinite and the relative timing of diagenetic events generally are not accounted for in such calculations. Complications in calculating mass balance arise if a given element is lost from the rock at one time, but gained at another (Milliken *et al.*, 1994). Apart from timing considerations, the microscale spatial distribution of authigenic phases also has implications for the nature of the transport system responsible for element mobilization (Milliken, 1995). This paper presents detailed observations on distribution of kaolinite (and/or dickite, but referred to subsequently only as kaolinite) at the thin-section scale in foreland basin sandstones of the Pennsylvanian Breathitt Formation in the southern Appalachian Basin and illustrates the integration of this information into a simple mass-balance calculation.

Microscale distribution of kaolinite in these rocks reveals a situation analogous to that described above, kaolinite being precipitated prior to most of the detrital feldspar dissolution. Interpretation of patchy kaolinite distribution as evidence of a feldspar-replacement origin does not fully withstand detailed scrutiny of rock textures. Published observations for sandstones of diverse ages in other basins suggest that the microscale distribution of kaolinite seen in Breathitt Formation sandstones is not unique.

SAMPLING AND METHODS

Sampling was accomplished during a regional study of sandstones of the Breathitt and Lee formations (lower and middle Pennsylvanian) in the southern Appalachian Basin around the Pine Mountain Overthrust (PMO; Milliken, 1992, 1998; Boettcher & Milliken, 1994; Fu *et al.*, 1994) (Fig. 1). Samples ($n = 548$) were obtained from 317 localities, above and below the PMO; 273 sandstone samples were point-counted (200 points). Among the point-counted samples are 147 samples of Breathitt sandstone from the undeformed region below and in front of the PMO and 52 samples from the synclinally deformed sandstones of equivalent age above the PMO.

For this study, additional petrographic observations on kaolinite distribution were made using conventional plane- and cross-polarized light microscopy (PPL, X-POL), secondary electron imaging (SE-I), scanned cathodoluminescence imaging (CL-I), and backscattered electron imaging (BSE-I) using standard thin-sections impregnated with blue-dyed epoxy. Thin-sections were polished and carbon-coated prior to observations with electron beam imaging. One additional kaolinite-rich sample (KY92A, from below the PMO) was selected for detailed point counting (1000 points) using light microscopy. This sample is coarser grained than average for the regional sample set, a choice made explicitly to reduce the uncertainty of rock-component identification. The paragenetic relationships observed in this sample are, however, representative of those observed in the regional sample set as a whole.

All thin-sections were stained with Na-cobalt-initrite for potassium feldspar identification (technique of Houghton, 1980). Ca-plagioclase is highly sodic in these samples, so staining for these grains was not attempted; plagioclase, much of it untwinned, was identified by included alteration products and also by cleavage traces brought into relief during the staining procedure for potassium feldspar.

In point counting, authigenic components are tabulated specifically as either grain-replacements (in space formerly occupied by a detrital grain) or as cements (filling primary pores). Dissolution and replacement of feldspar and lithic fragments were tabulated so as to preserve (to the extent possible) information about both present texture and mineralogy and the initial character of the primary detrital grain. A portion of both secondary pores and grain replacements are tabulated as such, even though unambiguous information on the identity of the original grain was not preserved.

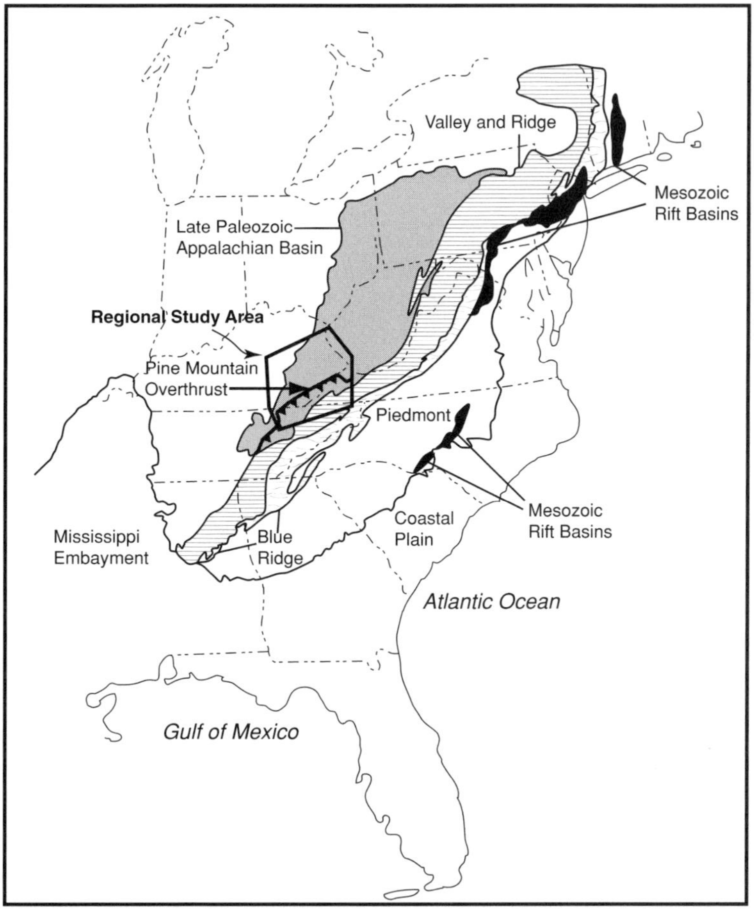

Fig. 1 Map of the southern Appalachian Basin. Modified from Boettcher & Milliken (1994). Breathitt Formation sandstones were sampled in the region both northwest and southeast of the Pine Mountain Overthrust.

DETRITAL MINERALOGY AND GENERAL DIAGENETIC HISTORY

Pennsylvanian siliciclastic units in eastern Kentucky, including the middle Pennsylvanian Breathitt Formation, are primarily fluvio-deltaic deposits shed from thrust sheets to the southeast during the late Palaeozoic Alleghanian orogeny (Smith *et al.*, 1971; Cobb *et al.*, 1981; Tankard, 1986). Maximum burial of 3 to 4 km occurred in this part of the Appalachian Basin foreland during the final stages of the orogeny (Quinlan & Beaumont, 1984; Tankard, 1986; Beaumont *et al.*, 1987). The thermal history determined by apatite fission-track geothermometry for sandstones in the immediate vicinity of the PMO suggests maximum temperatures above the annealing temperature for apatite (*c.* 110°C) and a long period of cooling that ended with rapid uplift in the Miocene (Boettcher & Milliken, 1994).

Detrital compositions of most Breathitt sandstones plot in the field of feldspathic litharenites (average = $Qt_{64}F_{13}L_{23}$) (classification of Folk, 1980). Lithic fragments are dominated by pelitic metasedimentary rocks and include minor amounts of sedimentary and volcanic grains.

The general diagenetic history, similar above and below the PMO, includes early calcite and siderite cements, mostly as concretions, and cementation by quartz, kaolinite, ankerite and late calcite (Milliken, 1992, 1998; Fu *et al.*, 1994). The detrital feldspar assemblage has undergone extensive alteration by some unknown combination of dissolution and replacement (by albite, calcite and ankerite). Detrital K-feldspar is present except in the far southeastern region of the larger study area, where albitization has gone to near-completion. Compactional porosity loss is a prominent aspect of the diagenetic history, accentuated by the presence of ductile lithic fragments and also by abundant micaceous debris, which has promoted pressure dissolution at grain margins (Milliken, 2001). Fracturing of brittle grains, as revealed by CL-I, is an additional contributor to the compaction.

Minor differences in the diagenesis of sandstones above and below the PMO probably are related to the greater burial and higher temperatures experienced on the southeastern side of the basin and a history of late meteoric recharge that apparently had a greater affect on the undeformed region in front of the PMO. Above the PMO there is a greater degree of quartz cementation, less kaolinite and greater compaction (Milliken, 1998, table 1, p. 90).

KAOLINITE DISTRIBUTION IN THIN-SECTION

Kaolinite in Breathitt sandstones is exclusively authigenic and occurs in two morphological types in thin-section. A volumetrically minor portion of kaolinite (estimated < 1% of total kaolinite) is affiliated with detrital muscovite (Fig. 2). This muscovite-associated kaolinite has relatively low microporosity. It does not appear to replace muscovite, but rather is localized between 001 surfaces that have been widely separated by expansion. The dramatic degree of mica inflation suggests a largely precompaction timing for this kaolinite. Not all muscovite grains in Breathitt sandstones have associated kaolinite, even in samples that contain kaolinite in significant amounts.

The volumetrically significant form of kaolinite in Breathitt sandstones is microporous masses of vermiform and blocky stacks of pseudohexagonal crystals (typically in the range of 5–10 μm across) (Fig. 3). This form of kaolinite has a highly irregular distribution at the thin-section scale. Because kaolinite spatial distribution is so discontinuous and also not strongly correlated with a particular mode of localization (as, for instance, quartz cement is affiliated with clean surfaces on detrital quartz), observations relevant to assessing kaolinite genesis relate not only to where the microporous kaolinite *is* observed, but also to where it is *not* observed.

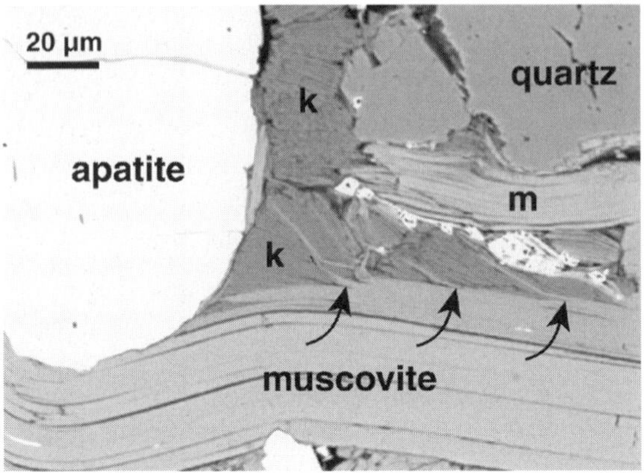

Fig. 2 Authigenic (pore-filling) kaolinite (k) affiliated with a detrital muscovite (m), one of the two petrographic types of authigenic kaolinite observed in this study. Other detrital grains in the field of view include quartz and apatite. Very bright crystals are former authigenic siderite, now replaced by Fe-oxide. Arrows indicate mica flakes that have been separated from the mother grain by expansion related to the growth of kaolinite. Pores are black. Backscattered electron imagery.

The remainder of this section describes the salient petrographic constraints that bear on interpretation of the timing of microporous kaolinite precipitation and, in particular, the mass balance of Al at the thin-section scale.

Microporous kaolinite occurs in primary pores (Fig. 4) adjacent to all of the common detrital grain types that are observed in the Breathitt. The irregularity of the kaolinite distribution is such that many primary pores do not contain kaolinite, however. Most kaolinite precipitation post-dates most quartz precipitation in the Breathitt, although ragged edges

Fig. 3 Microporous kaolinite (possibly dickite) showing typical vermicular habit.

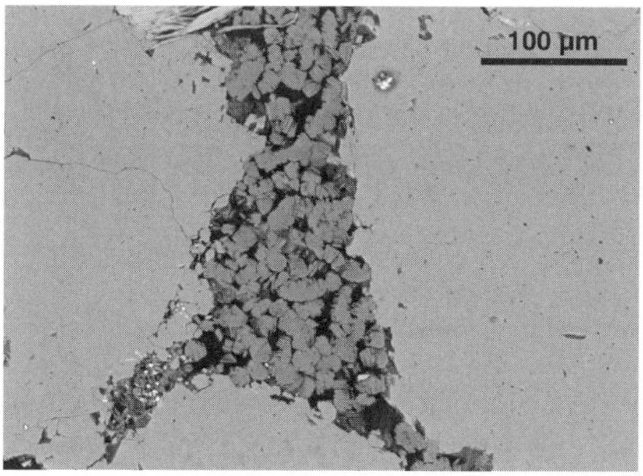

Fig. 4 Microporous kaolinite filling a primary pore. Vermiform kaolinite crystals are surrounded by quartz, a portion of which is authigenic. Pores are black. Backscattered electron imagery.

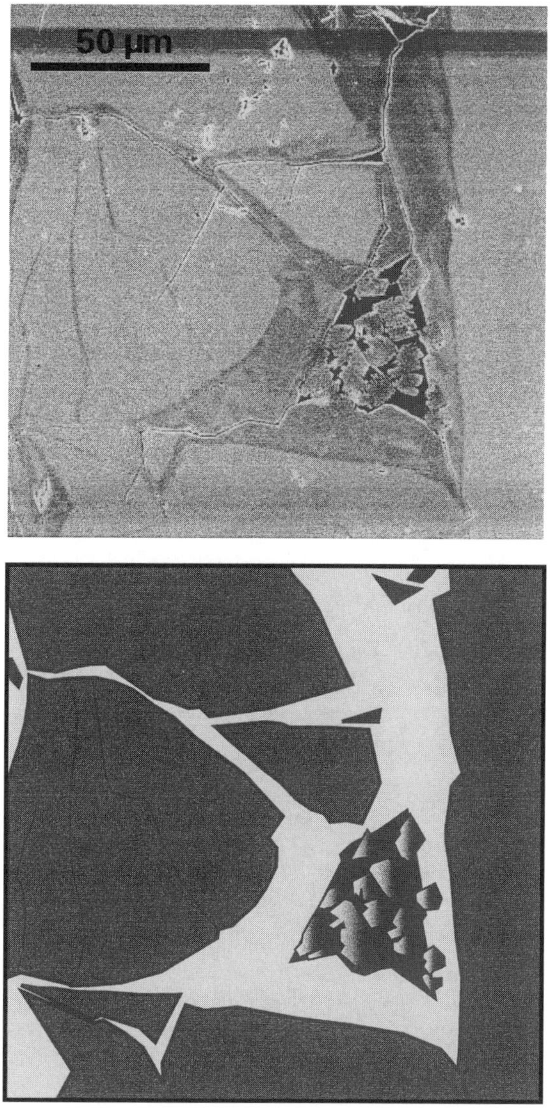

Fig. 5 Slightly overlapping timing of quartz and kaolinite precipitation is documented in this multichannel monochrome image made from combined secondary electron and cathodoluminescence signals. In the accompanying line drawing, detrital quartz is dark grey, authigenic quartz is white, kaolinite is shaded grey and surviving primary porosity is black. Most of the quartz cement emplacement preceded growth of the kaolinite. Kaolinite crystals are, however, partially engulfed by the later stages of quartz cement, indicating that the timing of these two phases overlapped.

on some overgrowths and kaolinite booklets engulfed by the outer zones of quartz cement attest to some overlap in the timing of these phases (Fig. 5). Based on the observation that quartz cement generally does not occur along contacts between detrital quartz and deformed lithic fragments, quartz cementation is interpreted to be largely (although not entirely) post-compaction (Milliken, 2001). Because kaolinite precipitation largely post-dates quartz cementation, it must be post-compaction as well.

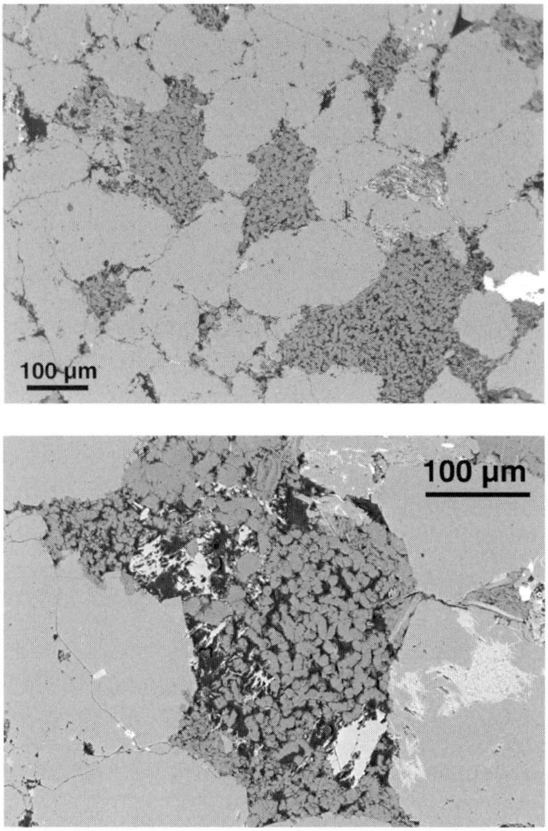

(A)

(B)

Fig. 6 Microporous kaolinite distributed as discrete grain-sized patches, both with (A) and without (B) evidence of associated detrital feldspar remnants. (A) Three such patches of kaolinite in a sample that is otherwise dominated by detrital quartz. Pores are black. Backscattered electron imagery. (B) Microporous kaolinite replacing a dissolved K-feldspar (bright remnants). Kaolinite also occurs within adjacent primary pores. Surrounding grains are detrital quartz and lithic fragments. A portion of the quartz in this field of view is authigenic. Pores are black. Backscattered electron imagery.

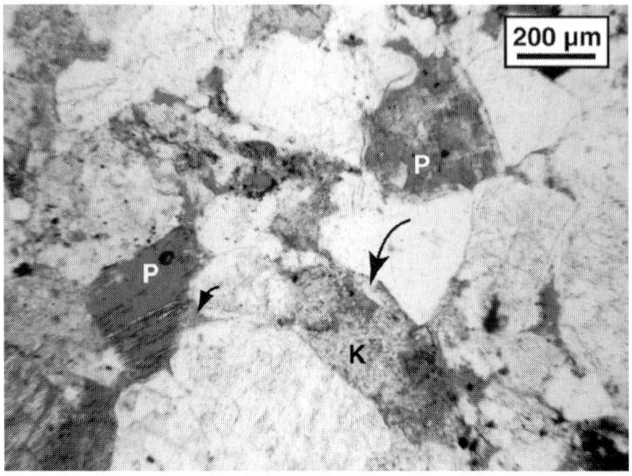

Fig. 7 Three skeletal feldspars. The grain lower centre is largely filled by kaolinite (K), the other two are not and display significant secondary porosity (P). Minor kaolinite is seen in a primary pore adjacent to the skeletal grain on the left (small arrow); no discernible kaolinite is associated with the skeletal grain on the upper right. Minor amounts of quartz cement (large arrow) invade the kaolinized grain, supporting the interpretation that dissolution of this grain occurred relatively early in the diagenetic history, pre-dating some of the quartz precipitation. Plane-polarized light.

Locally, microporous kaolinite occurs in discrete masses that are roughly the size and dimensions of detrital grains (Fig. 6A). If these discrete patches are fillings of primary pores (i.e. cements), their formation necessarily predates the major portion of the compaction that has affected these rocks because present intergranular volumes (IGVs) are sufficiently reduced so that intergranular spaces do not approach the size of these kaolinite patches. If, however, as seems most likely, these discrete patches are grain-replacements there is little or no evidence remaining as to the nature of the grains that were replaced. It is likely that discrete, grain-sized kaolinite patches include portions of both cement and grain-replacement. However, without clear evidence as to the former extent of the replaced grain (e.g. grain remnants) the relative proportion of grain replacement versus cement in these discrete grain-sized patches cannot be assessed very accurately.

Kaolinite also is observed within former secondary pores that are clearly related to feldspar dissolution (Fig. 6B), inasmuch as these spaces still contain skeletal remnants of the partially dissolved feldspar. Both K-feldspar and plagioclase are observed to be replaced by kaolinite. Detrital feldspar replacement by kaolinite typically is accompanied by additional kaolinite in adjacent primary pores (Figs 6B & 7).

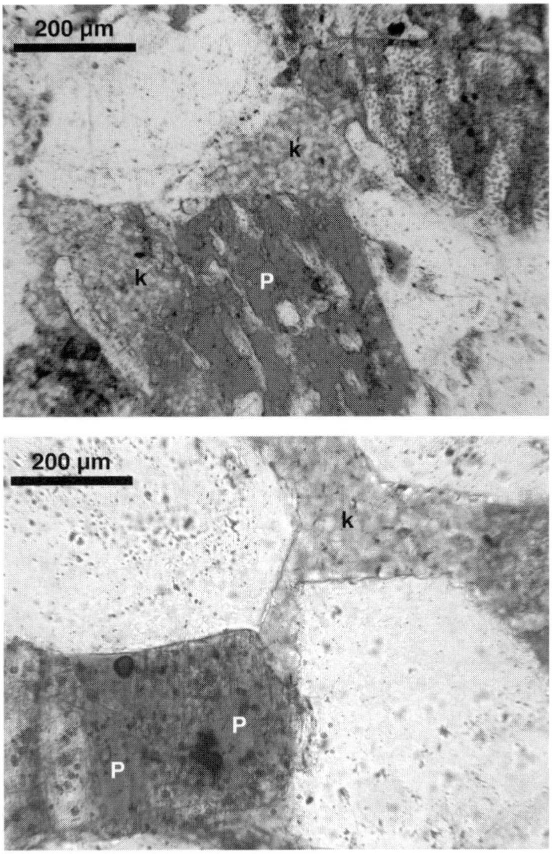

(A)

(B)

Fig. 8 Kaolinite (k) filling primary porosity immediately adjacent to extensively leached (skeletal) detrital feldspars. The precise limit of kaolinite distribution at the former grain margins argues for a chemical history in which kaolinite precipitation largely pre-dates the dissolution of the adjacent feldspar. (A) Grain remnants associated with the large secondary pore (P) are albitic. Minor kaolinization (k) of the feldspar is observed in the upper left corner of the grain, suggesting that dissolution of this portion of the grain occurred either prior to or synchronous with kaolinite precipitation. The partially dissolved grain on the upper right is K-feldspar (spotted stain). Plane-polarized light. (B) Apparent pore-throat sizes of secondary pores (P) within some skeletal feldspars exceed the size of some adjacent kaolinite-filled primary pores, suggesting that pore size and connectivity are not primary controls on kaolinite distribution. This kaolinite (k) fills a narrow primary pore immediately adjacent to an extensively leached potassium feldspar. Plane-polarized light.

Many, and perhaps most, partially dissolved detrital feldspars contain little or no kaolinite, however (Fig. 7). The occurrence of kaolinite in primary pores immediately adjacent to open secondary pore spaces in feldspars is suggestive of a timing sequence in which kaolinite precipitation precedes most feldspar dissolution (Fig. 8A & B). The occurrence of kaolinite in primary pores of relatively small diameter adjacent to markedly larger diameter secondary pores containing no kaolinite (Fig. 8B) argues against a simple pore-size or pore-connectivity limit on kaolinite distribution. In any case, the locally extensive replacement of some grains by kaolinite (examples cited above) clearly documents the potential for kaolinite to precipitate within secondary pores, if the timing sequence is appropriate.

Other than a marked tendency for kaolinite to occur in primary rather than secondary pores, no evidence of systematic spatial distribution of kaolinite with respect to detrital feldspars (partially dissolved or not) is observed (e.g. halo textures or systematic occurrence of kaolinite to one particular side of detrital feldspars or secondary pores). Thus, synchronous feldspar dissolution and kaolinite emplacement is not supported except for the case of kaolinite that occurs within secondary pores. Although it is true in a sense that kaolinite always occurs in proximity to feldspars in these rocks, it is equally true that kaolinite always occurs in proximity to detrital quartz, lithic fragments, quartz cement and primary pores. Thus, kaolinite displays no discernible selectivity for feldspar or other framework grains.

Multiple alteration textures also are widely observed in detrital feldspars. For example, many feldspars contain both kaolinite and intragranular (secondary) pores (Figs 6A, 7, 8A & 9). Complex replacements of feldspars by multiple authigenic phases, including kaolinite and secondary porosity are also observed (Fig. 9).

Kaolinite is observed locally to be engulfed, and therefore post-dated, at least in part, by ankerite (both a grain replacement and a cement) and a late phase of calcite cementation.

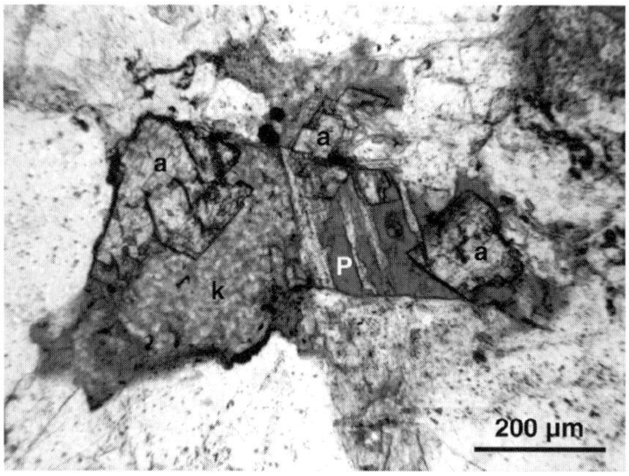

Fig. 9 Complex and multiphase alteration of this feldspar includes replacement by kaolinite (k) and ankerite (a) in addition to dissolution not accompanied by replacement (secondary porosity, P). Surviving feldspar remnants are albite and may represent a third replacement phase. Plane-polarized light.

Both of these late carbonates clearly post-date all of the quartz cementation (Milliken, 1998, 2001).

REGIONAL DATA FOR KAOLINITE DISTRIBUTION

Kaolinite abundance relative to the PMO is shown in Table 1. Below the PMO a greater proportion of samples contain kaolinite and, on average, kaolinite content in these samples is greater. Approximately 70% of kaolinite is distributed in primary porosity below the PMO, approximately 51% above. Regionally, there is no significant correlation between kaolinite content and the initial content of detrital feldspar, present feldspar content, plagioclase/K-feldspar ratio, or secondary porosity (Fig. 10), parameters that plausibly might be expected to correlate with kaolinite content in the case of a direct temporal relationship between feldspar dissolution and kaolinite precipitation.

DETAILED POINT-COUNT RESULTS: AN APPROACH TO INTEGRATING DIAGENETIC TIMING INTO MASS BALANCE CALCULATIONS

Is kaolinite spatially associated with dissolved detrital grains and to what degree? As illustrated in the previous section, kaolinite occurs in both primary and secondary pores, and conversely, kaolinite-free examples of both of these pore types are observed. Kaolinite occurs dominantly within primary pores and no systematic spatial affiliation of kaolinite with primary pores immediately adjacent to detrital feldspars is apparent. Detailed point counting (1000 points) of a single kaolinite-rich coarse-grained sample from the undeformed region in front of the PMO was undertaken to assess more quantitatively the distribution of kaolinite in different types of pores (Table 2).

Table 3 displays data pertinent to an examination of kaolinite distribution. The sample contains 6.4% kaolinite in primary and secondary pores combined. As in the regional sample set, kaolinite occurs mostly in primary pores

Table 1 Averaged point count data for kaolinite content in Breathitt sandstones above and below the PMO.

	n	Samples with kaolinite (%)	Average content* (%)	Average content† (%)	Range (%)	Cement‡ (% of total)
Below the PMO	147	88.4	1.7	2.3	0–10.3	70.2
Above the PMO	52	51.9	0.9	1.4	0–4.0	50.8

* Overall average volume per cent of whole rock.
† Average of samples containing kaolinite.
‡ Percent of total kaolinite that occurs in primary pores, averaged only for samples containing kaolinite.

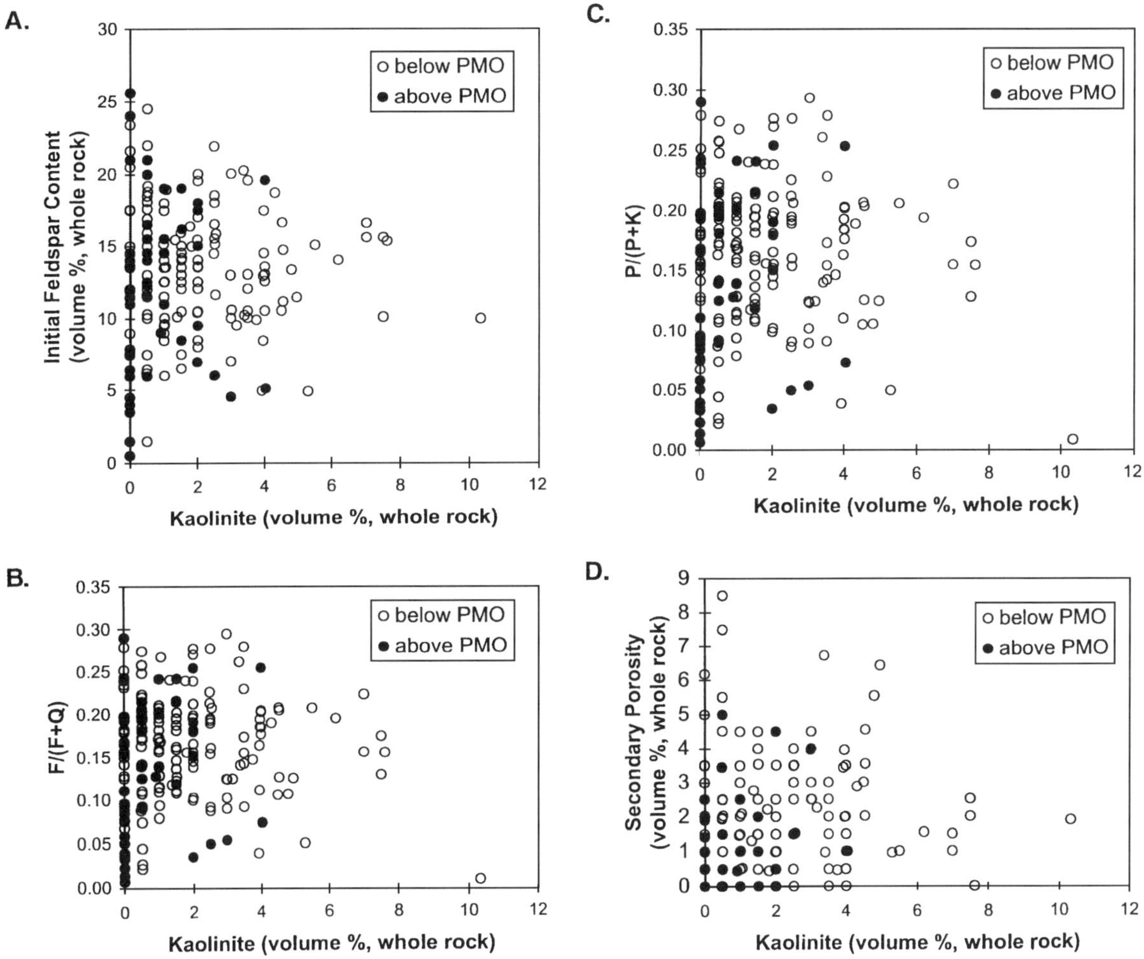

Fig. 10 Point count results from the regional set of Breathitt sandstones above and below the Pine Mountain Overthrust (PMO). Plots A–D all illustrate parameters that plausibly could be expected to manifest a correlation if kaolinite precipitation is temporally related to local feldspar dissolution at the thin-section scale. The absence of significant correlations in these plots lends support to the detailed petrographic observations that suggest that much of the observed feldspar dissolution did not result in kaolinite precipitation. (A) Restored initial detrital feldspar content versus total kaolinite (cement + grain replacement). (B) Detrital feldspar /(detrital feldspar + detrital quartz), a measure of present feldspar content that is somewhat corrected for variations in grain size, versus total kaolinite. (C) Plagioclase/(plagioclase + K-feldspar) versus total kaolinite. (D) Secondary porosity versus total kaolinite.

(75% of total kaolinite, including 3% as mica-associated kaolinite); grain replacement kaolinite (both within clearly identified skeletal feldspars and in discrete, grain-shaped masses) constitutes 25% of the total kaolinite. Primary and secondary pores currently constitute subequal portions of the total porosity, 43% and 57%, respectively.

Thus, assuming that pore types in the present rock are characteristic of those at the time of kaolinite emplacement, kaolinite has an apparent preference for emplacement in primary pores, because the percentage of total kaolinite occurring in primary pores exceeds the percentage of primary porosity in total porosity.

A more pertinent apportionment of primary

Table 2 Raw point count data for sample KY92A. The calculated parameters include ones that are relevant to a general characterization of this sample and also ones that are particularly useful for understanding kaolinite distribution.

	1	2
Detrital components:		
quartz		
monocrystalline	469	
polycrystalline	32	
lithic fragments		
metamorphic	101	
microquartz	4	
siltstone/mudstone	15	
volcanic	3	
plutonic	8	
unknown	33	
grain alteration		
secondary porosity		7
replacement		0
feldspars		
K-feldspar		
sericitized/unaltered	3	
kaolinite-replaced (feldspar)	2	
ankerite-replaced (feldspar)	0	
kaolinite-replaced (kaolinite)		2
ankerite-replaced (ankerite)		2
leached (feldspar)	16	
leached (secondary porosity)		20
plagioclase		
sericitized/unaltered	30	
kaolinite-replaced (feldspar)	2	
ankerite-replaced (feldspar)	0	
kaolinite-replaced (kaolinite)		3
ankerite-replaced (ankerite)		1
leached (feldspar)	15	
leached (secondary porosity)		16
non-QFR grains		
heavy mineral	3	
muscovite	5	
compaction seams	2	
unknown	0	

	1	2
Authigenic components:		
cements		
quartz	91	
kaolinite	48	
ankerite	11	
haematite	2	
grain replacements (grain unknown)		
kaolinite		11
ankerite		7
Porosity:		
primary	34	
secondary (grain unknown)		2
Total points	1000	71
Calculated parameters:		
QFR (present)	$Q_{68.3}F_{9.3}R_{22.4}$	
QFR (initial)	$Q_{64}F_{14}R_{22}$	
F/(Q + F) (present)	0.12	
F/(Q + F) (initial)	0.18	
P/(P + K) (present)	0.69	
P/(P + K) (initial)	0.60	
IGV	18.6	
kaolinite (whole rock)	6.4	
kaolinite cement	4.8	
kaolinite grain replacement	1.6	
ankerite cement	1.1	
ankerite grain replacement	1.0	
total porosity	7.9	
primary porosity	3.4	
secondary porosity	4.5	
cumulative loss of detrital components	7.1	

1. Components, detrital and authigenic, that serve an invariant role in calculations.
2. Components that can be variably accounted as either authigenic or detrital.

and secondary porosity is that which characterized the sample at the time of kaolinite precipitation. Assuming that all post-quartz cement porosity (present primary and secondary porosity + post-quartz cements + post-quartz grain replacements) was available at the time of kaolinite precipitation, the proportion of pore types would have been approximately 57% primary and 43% secondary. Under this second scenario, kaolinite also has an apparent preferential distribution into primary pores (Table 3).

A third scenario shown in Table 3 assumes that the existing secondary porosity post-dates kaolinite emplacement (and hence was

Table 3 Results for kaolinite, ankerite, and porosity in the detailed point count of sample KY92A. Porosity types are recalculated for different times in the diagenetic history, utilizing different assumptions about the timing of kaolinite precipitation relative to feldspar dissolution. See text for discussion.

	Kaolinite (%)			Ankerite (%)			Porosity (%)		
	Cement	Replacement	Total	Cement	Replacement	Total	Primary	Secondary	Total
Whole rock:									
today	4.8	1.6	6.4	1.1	1	2.1	3.4	4.5	7.9
total post-quartz*	0	0	0	0	0	0	9.3	7.1	16.4
post-quartz, pre-kaolinite†	0	0	0	0	0	0	9.3	2.6	11.9
Component:									
today—scenario 1	75	25		52	48		43	57	
scenario 2*	75	25		52	48		57	43	
scenario 3†	75	25		52	48		78	22	

* Assumes porosity available at the time of kaolinite precipitation is the sum of modern porosity + grain replacements + post-quartz cements.
† Assumes that all of modern secondary porosity is post-kaolinite, and that only pre/syn-kaolinite seconday porosity (represented by grain replacements) existed at the time of kaolinite precipitation.

unavailable for kaolinite precipitation). In this case only kaolinite-replaced and ankerite-replaced grains are summed to represent the secondary porosity prior to kaolinite emplacement; kaolinite cement and ankerite cement are added to the existing primary porosity as in the previous calculation. In this reconstruction primary pores make up approximately 78% of the total post-quartz cement porosity; secondary pores 22%. These totals match closely the observed proportional distribution of kaolinite within the different pore types (75% in primary; 25% in secondary), and suggest that kaolinite was not preferentially emplaced into any particular pore type, and instead, simply filled the different pore types in the proportion in which those pore types (primary and secondary) existed at the time of kaolinite precipitation.

Interestingly, ankerite in these rocks does have an apparent affinity for secondary porosity, 48% of the ankerite occurring in replacement mode, despite the fact that only 22% of the total porosity was secondary prior to ankerite precipitation.

Mass-balance calculations for Al, based on data from Tables 2 & 3, are performed for the end-member scenarios, scenario 2 (timing constraints are ignored and all silicate dissolution is construed as a local Al source) and scenario 3 (only pre/syn-kaolinite silicate dissolution available as a local Al source) (Table 4).

Calculations, referenced to 1 cm^3 of sandstone, utilize the following assumptions. Secondary pores in lithic fragments and unknown grains (oversized pores) are included in the calculation and are construed as feldspars with a K-feldspar/plagioclase ratio equivalent to that in the rock as a whole prior to feldspar dissolution ($P/(P + K) = 0.60$). Microporous kaolinite is assumed to have a porosity of 50% (Nadeau & Hurst, 1991). K-feldspar has the composition of end-member $KAlSi_3O_8$; plagioclase is oligoclase, $Na_{0.8}Ca_{0.2}Al_{1.2}Si_{2.8}O_8$.

Kaolinite precipitated in primary and secondary pores presently constitutes around 8.1×10^{-4} mol Al cm^{-3} in the sandstone. This estimate is sensitive to the assumption concerning the amount of microporosity in kaolinite; 6.5×10^{-4} mol Al cm^{-3} are present in kaolinite if 60% porosity in microporous kaolinite is assumed. The Al available from the sum of existing secondary porosity and grain replacements (dissolution scenario 2) amounts to

Table 4 Mass balance of Al with respect to feldspar dissolution and kaolinite precipitation in sample KY92A. Scenario 2 (dissolution) is calculated using the assumption that the sum of present secondary porosity and grain replacements represents the volume of the Al-source. Scenario 3 (dissolution) is based on the observed relative timing of kaolinite precipitation (post-dating most quartz cement but pre-dating most feldspar dissolution). See text for further discussion.

Scenario	Precipitation/ dissolution	Step 1: volume fraction of mineral precipitated/dissolved (vol. %* × 1 − (microporosity)† = vol. fraction‡)	Step 2: moles of mineral precipitated/dissolved (vol. fraction‡ × (density§/mol. wt¶) = mol. mineral cm^{-3}‡)	Step 3: moles Al precipitated/dissolved (mol. mineral cm^{-3}‡ × mol. Al/mol. mineral = mol. Al cm^{-3}‡)
	Precipitation:			
	$Al_2Si_2O_5(OH)_4$	6.4 × 0.5 × 0.01 = 0.032	0.032 × 2.65/210.2 = 4.0E-04	4.0E-04 × 2.0 = 8.1E-04
2: all feldspar loss serves as local Al source	Dissolution:			
	$KAlSi_3O_8$	2.8 × 1.0 × 0.01 = 0.028	0.028 × 2.57/278.3 = 2.6E-04	2.6E-04 × 1.0 = 2.6E-04
	$Na_{0.8}Ca_{0.2}Al_{1.2}Si_{2.8}O_8$	4.3 × 1.0 × 0.01 = 0.043	0.043 × 2.65/393.4 = 2.9E-04	2.9E-04 × 1.2 = 3.4E-04
	Total	7.1		6.1E-04
3: only pre/syn-kaolinite feldspar loss serves as Al source	Dissolution:			
	$KAlSi_3O_8$	1.0 × 1.0 × 0.01 = 0.010	0.010 × 2.57/278.3 = 9.2E-05	9.2E-05 × 1.0 = 9.2E-05
	$Na_{0.8}Ca_{0.2}Al_{1.2}Si_{2.8}O_8$	1.6 × 1.0 × 0.01 = 0.016	0.016 × 2.65/393.4 = 1.1E-04	1.1E-04 × 1.2 = 1.3E-04
	Total	2.6		2.2E-04

* In whole rock. † Fraction of mineral. ‡ In 1 cm^3 sandstone. § g cm^{-3}. ¶ g $mol.^{-1}$.

6.1×10^{-4} mol Al cm^{-3}; Al available from that portion of the silicate dissolution that preceded or was synchronous with kaolinite precipitation (dissolution scenario 3) is around 2.2 $\times 10^{-4}$ mol Al cm^{-3}.

Thus, petrographic constraints on the timing of kaolinite emplacement have significant ramifications for the calculation of mass balance. If apparent timing constraints are ignored (dissolution scenario 2), import of an additional 2×10^{-4} mol Al cm^{-3} is required to accomplish kaolinite emplacement; if timing constraints are utilized (dissolution scenario 3), the deficiency approaches 6×10^{-4} mol Al cm^{-3} for kaolinite precipitation. Additionally, application of the timing constraints implies the export of around 4×10^{-4} mol Al cm^{-3} as a result of late feldspar dissolution that was not accommodated by local kaolinite precipitation.

DISCUSSION

Petrographic associations

The pattern of microscale kaolinite distribution reported here (summarized in Table 5) is not unique to the Breathitt Formation. Authigenic kaolinite, comparable in many details to that reported here, has been described for sandstones from a wide variety of geological settings, and for samples from both outcrop and the subsurface. Early papers (despite the disadvantage of not having coloured impregnation media) documented the multiple modes of occurrence of kaolinite at the thin-section scale, recognizing cements, grain replacements and fracture fillings (e.g. Shelton, 1964, and papers cited therein).

Table 5 The salient petrographic aspects of kaolinite (and, possibly, associated dickite) in Breathitt sandstones.

Mica-associated kaolinite pre-dates much of the compaction
Discontinuous (patchy) masses of microporous kaolinite in both primary and secondary pores
Discrete grain-sized and grain-shaped masses of microporous kaolinite
Overlapping timing of quartz and kaolinite precipitation in which most quartz precipitation pre-dates most kaolinite (i.e. kaolinite emplacement is synchronous with only the latter stages of quartz precipitation)
Overlapping timing of kaolinite precipitation and detrital feldspar dissolution, most dissolution post-dating kaolinite precipitation

The association of low-porosity kaolinite with muscovite grains has been reported previously (e.g. Rossel, 1982; Kantorowicz, 1985; Ehrenberg & Nadeau, 1989; Pevear *et al.*, 1991; Burley & MacQuaker, 1992; Morad *et al.*, 1994; Osborne *et al.*, 1994), although some of these papers interpret the kaolinite as a mica replacement rather than as a precipitate between expanded mica flakes.

The occurrence of kaolinite as both a cement and a grain replacement having patchy distribution within a single sample is also a common observation (e.g. Stoessell, 1987; Hayes & Boles, 1992; Houseknecht & Ross, 1992; Platt, 1993; Matyas & Matter, 1997), as is the apparent preference of kaolinite for primary pores (e.g. Siebert *et al.*, 1984; Land *et al.*, 1987; Stoessell, 1987; Hayes & Boles, 1992; Houseknecht & Ross, 1992).

Rather diverse timing of kaolinite precipitation in sandstones has been reported, ranging from relatively early (pre-quartz cement), to early, yet at elevated temperature, to synchronous with quartz cementation, to relatively late kaolinite that post-dates quartz cementation (e.g. Glasmann *et al.*, 1989; Baker & Golding, 1992; Bjørlykke & Aagaard, 1992; Longstaffe *et al.*, 1992; Macaulay *et al.*, 1993; Morad *et al.*, 1994; Stewart *et al.*, 1994). Kaolinite precipitation that pre-dates nearby feldspar dissolution has been noted specifically in several previous studies (Siebert *et al.*, 1984; Land *et al.*, 1987; Stoessell, 1987).

As in the Breathitt example, halo textures or other specific spatial associations with detrital feldspars are not reported and previous authors also have commented on this fact (e.g. Stoessell, 1987).

Several studies have noted a tendency for

kaolinite precipitation to occur preferentially within units of greater porosity and permeability (e.g. Bayliss *et al.*, 1965; Huggett, 1984; Osborne *et al.*, 1994; Khanna *et al.*, 1997; Matyas & Matter, 1997), a trend that has not been examined specifically in the present study, although it is true that the most notable examples of kaolinite in the Breathitt occur in some of the coarsest sandstones in the regional sample set, which is dominated by well-sorted channel sandstones.

Acidic fluid, either meteoric or coal-related or both, is perhaps the most ubiquitous interpretation related to the occurrence of kaolinite (Burley, 1984; Kantorowicz, 1984; Glasmann *et al.*, 1989; Baker & Golding, 1992; Bjørlykke & Aagaard, 1992; Burley & MacQuaker, 1992; Longstaffe *et al.*, 1992; Macaulay *et al.*, 1993; Platt, 1993; McAulay *et al.*, 1994; Morad *et al.*, 1994; Stewart *et al.*, 1994; Lanson *et al.*, 1996; Matyas & Matter, 1997; De Ros & Morad, 1998; Cookenboo & Bustin, 1999). Recharge of meteoric water into the undeformed regions in front of the PMO during late diagenesis is a likely fluid flow model that has been proposed for other foreland basins (e.g. Deming *et al.*, 1990) and, of course, coals are an important component of the Breathitt Formation (e.g. Smith *et al.*, 1971). The issue of fluid composition is not addressed in this petrographic study, but Breathitt Formation sandstones are plausible as an example of this most commonly reported kaolinite association. In particular, it is likely that late, topographically driven meteoric recharge during basin uplift is responsible for the greater abundance of kaolinite in the undeformed portions of the Breathitt Formation below the PMO. It is interesting, however, that the petrographic associations reported here also conform to the reported characteristics of kaolinite in formations that have not encountered meteoric fluids because of rapid early burial and overpressuring (Siebert *et al.*, 1984; Land *et al.*, 1987; Lynch, 1996).

The reason for a greater proportion of kaolinite as a grain replacement above the PMO (near 50% as opposed to 30% below the PMO, Table 1) is uncertain, but may relate to the greater degree of compaction (pre-kaolinite) in the more deeply buried southeastern parts of the basin.

With the exception of the wide range reported in other studies for the apparent timing of kaolinite relative to quartz cementation, the assemblage of petrographic characteristics observed for kaolinite in Breathitt sandstones (Table 5) constitutes a sort of syndrome of petrographic associations for this mineral that has widespread applicability.

Aluminium source(s), transport and localization of kaolinite

With respect to the mass balance of Al, kaolinite is typically construed to be derived directly from local feldspar dissolution without a rigorous petrographic assessment of kaolinite distribution and timing (e.g. Boles, 1984; Giles, 1987; Stoessell, 1987; Giles & De Boer, 1990; Hayes & Boles, 1992; Houseknecht & Ross, 1992). Other authors have noted apparent Al-imbalances in diagenesis, both deficits (McLaughlin *et al.*, 1994) and excesses (Platt, 1993).

The imbalances of Al observed in this study, together with the lack of halo textures around existing and former feldspars, supports advection as the primary mechanism of Al transport. Other lines of evidence bear on the plausibility of this interpretation. For example, kaolinite commonly is reported as vein-filling, suggesting the emplacement of this phase under conditions of highly open fluid flow (Shelton, 1964; Woronick & Land, 1985; Guscott *et al.*, 1997). Extensive kaolinite cement also is reported in highly quartz-rich sandstones that bear no evidence of feldspar dissolution (e.g. Minster & Fraser, 1982; Loughnan & Roberts, 1986). Kaolinite authigenesis in carbonate rocks is particularly compelling evidence of Al mobility (Maliva *et al.*, 1999) and is much more widespread than commonly recognized (e.g. Dunham, 1972; Rudolph, 1978; Naiman, 1982; Leary, 1984; Woronick & Land, 1985; Hill, 1996).

Despite the marked Al imbalances observed in this study, feldspar dissolution on some scale, in both sandstones and associated shales, is still very attractive as a potential Al source. Given the complexity of kaolinization history, however, it is reasonable that other sources should be considered as well. For example, dissolution of detrital clay minerals in both sandstones and shales and amorphous grain coatings are examples of reactions that may contribute Al for authigenic phases. Land's (1997) observation that particulate or colloidal transport of Al has not been considered is still true.

Patchy distribution may be a key constraint on the type of model that will explain kaolinite genesis. For example, a microbial control on clay-mineral precipitation (Folk & Lynch, 1997) could serve as a mechanism of patchy kaolinite localization. An intimate association between coarsely crystalline kaolinite and bacterial and fungal remains has been reported in tonsteins (Moore, 1964a,b). Considering the extensive distribution of microbial communities in the deep subsurface (e.g. Sinclair & Ghiorse, 1989; Spark *et al.*, 2000), it would be reasonable to search for a similar association between microbes and the precipitation of patchy subsurface kaolinite cements.

An alternative that cannot be assessed by conventional petrography is that kaolinite forms haloes only around certain feldspars (perhaps the ones that were dissolving at the same time kaolinite was precipitating), but these feldspars are outside the plane of the thin-section. This scenario is not supported by any observations of halo textures anywhere in the thin-section, but cannot be firmly discounted. Assessment by three-dimensional imaging methods, for example X-ray computed tomography or laser confocal microscopy, offer the possibility of resolving this issue definitively.

Whatever the mechanism of kaolinite localization, timing constraints suggest that a model for Al mass balance in sandstones will not be a simple one. As has been demonstrated, different timing scenarios lead to very different conclusions with respect to mass balance. Constructing a correct quantitative mass balance applicable to a diverse assemblage of phases reacting in a complex temporal sequence within a common pore fluid is a formidable challenge. Petrographic relationships among grains, cements and porosity record the timing and spatial affiliations between chemically active components in sandstones and constitute a powerfully informative line of evidence that should not be ignored. For now, the long-standing paradox between petrographic evidence for Al mobility in rocks and other evidence relating to the limited transportability of Al in subsurface pore fluids remains with no clear resolution in sight. Perhaps new avenues of investigation (e.g. into biogenic factors) are needed in order to make progress on this compelling issue.

CONCLUSIONS

1 Kaolinite in Breathitt Formation sandstones occurs in two petrographic forms, one a low-porosity form spatially associated with detrital muscovites and the other a microporous aggregate of kaolinite crystals with patchy distribution in both primary and secondary pores.

2 Kaolinite precipitation overlapped the latter stages of quartz precipitation, but post-dates most of the quartz cementation.

3 Today, kaolinite is distributed dominantly in primary pores and largely pre-dates the formation of secondary porosity.

4 Kaolinite is distributed randomly among the primary and secondary pores that existed at the time of kaolinite precipitation. No control on the thin-section-scale localization of microporous kaolinite patches is apparent.

5 Mass-balance calculations based on the relative timing of kaolinite precipitation and feldspar dissolution suggest that Al was imported into the sandstone during kaolinite precipitation, but lost from the sandstones during later feldspar dissolution.

6 The lack of halo textures in kaolinite and the imbalances in Al during Breathitt diagenesis suggest that Al transport was effected largely by advection.

7 The greater content of kaolinite in undeformed Breathitt sandstones below the PMO is probably related to the greater degree of late meteoric recharge in this portion of the formation.

ACKNOWLEDGEMENTS

This study was supported by the Donors of the Petroleum Research Fund of the American Chemical Society (ACF-PRF 22805-AC8). Partial support of publication costs was provided by the Owen-Coates Fund of the Geology Foundation, University of Texas at Austin. An early version of this manuscript benefited from comments by Earle McBride and Rogerio Souza. Constructive reviews by Richard Worden, Dominic Strogen and an anonymous reviewer also contributed to the clarity of the manuscript.

REFERENCES

Baker, J.C. & Golding, S.D. (1992) Occurrence and palaeohydrological signficance of authigenic kaolinite in Aldebaran sandstone, Denison Trough, Queensland, Australia. *Clay. Clay Mineral.*, **40**, 273–279.

Bayliss, P., Loughnan, F.C. & Standard, J.C. (1965) Dickite in the Hawkesbury Sandstone of the Sydney Basin, Australia. *Am. Mineral.*, **50**, 418–300.

Beaumont, C., Quinlan, G.M. & Hamilton, J. (1987) The Alleghanian orogeny and its relationship to the evolution of the eastern interior, North America. In: *Sedimentary Basins and Basin-forming Mechanisms*, Canadian Society of Petroleum Geologists Memoir 12 (Ed. by C. Beaumont and A.J. Tankard), Calgary, pp. 425–445.

Bjørlykke, K. (1979) Cementation of sandstones, discussion. *J. Sed. Petrol.*, **49**, 1358–1359.

Bjørlykke, K. & Aagaard, P. (1992) Clay minerals in North Sea Sandstones. In: *Origin, Diagenesis, and Petrophysics of Clay Minerals in Sandstones*, Society for Economic Paleontologists and Mineralogists Special Publication 47 (Ed. by D.W. Houseknecht and E.D. Pittman), Tulsa, OK, pp. 65–80.

Boettcher, S.S. & Milliken, K.L. (1994) Mesozoic–Cenozoic unroofing of the southern Appalachian Basin: apatite fission track evidence from middle Pennsylvanian sandstones. *J. Geol.*, **102**, 655–663.

Boles, J.R. (1984) Secondary porosity reactions in the Stevens Sandstone, San Joaquin Valley, California. In: *Clastic Diagenesis* (Ed. by D.A. MacDonald and R.C. Surdam). *AAPG Mem.*, **37**, 217–224.

Boles, J.R. & Franks, S.G. (1979) Cementation of sandstones, reply. *J. Sed. Petrol.*, **49**, 1362.

Burley, S.D. (1984) Patterns of diagenesis in the Sherwood Sandstone Group (Triassic), United Kingdom. *Clay Mineral.*, **19**, 403–440.

Burley, S.D. & MacQuaker, J.H.S. (1992) Authigenic clays, diagenetic sequences and conceptual diagenetic models in contrasting basin-margin and basin-center North Sea Jurassic sandstones and mudstones. In: *Origin, Diagenesis, and Petrophysics of Clay Minerals in Sandstones*, Society for Economic Paleontologists and Mineralogists Special Publication 47 (Ed. by D.W. Houseknecht and E.D. Pittman), Tulsa, OK, pp. 81–110.

Cobb, J.C., Chesnut, D.R., Hester, N.C. & Hower, J.C. (1981) *Coal and Coal-bearing Rocks of Eastern Kentucky*. Guidebook for Annual, Geological Society of America Coal Division Field Trip on 5–8 November, Kentucky Geological Survey, Lexington, KY, 169 pp.

Cookenboo, H.O. & Bustin, R.M. (1999) Pore water evolution in sandstones of the Groundhog Coalfield, northern Bowser Basin, British Columbia. *Sed. Geol.*, **123**, 129–146.

De Ros, L.F. & Morad, S. (1998) Meteoric-water influx and porosity evolution in turbidite sandstones; evidence from Late Cretaceous/early Tertiary reservoirs of Espirito Santo Basin, eastern Brazilian margin. *AAPG Bull.*, **82**, 1909.

Deming, D., Nunn, J.A. & Evans, D.G. (1990) Thermal effects of compaction-driven groundwater flow from overthrust belts. *J. Geophys. Res.*, **95**, 6669–6683.

Dunham, R.J. (1972) *Capitan Reef, New Mexico and Texas: Facts and Questions to Aid Interpretation and Group Discussion.* Permian Basin Section, Society for Economic Paleontologists and Mineralogists, Tulsa, OK.

Ehrenberg, S.N. & Nadeau, P.H. (1989) Formation of diagenetic illite in sandstones of the Garn Formation, Haltenbanken area, mid-Norwegian continental shelf. *Clay Mineral.*, **24**, 233–253.

Folk, R.L. (1980) *Petrology of Sedimentary Rocks.* Hemphill Publishing Company, Austin, TX.

Folk, R.L. & Lynch, F.L. (1997) The possible role of nannobacteria (dwarf bacteria) in clay mineral diagenesis and the importance of careful sample preparation in high magnification SEM study. *J. Sed. Petrol.*, **67**, 583–589.

Fu, L., Milliken, K.L. & Sharp, J.M., Jr. (1994) Porosity and permeability variations in fractured and liesegang-banded Breathitt sandstones (Middle Pennsylvanian), eastern Kentucky: diagenetic controls and implications for modeling dual-porosity systems. *J. Hydrol.*, **154**, 351–381.

Giles, M.R. (1987) Mass transfer and problems of secondary porosity creaton in deeply buried hydrocarbon reservoirs. *Mar. Petrol. Geol.*, **4**, 188–204.

Giles, M.R. & De Boer, R.B. (1990) Origin and significance of redistributional secondary porosity. *Mar. Petrol. Geol.*, **7**, 378–397.

Glasmann, J.R., Clark, R.A., Larter, S., Briedis, N.A. & Lundegard, P.D. (1989) Diagenesis and hydrocarbon accumulation, Brent Sandstone (Jurassic), Bergen High Area, North Sea. *AAPG Bull.*, **73**, 1341–1360.

Guscott, S.C., Burley, S.D. & Marshall, J.D. (1997) The role of faulting in the localisation of diagenetic cements and porosity, a UK North Sea oil field. In: *American Association of Petroleum Geologists Annual Convention*, Vol. 6, Dallas, TX, 6–9 April, p. 45.

Hayes, M.J. & Boles, J.R. (1992) Volumetric relations between dissolved plagioclase and kaolinite in sandstones: implications for aluminum mass transfer in the San Joaquin Basin, California. In: *Origin, Diagenesis, and Petrophysics of Clay Minerals in Sandstones*, Society for Economic Paleontologists and Mineralogists Special Publication 47 (Ed. by D.W. Houseknecht and E.D. Pittman), Tulsa, OK, pp. 111–124.

Hill, C.A. (1996) *Geology of the Delaware Basin, Guadalupe, Apache, and Glass Mountains, New Mexico and West Texas*. Permian Basin Section, Society for Economic Paleontologists and Mineralogists, Tulsa, OK, 480 pp.

Houghton, H.F. (1980) Refined techniques for staining plagioclase and alkali feldspars in thin section. *J. Sed. Petrol.*, **50**, 629–631.

Houseknecht, D.W. & Ross, L.M., Jr. (1992) Clay minerals in Atokan deep-water sandstone facies, Arkoma Basin: Origins and influence on diagenesis and reservoir quality. In: *Origin, Diagenesis, and Petrophysics of Clay Minerals in Sandstones*, Society for Economic Paleontologists and Mineralogists Special Publication 47 (Ed. by D.W. Houseknecht and E.D. Pittman), Tulsa, OK, pp. 227–240.

Huggett, J.M. (1984) Controls on mineral authigenesis in coal measures sandstones of the East Midlands, UK. *Clay Mineral.*, **19**, 343–357.

Kantorowicz, J. (1984) The nature, origin, and distribution of authigenic clay minerals from Middle Jurassic Ravanscar and Brent Group sandstones. *Clay Mineral.*, **19**, 359–375.

Kantorowicz, J.D. (1985) The petrology and diagenesis of Middle Jurassic clastic sediments, Ravenscar Group, Yorkshire. *Sedimentology*, **32**, 833–853.

Khanna, M., Saigal, G.C. & Bjørlykke, K. (1997) Kaolinization of Upper Triassic–Lower Jurassic sandstones of the Tampen Spur Area, North Sea: implications for early diagenesis and fluid flow. In: *Basin-wide Diagenetic Patterns: Integrated Petrologic, Geochemical, and Hydrologic Considerations*, Society for Economic Paleontologists and Mineralogists Special Publication 57 (Ed. by I. Montañez, J.M. Gregg and K.L. Shelton), Tulsa, OK, pp. 253–268.

Land, L.S. (1997) Mass-transfer during burial diagenesis in the Gulf of Mexico sedimentary basin. In: *Basin-wide Diagenetic Patterns: Integrated Petrologic, Geochemical, and Hydrologic Considerations*, Society for Economic Paleontologists and Mineralogists Special Publication 57 (Ed. by I. Montañez, J.M. Gregg and K.L. Shelton), Tulsa, OK, pp. 29–39.

Land, L.S. & Dutton, S.P. (1978) Cementation of a Pennsylvanian deltaic sandstone: isotopic data. *J. Sed. Petrol.*, **48**, 1167–1176.

Land, L.S. & Dutton, S.P. (1979) Cementation of sandstones, reply. *J. Sed. Petrol.*, **49**, 1359–1361.

Land, L.S., Milliken, K.L. & McBride, E.F. (1987) Diagenetic evolution of Cenozoic sandstones, Gulf of Mexico sedimentary basin. *Sed. Geol.*, **50**, 195–225.

Lanson, B.L., Beafort, D., Berger, G., Baradat, J. & La Charpagne, J.-C. (1996) Illitization of diagenetic kaolinite-to-dickite conversion series: Late stage diagenesis of the Lower Permian Rotliegende sandstone reservoir, offshore of the Netherlands. *J. Sed. Res.*, **66A**, 501–518.

Leary, D.A. (1984) *Diagenesis of the Permian (Guadalupian) San Andres and Grayburg Formations, Central Basin Platform, West Texas*. MA, The University of Texas at Austin, 129 pp.

Longstaffe, F.J., Tilley, B.J. & Ayalon, A. (1992) Controls on porewater evolution during sandstone diagenesis, Western Canada sedimentary basin: an oxygen isotope perspective. In: *Origin, Diagenesis, and Petrophysics of Clay Minerals in Sandstones*, Society for Economic Paleontologists and Mineralogists Special Publication 47 (Ed. by D.W. Houseknecht and E.D. Pittman), Tulsa, OK, pp. 13–34.

Loughnan, F.C. & Roberts, F.I. (1986) Dickite- and kaolinite-bearing sandstones and conglomerates in Illawarra coal measures of the Sydney Basin, New South Wales. *Aust. J. Earth Sci.*, **33**, 325–332.

Lynch, F.L. (1996) Mineral/water interaction, fluid flow, and Frio sandstone diagenesis: Evidence from the rocks. *AAPG Bull.*, **80**, 486–504.

Macaulay, C.I., Fallick, A.E. & Haszeldine, R.S. (1993) Textural and isotopic variations in diagenetic kaolinite from the Magnus oilfield sandstones. *Clay Mineral.*, **28**, 625–639.

Maliva, R.G., Dickson, J.A.D. & Fallick, A.E. (1999) Kaolin cements in limestones: potential indicators of organic-rich pore waters during diagenesis. *J. Sed. Res.*, **69**, 158–163.

Matyas, J. & Matter, A. (1997) Diagenetic indicators of meteoric flow in the Pannonian Basin, southeast Hungary. In: *Basin-wide Diagenetic Patterns; Integrated Geochemical and Hydrologic Considerations*, Society for Economic Paleontologists and Mineralogists Special Publication 57 (Ed. by I. Montañez, J.M. Gregg and K.L. Shelton), Tulsa, OK, pp. 281–296.

McAulay, G.E., Burley, S.D., Fallick, A.E. & Kusznir, N.J. (1994) Palaeohydrodyanmic fluid flow regimes during diagenesis of the the Brent Group in the Hutton–NW Hutton reservoirs: constraints from oxygen isotope studies of authigenic kaolin and reverse flexural modelling. *Clay Mineral.*, **29**, 609–629.

McLaughlin, O.M., Haszeldine, R.S., Fallick, A.E. & Rodgers, G. (1994) The case of the missing clay, aluminum loss and secondary porosity, South Brae oilfield, North Sea. *Clay Mineral.*, **29**, 651–663.

Milliken, K.L. (1992) Regional diagenetic variations in Middle Pennsylvanian foreland basin sandstones of the southern Appalachians. In: *Geological Society of America Annual Meeting, Abstr. Programs Geol. Soc. Am.*, **24**, 57–58.

Milliken, K.L. (1995) Petrographic evidence of open-system chemical reactions in mudrocks in the Gulf of Mexico sedimentary basin. In: *Geological Society of American Annual Meeting, Abstr. Programs Geol. Soc. Am.*, **27**, A461.

Milliken, K.L. (1998) Carbonate diagenesis in non-marine foreland sandstones at the western edge of the Alleghanian overthrust belt, southern Appalachians. In: *Carbonate Cementation in Sandstones. Distribution Patterns and Geochemical Evolution*, International Association of Sedimentologists Special Publication 26 (Ed. by S. Morad), Blackwill Science, Oxford, pp. 87–105.

Milliken, K.L. (2001) Diagenetic heterogeneity in sandstone at the outcrop scale, Breathitt Formation (Pennsylvanian), eastern Kentucky. *AAPG Bull.*, **84**, 795–815.

Milliken, K.L., Mack, L.E. & Land, L.S. (1994) Elemental mobility in sandstones during burial: Whole-rock chemical and isotopic data, Frio Formation, South Texas. *J. Sed. Res.*, **A64**, 788–796.

Minster, T. & Fraser, A.G. (1982) Authigenic kaolinite in the Amir Formation, southern Negev, Israel. *Israel J. Earth Sci.*, **31**, 1–7.

Moore, L.R. (1964a) The *in situ* formation and development of some kaolinite macrocrystals. *Clay Mineral. Bull.*, **5**, 338–352.

Moore, L.R. (1964b) The microbiology, mineralogy, and genesis of a tonstein. *Proc. Yorkshire Geol. Soc.*, **34**, 235–285.

Morad, S., Ben Ismail, H., De Ros, L.F., Al-Aasm, I.S. & Serrihini, N.E. (1994) Diagenesis and formation waters chemistry of Triassic reservoir sandstones from southern-Tunisia. *Sedimentology*, **41**, 1253–1272.

Nadeau, P.H. & Hurst, A. (1991) Application of back-scattered electron microscopy to the quantification of clay mineral microporosity in sandstone. *J. Sed. Petrol.*, **61**, 921–925.

Naiman, E. (1982) *Sedimentation and diagenesis of a shallow marine carbonate and siliciclastic shelf sequence: the Permian (Guadalupian) Grayburg Formation, southeastern New Mexico.* MA, The University of Texas at Austin, 197 pp.

Osborne, M., Haszeldine, R.S. & Fallick, A.E. (1994) Variation in kaolinite morphology with growth temperature in isotopically mixed pore-fluids, Brent Group, UK North Sea. *Clay Mineral.*, **29**, 591–608.

Pevear, D.R., Klimentidis, R.E. & Robinson, G.A. (1991) Petrogenetic significance of kaolinite nucleation and growth on pre-existing mica in sandstones and shales. In: *Annual Clay Minerals Conference*, Vol. 28, Clay Minerals Society, Houston, TX, p. 125.

Platt, J.D. (1993) Controls on clay mineral distribution and chemistry in the Early Permian Rotliegend of Germany. *Clay Mineral.*, **28**, 393–416.

Quinlan, G.M. & Beaumont, C. (1984) Appalachian thrusting, lithosphere flexure, and the Paleozoic stratigraphy of the eastern interior of North America. *Can. J. Earth Sci.*, **21**, 973–996.

Rossel, N.C. (1982) Clay mineral diagenesis in Rotliegend aeolian sandstones of the Southern North Sea. *Clay Mineral.*, **17**, 69–77.

Rudolph, K.W. (1978) *Diagenesis of back-reef carbonates: an example from the Capitan complex.* MA, The University of Texas.

Shelton, J.W. (1964) Authigenic kaolinite in sandstones. *J. Sed. Petrol.*, **34**, 102–111.

Siebert, R.M., Moncure, G.K. & Lahann, R.W. (1984) A theory of framework grain dissolution in sandstones. In: *Clastic Diagenesis* (Ed. by D.A. MacDonald and R.C. Surdam). *AAPG Mem.*, **37**, pp. 163–175.

Sinclair, J.L. & Ghiorse, W.C. (1989) Distribution of aerobic bacteria, protozoa, algae, and fungi in deep subsurface sediments. *Geomicrobiology J.*, **7**, 15–31.

Smith, G.E., Dever, G.R., Jr., Horne, J.C., Ferm, J.C. & Whaley, P.W. (1971) Depositional environments of eastern Kentucky coals. In: *Annual Field Conference Coal Division*, Kentucky Geological Survey, Lexington, KY, p. 22.

Spark, I., Patey, I., Duncan, B., Hamilton, A., Devine, C. & McGovern-Traa, C. (2000) The effects of indigenous and introduced microbes on deeply buried hydrocarbon reservoirs, North Sea. *Clay Mineral.*, **35**, 5–12.

Stewart, R.N.T., Fallick, A.E. & Haszeldine, R.S. (1994) Kaolinite growth during pore-water mixing: isotopic data from Palaeocene sands, North Sea, UK. *Clay Mineral.*, **29**, 627–636.

Stoessell, R.K. (1987) Mass transport in sandstones around dissolving plagioclase grains. *Geology*, **15**, 295–298.

Stoessell, R.K. & Pittman, E.D. (1990) Secondary porosity revisited; the chemistry of feldspar dissolution by carboxylic acids and anions. *AAPG Bull.*, **74**, 1795–1805.

Tankard, A.J. (1986) Depositional response to foreland deformation in the Carboniferous of eastern Kentucky. *AAPG Bull.*, **70**, 853–868.

Woronick, R.W. & Land, L.S. (1985) Late burial diagenesis, Lower Cretaceous Pearsall and Lower Glen Rose formations, South Texas. In: *Carbonate Cements*, Society of Economic Paleontologists and Mineralogists Special Publication 36 (Ed. by N. Schneidermann and P.M. Harris), Tulsa, OK, pp. 265–275.

Int. Assoc. Sedimentol. Spec. Publ. (2003) **34**, 361–382

The role of the Cimmerian Unconformity (Early Cretaceous) in the kaolinitization and related reservoir-quality evolution in Triassic sandstones of the Snorre Field, North Sea

J.M. KETZER[1], S. MORAD[1], J.P. NYSTUEN[2,*] *and* L.F. DE ROS[3]

[1] *Department of Earth Sciences, Uppsala, Villa vägen 16, S-752 36, Uppsala, Sweden, e-mail: marcelo.ketzer@geo.uu.se, sadoon.morad@geo.uu.se*

[2] *Saga Petroleum ASA, 1301 Sandvika, Norway*

[3] *Universidade Federal do Rio Grande do Sul, Instituto de Geociências, Av. Bento Gonçalves, 9500, CEP 91501-970, Porto Alegre, Brazil, e-mail: lfderos@pop.inf.ufrgs.br*

ABSTRACT

Modal petrographic analysis, textural criteria and petrophysical data were used to unravel the origin and timing of kaolinite formation and its impact on reservoir properties of Triassic sandstones of the Lunde Formation, Snorre Field, northern North Sea. Kaolinite formation occurred by the replacement of detrital feldspar, mica, rock fragments, mud intraclasts and pseudomatrix. Some of the kaolinitization occurred shortly after deposition, but most occurred during the Early Cretaceous regional uplift and formation of the late Cimmerian Unconformity. Kaolinite stable isotopic data ($\delta^{18}O_{SMOW} = +13.9‰$ to $+18.5‰$, and $\delta D_{SMOW} = -83‰$ to $-69‰$) support kaolinite formation from Early Cretaceous meteoric waters. Kaolinite content in the sandstones increases from less than 5 to up to 20 vol.% within the first 200 m below the unconformity and average total porosity was enhanced by about 5%. Where mudstones were present between the unconformity and the Triassic sandstones, meteoric water flushing and, hence, kaolinite formation, were limited. However, in some cases, kaolinitization in sandstones buried under thick mudstones was aided by major normal faults that were hydraulically connected to the unconformity surface. The lack of a negative correlation between kaolinite and feldspar content is attributed to: (i) the presence of kaolinite sources other than feldspar (pseudomatrix, mica, mud intraclast and rock fragments), (ii) the strong variations in the initial detrital mineralogical composition of the sandstones, and (iii) to mass transfer of Si and Al ions on scales greater than that of the thin-section.

INTRODUCTION

Diagenetic modifications that occur after deposition at near-surface conditions (i.e. low T, P and surface-related fluids) and before significant burial (e.g. < 2 km; Morad *et al.*, 2000) are referred to here as eodiagenesis, whereas modifications occurring after significant burial (e.g. > 2 km; Morad *et al.*, 2000), under the influence of increasing P and T, and by modified diagenetic fluids, are referred to as mesodiagenesis

* Present address: Department of Geology, Oslo University, N-0316 Oslo, Norway.

(cf. Choquette & Pray, 1970). Modifications related to the exposure of previously buried sedimentary sequences to near-surface meteoric conditions below unconformities and/or along faults are referred to as telodiagenesis (Choquette & Pray, 1970). Unconformities often are considered as important features for petroleum exploration because they may promote telogenetic enhancement of reservoir quality of siliciclastic and carbonate rocks and act as avenues for hydrocarbon migration (Shanmugan, 1988, 1990; Shanmugan & Higgins, 1988; Giles & De Boer, 1990; Ghosh, 1997). Reservoir-quality enhancement in sandstones below unconformities occurs owing to the dissolution of detrital silicates and cements by undersaturated meteoric water (Larese *et al.*, 1983). Silicate dissolution is most extensive under semi-humid to humid climatic conditions and typically is accompanied by kaolinite precipitation. As the scale and magnitude of diagenetic modifications below unconformities are still matters of debate (Bjørkum *et al.*, 1990; Bloch, 1994), precise reservoir-quality prediction in such settings remains problematic.

Although kaolinite is one of the most important diagenetic clay minerals in Triassic and Jurassic sandstone reservoirs of the North Sea, its origin is still a controversial topic. Several authors have suggested that meteoric-water flushing and kaolinitization took place shortly after deposition of the sand, i.e. during near-surface eodiagenesis (Nedkvitne & Bjørlykke, 1992; McAulay *et al.*, 1994; Khanna *et al.*, 1997). Other authors argued that kaolinitization occurred during telodiagenesis below the late Cimmerian unconformity (hereafter referred to as the Cimmerian Unconformity), during the Early Cretaceous regional uplift (Hancock & Taylor, 1978; Sommer, 1978; Emery *et al.*, 1990; Macaulay *et al.*, 1993). Emery *et al.* (1990) and Macaulay *et al.* (1993) concluded that the zone of telogenetic kaolinitization in the marine Magnus Sandstones Member (Upper Jurassic) below the Cimmerian Unconformity in the Magnus Field was of minor importance and less than 10 m thick. Bjørkum *et al.* (1990, 1993) argued that no telogenetic kaolinite should be preserved in the Jurassic and Triassic North Sea sandstones because the rates of uplift and erosion were faster than the rates of meteoric-water alteration and kaolinite formation below unconformities. Bjørkum *et al.* (1990) suggested instead that kaolinite in these sandstones is detrital in origin. In some papers, no distinction has been made between eogenetic and telogenetic kaolinite (Bjørlykke & Brensdal, 1986; Lønøy *et al.*, 1986; Glasmann *et al.*, 1989a; McAulay *et al.*, 1993). Additionally, mesogenetic formation of kaolinite by the interaction of sandstones with organic acids and CO_2 derived from the maturation of organic matter was suggested by Burley (1986).

The aim of this paper is to elucidate the origin and timing of kaolinite formation and its impact on the modification of reservoir quality in sandstones of the Lunde Formation (Upper Triassic) from the Snorre Field in the northern North Sea. Within this context, particular emphasis is placed upon deciphering the role of telodiagenesis below the Cimmerian Unconformity.

GEOLOGICAL SETTING

The Snorre Field is located in the northwestern shoulder of the Viking Graben, in the Tampen Spur area, northern North Sea (Fig. 1). The reservoirs are fluvial sandstones of the Lunde (Upper Triassic) and the Statfjord (Upper Triassic to Lower Jurassic) formations, which are positioned on the crests of uplifted faulted blocks, presently buried to depths between 2300 and 3800 m. Deposition of these formations occurred during thermal subsidence subsequent to a Permian–Triassic rifting episode (Fig. 2). This thermal subsidence phase persisted throughout the Middle Jurassic, when marine sediments of the Dunlin Group and deltaic sediments of the Brent Group were deposited. A new rifting event occurred during the Middle to Late Jurassic, which resulted in the deposition of marine sediments of the Viking Group, which contains the most important hydrocarbon source rocks, namely

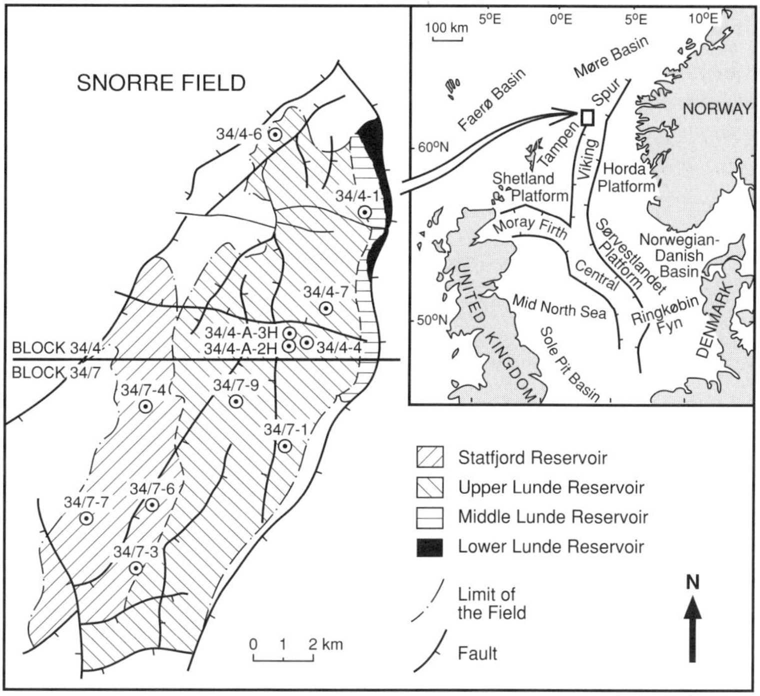

Fig. 1 Map of the northern North Sea area, showing location of the Snorre Field and the wells studied.

the Draupne Formation, which is equivalent to the Kimmeridge Clay Formation (Fig. 2). This rifting event was accompanied by block rotation and uplifting during the Late Jurassic to Early Cretaceous, which resulted in subaerial exposure and partial erosion of the Triassic and Jurassic units, and thus formation of the Cimmerian Unconformity. These units subsequently were covered by marine deposits of the Lower Cretaceous Cromer Knoll and Upper Cretaceous Shetland groups (Fig. 2). The relief created by block rotation and uplift in the Tampen Spur area is believed to have been in the order of hundred of metres above sea-level (Yielding, 1990; Brun & Tron, 1993). Presently, the Cimmerian Unconformity occurs in the Snorre Field at depths of about 2400 m ± 150 m.

The Lunde Formation in the Snorre Field varies in thickness between 300 and 1200 m, and is composed of sandstones and mudstones deposited in low- and high-sinuosity fluvial systems, and subordinate lacustrine environments. The formation is subdivided into three informal members. The lower member is composed of reddish brown to greenish grey mudstones, siltstones and very fine- to fine-grained sandstones deposited in a distal alluvial plain dominated by sheet-flood and channel deposits, showing a coarsening upward pattern. The middle member is composed of greyish, slightly bioturbated mudstones, commonly with desiccation cracks, and siltstones and very fine- to fine-grained sandstones deposited on a distal alluvial plain. Ephemeral sheet-flood deposits are intercalated with shallow-water lacustrine deposits and palaeosols. The upper member is composed of repetitive fining upward motifs, representing single and multistorey channel deposits and sheet-flood fluvial deposits, comprising fine to coarse sandstones, intercalated with reddish brown floodplain mudstones with common palaeosol horizons (Nystuen & Fält, 1995). The upper member originally was divided into five reservoir units, A, B/C, D, E and F, based on sand/mud and net/gross ratios (Fig. 3; Hollander, 1987). In the present reservoir model, 12 reservoir zones are defined as

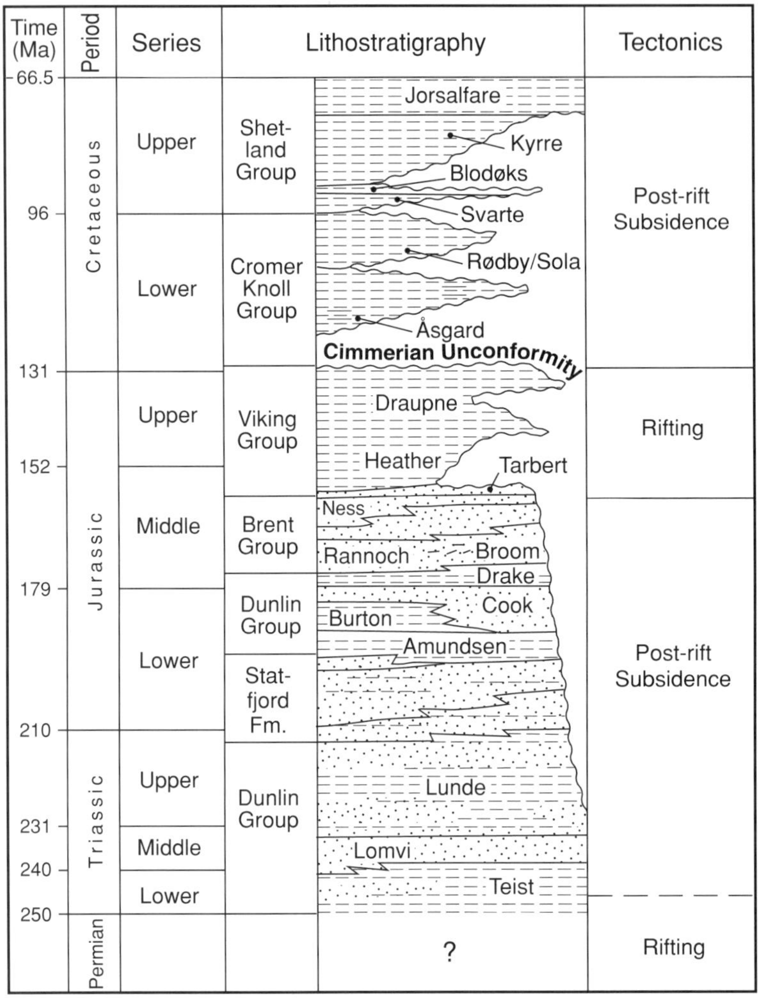

Fig. 2 Stratigraphic and tectonic features of the Tampen Spur area. Modified from Campbell & Ormaasen (1987), with lithostratigraphic nomenclature from Vollset & Doré (1984), and Isaksen & Tonstad (1989).

allostratigraphic units, corresponding to high-frequency sequences (L01–L12, Fig. 3; Diesen *et al.*, 1995). Owing to the lower sand/mud ratio and stronger carbonate cementation, both the lower and the middle Lunde members have poorer reservoir quality compared with the upper member, which represents the main reservoir of the Snorre Field.

SAMPLES AND METHODS

Samples and data from 12 wells located in the Snorre Field (Fig. 1), representing the entire thickness of the Lunde Formation, were used for this study. Helium porosity and permeability (not corrected for the Klinkenberg effect) were measured from 3.8-cm diameter core plugs in a confining pressure chamber capable of a load of up to 10 000 p.s.i. Around 600 thin-sections from conventional cores and side-wall cores, located between 2400 and 2800 m below sea floor, were prepared after impregnation of the samples with coloured epoxy resin. The modal compositions of the sandstones were obtained by counting 300 points in each thin-section. Sixty-five samples were covered by a thin layer of gold and examined using a Philips XL30 scanning electron microscope (SEM), with an acceleration voltage of 10 kV, equipped with an EDAX system. The clay mineral composition in 55 mudstones and siltstones samples

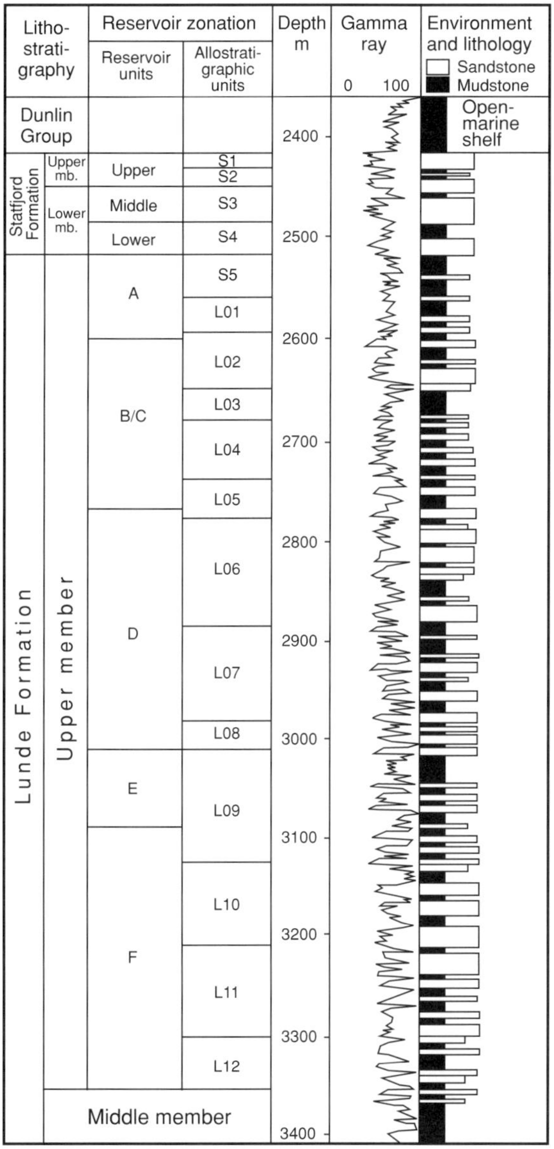

Fig. 3 Stratigraphy and reservoir zonation of the upper member of the Lunde Formation in the Snorre Field, represented by well 34/7-3 (modified from Nystuen & Fält, 1995).

(fraction < 2 μm) were analysed in a Philips X-ray diffractometer equipped with a Cu (Kα) radiation and Ni-filter. Abundances of individual clay minerals were calculated on the basis of areas below the diffraction peaks.

Oxygen and hydrogen isotope analyses were performed on eight kaolinite samples from wells 34/4-1, 34/4-A-2H and 34/4-A-3H (Fig. 1), between 2500 and 2700 m of depth. Clay fractions smaller than 5 μm were separated by centrifugation and their purity was assessed by means of SEM and X-ray diffraction (XRD) analyses. These analyses indicated the presence of < 10% illitic clay minerals. The samples were firstly degassed for 2 h under vacuum at 150°C. The oxygen isotopic composition of the kaolinite was measured by extracting oxygen after reacting the samples with bromine pentafluoride (BrF_5) at 600°C, and collection of O_2 via a liquid nitrogen cold-trap, according to the method described by Clayton & Mayeda (1963). Quantitative oxygen isotope analysis was made by conversion to CO_2 over red-hot graphite and measured in a VG-Micromass 602C mass spectrometer. A standard with an oxygen isotope fractionation factor of 1.0412 was used to calibrate the reference gas. A $\delta^{18}O$ value of +9.7‰ was obtained for the silica standard NBS-28 during these experiments. Duplicate analyses were performed with a maximum error of ±0.28‰. Hydrogen for hydrogen isotope analyses was obtained from 50-mg samples of pure kaolinite using a procedure modified after Bigeleisen *et al.* (1952) and Godfrey (1962). Samples were degassed at 150°C for 2 h under vacuum, and then maintained under vacuum at room temperature for 12 h. The samples were then reheated under vacuum for 30 min at 200°C prior to extraction of the hydroxyl group and hydrogen. During extraction, the samples were heated to 1000°C for 30 min and the resulting gases passed first over CuO at 500°C, and then metallic U at 800°C. The hydrogen gas was analysed using a PRISM II dual inlet, gas-source mass spectrometer. Precision of the kaolinite δD measurements was ±2‰. Isotopic results for both hydrogen and oxygen are reported relative to the Vienna Standard Mean Ocean Water (V-SMOW).

DETRITAL COMPOSITION AND DIAGENETIC HISTORY

The sandstones are dominantly subarkoses and arkoses, and less commonly lithic arkoses and

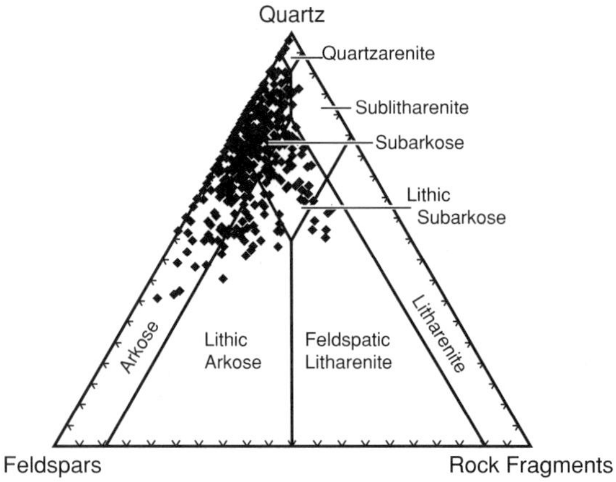

Fig. 4 Detrital composition of the Lunde Formation sandstones plotted on McBride's (1963) sandstone classification diagram.

lithic subarkoses (average $Q_{71}F_{23}L_{06}$; Fig. 4). The quartz content varies between 30 and 78% (average 44% of bulk rock volume) and is dominantly monocrystalline. Feldspar content varies between 2 and 42% (average 14 vol.%) with plagioclase dominating over K-feldspar. Rock fragments (0–23%, average 4 vol.%), include granites, low-grade metamorphic rocks and rarely siliciclastic sedimentary rocks. Other framework grains include mica (0–28 vol.%; average 5 vol.%), mud intraclast (0–30 vol.%; average 5 vol.%), and trace amounts of heavy minerals (Fe–Ti oxides, zircon, epidote, tourmaline, apatite, garnet, rutile, staurolite, kyanite, spinel). In most of the sandstones, mud intraclasts are deformed into pseudomatrix as a result of mechanical compaction. The XRD analyses of mudstones and siltstones revealed that the clay minerals (< 2 μm) are dominated by mixed-layer illite–smectite and chlorite–smectite (I/S and C/S; average 83 vol.%), with lesser amounts of kaolinite (average 6 vol.%), and chlorite (average 11 vol.%).

The diagenetic history of the Lunde Formation is complex, being related to two burial phases that were separated by an uplift phase (Fig. 5; Morad *et al.*, 1990, 1998). The first burial phase was characterized by a fast subsidence rate during the Middle Triassic and a much slower rate during the Late Jurassic. The maximum burial depth of the base of the Lunde Formation was approximately 1700 m. Diagenetic processes during this burial phase include: (i) mechanical infiltration of smectitic clays and precipitation of vadose and phreatic calcrete, dolocrete, siderite and small amounts of kaolinite at near-surface conditions; and (ii) precipitation of calcite, siderite and dolomite during burial to 1700 m (Morad *et al.*, 1998).

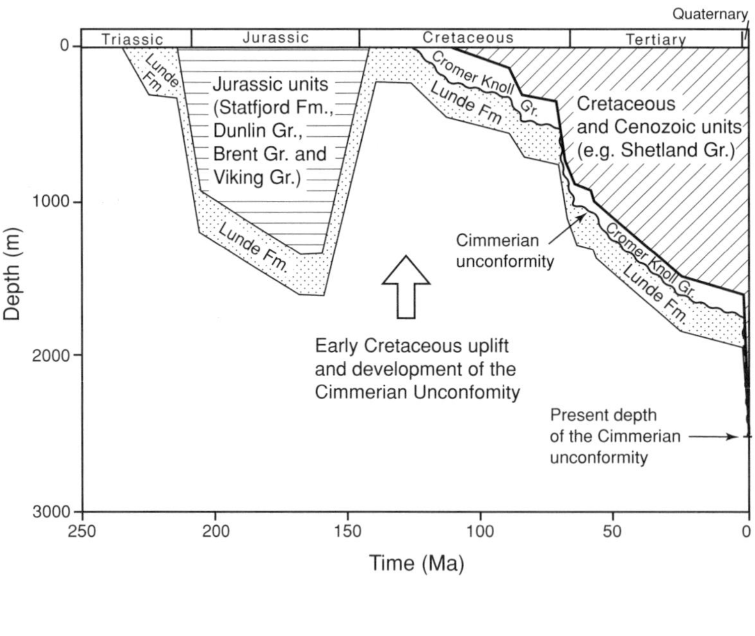

Fig. 5 Burial history curve of the Snorre Field area as represented by well 34/4-1. Note the first burial phase (Triassic to Late Jurassic) was faster than the second burial phase (Early Cretaceous to late Tertiary). Accelerated burial rates prevailed during the late Tertiary and Quaternary. Duration of subaerial exposure of the Lunde Formation during Early Cretaceous uplift was about 10–15 Myr.

The second burial phase to depths of 2400–2800 m occurred subsequent to the Early Cretaceous uplift and was characterized by a slow but continuous subsidence rate during Cretaceous and early Tertiary times, followed by a much higher subsidence rate during late Tertiary and Quaternary times (Fig. 5). The second burial phase has resulted in partial recrystallization of kaolinite, albitization of detrital plagioclase and precipitation of quartz overgrowths and small amounts of chlorite, calcite, ankerite, dickite and illite.

PETROGRAPHY AND ISOTOPIC COMPOSITION OF KAOLIN

Kaolinite occurs as patches composed of thin (≤ 1 μm), pseudohexagonal crystals arranged as booklet and vermicular aggregates with high intercrystalline microporosity (visual estimation of about 40–50%; Fig. 6a). The kaolinite patches are distributed irregularly and formed mainly by the replacement of pseudomatrix, micas, feldspars and rock fragments. Small amounts of kaolinite also occur as pore-filling cement. Kaolinite patches that have replaced pseudomatrix (Fig. 6b) are common in medium- and coarse-grained sandstones, being distinguished by the presence of small amounts of remnant I/S clays and, in some cases, silt-size quartz, feldspars, micas, and Fe- and Ti-oxides. These patches have lower intercrystalline microporosity compared with kaolinite that fills the intergranular pores and replaces feldspars (Fig. 6b). Kaolinitized micas occur in fine- to coarse-grained sandstones and display the typical expanded texture into open pores. In some cases, kaolinitized biotite displays small (≈ 20 μm) siderite crystals between the expanded lamellae.

The SEM examination revealed that kaolinite platelets in about 5% of the samples are intercalated with thicker crystals (2–3 μm) of well-ordered kaolinite and dickite, similar to those described by Beaufort *et al.* (1998; Fig. 6c). Visual estimation using SEM images indicates that the average proportion of such thicker

(a) (b) (c)

Fig. 6 Scanning electron microscope images showing: (a) the delicate habit of pore-filling kaolinite patches with booklet and vermicular habit, and the high intercrystalline microporosity, indicative of an *in situ*, diagenetic origin; (b) kaolinite patches replacing pseudomatrix with lower intercrystalline microporosity than pore-filling kaolinite (Fig. 6a); (c) kaolinite aggregate in which thinner kaolinite platelets (K) alternate with thicker dickite (D).

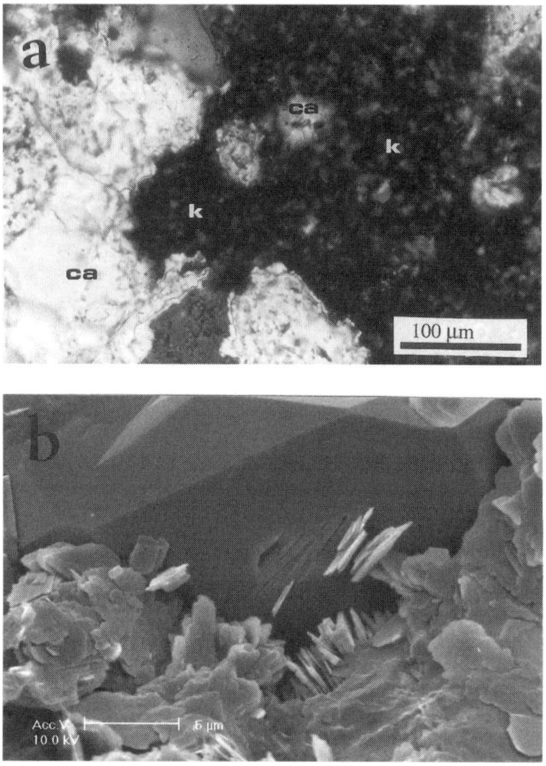

Fig. 7 (a) Optical photomicrograph (x-nicols) showing kaolinite patches (k) filling dissolution voids in eogenetic calcite (ca). Note the corroded borders of calcite and the presence of calcite remnants in the centre of the photo. (b) Scanning electron microscope image showing kaolinite platelets engulfed by quartz overgrowth.

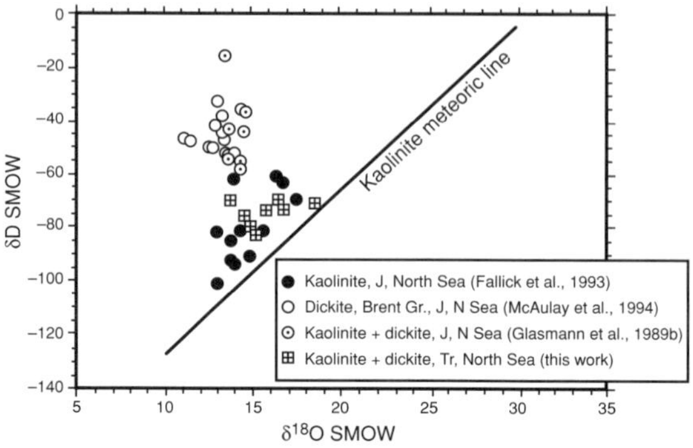

Fig. 8 Plot of $\delta^{18}O$ versus δD of kaolinite from the Lunde Formation sandstones and from Jurassic sandstones in other North Sea Fields. Note that the positions of the Lunde samples (this work) and some of the other North Sea samples are close to the 'meteoric kaolinite line'. The overall, slight to considerable deviation from this line (e.g. samples of Glasmann *et al.* (1989b) and McAulay *et al.* (1994)) is attributed to isotopic resetting (particularly of H isotopes) owing to recrystallization of kaolinite and formation of dickite during burial.

crystals is minimal and rarely exceeds 10 vol.%. Aggregates composed of kaolinite and dickite typically have higher intercrystalline microporosity than aggregates composed exclusively of kaolinite. The rate of burial transformation of kaolinite into well-ordered kaolinite and dickite with increasing burial depths (Ehrenberg *et al.*, 1993; McAulay *et al.*, 1994; Beaufort *et al.*, 1998) cannot be evaluated here owing to the narrow depth range of the core samples available (2300–2800 m). Kaolinite is, in a few samples, engulfed by displacive, eogenetic calcrete and dolocrete, but commonly it fills dissolution voids in calcrete (Fig. 7a). Kaolinite is engulfed by, and hence pre-dates, mesogenetic quartz overgrowths (Fig. 7b).

The $\delta^{18}O$ values of kaolinite vary between +13.9‰ and +18.5‰, and δD between −83‰ and −69‰ relative to V-SMOW. These isotopic values display various degrees of deviation from the 'meteoric kaolinite line' (cf. Savin & Epstein, 1970; Fig. 8) mainly owing to enrichment in D. The meteoric kaolinite line is constructed based on the fractionation of oxygen and hydrogen isotopes between kaolinite and meteoric water at various latitudes. With increasing latitude, meteoric waters, and consequently kaolinite, become enriched with ^{16}O and ^{1}H isotopes relative to ^{18}O and D, respectively.

RESERVOIR PROPERTIES

The sandstones reveal a wide range of permeability and porosity. Permeability varies between 0.1 and 2800 mD (average 273 mD) and shows a positive correlation with He porosity (5 and 32%, average 23%; Fig. 9). Porosity and permeability generally are higher in the coarse-grained sandstones, such as those deposited between the base and middle of upward-fining

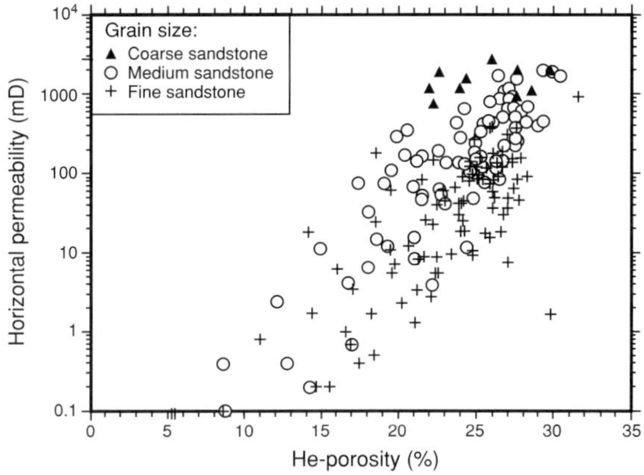

Fig. 9 A scatter plot showing a positive correlation ($r = +0.4$) between He porosity and permeability of fine-, medium- and coarse-grained sandstones of the Lunde Formation. Variation in permeability for each porosity value is attributed mainly to the presence of variable amounts of kaolinite, clay pseudomatrix and isolated intragranular pores.

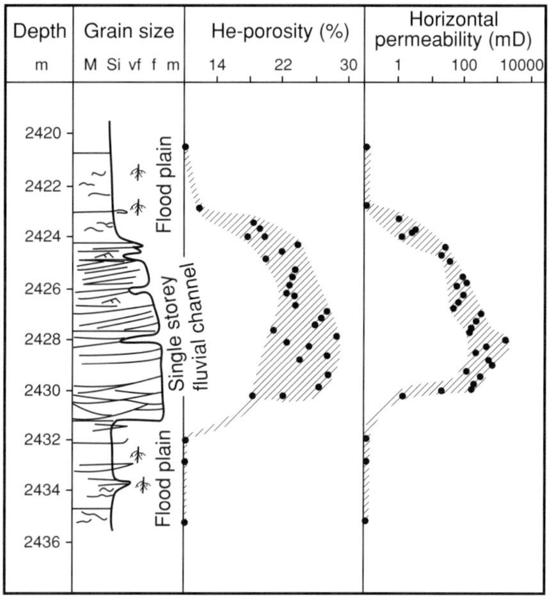

Fig. 10 Variations in He porosity and permeability of sandstone lithofacies in a single fining upward single-storey fluvial channel unit, showing higher values in the middle of the channel and lower values at the base and top of the channels and floodplain. Upper member of the Lunde Formation (reservoir unit D), well 34/7-1 (modified from Nystuen & Fält, 1995).

single-storey fluvial channel deposits, than in the medium- and fine-grained sandstones, such as those deposited in the upper parts of the channels and in crevasse splays (Figs 9 & 10).

Porosity of the sandstones includes: (i) primary intergranular macropores, highly controlled by sandstone texture, and thus by the depositional facies; (ii) secondary intragranular macro- and micropores formed as a result of partial to complete dissolution of feldspar and, less commonly, mica, mud intraclast and rock fragments; (iii) secondary intergranular macropores and micropores formed by the dissolution of calcite cement; and (iv) intercrystalline micropores between kaolinite and chlorite platelets, and within pseudomatrix. Modal porosity (0–30%; average 10%) has a positive correlation with He porosity ($r = +0.6$; Fig. 11). The higher He-porosity values are attributed to the presence of micropores in diagenetic clay minerals and pseudomatrix. This particularly is evident in samples that have low modal porosity (< 2%) but high He porosity (> 25%). Such samples have ≥ 10 vol.% pseudomatrix and/or kaolin (Fig. 11). When the volume of pseudomatrix is close to 30% or when the samples are tightly cemented by calcite (≥ 20

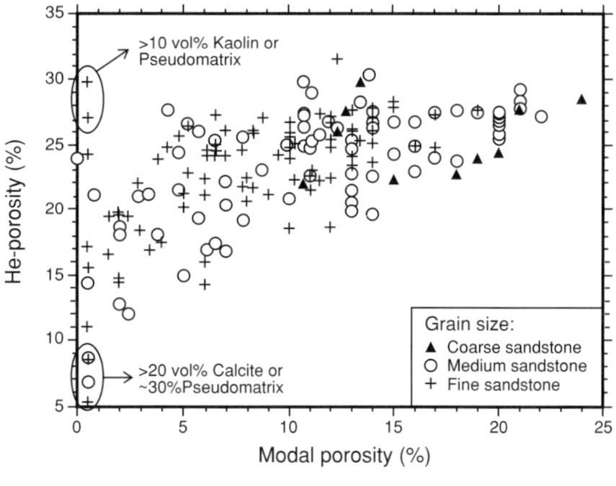

Fig. 11 A scatter plot showing a positive correlation ($r = +0.6$) between He porosity and modal porosity in the Lunde Formation. Samples with modal porosity < 1% and He-porosity < 10% have 30 vol.% of pseudomatrix or more than 20 vol.% of calcite. Samples with modal porosity < 1% and He-porosity between 25 and 30% have more than 10 vol.% of pseudomatrix and/or kaolinite.

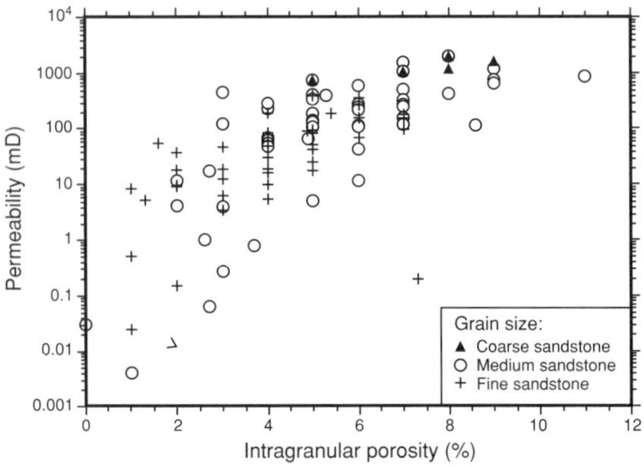

Fig. 12 A scatter plot showing a positive correlation ($r = +0.7$) between permeability and intragranular porosity in the sandstones. The intragranular pores were formed by the dissolution of feldspar, mica and rock fragments.

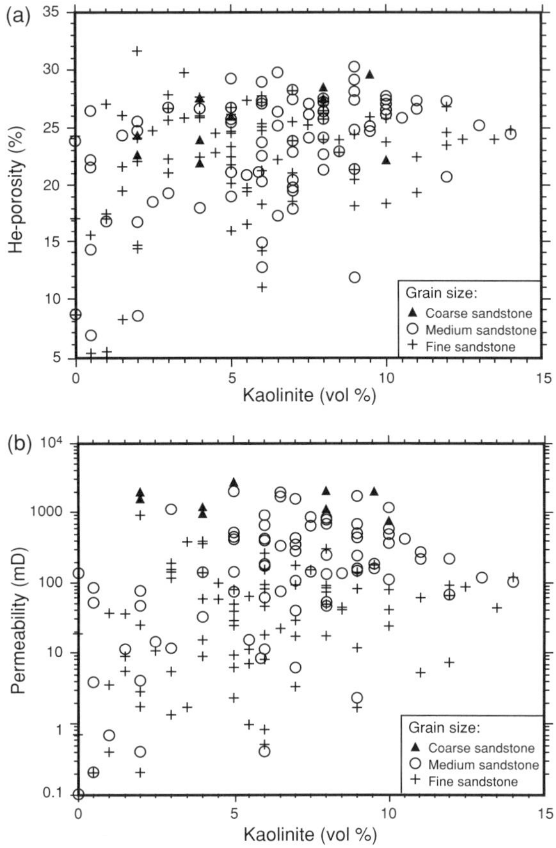

Fig. 13 Scatter plots showing: (a) positive correlation ($r = +0.4$) between kaolin content and He porosity; and (b) the lack of correlation between the overall kaolinite content and permeability ($r = +0.1$). However, note that there is a noticeable decrease in permeability of sandstones with kaolinite content exceeding 10 vol.%.

vol.%), both modal and He porosity decrease to < 10% (Fig. 11). Intragranular porosity (1–11%) shows a positive correlation with permeability ($r = 0.7$; Fig. 12).

Kaolinite contents also have a positive correlation with He porosity ($r = +0.4$; Fig. 13a), but no correlation with permeability ($r = +0.1$; Fig. 13b). However, high permeability (≥ 500 mD) and porosity (> 28%) values are encountered in sandstones that are devoid of carbonate cement and pseudomatrix, and with kaolin content < 10 vol.%. Low porosities and associated low kaolin contents are encountered in sandstones with a pseudomatrix content close to 30 vol.% and in samples tightly cemented by calcite (≥ 20 vol.%; Fig. 11).

DISTRIBUTION OF KAOLIN, FELDSPAR, MICA AND POROSITY BELOW THE UNCONFORMITY

Three main patterns of vertical distribution of kaolinite and reservoir properties were recognized below the Cimmerian Unconformity in the wells studied. No systematic variations in the amounts of feldspar were recognized. The lack of systematic permeability data from the intervals studied precluded inclusion of this parameter in the plots. The three distribution patterns of kaolinite, best illustrated in wells for which samples and modal analyses of sandstones could be systematically obtained for at least the first 500 m below the Cimmerian Unconformity, include:

1 Fairly systematic increase in kaolinite content towards the Cimmerian Unconformity in wells 34/4-1, 34/4-4, 34/4-6 and 34/7-1 (Fig. 14a) in which the Lunde Formation is cut by the unconformity. Within the first 200 m immediately below the unconformity, the kaolinite content reaches values close to 20 vol.% (average 4 vol.%), but decreases to 0–3 vol.% (average

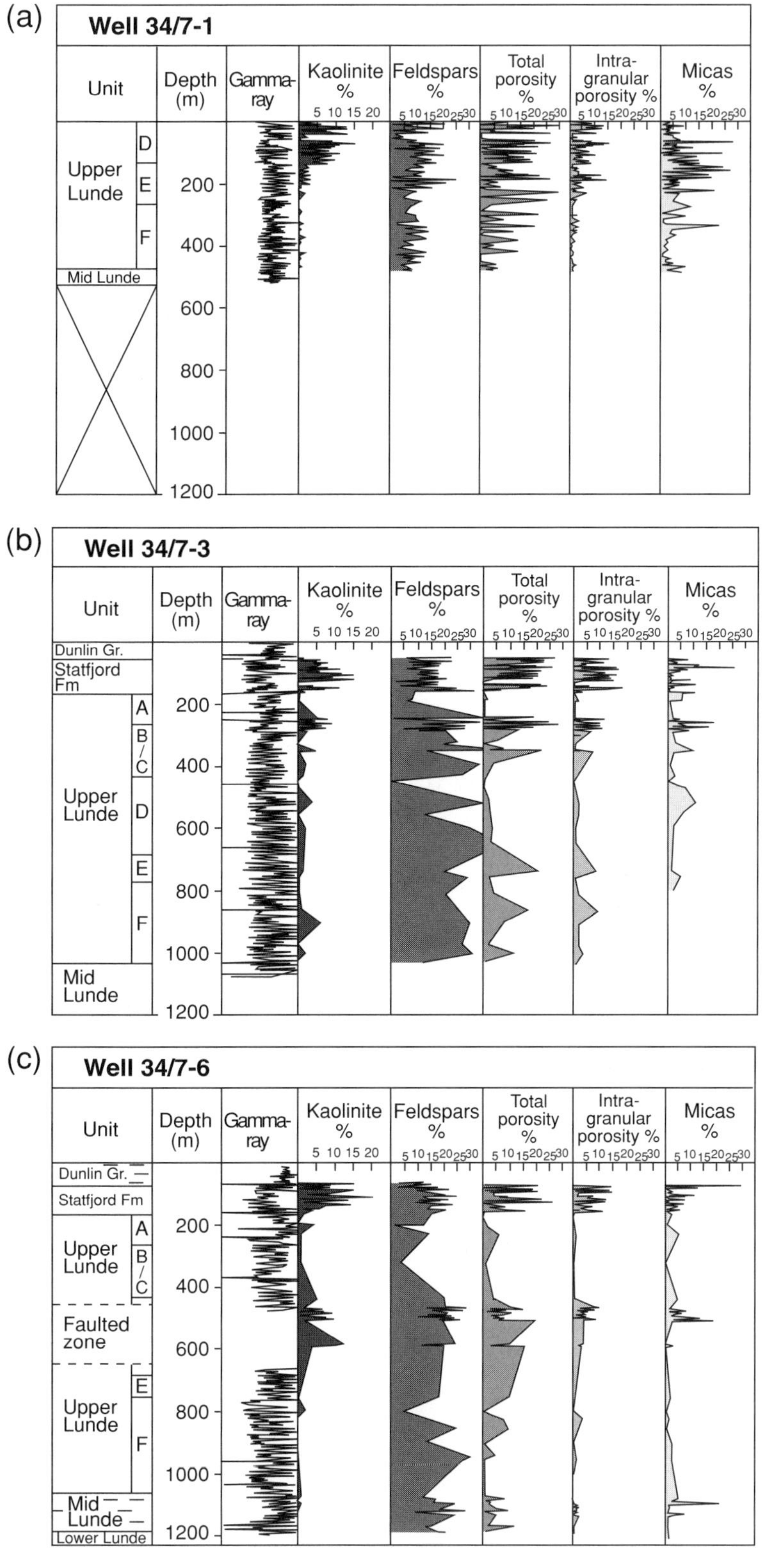

Fig. 14 Vertical distribution patterns of kaolinite, feldspars, micas and porosity below the Cimmerian Unconformity in the wells studied. (a) Kaolinite content increases systematically towards the unconformity. (b) Absence of systematic distribution pattern of kaolinite with respect to the unconformity, which is typical for wells in which the Lunde Formation is covered by thick mudstones of the Dunlin Group that prevented effective percolation of meteoric waters. (c) Kaolinite content increases in the vicinity of a fault plane that connected to the Cimmerian Unconformity, and thus enhanced meteoric-water incursion. See Fig. 1 for well locations.

< 1 vol.%) below 200 m. Intragranular porosity increases from < 5 vol.% (average 1 vol.%) to 15 vol.% (average 4 vol.%) within the 200 m below the unconformity. Average total modal porosity increases from 6% within 200 m below the unconformity to 11% at greater depths (Fig. 14a). Feldspar content (2–25 vol.%) does not change systematically in relationship to the unconformity. Conversely, the decrease in mica content is fairly concomitant with the increase in kaolinite towards the unconformity (Fig. 14a).

2 No obvious systematic distribution of kaolinite content (0–9 vol.%, average 3 vol.%) relative to the Cimmerian Unconformity (wells 34/7-4, 34/7-7 and 34/7-3; Fig. 14b). The feldspar (5–42 vol.%, average 20 vol.%) and mica (0–18 vol.%, average 5 vol.%) contents and porosity (0–28%, average 7%) also display no systematic change in relation to the unconformity. In these wells, mudstones of the Dunlin Group and rocks of the Statfjord Formation occur between the Cimmerian Unconformity and the Lunde Formation (Fig. 14b). Kaolinite content increases considerably only in the Jurassic Statfjord Formation (0–15 vol.%, average 5 vol.%).

3 Close inspection of the structural setting of some of the wells (e.g. 34/7-6) reveals that kaolinite content increases from less than 5 vol.% to 12 vol.% in sandstones lying in the vicinity of major normal faults that are connected to the Cimmerian Unconformity (Fig. 14c). Such faults are related to the Late Jurassic to Early Cretaceous rifting and footwall uplift of the Viking Graben (Yielding, 1990). In well 34/7-6, feldspar and mica contents do not change in the fault zone, but the intragranular porosity increases from less than 5% to about 10% (Fig. 14c).

DISCUSSION

Origin of kaolinite

There are several lines of evidence indicating an *in situ*, diagenetic origin for kaolinite, including: (i) delicate euhedral booklets and vermicular textures; (ii) kaolinitized mica flakes expanded into open pores, and (iii) high intercrystalline microporosity within patches of pore-filling kaolin, which typically is encountered in diagenetic (as opposed to detrital) clay aggregates (Hurst & Nadeau, 1995). Additionally, a detrital intraclastic origin related to the erosion and redeposition of floodplain deposits during channel migration is unlikely considering the low kaolinite content (average 6 vol.%) as well as the relative absence of kaolinitized mud in the floodplain mudstones. Instead, clay minerals in these mudstones are dominated by I/S and C/S, evolved from smectite formed under the semi-arid climate that prevailed during the Triassic.

It is unlikely that kaolinite formation occurred during mesodiagenesis by circulation of organic acids and CO_2, produced during thermal maturation of organic matter because of (i) the lack of pervasive dissolution of mesogenetic carbonate cements (Morad *et al.*, 1998) and because (ii) mesogenetic kaolin is characterized by thick booklets of well-ordered kaolinite and blocky dickite, rather than by the observed dominantly thin, stacked pseudohexagonal crystals (cf. Beaufort *et al.*, 1998). Nevertheless, small amounts of mesogenetic kaolin may have precipitated as a by-product of the albitization of detrital calcic plagioclase (Morad *et al.*, 1990).

Timing of kaolinitization

The complex burial history of the Triassic sandstones (Fig. 5) renders a precise determination of the timing of kaolinitization difficult. The paragenetic relationships of kaolinite with other cements indicate that kaolinite was formed during near-surface eodiagenesis (Late Triassic) and following uplift during telodiagenesis below the Cimmerian Unconformity. Petrographic criteria indicate that kaolinite was formed dominantly under telogenetic conditions as a result of Early Cretaceous exposure. These criteria include: (i) abundant kaolinite fills dissolution voids in grain-displacive, eogenetic calcrete (Fig. 7a), (ii) detrital silicates (feldspars, mud intraclast and micas), which are totally enclosed by eogenetic calcrete and dolocrete, show little evidence of kaolinitization, and (iii) sandstones in which the pore systems are totally blocked by smectitic pseudomatrix (formed as a result of mechanical compaction of mud intraclasts) show only minor kaolinitization owing to their very low permeability, which inhibited meteoric water flow. Note that in cases in which kaolinitization of pseudomatrix had occurred during eodiagenesis, then these low-permeability sandstones should contain abundant pseudomatrix of kaolinitic and not smectitic composition.

The distribution pattern of kaolinite is another key piece of evidence to unravel the timing of kaolinitization. The amounts of kaolinite formed during telodiagenesis are expected to increase towards unconformity surfaces owing to the greater degree of undersaturation (particularly with aK^+) and lower pH of descending meteoric waters. With increasing depth below the unconformity, meteoric water becomes progressively less undersaturated and has a higher pH, and hence, loses its capacity to dissolve common framework grains such as feldspars and micas. Thus, the increase in the overall abundance of kaolinite (from less than 5 to about 20 vol.%), and in intragranular porosity (from < 5 to > 10 vol.%), in the first 200 m below the Cimmerian Unconformity (Fig. 14a), is attributed to meteoric telodiagenesis. The increase in kaolinite content is, however, not accompanied by a concomitant decrease in the feldspar content, which is anticipated to be a common source of aluminium and silica for kaolinite precipitation. This lack of correlation, which has been observed in other North Sea reservoirs (e.g. Nedkvitne & Bjørlykke, 1992), is attributed to: (i) high initial variability of feldspar content (5–25 vol.%); (ii) kaolinitization of other silicate minerals, such as smectite-rich pseudomatrix, micas and siliciclastic rock fragments. Kaolinitization of mica, for instance, is indicated by the systematic decrease in mica content in the first 100 m approaching the Cimmerian Unconformity (Fig. 14a); and (iii) mobilization of Al and Si on scales greater than that of the thin-section.

We infer that normal faults that were developed or reactivated during the tectonic uplift and connected to the Cimmerian exposure surface have allowed the gravitationally induced downward flow of meteoric waters. Such a circulation pattern has promoted the telogenetic dissolution of grains and precipitation of kaolinite in sandstone beds that were neither directly exposed nor connected to the unconformity surface via porous sand-rich strata. At depths less than a few hundred metres, faults can focus meteoric water flow and thus silicate alteration (Herczeg & Payne, 1992; Parnell *et al.*, 2000). This is manifested by some of the wells studied (e.g. well 34/7-6; Fig. 14c), where sandstone beds that are cut by major faults display an increase in both kaolinite content (from < 5 to > 10 vol.%) and intragranular porosity (from < 5 to 12%). The close association of elevated amounts of kaolinite and intragranular porosity with such normal faults is thus attributed to telodiagenesis.

In wells where the Lunde Formation sandstones are neither cut by the unconformity nor by faults, no upward increase in kaolinite content is recognized (e.g. well 34/7-3; Fig. 14b). We anticipate that the presence of thick (> 50 m) mudstone layers of the Dunlin Group in such wells inhibited meteoric-water infiltration and telogenetic alterations in these reservoir sandstones (Fig. 15).

The interpreted, dominantly telogenetic rather than eogenetic origin of the kaolinite is further supported by differences in climatic conditions encountered during these diagenetic phases (Nystuen *et al.*, 1989). During deposition and eodiagenesis (Late Triassic), the climate was semi-arid, which favoured the formation of calcrete and dolocrete rather than of kaolinite (Morad *et al.*, 1998). Conversely, the humid climate that prevailed during telodiagenesis (Early Cretaceous) favoured pervasive kaolinitization of detrital silicates owing to the more efficient leaching and removal of ions (K, Ca, Mg) from the sandstones. This postulation is supported by the scarcity of kaolinite within eogenetic carbonate concretions, in contrast to its abundance in the uncemented host sediments.

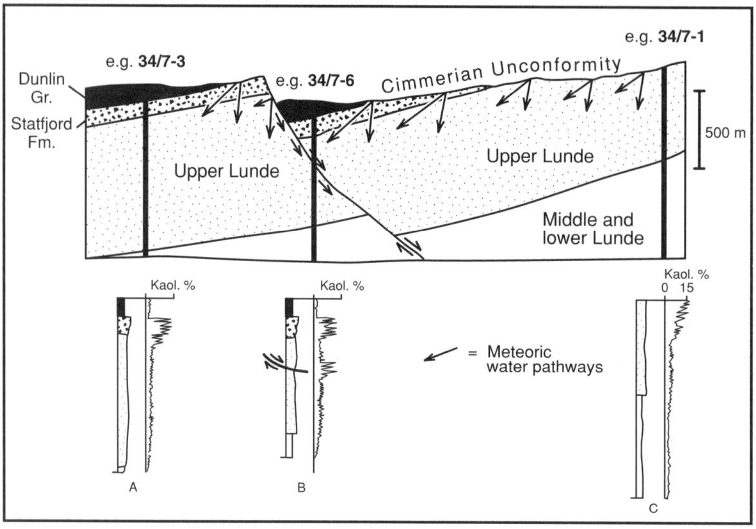

Fig. 15 Schematic section representing meteoric-water pathways below the Cimmerian Unconformity and three possible distribution patterns of kaolinite. In well A, kaolinite content is not related to the unconformity because meteoric flow is inhibited by mudstones of the Dunlin Group. In well B, meteoric flow occurs through a fault hydraulically connected to the Cimmerian Unconformity. In well C, kaolinite content is spatially related to the Cimmerian Unconformity because the Lunde Formation is cut by the unconformity.

The climate was more humid during the deposition and eodiagenesis of the Statfjord Formation (Lower Jurassic) than during deposition and eodiagenesis of the Lunde Formation. Consequently, in wells where both the Statfjord and the Lunde formations are protected from intense telogenetic alteration by the overlying mudstones of the Dunlin Group, kaolinite is more abundant in the Statfjord than in the Lunde sandstones (Fig. 14b).

Isotopic composition of kaolinite and meteoric waters

The texture of kaolinite indicates a dominantly low-temperature, meteoric-water origin. Thus, the oxygen and hydrogen isotopic composition of kaolinite can be used to unravel the origin and timing of kaolinitization, because such isotopes reflect the palaeolatitude, and hence palaeoclimatic control on the composition of meteoric waters involved. The palaeolatitude and palaeoclimate, and hence isotopic composition of meteoric waters, were different during eo- and telodiagenesis. Triassic meteoric waters at the interpreted palaeolatitude (20–26°N; Morad *et al.*, 1998) had an isotopic composition of $\delta^{18}O \approx -5‰$ (Morad *et al.*, 1998), whereas Late Jurassic to Early Cretaceous meteoric water had $\delta^{18}O \approx -7$ to $-9‰$ (Hamilton *et al.*, 1987; Hudson & Andrews, 1987; McAulay *et al.*, 1994). Using the fractionation equation between kaolinite and water (Land & Dutton, 1978), the range of isotopic values of kaolinite ($\delta^{18}O$ between +13.8‰ and +18.5‰) and assuming a near-surface temperature range of 15–25°C for precipitation of kaolinite, the isotopic composition of water varies between $\delta^{18}O = -13.5$ and $-6.8‰$, respectively (Fig. 16). The uppermost value ($-6.8‰$) is thus similar to the isotopic composition of the Early Cretaceous meteoric water, and is thus indicative of precipitation during telodiagenesis. However, lower values may represent more ^{18}O-depleted meteoric waters, suggesting partial resetting of the isotopic composition as a result of recrystallization of kaolinite into ordered kaolinite and disordered dickite during the subsequent burial phase. Such a partial recrystallization and its transformation of kaolinite into disordered dickite would explain the range of $\delta^{18}O$ values verified in the kaolinite samples from the Lunde Formation (Fig. 16). Variation in $\delta^{18}O$ values of meteoric waters, and hence of precipitated diagenetic minerals, also may be the result of changes in the amounts and source of rainfall in the area over a time-scale of several thousand years (Burns & Matter, 1995). The $\delta^{18}O$ values of meteoric waters are inversely related to the amounts of rainfall.

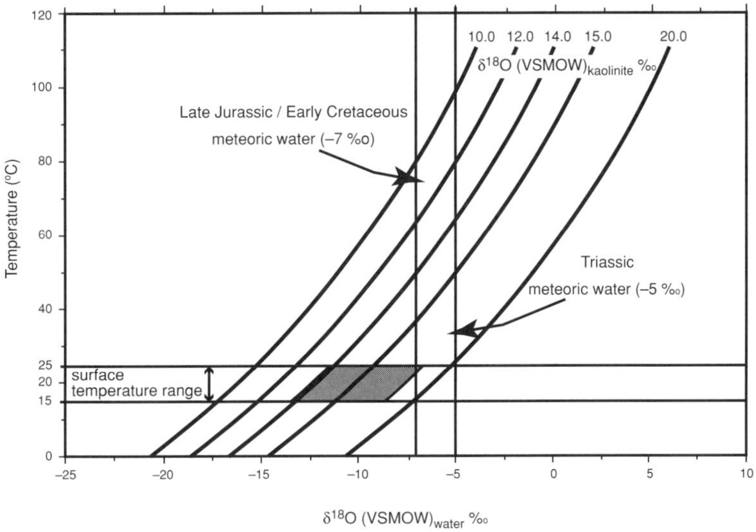

Fig. 16 Range of $\delta^{18}O$ values of the pore fluids (–13.5 to –6.8‰) constrained for the precipitation of the kaolinite analysed ($\delta^{18}O$ between +13.9 to +18.5‰), assuming a near-surface temperature of 15–25°C. Note that the lowest value is similar to the isotopic composition of the Jurassic meteoric water for the North Sea. Curves represent the temperature-related oxygen isotopic equilibrium between kaolinite and water.

The hydrogen isotopic signatures of the kaolinite analysed ($\delta D = -83‰$ to $-69‰$) are enriched in the light, ^1H isotope, which is typical for meteoric waters that are enriched in ^{16}O isotope. However, as is the case with kaolinite from other North Sea reservoirs, the δD values of the kaolinite analysed are relatively more enriched in deuterium relative to the meteoric kaolinite line (Fig. 16; see Morad *et al.*, this volume, pp. 63–91). The deviation of δD values in North Sea kaolinite from this line probably results from differences in the relative rates of equilibrium of kaolinite oxygen and hydrogen isotopes with pore waters. This is in agreement with the findings made by several authors, that the low-temperature, hydrogen isotope exchange in clay minerals might occur without a concomitant alteration of the oxygen isotopes (O'Neil & Kharaka, 1976; Bird & Chivas, 1988; Longstaffe & Ayalon, 1990).

Reservoir heterogeneity and reservoir-quality evolution below the Cimmerian Unconformity

Despite the impact of various types of diagenetic alterations, reservoir quality is still strongly influenced by depositional facies. This is clearly verified by the impact of sandstone texture on porosity and permeability (Fig. 9). The coarse-grained sandstones, such as those deposited at the base to the middle part of fining upward, fluvial single-storey channel bodies still reveal a higher permeability than finer grained sandstones deposited on top of channels or in crevasse splays (Fig. 10). Most deterioration in both porosity and permeability results from the formation of abundant pseudomatrix and eogenetic carbonate cement. The wide range of permeability encountered for each porosity value (Fig. 9) is attributed to variations in the amounts and distribution pattern of interstitial clay minerals (mainly kaolinite and pseudomatrix), variations in grain size and, to a lesser extent, to the presence of intragranular pores that are poorly connected to the intergranular pore conduits in the sandstones.

Reservoir quality displays a considerable vertical and lateral heterogeneity that varies in magnitude at different scales (Fig. 17). On a scale of tens to hundreds of metres, the controls on reservoir heterogeneity are related to the geometry, frequency, dimension and orientation of sandstone bodies, as well as on sand/mud and net/gross ratios (Diesen *et al.*, 1995; Nystuen & Fält, 1995). At the same scale, further reservoir heterogeneity was introduced by telodiagenesis below the Cimmerian Unconformity. The precise extent of reservoir-quality evolution below the unconformity is difficult to

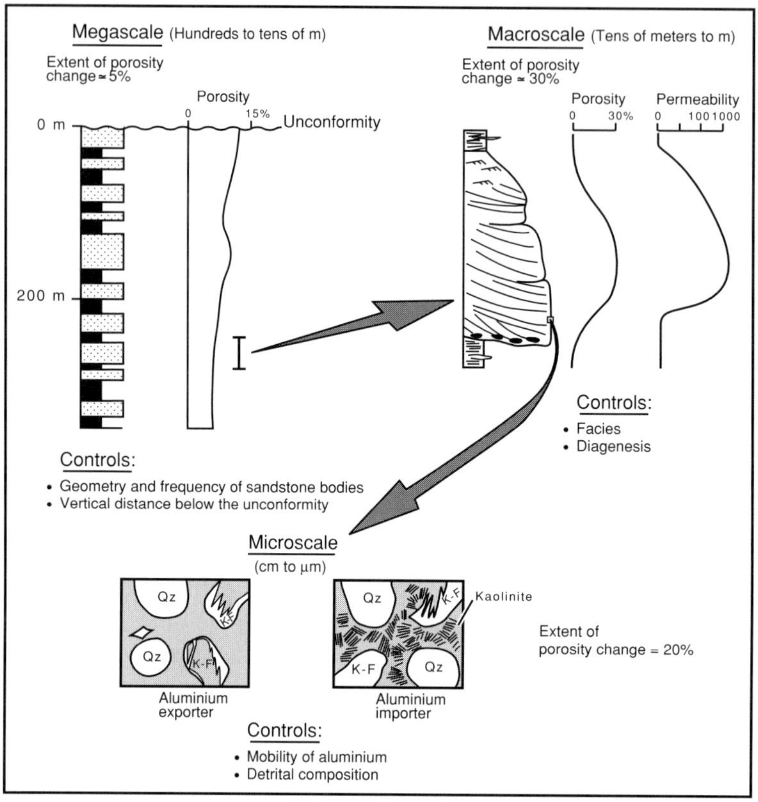

Fig. 17 A sketch showing the scales of heterogeneity in reservoir porosity and permeability owing to the impact of telogenetic alterations in the Lunde Formation.

evaluate, partly because it was overprinted to variable extents by mesodiagenesis during the second burial phase (Fig. 5). Intragranular porosity formed during telodiagenesis was, however, better preserved than that formed during eodiagenesis. Telogenetic dissolution occurred in sandstones that were already compacted and partially cemented, and thus possessed a more stable framework that prevented a pervasive collapse of secondary pores. There is a recognizable trend of an overall increase in intragranular porosity from < 5 to 15 vol.% and in average modal porosity by about 5% within the first 200 m below the unconformity (Fig. 14a).

Reservoir heterogeneity on the scale of metres to tens of metres is related to facies (Fig. 17) and to alterations induced during eodiagenesis and telodiagenesis. Diagenetic processes that had an important impact on porosity and/or permeability of the Lunde Formation sandstones (including carbonates, quartz and clay-mineral cements and pseudomatrix) vary widely in abundance and distribution pattern. On scales less than 1 m, important controls on reservoir heterogeneity include detrital composition and diagenesis. Considering telodiagenesis, the mobility of aluminium is of particular importance. Mobility of aluminium on distances greater thin-section scale may partly explain the poor positive correlation between kaolinite content and porosity and permeability (Fig. 13), and between kaolinite and feldspar contents below the unconformity (Fig. 14). Similar observations have been made for other North Sea sandstones reservoirs, such as in the Brent Group (Middle Jurassic) of the Hukdra Field (Nedkvitne & Bjørlykke, 1992). These authors attributed the presence of trace amounts of kaolinite in sandstones with abundant feldspar-dissolution, and intragranular porosity to advective mass transfer along laterally continuous, permeable beds.

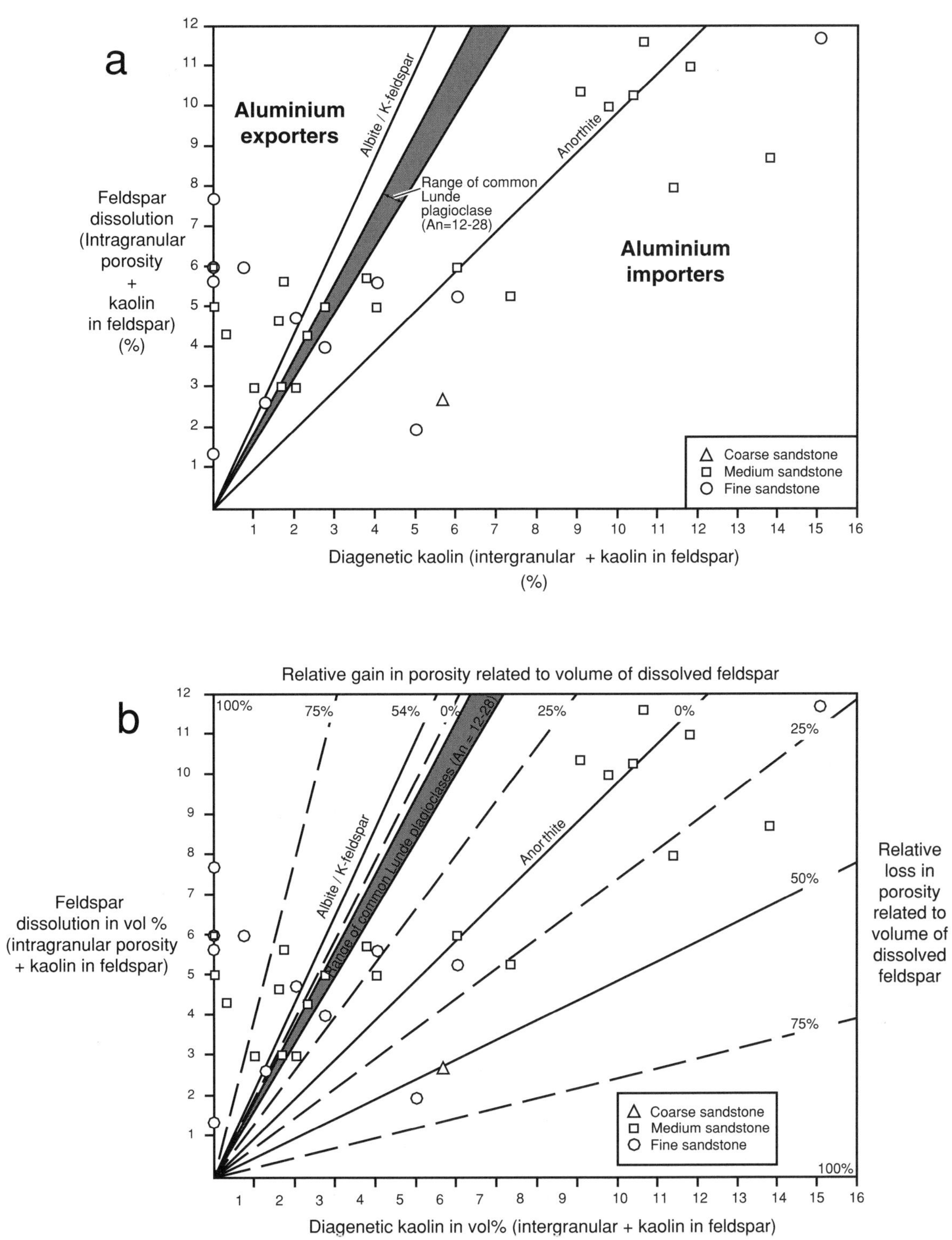

Fig. 18 Plot of: (a) sources of aluminium versus kaolinite showing samples that have 'exported' and 'imported' aluminium, and (b) concomitant relative gains in porosity.

Table 1 Mass-balance reactions of kaolinitization of common detrital grains in the Lunde Formation and the respective calculated volume of products.

Reactions*	Volume of products: Kaolinite (%)	Volume of products: Porosity (%)
$2KAlSi_3O_8 + 2H^+ + 9H_2O =$ $Al_2Si_2O_5(OH)_4 + 4H_4SiO_4 + 2K^+$ (1) **microcline** (mol. vol. = 2 × 108.7 cm³) kaolinite (mol. vol. 99.5 cm³)	46	54
$1.61Na_{0.74}Ca_{0.27}Al_{1.24}Si_{2.75}O_8 + 5.84H_2O + 2.04H^+ =$ **plagioclase**† (mol. vol. = 1.6 × 100.5 cm³) $Al_2Si_2O_5(OH)_4 + 1.19Na^+ + 0.43Ca^{+2} + 2.43H_4SiO_4$ (2)	60	40
$0.73K_{0.85}Na_{0.03}(Fe_{0.10}Mg_{0.09}Al_{1.88})(Al_{0.87}Si_{3.13})O_{10}(OH)_2 + 1.36H_2O + 0.94H^+ =$ **muscovite**‡ (mol. vol. = 0.73 × 140.7 cm³) $Al_2Si_2O_5(OH)_4 + 0.62K^+ + 0.02Na^+ + 0.07\,Fe^{+2} + 0.06Mg^{+2} + 0.28H_4SiO_4$ (3) kaolinite (mol. vol. = 99.5 cm³)	97	3
$1.38K_{0.89}Na_{0.11}(Fe_{1.21}Mg_{1.0}Al_{0.33}Ti_{0.19})(Al_{1.12}Si_{2.88})O_{10}(OH)_2 + 0.32H_2O + 8.48H^+ =$ **biotite**‡ (mol. vol. = 1.38 × 149.3 cm³) $Al_2Si_2O_5(OH)_4 + 1.23K^+ + 0.15Na^+ + 1.67Fe^{2+} + 1.38Mg^{2+} + 0.26Ti^{4+} + 1.97H_4SiO_4$ (4) kaolinite (mol. vol. = 99.5 cm³)	48	52
$1.18Na_{0.27}(Al_{1.54}Mg_{0.32}Fe_{0.22})(Al_{0.16}Si_{3.84})O_{10}(OH)_2 + 4.5H_2O + 1.84H^+ =$ **smectite**§ (mol. vol. = 1.18 × 218.8 cm³) $Al_2Si_2O_5(OH)_4 + 0.32Na^+ + 0.38Mg^{2+} + 0.26Fe^{2+} + 2.53H_4SiO_4 + 0.35H_2O$ (5) kaolinite (mol. vol. = 99.5 cm³)	39	61

* Molar volumes (at 25°C and 1 atm) from Robie & Waldbaum (1968) and Ranson & Hegelson (1994).
† Structural formula from Morad *et al.* (1990).
‡ Structural formula from Morad (1990).
§ Structural formula from Kerr *et al.* (1950).

Hence, sandstones behave either as Al^{3+} 'importers' or as Al^{3+} 'exporters'. A plot of the amounts of potential aluminium sources (e.g. feldspars) versus kaolinite content in a sandstone sample allows the distinction between samples that are Al^{3+} exporters and Al^{3+} importers (Fig. 18a). Aluminium importers undergo a relative loss in porosity, whereas aluminium exporters gain (intragranular) porosity and permeability (Fig. 18b). As the importation of aluminium is manifested by the precipitation of intergranular kaolinite, it also results in a net loss of permeability. This effect is most evident in samples with kaolin content greater than 10 vol.% (Fig. 13b).

The detrital composition of the sandstones also exerted an important control on reservoir heterogeneity at thin-section scale owing to variations in the amount and type of unstable grains prone to meteoric dissolution. Variations in the amounts of feldspars, mud intraclasts, unstable rock fragments, micas and pseudomatrix, also explain the weakly positive (or lack of) correlation between kaolinite and total porosity and permeability ($r = 0.4$ and $r = 0.1$, respectively; Fig. 13). Near-surface kaolinitization of K-feldspar induces a decrease in molar volume (i.e. porosity gain) by about 54% of the total volume of altered feldspars, as the excess of silica does not precipitate as quartz under semi-humid to humid climatic conditions (McBride, 1989; Table 1). Kaolinitization of plagioclase (An = 12–28; Morad *et al.*, 1990) induces a molar volume reduction by about 40% (Table 1). The kaolinitization of muscovite is not accompanied by significant changes in porosity (Table 1), whereas kaolinitization of biotite is accompanied by a 50% molar-volume reduction (Table 1), yet the formation of Fe and Ti oxides as a by-product of kaolinitization of biotite may reduce the porosity gained to 25%. Despite such gain in porosity, eogenetic kaolinitization of micas can cause major deterioration of permeability owing to expansion of mica flakes into open pores.

Kaolinitization of I/S pseudomatrix may have induced a net gain in porosity that is proportional to the smectite content, as smectite kaolinitization induces a molar-volume reduction of 61%, whereas kaolinitization of illite causes no molar volume change (Table 1). For example, if the I/S contained 25% smectite interstratification, then the kaolinitization of mud intraclasts would result in a microporosity value gain of 15%, which is similar to that observed in detrital clay aggregates (≈ 10%; Hurst & Nadeau, 1995; Fig. 6b). The overall porosity modification owing to telodiagenesis can be estimated based on the proportion of detrital minerals that have undergone kaolinitization. The kaolinitization of a Lunde sandstone with 8% feldspars (5% plagioclase and 3% K-feldspar), 3% pseudomatrix, 3% muscovite and 2% biotite results in a molar volume reduction (i.e. enhanced porosity) of 6.6%.

CONCLUSIONS

1 The recognition of the precise timing and extent of kaolinitization and related modifications of reservoir quality in sandstone sequences that have undergone multiple phases of burial and uplift is difficult. Nevertheless, this study has demonstrated that the incursion of meteoric water below unconformity surfaces has an important impact on the diagenetic and reservoir-quality evolution of siliciclastic sequences.

2 Telodiagenesis below the Cimmerian Unconformity (Early Cretaceous) has resulted in considerable amounts of silicate dissolution and kaolinite formation in Triassic sandstones of the Lunde Formation, Snorre Field, northern North Sea. The main telogenetic alterations are encountered in the first 200 m below the unconformity.

3 Porosity enhancement in the sandstones induced by telogenetic kaolinitization is still prevalent despite subsequent burial and mesodiagenesis, partly because telodiagenesis occurred in already compacted and partially cemented sandstones. Average porosity of the sandstones increased by about 5% in the first 200 m immediately below the unconformity.

4 The dominance of telogenetic over eogenetic kaolinite formation is recognized, based on the following criteria: (i) systematic increase in kaolinite content, from < 5 to 20 vol.%, towards the unconformity. This pattern is less evident when thick (> 50 m) mudstone layers are present between the sandstones and the unconformity. (ii) Increase in kaolinite content in the vicinity of major faults connected to the unconformity surface. (iii) Absence of kaolinite in eogenetic carbonates, except within secondary dissolution pores.

5 The oxygen isotopic composition of kaolinite ($\delta^{18}O$ = +13.9 to +18.5‰) is compatible with Early Cretaceous telogenetic, rather than with Triassic eogenetic, meteoric waters ($\delta^{18}O$ = −7 to −9‰ and −5‰, respectively), assuming near-surface temperature ranges (15–25°C). However, the lowermost $\delta^{18}O$ values of kaolinite probably reflect isotopic resetting during partial recrystallization and dickitization of kaolinite.

6 The telogenetic rather than eogenetic formation of kaolinite presumably was enhanced by more humid climatic conditions that prevailed during the Early Cretaceous compared with the semi-arid climate during the Triassic.

7 The lack of correlation between diagenetic kaolinite and feldspar, which is characteristic of other North Sea reservoirs, is attributed to Al^{3+} transfer over scales larger than that of the thin-section, and to the presence of other kaolinite precursors (mica, mud intraclast, pseudomatrix and rock fragments).

ACKNOWLEDGEMENTS

We are grateful to Saga Petroleum ASA for access to samples and data. J.M. Ketzer thanks the Brazilian Research Council, CNPq, for providing a PhD scholarship (grant no. 200059/98-0). S. Morad thanks the Swedish Natural Sciences Research Council (NFR) and J.P. Nystuen thanks the colleagues in the former

Saga Petroleum ASA for many stimulating discussions on the geology of the Snorre Field. L.F. De Ros acknowledges the support by the Brazilian Research Council, CNPq. We all thank Richard Worden and Susanne Schmid for reviewing the manuscript and Stuart Haszeldine for comments and suggestions.

REFERENCES

Beaufort, D., Cassagnabere, A., Petit, S. *et al.* (1998) Kaolinite-to-dickite reaction in sandstone reservoirs. *Clay Mineral.*, **33**, 297–316.

Bigeleisen, J., Perlman, M.L. & Prosser, H.C. (1952) Conversion of hydrogenic materials to hydrogen for isotope analysis. *Anal. Chem.*, **24**, 1356–1357.

Bird, M.I. & Chivas, A.R. (1988) Stable isotope evidence for low temperature kaolinite weathering and post-formational hydrogen-isotope exchange in Permian kaolinites. *Chem. Geol.*, **72**, 249–265.

Bjørkum, P.A., Mjos, R., Walderhaug, O. & Hurst, A. (1990) The role of the late Cimmerian unconformity for the distribution of kaolinite in the Gullfaks Field, northern North Sea. *Sedimentology*, **37**, 395–406.

Bjørkum, P.A., Knarud, R. & Bergan, M. (1993) How important is the late Cimmerian unconformity in controlling the formation of kaolinite in sandstones of the North Sea? (examples from the Snorre Field). In: *Diagenesis and Basin Development*, American Association of Petroleum Geologists Studies in Geology 36 (Ed. by A.D. Horbury and A.G. Robinson), Tulsa, OK, pp. 261–269.

Bjørlykke, K. & Brendsdal, A. (1986) Diagenesis of the Brent Sandstone in the Statfjord Field, North Sea. In: *Roles of Organic Matter in Sediment Diagenesis*, Society for Economic Paleontologists and Mineralogists Special Publication 38 (Ed. by D.L. Gautier), Tulsa, OK, pp. 157–167.

Bloch, S. (1994) Secondary porosity in sandstones: significance, origin, relationship to subaerial unconformities, and effect on predrill reservoir quality prediction. In: *Reservoir Quality Assessment and Prediction in Clastic Rocks*, Society for Economic Paleontologists and Mineralogists Short Course 30, (Ed. by M.D. Wilson), Tulsa, OK, pp. 137–161.

Brun, J.P. & Tron, V. (1993) Development of the North Viking Graben: inferences from laboratory modelling. *Sed. Geol.*, **86**, 31–51.

Burley, S.D. (1986) The development and destruction of porosity within Upper Jurassic reservoir sandstones of the Piper and Tartan fields, Outer Morey Firth, North Sea. *Clay Mineral.*, **21**, 649–694.

Burns, S.J. & Matter, A. (1995) Geochemistry of carbonate cements in surficial alluvial conglomerates and their paleoclimatic implications, Sultanate of Oman. *J. Sed. Res.*, **A65**, 170–177.

Campbell, C.J. & Ormaasen, E. (1987) The discovery of oil and gas in Norway: an historical synopsis. In: *Geology of the Norwegian Oil and Gas Fields* (Ed. by A.M. Spencer), Graham & Trotman, London, pp. 1–37.

Choquette, P.W. & Pray, L.C. (1970) Geologic nomenclature and classification of porosity in sedimentary carbonates. *AAPG Bull.*, **54**, 207–250.

Clayton, R.N. & Mayeda, T.K. (1963) The use of bromine pentafluoride in the extraction of oxygen from oxides and silicates for isotopic analysis. *Geochim. Cosmochim. Acta*, **27**, 43–52.

Diesen, G.W., Edvardsen, A., Nystuen, J.P., Sverdrup, E. & Tollefsrud, J.I. (1995) *Geophysical and Geological Tools and Methods used for Reservoir Characterisation and Modelling of the Snorre Field—North Sea, Norway.* Technology Research Center Special Publication 5, Japan National Oil Company, Chiba, pp. 69–90.

Ehrenberg, S.N., Aagaard, P., Wilson, M.J., Fraser, A.R. & Duthie, D.M.L. (1993) Depth-dependent transformation of kaolinite to dickite in sandstones of the Norwegian continental shelf. *Clay Mineral.*, **28**, 325–352.

Emery, D., Myers, R.J. & Young, R. (1990) Ancient subaerial exposure and freshwater leaching in sandstones. *Geology*, **18**, 1178–1181.

Fallick, A.E., Macaulay, C.I. & Haszeldine, R.S. (1993) Implications of linearly correlated oxygen and hydrogen isotopic compositions for kaolinite and illite in the Magnus Sandstone, North Sea. *Clay. Clay Mineral.*, **41**, 184–190.

Ghosh, S.K. (1997) Porosity distribution in relation to unconformities and its implications. *American Association of Petroleum Geologists and Society for Sedimentary Geology Annual Meeting Abstracts*, Vol. 6, Dallas, 6–9 April, p. 40.

Giles, M.R. & De Boer, R.B. (1990) Origin and significance of redistributional secondary porosity. *Mar. Petrol. Geol.*, **7**, 378–397.

Glasmann, J.R., Clark, R.A., Larter, S., Briedis, N.A. & Lundegard, P.D. (1989a) Diagenesis and hydrocarbon accumulation; Brent Sandstone (Jurassic), Bergen High area, North Sea. *AAPG Bull.*, **73**, 1341–1360.

Glasmann, J.R., Lundegard, P.D., Clark, R.A., Penny, B.K. & Collins, I.D. (1989b) Geochemical evidence for the history of diagenesis and fluid migration: Brent Sandstone, Heather Field, North Sea. *Clay Mineral.*, **24**, 255–284.

Godfrey, J.D. (1962) The deuterium content of hydrous minerals from the East-Central Sierra Nevada and Yosemite National Park. *Geochim. Cosmochim. Acta*, **26**, 1215–1245.

Hamilton, P.J., Fallick, A.E., MacIntyre, R.M. & Elliot, S. (1987) Isotopic tracing of the provenance and diagenesis of Lower Brent Group sands, North Sea. In:

Petroleum Geology of NW Europe (Ed. by J. Brooks and K. Glennie), Graham & Trotman, London, pp. 939–949.

Hancock, N.J. & Taylor, A.M. (1978) Clay mineral diagenesis and oil migration in the Middle Jurassic Brent Sand Formation. *J. Geol. Soc. London*, **135**, 67–72.

Herczeg, A.L. & Payne, T.E. (1992) Recharge and weathering processes in fractured rock aquifers in Northern Australia. In: *Water Rock Interaction—Low Temperature Environments* (Ed. by Y.K. Khahara and A.S. Maest), Balkema, Rotterdam, pp. 561–564.

Hollander, N.B. (1987) Snorre. In: *Geology of the Norwegian Oil and Gas Fields* (Ed. by A.M. Spencer), Graham & Trotman, London, pp. 307–318.

Hudson, J.D. & Andrews, J.E. (1987) The diagenesis of the Great Estuarine Group, Middle Jurassic, Inner Hebrides, Scotland. In: *Diagenesis of Sedimentary Sequences* (Ed. by J.D. Marshall). *Geol. Soc. London Spec. Publ.*, **36**, 259–276.

Hurst, A. & Nadeau, P.H. (1995) Clay microporosity in reservoir sandstones: an application of quantitative electron microscopy in petrophysical evaluation. *AAPG Bull.*, **79**, 563–573.

Isaksen, D. & Tonstad, D. (1989) A revised Cretaceous and Tertiary lithostratigraphic nomenclature for the Norwegian North Sea. *Norsk Petrol. Direktorat Bull.*, **5**, 59 pp.

Kerr, P.F., Hamilton, P.K. & Pill, R.J. (1950) *Analytical Data on Reference Clay Minerals*. Project 49: Clay Minerals Standards, Preliminary Report 7, American Petroleum Institute, pp. 1–60.

Khanna, M., Saigal, G.C. & Bjørlykke, K. (1997) Kaolinitization of the Upper Triassic–Lower Jurassic sandstones of the Tamper Spur area, North Sea: implications for early diagenesis and fluid flow. In: *Basin-wide Diagenetic Patterns: Integrated Petrologic, Geochemical and Hydrological Considerations*, Society for Economic Paleontologists and Mineralogists Special Publication 57, (Ed. by P. Montañez, J.M. Gregg and K.L. Shelton), Tulsa, OK, pp. 253–268.

Land, L.S. & Dutton, S.P. (1978) Cementation of a Pennsylvanian deltaic sandstone: isotopic data. *J. Sed. Petrol.*, **48**, 1167–1176.

Larese, R.E., Haskell, N.L., Prezbindsowski, D.R. & Beju, D. (1983) Porosity development in selected Jurassic sandstones from the Norwegian and North Seas, Norway—an overview. In: *Petroleum Geology of the North European Margin* (Ed. by A.M. Spencer, E. Holter, S.O. Johnen, *et al.*), Graham & Trotman, London, pp. 81–95.

Longstaffe, F.J. & Ayalon, A. (1990) Hydrogen-isotope geochemistry of diagenetic clay minerals from Cretaceous sandstones, Alberta, Canada: evidence for exchange. *Appl. Geochem.*, **5**, 657–668.

Lønøy, A., Akselsen, J. & Rønning, K. (1986) Diagenesis of a deeply buried sandstones reservoirs: Hild Field, northern North Sea. *Clay Mineral.*, **21**, 497–511.

Macaulay, C.I., Fallick, A.E. & Haszeldine, R.S. (1993) Textural and isotopic variations in diagenetic kaolinite from the Magnus Oilfield sandstones. *Clay Mineral.*, **28**, 625–639.

McBride, E.F. (1963) A classification of common sandstones. *J. Sed. Petrol.*, **33**, 664–669.

McBride, E.F. (1989) Quartz cement in sandstones: a review. *Earth Sci. Revi.*, **26**, 69–112.

McCaulay, G.E., Burley, S.D. & Johnes, L.H. (1993) Silicate mineral authigenesis in the Hutton and NW Hutton fields: implications for sub-surface porosity development. In: *Petroleum Geology of the Northwest Europe: Proceedings of the 4th Conference* (Ed. by J.R. Parker), Geological Society Publishing House, Bath, pp. 1377–1394.

McCaulay, G.E., Burley, S.D., Fallick, A.E. & Kusznir, N.J. (1994) Paleohydrodynamic fluid flow regimes during diagenesis of the Brent Group in the Hutton–NW Hutton reservoirs: constraints from oxygen isotopic studies of authigenic kaolin and reverse flexural modelling. *Clay Mineral.*, **29**, 609–626.

Morad, S. (1990) Mica alteration reactions in Jurassic reservoir sandstones from the Haltenbanken area, offshore Norway. *Clay. Clay Mineral.*, **38**, 584–590.

Morad, S., Bergan, M., Knarud, R. & Nystuen, J.P. (1990) Albitization of detrital plagioclase in Triassic reservoir sandstones from the Snorre Field, Norwegian North Sea. *J. Sed. Petrol.*, **60**, 411–425.

Morad, S., De Ros, L.F., Nystuen, J.P. & Bergan, M. (1998) Carbonate diagenesis and porosity evolution in sheetflood sandstones: evidence from the Middle and Lower Lunde Members (Triassic) in the Snorre Field, Norwegian North Sea. In: *Carbonate Cementation in Sandstones*, International Association of Sedimentologists Special Publication 26 (Ed. by S. Morad), Blackwell Science, Oxford, pp. 53–85.

Morad, S., Ketzer, J.M. & De Ros, L.F. (2000) Spatial and temporal distribution of diagenetic alterations in siliciclastic rocks: implications for mass transfer in sedimentary basins. *Sedimentology*, **47** (Supplement 1), 95–120.

Morad, S., Worden, R.H. & Ketzer, J.M. (2003) Oxygen and hydrogen isotopic composition of diagenetic clay minerals in sandstones: a review of the data and controls. In: *Clay Mineral Cements in Sandstones*, International Association of Sedimentologists Special Publication 34 (Ed. by R.H. Worden and S. Morad), Blackwell Publishing, Oxford, pp. 63–91.

Nedkvitne, T. & Bjørlykke, K. (1992) Secondary porosity in the Brent Group (Middle Jurassic), Hundra Field, North Sea: implication for predicting lateral continuity of sandstones. *J. Sed. Petrol.*, **61**, 23–34.

Nystuen, J.P. & Fält, L.M. (1995) Upper Triassic–Lower Jurassic reservoir rocks in the Tampen Spur area, Norwegian North Sea. In: *Petroleum Exploration and Exploitation in Norway*, Norwegian Petroleum Society Special Publication 4 (Ed. by S. Hanslien), Elsevier, Amsterdam, pp. 135–179.

Nystuen, J.P., Knarud, R., Jorde, K. & Stanley, K.O. (1989) Correlation of Triassic to Lower Jurassic sequences, Snorre Field and adjacent areas, northern North Sea. In: *Correlation in Hydrocarbon Exploration* (Ed. by J.D. Collinson), Norwegian Petroleum Society, Graham & Trotman, London, pp. 273–289.

O'Neil, J.R. & Kharaka, Y.F. (1976) Hydrogen and oxygen isotope exchange reactions between clay minerals and water. *Geochim. Cosmochim. Acta*, **40**, 241–246.

Parnell, J., Baron, M. & Boyce, A. (2000) Controls on kaolinite and dickite distribution, highland Boundary Fault Zone, Scotland and Northern Ireland. *J. Geol. Soc. London*, **157**, 635–640.

Ranson, B. & Helgeson, H.C. (1994) Estimation of the standard molal heat capacities, entropies, and volumes of 2:1 clay minerals. *Geochim. Cosmochim. Acta*, **58**, 4537–4547.

Robie, R.A. & Waldbaum, D.R. (1968) Thermodynamic properties of minerals and related substances at 298.15°K (25°C) and one atmosphere (1.013 Bars) pressure and at higher temperatures. *US Geol. Surv. Bull.*, **1259**, 256 pp.

Savin, S.M. & Epstein, S. (1970) The oxygen and hydrogen isotopic geochemistry of clay minerals. *Geochim. Cosmochim. Acta*, **49**, 57–66.

Shanmugan, G. (1988) Origin, recognition, and importance of erosional unconformities in sedimentary basins. In: *New Perspectives in Basin Analysis* (Ed. by K.L. Kleinspehn and C. Paola), Springer-Verlag, New York, pp. 83–108.

Shanmugan, G. (1990) Porosity prediction in sandstones using erosional unconformities. In: *Prediction of Reservoir Quality Through Chemical Modeling* (Ed. by I.D. Meshri and P.J. Ortoleva), *AAPG Mem.*, **49**, 1–23.

Shanmugan, G. & Higgins, J.B. (1988) Porosity enhancement from chert dissolution beneath Neocominian unconformity: Ivishak Formation, North Slope, Alaska. *AAPG Bull.*, **72**, 523–535.

Sommer, F. (1978) Diagenesis of Jurassic sandstones in the Viking Graben. *J. Geol. Soc.*, **135**, 63–67.

Vollset, J. & Doré, A.G. (1984) A revised Triassic and Jurassic lithostratigraphic nomenclature for the Norwegian North Sea. *Norsk Petrol. Direktorat Bull.*, **3**, 53 pp.

Yielding, G. (1990) Footwall uplift associated with Late Jurassic normal faulting in the northern North Sea. *J. Geol. Soc.*, **147**, 219–222.

Int. Assoc. Sedimentol. Spec. Publ. (2003) **34**, 383–408

The formation and stability of kaolinite in Brent sandstone reservoirs: a modelling approach

ÉTIENNE BROSSE[1], THOMAS MARGUERON[2,*], CÉDRIC CASSOU[1], BERNARD SANJUAN[3], ANDREW CANHAM[4], JEAN-PIERRE GIRARD[3], JEAN-CLAUDE LACHARPAGNE[5] *and* FRÉDÉRIC SOMMER[2]

[1]*Institut Français du Pétrole, Rueil-Malmaison, France, e-mail: etienne.brosse@ifp.fr*
[2]*Total-Fina, Saint-Rémy-lès-Chevreuse, France*
[3]*Bureau de Recherches Géologiques et Minières, Orléans, France*
[4]*IreS, Chester, United Kingdom*
[5]*Elf, Pau, France*

ABSTRACT

The precipitation of kaolinite in reservoir sandstones is a relatively early diagenetic reaction, which depends on the initial proportions of mica and feldspar varieties in the sand, and on the nature and flow rate of the interstitial water. A water–rock interaction code was used to explore the influence of the mineralogical composition and of the water characteristics, on the formation and stability of kaolinite in subarkosic sandstones. The water compositions tested were meteoric or, alternatively, marine. The detrital aluminosilicates considered were plagioclase (anorthite and low-T albite), K-feldspar, phengite and quartz. From this assemblage, below 75°C plagioclase was the main control for kaolinite formation. Anorthite in all cases acts as a precursor of kaolinite, whereas low-T albite was destabilized only by influx of marine or meteoric water. Phengite dissolved only in the marine case, which also induced some K-feldspar overgrowth. After burial, in more and more confined conditions, K-feldspar and some kaolinite were altered, and consequently illite precipitated. The assemblage of kaolinite, K-feldspar and residual phengite was destabilized at *c.* 95°C, and illite began to form. If the precipitated clay mineral was an Fe, Mg-illite and if the system was closed, this reaction was of limited extent at this temperature. A more efficient illitization process occurred between 120 and 130°C.

The preceding scenario was compared with petrographical and geochemical data obtained from samples of the Brent Group reservoirs cored in the Hild Field. The main discrepancy between the simulation results and the data is a generation of kaolin-mineral polytype (seemingly dickite) that according to the petrographical and geochemical constraints occurred in the 105–125°C range. The conditions explored in the modelling approach were not able to reproduce the formation of this mineral above 90°C.

* Present address: Institut Français du Pétrole, Rueil-Malmaison, France.

INTRODUCTION

Many papers have described diagenetic polytypes of the kaolin-group minerals in reservoir sandstones. Particular attention has been paid to the Brent Group (Middle Jurassic) from the northern North Sea. From petrographical, mineralogical and geochemical features it has been discussed at what stage of diagenesis kaolin developed, and from what dissolved precursor minerals (namely detrital feldspars and micas) it formed (among others: Bjørlykke & Brendsdal, 1986; Scotchman *et al.*, 1989; Bjørlykke & Aagaard, 1992; Bjørlykke *et al.*, 1992; Giles *et al.*, 1992; Ehrenberg *et al.*, 1993; Emery *et al.*, 1993; Macaulay *et al.*, 1993; McAulay *et al.*, 1993, 1994; Hogg *et al.*, 1995; Khanna *et al.*, 1997; Potdevin & Hassouta, 1997; Beaufort *et al.*, 1998; Cassagnabère, 1998; Morad *et al.*, 1998; Hassouta, 1999; Maréchal, 2000). However, neither all feldspar varieties nor probably all mica grains present the same stability range and consequently the same alteration potential. Accordingly, diagenetic processes dissolve feldspar and mica grains selectively. Whereas it is rare to observe Ca-rich plagioclase, even at shallow burial depth, it is common to observe well-preserved K-feldspar or muscovite even at a depth greater than about 2 km. A related aspect is that the intensity of the diagenetic transformations is determined by the mineralogical composition of the sediment inherited from the source province, and the depositional processes. Different proportions of the detrital constituents result in different assemblages of the diagenetic by-products. In addition, the stability of feldspars and micas, and their inclination to dissolve and deliver kaolin-producing components, depends on the geological and geochemical context, particularly on the composition and movement of interstitial waters. Whatever the context invoked for explaining the formation of kaolinite *sensu lato*, the need for an active fluid flow has been stressed because some elements contained in detrital feldspars or micas (K, Na, possibly also Ca) must be exported out of the sandstone (e.g. Khanna *et al.*, 1997).

The preceding parameters, i.e. firstly intrinsic stability and proportions of the minerals, and secondly water composition and water-flow velocity, can combine in a complex, non-linear way. Discussing their respective role can lead to confusing conclusions, especially if a quantitative evaluation of the diagenetic effects on porosity and permeability distribution is required for engineering purposes. In that respect, numerical modelling of water–rock interactions, because it explicitly takes into account the complexity of the phenomena, can provide arguments in favour of one or other of the hypotheses that arise from petrographical, mineralogical and geochemical analyses. Such a modelling approach is proposed here (following Brosse *et al.*, 2000), to assess the origin of kaolinite in Brent Group reservoirs, and to discuss its stability.

The range of temperatures investigated in the simulations was 25 to 150°C. It was thus possible to reach the domain where kaolinite begins to dissolve, and illite to form. The precipitation of illite in the Brent sandstones has been addressed in a preceding paper (Brosse *et al.*, 2000). The present work integrates a more detailed composition of illite, and provides new arguments to understand the conditions of illitization.

CONSTRAINTS FROM THE HILD FIELD RESERVOIRS

An original data set was generated for the Brent Group sandstones of the Hild field (Lacharpagne *et al.*, 1999; Girard *et al.*, 2001). The Hild Field is located on the western side of the North Viking Graben, southeast of the Greater Alwyn area and west-southwest of the Oseberg Field (Fig. 1). The structure of the field is determined by block faulting and tilting associated with the Jurassic rift that affected the area. The absence of the uppermost Tarbert units, in crestal locations of the tilted blocks, was interpreted to be the result of erosion (Fig. 2, well 30/4-2; see also fig. 11 in Rønning *et al.* (1986), where 30/4-2 is the well to the

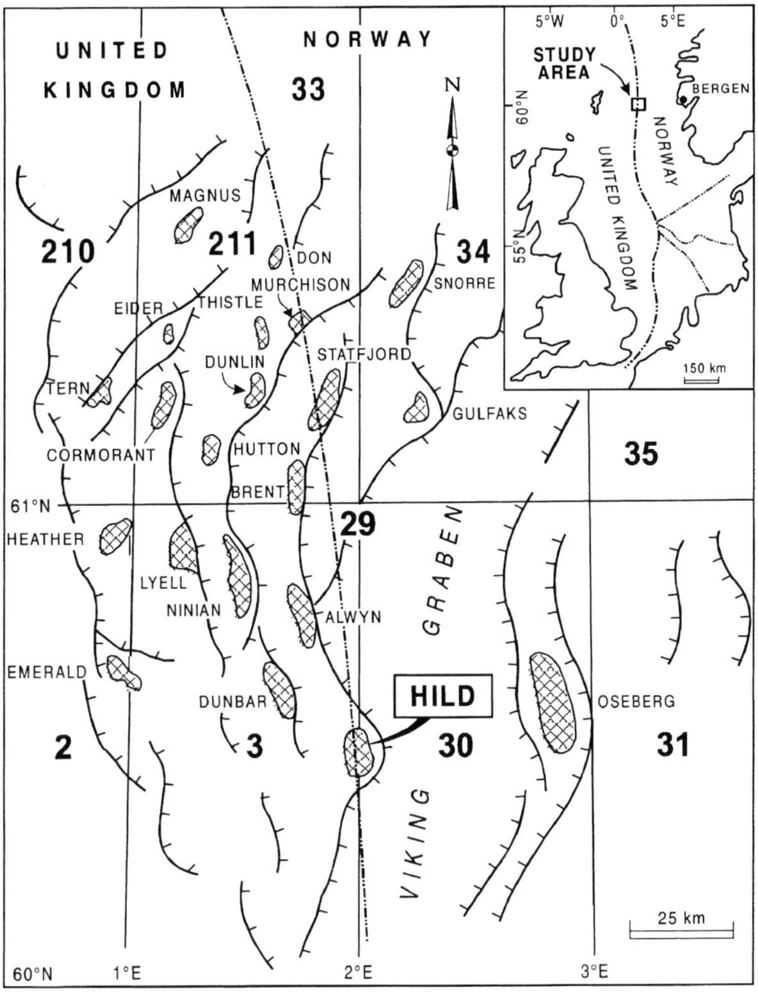

Fig. 1 Situation of the Hild Field in the North Sea.

right, 29/6-1 and 30/7-8 the wells to the left). The question of erosion and possible emersion is examined in the discussion below.

Analytical data were acquired from core samples from three wells, namely 29/6-1, 30/7-8 and 30/4-2 (Fig. 2). The reservoir sandstones studied are from the Tarbert Formation and from the upper part of the Ness Formation, between *c.* 3800 and 4300 m burial depth. Present temperature in the Tarbert Formation is 153°C for well 29/6-1, 145°C for 30/-8 and 118°C for 30/4-2. Compared with hydrostatic pressures (*c.* 38 MPa at 4000 m), 35 to 40 MPa overpressure values have been measured. At this depth fluid pressure twice the value of hydrostatic pressure is consistent with the pressure values of Jurassic reservoirs in the Viking Graben (Buhrig, 1989).

In the cored sections studied the interpretation of sedimentary facies proposed by Rønning *et al.* (1986) and Lacharpagne *et al.* (1999) are consistent, but unfortunately lack integration with the regional context. Both interpretations recognized environments ranging from a delta plain to the proximal offshore area of a delta complex. Sandstones of the Ness Formation are alluvial (distributary channel and crevasse splay) and organized in relatively thin, isolated

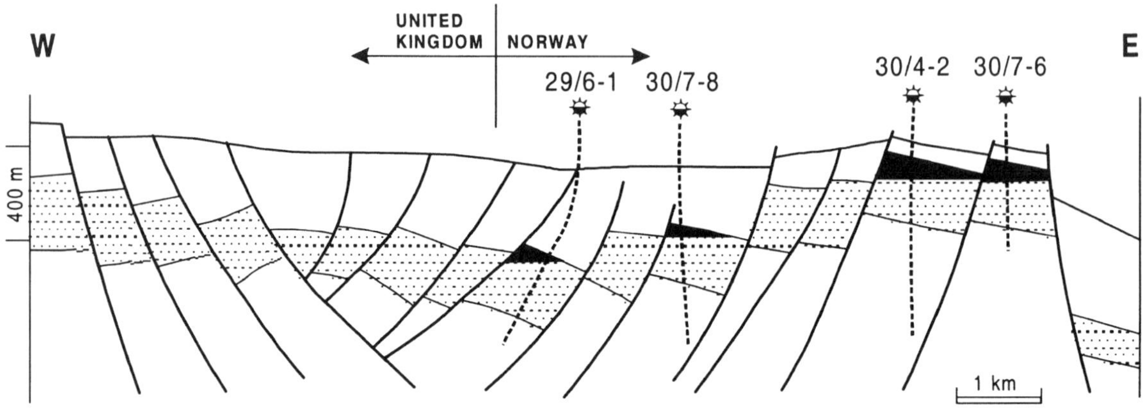

Fig. 2 Geological cross-section over the Hild Field. Data from the cored wells 29/6-1, 30/7-8 and 30/4-2 were used for constraining and discussing the simulations.

sequences; those of the Tarbert Formation are lagoonal or constitute part of barrier–bar complexes.

Quantification of the mineral phases and diagenetic trends versus burial depth

Ninety-four samples of sandstone reservoir were selected to represent the various sedimentary facies recognized from cores. They were studied petrographically using 500-point modal analysis of thin-sections. The precision increases with the value, e.g. 5 ± 2% and 30 ± 4% (Van der Plas & Tobi, 1965). Details on the preparation, staining and counting techniques, together with the comprehensive table of results, are given by Lacharpagne *et al.* (1999). This data set has been used to define and discuss the modelling approach of the present paper.

Detrital minerals are mainly monocrystalline quartz (> 50%, proportions are expressed in volumetric per cent of the whole rock, including porosity), mica with largely dominant muscovite-type minerals (0 to 10%, average 1.9%), rare feldspars almost exclusively composed of albite (always < 2%, only *c.* 0.1% on average), and lithic fragments (in general 10 to 20%, dominated by polycrystalline quartz). According to the classification proposed by McBride (1963), if polycrystalline quartz is counted in the lithic component, most of these sandstones are presently sublitharenites.

Diagenetic minerals include quartz overgrowth (up to 25%, 8.5% on average), kaolin-group minerals (0 to 16%, most < 12%, 2.7% on average), illite (1 to 15%, 5.4% on average), and various carbonates (non-ferroan calcite, ferroan calcite, dark blue stained carbonate, ferroan dolomite).

Additional to the identified phases, there is a proportion of optically unresolvable clay (up to 25%, 5.7% on average), which could be partly detrital, partly diagenetic. The proportion of mica increases with the proportion of unresolvable clay. Unresolvable clay could contain some detrital illite or kaolinite, however, the illite and the kaolinite phases presented in Figs 3 & 5 were recognized as strictly diagenetic.

The more mature (i.e. quartz-rich) compositions of sandstone were observed in the Tarbert Formation coarsening-upward sequences, interpreted as prograding barriers (Lacharpagne *et al.*, 1999). These sediments are also the coarsest and those that contain the highest proportion of polycrystalline quartz. Conversely, the back-barrier sandstones are richer in detrital clay and mica, and are finer grained. In the framework of the existing sedimentological description this back-barrier clay is the only significant difference in mineral composition that could be

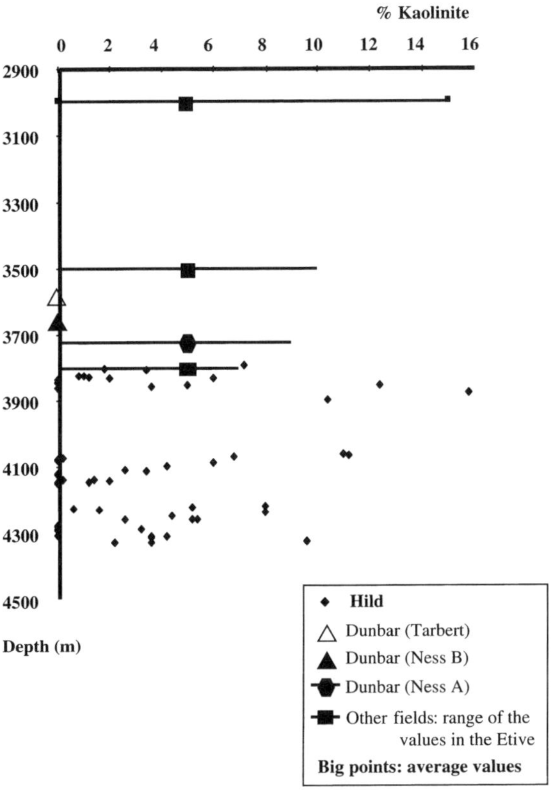

Fig. 3 Distribution of diagenetic kaolinite in the Brent sandstone reservoirs of the Hild Field. Comparison with ranges of values obtained from the Brent sandstones of other fields in the same province (Bjørlykke *et al.*, 1992; Giles *et al.*, 1992; Durand *et al.*, 2000).

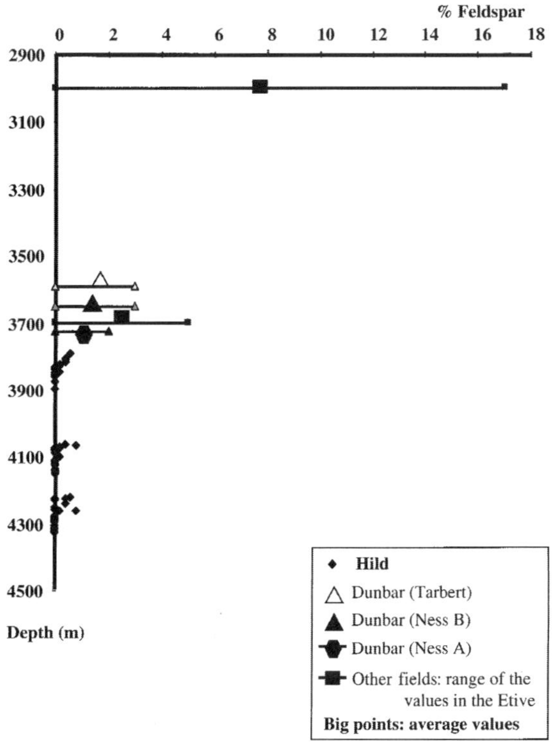

Fig. 4 Distribution of K-feldspar in the Brent sandstone reservoirs of the Hild Field. Comparison with ranges of values obtained from the Brent sandstones of other fields in the same province (Bjørlykke *et al.*, 1992; Giles *et al.*, 1992; Durand *et al.*, 2000).

related to the depositional facies. In particular, relationships between the kaolinite abundance and the facies type, as those noted by Osborne *et al.* (1994) or Khanna *et al.* (1997), were not observed here.

The amount of kaolinite exhibits a larger range of values in the Hild Field than in many other fields of the North Viking Graben area, nevertheless, the average values are similar in the various sets of samples considered here for comparison (Fig. 3). In contrast, from the distribution versus depth of the kaolinite content, the Brent reservoirs of Hild turn out to be unusual. Even at 4300 m a high kaolinite proportion is still present. A decrease of kaolinite proportion with increasing burial depth was observed by Bjørlykke *et al.* (1992) (their fig. 3c) and by Giles *et al.* (1992) (their fig. 15) in a range of depth values (2000 to 4000 m) different from the range investigated here. From the data set presented by Chuhan *et al.* (2000) kaolinite is not significantly present at a depth greater than 4000 m, except at Hild (their table 4). In the Tarbert Formation and in the Ness B unit from the Dunbar Field, which, like Hild, belong to the Greater Alwyn area, there is no kaolinite (Fig. 3). In this case it was proposed that kaolinite completely dissolved to contribute to illite formation (Brosse *et al.*, 2000). The observations made at Hild suggest that kaolinite was abnormally resistant to diagenetic alteration.

Compared with the Brent sediments from the other cited examples, the most salient feature at Hild is that the sandstones are very poor in

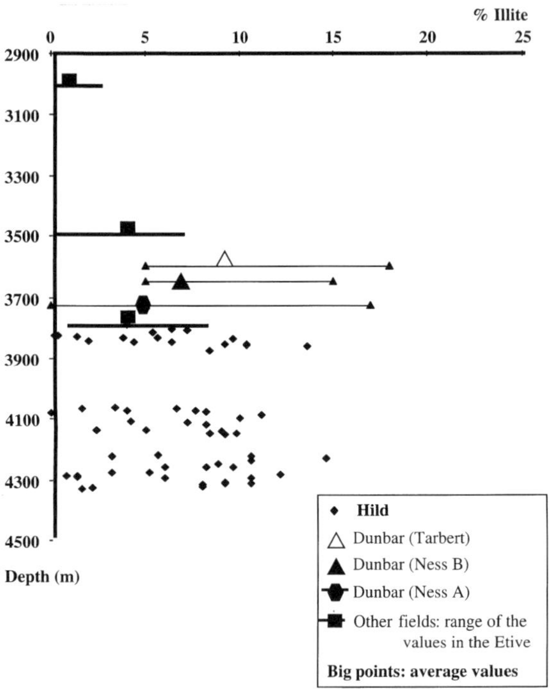

Fig. 5 Distribution of diagenetic illite in the Brent sandstone reservoirs of the Hild Field. Comparison with ranges of values obtained from the Brent sandstones of other fields in the same province (Bjørlykke *et al.*, 1992; Giles *et al.*, 1992; Durand *et al.*, 2000).

feldspar (Fig. 4). Figure 4 is comparable to fig. 3b of Bjørlykke *et al.* (1992). As the Hild reservoirs are deeper than those from other reported fields it remains to infer the initial feldspar amount in the sediments. A hypothesis discussed by Lønøy *et al.* (1986) is that the Brent sandstones at Hild have never contained a significant amount of K-feldspar. If this was true, the provider of K contained in diagenetic illite could only be detrital mica and/or interstitial water transporting elements from other formations. However, the illite percentage is consistent with other data from the Brent reservoirs presented by Giles *et al.* (1992) (their fig. 17) or Giles (1997) (his fig. 15.10). It exhibits a marked increase at depths greater than 3500 m (Fig. 5). The amount of diagenetic illite increases from 1 or 2% at 3000 m to more than 5% on average at 4000 m. At Hild it can reach 15%.

Formation of illite requires, at least, a source of Al, Si and K. Alteration of both K-feldspar and kaolinite at depth are adequate mechanisms to provide these elements. Nevertheless the trends exhibited by K-feldspar and illite correspond to the 3000 to 4000 m depth range, where no significant variation of the kaolinite quantity is observed. It is remarkable that, although decreasing with depth, the proportion of kaolin-group minerals locally remains relatively high at 4300 m in the reservoir sandstones cored from Hild. These observations suggest that illitization was more directly related to K-feldspar dissolution than to kaolinite dissolution.

Chronology and context of the main diagenetic events

Other analyses on the samples from Hild were available for the present modelling study. They included cold-cathodoluminescence microscopy, scanning electron microscopy (SEM; secondary electron mode), X-ray diffractometry (XRD) of both whole-rock and < 2 μm size fractions, fluid inclusion geothermometry, K–Ar datings of illite-rich separated fractions, and analyses of isotopic compositions ($\delta^{18}O$ and $\delta^{13}C$) of both authigenic kaolinite- or illite-rich fractions (using LASER microprobe ablation, ion microprobe, or conventional analyses from separated fractions). Detailed results of these investigations have been reported by Lacharpagne *et al.* (1999) and Girard *et al.* (2001). Only the contributions of these papers to the understanding of kaolinite formation and stability are summarized in the present section. Figure 6 shows the paragenetic sequence inferred from the various analyses mentioned.

Cold-cathodoluminescence microscopy was helpful to reveal the existence of: (i) skeletal, degraded feldspar cores inside kaolinite (or dickite) patches; (ii) two distinct silicification phases; (iii) various carbonate cements.

Two types of kaolin-group minerals, K1 and K2, were recognized. Type K1 presents a vermicular texture and replaces detrital grains, notably mica. Some dissolution of calcite could

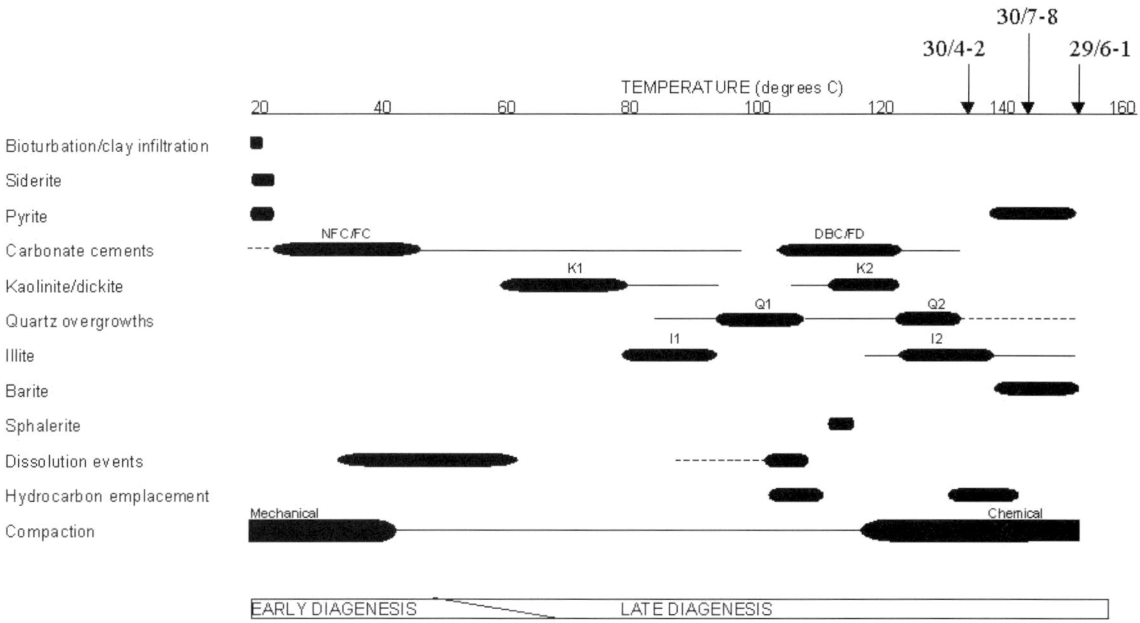

Fig. 6 Paragenetic sequence inferred from petrographic and geochemical analyses (from Lacharpagne *et al.*, 1999): DBC, dark-blue stained carbonate; NFC, non-ferroan calcite; FC, ferroan calcite; FD, ferroan dolomite.

have occurred during K1 formation. Type K2 presents a blocky texture and generally is developed inside secondary porosity. According to morphologies described by Ehrenberg *et al.* (1993) or Cassagnabère (1998) for rigorously defined polytypes, K1 should be kaolinite and K2 dickite. Twelve samples were processed to separate a 5–10 μm fraction containing almost exclusively kaolin. From XRD and infra-red spectrometry of this fraction, proportions of kaolinite and dickite were derived. Dickite is the major constituent. The isotopic signature of the kaolinite polytype, although obtained on a limited number of samples, was better explained by a meteoric than by a marine influence. On the other hand, detailed petrographic observations did not show any direct genetic link from K1 to K2.

Type K2 post-dates a first silicification episode, characterized by a phase (Q1) of quartz overgrowth, seemingly formed (at least in part) at the expense of feldspar, and comparable to the Z0 zone defined by Hogg *et al.* (1992) at Dunbar. Some dissolution of calcite may have occurred during K2 formation. According to fluid inclusions, Q1 formed between 95 and 110°C, whereas a second diagenetic overgrowth of quartz (Q2) began to precipitate between 125 and 135°C and probably continued to form, up to the present-day reservoir conditions (Girard *et al.*, 2001). Unfortunately, it was impossible to quantify separately the amounts of the two phases of quartz overgrowth.

Finally, two types of diagenetic illite were identified and called I1 and I2. The former is a relatively minor constituent. It can coat grains or, alternatively, it can present a platy texture developed mainly at the expense of K1 before any silicification step. The latter presents a pore-bridging, fibrous texture, developed notably as extensions on the edges of I1 crystals. It is coeval with Q2 or slightly post-dates Q2, and was formed at the expense of I1 or detrital clays, but not at the expense of K2, which is neither altered by illitization nor stained by hydrocarbons.

In terms of heritage, chronology and temperature, available constraints derived from petrography and geochemistry are the following (Fig. 6):

- K1 precipitated first, particularly from mica dissolution, at temperatures lower than 95°C, possibly under meteoric influence (isotopic data remain ambiguous here);
- I1, mainly replacing K1, formed also at temperatures lower than 95°C;
- Q1, of minor abundance, precipitated between 95 and 110°C (early Tertiary, according to the subsidence curve);
- K2 precipitated between 105 and 125°C, possibly from an alteration of feldspars (according to petrographical criteria);
- Q2, the major phase of quartz overgrowth, precipitated above 125°C, partly as a consequence of pressure solution (since Eocene–Oligocene);
- I2, derived from the final alteration of K-feldspar, was coeval with Q2 or slightly later; isotopic data indicated a 120 to 140°C range of temperature for its formation, and K–Ar dating suggested that illite mainly developed during, and maybe also after, Eocene times (considering possible contamination by mica-type minerals of the 0.2 μm dated fraction, the 45 Ma age is given by Girard *et al.* (2001) as an upper limit).

The amounts of K1, K2, I1 and I2 are difficult to quantify from thin-section. An estimation was attempted, less precise than those obtained for the whole kaolin-group mineral and illite fractions. In wells 30/7-8 and 29/6-1 the K2/K1 ratio increases with depth and with the water-to-hydrocarbon ratio in the pore space. The I2/I1 ratio is much higher in samples from 29/6-1, the deepest well, than in samples from the two other wells.

Non-Fe calcite is the earliest carbonate cement. It is notably confined to the top reservoir interval of the crestal 30/4-2 well. It pre-dates at least Q1, and, according to the quantities reached (up to 38.6 vol.% of the whole rock), may also pre-date significant compaction. Nevertheless, the isotopic data are not unambiguous enough to attribute its formation to a meteoric or, conversely, a marine influence. The calcite cementation did not preserve a pristine initial mineralogy, including anorthite, as was observed for instance in the neighbouring Ellon Field (Potdevin & Hassouta, 1997; Hassouta, 1999). On the other hand, an important dissolution episode seems to have affected the carbonates during K2 formation. In conclusion, the carbonate cementation episodes were difficult to date in the diagenetic history. Unfortunately, they were not able to provide a clue for better constraining the diagenetic evolution of aluminosilicates.

Composition of mica- and illite-type minerals

The mineral phases chosen to represent a system in the geochemical modelling, determine both the set of elements taken into account in the calculations and the thermodynamic constants that constrain the heterogenous reactions. For most of the detrital phases considered here, namely quartz, K-feldspar, low-T albite, anorthite, calcite and kaolinite, the thermodynamic constants were chosen from those available in the Centre de Géochimie de la Surface (CGS) data base (Fritz, 1981). However, for mica and illite, which can exhibit large variations in elemental composition that induce strong differences in thermodynamic constants, neither the CGS data base nor SUPCRT (Johnson *et al.*, 1992) contain relevant data. In this case the thermodynamic constants were calculated in a way that took the composition of the mineral into account.

Sets of illites and detrital micas were analysed from two wells (29/6-1 and 30/7-8) using SEM and energy dispersive spectrometry (EDS). Both types of mineral contain significant quantities of iron and magnesium (Fig. 7). Relative to illite, Fe and Mg are relatively more abundant in the muscovite grains. Such muscovite-type minerals are called phengite. For both mica and illite, the highest proportions of magnesium and iron were observed in the samples from the well 29/6-1, where the reservoir is the deepest. In Fig. 7 these analyses can be compared with compositions of other samples from the Brent Group reservoirs. For the simulations of the diagenetic evolution at Hild the following compositions were considered.

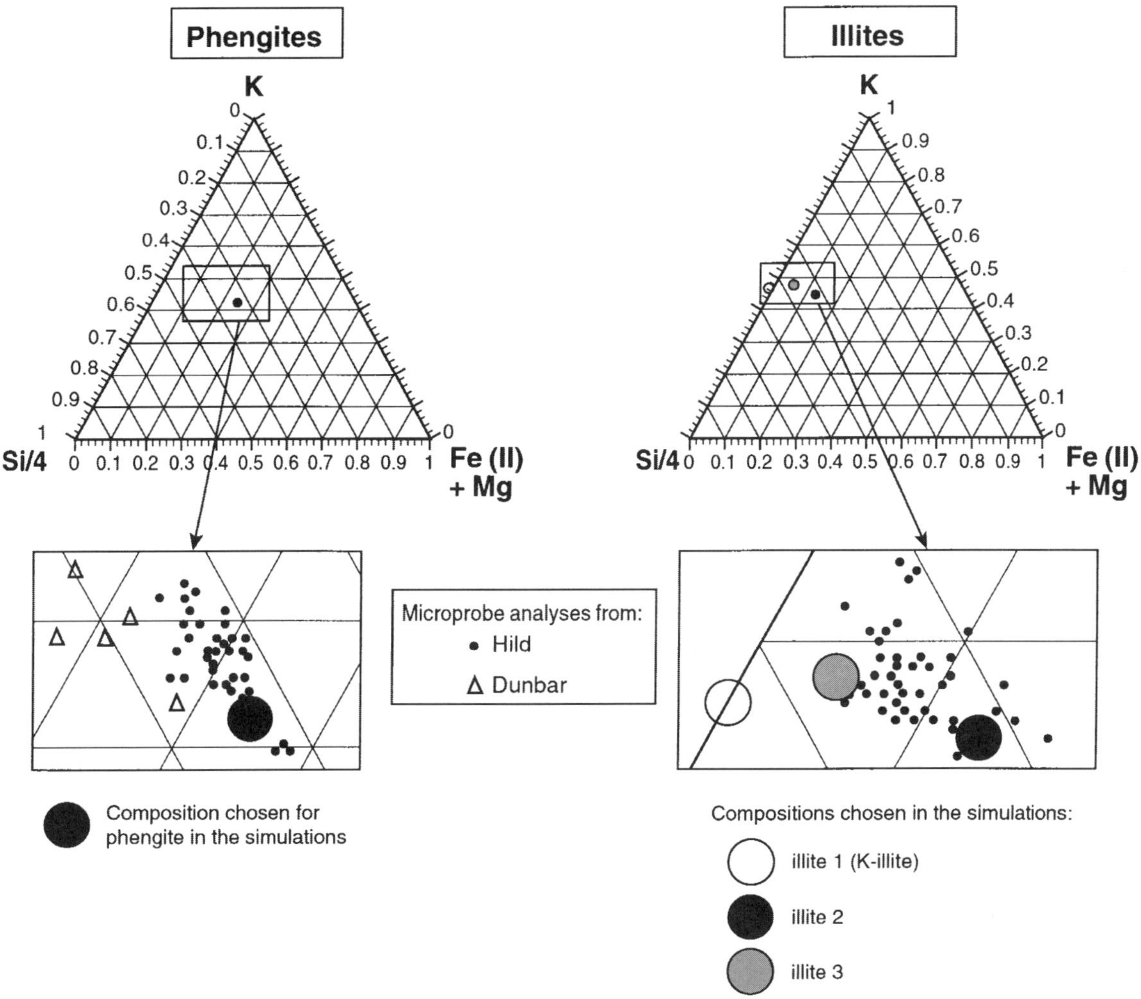

Fig. 7 Composition of the detrital muscovite-type (phengite) and illite-type minerals. Data deduced from local scanning electron microscopy and energy-dispersive spectrometry analyses assuming Fe as Fe(II). Samples are from wells 29/6-1 and 30/7-8 of Hild (J.-P. Girard's analyses), and from the Dunbar field (Durand *et al.*, 2000).

1 For mica, a phengite: $K(Al_{1.54}Fe_{0.345}Mg_{0.245})(Si_{3.2}Al_{0.8})O_{10}(OH)_2$.

2 A reference K-illite (called illite 1) $K_{0.7}(Al_2)(Si_{3.3}Al_{0.7})O_{10}(OH)_2$ and two types of Fe, Mg-illite with distinct proportions of Fe and Mg, namely an illite 2 corresponding to the structural formula $K_{0.85}(Al_{1.665}Fe_{0.108}Mg_{0.149})(Si_{3.293}Al_{0.823})O_{10}(OH)_2$ and an illite 3 with the structural formula $K_{0.81}(Al_{1.93}Fe_{0.05}Mg_{0.05})(Si_{3.2}Al_{0.8})O_{10}(OH)_2$.

The formulae above were calculated as closely as possible to the analytical values in order to respect electrical neutrality. All Fe was assumed to be Fe(II). From the analytical data set the phengite composition was chosen to be relatively rich in Fe and Mg, and the compositions of illite 2 and illite 3, respectively, as being richer and less rich in Fe and Mg. In that way, the influence of the illite composition could be tested through modelling, without changing the composition of phengite. The obligation to make such choices can be considered as a drawback of the modelling approach presented here. It is unavoidable with lower temperature geochemical models that do not include a formalism for the precipitation of solid solutions.

CONSTRUCTION OF THE MODELLING SYSTEMS

Basic principles of the DIAPHORE (diagenetic processes in heterogeneous oil reservoirs) numerical model were explained by Le Gallo *et al.* (1998) and Brosse *et al.* (2000). Because compaction is not taken into consideration, DIAPHORE must be applied to short-term diagenetic episodes during which compaction is not a dominant phenomenon compared with the geochemical reactions. The version of DIAPHORE used for the present study permitted consideration of a geothermal gradient within the system. To account for feedback of diagenetic reactions on reservoir properties, minerals were approximated as spheres of variable size (Le Gallo *et al.*, 1998). The size of the spheres representing the detrital grains increased during precipitation and decreased during dissolution. By contrast the spheres representing the diagenetic cements likely to modify the permeability, e.g. illite, were kept the same, small size, and the number of spheres was increased proportionally to the volume of the mineral formed. A coefficient could be applied to the surface of a specific mineral to account for the difference between the geometric and the reactive surface area. In the present case, an adaptation of the Kozeny–Carman formalism was sufficient to account for the reduction of permeability from 100 mD to the few millidarcys observed in the reservoirs considered.

Hypotheses to be tested for a better understanding of kaolinite formation and stability

A diagenetic scenario was proposed by Lacharpagne *et al.* (1999). The first important diagenetic step was the precipitation of kaolinite *sensu stricto* (K1), from plagioclase dissolution and to some extent from mica dissolution, and the precipitation of non-Fe calcite. The transformation took place either as a result of water mixing related to eustatic changes during depositional times (Fig. 8), or as a result of water flow related to the tilting of structural blocks during the Cimmerian tectonic event (Fig. 9).

The subsequent diagenetic steps were coeval with compaction and progressive closure of the system (Fig. 10). Kaolinite was partly altered to illite-type minerals (at least a part of I1) before temperatures reached 95°C. Then a moderate quantity of quartz overgrowth (Q1) formed. Type K2 (probably dickite) formed between 105 and 125°C, possibly from K-feldspar dissolution. Finally Q2 and I2 formed above 125°C,

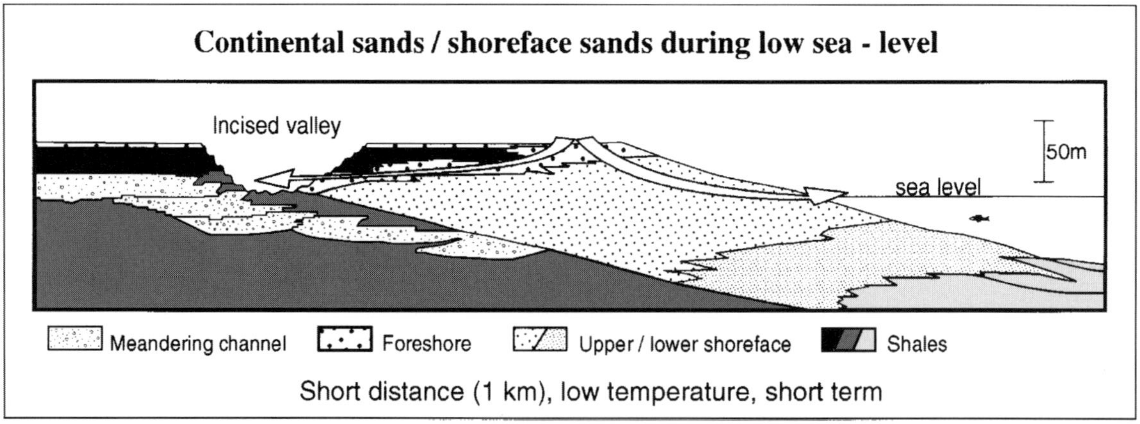

Fig. 8 Hydrodynamic pattern through continental and shoreface sandstones during sea-level lowstand.

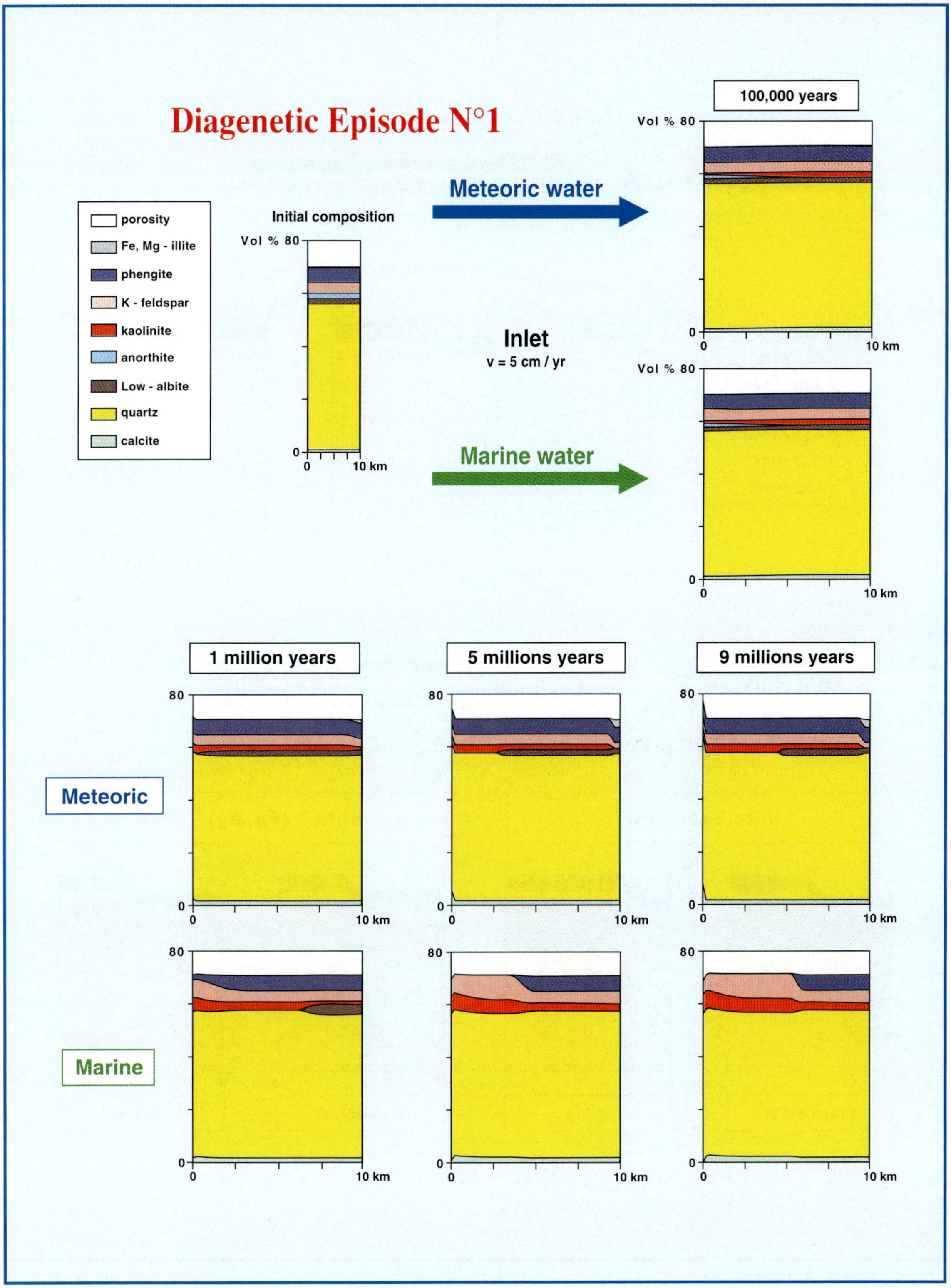

Plate I Diagenetic transformations computed by the DIAPHORE model during episode 1 defined in the text. Comparison between a meteoric and a marine regime.

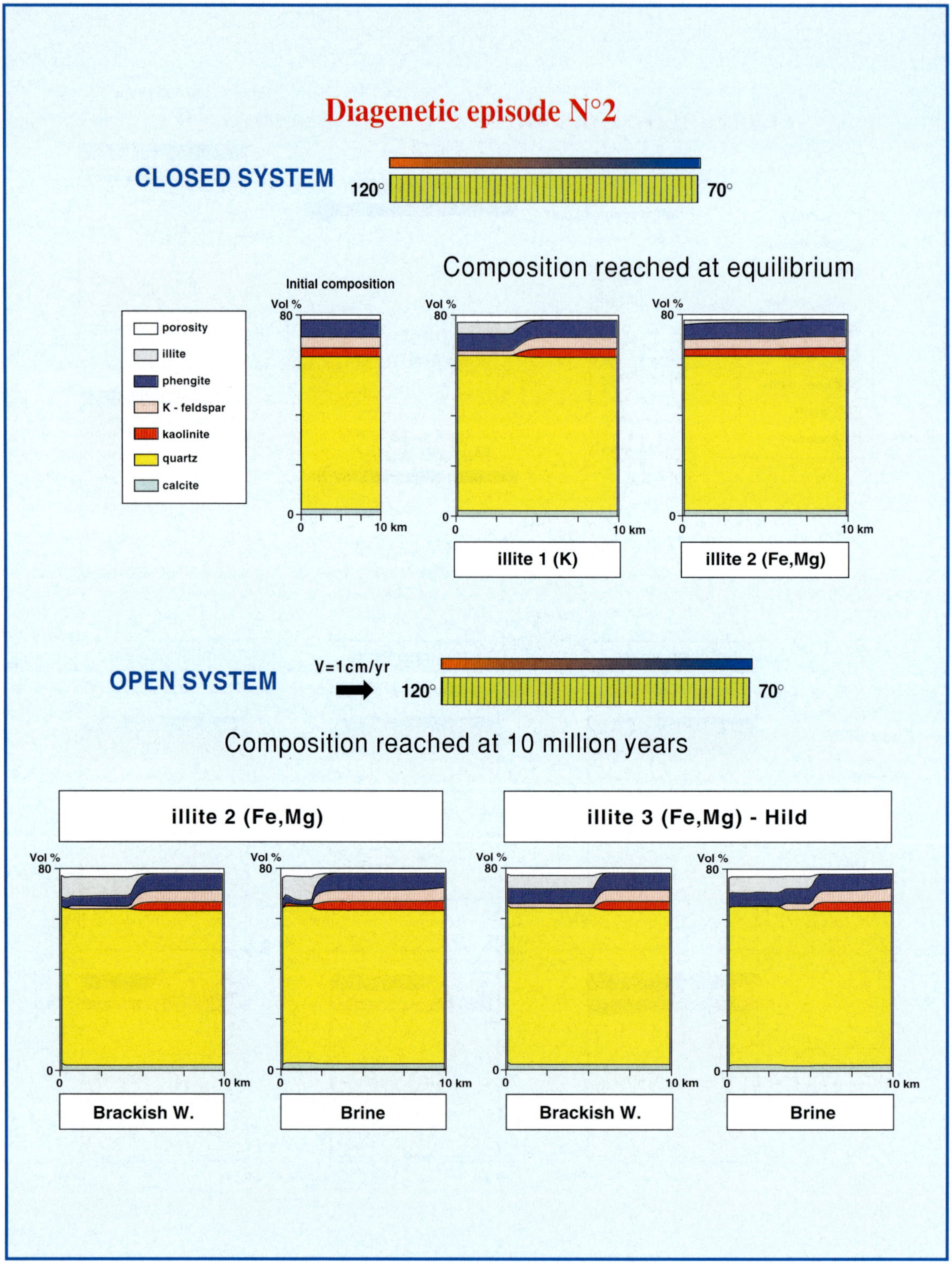

Plate 2 Diagenetic transformations computed by the DIAPHORE model during episode 2 defined in the text. Closed system (upper part) and open system (lower part). Differences between the results were induced by various types of illite and various water compositions.

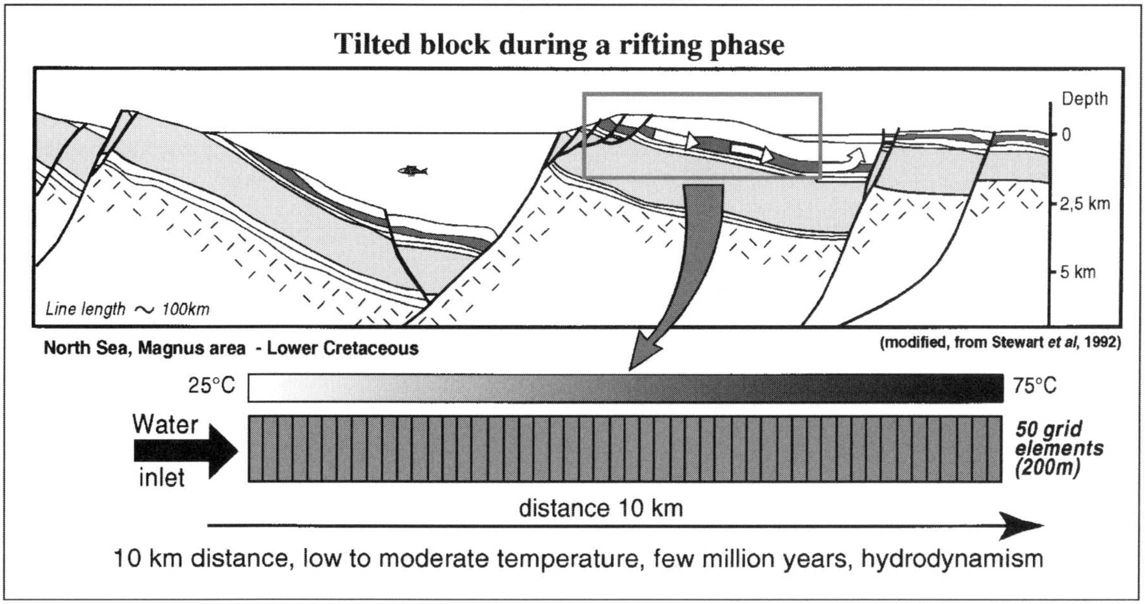

Fig. 9 Geological situation and hydrodynamic pattern induced by the tilting of faulted blocks during a rifting phase. A grid of 50 elements was designed to take into account a 10 km long reservoir system in the Brent Group sandstones.

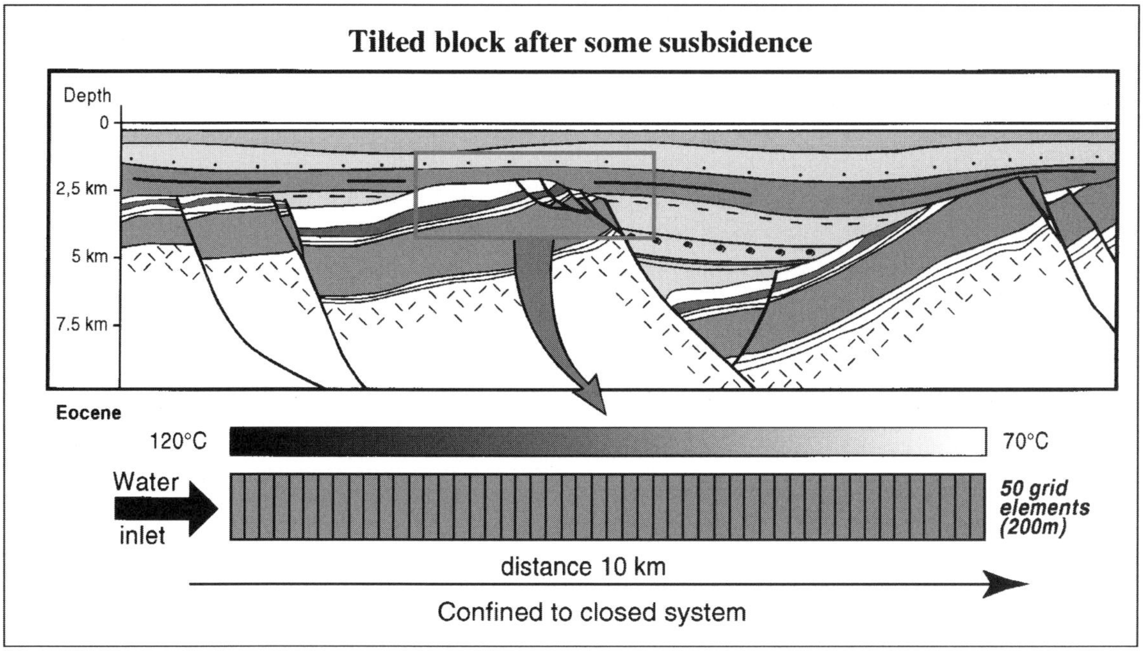

Fig. 10 Geological situation induced by the subsidence of a tilted block. Hydrodynamic regimes progressively decrease in intensity, whereas temperature increases. Compactional waters can flow upwards, from deeper, warmer zones of the basin.

Table 1 Water compositions used in the simulations (total concentrations of the elements). The concentrations are expressed in grams per litre: TIC, total inorganic carbon. The data were provided by the oil company operating at Hild.

	Concentrations ($g\ l^{-1}$)								
	Al	K	Na	Ca	Mg	Si	TIC	Cl	Fe
Meteoric (episode 1)	$2.7\ 10^{-9}$	$7.8\ 10^{-4}$	$4.36\ 10^{-3}$	0.1158	$7.29\ 10^{-3}$	$5.61\ 10^{-3}$	0.06	0.0124	$2.89\ 10^{-3}$
Marine (episode 1)	$2.7\ 10^{-9}$	0.3927	10.6511	0.4064	1.2773	$1.98\ 10^{-3}$	0.06	21.0646	$2.89\ 10^{-3}$
Hild brine (episode 2)	$2.5\ 10^{-7}$	0.2191	22.7	2.825	0.1632	$1.3\ 10^{-3}$	0.183	40.5	$1.3\ 10^{-3}$

preserving kaolin-group minerals at least in part.

During their deposition the Ness-type alluvial sediments contained river water, whereas Tarbert-type lagoonal/deltaic to shoreface sediments contained brackish to marine water. Any sea-level lowstand could eventually renew the interstitial water with fresh water (Fig. 8). During Late Jurassic times and eventually part of Early Cretaceous times a tilted block could offer a sufficient hydraulic head for meteoric water, or even a mixing of meteoric and marine waters, to recharge the sands (Fig. 9). Finally, from Cretaceous times, progressively more basinal-type waters invaded the sandstones (Fig. 10). This geological and hydrodynamic scenario is very similar to the schematic evolution proposed by McAulay *et al.* (1993) from data on the Hutton Field area (see their fig. 10). With 70 g l^{-1} the present-day formation water is more saline at Hild than in most other fields of the North Viking Graben (Warren & Smalley, 1994; Bazin *et al.*, 1997b). An analysis of this water is provided in Table 1. More detail on the composition of Hild formation water is given by Sanjuan *et al.* (this volume, pp. 425–452).

The following questions were tested through geochemical modelling.

1 What is the capacity to form calcite and kaolinite at relatively low temperature? What is the need for hydrodynamism? What is the nature of the water?

2 Is it possible to generate kaolin-group minerals (kaolinite or dickite) up to 125°C, in particular from K-feldspar dissolution, in a flow system much more closed than the preceding low-temperature one, and during carbonate dissolution? What is the temperature of destabilization of these kaolin-group minerals?

Numerical modelling was applied successively to two reservoir systems built from the diagenetic scenario in order to represent the conditions of kaolinite formation (episode 1) and the conditions of illite formation (episode 2). The system represented in the model is a one-dimensional reservoir 10 km long, and homogeneous in terms of mineralogical composition at the initial time of any new diagenetic episode. The water that infiltrates the system is characterized by an imposed composition at the inlet (Table 1) and by a flow-rate value. The system may be subjected to a temperature gradient along the flow axis. During the simulated episode, water is injected continuously at the temperature of the inlet and its temperature progressively changes as it flows throughout the system. At water-flow velocities lower than a few tens of centimetres per year, the convective component of heat transport is negligible compared with the conductive part (Giles, 1997).

First simulated diagenetic episode: kaolinite formation between 25°C and 75°C

A first diagenetic episode was simulated in a situation where the reservoir, included in a tilted block, could admit a water recharge at low temperature (Fig. 9). The geometry of the system was inspired by the reconstruction proposed by Stewart *et al.* (1992) for the Magnus zone during the Cimmerian tectonic event. This assumption is discussed below. As a first hypothesis to be tested, a hydraulic head strong enough to induce a significant water flow in all the system was assumed. It means that the emergent landscape was elevated several hundred metres above sea-level (consider the Suez Rift as a modern analogue, e.g. Patton *et al.*, 1994) and thus the infiltrating water was dominated by meteoric recharge. An opposite alternative to this hypothesis was to consider that the sandstone had been infiltrated by marine-derived water. By testing such different hypotheses, a difference in diagenetic evolution induced by the nature of water could be investigated.

In the system built for the simulations, water enters at 25°C at the surface. The water-flow rate is between 5 cm yr^{-1} and 25 cm yr^{-1} (pore velocity, i.e. Darcy's velocity divided by porosity). A TEMISPACK calculation made in the context of the Mahakam Delta basin (Burrus *et al.*, 1992) showed that such values can be reached easily in an aquifer formation recharged by meteoric water from a hydraulic head of a few hundred metres, provided that the water can escape from the deep part of the tilted block, for instance laterally in another aquifer, or vertically in a fault. Finally, the concentration of dissolved mineral carbon was fixed in order that the pCO_2 value was 2.5×10^{-2} bar at 50°C, a value that conforms to the trend described by Smith & Ehrenberg (1989). The system was subjected to a temperature gradient of 5°C km^{-1}, increasing along the flow axis.

Second simulated diagenetic episode: kaolinite destabilization and illite formation between 70°C and 120°C

A second diagenetic episode was simulated assuming that the top of the tilted block was buried at *c.* 1000 m (Fig. 10). The same temperature gradient, 5°C km^{-1}, was maintained along the 10 km system. Two regimes of water flow were tested: a closed system and an open system infiltrated by basin-derived brine (twice sea-water salinity) (Table 1) at a 1 cm yr^{-1} flow rate. The dissolved total carbon value was fixed in the same way as previously, in order that the pCO_2 value was 0.236 bar at 100°C. The initial composition, assumed to be homogeneous along the system in order to facilitate the discussion, was adapted from the composition reached at the end of episode 1.

The mineralogical and geochemical system considered in the simulations

Detrital feldspar is the main precursor for diagenetic kaolinite and illite. From petrographical criteria a hypothetical reconstitution of the initial average feldspar content was proposed by Lacharpagne *et al.* (1999): 7.5 vol.% (range 1 to 18%), equally distributed between K-feldspar and plagioclase. This assumption was used to define the starting composition of the sediment to be considered in the present modelling approach. This point is discussed below.

The following geochemical system and initial mineral proportions for episode 1 were considered (30% was chosen as an initial porosity value): quartz SiO_2 (55 vol.%); K-feldspar $KAlSi_3O_8$ (4 vol.%); anorthite $CaAl_2Si_2O_8$ (2 vol.%); low-*T* albite (or low-albite) $NaAlSi_3O_8$ (2 vol.%); phengite-type mica $K(Al_{1.54}Fe_{0.345}Mg_{0.245})(Si_{3.2}Al_{0.8}O_{10})(OH)_2$ (5.8 vol.%); kaolinite $Si_2Al_2O_5(OH)_4$ (0.1 vol.%); illite 1, 2 or 3 as defined above (0.1 vol.%); calcite $CaCO_3$ (1 vol.%). Kaolinite and illite were introduced in the system with a trivial volume in order to serve as a nucleus for their precipitation.

A basic philosophy of our modelling approach is that minerals not included in the geochemical system cannot precipitate. In other words, the determination of the geochemical system agrees with observations made on the core samples of the reservoir studied, or on samples from reservoirs considered as equivalent (e.g. for Brent, Ellon or Dunbar in the case of Hild). The minerals in the list above include all the major phases observed from core samples of the Ness and Tarbert reservoirs known in the Hild and Greater Alwyn areas (Lønøy *et al.*, 1986; Jourdan *et al.*, 1987; Hogg *et al.*, 1992; Potdevin & Hassouta, 1997; Lacharpagne *et al.*, 1999; Brosse *et al.*, 2000; Chuhan *et al.*, 2000; Durand *et al.*, 2000). In particular it is remarkable that in the Ellon Field, where samples including zones cemented early by calcite could be studied (Potdevin & Hassouta, 1997), no other major detrital phase could be detected. Finally, one important drawback of the geochemical model used here is that the precipitation of a solid solution with an evolving composition still cannot be represented. This is an important limitation for representing diagenetic clay minerals.

According to the minerals present in the geochemical system, the 11 elements taken in consideration in the calculations were Si, Al, K, Ca, Na, Fe, Mg, C, O, H and Cl. The aqueous species taken into account were the following: H^+, $Al(OH)_4^-$, $Al(OH)_3^0$, $Al(OH)_2^+$, $Al(OH)^{2+}$, Al^{3+}, K^+, KCl^0, Na^+, $NaCl^0$, Ca^{2+}, $CaCO_3^0$, $CaHCO_3^+$, $Ca(OH)^+$, Mg^{2+}, $MgCO_3^0$, $MgHCO_3^+$, $Mg(OH)^+$, Fe^{2+}, $FeCO_3^0$, $FeHCO_3^+$, $Fe(OH)^+$, $H_4SiO_4^0$, $H_3SiO_4^-$, CO_3^{2-}, HCO_3^-, $H_2CO_3^0$, Cl^-, HCl^0, CO_2 and OH^-. As no redox reaction was expected, the Fe(III) species were discarded. No data were available on the dissolved organic carbon. According to the calculations presented by Bazin *et al.* (1997a) and Giles (1997), its concentration would have to reach very high values before shifting the pH value, buffered by aluminosilicates. Berger *et al.* (1995) came to the same conclusion. However, it seems from experimental work that the presence of organic acids has a limited effect (less than half an order of magnitude) on the dissolution rate of K-feldspar (Bevan & Savage, 1991).

Thermodynamic constants and kinetic parameters

The thermodynamic constants for the aqueous species and most of the mineral phases considered are listed in Fritz (1981). The thermodynamic constants to be considered for the muscovite- and illite-type minerals depend on their composition. Sanjuan *et al.* (this volume, pp. 425–452) demonstrated the difficulties encountered in obtaining reliable thermodynamic constants for illites and phengites from the literature. Calculating thermodynamic constants from a theoretical approach, or in contrast, deriving them from experimental results, produces differences in the values.

For the simulations presented here, the thermodynamic constants required were calculated using the method published by Tardy & Garrels (1974). According to Aja (1995), such theoretical values could overestimate the stability of clay minerals, so that the precipitation of illite predicted in the simulations reported below could occur too early compared with natural illites. Although this is a serious limitation, it will be shown that the availability, in the system, of Fe and Mg present in the mineral, is at least an equally important control on illitization.

The so-called polyhedra method estimates the Gibbs free energy of formation, ΔG, of silicates (Tardy & Garrels, 1974). The method of calculation is based on the assumption that a silicate can be represented by a set of oxide and hydroxide components. Tardy & Garrels (1974) provided the ΔG values of oxides and hydroxides engaged in the silicated bonds. For a given crystallochemical environment the Gibbs free energy of formation, ΔG_f, of an oxide or a hydroxide, and particularly the standard one, ΔG_f^0, at 298.15 K and 1 bar, are tabulated (Chermak & Rimstidt, 1989, 1990). According to its type and structural formula, the mineral considered presents a variety of crystallographic sites with specific coordination numbers, each site being occupied by a given cation. A cation is coordinated to hydroxyl groups (–OH) and to oxygen atoms (=O). The proportion of oxides and hydroxides in the polyhedra decomposition represents the proportion of coordinations with, respectively, oxygen and hydroxyl. In addition to the hypotheses explained by Tardy & Garrels (1974), it was assumed that: (i) Fe is only present as Fe(II); and (ii) in the octahedral sites the elements Al, Mg and Fe share the hydroxyl and oxygen coordinations proportionally to their total abundance in these sites. Finally, in order to determine the thermodynamic equilibrium constant for the hydrolysis of the mineral considered, the Gibbs free energy of formation of the corresponding aqueous species had to be included in the calculation,

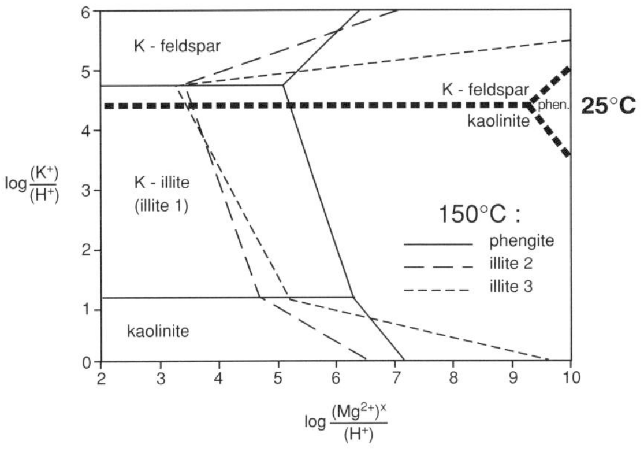

Fig. 11 Combined activity diagram with superposition of stability fields for: (i) K-feldspar, kaolinite and phengite at 25°C (K-illite is not stable at this temperature); (ii) K-feldspar, kaolinite, phengite, K-illite, illite 2 and illite 3 at 150°C. In $\log \frac{(Mg^{2+})^x}{(H^+)}$ the value of x changes with the phase composition: 0.21 for phengite; 0.29 for illite 2; 0.25 for illite 3. No pressure effect.

using the same standard reference state. These values were taken from Jacquot (2000).

When considering the system Si, Al, K, Ca, Na, C, O, H and Cl, and the thermodynamic control only, the thermodynamic data available predict that K-muscovite (which represents illite in this system) forms at low temperature from the reaction of K-feldspar and kaolinite (e.g. Bjørkum & Gjelsvik, 1988; Berger *et al.*, 1995; Brosse *et al.*, 2000). Of course it is never observed from core samples. Conversely, when Fe and Mg are added to the geochemical system and when a phengite composition is considered to represent mica, the domains of mineral-phase stability are modified (Fig. 11), in such a way that the temperature threshold, below which the association of K-feldspar and kaolinite is stable, increases.

For dickite, calculations were made using the experimental data provided by Zotov *et al.* (1998), and discussed below. The thermodynamic constants of the hydrolysis reactions relevant to the presented simulations are reported in Table 2. No pressure correction was applied.

The kinetic rates of mineral dissolution were taken from the literature. They are reported in Table 2. The DIAPHORE numerical model represents the pH dependence on the dissolution rate, in the way explained by Madé *et al.* (1994) or Le Gallo *et al.* (1998). However, the reactions considered here took place in the neutral domain only, in which the pH dependence is considered nil.

Using the microreversibility principle it was assumed that a rate of precipitation is similar to the rate of dissolution, with the opposite sign. We are fully aware that this formalism is not well adapted to represent the precipitation rate of some minerals, studied by laboratory experiments (e.g. Oelkers (1996) for a review). Nevertheless, it has the advantage of preventing numerical problems, not yet solved when assuming more complex formalisms. Furthermore, errors resulting from the poor control of reactive surface areas penalizes kinetic effects even more severely. On the other hand, an oversaturation threshold can be introduced to prevent the precipitation of a mineral as long as its activity product has not reached a given value. This parameter was used for phengite. Because secondary phengite was never detected from core samples, it was assumed that the mineral could not precipitate unless an oversaturation level of 5 (log unit) was reached. Such a hypothesis is the result of the imperfection of the present-day codes and should disappear as soon as geochemical modelling can calculate the evolving composition of solid solutions during their precipitation (see Michau (1997) as a first approach).

Initial geometric surface areas of the minerals were estimated, through the sphere model described above, from the average size of the mineral grains (Table 2), quantified by petrography (Lacharpagne *et al.*, 1999). A coefficient was also roughly estimated from petrography and applied to the geometric surface area to derive the reactive surface area (Table 2). This parameter is poorly constrained.

Diffusion–dispersion

The status of diffusion–dispersion with respect to advection as a governing process in the

Table 2 Data used in the simulations. Thermodynamic data from the Centre de Géochimie de la Surface (CGS) data base (Fritz, 1981) or, for phengite and illites, calculated from the method of Tardy & Garrels (1974). See text for the composition of illites. Kinetic coefficients in the range of neutral pH at 25°C and 120°C, from: [1] Brady & Walther (1990); [2] Helgeson *et al.* (1984); [3] Stumm & Wieland (1990); [4] Knauss & Wolery (1988); [5] Carroll-Webb & Walther (1988); [6] Sjöberg & Rickard (1984).

	Quartz	Anorthite	Low-albite	K-feldspar	Phengite	Kaolinite	Illite 1	Illite 2	Illite 3	Calcite
Initial state (episode 1):										
Proportion (porosity = 0.3)	0.55	0.02	0.02	0.04	0.058	0.001	0.001	0.001	0.001	0.01
Grain size (diameter, mm)	0.12	0.1	0.1	0.1	0.02	0.0005	0.0005	0.0005	0.0005	0.006
Coefficient for reactive surface area	0.08	0.08	0.08	0.08	0.02	0.02	0.06	0.06	0.06	$2\ 10^{-6}$
Molar volume ($cm^3\ mol^{-1}$)	22.69	100.07	100.07	108.72	146.90	99.52	133.40	133.40	133.40	36.93
Thermodynamics:										
25°C log K (for hydrolysis)	−3.999	−19.488	−20.178	−23.192	−35.692	−39.14	−51.238	−44.777	−49.705	−8.37
150°C log K (for hydrolysis)	−2.718	−16.081	−14.076	−15.715	−28.614	−28.164	−38.936	−34.467	−38.105	−10.25
Reference	CGS	CGS	CGS	CGS	Calculated	CGS	Calculated	Calculated	Calculated	CGS
Kinetics:										
25°C rate ($mol\ m^{-2}\ yr^{-1}$)	$6.74\ 10^{-7}$	$1.05\ 10^{-5}$	$1.05\ 10^{-5}$	$4.18\ 10^{-5}$	$1.21\ 10^{-6}$	$1.57\ 10^{-6}$	$1.21\ 10^{-6}$	$1.21\ 10^{-6}$	$1.21\ 10^{-6}$	$1.396\ 10^{3}$
120°C rate ($mol\ m^{-2}\ yr^{-1}$)	$9.16\ 10^{-5}$	$2.18\ 10^{-4}$	$2.18\ 10^{-4}$	$8.69\ 10^{-4}$	$6.23\ 10^{-6}$	$7.99\ 10^{-6}$	$6.23\ 10^{-6}$	$6.23\ 10^{-6}$	$6.23\ 10^{-6}$	$5.454\ 10^{3}$
Reference	[1]		[2]	[2]	[3],[4]	[3],[5]				[6]

transfer of chemical elements can be approached by the Peclet number P_e, which refers to a distance, L (Oelkers, 1996). If L is the scale of interest for the problem investigated, e.g. 200 m in the present case (size of a grid element in the modelled system, and relevant distance for the characterization of a reservoir), it is possible to define P_e from a variable change in the (simplified) conservation equation of an element:

$$P_e = \frac{uL}{D}$$

where u is the Darcy velocity (m s^{-1}) and D is the diffusion–dispersion coefficient (m^2 s^{-1}). It can be shown that when $P_e >> 1$ the diffusion–dispersion mechanism has a negligible effect on the element's transfer compared with advection. An approximate calculation was made involving a 200 m value for L, a 1 cm yr^{-1} value for u and a 10^{-5} cm^2 s^{-1} value for D, that resulted in a P value of 63. As this calculation was very approximate, a DIAPHORE calculation was also made involving diffusion–dispersion at a flow rate of 5 cm yr^{-1}. There was no detectable difference with the case of pure advection. Accordingly, diffusion was neglected. For the simulations in closed systems it was justified by the very small difference of water composition between the different parts of the reservoir. Possible exchanges between shale and sandstone were not investigated.

MODELLING RESULTS

The results of a simulation are expressed as the composition of rock and water versus distance. In computer simulations, water–rock interaction in a system open to water flow often results in the formation of mineral zones characterized by a mineralogical composition and separated by reaction fronts. In nature, such fronts are known, e.g. dolomitization (up to 100% of the minerals are transformed) and redox reactions (sometimes very easy to observe from colour change). In subarkosic sandstones, because of the nature of the minerals involved (feldspars, micas, diagenetic clay minerals), because of their proportion in the rock, and owing to the possible overprinting of detrital and diagenetic heterogeneities, such fronts, if they exist, should be much less easy to detect.

When interpreting the mineralogical composition computed at a given time somewhere from the inlet, the characteristics of the corresponding zone and of the simulation must be integrated: initial composition of the rock, fixed temperature range, fixed geometry and infiltration of the sandstone by water evolved throughout the upstream part of the system.

Episode 1, kaolinite formation

In Plate 1 (facing p. 392) the mineralogical compositions computed in the presence of meteoric or, conversely, of a marine-derived interstitial water can be compared. Results are presented for 100 000 yr, 1 Myr, 5 Myr and 9 Myr. For both meteoric and marine hypotheses the dissolution of anorthite is a relatively short-term event, achieved in approximately 100 000 yr. For kinetic and thermodynamic reasons its efficiency increases with temperature. The dissolution of anorthite delivers solutes to the system, which allow the formation of calcite and kaolinite. There is no influence of the fluid flow on the dissolution of anorthite. In a closed system, the minerals are at equilibrium after anorthite dissolution.

If the water is mobile the dissolution of anorthite is followed by the dissolution of low-T albite and by concomitant formation of additional kaolinite. A reaction front develops and crosses the system from the inlet to the outlet. The new reaction is much slower than the dissolution of anorthite. The process is not controlled by a difference of kinetics (Table 2) but by the advective water flow. In the meteoric case the reaction front after 9 Myr has only affected a 5000 m distance from the inlet, i.e. the front velocity is *c.* 0.6 mm yr^{-1}. It is 12 times faster in the presence of the marine-derived water. This result seems paradoxical. We checked that, in the part of the system where albite is still present (water there has a different composition from water infiltrating the rock at the inlet), both marine-derived and meteoric-derived aqueous solutions are undersaturated with respect to low-T albite. We attribute the higher efficiency of dissolution, in the case of marine-derived water, to the chlorinity value, which allows a higher dissolved load to be transported. Although thermodynamics promotes low-T albite stability as temperature increases (in contrast with what occurs for anorthite), neither marine nor meteoric influence can stabilize albite in the long term. However, the effect of alkalinity could be important. It was beyond the scope of the present work to explore albite stability, but other modelling tests showed that the stability and the dissolution rate of albite in the warmest part of the reservoir considered here (75°C) are extremely sensitive to the combination of water composition and water-flow velocity.

Finally, after more than 10 Ma at a 5 cm yr^{-1} flow rate, or in the presence of a more active water flow (25 cm yr^{-1} was tested), the diagenetic effects of the marine and meteoric regimes begin to diverge in a more radical way. With a meteoric influx, the assemblage of aluminosilicates (still in place after the complete dissolution of albite) is stable. Nevertheless, in the long term a carbonate cement can develop close to the inlet and may occlude the pore space. With a marine influx, phengite completely dissolves, whereas K-feldspar and additional kaolinite precipitate (see Plate 1, facing p. 392).

Episode 2, kaolinite destabilization and illite formation

The initial mineralogical composition for the simulation of the second diagenetic episode was chosen to be homogeneous in the 10-km-long reservoir. We decided to start from the composition reached just after low-albite dissolution, during episode 1. An initial porosity value of *c.* 20% was assumed.

When the system is closed, the mineralogical composition reaches equilibrium before 1 Ma (see Plate 2, facing p. 392, upper part). Some illite forms in the warmest part of the reservoir (120°C, to the left), nevertheless, in the reaction that occurs, the temperature of incipient illitization, and the amount of illite formed, differ with the type of illite considered. For illite 1 (K-illite) the process is a strong destabilization of the kaolinite/K-feldspar assemblage, whereas phengite remains stable. This transformation has been discussed in detail by Brosse *et al.* (2000). In the present case it occurs at *c.* 100°C and is achieved as soon as one of the two minerals, kaolinite or K-feldspar according to their initial abundance, has disappeared. The quantity of illite formed depends on the quantity of the (K-feldspar + kaolinite) precursor assemblage. In contrast, for an Fe, Mg-illite the process is destabilization of the three aluminosilicates initially present in the system, namely kaolinite, K-feldspar and phengite. It occurs above 90°C, but the amount of illite formed is very limited because in this case the geochemical system, controlled by Fe and Mg exchanges, can reach an equilibrium without requiring the complete dissolution of a mineral. However, at a higher temperature, the preceding assemblage of minerals is destabilized. A closed-system simulation was made in a 150 to 100°C temperature range in the presence of the Hild brine. It revealed the existence of a temperature threshold between 120 and 130°C, above which the precipitation of Fe, Mg-illite and the dissolution of kaolinite, K-feldspar and phengite are more active. After 6 Myr in the warmest part of the system (up to 150°C in this simulation), the proportion of additional illite reaches 1.5%, whereas the amounts of dissolved kaolinite, K-feldspar and phengite reach *c.* 0.5% for each mineral. It is likely that this reaction ends only when one of the three minerals is exhausted. Moreover, the presence of K-feldspar at this step is a prerequisite of the illitization process. A simulation made in the same conditions (closed system, brine, 150°C–100°C) showed that the assemblage of phengite and kaolinite is stable and cannot lead to illitization.

When the system is open to water circulation (1 cm yr^{-1}) the simulated diagenetic evolution divides the reservoir into three zones separated by reaction fronts (see Plate 2, facing p. 392, lower part, situation after 10 Myr):

1 the shallowest, coldest zone (70°C at the outlet), where the initial composition is preserved in the long term (10 Myr), even if very limited dissolution of kaolinite and K-feldspar is observed in the presence of saline water (Hild brine) to the benefit of phengite precipitation;
2 the intermediate zone, where one mineral, kaolinite or, alternatively, K-feldspar, is completely dissolved and where some illite and a minor amount of quartz are formed;
3 the deepest, warmest zone (120°C at the inlet), where residual K-feldspar (if any) is dissolving and where additional illite, quartz and calcite form (in Plate 2 this zone is generally very narrow, and it appears only in the deepest part of the system).

The position of the reaction front with respect to the thermal gradient and to the distance in the flow direction depends not only on the temperature and water-flow velocity but also on the nature of the illite involved, and on the water composition. Illite 2, the richest in Fe and Mg, and the closest to phengite in terms of composition (Fig. 7), corresponds to dissolution of phengite that occurs at a higher temperature (95 to 110°C, depending on water salinity) relative to illite 3 (90°C). On the other hand, a brackish water better preserves K-feldspar than the brine derived from the present Hild water. The best preservation of kaolinite occurs with illite 2, in the presence of brine. Nevertheless, in none of the cases examined would it be possible to preserve large quantities of kaolinite, except if larger quantities than those considered in the simulations were present initially.

DISCUSSION

Although significantly improved by recent progress, computer-aided geochemical modelling still cannot pretend to provide an accurate replication of all the processes that occur during diagenetic reactions. It should be a tool for the petrographer and the geochemist, useful to test the effect of parameters as important as temperature, nature and flow rate of the interstitial water, or mineralogical composition. At least some of the thermodynamic and kinetic data must be considered with great care. The polyhedra method that was used here to fix values of log K for illite, was not specifically validated for this mineral. The microreversibility principle that is often considered for deducing kinetic rates of precipitation from rates experimentally driven for dissolution, is pragmatic but was proven unable to account for the precipitation of several minerals addressed in experimental work (e.g. Schott & Oelkers, 1995; Oelkers, 1996; Giles, 1997). The appraisal of reactive surface areas is such a difficult problem that this parameter could be presented as an empirical way to calibrate the model (Holstad, 1994). On the other hand, a modelling tool such as DIAPHORE, because it does not integrate compaction, requires the truncation of geological history in short steps (few million years) so that the impact of the whole diagenetic evolution is difficult to evaluate.

In spite of these limitations, the preceding results managed to capture some major diagenetic transformations that have been observed and analysed from core samples. They are discussed now with respect to these data. Then two particular aspects are discussed with reference to the literature available on the Brent province: (i) the initial composition of the sand; and (ii) the character of the interstitial water (composition and flow rate).

Comparison of the modelling results and the data obtained from the Hild Field

The simulations presented above were based on an average value of the mineralogical composition analysed in the Ness and Tarbert formations of the Hild Field. They showed several successive transformations.

1 Starting with 4% K-feldspar and 4% plagioclase (half anorthite, half low-T albite), it was possible to form *c.* 3% kaolinite before 75°C only from the alteration of the bulk of the plagioclase, an alteration that was observed in the core samples. This kaolinite is equivalent to the kaolinite K1 described by Lacharpagne *et al.* (1999). As up to 9% plagioclase was possible initially (Lacharpagne *et al.*, 1999), up to *c.* 7% diagenetic kaolinite could be formed in this way. It is less than the extreme values (12 to 16%) sometimes observed (see above). At this stage, the sand has kept its original content of muscovite (considered as 6% in the simulations) and K-feldspar. Under the influence of meteoric, fresh water and relatively moderate hydrodynamism, this situation could remain stable in the long term.

2 At a higher temperature, muscovite and K-feldspar could produce illite in a two-step process. Between 90 and 100°C in a closed system, a small amount of illite, equivalent to the illite I1 described by Lacharpagne *et al.* (1999), formed from the partial dissolution of muscovite, K-feldspar and kaolinite. In an open system the transformation could be more rapid, leading to the dissolution of the less abundant mineral in the K-feldspar/kaolinite assemblage. Between 120 and 130°C, in a closed system and in the presence of the Hild brine, a larger amount of illite was slowly produced, from the progressive dissolution (not completed at 150°C) of the same three residual aluminosilicates. Up to 150°C, which is the maximum present-day temperature at Hild, the presence of K-feldspar controlled the formation of illite. Compared with the average values obtained for the 94 samples in the data set, these results explain, in any of the tested conditions, that illitization could induce the total dissolution of K-feldspar. Only if the system was closed (or very confined) at depths greater than those approached when temperature reached *c.* 90°C, and if the K-feldspar quantity at this step was smaller than the quantity of kaolinite, the modelling results explain also the assemblage of aluminosilicates observed today, i.e. 2.8% kaolinite, 5.2% illite and no K-feldspar on average. The model reproduced also a two-step illitization process that can explain the two phases of illite, I1 and I2, observed by Lacharpagne *et al.* (1999). According to these figures, and assuming that the meteoric influence was dominant during the first part of diagenesis, the initial K-feldspar proportion would have to be in the range 2–2.5% and the initial plagioclase proportion in the range 7–7.5% (average values, proportions in the whole mineral fraction excluding porosity).

3 A marine influence during the first part of diagenesis would allow a plagioclase value slightly lower than the value mentioned above, because kaolinite could be produced also from a partial dissolution of muscovite. However, this would not be sufficient to invert the ratio required between the two types of feldspar (K-feldspar/plagioclase < 1).

The simulations dedicated to episode 1 showed that kaolinite formation, between 25 and 75°C, derived mainly from the dissolution of plagioclase. However, the dissolution of Na-rich plagioclase occurred only if the interstitial water was part of a flowing hydrodynamic regime. In that case, both marine and meteoric conditions allow albite dissolution. In the case of seawater flow, diagenetic kaolinite also formed as a result of the dissolution of phengite, accompanied by the precipitation of some K-feldspar.

An additional modelling test (not illustrated here) demonstrated that a strong meteoric-derived water-flow regime (25 cm yr^{-1}), if maintained over a long enough period of time (> 2 Myr), resulted in massive calcite precipitation near the system inlet. Such a pattern can be compared, at Hild, with the great abundance of non-Fe calcite localized in the upper part of the reservoir, in the crestal well 30/4-2. Nevertheless, sensitivity of this diagenetic behaviour to the value of pCO_2, a poorly constrained parameter, was not examined. On the other hand, there is no indisputable argument to infer the nature, meteoric or marine, of the water recharge. Isotopic arguments are more in favour of a meteoric influence. According to the modelling results the local presence of calcite cement argues slightly more in favour of a freshwater-flow regime. The partial replacement of muscovite by kaolinite observed by petrographers (Lacharpagne *et al.*, 1999), however, is better explained by marine-derived water.

In the 70–120°C range of temperature it was impossible in the various simulation tests to obtain both precipitation of kaolinite and dissolution of K-feldspar. This is the only major discrepancy between the scenario proposed by petrographers (Fig. 6) and the indications provided by modelling. Compared with kaolinite, the stability field of dickite is slightly shifted towards higher temperature, nevertheless, according to the thermodynamic data published by Zotov *et al.* (1998), the shift would be too small to explain dickite formation at this stage. Petrographic observations indicated also that calcite could be dissolved during precipitation of the kaolin-group mineral K2 (probably dickite). This effect was not reproduced by simulation. Variable pH conditions were not tested here, but could be a way to achieve such a result.

Another discrepancy between the model used and the observations concerns quartz cementation. A common feature of all the simulated mineral transformations is that they release silica. The modelled amount of quartz formed, in the 1–2 vol.% range for each major diagenetic episode, however, does not account in total for the observed 8.5%. As discussed by Brosse *et al.* (2000) the main reason for this discrepancy is probably that the model does not include pressure solution, a major agent of quartz cementation in sandstones at depths greater than *c.* 3000 m (e.g. Walderhaug, 1994).

Initial mineralogical composition

Dissolution of both feldspar and mica frequently has been mentioned as the main cause of kaolinite formation in the Brent reservoirs (e.g. Bjørlykke & Brendsdal, 1986; Kantorowicz, 1990; Ehrenberg *et al.*, 1993; Emery & Robinson, 1993). Nevertheless, it has been very difficult to quantify from petrography the contribution of each type of mineral to the resulting amount of kaolinite. Girard (1995) observed, in Middle Jurassic sandstone reservoirs from the Oseberg field, that kaolinite is related to mica alteration, but not to K-feldspar alteration. The stability of K-feldspar with respect to kaolinitization was also noted by Bjørlykke & Brendsdal (1986), in the case of the Brent sandstones from the Brent and the Statfjord fields. The initial average K-feldspar proportion considered in the present work (4%) is lower than the minimum average proportion suggested by Bjørlykke *et al.* (1992) for the Brent province (10%), and much lower than in some other fields (Glasmann, 1992; Harris, 1992). The Brent sandstones of the Hild field, deposited in the south of the Brent province, represent the more continental facies.

The low K-feldspar/plagioclase ratio reconstructed from the modelling results is unusual for sandstones (Pettijohn *et al.*, 1972). Bjørlykke & Brendsdal (1986), Harris (1989), Bjørlykke *et al.* (1992) and Chuhan *et al.* (2000) suggested that the ratio between K-feldspar and plagioclase varies to a large extent in the Brent sandstones, from one formation to the other, and from one area to the other. The variation is not related in a simple way to sedimentary facies (e.g. Giles *et al.*, 1992). Unfortunately, this relationship has not been explored in a systematic way by sedimentologists. The parity assumed between the initial amounts of K-feldspar and plagioclase does not seem valid for the Hild field. Chuhan *et al.* (2000) have remarked on the specific composition of the reservoirs at Hild, and have suggested that initially they were poor in feldspar. Our modelling results are more in favour of the alternative hypothesis formulated by Lønøy *et al.* (1986), that plagioclase could have been more abundant than K-feldspar in the Brent sandstones of this particular field. A second alternative hypothesis, not tested in the modelling approach, is that a given amount of detrital biotite would compensate for a lack of plagioclase and would be altered at the same point of the kaolinitization process. Only 0.05% (0 to 0.8%) biotite was detected in thin-section, and the observations are not sufficient to derive an initial content.

A way to approach the initial mineralogy is provided by early cemented carbonate zones. Hassouta (1999) quantified by point counting (1000 points) the proportions of minerals inside and outside calcite-cemented zones of the Tarbert Formation cored from the Ellon field (Greater Alwyn area) (see also Potdevin &

Hassouta, 1997). According to its volumetric proportion and isotopic signature, calcite was interpreted as a cement formed before burial depth reached *c.* 1000 m. The mineralogical composition preserved in the cemented zones of the sandstone showed that 3% kaolinite was present, whereas plagioclase amount was lower than 0.5%, with evidence of dissolution. Even assuming that kaolinite was derived only from plagioclase, the initial abundance of plagioclase in this case is clearly lower than that of K-feldspar, which disagrees with the conclusion derived above for Hild.

Composition and rate of water flow

The simulations showed that a meteoric character is not required for the water-flow regime to form kaolinite at low temperature. The isotopic composition of diagenetic kaolinite often has been used to investigate the origin of the water present during the precipitation of kaolinite. Whereas the isotopic signature is meteoric at Ninian (Emery & Robinson, 1993) and Oseberg (Girard, 1995), it is marine at Lyell (Emery & Robinson, 1993), Magnus (Fallick *et al.*, 1993) and Dunbar (Hogg *et al.*, 1995).

According to stratigraphic data only the upper part of the Tarbert Formation was eroded from the crest of the tilted block located at the top of the structure (well 30/4-2) (Rønning *et al.*, 1986). This feature, comparable with what has been described from Ninian (Van Vessem & Gan, 1991), suggests that emersion occurred (see also Lee & Hwang, 1993). The existence of a perennial elevated topography comparable to what is observed today in the Rift of Suez, would have produced a high degree of erosion. Such a situation has been described in the Jurassic context of the North Viking Graben, e.g. at Magnus (Stewart *et al.*, 1992) and Snorre (Dahl & Solli, 1993) (see also Yielding & Roberts, 1992). It does not seem highly realistic here. Accordingly, we think that meteoric-derived hydrodynamic flow would have been of limited extent, or duration, at Hild.

Another result of geochemical modelling concerns the link between feldspar stability and hydrodynamism. Water flow is required for obtaining the alteration of albite. As detrital albite is rare or absent in the Brent sandstones at Hild, and as plagioclase abundance is required by the formation of kaolinite in place, it can be inferred that these sediments systematically suffered hydrodynamic regimes at the beginning of the diagenetic evolution.

A third point of discussion is K-feldspar overgrowth. At low temperature (< 75°C), K-feldspar overgrowth is possible only when the water flow has a marine signature. The presence of K-feldspar overgrowth frequently has been observed in the Brent and other Middle Jurassic sandstones of the North Sea: from the Brent and Statfjord fields (Bjørlykke & Brendsdal, 1986; Harris, 1992), the Magnus Field (Emery *et al.*, 1993), the Oseberg Field (Girard, 1995) and the Ellon field (Potdevin & Hassouta, 1997; Hassouta, 1999).

Finally, the flow-rate values are also a question for debate. The simulations showed that if very active hydraulic flow (of several tens of centimetres, or even several metres, per year) is maintained for long periods of time, very strong diagenetic alterations are induced. As the complete dissolution of muscovite was not observed in the core samples examined, it can be argued that such hydrodynamism was not possible, or alternatively that the rates of the processes are too high in the model.

CONCLUSIONS

Geochemical modelling was used to address the problem of kaolinite formation and stability in sandstones of the Brent Group. The results were compared with analytical data acquired from core samples from the Hild Field. The analytical and the modelling results were similar for the formation of both vermicular kaolinite and illite. Below 75°C the kaolinite could easily be derived from the alteration of, at least, Ca-rich plagioclase. However, fluid movement was required to dissolve Na-rich plagioclase. At higher temperatures, a two-step destabilization of kaolinite and illitization, well constrained by the data base, was reproduced by the model. In contrast, two major differences between the model and the data concerned the blocky kaolin-group mineral (probably dickite) and the total amount of quartz. The stability of kaolin-group minerals at depth is lower in the simulations than it seems to be in nature. It was not possible with the assumed modelling hypotheses to simulate the precipitation of kaolinite or dickite in the 105–125°C range of temperature. Conversely, geochemical modelling showed that kaolin-group minerals are very prone to dissolution above a 90 to 100°C threshold, and that factors favourable to the stability of kaolin are: (i) a completely closed system; (ii) a marked ferro-magnesian character of illite; and (iii) the presence in the pore space of a brine sufficiently concentrated in salts, i.e. of composition close to the present-day Brent formation water at Hild. Moreover, the interpretation of the modelling results showed that the Ness and Tarbert sandstones of the Hild area were probably characterized by a specific composition, particularly rich in plagioclase (or perhaps biotite) and relatively poor in K-feldspar. Nevertheless, modelling showed that K-feldspar is a prerequisite precursor of illite formation. The particular composition of the sands at Hild, i.e. plagioclase more abundant than K-feldspar, could explain why K-feldspar disappeared, whereas kaolin-group minerals are still present.

The numerical modelling approach still cannot pretend to be an exact reconstitution of the diagenetic evolution. Nevertheless it allows the testing of hypotheses formulated from the petrographic and geochemical data, and accordingly might be considered as a tool to monitor the interpretation of the data. Many other hypotheses suggested by the data have not been tested yet, for instance the transformation of biotite into kaolinite, or the impact of pH variations on the formation of late kaolinite, and of pCO_2 variations on the formation of carbonate cements.

ACKNOWLEDGEMENTS

The application of DIAPHORE to the Hild case study was partly funded by the Thermie Programme of the EEC (*Diamap* Project, Contract N. OG/233/94-FR/UK, 1997–1999). E. Jacquot (CGS, Strasbourg University) helped to achieve several thermodynamic calculations. The corrections, remarks and questions formulated by R. Worden, and the reviews made by G. Berger and S. Haszeldine, greatly improved the manuscript.

REFERENCES

Aja, S.U. (1995) Thermodynamic properties of some 2 : 1 layer clay minerals from solution-equilibration data. *Eur. J. Mineral.*, **7**, 325–333.

Bazin, B., Brosse, E. & Sommer, F. (1997a) Chemistry of oil-field brines in relation to diagenesis of reservoirs. 1. Use of mineral stability fields to reconstruct *in situ* water composition. Example of the Mahakam basin. *Mar. Petrol. Geol.*, **14**(5), 481–495.

Bazin, B., Brosse, E. & Sommer, F. (1997b) Chemistry of oil-field brines in relation to diagenesis of reservoirs. 2. Reconstruction of palaeo-water composition for modelling illite diagenesis in the Greater Alwyn area (North Sea). *Mar. Petrol. Geol.*, **14**(5), 497–511.

Beaufort, D., Cassagnabère, A., Petit, S., *et al.* (1998) Kaolinite-to-dickite reaction in sandstone reservoirs. *Clay Mineral.*, **33**, 297–316.

Berger, G., Lacharpagne, J.C., Velde, B., Beaufort, D. & Lanson, B. (1995) Mécanisme et contraintes cinétiques des réactions d'illitisation d'argiles sédimentaires, déduits de modélisations d'interactions eau-roche. *Bull. Centres Rech. Prod. Elf-Aquitaine*, **19**, 225–234.

Bevan, J. & Savage, D. (1991) The effect of organic acids on the dissolution of K-feldspar under conditions relevant to burial diagenesis. *Mineral. Mag.*, **53**, 415–425.

Bjørkum, P.A. & Gjelsvik, N. (1988) An isochemical model for formation of authigenic kaolinite, K-feldspar and illite in sediments. *J. Sed. Petrol.*, **58**(3), 506–511.

Bjørlykke, K. & Aagaard, P. (1992) Clay minerals in North Sea sandstones. In: *Diagenesis and Petrophysics of Clay Minerals in Sandstones*, Society of Economic Paleontologists and Mineralogists Special Publication 47 (Ed. by D.W. Houseknecht and D.W. Pittman), Tulsa, OK, pp. 65–80.

Bjørlykke, K. & Brendsdal, A. (1986) Diagenesis of the Brent sandstone in the Statfjord field, North Sea. In: *Role of Organic Matter in Sediment Diagenesis*, Society of Economic Paleontologists and Mineralogists Special

Publication 38 (Ed. by D.L. Gautier), Tulsa, OK, pp. 157–166.

Bjørlykke, K., Nedkvitne, T., Ramm, M. & Saigal, G.C. (1992) Diagenetic processes in the Brent group. In: *Geology of the Brent Group* (Ed. by A.C. Morton, R.S. Haszeldine, M.R. Giles and S. Brown). *Geol. Soc. London Spec. Publ.*, **61**, 263–288.

Brady, P.V. & Walther, J.V. (1990) Kinetics of quartz dissolution at low temperatures. *Chem. Geol.*, **82**, 253–264.

Brosse, E., Matthews, J., Bazin, B., Le Gallo, Y. & Bildstein, O. (2000) Related quartz and illite cementation in the Brent sandstones: a modelling approach. In: *Quartz Cementation in Sandstones*, International Association of Sedimentologists Special Publication 29 (Ed. by R.H. Worden and S. Morad), Blackwell Science, Oxford, pp. 51–66.

Buhrig, C. (1989) Geopressured Jurassic reservoirs in the Viking Graben: modelling and geochemical significance. *Mar. Petrol. Geol.*, **6**, 31–48.

Burrus, J., Brosse, E., Choppin de Janvry, G., Grosjean, Y. & Oudin, J.-L. (1992) Basin modelling in the Mahakam delta based on the integrated 2D model TEMISPACK. In: *21st Annual Convention Indonesian Petroleum Association Proceedings*, 11 April, pp. 23–43.

Carroll-Webb, S.A. & Walther, J.V. (1988) A surface complex reaction model for pH-dependence of corundum and kaolinite dissolution rates. *Geochim. Cosmochim. Acta*, **52**, 2609–2623.

Cassagnabère, A. (1998) *Caractérisation et interprétation de la transition kaolinite-dickite dans les réservoirs à hydrocarbures de Frøy et Rind (Mer du Nord, Norvège)*. Thèse Université Poitiers, 238 pp.

Chermak, J.A. & Rimstidt, J.D. (1989) Estimating the thermodynamic properties of silicate minerals at 298 K from the sum of polyhedral contributions. *Am. Mineral.*, **74**, 1023–1031.

Chermak, J.A. & Rimstidt, J.D. (1990) Estimating the free energy of formation of silicate minerals at high temperatures from the sum of polyhedral contributions. *Am. Mineral.*, **75**, 1376–1380.

Chuhan, F.A., Bjørlykke, K. & Lowrey, C. (2000) The role of provenance in illitization of deeply buried reservoir sandstones from Haltenbanken and north Viking Graben, offshore Norway. *Mar. Petrol. Geol.*, **17**, 673–689.

Dahl, N. & Solli, T. (1993) The structural evolution of the Snorre Field and surrounding areas. In: *Petroleum Geology of Northwest Europe: Proceedings of the 4th Conference* (Ed. by Parker, J.R.), Geological Society Publishing House, Bath, pp. 1159–1166.

Durand, C., Matthews, J.C., Le Gallo, Y., Brosse, E. & Sommer, F. (2000) Illitisation and silicification in Greater Alwyn: I. Assessing and synthesising experimental data. *Clay Mineral.*, **35**, 215–229.

Ehrenberg, S.N., Aagaard, P., Wilson, M.J., Fraser, A.R. & Duthie, D.M.L. (1993) Depth-dependent transformation of kaolinite to dickite in sandstones of the Norwegian continental shelf. *Clay Mineral.*, **28**, 325–352.

Emery, D. & Robinson, A. (1993) *Inorganic Geochemistry: Applications to Petroleum Geology*. Blackwell Scientific Publications, Oxford, 254 pp.

Emery, D., Smalley, P.C. & Oxtoby, N.H. (1993) Synchronous oil migration and cementation in sandstone reservoirs demonstrated by quantitative description of diagenesis. *Phil. Trans. Roy. Soc. London Ser. A*, **344**, 115–125.

Fallick, A.E., Macaulay, C.I. & Haszeldine, R.S. (1993) Implication of linearly correlated oxygen and hydrogen isotopic compositions for kaolinite and illite in the Magnus Sandstone, North Sea, *Clay. Clay Mineral.*, **41**, 184–190.

Fritz, B. (1981) *Etude thermodynamique et modélisation des réactions hydrothermales et diagénétiques*. Mémoires des Sciences Géologiques 65, Strasbourg University, 198 pp.

Giles, M.R. (1997) *Diagenesis: a Quantitative Perspective*. Kluwer Academic Publishers, Dordrecht, 526 pp.

Giles, M.R., Stevenson, S., Martin, S.V., *et al.* (1992) The reservoir properties and diagenesis of the Brent group: a regional perspective. In: *Geology of the Brent Group* (Ed. by A.C. Morton, R.S. Haszeldine, M.R. Giles and S. Brown). *Geol. Soc. London Spec. Publ.*, **61**, 289–328.

Girard, J.-P. (1995) *Diagenesis of the Oseberg Reservoir Formation (Middle Jurassic), Oseberg Field, Northern Nborth Sea: a Core Study*. BRGM Report N. R38556, Bureau de Recherches Gédogiques et Minières, Orleans, 97 pp, 6 plates.

Girard J.-P., Munz I.A., Johansen H., Hill S.J. & Canham A.C. (2001) Conditions and timing of quartz cementation in Brent reservoirs, Hild field, North Sea: Constraints from fluid inclusions and SIMS oxygen isotope microanalysis. *Chem. Geol.*, **176**, 73–92.

Glasmann, J.R. (1992) The fate of feldspars in Brent group reservoirs, North Sea: a regional synthesis of diagenesis in shallow, intermediate and deep burial environments. In: *Geology of the Brent Group* (Ed. by A.C. Morton, R.S. Haszeldine, M.R. Giles and S. Brown). *Geol. Soc. London Spec. Publ.*, **61**, 329–350.

Harris, N.B. (1989) Diagenetic quartzarenite and destruction of secondary porosity: an example from the Middle Jurassic Brent sandstone of Northwest Europe. *Geology*, **17**, 361–364.

Harris, N.B. (1992) Burial diagenesis of Brent sandstones: a study of Statfjord, Hutton and Lyell fields. In: *Geology of the Brent Group* (Ed. by A.C. Morton, R.S. Haszeldine, M.R. Giles and S. Brown). *Geol. Soc. London Spec. Publ.*, **61**, 351–375.

Hassouta, L. (1999) *La comparaison de grès cimentés et de grès non cimentés par la calcite du groupe du Brent (Zone d'Alwyn, Mer du Nord). Une clé pour l'établissement de bilans de matière et la compréhension des processus de formation du quartz et des argiles (illite, kaolinite, dickite).* Thèse Université des Sciences et Technologies, Lille, 167 pp.

Helgeson, H.C., Murphy, W. & Aagaard, P. (1984) Thermodynamic and kinetic constraints on reaction rates among minerals and aqueous solutions: II. Rate constants, effective surface area and hydrolysis of feldspars. *Geochim. Cosmochim. Acta*, **48**, 2405–2432.

Hogg, A.J.C., Sellier, E. & Jourdan, A.J. (1992) Cathodoluminescence of quartz cements in Brent group sandstones, Alwyn South, UK North Sea. In: *Geology of the Brent Group* (Ed. by A.C. Morton, R.S. Haszeldine, M.R. Giles and S. Brown). *Geol. Soc. London Spec. Publ.*, **61**, 421–440.

Hogg, A.J.C., Pearson, M.J., Fallick, A.E. & Hamilton, P.J. (1995) An integrated thermal and isotopic study of the diagenesis of the Brent Group, Alwyn South, U.K. North Sea. *Appl. Geochem.*, **11**, 531–546.

Holstad, A. (1994) Diagenetic processes in sedimentary basins—a DAE application. *Ann. Numerical Math.*, **1**, 461–482.

Jacquot, E. (2000) *Modélisations thermodynamiques et cinétiques des réactions géochimiques dans les réservoirs profonds: application au site européen de recherche en géothermie profonde de Soultz-sous-Forêt (Bas Rhin, France).* Thèse Université de Strasbourg, 200 pp.

Johnson, J.W., Oelkers, E.H. & Helgeson, H.C. (1992) SUPCRT92: software package for calculating the standard molal thermodynamic properties of minerals, gases, aqueous species, and relations among them as functions of temperature and pressure. *Comput. Geosci.*, **18**, 899–947.

Jourdan, A., Thomas, M., Brevart, O., *et al.* (1987) Diagenesis as the control of the Brent sandstones reservoir properties in the Greater Alwyn area (East Shetland Basin). In: *Petroleum Geology of North West Europe* (Ed. by J. Brooks and K. Glennie), pp. 951–961.

Kantorowicz, J.D. (1990) The influence of variations in illite morphology on the permeability of Middle Jurassic Brent Group sandstones, Cormorant Field, UK North Sea. *Mar. Petrol. Geol.*, **7**, 66–74.

Khanna, M., Saigal, G.C. & Bjørlykke, K. (1997) Kaolinitization of Upper Triassic – Lower Jurassic sandstones of the Tampen Spur area, North Sea: implications for early diagenesis and fluid flow. In: *Basin-wide Diagenetic Patterns*, Society for Economic Paleontologists and Mineralogists Special Publication 57 (Ed. by I.P. Montanez, J.M. Gregg and K.L. Shelton), Tulsa, OK, pp. 253–268.

Knauss, K.G. & Wolery, T.J. (1988) The dissolution kinetics of quartz as a function of pH and time at 70°C. *Geochim. Cosmochim. Acta*, **52**, 43–53.

Lacharpagne, J.C., Girard, J.P., Sanjuan, B., *et al.* (1999) *DIAMAP—Reservoir Quality Mapping through Numerical Prediction of Mineral Diagenesis.* Final Technical Report, EEC Thermie Contract OG/165/96. 106 pp.

Le Gallo, Y., Bildstein, O. & Brosse, E. (1998) Modeling diagenetic changes in reservoir permeability, porosity and mineral compositions with water flow. In: *Reactive Transport Modeling of Natural Systems* (Ed. by C. Steefel). *J. Hydrol.*, **209**, 366–388.

Lee, M.J. & Hwang, Y.J. (1993) Tectonic evolution and structural styles of the East Shetland Basin. In: *Petroleum Geology of Northwest Europe: Proceedings of the 4th Conference* (Ed. by J.R. Parker), Geological Society Publishing House, Bath, pp. 1137–1149.

Lønøy, A., Akselsen, J. & Rønning, K. (1986) Diagenesis of a deeply buried sandstone reservoir: Hild field, Northern North Sea. *Clay Mineral.*, **21**, 497–512.

Macaulay, C.I., Fallick, A.E. & Haszeldine, R.S. (1993) Textural and isotopic variations in diagenetic kaolinite from the Magnus oilfield sandstones. *Clay Mineral.*, **28**, 625–639.

Madé, B., Clément, A. & Fritz, B. (1994) Modélisation thermodynamique et cinétique des réactions diagénétiques dans les bassins sédimentaires. *Rev. Inst. Fr. Pétrol.*, **49**(6), 569–602.

Maréchal, B. (2000) *Caractérisation et interprétation de la variabilité chimique et minéralogique des grès réservoirs. Implications pour la modélisation numérique de la diagenèse minérale des grès du Ravenscar Group (Yorkshire, Royaume-Uni).* Thèse Ecole des Mines de Saint-Etienne, 272 pp + annexes.

McAulay, G.E., Burley, S.D. & Johnes, L.H. (1993) Silicate mineral authigenesis in the Hutton and NW Hutton fields: implications for subsurface porosity development. In: *Petroleum Geology of Northwest Europe* (Ed. by J.R. Parker), Geological Society Publishing House, Bath, pp. 1377–1394.

McAulay, G.E., Burley, S.D., Fallick, A.E. & Kusznir, N.J. (1994) Palaeohydrodynamic fluid flow regimes during diagenesis of the Brent Group in the Hutton–NW Hutton reservoirs: constraints from oxygen isotope studies of authigenic kaolin and reverse flexural modelling. *Clay Mineral.*, **29**, 609–626.

McBride, E. (1963) A classification of common sandstones. *J. Sed. Petrol.*, **33**, 664–669.

Michau, N. (1997) *Modélisation cinétique de la formation des argiles en fonction de leur composition et de leur morphologie.* Thèse Université Louis Pasteur, 184 pp.

Morad, S., De Ros, L.F., Nystuen, J.P. & Bergan, M. (1998) Carbonate diagenesis and porosity evolution in sheet-flood sandstones: evidence from the Middle and Lower Lunde Members (Triassic) in the Snorre Field, Norwegian North Sea. In: *Carbonate Cementation in Sandstones*, International Association of Sedimentologists Special Publication 26 (Ed. by S. Morad), Blackwell Science, Oxford, pp. 53–85.

Oelkers, E. (1996) Physical and chemical properties of rocks and fluids for chemical mass transport calculations. In: *Reactive Transport in Porous Media*,

Reviews in Mineralogy 34 (Ed. by P.C. Lichtner, C.I. Steefel and E.H. Oelkers), Mineral Society of America, Washington, DC, pp. 131–191.

Osborne, M., Haszeldine, S.R. & Fallick, A.E. (1994) Variation in kaolinite morphology with growth temperature in isotopically mixed pore-fluids, Brent Group, UK North Sea. *Clay Mineral.*, **29**, 591–608.

Patton, T.L., Moustafa, A.R., Nelson, R.A. & Abdine, S.A. (1994) Tectonic evolution and structural setting of the Suez Rift. In: *Interior Rift Basins* (Ed. by S.M. Landon). *AAPG Mem.*, **59**, 9–55.

Pettijohn, F.J., Potter, P.E. & Siever, R. (1972) *Sand and Sandstone*. Springer-Verlag, Berlin, 618 pp.

Potdevin, J.-L. & Hassouta, L. (1997) Bilan de matière des processus d'illitisation et de surcroissance de quartz dans un réservoir pétrolier du champ d'Ellon. *Bull. Soc. Geol. Fr.*, **168**(2), 219–230.

Rønning, K., Johnston, C.D., Johnstad, D.S.E. & Songstad, P. (1986) Geology of the Hild field. In: *Habitat of Hydrocarbons on the Norwegian Continental Shelf*. Norwegian Petroleum Society, Graham & Trotman, London, pp. 199–206.

Sanjuan, B., Girard, J.-P., Lanini, S., Bourguignon, A. & Brosse, E. (2003) Geochemical modelling of diagenetic illite and quartz cement formation in the Brent Group sandstone reservoirs: example of the Hild Field, Norwegian North Sea. In: *Clay Mineral Cements in Sandstones*, International Association of Sedimentologists Special Publication 34 (Ed. by R.H. Worden and S. Morad), Blackwell Publishing, Oxford, pp. 425–452.

Schott, J. & Oelkers, E.H. (1995) Dissolution and crystallisation rates of silicate minerals as a function of chemical affinity. *Pure Appl. Chem.*, **67**, 903–910.

Scotchman, I.C., Johnes, L.H. & Miller, R.S. (1989) Clay diagenesis and oil migration in Brent Group sandstones of NW Hutton field, UK North Sea. *Clay Mineral.*, **24**(2), 339–374.

Sjöberg, E.L. & Rickard, D.T. (1984) The influence of experimental design on the rate of calcite dissolution. *Geochim. Cosmochim. Acta*, **47**, 2281–2285.

Smith, J.T. & Ehrenberg, S.N. (1989) Correlation of carbon dioxide abundance with temperature in clastic hydrocarbon reservoirs: relationship to inorganic chemical equilibrium. *Mar. Petrol. Geol.*, **6**, 129–135.

Stewart, I.J., Rattey, R.P. & Vann, I.R. (1992) Structural style and the habitat of hydrocarbons in the North Sea. In: *Structural and Tectonic Modelling and its Application to Petroleum Geology* (Ed. by B.T. Larsen and R.M. Larsen), Norwegian Petroleum Society, Graham & Trotman, London, pp. 197–220.

Stumm, W. & Wieland, E. (1990) Dissolution of oxides and silicate minerals: rates depend on surface speciation. In: *Aquatic Surface Chemistry* (Ed. by W. Stumm), Wiley Interscience, New York, pp. 367–400.

Tardy, Y. & Garrels, R.M. (1974) A method of estimating the Gibbs energies of formation of layer silicates. *Geochim. Cosmochim. Acta*, **38**, 1101–1116.

Van der Plas, L. & Tobi, A.C. (1965) A chart for judging the reliability of point counting results. *Am. J. Sci.*, **263**, 87–90.

Van Vessem, E.J. & Gan, T.L. (1991) The Ninian Field, Blocks 3/3 & 3/8, UK North Sea. In: *United Kingdom Oil and Gas Fields, 25 Years Commemorative Volume*, Geological Society Memoir 14 (Ed. by I.L. Abbotts), Geological Society Publishing House, Bath, pp. 175–182.

Walderhaug, O. (1994) Precipitation rates for quartz cement in sandstones determined by fluid-inclusion microthermometry and temperature-history modeling. *J. Sed. Res.*, **A64**, 324–333.

Warren, E.A. & Smalley, P.C. (1994) *North Sea Formation Waters Atlas*. Geological Society Memoir 15, Geological Society Publishing House, Bath, 208 pp.

Yielding, G. & Roberts, A. (1992) Footwall uplift during normal faulting—implications for structural geometries in the North Sea. In: *Structural and Tectonic Modelling and its Application to Petroleum Geology*, Norwegian Petroleum Society Special Publication 1 (Ed. by R.M. Larsen and B.T. Brekke), Elsevier, Amsterdam, pp. 289–304.

Zotov, A., Galev, A.M. & Schott, J. (1998) An experimental study of kaolinite and dickite relative stability at 150–300°C and the thermodynamic properties of dickite. *Am. Mineral.*, **83**, 516–524.

Illite case studies

Int. Assoc. Sedimentol. Spec. Publ. (2003) **34**, 411–424

Illite fluorescence microscopy: a new technique in the study of illite in the Merrimelia Formation, Cooper Basin, Australia

N.M. LEMON* *and* C.J. CUBITT†

National Centre for Petroleum Geology and Geophysics (NCPGG), Thebarton Campus, University of Adelaide, South Australia 5005, Australia, e-mail: nlemon@ncpgg.adelaide.edu.au

ABSTRACT

Illite from the Palaeozoic glacial succession (Merrimelia Formation and Tirrawarra Sandstone) of the Cooper Basin in central Australia fluoresces when viewed under ultraviolet light. A series of tests indicate that the illite itself fluoresced and the phenomenon was not an artefact of remnant hydrocarbons or interference from the resin used to prepare thin-sections. Illite generally is very fine grained and difficult to discriminate using standard transmitted light microscopy, particularly in mixtures of other clays. Illite fluorescence microscopy is an effective tool for observing relationships between diagenetic illite and other authigenic phases and to quickly and cost-effectively assess illite proportions from normally prepared, unpolished thin-sections. Illite fluorescence microscopy analysis, combined with video imaging, was used to assist in estimating illite percentages in porous but low permeability glacio-aeolian sandstone to provide a better assessment of reservoir quality. The results of this study are presented here to encourage others to look for fluorescent illite to see if the technique is more widely applicable.

INTRODUCTION

The normal authigenic habit of illite, fine fibres and bladed platelets, commonly is found fringing and bridging the pore spaces between framework grains, and in doing so adversely influences the deliverability of hydrocarbons from reservoir rocks. The fine grain size of authigenic illite makes it difficult to estimate proportions or see diagenetic relationships using standard transmitted light microscopy, particularly when illite is mixed with other clay phases. X-ray diffraction (XRD) and scanning electron microscopy (SEM) usually have been used to identify illite and to depict its distribution, but this requires the application of two additional preparations and techniques beyond standard thin-section description. Fluorescence microscopy appears to be a suitable new, cost-effective tool using the standard petrological preparation to better constrain illite proportions in reservoir and non-reservoir sediments, provided that the illite responds to UV light.

Illite in samples from several basins throughout Australia, including samples from the Merrimelia Formation and Tirrawarra Sandstone in the Permian Cooper Basin in central Australia, had been observed by one of the

* Current address: SANTOS Ltd, 91 King William St, Adelaide 5000, SA, Australia.
† Current address: Woodside Ltd., 1 Adelaide Terrace, Perth 6000, WA, Australia.

authors (NML) to fluoresce under ultraviolet light. It since has been found that almost all samples from the Merrimelia Formation and associated Tirrawarra Sandstone across the Cooper Basin contain fluorescent illite. A series of tests were conducted to prove that the illite itself fluoresces and that the effect is not an artefact of section preparation or some associated mineral or organic matter.

The fluorescent properties of illite might prove more than just an oddity and, if widespread, could be used to identify and quantify the mineral, particularly in hydrocarbon reservoirs where the mineral can profoundly and adversely influence reservoir properties.

GEOLOGICAL SETTING

All samples used in this study are from the upper Carboniferous to Lower Permian glacial succession of the Cooper Basin. The Cooper Basin is Australia's largest onshore gas basin, with subordinate oil reserves in some fields. The basin is situated in central Australia and extends from the northeast corner of South Australia into southwest Queensland (Fig. 1). The southern Cooper Basin occupies an area of 60 000 km^2 and it preserves an infrabasinal suite of up to 1200 m of non-marine sandstones, shales and coals, which form the Gidgealpa Group of Late Carboniferous to Permian age. The basin unconformably overlies the Precambrian to mid-Palaeozoic Warburton Basin succession and is in turn overlain unconformably by the Mesozoic–Tertiary Eromanga Basin (Fig. 2).

The glacigenic Merrimelia Formation was deposited unconformably on the tectonically deformed Warburton Basin. The Tirrawarra Sandstone, one of the main hydrocarbon-producing reservoirs in the basin, in part overlies and in part is a lateral equivalent of the Merrimelia Formation. In the past, hydrocarbon explorers regarded the Merrimelia Formation as non-economic or 'basement' and the contact with the Tirrawarra Sandstone was perceived as an unconformity. This stratigraphical relationship was redefined by Williams & Wild (1984) as an interfingering, gradational boundary (Fig. 3). This relationship also holds true petrologically, with the same rock fragments and associated authigenic clay phases observed in both units (Chaney *et al.*, 1997). It has been suggested by Chaney *et al.* (1997) that the Tirrawarra Sandstone (*sensu stricto*) be regarded as a 'facies type' with the Merrimelia Formation glacial complex. Accordingly, in this paper, the Merrimelia sediments are considered as those derived directly from glacial deposition and the Tirrawarra Sandstone comprises those sediments deposited by braid plain processes.

Sedimentology

Sedimentological analysis of the Merrimelia Formation by several workers (Williams & Wild, 1984; Williams *et al.*, 1985, 1987; Chaney *et al.*, 1997) revealed a complex glacigenic environment, including glacio-lacustrine, deltaic and shorezone sediments, diamictites, fluvial lithologies and glacio-aeolian sandstones. These facies are observed within terminoglacial and proglacial environments and interfinger regularly (laterally and vertically) throughout the Merrimelia Formation, exhibiting rapid environmental change related to the position of the ice sheet (Chaney *et al.*, 1997). The Tirrawarra Sandstone was deposited in a braided river system which reworked the glacial sediments away from the terminoglacial environment.

Petrography

Rock fragments and detrital quartz grains constitute 25–30% and 40–45% respectively of the total Merrimelia rock component. Fine-grained matrix, silica, carbonate and authigenic clay phases usually constitute up to one-third of each thin-section sampled, thereby leaving the sample with no visible porosity. The proportions of quartz framework grains with straight and undulose extinction within the more mineralogically mature Merrimelia facies were approximately the same. Feldspar grains and micas constitute approximately 1% and 2%,

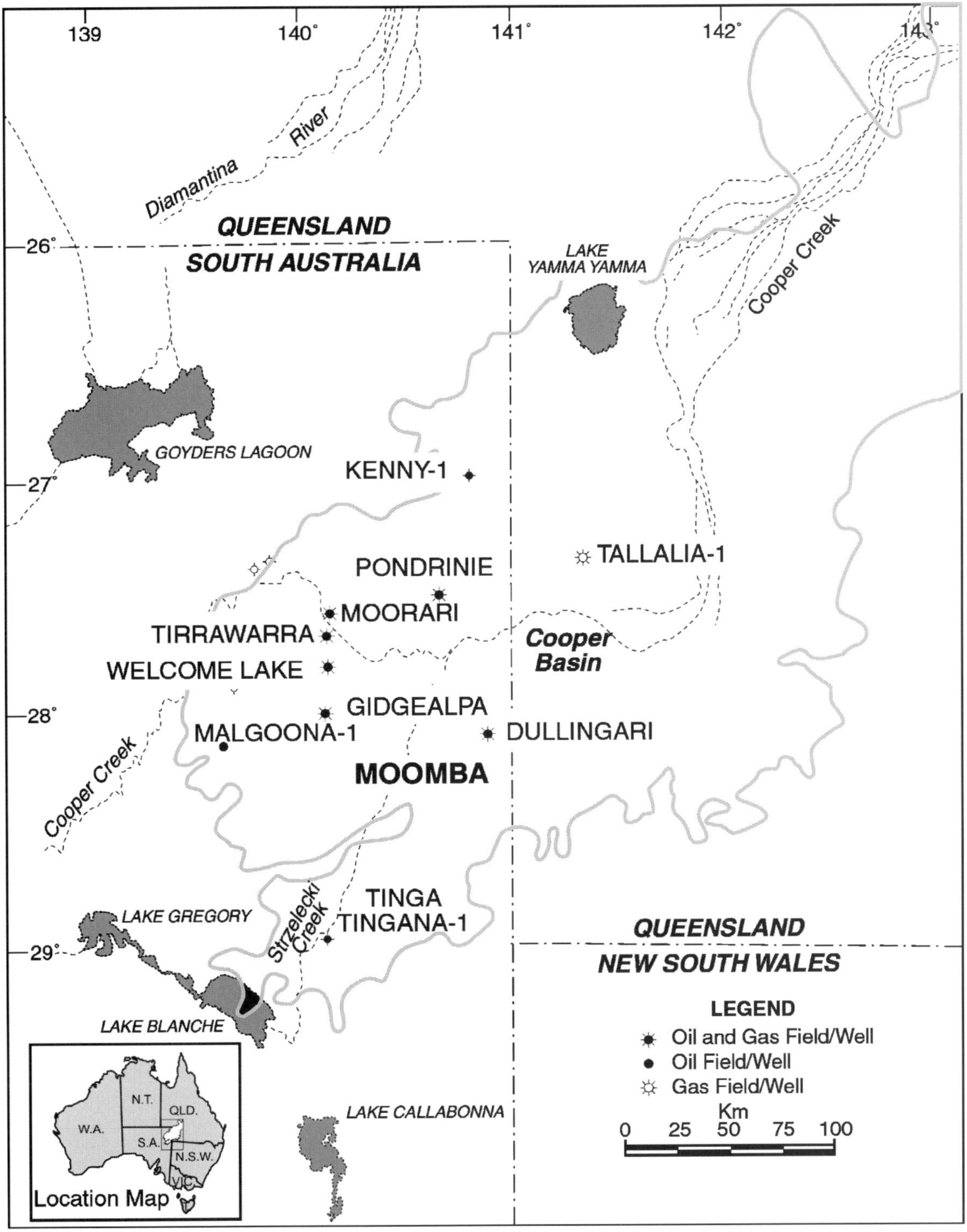

Fig. 1 Location map of the Cooper Basin (modified from Apak *et al.*, 1997).

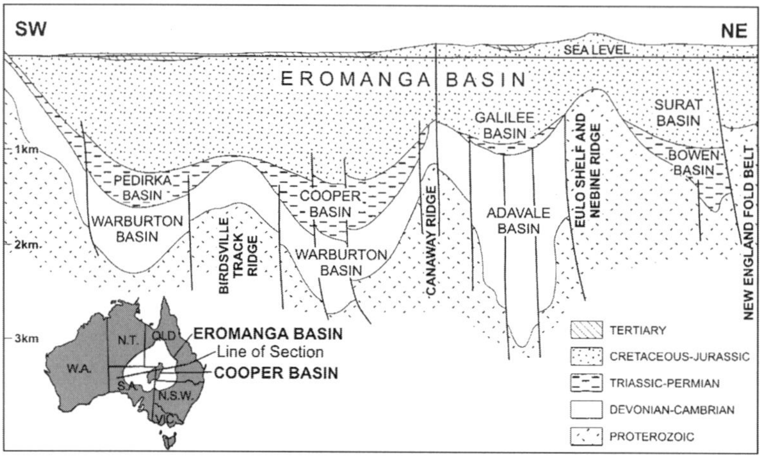

Fig. 2 A regional cross-section showing the Carboniferous to Triassic Cooper Basin completely covered by the Mesozoic Eromanga Basin and above the Cambro-Devonian Warburton Basin (modified from Carne, 1986).

Age	Palyn Zone (mod from Price et al., 1985)	Stratigraphy of Lower Gidgealpa Group
Early Permian	PP2.1	lower Patchawarra Formation
	PP1.2.2	Tirrawarra Sandstone
Late Carboniferous	PP1.2.1	Merrimelia Formation
Cambrian - Devonian		Warburton Basin

Fig. 3 Stratigraphical relationships in the Lower Gidgealpa Group of the Cooper Basin (after Chaney *et al.*, 1997). Palyn = palynological zones.

respectively, of the detrital total in the Merrimelia Formation.

Merrimelia Formation sediments contain up to 76 different rock fragment species. Sedimentary rock fragments comprise 42%, igneous 29%, metamorphic 26% and pyroclastic rock fragments 4% of the total amount observed. Shale rock fragments were the most commonly observed lithic clast type, accounting for 20% of lithic grains identified. Rhyodacite porphyry, which accounts for 13% of the rock fragment total, was the most common igneous type. There is a range of metamorphic types covering 25 differing lithologies with no one species dominant (Chaney *et al.*, 1997).

Figure 4 shows that illite is the most common diagenetic mineral observed in Merrimelia sediments. The majority of this clay is observed as part of the matrix component with replacement of unstable rock fragments the next most commonly observed occurrence.

METHODS

Petrographic analysis

In total, 160 core samples were collected on a basin-wide scale, embracing the full range of stratigraphical and facies variations possible within the Merrimelia Formation. A total of 160 thin-sections were prepared with blue epoxy resin impregnation to facilitate porosity identification. Selected sections were polished and examined using microprobe and fluorescence microscopy techniques. Quantification of thin-section mineralogy (200 counts per slide) was obtained using a combination of visual percentage estimation charts (Terry & Chilingar, 1955) and a video image analysis system (VIDEO PRO 32©) linked to an Olympus BHT microscope. Sandstone compositions were calculated using TRIPLOT 3.0™ ternary diagram software and classified according to Dott (1964).

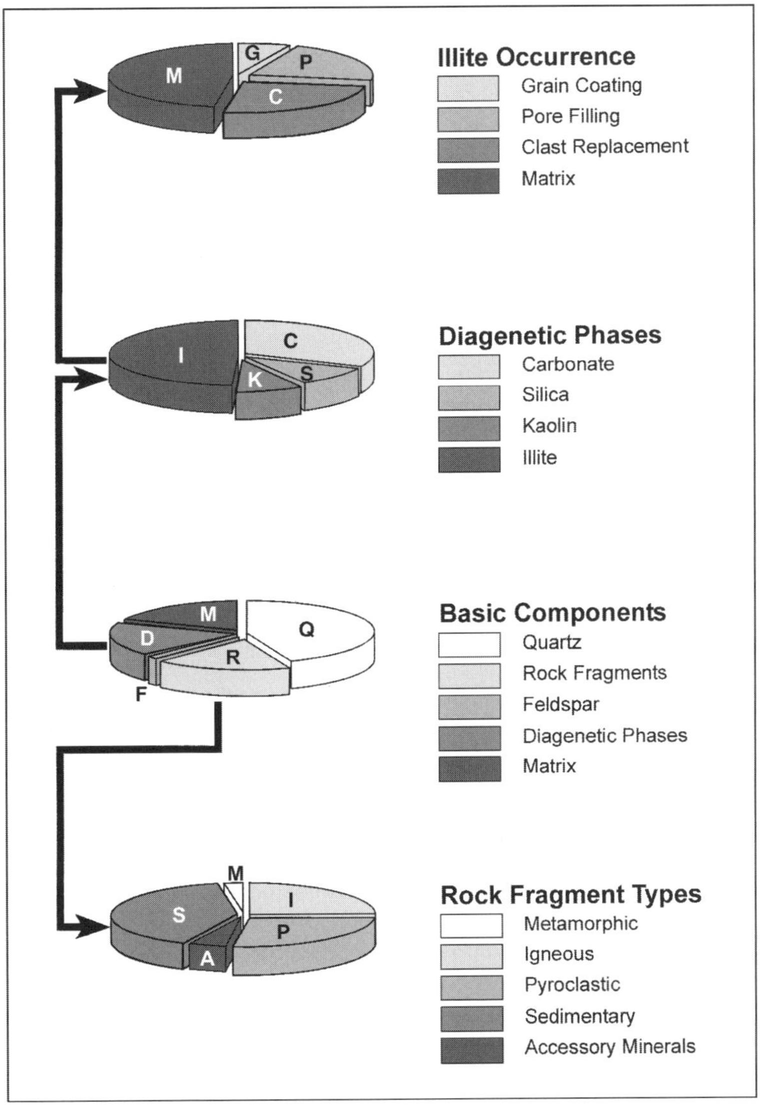

Fig. 4 Compositional subdivision of Merrimelia sediments.

Fluorescence microscopy (FM)

A total of 24 thin-sections from 23 wells were selected for fluorescence analysis. The light source consisted of a 100 W Olympus (BH2-RFL-T3) mercury vapour lamp, which was connected to an Olympus BHSP binocular petrological microscope. The light from the mercury vapour lamp was not filtered but passed straight through an Olympus 'B' excitation cube, which gives blue and UV light from 510 to 490 μm. A return light filter in the microscope removes all light of wavelengths shorter than 515 μm, presenting an image with only red, yellow and green colours. This is a similar set-up to most coal petrology microscopes with a fluorescence facility. Excitation of the illite crystal lattice is seen as yellow-green fluorescence, providing a valuable tool for the optical characterization of authigenic illite. Photomicrographs were taken of all relationships recorded, using Kodak Ektapress 1600 ASA film.

Electron microprobe analysis (EM)

Elemental compositions of all authigenic phases were determined on a CAMECA SX 51 microprobe operating at 15 kV to 20 kV, using a 20 nA beam current and a 0.2 μm beam diameter.

Scanning electron microscopy (SEM)

The SEM studies were undertaken on freshly broken core samples, as well as on < 2 μm and < 4 μm clay separates. Both sample types were coated with carbon and combined gold–palladium. Scanning electron microscopy and semi-quantitative compositional analyses were performed on a XL20 Philips electron microscope, which was connected to an energy dispersive X-ray (EDX) analysis system.

Transmission electron microscopy (TEM)

Clay samples settled from suspension were inspected under a Philips CM 200 transmission electron microscope.

X-ray diffraction (XRD) analysis

As part of routine mineralogical investigations, bulk samples were prepared as air-dried XRD smears and analysed for each thin-section core sample that was obtained. Clay separates at < 4 μm and < 2 μm were prepared and run for selected illite-rich samples identified from bulk XRD analyses. Clay fractions were run as air-dried and glycolated, orientated samples, prepared by the filter transfer method outlined in Moore & Reynolds (1989). The amount of expandable layers in the illitic I/S (illite–smectite mixed layer) was estimated from broad XRD peaks using peak positions given in Moore & Reynolds (1989) and by comparison with calculated XRD profiles generated using NEWMOD (Reynolds, 1980). Mineral identification was performed using TRACES 3.0© software. Samples were run on a Philips PW1050 X-ray diffractometer at 50 kV and 35 mA, using Co–Kα-radiation, at a scan speed of 2° min^{-1}.

THE NATURE OF ILLITE IN THE MERRIMELIA FORMATION

Characterization

Illite samples prepared from whole-rock samples of Merrimelia/Tirrawarra sediments show an interstratified illite–smectite (I/S) clay with about 90% illite. Based on the XRD character of orientated samples, this I/S clay has an ordered interstratification (Moore & Reynolds, 1989) with very little intermixed discrete illite. Samples prepared as size separates show that both 2M1 and 2M2 polytypes are present, but at this stage the distribution of each with regard to cement, matrix and alteration products is not known.

Illite is composed of laths and flakes. The laths vary in length from < 1 μm to 7 μm, whereas the flakes are approximately 2 μm in diameter (Fig. 5A). The thick illite flakes also may be interpreted as illite laths that have been irregularly stacked, possibly by the collapse of laths during air drying of the sample (McHardy *et al.*, 1982).

Illite morphologies, textures and authigenesis

Authigenic illite in the Merrimelia/Tirrawarra glacial complex forms three distinct building-block morphologies: filamentous (Fig. 5B), platy (Fig. 5C) and boxwork (Fig. 5D). These three basic units combine to form a number of illite textures, which have been grouped into five categories: isolated illite fibres, flame boxwork/filamentous, honeycomb boxwork/filamentous, grain coating illite and filamentous after feldspar dissolution.

The complex interplay of different illite textures suggests that illite was formed and is continuing to be formed via a number of diagenetic processes. Observational evidence suggests that there are four main diagenetic processes that led to the formation of illite in Merrimelia/Tirrawarra sediments. These processes are:

1 recrystallization of grain-coating and pore-lining smectite;

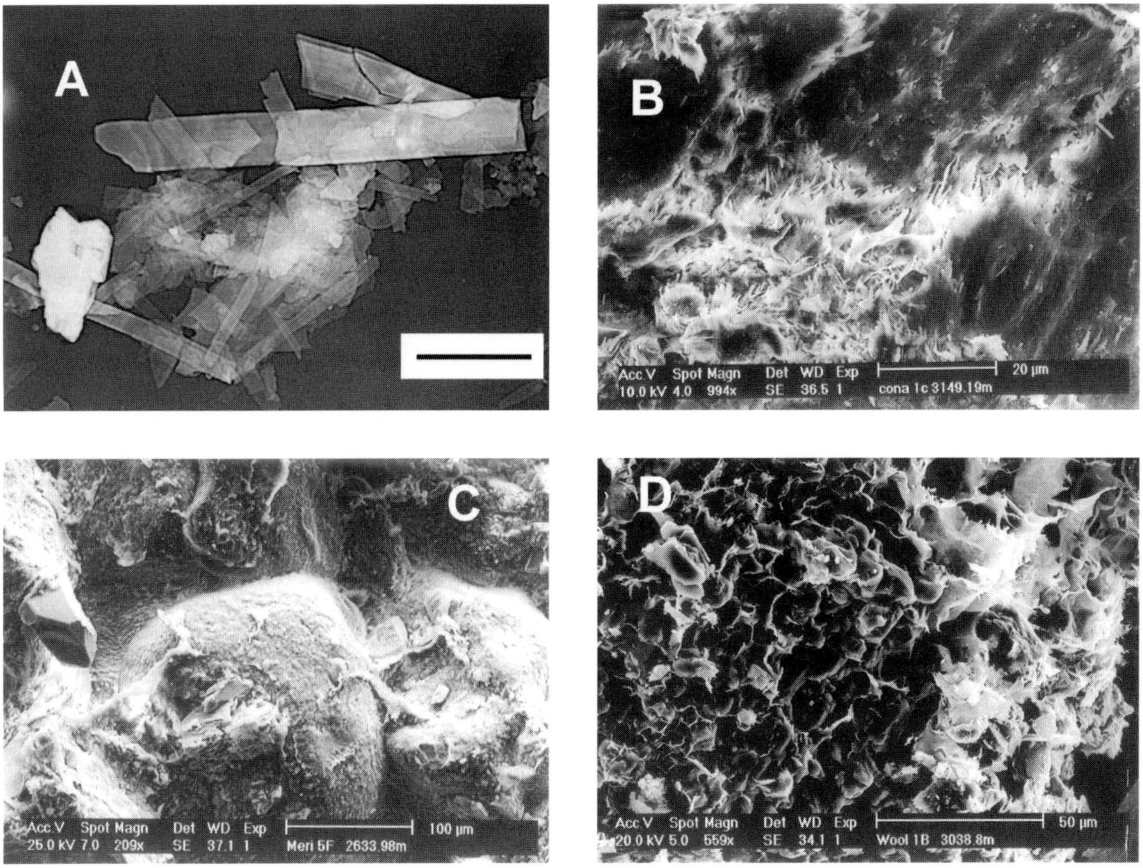

Fig. 5 (A) Merrimelia 18 (2548.03 m). A transmission electron microscopy (TEM) photomicrograph illustrating flake and lath illite particles (scale bar is 1 µm). (B) Coonatie 1 (3149.19 m). A scanning electron microscopy (SEM) photomicrograph of filamentous illite. (C) Merrimelia 5 (2633.98 m). A SEM photomicrograph of aeolian sandstone, illustrating platy illite coating framework grains. (D) Woolkina 1 (3038.8 m). A SEM photomicrograph illustrating honeycomb boxwork illite surrounding kaolin and filamentous illite extending from the boxwork substrate.

2 precipitation of pore-filling illite;
3 partial or total replacement of unstable feldspars/unstable rock fragments;
4 transformation and recrystallization of smectitic matrix to illite with depth.

ILLITE FLUORESCENCE: DISCUSSION

The uses of fluorescence

The uses of fluorescence microscopy in the earth sciences are largely concerned with organic compounds as they give a reliable fluorescence response (Van Gijzel, 1979). Fluorescence microscopy is the main observational tool in micro-organic and microfossil observations, hydrocarbon fluid inclusion identification and interpretation, coal-characterization (Quick, 1992; Teichmuller & Teichmuller, 1982), organic maturity observations (Teichmuller & Durand, 1983) and palynology investigations. Used in combination with other mineralogical identification techniques, commercial tungsten and uranium ore minerals are unequivocally

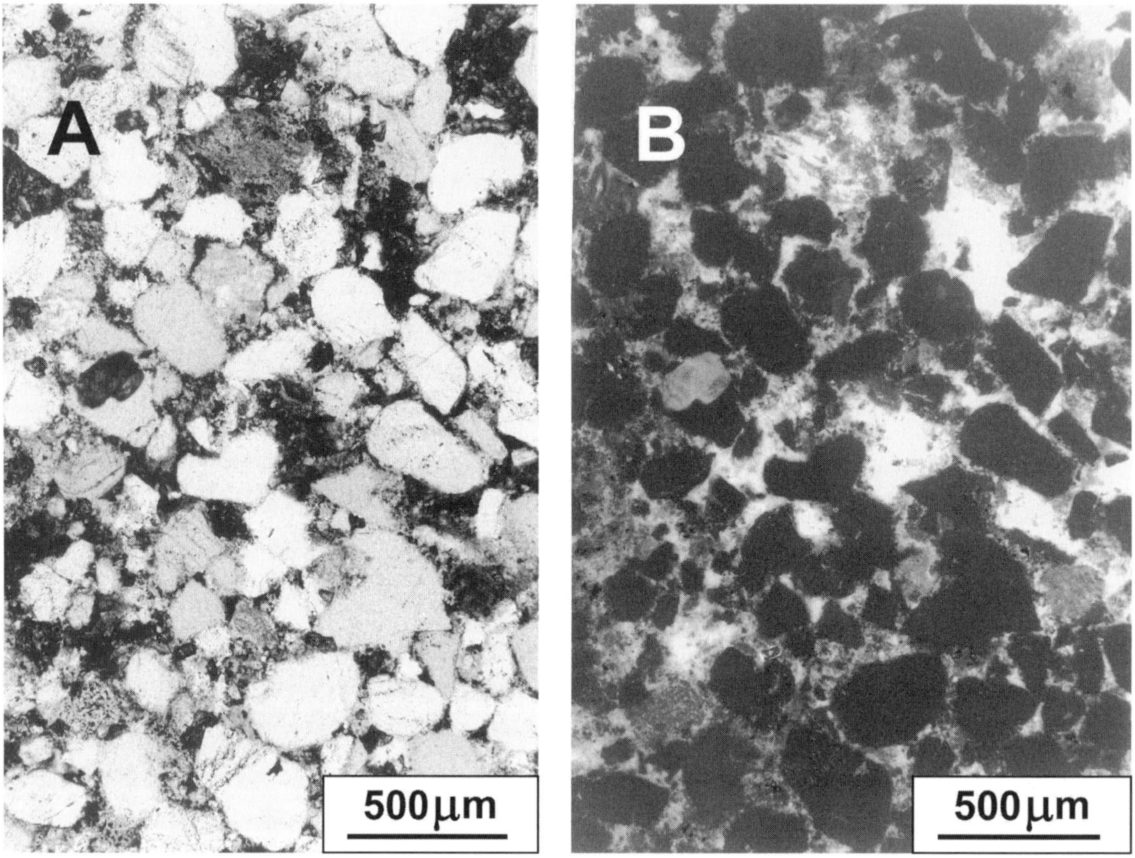

Fig. 6 (A) Gidgealpa 5 (2298.57 m) (plane polarized light). Photomicrograph of a typical litharenite from the Merrimelia Formation. White grains are quartz, grey grains are mainly rock fragments altered to a mixture of clays, mainly illite and kaolin. Dark areas are mainly matrix, coloured by organic matter. (B) The same view as (A) viewed in fluorescence mode. The bright areas are patches of relatively pure illite luminescing greenish yellow. The mid-grey grains are altered rock fragments and matrix, now composed of a mixture of clays (illite and kaolin).

identified by their fluorescing emissions (Duda & Rejl, 1986).

The use of fluorescence microscopy in sedimentary thin-section petrography largely has been limited to assessment of the organic content of these rocks. Soeder (1990), however, has adapted common medical staining techniques, using rhodamine-B dye, to study pore structures in polished thin-sections with the view to better understand pore networks in tight gas reservoirs. Fluorescence microscopy also has been used by Yurewicz & Dravis (1984) to augment standard carbonate petrography to aid the recognition of depositional facies in dolomitized and recrystallized limestones. To date, no mention of illite fluorescence has been found in the published literature with regard to siliciclastic rocks.

Nature of illite fluorescence

Figure 6A illustrates the 'typical' appearance of illite in plane polarized light. Illite occurs as both matrix and altered rock fragments compacted to pseudomatrix. The illite commonly is mixed with other clays (mainly kaolin), but may be a relatively pure phase. The same field of view illuminated with UV light (Fig. 6B) shows illite fluorescing in grey and white tones and in greater proportions than that distinguished in Fig. 6A. Illite fluorescence colours range from green-yellow to yellow. The brightest areas are those where masses of relatively pure illite are present, whereas those zones where illite is diluted by other clays show a lower intensity of fluorescence. Illite is the main fluorescent mineral in the samples studied, although other fluorescing phases were observed in the thin-sections, including the epoxy resin mounting media used in thin-section manufacture (smooth green fluorescence) (see Plate 1A, facing p. 424) and zoned carbonate crystals (orange fluorescence) (see Plate 1B, facing p. 424). Plate 1A illustrates the differences between illite and epoxy fluorescence. Illite fluorescence is yellow with an irregular, patchy appearance whereas the mounting medium fluoresces an homogeneous green colour in keeping with the amorphous nature of the resin. There is a weak edge effect associated with the mounting resin. Plate 1A shows a bright edge to the thick band of resin across the lower right of the image.

A series of tests was conducted to show that the fluorescence observed did indeed emanate from illite and was not an artefact of edge effects combined with fluorescence of the mounting/impregnation epoxy resin. It could be argued that the fluorescence is produced in the resin or in a carbonate mineral and the light is transmitted through transparent minerals to be focused at a boundary where refractive indices change. The filamentous and platy habits of illite might provide a vastly increased surface area where such edge effects are concentrated. Even the smooth edge of a quartz grain might respond in the same manner, thereby falsely attributing an illite rim to such edges. The colour of the response under UV illumination enables some discrimination of the fluorescent phase. Illite fluoresces yellow, Araldite resin is lime to emerald green and the carbonate phases from the Merrimelia Formation that fluoresce are orange yellow under UV light.

Crushed illite-rich samples (< 2 μm and < 4 μm clay separates) were prepared as strewn mounts on glass slides, one dispersed in water and one in paraffin. Both samples fluoresced (Fig. 7A & B), indicating that fluorescence is not a function of the epoxy resin.

The thin fluorescing rims observed in some samples (i.e. Plate 1C, facing p. 424) could well be confused with an epoxy resin edge effect. Grain-coating platy illite is known from the Merrimelia Formation (Figs 5C & 7C), and the sample illustrated in Plate 1C was checked for such rims. The SEM images (Fig. 7C) confirm that there is an illite coating on the grains of this sample.

Another case where confusion over edge effects might arise is illustrated in a sample from the well Kenny 1. The Araldite has effectively penetrated the micropores between loosely packed, pore-filling kaolin booklets and platelets (Fig. 7D). A fluorescence image of a similar area of kaolin-filled pores (see Plate 1D, facing p. 424) suggests that the microporosity between platelets also contains illite, as the colour of fluorescence is yellow rather than the dull green expected for epoxy alone. The SEM image (Fig. 5C) confirms the presence of illite, showing that the kaolin booklets are a substrate for illite fibres arranged as a boxwork. The detail of Plate 1D shows the pseudohexagonal kaolin platelets as dark patches within the fluorescing illite mass.

It also is possible that the fluorescent response attributed to illite could be coming from hydrocarbons trapped between the illite fibres. The extensive surface area and complex interplay of boxwork and filamentous illite could have ensnared and bound hydrocarbon phases within micropores. The areas of hydrocarbon fluorescence might then largely correspond with areas of illite. The consistency of illite fluorescence across the basin, however, in both permeable and poorly permeable Merrimelia and Tirrawarra sediments, is not mirrored in hydrocarbon distribution in this unit across the Cooper Basin.

The sample illustrated in Fig. 7A was prepared by dispersing crushed core in paraffin. If the sample contained fluorescent hydrocarbons, the mounting medium should dissolve

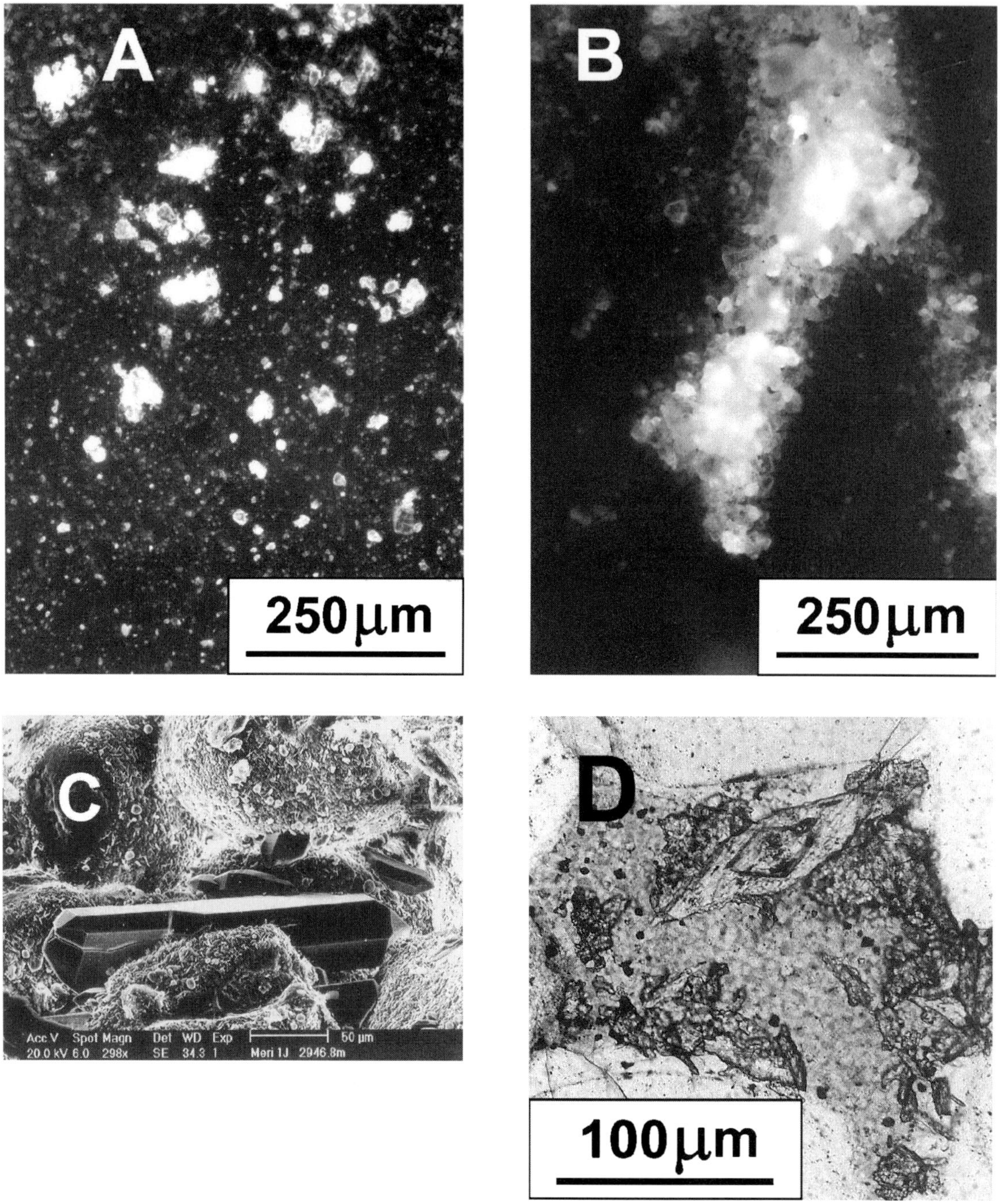

Fig. 7 (A) Photomicrograph in fluorescence mode of crushed sample from Merrimelia 18 (2548.03 m) strewn on a glass slide dispersed in paraffin. Clumps of illite fluoresce with a strong greenish yellow colour. (B) The same sample as used in (A) dispersed in water. Under UV illumination, illite particles fluoresce with the same greenish yellow colour as when dispersed in paraffin or araldite resin. (C) Merrimelia 1 (2946.80 m). A scanning electron microscopy photomicrograph illustrating the platy nature of illite, which coats framework grains. Illite is absent where framework grains touch (under the C label). Late authigenic quartz also has no illite coating. (D) Woolkina 1 (3038.80 m) (plane polarized). Detail of a kaolin-filled pore. The mottled texture shows porosity between kaolin booklets. There is no suggestion of illite between kaolin booklets viewed in this way but it is clearly seen in fluorescence mode, as in Plate 1D (facing p. 424).

Table 1 Average microprobe elemental compositions of illite morphologies.

Morphology	MgO	Al_2O_3	SiO_2	K_2O	TiO_2	FeO	Na_2O	CaO	MnO
Boxwork	2.20	26.83	56.23	8.58	0.00	6.15	0.00	0.00	0.00
Filamentous	0.12	32.93	55.43	8.55	1.10	1.78	0.04	0.00	0.00
Platy	1.07	23.86	42.54	6.83	0.06	1.49	0.13	0.31	0.02

some of the hydrocarbon and induce fluorescence in the medium. This is not the case, thereby supporting the concept that it is indeed the illite itself that is fluorescing.

Cause of illite fluorescence

Having discounted the fluorescence response from the mounting media, other mineral phases and hydrocarbons, one is drawn to the conclusion that the illite itself fluoresces. No work has yet been done that identifies the cause of the fluorescent response. Some microprobe analyses were collected from various illite morphologies (Table 1) but there was no consistent pattern to the trace elements that could be correlated with the fluorescence response.

Determination of the cause of the response is not the subject of this paper but could well be taken up by other workers. A lack of understanding of the cause does not diminish the potential value of fluorescence to identification and mapping the distribution of illite.

ESTIMATING ILLITE PROPORTIONS: FLUORESCENCE ILLITE TECHNIQUE

Figure 8 illustrates an example where image analysis was used to estimate the illite proportion from a fluorescent image of the same thin-section illustrated in Plate 1C (facing p. 424). Illite surrounds the framework grains in this glacio-aeolian sandstone. The image analysis system used both colour and intensity discrimination, thereby allowing for separation of the yellow illite response from the green resin and edge response. The binary images in Fig. 8 were matched to the thicknesses of the illite rims observed under SEM, then area counts were done on the resulting images. This process of image analysis was then repeated randomly

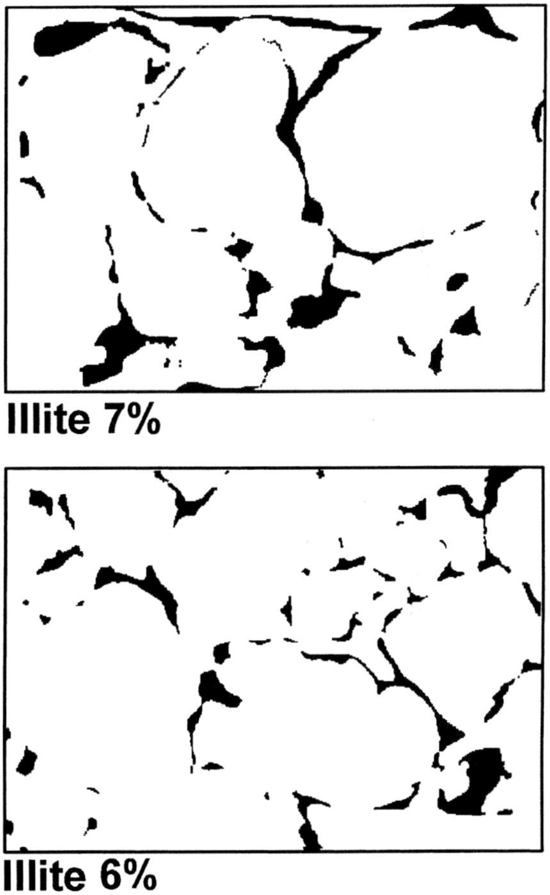

Fig. 8 Merrimelia 5 (2627.98 m). Binary images produced from the fluorescence image of parts of the same thin-section as Plate 1C (facing p. 424). Illite (in black) is estimated as 7% and 6%. The binary images were created to match the illite distributions seen in scanning electron microscopy.

across each slide to expand the area of illite estimation. The illite percentage quoted under each of these images is an average of several readings.

This type of image analysis can be done on a standard thin-section without the need for polishing. The response is better with no cover slip in place.

Illite image analysis indicates that platy grain-coating illite in glacio-aeolian sandstones constitutes 6–10% of the total rock volume. These percentages show good agreement with the corresponding XRD spectra, so the authors accepted this technique as an accurate estimate of the proportion of illite in samples from the Merrimelia Formation. It should be noted that SEM images were required to confirm the illite distribution patterns.

RESERVOIR QUALITY ASSESSMENT: ILLITE FLUORESCENCE

Illite is well documented in the literature, particularly with regard to reservoir quality. An understanding of illite proportions and authigenesis is crucial in describing the reservoir quality in sandstones that contain illite. Merrimelia and Tirrawarra sediments fall into this category, often containing up to 20% illite. The lowest illite proportions observed in this glacigenic sequence were in glacio-aeolian sandstones in the Merrimelia Field. This sequence was first recognized and described as a potential reservoir interval by Williams *et al.* (1985, 1987).

Glacio-aeolian sandstones in the Merrimelia Field are the best sorted of all Merrimelia sediments, with sorting values ranging from 0.25 to 0.51 ϕ (very well to well sorted (Folk & Ward, 1957)). The sizes of framework grains range from medium to very fine grained (0.25 –0.125 mm). Rock fragments are rare (average 3% of the total grains counted) and generally are altered to kaolin and illite. Feldspars (microcline and plagioclase) also are very rare, as most have been partially or almost totally altered to illite with minor kaolin. Very few squeezed ductile rock fragments were observed in these sandstones. The rounding of grains is high, with some quartz grains exhibiting almost perfect sphericity.

Image analysis of thin-sections through the aeolian sandstone complex indicates porosities ranging from 10% to 20%. Two-thirds of pore spaces are primary, and compaction indices indicate that approximately 30% to 50% of primary porosity has been removed. Secondary porosity has been created by the dissolution of rare feldspars and fine-grained siderite cement.

Illite fluorescence and SEM images show that illite forms a very thin veneer on all framework grains (Plate 1C & Fig. 7C). This illite phase accounts for a small proportion of the rock volume but diminishes reservoir quality by partially occluding pore throats (Plate 1C, facing p. 424) and increasing water saturation.

Along with illite, primary porosity is also occluded by authigenic silica druse (Fig. 7C) and in rare cases, siderite/calcite cementation. Fine-grained, stoss-side, grain-flow laminations, typical of aeolian sediments, have been preferentially silica cemented, producing thin high and low porosity zones where the silica-cemented zones form permeability baffles. Similar aeolian sandstones are reported by Seeman (1982) in the southern North Sea Gas Basin.

Reservoir quality in this glacio-aeolian sequence is poor despite the well-sorted, porous nature. The role of silica cement barriers was identified via normal thin-section petrology. The role of illite in these sediments was found by using a combination of illite fluorescence and SEM images. The combination of petrology techniques identified the role of all diagenetic phases in these sediments and reservoir quality assessment was made with more confidence.

CONCLUSIONS AND RECOMMENDATIONS

Illite from the Merrimelia Formation of the

Cooper Basin fluoresces yellow-green. Our experiments show that the fluorescence is from the mineral itself, is not from trapped hydrocarbons and can be distinguished from fluorescence associated with slide preparation resins and other minerals. Image analysis under UV illumination can provide a quick, low-cost illite assessment technique using normally prepared thin-sections. Confidence in the results is raised when SEM images are used to confirm the identity and mode of occurrence of the illite.

Confirmation of the value of this technique will be achieved if other workers apply it to a wide range of illite samples.

ACKNOWLEDGEMENTS

The authors would like to thank G. Carne, S. Jones, S. Tiainen and especially J. Kaldi for proof-reading and improving the original manuscript. Their suggestions and critical assessment of the technical aspects of this paper were invaluable. The paper benefited greatly from the reviews by Norman Oxtoby and Cathal Dillon, both of whom provided valuable comments and suggestions. Editorial comment from Richard Worden is gratefully acknowledged. We also would like to thank the NCPGG, for providing facilities and funding for this study. We sincerely thank the staff at the University of Adelaide Centre for Electron Microscopy and MicroStructure Analysis (CEMMSA) for help with the TEM, microprobe and SEM. G. Kraishan and P. Grech drew many of the diagrams in this paper. The authors acknowledge the support of the Cooper Basin Joint Venture partners and thank them for their permission to publish this work.

REFERENCES

Apak, S.N., Stuart, W.J., Lemon, N.M. & Wood, G. (1997) Structural evolution of the Permian–Triassic Cooper Basin, Australia: relation to hydrocarbon trap styles. *AAPG Bull.*, **81** (4), 533–555.

Carne, G. (1986) *Cooper/Eromanga Basin Study. A Detailed Analysis of the Cooper–Eromanga Basins.* Unpublished Report, Santos Ltd, Adelaide.

Chaney, A.J., Cubitt, C.J. & Williams, B.P.J. (1997) Reservoir potential of glacio-fluvial sandstones: Merrimelia Formation, Cooper Basin, South Australia. *Aust. Petrol. Prod. Explor. Assoc. J.*, **37** (1), 154–176.

Dott, R.H. Jr. (1964) Wacke, greywacke and matrix—what approach to immature sandstone classification? *J. Sed. Petrol.*, **34**, 625–632.

Duda, R. & Rejl, L. (1986) *Minerals of the World.* Spring Books, Twickenham.

Folk, R.L. & Ward, W. (1957) Brazos River Bar: a study in the significance of grain size parameters. *J. Sed. Petrol.*, **27**, 3–26.

McHardy, W.J., Wilson, M.J. & Tait, J.M. (1982) Electron microscope and X-ray diffraction studies of filamentous illitic clay from sandstones of the Magnus Field. *Clay Mineral.*, **17**, 23–40.

Moore, D.M. & Reynolds, R.C. (1989) *X-Ray Diffraction and the Identification and Analysis of Clay Minerals.* Oxford University Press, Oxford.

Quick, J.C. (1992) *Fundamental characterization of New Zealand bituminous coal for prediction of carbonization behaviour—with special emphasis on fluorometric analysis.* Unpublished PhD Thesis, University of Canterbury, NZ.

Price, P.I., Filatoff, J., Williams, A.J., Pickering, S.A. & Wood, G.R. (1985) *Late Palaeozoic and Mesozoic Palynostratigraphical Units, Cooper and Eromanga Basins.* Unpublished Report, CSR Oil and Gas Division, Brisbane.

Reynolds, R.C. Jr. (1980) Interstratified clay minerals. In: *Crystal Structures of Clay Minerals and their X-ray Identification*, Mineralogical Society Monograph 5 (Ed. by G.W. Brindley and G. Brown), London, pp. 249–304.

Seeman, U. (1982) Depositional facies, diagenetic clay minerals and reservoir quality of Rotliegend sediments in the southern Permian Basin (North Sea): a Review. *Clay Mineral.*, **17** (1), 58–68.

Soeder, D.J. (1990) Applications of fluorescence microscopy to study of pores in tight rocks. *AAPG Bull.*, **74** (1), 30–40.

Teichmuller, M. & Durand, B. (1983) Fluorescence microscopial rank studies on liptinites and vitrinites in peat and coals, and comparisons with results of the Rock-Eval Pyrolysis. *Int. J. Coal Geol.*, **2**, 197–230.

Teichmuller, M. & Teichmuller, R. (1982) Fundamentals of coal petrology. In: *Stachs Textbook of Coal Petrology*, 3rd edn (Ed. by E. Stach, M-Th. Mackowsky, M. Teichmuller, G.H. Taylor., D. Chandra and R. Teichmuller), Gebruder Borntebarth, Stuttgart, pp. 5–82.

Terry, R.D. & Chilingar, G.V. (1955) Summary of 'Concerning some additional aids in studying sedimentary formations,' by Shvetsov M.S. *J. Sed. Petrol.*, **25**, 229–234.

Van Gijzel, P. (1979) Manual of the techniques and some geological applications of fluorescence microscopy. *Workshop Manual, 12th Annual Meeting of American Association of Stratigraphic Palynologists, Dallas.*

Williams, B.P.J. & Wild, E.K. (1984) The Tirrawarra Sandstone and Merrimelia Formation of the Southern Cooper Basin, South Australia: the sedimentation and evolution of a glacio-fluvial system. *Aust. Petrol. Explor. Assoc. J.*, **24** (1), 377–392.

Williams, B.P.J., Wild, E.K. & Suttill, R.J. (1985) Paraglacial aeolianites: potential new hydrocarbon reservoirs, Gidgealpa Group, Southern Cooper Basin. *Aust. Petrol. Explor. Assoc. J.*, **35**, 291–310.

Williams, B.P.J., Wild, E.K. & Suttill, R.J. (1987) Late Proterozoic cold-climate aeolianites, Southern Cooper Basin, South Australia. In: *Desert Sediments: Ancient and Modern* (Ed. by L. Frostick and I. Reid). *Geol. Soc. London Spec. Publ.*, **35**, pp. 233–249.

Yurewicz, D.A. & Dravis, J.J. (1984) Improved recognition of sedimentary fabrics using fluorescence microscopy—implications for interpreting carbonate facies and diagenetic history. *AAPG Bull.*, **68**, 542.

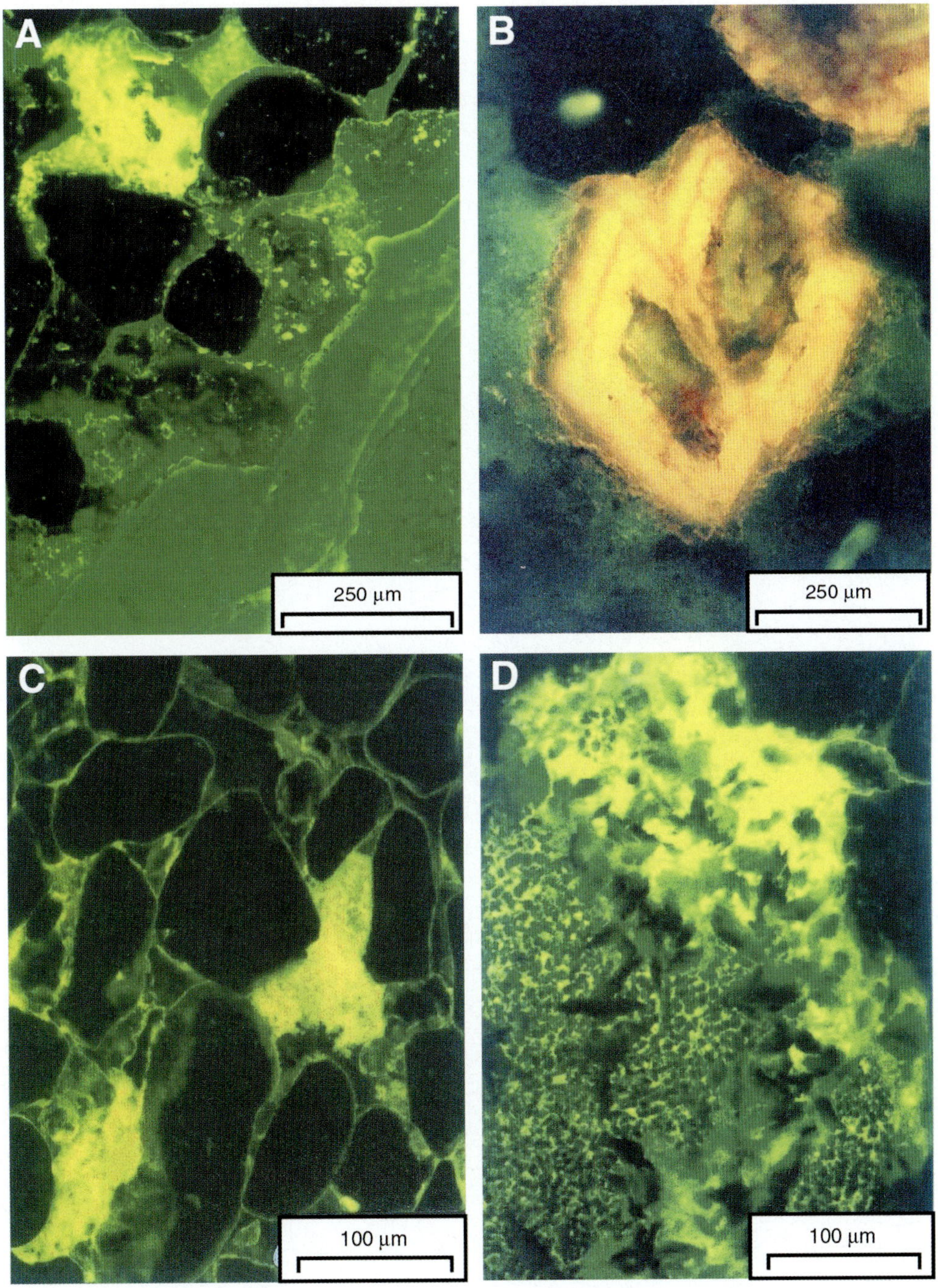

Plate 1 (A) Lake Hope 1 (2493.70 m) (fluorescence mode). The photomicrograph illustrates the difference between the bright patchy fluorescence response of illite (yellow colour) and the smooth, homogeneous mounting medium (lime green in colour). (B) Paning 1 (3129.9 m) (fluorescence mode). Detail of siderite cement crystals, fluorescing orange yellow colour). (C) Merrimelia 5 (2627.98 m) (fluorescence mode). Thin rims of illite around framework grains and altered rock fragments fluoresce strongly (greenish yellow in colour). (D) Kenny 1 (3108.47 m) (fluorescence mode). The black spotted area at the centre and left of the image contains kaolin booklets overgrown by wispy illite. The bright yellow zone is illite alteration of a rock fragment. Laths of siderite grow in both the kaolin and illite areas.

Int. Assoc. Sedimentol. Spec. Publ. (2003) **34**, 425–452

Geochemical modelling of diagenetic illite and quartz cement formation in Brent sandstone reservoirs: example of the Hild Field, Norwegian North Sea

BERNARD SANJUAN[1], JEAN-PIERRE GIRARD[1], SANDRA LANINI[1], ANNE BOURGUIGNON[1] *and* ÉTIENNE BROSSE[2]

[1]*BRGM-3, Av. Claude Guillemin, 45060 Orleans Cedex 2, France, e-mail: b.sanjuan@brgm.fr*

[2]*IFP-1, Av. Bois-Préau, 92500 Rueil-Malmaison Cedex, France*

ABSTRACT

Deep burial diagenesis in the Brent reservoir sandstones of the Hild Field, Norwegian North Sea, was responsible for the development of illite and quartz, and the extensive dissolution of kaolinite and K-feldspar at $T > 100°C$. Geochemical modelling and numerical simulations were conducted in an attempt to reproduce these diagenetic processes.

Present-day formation water in Hild is a Na–Cl brine (total dissolved solids ≈ 70 g l^{-1}) with a calculated pH of about 5.2 in reservoir conditions. Saturation indices (150°C, 800 bar) indicate equilibrium with kaolinite, illite, paragonite and calcite, near-equilibrium with quartz and disordered dolomite, and marked undersaturation with respect to anorthite, albite, K-feldspar and phengite. Numerical simulations performed in closed- and open-system conditions, using sea water, present-day formation water and fresh water as pore-water, indicate that illitization in Hild is controlled by the closed-system reaction: K-feldspar + kaolinite ⇒ illite + 2 quartz + H_2O. The results indicate further that the initial proportion of K-feldspar and kaolinite in the reacting rock constitutes the primary factor controlling the amount of diagenetic quartz and illite generated in numerical models. Variables such as T (between 100 and 150°C), pCO_2 and aqueous acetate concentration have little influence on simulation outcomes.

The amount of diagenetic illite (average = 6%) present in the reservoir can readily be reproduced numerically by adjusting the K-feldspar content and K-feldspar/kaolinite molar ratio (which must be < 1) in the initial assemblage reconstructed from petrography. Spatial variation in the present-day diagenetic illite content most likely reflects primarily variation in K-feldspar abundance at the time of illitization. In contrast, the amount of diagenetic quartz observed in the reservoir (average = 10%) cannot be achieved numerically by the reaction described, suggesting that a contribution by pressure-solution was significant. Numerical simulation further illustrates that the illitization process is not responsible for the loss of porosity that affected the Hild Brent sandstones. Mechanical compaction and quartz cementation are more likely causes.

This study illustrates that numerical modelling in closed-system or open-system conditions is useful in simulating diagenetic transformations observed in illitized Brent reservoirs provided that sufficient constraints can be placed on mineralogy, fluid chemistry and pressure–temperature conditions of the system.

INTRODUCTION

Clay mineral reactions during burial diagenesis are controlled thermodynamically by changes in mineral stability mainly in response to variation in temperature and fluid chemistry, and kinetically by reaction rates in relation to burial rate, temperature and fluid flow. Geochemical modelling therefore constitutes a valuable tool to investigate, reproduce and hence predict diagenetic transformations that affect hydrocarbon reservoirs. Because diagenesis in general, and clay diagenesis in particular, can have a significant impact on reservoir quality, the petroleum industry has shown a growing interest in the development of numerical simulators of water–rock interaction in sediments over the past decade (Wilson, 1994).

The aim of the work reported here was to reproduce, by use of numerical modelling, the diagenetic evolution of Brent reservoirs in the Hild Field, Norwegian sector of the northern North Sea (Fig. 1). A comprehensive investigation of more than 110 core samples showed that Hild Brent reservoir sandstones underwent intense diagenetic alteration (Lacharpagne *et al.*, 1999). The main features are the development of non-ferroan calcite and kaolinite cements during shallow, early diagenesis (< 100°C), and quartz, illite, dickite and ferroan calcite during deep, late diagenesis (> 100°C). A summary of the diagenetic history of the Hild Brent sandstones is given in Brosse *et al.* (this volume, pp. 383–408), and details can be found in Lacharpagne *et al.* (1999) and Girard *et al.* (2001, 2002).

Here we report the results of numerical modelling efforts conducted in an attempt to reproduce the development of quartz and illite cements, and the dissolution of K-feldspar and kaolinite, observed in the Hild Brent sandstones at temperatures of 100–150°C. Among late diagenetic cements (> 100°C), dickite (Girard *et al.*, 2002) was not considered in our study because of the great discrepancy between thermodynamic data (Zotov *et al.*, 1998) and field observations (Ehrenberg *et al.*, 1993) for this mineral. The precipitation of diagenetic kaolinite at low and intermediate depth and temperature in Hild was modelled and the results are presented in Brosse *et al.* (this volume, pp. 383–408).

The geochemical modelling was conducted in two stages, using commercial as well as numerical codes developed in house. In the first stage, closed-system simulations were performed in order to investigate the effects of a number of key variables and evaluate which are most influential in the reconstruction of deep-burial diagenetic transformations in Hild. The objective was to better understand why different amounts of illite are found in different wells. In the second stage, coupled chemistry–transport simulations were performed to investigate whether open-system conditions would produce significantly different outcomes.

Closed-system simulations were undertaken by the use of EQ3/6 (Wolery, 1995) and by the use of a specific chemical simulator named HILDSIM, which was specifically developed within the Allan™/Neptunix environment for this application (Czernichowski *et al.*, 1996; Kervévan & Baranger, 1998). Simulations also were performed with DIAPHORE (Le Gallo *et al.*, 1998), i.e. the geochemical code developed by Institut Français du Pétrol (IFP), for comparison. Comparative tests between these three geochemical codes indicate that similar results are obtained when the same thermodynamic data base is used. Coupled chemistry–transport simulations were performed by coupling HILDSIM with the MARTHE hydrodynamic

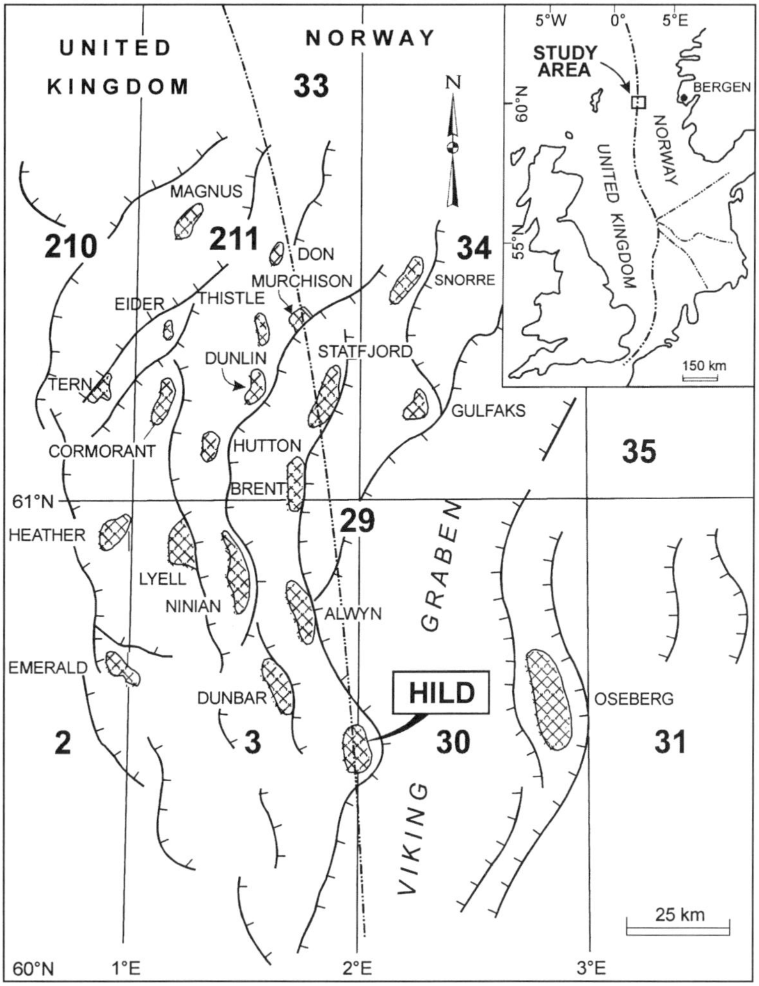

Fig. 1 Location map of the Hild Field (reproduced after Girard *et al.*, 2001).

code developed by Bureau de Recherches Géologiques et Minières (France) (BRGM; Thiery, 1994; Kervévan *et al.*, 1998).

CONSTRAINTS ON SIMULATION CONDITIONS

Reservoir mineralogy

The mineralogical compositions of the Tarbert and Ness Formation sandstones, and the nature of diagenetic transformations and their relative chronology were reconstructed from existing data (Total Norge, 1993) and from new data acquired in the course of this project from wells 29/6-1, 30/4-2 and 30/7-8 (Lacharpagne *et al.*, 1999). The main mineral constituents observed in core samples are quartz (major component), feldspar, detrital muscovite and biotite, kaolinite, dickite, calcite, dolomite and illite in variable proportions. Because mineralogical composition changes with sedimentary facies and because initial detrital mineralogy is not known precisely (Lacharpagne *et al.*, 1999), five different mineralogical assemblages were considered for geochemical modelling (Table 1). Present porosity, which includes both primary

Table 1 Different reconstructed depositional compositions of the Hild Brent sandstones used for numerical modelling.

Mineral	Composition 1 (% vol.)	Composition 1b (% vol.)	Composition 1c (% vol.)	Composition 2 (% vol.)	Composition 3 (% vol.)
Quartz	46	47	47	58	49
K-feldspar	10	8	14	3	3
Albite	5	1	0	0	0.5
Anorthite	0	5	0	0	5.5
Detrital phengite (+ clay matrix)	4	4	4	4	1
Kaolinite	0	0	0	0	3
Calcite	0	0	0	0	4
Disordered dolomite	0	0	0	0	2
Porosity	35	35	35	35	32
Total	100	100	100	100	100

and secondary porosity, ranges from 5 to 30% and averages 10%. Initial porosity values were considered to be around 30–35%.

Clays and thermodynamic data

Electron microprobe (EM) analyses of detrital muscovite and diagenetic illite were conducted on two samples (29/6-1, 4230 m and 30/7-8, 4121 m) in order to document any major difference in chemical composition that would bear on thermodynamic properties of the two types of 2:1 phyllosilicates. The results of 83 individual EM analyses indicate that detrital micas are phengitic ($Fe + Mg = 0.5$) in composition, whereas diagenetic illite is aluminous ($Fe + Mg = 0.1$) and highly charged ($K = 0.8$, based on 11 oxygens per formula) (Lacharpagne *et al.*, 1999; Brosse *et al.*, this volume, pp. 383–408). In theory, such compositional differences should be taken into account in thermodynamic modelling.

A comparison of measured and predicted Gibbs free energy of formation (ΔG_f°) for two different illite compositions is shown in Table 2. It illustrates the great variability of values found in the literature. The same applies to published values of ΔG_f° for pure muscovite ($KAl_3Si_3O_{10}(OH)_2$) (Table 2). It generally is admitted that ΔG_f° must be known at ±0.7 kcal mol^{-1} or better for acceptable thermodynamic modelling (in particular, saturation state calculations; Helgeson *et al.*, 1978; Michard, 1989). Current knowledge of this parameter for clays such as illite and muscovite is not accurate enough. The uncertainty on experimentally determined values of ΔG_f° for clays is the result of complications related to experimental techniques and to the nature of clay minerals (May *et al.*, 1986). The large uncertainty of values of ΔG_f° estimated by theoretical methods (Tardy & Garrels, 1974; Tardy & Duplay, 1992) result from approximations made in the theoretical models themselves. Theoretical models often overestimate the stability of clay minerals (Table 2), possibly in relation to differences in hydration states (among other things) between clays and well-crystallized, high-temperature micas often used as end-member components in the calculations (Aja, 1995).

An additional uncertainty arises from the poor knowledge of the standard enthalpy of formation (ΔH_f°) of clay minerals, a parameter involved in the calculation of action mass law constants for dissolution–precipitation reactions at $T > 25°C$. Values of ΔH_f° encountered in the literature can be highly variable, as illustrated in Table 2 for muscovite. In particular, the value derived from the prediction model based on refined crystal structures of multisite compounds (Vieillard, 1994) is associated with a very high uncertainty.

Table 2 Comparison of measured and predicted ΔG_f° (kcal mol^{-1}) and ΔH_f° (kcal mol^{-1}) for illite and muscovite.

Source	Determination	ΔG_f° illite 1 ($K_{0.69}Mg_{0.16}Al_{2.37}Si_{3.47}O_{10}(OH)_2$)	ΔG_f° illite 2 ($K_{0.85}Mg_{0.12}Al_{2.61}Si_{3.27}O_{10}(OH)_2$ (Aja, 1995))
Aja (1995)	Measured	−1303.1 ± 1.6	−1315.1 ± 2.0
Chermak & Rimsdit (1989)	Measured	−1303.1 ± 5.8	−1321.8 ± 12.4
Tardy & Duplay (1992)	Predicted	−1311.3	−1323.9
Tardy & Garrels (1974)	Predicted	−1309.4	−1323.3

Source	Determination	Muscovite ($KAl_3Si_3O_{10}(OH)_2$) ΔG_f°	ΔH_f°
Robie *et al.* (1978)	Measured	−1338.59	−1428.48
Michard (1983)	After Montoya & Hemley (1975)	−1335.16	−1426.27
Helgeson *et al.* (1978)	estimated (values selected in EQ3/6 data0.sup.R2 data base)	−1336.30	−1427.41
Vieillard (1994)	Predicted		−1423.25 ± 4.79

In light of the uncertainty associated with available values for ΔG_f° and ΔH_f° of muscovite and illite, no attempt was made to calculate specific values of these parameters based on the measured chemical compositions of Hild detrital phengite and diagenetic illite. This approach was, however, attempted by Brosse *et al.* (this volume, pp. 383–408). In the geochemical simulations reported here and on the basis of early simulation tests, it was decided to use the thermodynamic data for pure 'muscovite' provided in the data0.com.R2 data base (EQ3/6; Wolery, 1995) to represent Hild diagenetic illite, which essentially is free of Fe and Mg. The thermodynamic data for the phase termed 'Mg illite' in the EQ3/6 data base was used to represent Hild detrital phengite. In this paper, only the terms diagenetic illite and detrital phengite (or illite and phengite) are used in the text.

Formation fluid: chemical composition

Two chemical analyses of present Brent formation waters in the Hild Field are available in the literature (Warren & Smalley, 1994) for wells 29/6-1 (4294 m) and 25/6-1 (4258 m). One additional chemical composition was provided by the IFP for well 29/6-1. All three compositions are similar (Table 3) and indicate that Hild Brent formation water is a Na–Cl brine with a TDS (total dissolved solids) around 70 g l^{-1}, i.e. twice that of sea water.

Measurements of pH provided with water chemical analyses were done at the well-head, and therefore are affected by CO_2 degassing and temperature decrease between down-hole and surface. Another difficulty for speciation calculations arises from alkalinity, which generally is considered to represent HCO_3 only. However, it also can include a number of other species such as organic acids, which were not analysed in Hild formation waters. Because reported pH and bicarbonate concentration are unlikely to be representative of reservoir conditions and because aqueous Al concentration was not measured, these parameters had to be calculated (Table 3). Aqueous bicarbonate and Al concentrations were determined assuming equilibrium with calcite and kaolinite, respectively. pH was calculated by fixing fCO_2 (CO_2 fugacity) to values in agreement with those of Smith & Ehrenberg (1989) for North Sea fields at 150°C (pCO_2 values from 9 to 25 bar).

Table 3 Chemical and isotopic composition of Hild formation water.

Hild water	Sampling date	T (°C)	TDS mg l^{-1}	Specific gravity	pH (25°C)	Na mg l^{-1}	K mg l^{-1}	Ca mg l^{-1}	Mg mg l^{-1}	Cl mg l^{-1}	HCO_3 mg l^{-1}	SO_4 mg l^{-1}	SiO_2 mg l^{-1}	Al µg l^{-1}	CH_3COOH mg l^{-1}	δD (‰)	$\delta^{18}O$ (‰)
Well 29/6-1 †	26/04/1982	?	69928	1.047	7.40?	23800	260	2650	170	41850	850?	0	58				
Well 29/6-1 ‡		152	72666		6.50?	24403	235	3030	175	43745	525?	0				−46.0	2.2
Well 25/6-1 ‡		151	71399		6.27?	24334	262	2705	175	42753	866?	0				−32.0	2.6
*Water a		150	70000	1.05	4.98	23800	260	2650	170	41850	650	0	58	59	300		
*Water b		150			5.09						490			66	300		
*Water c		150			5.20						386			82	300		
*Water d		150			5.30						295			94	300		

† Analysis provided by IFP.
‡ Analysis from Warren & Smalley (1994).
* pH, HCO_3 and Al concentrations were calculated using the following assumptions. Water a: $fCO_2 = 25$ bar and water in equilibrium with calcite and kaolinite; water b: $fCO_2 = 15$ bar and water in equilibrium with calcite and kaolinite; Water c: $fCO_2 = 10$ bar and water in equilibrium with calcite and kaolinite; water d: $fCO_2 = 5.5$ bar and water in equilibrium with calcite and kaolinite.
? = questionable value.

Formation fluid: saturation state

Saturation indices (SI: the log ratio of solubility product and the equilibrium saturation state) calculated under various conditions are presented in Fig. 2. In recognition of the inherent uncertainty involved in equilibrium thermodynamic calculations, it usually is considered that equilibrium is reached when $SI = 0 \pm 0.3$. Water is oversaturated with respect to a mineral when $SI > 0.3$ and undersaturated when $SI < -0.3$.

All calculations were done using the EQ3NR code and the following data bases:

- data0.sup.R2 data base of the EQ3/6 software package (Wolery, 1995);
- data0.sup.V8.R6 data base (T.J. Wolery, personal communication, 1997), which accounts for the effect of pressure on thermodynamic constants.

Both data bases rely heavily on the SUPCRT92 data base (Johnson *et al.*, 1992), a reference software package for calculating the standard molal thermodynamic properties of minerals, gases, aqueous species and reactions. Because thermodynamic constants for dissolution–precipitation reactions of Mg-rich illite and Na–K–Ca–Mg montmorillonites are not available in the data0.sup.R2 data base, values from the data0.com.R2 data base of the EQ3/6 software package (Wolery, 1995) were used. However, the influence of pressure on thermodynamic constants of these minerals is not known.

Given the water ionic strengths ($I < 1$ M), the extended Debye–Hückel formalism (B-dot equation) was used to calculate the ion activity coefficients for electrically charged species. In the case of polar neutral aqueous species, the activity coefficients were set to unity. The activity of aqueous CO_2 was calculated from the expression after Drummond (1981). The fugacity coefficient for CO_2 gas was determined according to the equations given by Duan *et al.* (1992).

Figure 2a shows the results of calculations at 150°C, 800 bar and $fCO_2 = 5.5$ bar (corresponding to a pCO_2 of 9 bar). It illustrates that Hild formation water is in equilibrium with calcite, kaolinite (because the water initially was assumed to be controlled by these minerals), illite and paragonite, close to equilibrium with quartz and disordered dolomite, and very undersaturated with respect to all feldspars (anorthite, albite, K-feldspar). The slight undersaturation with respect to quartz most likely results from inaccuracies in measured dissolved silica concentration, possibly owing to precipitation of amorphous silica during cooling prior to measurement. The slight oversaturation with respect to disordered dolomite probably is caused by the large variability of chemical composition and crystallinity for this type of mineral (different solid solutions), which makes the thermodynamic data inaccurate. It can be seen in Fig. 2b that if a higher CO_2 fugacity of 15 bar is used (pCO_2 of 25 bar), saturation indices are not affected significantly. Only pH is modified slightly.

If smectite (i.e. the Na, K, Ca and Mg montmorillonites and pyrophyllite from the EQ3/6 data base) and phengite are considered in the mineralogical assemblage, results indicate that Hild formation water is significantly undersaturated with respect to these two minerals (Fig. 2c & d; the effect of pressure is not considered). According to Kharaka *et al.* (1985) and Egeberg & Aagaard (1989), dissolved silica in formation water from clastic reservoirs generally is in equilibrium with quartz at $T > 70$°C. If equilibrium with quartz is imposed on the system (using a value of 115 mg l^{-1} for dissolved silica), calculations indicate that the fluid remains undersaturated with respect to phengite but close to equilibrium with respect to montmorillonites (Fig. 2e & f). Ignoring the effect of pressure has little influence on the water saturation index with respect to disordered dolomite, which becomes slightly negative (Fig. 2c–f).

Application of various aqueous cation geothermometers (Na/K, Na/K/Ca, K/Mg; Michard, 1989), all based partly on the control of K by K-feldspar, yielded temperatures of 80–120°C, which are significantly lower than measured reservoir temperatures (140–150°C).

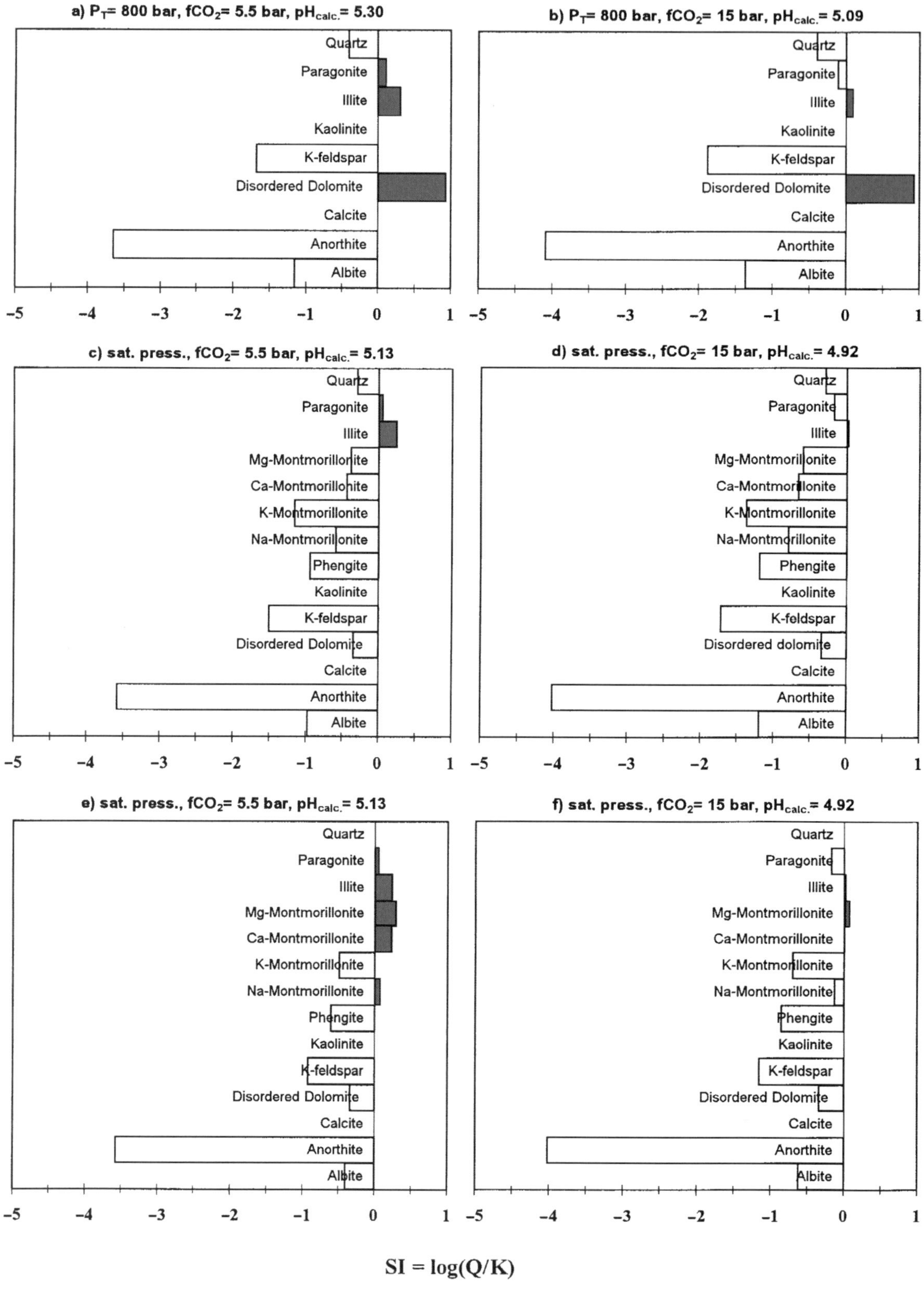

Fig. 2 Saturation indices (SI) for Hild formation water calculated at 150°C.

THE HILDSIM SIMULATOR

The dedicated geochemical simulator HILDSIM was designed as described in Kervévan & Baranger (1998). It calculates water solute speciation and water–rock–CO_2 interactions as a function of reaction progress (i.e. time) at temperatures ranging from 0 to 300°C and pressures from 0 to 5 kbar. The current version of HILDSIM takes into account nine minerals (calcite, dolomite, anorthite, albite, K-feldspar, kaolinite, illite, phengite and quartz) and 30 reactions involving 41 relevant dissolved species (11 chemical elements: O, H, Na, K, Ca, Mg, Al, Si, S, C, Cl), and CO_2 as a gas phase. Regarding organic species, only dissolved acetic acid, the acetate ion and complexes associated with Na, K, Ca, Mg and Al ions were included. Acetic acid is by far the dominant organic acid in oil-field waters of the North Sea (Barth, 1991). Redox equilibrium amongst carbon species (CO_2, CH_4, HCO_3, CH_3COO, etc.) was not considered. HILDSIM uses the same thermodynamic data bases as listed above (see 'Formation fluid: saturation state').

In order to make HILDSIM as efficient and robust as possible for subsequent coupling with the MARTHE transport code, activities of charged species were calculated according to the Davies equation. Activities of neutral species were considered equal to concentrations. Test calculations performed using the B-dot model and the Davies equation yielded similar results (Appendix 1). Only minor differences occur in the calculated pH, fCO_2 and the aqueous speciation. The activity of aqueous CO_2 and the fugacity coefficient for CO_2 gas were determined as in EQ3/6 (see above).

Generally, the effect of pressure on thermodynamic constants can be neglected up to 300 bar (Aggarwal *et al.*, 1990), but should be taken into account at higher pressures. Most of our closed-system simulations were conducted at a total pressure of 800 bar (measured reservoir pressure). Therefore, the influence of pressure on thermodynamic constants was accounted for, except for runs in which phengite and/or Na-montmorillonite were considered because the pressure effect for the minerals is not known. However, comparative tests performed in the course of this study suggest that saturation index (for silicate minerals) and speciation calculations are not greatly affected by neglecting the pressure effect at P and T conditions of the Hild reservoir (Appendix 2).

In HILDSIM, each of the individual modules describing mineral reactions contains a kinetic dissolution–precipitation law (derived from transition state theory) involving a kinetic constant, a reactive surface area and species concentrations. The value of the kinetic constant and the expression of the reactive surface can be readily adjusted as needed. The module for CO_2 gas also contains a kinetic law for dissolution in water, dependent on a kinetic constant and the CO_2 partial fugacity of the system. In order to simulate equilibrium conditions, kinetic constants were set to high values, making reactions essentially instantaneous.

All calculations performed with HILDSIM were done with reference to an initial mass of 1 kg of H_2O. Moles of minerals reacting with H_2O were calculated according to their initial volume per cent and to their molar volume. Porosity, as well as density and salinity (TDS) of water were taken into account.

GEOCHEMICAL MODELLING: CLOSED-SYSTEM CONDITIONS

Simulation conditions and influence of key variables

The main diagenetic reaction reported in Hild reservoirs is the extensive dissolution of K-feldspar and kaolinite, and the formation of diagenetic quartz and illite at elevated temperatures (100–150°C). These four minerals and their interactions with water at temperatures of 80 to 150°C were considered mainly in closed-system simulations. Simulations also were conducted in which albite and calcite were present.

Four different simplified mineralogical compositions (Table 4), modified after those reported in Table 1, were used in the simulations.

Table 4 Mineralogical compositions used in the closed-system simulations investigating the effect of relevant variables (initial mineralogy, fCO_2, temperature, aqueous acetate concentration).

Mineral	Assemblage 1 (% vol.)	Assemblage 2 (% vol.)	Assemblage 3 (% vol.)	Assemblage 4 (% vol.)
Quartz	52	62	52	52
K-feldspar	10	10	10	10
Albite	0	0	5	10
Illite (EQ3/6 muscovite)	3	3	3	3
Kaolinite	15	5	10	5
Porosity	20	20	20	20
Total	100	100	100	100

The initial chemical composition of water was derived from formation fluid data provided by the IFP (Table 3). Aqueous aluminium, silica and bicarbonate concentrations were determined assuming equilibrium at 80°C with quartz, kaolinite and calcite respectively. pH was calculated by fixing fCO_2 to a value of 0.2 bar in agreement with data compiled by Smith & Ehrenberg (1989) for North Sea fields. Numerical simulations conducted in this study indicate that the influence of water chemical composition on outcomes of closed-system simulations is small (Lacharpagne *et al.*, 1999).

Over 50 closed-system simulations were performed, using EQ3/6 and HILDSIM, in order to investigate the effect of relevant variables and evaluate which of those may be most important in the reconstruction of diagenetic illite and quartz formation at Hild. Particular emphasis was placed on assessing the effect of initial mineralogical composition, fCO_2, temperature and acetate ion concentration on the evolution of the mineralogical assemblage. Four different mineralogical compositions (Table 4 & Figs 3–6) were used in these simulations.

Influence of initial mineralogy

Outcomes of simulations performed at 150°C (fCO_2 = 10 bar) for the four different initial mineralogical compositions are shown in Fig. 3. In all cases, quartz and diagenetic illite are precipitated, and K-feldspar and kaolinite are dissolved. As the reaction progresses, the most abundant of the latter two minerals reaches equilibrium whereas the other is progressively lost and disappears from the system. At that time, the reaction stops.

The amount of illite and quartz formed at the end of the reaction directly reflects the total amount of K-feldspar and kaolinite dissolved in the process. Prevailing factors in the control of illite formation under closed-system conditions are K and Al availability.

In albite-free simulations, the final pH is lower for mineralogical assemblage 1 than 2 (4.74 compared with 5.88). This is because more K-feldspar is dissolved (lower consumption of protons) and less illite is formed in the first case (see dissolution–precipitation reactions). In simulations involving albite, the difference in final pH is less significant (5.52 compared with 6.08) because albite contributes to pH buffering.

Influence of fCO_2

The influence of fCO_2 was evaluated in 10 simulations at 150°C for different mineralogical assemblages (Fig. 4). Results indicate that variation in fCO_2 primarily affects pH, which decreases with increasing fCO_2, but in a very minor way. This is because pH is strongly buffered by dissolution–precipitation reactions. The small changes in pH are, however, sufficient to induce modifications in water speciation. In the absence of albite, the overall reactions and amounts of mineral dissolution–precipitation are little affected by fCO_2 variation. In the presence of albite (assemblages 3 and 4), the

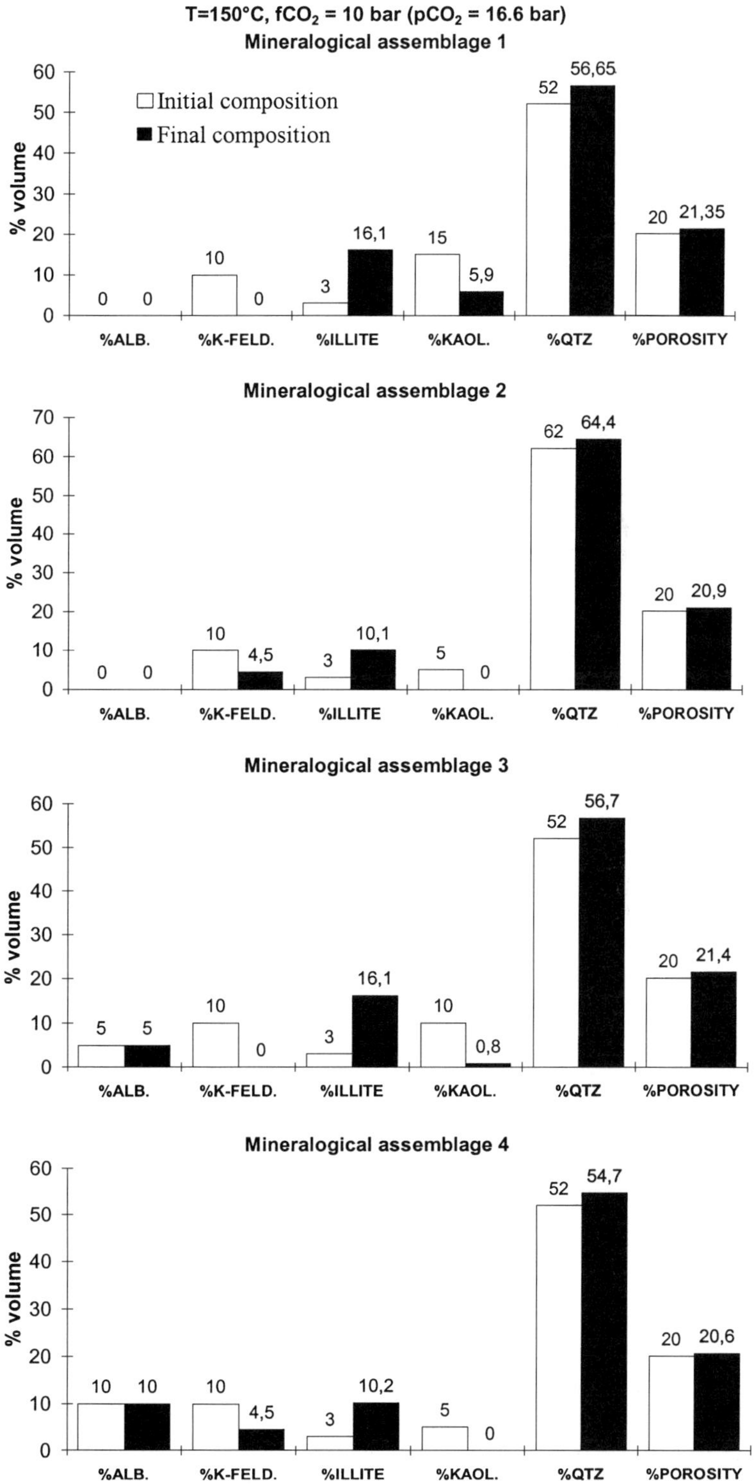

Fig. 3 Outcomes of closed-system simulations illustrating the influence of the initial mineral assemblage.

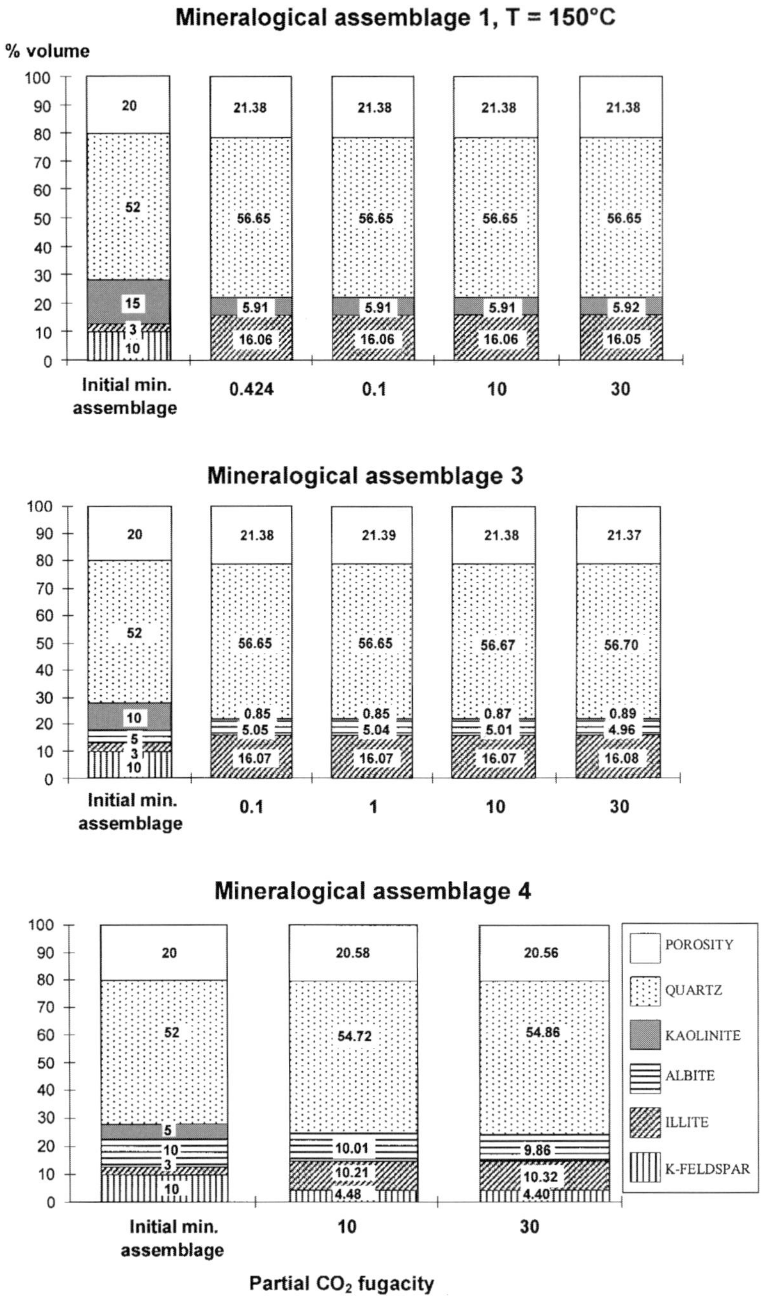

Fig. 4 Outcomes of closed-system simulations illustrating the influence of fCO_2.

plagioclase precipitates up to fCO_2 values of about 10 bar while it is unstable and dissolves at a value of 30 bar. Additional runs have shown that albite is dissolved at lower fCO_2 values when calcite is present in the system.

Influence of temperature

Six simulations were carried out at constant fCO_2 (= 10 bar) using initial mineralogy 1 and 3 at $T = 80$, 100 and 150°C (Fig. 5). No significant

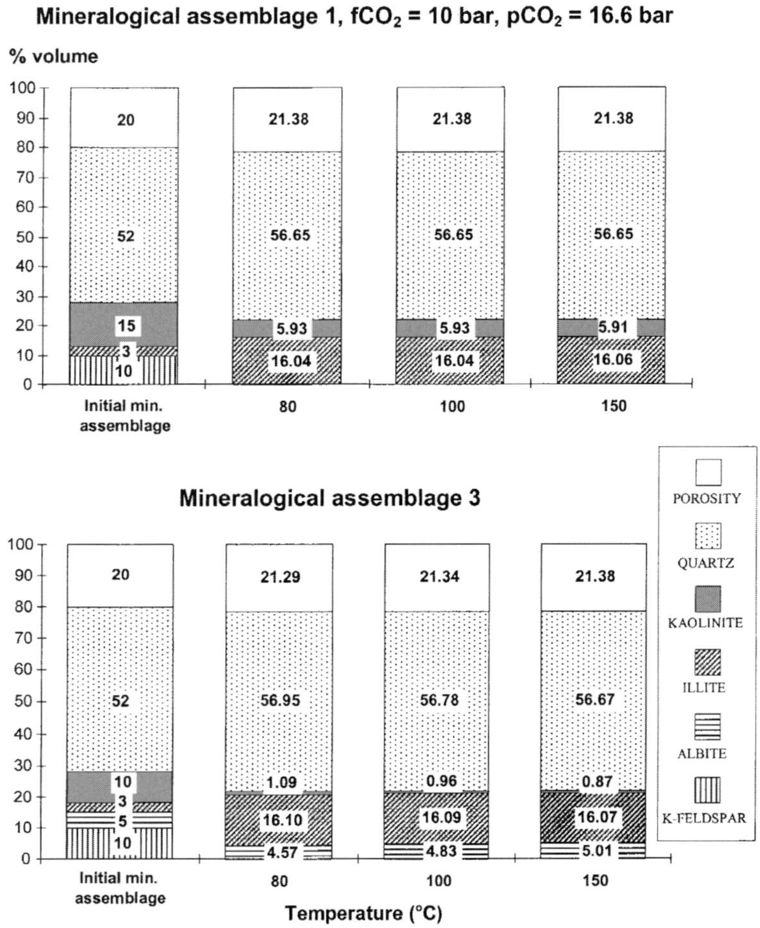

Fig. 5 Outcomes of closed-system simulations illustrating the influence of temperature.

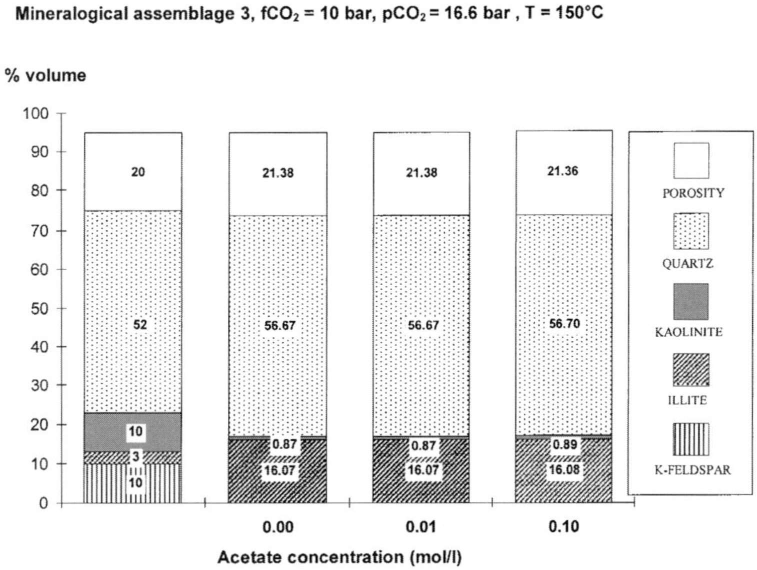

Fig. 6 Outcomes of closed-system simulations illustrating the influence of acetate ions.

effect of temperature is detected for the albite-free assemblage. In contrast, simulations performed in the presence of albite indicate that albite stability increases with temperature. Albite is slightly dissolved up to 100°C and precipitates at 150°C. In all simulations, pH decreases slightly with increasing temperature. The pH drops more markedly when albite precipitates.

Influence of aqueous acetate concentration

Three simulations were performed at 150°C and at $fCO_2 = 10$ bar using mineralogy 3 and for concentrations of dissolved acetate ranging from 0 to 0.1 M (note that reported acetate concentrations in waters from North Sea oilfields are lower than 0.02 M according to Barth, 1991). No significant effect is evident (Fig. 6). This primarily results from the fact that, under our simulation conditions, the dominant species for aqueous aluminium is AlO_2^-. These results agree with the findings of Harrison & Thyne (1991) that organic acid anions are ineffective at neutral to alkaline pH in modifying the stability of aluminosilicate minerals. In contrast, these authors predict that organic anions are variably effective in changing the stability of carbonates under a wide range of pH.

Overall, the results of closed-system simulations indicate that the illite + quartz assemblage is more stable than the kaolinite + K-feldspar association at 80°C and above. Depending on T and fCO_2 conditions, albite can be dissolved or precipitated but only in very small amounts (±0.5% in volume). Porosity systematically increases slightly, i.e. by 0.5 to 1.5%. Closed-system simulations as conducted in this study indicate that initial mineralogy is the primary parameter influencing simulation outcome. The effect of the other parameters considered above is small or insignificant. The amount of illite formed in closed-system simulations is directly proportional to the amount of K-feldspar present in the system when the K-feldspar/kaolinite molar ratio is no greater than 1.

Application to modelling Hild diagenesis

Reconstruction of the chemical composition of present formation water

In order to determine the mineralogical assemblage that allowed the reconstruction of the chemical composition of the present formation water, closed-system simulations were performed at 150°C and variable fCO_2 values. Chemical composition of the initial waters used in these calculations is shown in Table 3 (waters a, b, c and d) and the results are presented in Table 5. A mineralogical assemblage composed of quartz, K-feldspar, kaolinite, illite, calcite and disordered dolomite was considered. In some runs, phengite, albite, paragonite or Na-montmorillonite also were considered in the mineral assemblage. The effect of pressure on thermodynamic data was neglected. Direct comparison with runs where pressure was included suggests that the influence of pressure is relatively insignificant on the calculated chemical water composition (Table 5).

Simulations including albite or phengite yield formation water compositions that are very different from the analytical data. This is consistent with the results of saturation index calculations, indicating that albite and phengite are not in equilibrium with Hild formation water at 150°C. Control of dissolved Na by albite leads to higher aqueous concentrations of Na and, consequently, to lower concentrations of aqueous Ca, Mg and K compared with measured values. In the simulations that include phengite, the final dissolved Mg concentrations are very high, resulting in lower concentrations of dissolved Ca and K, compared with analytical data.

Simulations with no albite and with fCO_2 = 25 bar yield final water compositions that best approach analytical data. However, equilibrium of Hild formation water with respect to paragonite or Na-montmorillonite cannot be excluded totally. The concentration of dissolved Mg is slightly higher than analytical data. This may be related to uncertainty over the thermodynamic constant for disordered dolomite.

Table 5 Final chemical composition of pore-water resulting from closed-system simulations in comparison with measured composition (analytical data). Minimum mineralogical assemblage considered in simulations: quartz, K-feldspar, kaolinite, illite, calcite and disordered-dolomite. Additional minerals considered in some simulations are indicated in first column.

Water	Pressure influence	fCO_2	pH	Na mg l^{-1}	K mg l^{-1}	Ca mg l^{-1}	Mg mg l^{-1}	HCO_3 + acetate mg l^{-1}	Cl mg l^{-1}	SO_4 mg l^{-1}	SiO_2 mg l^{-1}	Al μg l^{-1}
Hild 29/6-1 (Analytical data)			?	23800	260	2650	170	850	41850	0	58	?
Hild 150°C, P_T = 800 bar	Yes	10	5.19	22952	138	2254	363	706	40451	0	147	32
Hild 150°C	No	25	4.84	22955	251	2216	389	737	40456	0	115	9
Hild 150°C, P_T = 800 bar	Yes	25	4.98	22954	228	2258	361	851	40455	0	147	12
Hild 150°C, albite	No	10	5.48	26004	58	310	54	820	40544	0	115	25
Hild 150°C, albite, P_T = 800 bar	Yes	16	5.60	26315	54	244	38	1410	40631	0	148	71
Hild 150°C, albite, P_T = 800 bar	Yes	25	5.60	26629	55	174	27	1952	40644	0	148	70
Hild 150°C, paragonite	No	9	5.08	23214	145	2051	361	564	40467	0	115	12
Hild 150°C, paragonite, P_T = 800 bar	Yes	10	5.19	22969	139	2243	361	711	40452	0	147	32
Hild 150°C, paragonite	No	25	5.04	25091	158	900	157	846	40584	0	115	12
Hild 150°C, Na-montmorillonite	No	10	4.87	19353	232	4512	798	567	40231	0	115	9
Hild 150°C, Na-montmorillonite	No	14	4.95	22822	194	2299	404	627	40446	0	115	13
Hild 150°C, Na-montmorillonite	No	25	5.15	25689	123	5562	97	950	40623	0	115	14
Hild 150°C, phengite (without disordered-dolomite)	No	10	6.15	22516	12	17	2387	3538	39684	0	117	100
Hild 150°C, phengite (without disordered-dolomite)	No	25	6.12	22540	13	10	3418	8671	39725	0	117	94

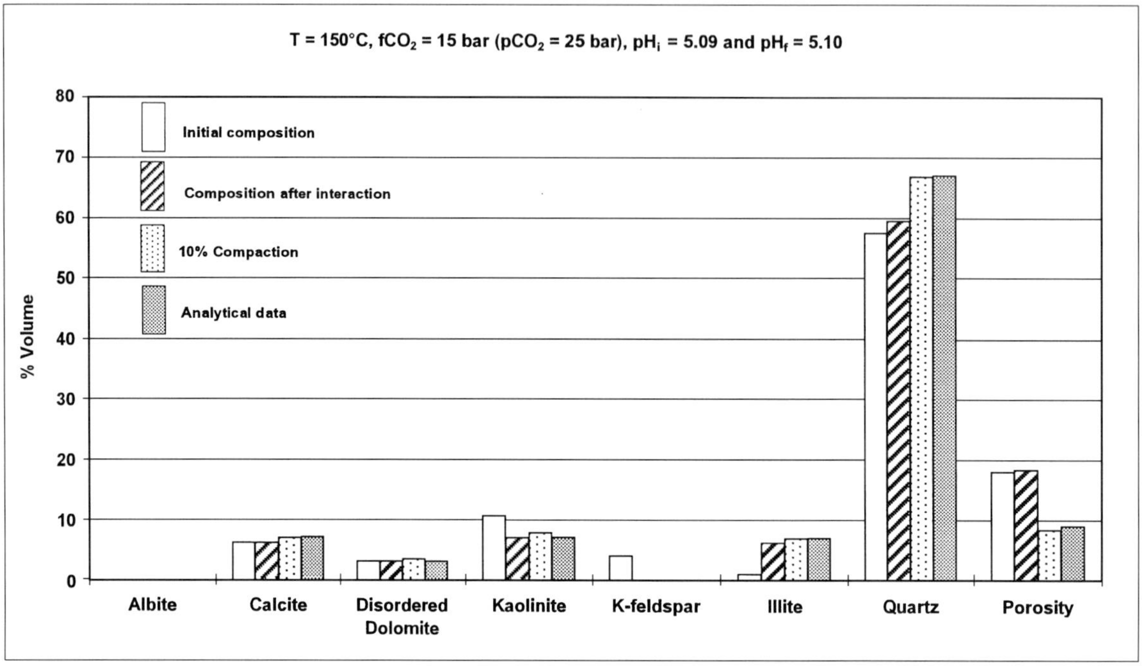

Fig. 7 Numerical modelling of water–sandstone interactions in closed-system conditions (T = 150°C, fCO_2 = 15 bar (pCO_2 = 25 bar), pH_i = 5.09 and pH_f = 5.10).

The results of closed-system simulations at 150°C lead to the following observations.

1 Aqueous Na appears to behave conservatively (same behaviour as chloride) or its concentration is controlled by smectite or paragonite. The latter scenario seems unlikely as no paragonite was reported in the Hild Brent sandstones.

2 Concentrations of aqueous K, Ca, Mg, Si and Al are controlled by diagenetic illite, calcite, disordered dolomite, quartz and kaolinite, respectively.

Geochemical modelling of diagenetic illite formation

A closed-system simulation at 150°C was conducted in an attempt to reproduce the observed average volumes of diagenetic illite and quartz in the reservoir. Chemical composition of water b in Table 3 was used (fCO_2 = 15 bar, pCO_2 = 25 bar). An initial mineralogical composition (Fig. 6) was determined so that the final quantities of kaolinite and K-feldspar dissolved and illite and quartz precipitated were in agreement with the amounts derived from petrographic observations. Based on burial history considerations (Lacharpagne *et al.*, 1999), a loss of 10% porosity owing to mechanical compaction was assumed.

Results of the geochemical simulation (Fig. 7) can be described by the following reaction, which has been used in other modelling studies (e.g. Berger *et al.*, 1997; Brosse *et al.*, 2000):

$$\underset{(1.1\%)}{\text{K-feldspar}} + \underset{(1\%)}{\text{kaolinite}} \Rightarrow \underset{(1.4\%)}{\text{illite}} + \underset{(0.5\%)}{2\ \text{quartz}} + H_2O \quad (1)$$

Percentages indicated are volume per cent. According to stoichiometry, 4.7% of K-feldspar and 4.3% of kaolinite are required to react in order to generate the average *c.* 6% diagenetic illite observed in thin-sections (Lacharpagne *et al.*, 1999). However, according to this reaction only 2.1% quartz would be produced, i.e. a much smaller amount than the average *c.* 10% observed petrographically. If the amount of diagenetic illite derived from point-counting is corrected for microporosity (*c.* 50% according to Nadeau & Hurst, 1991), only 3% illite would need to be produced and only 1% quartz would form. Consequently, the large volumes of diagenetic quartz observed in the Hild Brent sandstones cannot be produced following this reaction in closed-system numerical modelling.

Additional closed-system simulations were carried out at $T < 100°C$. They were performed to investigate whether a significant amount of quartz cement could be formed during early diagenetic stages, i.e. the formation of diagenetic kaolinite at the expense of detrital phengite and feldspar at $T < 100$ °C, prior to illitization processes. These low-temperature simulations systematically lead to precipitation of diagenetic illite at the expense of K-feldspar and kaolinite. In order to circumvent this problem, the precipitation of illite had to be prevented arbitrarily by excluding this mineral from the system (a similar approach was advocated by Berger *et al.* (1997) and Brosse *et al.* (2000)). Under such conditions, it was observed that low temperature simulations failed to generate the large amount of diagenetic quartz observed in Hild. In order to avoid an arbitrary prevention of illite precipitation as indicated above, it was attempted to replace quartz by chalcedony (a Si-phase more soluble than quartz). This did not help. To stop illite precipitation at $T < 100°C$, amorphous silica had to be considered in the system instead of quartz, but in this case, the amount of silica precipitated following aluminosilicate dissolution was even lower than that obtained with chalcedony or quartz (amorphous silica being more soluble than the two other silica phases). In addition, K-feldspar was precipitated in conjunction with kaolinite. All of these results are not in agreement with petrographic observations, and indicate that alternative sources of silica must be considered. This is discussed further below.

GEOCHEMICAL MODELLING: OPEN-SYSTEM CONDITIONS

Coupled chemistry–transport model (HILDSIM–MARTHE)

The objective of simulations in open-system conditions was to investigate whether significant differences would arise in comparison with closed-system simulations. In particular, it was interesting to explore mechanisms that could reproduce the large amounts of diagenetic quartz observed in the Hild Brent. The HILDSIM chemical simulator was coupled with the hydrodynamic transport code developed by BRGM and named MARTHE (Thiery, 1994; Kervévan *et al.*, 1998). The resulting coupled chemistry–transport code operates by sequential calls to the transport and the chemistry subroutines for a number of successive time-steps in each grid block.

Coupled simulations were carried out in one dimension, using an approach very similar to that described in Brosse *et al.* (2003, this volume, pp. 383–408). The geometry of the Hild reservoir was simplified as a succession of 50 contiguous cells, each of which was a cube of 100 m a side. Porosity was assumed to be saturated with water, and temperature was assumed constant throughout the reservoir. Water flow was considered unidirectional and constant through the system. That is, at any instant, an equivalent mass of fluid enters the reservoir at the inlet ($x = 0$ m) and leaves the reservoir at the outlet ($x = 5000$ m). A permeability (K) of 3.5×10^{-5} m s^{-1} ($\approx$ 3500 mD), i.e. in the upper range of permeabilities encountered in Hild reservoirs (range is 0.05–5000 mD), was used. The effect of precipitation–dissolution reactions on porosity and permeability was not taken into account. A hydraulic head (H) of zero was attributed to the outlet of the reservoir. At the inlet, an appropriate value of H was calculated to constrain the flow rate (Q) through the reservoir to the desired value (see below) according to:

$$Q = S \times K \times H/5000$$

where S is the cell cross-section ($100 \times 100 = 10^4$ m^2). A longitudinal dispersion of 500 m was applied arbitrarily to the transport of dissolved constituents (the value of this parameter is unknown but a good rule of thumb is to set it equal to one-tenth of field length).

Simulation conditions

Basin modelling of the Hild area (Séjourné, 1997) indicates that mean water flow during deep burial conditions at Hild is very slow, with calculated fluid velocities of the order of 0.1 to 0.5 mm yr^{-1}. These values are within the range of fluid velocities considered representative of compactionally driven flow (Giles, 1997). Therefore, Darcy fluid velocities of 0.1 and 0.5 mm yr^{-1}, corresponding to flow rates of 2.73×10^{-3} and 1.37×10^{-2} m^3 day^{-1}, were used in our simulations.

Temperature was considered uniform throughout the reservoir at any instant in time, but it was allowed to increase linearly with time from 100 to 150°C between the beginning and the end of the simulation. This was done in an attempt to simulate increased burial depth, hence temperature, of the reservoir with time during illitization processes.

According to the reconstruction of the diagenetic history (Girard *et al.*, 2001, 2002), the main illitization event (I2 of Girard *et al.*, 2002) is most likely to have occurred over the past 30 Myr. However, some illite (I1 of Girard *et al.*, 2002) may have formed as early as 70–80 Ma. Therefore, coupled simulations were performed over time durations of at least 60 Myr, and some were extended up to 120 Myr. It must be kept in mind that, other things being equal, simulation outcomes generated over 60 to 120 Myr would occur over shorter time spans if fluid velocities higher than 0.1–0.5 mm yr^{-1} were considered.

The following minerals were considered in the coupled modelling: quartz, K-feldspar, illite, phengite, calcite and dolomite. The initial mineralogical composition of the reservoir prior to coupled modelling at 100–150°C was taken as the final mineralogy resulting from coupled simulations at 60–100°C (not reported here) and using the reconstructed depositional mineralogy reported in Table 1 as the starting assemblage. In low-temperature simulations (see above and Brosse *et al.*, this volume, pp. 383–408), anorthite and albite are totally dissolved and kaolinite is formed. Consequently, anorthite and albite were excluded from the high-temperature (100–150°C) coupled simulations. Slightly variable initial proportions of the minerals indicated above were tested. They are shown in Figs 8–11 (first frame, upper left-hand side). Simulations reported in Figs 8 & 9 are based on a similar initial mineralogy (except for carbonates), in which K-feldspar is more abundant than kaolinite. In the simulation reported in Fig. 10, we considered the reverse situation, in which kaolinite initially is more abundant than K-feldspar. In the simulation presented in Fig. 11, the initial mineralogy is that used in closed-system simulations (Fig. 7).

The system is initialized by equilibrating Hild formation water with the selected initial rock assemblage throughout. Then, extraneous water (injection water) is introduced at the inlet of the reservoir at the selected velocity. Three types of injected fluids were considered: sea water, fresh water and Hild formation water. Chemical compositions used for these end-member fluids are given in Appendix 3. The temperature of the injected fluid was set equal to that of the pore fluid present in the reservoir.

In coupled chemistry–transport simulations, it is often practical to select calculation time-steps corresponding to the time necessary for the fluid to move through one cell, i.e. a Courant number of 1. In our study, it was necessary to use a time-step corresponding to a Courant number of 2.5 in order to optimize total calculation time and maintain a good accuracy of chemical calculations. This resulted in a calculation time-step significantly greater than the time required for mineral reactions to reach equilibrium. Consequently, the kinetics of reactions had little influence on simulation outcomes. This approximation was not considered to introduce much uncertainty because at high

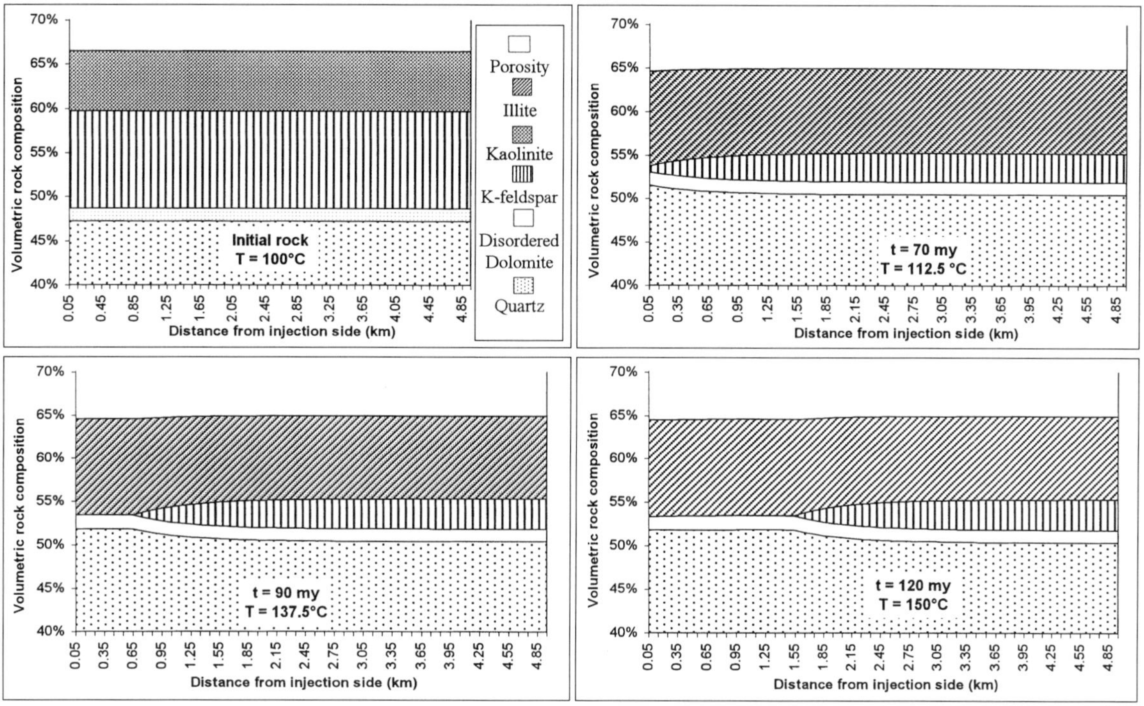

Fig. 8 Results of coupled modelling. Injected water: sea water; $T = 100$ to 150°C; fluid velocity = 0.1 mm yr^{-1}.

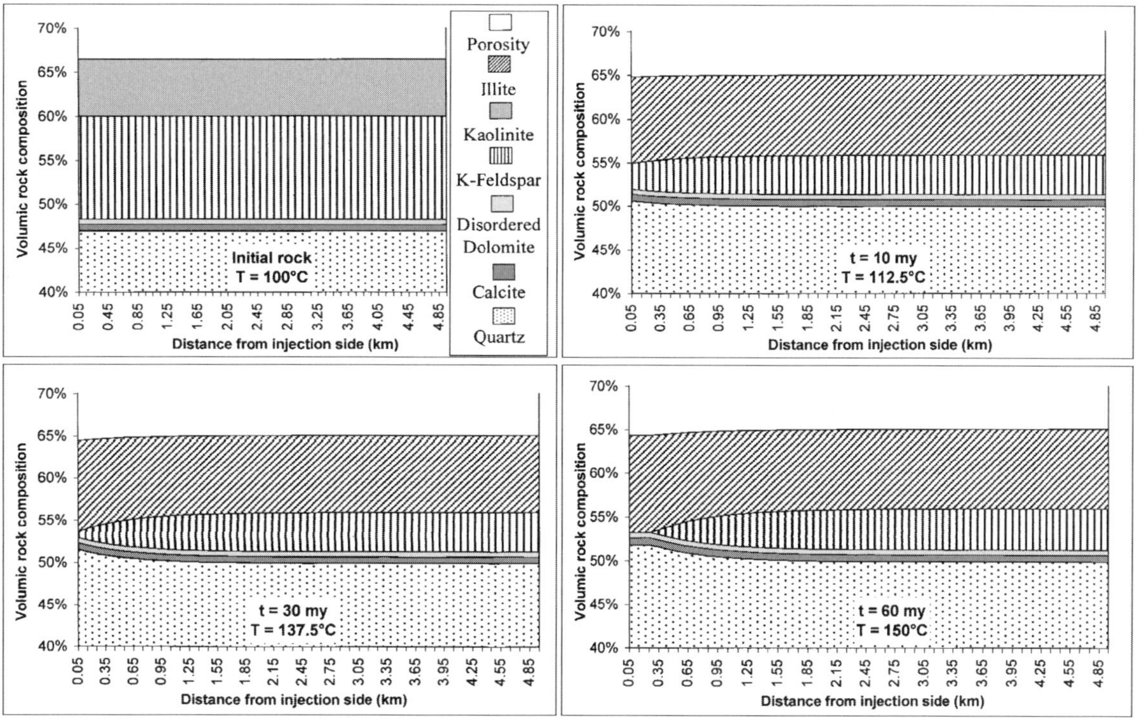

Fig. 9 Results of coupled modelling. Injected water: fresh water; $T = 100$ to 150°C; fluid velocity = 0.5 mm yr^{-1} ($Q = 0.01368$ m^3 day^{-1}).

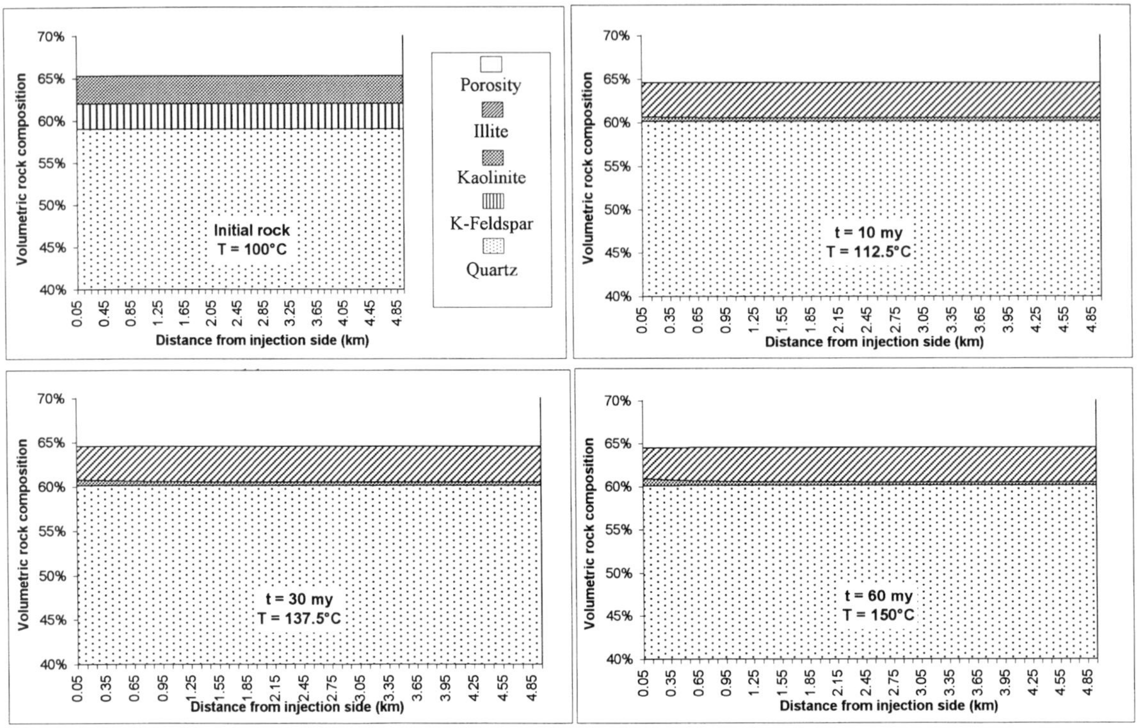

Fig. 10 Results of coupled modelling. Injected water: fresh water; $T = 100$ to $150°C$; fluid velocity = 0.5 mm yr^{-1} ($Q = 0.01368$ m^3 day^{-1}). Initial rock mineralogy differs from simulation run shown in Fig. 9 (no carbonates, K-feldspar/kaolinite = 1).

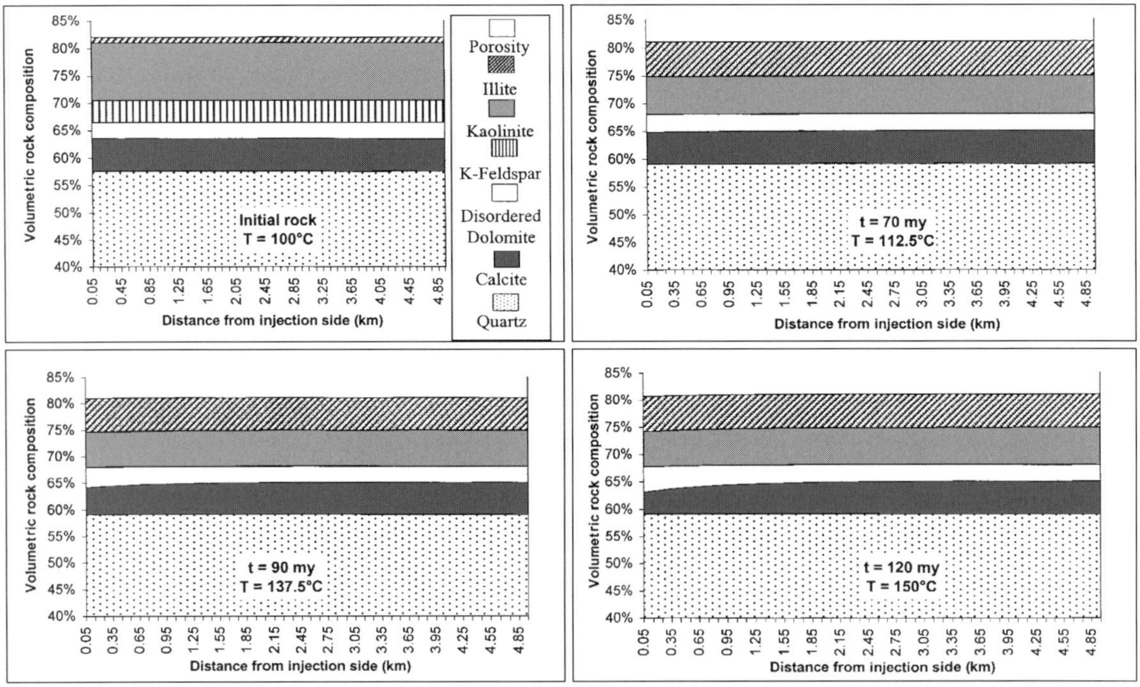

Fig. 11 Results of coupled modelling. Injected water: sea water; $T = 100$ to $150°C$; fluid velocity = 0.1 mm yr^{-1} ($Q = 0.00273$ m^3 day^{-1}).

temperatures reactions become essentially instantaneous (relative to the overall timescale being considered in these simulations) and local thermodynamic equilibrium can be assumed to be representative (Giles, 1997).

Results

A selection of illustrative examples of computation outcomes are shown in Figs 8–11. Variations in mineral abundance and porosity are reported as a function of distance from injection side at different simulation times.

Comparison of simulation outcomes (including those in Figs 8–11) indicates that the type of fluid injected into the reservoir—sea water, fresh water or Hild formation water—has relatively little influence on the final mineralogical composition of the system. In addition, it can be shown that, under the conditions considered, the progress and magnitude of simulated diagenetic transformations essentially are controlled by fluid velocity. The faster the fluid flow, the more rapid the diagenetic reaction.

Results of coupled simulations generally are comparable to those of closed-system simulations, although some significant differences exist. In all cases, illite and quartz are formed at the expense of K-feldspar and kaolinite. As in closed-system simulations, the final content of K-feldspar and kaolinite in the reservoir primarily is controlled by the initial proportions of these two minerals. The least abundant of the two minerals in the initial assemblage is rapidly consumed (essentially instantaneously in this modelling). However, in contrast with closed-system simulations, in the case of complete consumption of kaolinite, the reaction does not necessarily stop if some K-feldspar remains available (in cases in which the initial K-feldspar/kaolinite molar ratio is greater than 1). This is illustrated in Figs 8 & 9, where it can be seen that K-feldspar continues to dissolve, generating extra illite, after dissolution of kaolinite is completed. This differs from the results of closed-system simulations and indicates that in open-system conditions, provided that there is enough time, the amount of diagenetic illite formed is directly proportional to the amount of K-feldspar present initially. Figures 10 & 11 show that if K-feldspar is completely dissolved first (that is if the initial K-feldspar/kaolinite molar ratio is lower than 1), residual kaolinite remains stable and no additional illite is formed.

Simulation outcomes shown in Figs 10 & 11 are very close to those obtained in closed-system simulations, as well as to petrographic observations. About 4% diagenetic illite and 2% quartz are formed at the expense of 4% K-feldspar and 3% kaolinite. Other runs, using the same initial mineralogy as in Fig. 11, but performed with fresh water or Hild formation fluid as the injection water yielded very similar outcomes.

Considering open-system conditions, therefore, does not reproduce the large amount of diagenetic quartz observed in the reservoir. In an attempt to enhance quartz precipitation, coupled simulations were conducted using the same initial rock mineralogy as in Fig. 11 and a hypothetical Si-rich injection water. The Si-rich injection water was chemically identical to Hild formation water, with the difference that dissolved silica concentration was multiplied by two. This essentially had no effect on the final mineral abundances. It suggests that even the introduction of deep/hot Si-rich fluids in the reservoir would not be sufficient to explain the high volume of diagenetic quartz observed (10–20%).

DISCUSSION

The results of our simulations indicate that the proportion of K-feldspar and kaolinite present in the initial mineralogical assemblage is a determining factor in the progress and outcome of the reaction. In the Hild Brent sandstones, only traces (< 1%) of K-feldspar are found today, whereas up to 4% kaolinite remains visible (Lacharpagne *et al.*, 1999). This indicates that the K-feldspar/kaolinite molar ratio prior to the main illitization event was in favour of kaolinite, i.e. lower than 1. This is in good agreement with the reconstructed diagenesis of the Hild Brent sandstones, which includes an extensive kaolinite cementation event at intermediate depth and temperature (Girard *et al.*, 2002). As shown by geochemical modelling (Brosse *et al.*, this volume, pp. 383–408), the formation of kaolinite below 100°C essentially can be attributed to the dissolution of detrital plagioclase and phengite. The extent to which some dissolution of K-feldspar might have occurred at that time is unknown.

Consequently, in the light of the simulation results reported here, the amount of diagenetic illite formed in the Hild Brent sandstones was most likely controlled directly by the amount of K-feldspar present in the formation after the kaolinitization event, i.e. around 100°C. The amount of kaolinite present in the system did not constitute a limiting factor, even if closed-system conditions are considered. Closed-system simulations further indicated that other parameters such as fCO_2, temperature and aqueous acetate concentration were also not determining factors in the control of the amount of diagenetic illite formed.

A problematic question at the beginning of our simulations was the higher thermodynamic stability of diagenetic illite relative to kaolinite below 100°C. This is opposite to field observations, which indicate that diagenetic illite in Brent reservoirs mainly forms at a depth greater than about 3 km (Smith & Ehrenberg, 1989; Bjørlykke *et al.*, 1992; Giles, 1997). The cause of this discrepancy probably is related to thermodynamic or kinetic factors, including (i) inappropriate chemical formulae and thermodynamic data for detrital mica and diagenetic illite and (ii) a poorly defined precipitation rate and a nucleation energy barrier for diagenetic illite (Berger *et al.*, 1997).

In our attempts to overcome the poor knowledge of thermodynamic and kinetic data for clays (see above), we found that the 'Mg illite' end-member of the data0.com.R2 data base (Wolery, 1995) was a good thermodynamic and kinetic representative for detrital phengite occurring in the Hild Brent sandstones, whereas diagenetic illite was well represented by the 'muscovite' end-member of the EQ3/6 data base. Our results, however, indicate that below 100°C, the EQ3/6 'muscovite' (representing diagenetic illite) must be excluded from the system and only 'Mg illite' (representing detrital phengite) should be considered in order to prevent diagenetic illite precipitation during early to intermediate diagenesis (which is not supported by petrography). The 'Mg illite' is less stable than kaolinite and, therefore, dissolves readily to form kaolinite at low temperature. At temperatures greater than 100°C, both EQ3/6 'muscovite' and 'Mg illite' can be considered in geochemical simulations. This empirical simplification allows reconciliation between numerical modelling and field observations when it comes to illitization processes observed in Hild. It can be considered as a good substitute to approximate theoretical assumptions, and should be further validated in applications to other oil fields. A similar approach is used to account for metastability in the silica system for chemical geothermometry applications (Arnorsson, 1975; Michard, 1979).

Results of simulations performed in closed and open systems at 100–150°C show a small increase in porosity (< 1%), reflecting mainly the increase in rock density (very low molar volume of quartz compared with that of K-feldspar and kaolinite) and the expulsion of H_2O from solid kaolinite (illite water content is less than kaolinite). This increase in porosity does not necessarily induce an increase in the permeability of the reservoir. It is well known that pervasive formation of fibrous illite often results in a significant reduction of permeability because illite has a tendency to block pore throats, thereby decreasing pore connectivity (Ehrenberg, 1990). However, it is contrary to field observations, which indicate that porosity is significantly lower on average in the deeper, more diagenetically altered (more illitized) well 29/6-1 relative to the shallow well 30/4-2 (Lacharpagne *et al.*, 1999).

Consequently, in the Hild Field, the main diagenetic process responsible for the loss of porosity in carbonate-free sandstones is not illitization but more likely mechanical com-

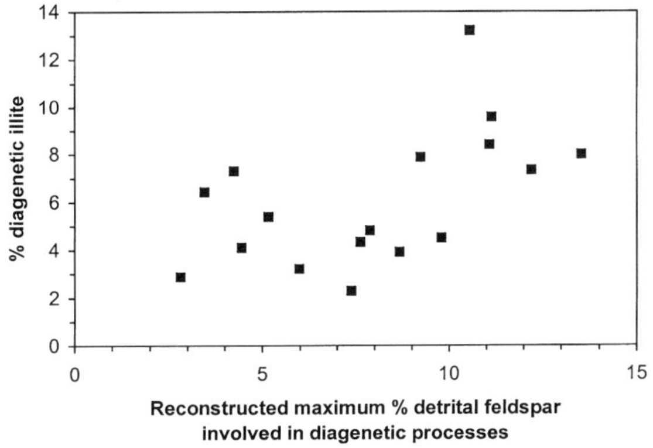

Fig. 12 Plot of percentage diagenetic illite in the Hild Brent sandstones versus maximum percentage detrital feldspar involved in diagenetic processes (reconstructed from point-counting data; Lacharpagne *et al.*, 1999). Data points represent stratigraphical unit averages.

paction and quartz cementation. The observed quantities of diagenetic quartz present in Hild could not be reproduced adequately by numerical modelling of the K-feldspar/kaolinite reaction as performed in this study. This indicates that the diagenetic quartz cement observed in Hild is not strictly sourced by illitization processes, i.e. dissolution of feldspar, nor by the introduction of putative Si-rich (possibly hot) extraneous fluids (see above). Other mechanisms must be considered, and in particular internal sources of Si such as pressure-solution (Lacharpagne *et al.*, 1999). Simple geochemical modelling of the illitization reaction is not sufficient on its own to predict the magnitude of porosity loss in the Hild reservoir as a result of diagenesis.

It is expected that the formation of illite will affect permeability (more so than porosity) and that part of the permeability loss recorded in deep parts of the Hild Field may indeed be related to illitization, as fibrous illite (I2) is more abundant in the deepest well (29/6-1). According to the results of geochemical simulations, diagenetic illite formation would be more significant in areas where detrital K-feldspar was initially more abundant. Figure 12 is a plot of the percentage of diagenetic illite in Hild Brent sandstones as a function of the reconstructed maximum amount of detrital K-feldspar possibly involved in diagenetic processes. The latter parameter corresponds to the difference between the reconstructed maximum percentage of detrital feldspar initially present in the sediment, as derived from point-counting data (Lacharpagne *et al.*, 1999), and the percentage of feldspar still present today. The percentage of illite is the percentage of total illite (I1 + I2) derived from point-counting data. The trend is that of an increasing percentage of illite with increasing percentage of feldspar. The correlation, however, is not good owing to the large uncertainty associated with the reconstruction of initial feldspar content in the sediment (and possibly to the fact that data points in Fig. 12 are for stratigraphical unit averages and not individual samples). In addition, the proportion of K-feldspar and plagioclase present in the initial sediment, and the quantity of detrital K-feldspar possibly dissolved to form kaolinite at $T < 100°C$ are not known. In spite of the large uncertainty, the trend shown in Fig. 12 supports our modelling results.

CONCLUSIONS

This study has attempted to reproduce numerically, illitization–quartzification diagenetic processes occurring in the Hild Field during deep burial at temperatures of 100–150°C. Closed-system simulations were performed using EQ3/6 and HILDSIM, a geochemical simulator specifically developed for this study. Open-system simulations were also performed by coupling HILDSIM with MARTHE, a transport code developed at BRGM. Variable initial mineralogical composition of the sandstone reservoir as well as different types of porewaters (sea water, fresh water, Hild formation water) were considered.

The results of geochemical modelling and numerical simulations performed in the course of this study lead to the following main remarks and conclusions.

1 The 'muscovite' end-member of the EQ3/6 data base (data0.sup.R2; Wolery, 1995) is a valid thermodynamic and kinetic representative of Hild diagenetic illite. The detrital muscovite, which exhibits a phengite chemical composition, is best represented by 'Mg illite' from the same data base.

2 The use of aqueous cation geothermometers to derive water–rock interaction temperatures, does not seem to be appropriate in reservoirs free of K-feldspar such as the Hild Brent sandstones.

3 Geochemical modelling results compared with the analytical mineral and fluid data suggest that illitization in Hild can be reproduced in closed-system conditions mainly following the reaction:

$$\text{K-feldspar} + \text{kaolinite} \Rightarrow \text{illite} + 2\ \text{quartz} + H_2O \quad (1)$$

4 In closed-system simulations, the quantity of diagenetic illite formed is controlled primarily by the amount of K-feldspar initially present. Optimum formation of illite is obtained when the molar K-feldspar/kaolinite ratio is lower or equal to 1. In the Hild Brent, this ratio is likely to have been lower than 1 by the time illitization started, owing to significant formation of kaolinite at an intermediate diagenetic stage. Variables such as temperature (between 100 and 150°C), pCO_2, and aqueous acetate concentration have little or no influence on the simulation outcomes.

5 Under conditions used in the simulations, coupled chemistry–transport modelling confirms the results obtained by closed-system modelling. Open-system simulations further indicate that illite can continue to precipitate even after kaolinite has been entirely consumed (as long as K-feldspar is available) according to the reaction:

$$3\ \text{K-feldspar} + 2H^+ \Rightarrow \text{illite} + 6\ \text{quartz} + 2K^+ \quad (2)$$

6 The amount of diagenetic illite (6% on average) present in the Hild reservoir can be readily reproduced by numerical modelling by adjusting initial mineralogy (K-feldspar and kaolinite content) appropriately. Petrographic evidence provides some support to the conclusion that variation in the amount of illite formed in the Hild Brent sandstones may have been controlled mainly by variation in the amount of K-feldspar present in the reservoir prior to illitization.

7 In constrast to illite, the large amount of quartz cement (10% on average) presently observed in the reservoir could not be generated by numerical modelling of the illitization processes as performed in this study. No more than 2–3% quartz cement can be produced, even when dissolved silica concentration in the pore fluid is increased arbitrarily. These results, combined with petrographic and mass balance evidence, suggest that silica was not sourced primarily by the dissolution of feldspar, nor by the introduction of Si-rich extraneous fluids. The contribution of other mechanisms, such as pressure solution, was most likely significant.

8 In the Hild Brent sandstones, the main diagenetic process responsible for the loss of porosity in carbonate-free sandstones is not illitization but more likely mechanical compaction and quartz cementation.

This study, together with other works, shows that numerical modelling in closed-system or open-system conditions is useful to simulate diagenetic transformations observed in reservoirs provided that sufficient constraints can be placed on mineralogy, fluid chemistry and pressure–temperature conditions of the system. Water–rock interaction modelling helped identify key parameters occurring during illitization and test the conditions of illite deposition (closed- or open-system conditions). A considerable advantage of numerical modelling is its ability to integrate all the constituents of a complex geochemical system. Global reactions such as (1) and (2) given above, the conditions of illite formation, and the reconstruction of the present chemical fluid composition could not be validated without modelling.

The lack of appropriate thermodynamic and kinetic data, or the large uncertainty associated with data available (especially for clays) may constitute a significant limitation to accurate prediction of diagenetic evolution by numerical modelling at the present time.

ACKNOWLEDGEMENTS

This work was conducted as part of the EU-funded DIAMAP project in the framework of the Thermie programme. Financial support was provided partly by the European Commission (DGXVII—contract OG-0165-96) and by the Research Division of BRGM. We are grateful to Norsk Hydro and Total-Fina for providing access to pre-existing data on Hild and to T.J. Wolery (Berkeley, USA) for providing his data0.sup.V8.R6 data base. Special thanks are expressed to A. Bariteau (IFP, France), who performed some of the early closed-system simulations during her postdoctorate at BRGM, and to J.C. Lacharpagne (Elf, France), F. Sommer (Total-Fina, France) and H. Johansen (IFE, Norway) for fruitful discussions and constructive criticism at different stages of the study. Helpful comments by reviewers D. Awwiller (Exxon), C. Rochelle (BGS) and R. Worden (University of Liverpool) greatly improved the clarity of the manuscript.

APPENDIX 1

Effect of activity models on water speciation, mineral saturation index (SI) and evolution of mineralogical assemblage

Activity model	T (°C)	fCO_2 (bar) Not fixed	pH	Molality (mol kg^{-1} of H_2O)								
				Na^+	K^+	Ca^{2+}	Mg^{2+}	Cl^-	SO_4^{2-}	HCO_3^-	SiO_2 (aq)	Al^{3+}
B-dot	150	0.485	5.12	0.848	3.81E-03	5.95E-02	3.85E-03	0.992	6.25E-05	1.78E-04	1.91E-03	7.13E-13
Davies	150	0.485	4.96	0.815	3.81E-03	5.75E-02	4.27E-03	0.957	4.66E-05	1.18E-04	1.91E-03	3.26E-12

Activity model	T (°C)	fCO_2 (bar)	log SI					
			Albite	K-feldspar	Illite	Kaolinite	Quartz	Calcite
B-dot	150	0.485	−0.43	−1.17	0	0	0	−1.04
Davies	150	0.485	−0.50	−1.17	0	0	0	−1.26

Activity model	Percentage					
	Albite	K-feldspar	Illite	Kaolinite	Quartz	Calcite
B-dot	0	0	16.10	5.90	56.65	0
Davies	0	0	16.10	5.90	56.65	0

APPENDIX 2

Effect of pressure on water speciation

				Molality (mol kg^{-1} of H_2O)								
T (°C)	P_{tot} (bar)	fCO_2 (bar)	pH	Na^+	K^+	Ca^{2+}	Mg^{2+}	Cl^-	SO_4^{2-}	HCO_3^-	SiO_2(aq)	Al^{3+}
150	4.75	9	4.99	0.9105	5.72E-03	6.25E-03	3.99E-03	1.662	6.74E-05	2.49E-03	1.91E-03	1.69E-12
150	800	9	5.15	0.9340	5.81E-03	6.41E-02	4.23E-03	1.093	7.22E-05	3.11E-03	2.45E-03	1.03E-12
150	4.75	25	4.77	0.9104	5.72E-03	6.17E-02	3.93E-03	1.067	6.75E-05	4.19E-03	1.91E-03	7.75E-12
150	800	25	4.92	0.9340	5.81E-03	6.30E-02	4.16E-03	1.093	7.25E-05	5.27E-03	2.45E-03	4.75E-12

APPENDIX 3

Initial chemical composition of the different waters used in open-system coupled simulations

Water	TDS mg l^{-1}	pH	Na mg l^{-1}	K mg l^{-1}	Ca mg l^{-1}	Mg mg l^{-1}	Cl mg l^{-1}	HCO_3 mg l^{-1}	SO_4 mg l^{-1}	SiO_2 mg l^{-1}	Al µg l^{-1}	CH_3COOH mg l^{-1}
Sea water (Nordstrom *et al.*, 1979)	35900	8.22	11019	408	422	1322	19805	145	2775	4.38	2.05	0
Fresh water (Sanjuan *et al.*, 1995)	543	7.20	4.37	0.78	116	7.29	2.13	391	9.61	12.02	0.0027	0
Hild water*	69088	5.30	23800	260	2650	170	41850	319	0	58	0.27	300

* pH, HCO_3 and Al concentrations were calculated assuming T = 100°C, fCO_2 = 3 bar (in agreement with data compiled by Smith & Ehrenberg (1989) for North Sea fields) and water in equilibrium with calcite and kaolinite.

REFERENCES

Aggarwal, P.K., Gunter, W.D. & Kharaka, Y.K. (1990) Effects of pressure on aqueous equilibria. In: *Chemical Modelling of Aqueous System II*, American Chemical Society, Washington, DC, pp. 87–101.

Aja, S.U. (1995) Thermodynamic properties of some 2 : 1 layer clay minerals from solution-equilibration data. *Eur. J. Mineral.*, **7**, 325–333.

Arnorsson, S. (1975) Application of the silica geothermometer in low temperature area in Iceland. *Am. J. Sci.*, **275**, 763–784.

Barth, T. (1991) Organic acids and inorganic ions in waters from petroleum reservoirs, Norwegian continental shelf: a multivariate statistical analysis and comparison with american reservoir formation waters. *Appl. Geochem.*, **6**, 1–15.

Berger, G., Lacharpagne, J.C., Velde, B., Beaufort, D. & Lanson, B. (1997) Kinetic constraints on illitisation reactions and the effects of organic diagenesis in sandstone/shale sequences. *Appl. Geochem.*, **12**, 23–35.

Bjørlykke, K., Nedkvitne, T., Ramm, M. & Girish, C. (1992) Diagenetic processes in the BRENT Group (Middle Jurassic) reservoirs of the North Sea: an overview. In: *Geology of the Brent Group* (Ed. by A.C. Morton, R.S. Haszeldine, M.R. Giles and S. Brown). *Geol. Soc. London Spec. Publ.*, **61**, 263–287.

Brosse, É., Matthews, J., Bazin, B., Le Gallo, Y. & Sommer, F. (2000) Related quartz and illite cementation in the Brent sandstones: a modelling approach. In: *Quartz Cementation in Sandstones*, International Association of Sedimentologists Special Publication 29 (Ed. by R.H. Worden and S. Morad), Blackwell Science, Oxford, pp. 51–66.

Brosse, É., Margueron, T., Cassou, C., *et al.* (2003) The formation and stability of kaolinite in Brent sandstone reservoirs: a modelling approach. In: *Clay Mineral Cements in Sandstones*, International Association of Sedimentologists Special Publication 34 (Ed. by R.H.

Worden and S. Morad), Blackwell Publishing, Oxford, pp. 383–408.

Chermak, J.A. & Rimstidt, J.D. (1989) Estimating the thermodynamic properties ($\Delta G°$f and $\Delta H°$f) of silicate minerals at 298.15°K. *Am. Mineral.*, **74**, 1023–1031.

Czernichowski, I., Sanjuan, B., Rochelle, C., *et al.* (1996) Analysis of the geochemical aspects of the underground disposal of CO_2. In: *Deep Injection Disposal of Hazardous and Industrial Waste* (Ed. by J.A. Apps and C.F. Tsang), Academic Press, San Diego, CA, pp. 565–583.

Drummond, S.E. Jr. (1981) *Boiling and mixing of hydrothermal fluids: chemical effects on mineral precipitation.* Unpublished PhD thesis, Pennsylvania State University. In: *SOLMINEQ.88: a Computer Program for Geochemical Modelling of Water–Rock Interactions* by Kharaka Y.K., Gunter, W.D., Aggarwal, P.R., Perkins, E.H. & De Brall, J.D. US Geological Survey Water Resources Investigations Report, 88–4227 (1988).

Duan, Z., Moller, N. & Weare, J.H. (1992) An equation of state for CH_4–CO_2–H_2O system: I. Pure systems from 0 to 1000°C and 0 to 8000 bar. *Geochim. Cosmochim. Acta*, **56**, 2605–2617.

Egeberg, P.K. & Aagaard, P. (1989) Origin and evolution of formation waters from oil fields on the Norwegian shelf. *Appl. Geochem.*, **4**, 131–142.

Ehrenberg, S.N. (1990) Relationship between diagenesis and reservoir quality in sandstones of the Garn Formation, Haltenbanken, mid-Norwegian continental shelf. *AAPG Bull.*, **74**, 1538–1558.

Ehrenberg, S.N., Aagaard, P., Wilson, M.J., Fraser, A.R. & Duthie, M.L. (1993) Depth-dependent transformation of kaolinite to dickite in sandstones of the Norwegian continental shelf. *Clay Mineral.*, **28**, 325–352.

Giles, M.R. (1997) *Diagenesis: a Quantitative Perspective—Implications for Basin Modelling and Rock Property Prediction.* Kluwer, Dordrecht, 526 pp.

Girard, J.-P., Munz, I.A., Johansen, H., Hill, S.J. & Canham, A. (2001) Conditions and timing of quartz cementation in Brent reservoirs, Hild Field, North Sea: constraints from fluid inclusions and SIMS oxygen isotope microanalysis. *Chem. Geol.*, **176**, 1–4.

Girard, J.-P., Munz, I.A., Johansen, H., Lacharpagne, J.-C. & Sommer, F. (2002) Temperature, timing and origin of fluids involved in the diagenesis of Hild Brent reservoirs, Norwegian North Sea: fluid inclusions, stable isotope and K/Ar dating. *J. Sed. Res.* (In press)

Harrisson, W.J. & Thyne, G.D. (1992) Predictions of diagenetic reactions in the presence of organic acids. *Geochim. Cosmochim. Acta*, **56**, 565–586.

Helgeson, H.C., Delany, J.M., Nesbitt, H.W. & Bird, D.K. (1978) Summary and critique of the thermodynamic properties of rock-forming minerals. *Am. J. Sci.*, **278-A**, 229 pp.

Johnson, J.W., Oelkers, E.H. & Helgeson, H.C. (1992) SUPCRT92: a software package for calculating the standard molal thermodynamic properties of minerals, gases, aqueous species and reactions from 1 to 5000 bar and 0 to 1000°C. *Comp. Geosci.*, **18**, 899–947.

Kervévan, C. & Baranger, Ph. (1998) SCS: specific chemical simulators dedicated to chemistry–transport modelling. Part I—design and construction of an SCS. Goldschmidt Conference, Toulouse, 29th August–3rd September. *Mineral. Mag.*, **62A**, 771–772.

Kervévan, C., Thiery, D. & Baranger, Ph. (1998) SCS: specific chemical simulators dedicated to chemistry–transport modelling. Part II—coupling of SCS with the hydro-transport modelling software MARTHE. Goldschmidt Conference, Toulouse, 29th August–3rd September. *Mineral. Mag.*, **62A**, 773–774.

Kharaka, Y.K., Hull, R.W. & Carothers, W. (1985) Water–rock interaction in sedimentary basins. In: *Relationship of Organic Matter and Mineral Diagenesis*, Society for Economic Paleontology and Mineralogy Special Publication 17 (Ed. by D.L. Gauthier, Y.K. Kharaka and R.C. Surdam), Tulsa, OK, pp. 79–174.

Lacharpagne, J.-C., Girard, J.-P., Sanjuan, B., *et al.* (1999) *DIAMAP—Reservoir Quality Mapping Through Numerical Prediction of Mineral Diagenesis.* Final Technical Report No. OG/165/96, Groupement Européen de Recherches Technologiques sur les Hydrocarbures, Rueil Malmaison, 106 pp + annexes.

Le Gallo, Y., Bildstein, O. & Brosse, E. (1998) Coupled reaction–flow modeling of diagenetic changes in reservoir permeability, porosity and mineral compositions. *J. Hydrol.*, **208**, 366–388.

May, H.M., Kinniburgh, P.A., Helmke, P.A. & Jackson, M.L. (1986) Aqueous dissolution, solubilities and thermodynamic stabilities of common aluminosilicate clay minerals: kaolinites and smectites. *Geochim. Cosmochim. Acta*, **50**, 1667–1677.

Michard, G. (1979) Géothermomètres chimiques. *Bull. Bur. Rech. Géol. Min. (2ème sér.), Sect. III*, **2**, 183–189.

Michard, G. (1983) *Recueil de données thermodynamiques concernant les équilibres eaux-minéraux dans les réservoirs hydrothermaux.* Report EUR 8590FR, Commission des Communautés Européennes, Bruxelles.

Michard, G. (1989) *Equilibres chimiques dans les eaux naturelles.* Editions Published, Paris, 357 pp.

Montoya, J.W. & Hemley, J.J. (1975) Activity relations and stabilities in alkali feldspar and mica alteration reactions. *Econ. Geol.*, **70**, 577–594.

Nadeau, P.H. & Hurst, A. (1991) Application of back-scattered electron microscopy to the quantification of clay mineral microporosity in sandstones. *J. Sed. Petrol.*, **61**, 921–925.

Nordstrom, D.K., Plummer, L.N., Wigley, T.M.L., *et al.* (1979) A comparison of computerized chemical models for equilibrium calculations in aqueous system. In: *Chemical Modeling in Aqueous Systems*, American Chemical Society Symposium Series, Vol. 93 (Ed. by E.A. Jenne), American Chemical Society, Washington, DC, 857–892.

Robie, R.A., Hemingway, B.S. & Fisher, J.R. (1978) Thermodynamic properties of minerals and related substances at 298.15°K and 1 bar (10^5 Pascal) pressure and at higher temperatures. *US Geol. Surv. Bull.*, **1452**, 456 pp.

Sanjuan, B., Girard, J.P. & Czernichowski, I. (1995) *Geochemical Modelling of the Main Diagenetic Processes in the Oseberg Sandstone Reservoir, Oseberg Field, Northern North Sea.* BRGM Report R38599, Bureau de Recherches Geologiques et Minières (France), Orleans, 167 pp.

Séjourné, C. (1997) *Hild Field—Temispack Diagenesis Study.* Internal Report No. 97-041RS, Elf EP, Pau, 10 pp. + figures.

Smith, J.T. & Ehrenberg, S.N. (1989) Correlation of carbon dioxide abundance with temperature in clastic hydrocarbon reservoirs: relationships to inorganic chemical equilibrium. *Mar. Petrol. Geol.*, **6**, 129–135.

Tardy, Y. & Duplay, J. (1992) A method of estimating the Gibbs free energies of formation of hydrated and dehydrated clay minerals. *Geochim. Cosmochim. Acta*, **56**, 3007–3029.

Tardy, Y. & Garrels, R.M. (1974) A method of estimating the Gibbs energies of formation of layer silicates. *Geochim. Cosmochim. Acta*, **38**, 1101–1116.

Thiery, D. (1994) *Logiciel Marthe (Modélisation d'Aquifère par un maillage Rectangulaire en régime Transitoire pour le calcul Hydrodynamique des Ecoulements)*, Version 5.4. Rapport BRGM R38085, Bureau de Recherches Geologiques et Minières (France), Orleans, 54 pp.

Total Norge (1993) *Sedimentology, Biostratigraphy, Diagenesis and Reservoir Properties (Scal) of the Tarbert Formation in Wells 30/4-2, 30/7-8 and 29/6-1, Hild Field, Norway.* Internal Report RL 5169, Total Exploration Production.

Vieillard, Ph. (1994) Prediction of enthalpy of formation based on refined crystal structures of multisite compounds: Part 1. Theories and examples. *Geochim. Cosmochim. Acta*, **58**, 4049–4063.

Warren, E.A. & Smalley, P.C. (1994) *North Sea Formation Waters Atlas.* Geological Society Memoir 15, Geological Society Publishing House, Bath.

Wilson, M.D. (1994) *Reservoir Quality Assessment and Prediction in Clastic Rocks.* Society of Economic Paleontologists and Mineralogists Short Course 30, Tulsa, OK, 432 pp.

Wolery, T.J. (1995) *EQ3/6, a Software Package for Geochemical Modeling of Aqueous Aystems: Package Overview and Installation Guide*, Version 7.2b. Laurence Livermore National Latoratory, Livermore, CA.

Zotov, A., Mukhamet-Galeev, A. & Schott, J. (1998) An experimental study of kaolinite and dickite relative stability at 150–300°C and the thermodynamic properties of dickite. *Am. Mineral.*, **83**, 516–524.

Int. Assoc. Sedimentol. Spec. Publ. (2003) **34**, 453–469

The effect of oil emplacement on diagenetic clay mineralogy: the Upper Jurassic Magnus Sandstone Member, North Sea

R.H. WORDEN[1] *and* S.A. BARCLAY[2]

[1] *Department of Earth Sciences, University of Liverpool, Brownlow Street, Liverpool L69 3GP, UK, e-mail: r.worden@liverpool.ac.uk*

[2] *Department of Geology and Geophysics, University of Edinburgh, EH9 3JW, UK*

ABSTRACT

The Upper Jurassic Magnus turbidite sandstone has higher illite/(illite + kaolin) and $K_2O/(K_2O + Al_2O_3)$ ratios in the oil zone than in the water zone. Both clay minerals grew during burial diagenesis at similar times as quartz and ankerite cementation and oil emplacement. Illite may have grown at the same time as, or slightly later than, kaolin. Both clay minerals probably grew as a result of the dissolution of aluminosilicate minerals (e.g. K-feldspar). Bulk chemical (X-ray fluorescence) data show that the oil zone has a greater amount of potassium than the water zone. Illite thus grew where the aqueous potassium could not escape from the sandstone, whereas kaolin grew where aqueous potassium was able to escape the host sandstone. The oil zone would have had many orders of magnitude slower transport of potassium than the water zone (by either advection or diffusion mechanisms). The clay-mineral distribution pattern is thus likely to result from the inability of the oil-zone sandstone to lose potassium relative to the water-zone sandstone. The pattern is not the result of primary facies or depositional environment variations because all samples come from the same facies and turbidite sands tend to have minimal lateral variation. Nor is the pattern the result of early diagenetic processes beneath the Cimmerian Unconformity because kaolin is found in greatest abundance furthest away from the unconformity. The prevalence of kaolin cannot be the result of minor differences in reservoir temperature between the crest (oil zone) and flank (water zone) of the structure because illite, and not kaolin, would be the favoured mineral at the greater depth and thus higher temperature of the water zone. The data presented here imply that diagenesis is modified (in terms of physicochemical mechanism and geochemistry) rather than stopped by oil emplacement. The data also imply that oil-zone mudstones may be slightly less illitized than water-zone mudstones, although appropriate data to test this hypothesis have not yet been collected.

INTRODUCTION

Oil emplacement has been shown to have a wide range of effects on diagenetic processes in sandstones. *In extremis*, oil has been attributed with the ability to totally stop diagenesis (e.g. Thomas, 1986; Robinson & Gluyas, 1992; Hogg & Pearson, 1995). Conversely, some have speculated that the addition of petroleum may have negligible or no impact upon inorganic diagenetic processes (e.g. Ehrenberg, 1990; Walderhaug, 1994; Darby *et al.*, 1997). However, an analysis of the physical effects of oil emplacement has demonstrated that oil will variably slow down cementation dependent on the source of the cement, the transport mechanism, wettability and the rock fabric (Worden *et al.*, 1998). Some effort has been spent examining the physicochemical impact of oil emplacement on quartz cementation but less attention has been given to the physical effect of oil emplacement on clay cementation.

Theoretically, the addition of oil to porous rocks can affect chemical as well as physical diagenesis. However, it is highly unlikely that oil will help transport aluminosilicate cementing species through complexation (e.g. organo-silica complexes are thought to be of no importance; Fein, 2000), but CO_2 entrained in the oil can cause carbonate cementation, increasing the formation water acidity and thus inducing silicate diagenesis (Worden & Barclay, 2000).

The addition of oil has a major impact upon transport of aqueous species. Advective flow of water in an oil reservoir is limited by the relative permeability scaling factor for two fluid phase systems. Diffusive transport is also inhibited, owing to increased tortuosity of the residual water in the pore network (Worden *et al.*, 1998). Grain surface reactions, such as dissolution and precipitation of minerals, theoretically will be unaffected by oil emplacement if the system is water-wet although surface processes will be halted in a totally oil-wet system (Worden *et al.*, 1998; Barclay & Worden, 2000a; Worden & Heasley, 2001).

In this paper, we report a systematic investigation of the distribution of clay minerals through an oil–water contact in order to assess the impact of oil emplacement on clay diagenesis in sandstones. A high-density sampling programme was used, resulting in the analysis of clay populations in more than 250 samples from five wells from one oil field.

General petroleum geology of the Magnus Field

The Magnus Field lies 160 km to the northeast of the Shetland Islands within UKCS blocks 211/12a and 211/7a. The field occurs in the Magnus province at the southern margin of the North Shetland Basin, which is typified by easterly dipping fault blocks and Upper Jurassic sandstone reservoirs (Fig. 1).

The reservoir occurs within the Late Jurassic Humber Group, which is subdivided into the Heather Formation and the overlying Kimmeridge Clay Formation (Barclay *et al.*, 2000). The main reservoir, the Magnus Sand Member (MSM), occurs within the Kimmeridge Clay Formation. The MSM was a submarine fan complex incorporating thick amalgamated channel deposits and thinly bedded turbidite channel sands and various fine grained facies including *sensu stricto* mudstones (De'ath & Schuyleman, 1981).

In late Volgian and Ryazanian times, the reservoir was tilted and eroded in the crest area of the field prior to the deposition of the Cretaceous Cromer Knoll Group. A second break in deposition occurred during Cenomanian and Turonian times, followed by burial during the rest of Cretaceous and Tertiary times. The field is a combined structural and stratigraphical trap; the seal is the unconformable Cretaceous mudstone that drapes the structure. The source rock was the Kimmeridge Clay Formation with the kitchen occurring off-structure in the basin. The structure was filled with oil at approximately 75 Ma (Shepherd *et al.*, 1990; Emery *et al.*, 1993). The oil–water transition zone occurs between about 3160 and 3180 m, although the local water saturation depends on a large number of factors including wettability, pore fabric (narrow pore throats and pore cul-de-sacs

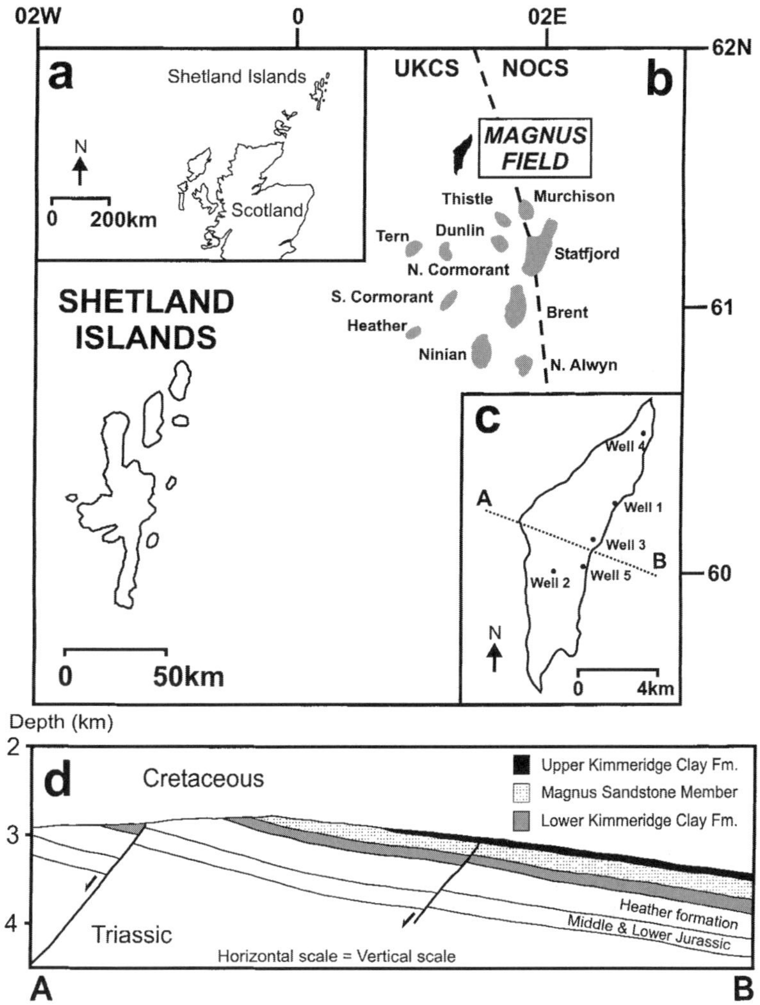

Fig. 1 Magnus petroleum geology: (a–c) Magnus Field location; (d) cross-section through the Magnus Field. Note inset map (c) of the field with the five wells added.

encouraging retention of water) and compartmentalization (Barclay *et al.*, 2000).

GEOCHEMISTRY OF THE MAGNUS SANDSTONE

Magnus is a subarkosic sandstone dominated by monocrystalline quartz and subordinate polycrystalline quartz, with K-feldspar, minor plagioclase and trace quantities of detrital muscovite (De'ath & Schuyleman, 1981). The massive structureless sandstones were relatively clean with no bioturbation or soft-sediment deformation introducing detrital clay into the sandstone. There are no signs of detrital kaolin or illite in these sandstones.

The diagenetic history of the Magnus Sandstone can be divided conveniently into three phases (Fig. 2):

1 initial shallow burial during the Late Jurassic;

2 minor uplift, erosion and telodiagenesis during the Early Cretaceous;

3 renewed burial towards the present-day depth during the Late Cretaceous through to the early Tertiary (Fig. 2).

The thermal history was determined using a one-dimensional heat flow model and calibrated to vitrinite reflectance (Emery *et al.*,

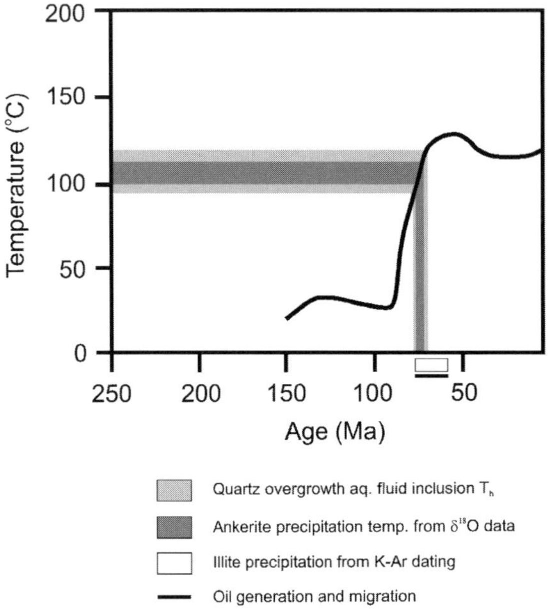

Fig. 2 Thermal history with the time–temperature information from mineral cements superimposed.

1993). The renewed burial diagenesis had the most important effect upon reservoir quality.

The shallow burial was associated with minor calcite cementation and physical compaction (Emery *et al.*, 1990). The transient telogenetic stage was responsible for localized kaolin growth at the expense of K-feldspar at the exposed crest of the field. The telogenetic K-feldspar decay and kaolin growth is thought to be restricted to within a few metres of the top of the Magnus sandstone at the structurally highest point within the field and is not widespread (Emery *et al.*, 1990).

The burial diagenetic cements include minor K-feldspar and more important quartz, ankerite, kaolin and illite (Shepherd *et al.*, 1990; Macaulay *et al.*, 1992; Emery *et al.*, 1993; Barclay & Worden, 2000b; Worden & Barclay, 2000). The carbonate cements within the MSM have $\delta^{13}C$ values of −7.7 to −12.8‰ PDB (a standard based on a Jurassic belemnite from the Pee Dee Formation) (Macaulay *et al.*, 1992) suggesting an influx of organically derived aqueous CO_2. Gaseous CO_2 is present in the associated gas in Magnus at a concentration of about 2% by volume and has $\delta^{13}C$ values of between −13 and −16‰ PDB, suggesting strongly that source-rock-derived CO_2 has been important in the diagenesis of the MSM (Worden & Barclay, 1999, 2000).

Oil fluid inclusions have been observed in diagenetically altered detrital feldspar grains, in feldspar cements, in healed fractures within quartz grains and in quartz cements. Rare feldspar overgrowths (less than 1% rock volume; Emery *et al.*, 1993), contain primary aqueous and hydrocarbon inclusions, the latter fluorescing yellow under ultraviolet illumination. Quartz overgrowths represent 3–12% of rock volume and contain primary hydrocarbon inclusions that fluoresce blue-green (i.e. slightly higher maturity than the oil in the petrographically earlier feldspar cement). Hydrocarbons within altered feldspar are a whitish-blue fluorescing oil, which appears to have a higher API° gravity than those in the cements. As the most volumetrically abundant oil inclusions are found within altered feldspars, processes of feldspar alteration (and concomitant aqueous and oil inclusion formation) may have been occurring since early Cretaceous times through to the present day. Clay diagenesis, the focus of this paper, will be dealt with in subsequent sections. The modelled timing of local source-rock maturation, distribution of hydrocarbon inclusions, and petrographic relationships between oil and mineral cements, all suggest that oil emplacement was at least partly synchronous with late burial diagenesis during the growth of quartz and ankerite cements (Emery *et al.*, 1993; Worden & Barclay, 1999, 2000).

SAMPLES AND METHODS

Samples were collected from five wells across the field covering the water zone, the oil zone and the transition zone. Over 500 samples were collected initially; at the same time the rock was characterized sedimentologically using conventional core description techniques. Only the main reservoir facies (structureless, massive sands of the Bouma-A facies) were considered in this study, thus removing any interfacies variability. Thus the data reported here include 257 samples; the remaining samples belong to other facies.

The paragenetic sequence and clay cement types were investigated using optical and scanning electron microscope (SEM) petrography. Clay types were identified in a JEOL 6400 SEM, operating at 20 kV, by observing morphology and by using semi-quantitative energy dispersive secondary X-ray analysis.

All samples were characterized mineralogically using X-ray diffraction (XRD). Raw data were quantified using the Rietveldt Method and Siroquant™ version 2.0 software (Sietronics Pty Ltd). The Rietveldt method iteratively computes mineral abundances by calculating a best-fit trace based upon an initial qualitative assessment of minerals identified from the raw data (Bish & Post, 1993). The software compares the raw trace to the calculated best-fit trace and determines an overall error (χ^2). The closer this χ^2 value is to unity, the better is the match between the traces. Values less than 6 have been deemed acceptable for bulk mineral quantification in simple systems (Bish & Post, 1993) and those obtained in this study were all < 4. This corresponds to estimated < 1% errors in quantification of the major mineral phases. Siroquant™ cannot differentiate between ferroan and non-ferroan calcite, although ferroan and non-ferroan dolomites have sufficient peak shifts to be discriminated.

Major and trace element geochemistry were determined for each sample using routine wavelength dispersive X-ray fluorescence (XRF) techniques. Rock samples were crushed, ground and made into pellets using a borax holder and a manual press. The XRF data were produced on a Siemens SRS 303 A5 automatic X-ray fluorescence spectrometer (with an end-window rhodium tube) calibrated against international standards.

The formation water conditions that favoured kaolin or illite precipitation were investigated using equilibrium thermodynamics. Although diagenesis is characterized by systems not being at equilibrium, reaction pathway modelling (equilibrium thermodynamics) is a useful approach because it shows how water–rock systems will evolve (given sufficient time). Phase diagrams were produced using the computer programs Act2™ and Tact™ (Bethke, 1994), and a set of thermodynamic data compiled and distributed by Lawrence Livermore National Laboratory (Delany & Lundeen, 1990). The modelling input parameters were taken from Barclay & Worden (2000b).

RESULTS

The results presented here all come from the same sedimentary facies (massive structureless, Bouma-A sands). None of the differences between samples reported in this section are the result of primary sedimentary differences in depositional facies.

The bulk geochemical analysis of the sandstones revealed a modest range of potassium contents from a minimum of about 1.5 wt% (in terms of the oxide) and a maximum of about 4.0 wt%. Although there was scatter in the data, the lowest values occur below and the highest values occur above the oil–water transition zone at about 3170 m (Fig. 3a). Comparison of the relatively labile potassium to an oxide that is considered immobile is a useful way to ensure that the comparison between samples is truly valid. Also, 'dilution' of the potassium-bearing minerals (clays, feldspars, etc.) by other (especially post-depositional) allogenic minerals such as carbonate cements can be corrected by this approach. The fractional parameter $K_2O/(K_2O + Al_2O_3)$ has been derived for each sample and plotted versus depth (Fig. 3b). This fraction has lower values below and higher values above the oil–water transition zone. The reduction in scatter relative to the simple K_2O data presumably reflects correction for diluting minerals (mentioned above).

Ultraviolet fluorescence microscopy confirmed that quartz cement contains primary oil inclusions. This proves that quartz cementation occurred when oil was present in the reservoir (Fig. 4a & b).

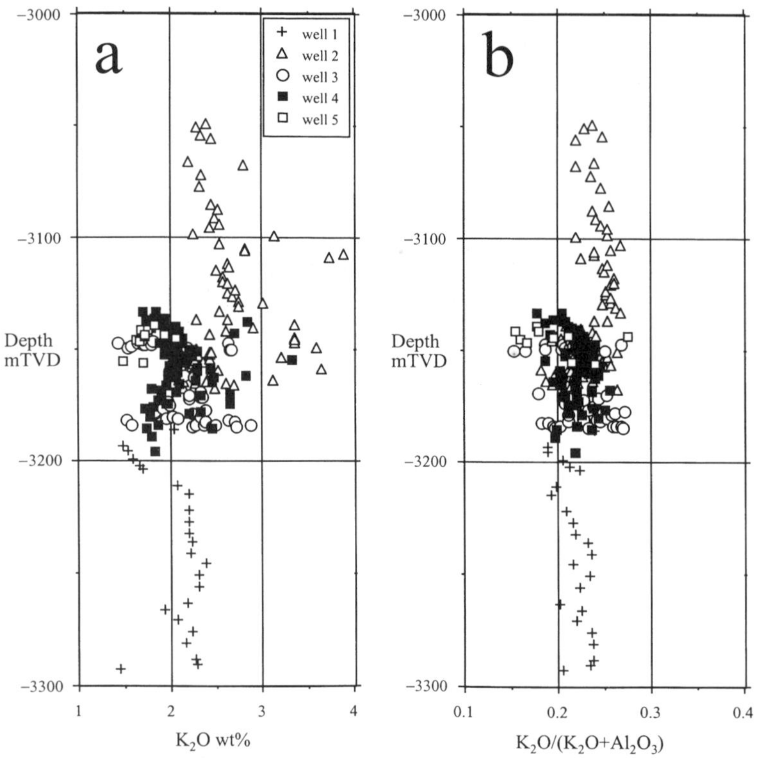

Fig. 3 X-ray fluorescence (XRF) data from the massive structureless sandstones of the Magnus Sandstone Member. (a) Simple plot of potassium oxide (as reported by the XRF) versus depth. (b) Plot of potassium oxide normalized to aluminium oxide to remove the effects of dilution by post-depositional minerals such as calcite and dolomite. The oil–water transition zone is reported to lie between 3160 and 3180 m although locally there may be elevated water saturations owing to local wettability and fabric heterogeneity (Barclay *et al.*, 2000). TVD = total vertical depth.

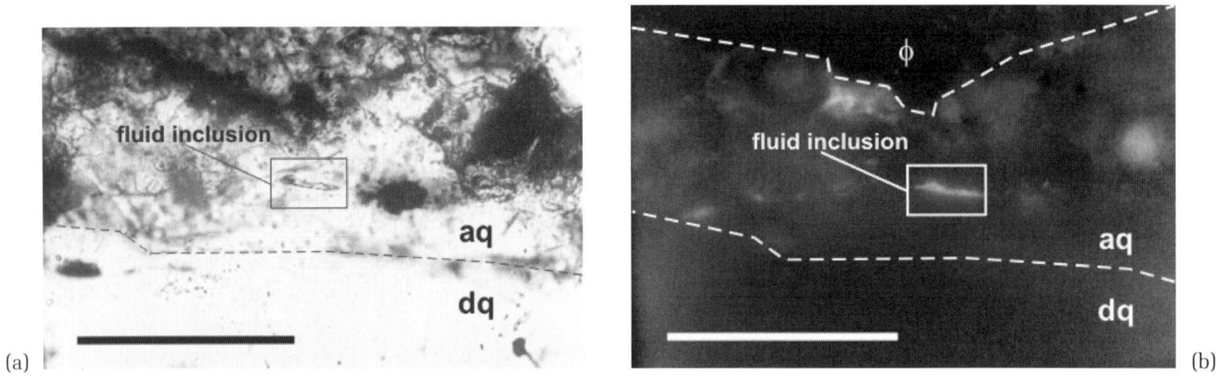

Fig. 4 (a) Fluid inclusion wafer photomicrograph of a primary, two-phase (liquid + gas) petroleum fluid inclusion in the middle of a quartz overgrowth (aq);dq = detrital quartz (scale bar = 100 μm). (b) Same view as in (a) but illuminated by UV light causing the petroleum in the inclusion to fluoresce. Boundaries of authigenic quartz overgrowth are indicated by white dotted lines; ϕ = porosity.

The SEM petrographic analysis showed that quartz and discrete booklets of kaolin are locally intergrown, suggesting that they grew at much the same time (Fig. 5a) and suggesting that kaolin grew at a similar time to oil emplacement. The kaolin also forms dense monomineralic patches where the relative timing of growth is less easily established (Fig. 5b). K-feldspar has been dissolved locally, leaving some secondary intragranular porosity. Trace

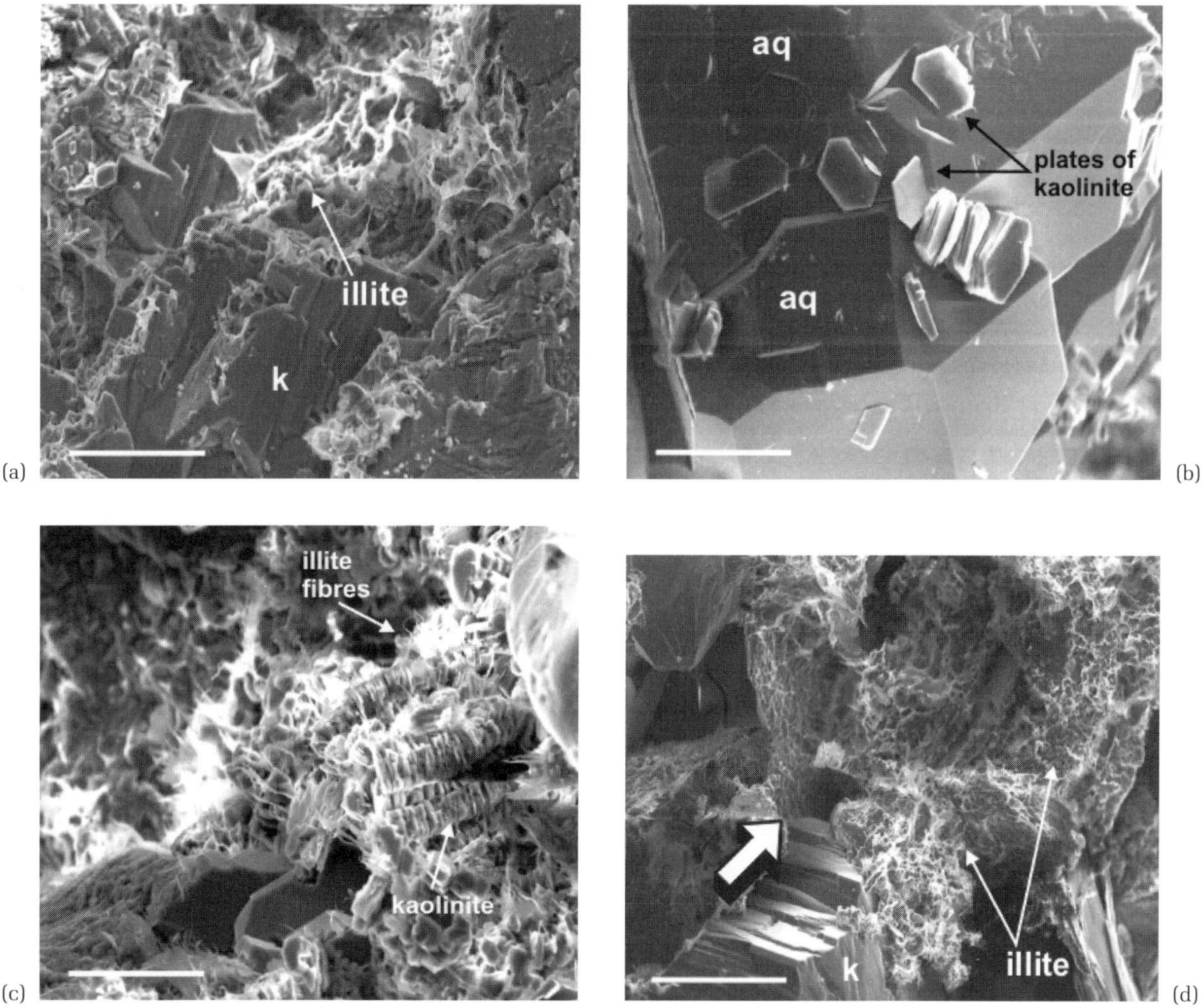

Fig. 5 (a) Scanning electron microscopy micrographs of diagenetic features from the Magnus Sandstone. Partially decomposed detrital K-feldspar grain (k) with authigenic illite growing on the surface (scale bar = 20 μm). (b) Quartz cement (aq) intergrown with plates of kaolinite (scale bar = 20 μm). (c) Illite cement that has grown on pre-existing kaolinite cement (scale bar = 15 μm). Illite may result from kaolinite decomposition or may simply have used it as an energetically favourable nucleation site. (d) Extensive growth of illite in pores (scale bar = 50 μm; arrow indicates direction of pore throat). This form of illite would seriously damage reservoir quality.

amounts of detrital muscovite were observed, although this generally was below detection limits using point counting (on 400 counts per section). Detrital muscovite therefore represents less than 0.25% of the rock. Fine filaments of illite grow on top of, or out of, kaolin plates, suggesting that the illite grew at the same time as or after the kaolin (Fig. 5c). In samples where illite is abundant, it effectively blocks pores and thus is highly detrimental to reservoir quality (Fig. 5d).

The amount of kaolin varies with depth within the same facies across the field (Fig. 6a). Above the transition zone, there is minimal kaolin, with many samples displaying no kaolin on XRD traces. Unlike kaolin, illite is relatively more abundant above the oil–water transition zone than it is below (Fig. 6b). Many

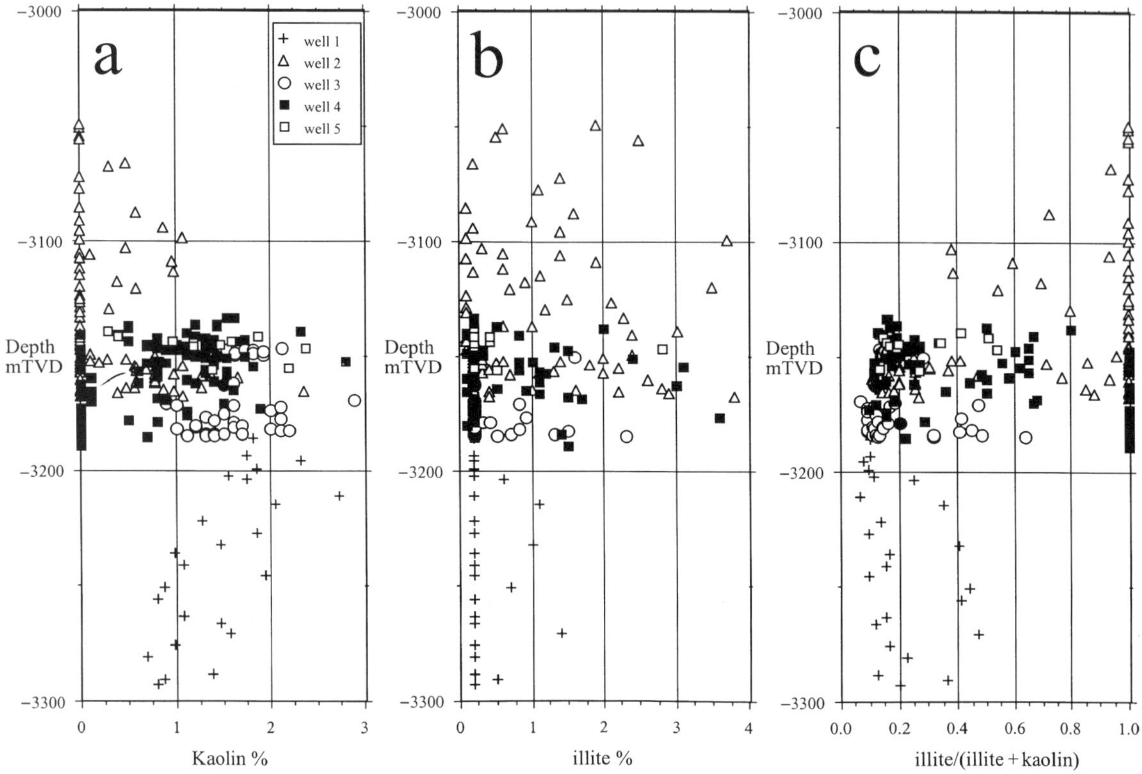

Fig. 6 Variations of absolute abundance of (a) kaolin with depth, (b) illite with depth and (c) the variation of the relative proportions of illite and kaolin with depth. The oil–water transition zone is reported to lie between 3160 and 3180 m although locally there may be elevated water saturations owing to local wettability and fabric heterogeneity (Barclay *et al.*, 2000). TVD = total vertical depth.

samples from the transition zone into the water zone display very small amounts of illite. The relative magnitude of the clay-mineral variations is best represented as a fraction. Thus, a plot of illite/(illite + kaolin) (Fig. 6c) displays higher values above than below the transition zone. Although the separation of illite and kaolin in the Magnus Sandstone is not perfect, there is a clear element of heterogeneity in the distribution of aluminosilicate clay minerals, with the oil zone favouring illite and the water zone favouring kaolin. The pattern of kaolin distribution (increasing down the structure) was suggested previously by Macaulay *et al.* (1993b), but the pattern of illite distribution has not been reported previously. Detrital muscovite (and any sericitic alteration of feldspars) and diagenetic illite are difficult to resolve using XRD data because they are mineralogically very similar. However, there will be a relatively minor contribution of detrital muscovite to these data because the less than 0.25 vol.% (reported above) is typically much less than the percentages reported for illite in Fig. 6b. There is no link or correlation between the amounts of detrital plagioclase and diagenetic kaolin, suggesting that localization of the kaolinization of calcic plagioclase is not a significant process (e.g. Morad *et al.*, 1990).

DISCUSSION

The key issues that need to be addressed to understand the clay mineral distribution include: (i) the timing of kaolin and illite growth relative to oil emplacement and other diagenetic phenomenon, (ii) the geochemical cause of the distribution pattern and (iii) the petrophysical origin of the geochemical controls (i.e. controls on material flux).

Clay-mineral diagenesis in the Magnus Sandstone Member: when did the clay minerals grow?

The paragenetic sequence for burial diagenetic mineral cements in the Magnus Sandstone Member includes late diagenetic quartz and ankerite cement as well as illite and kaolin. Emery *et al.* (1993) and Worden & Barclay (1999, 2000) reported fluid inclusion homogenization temperatures (converted into dates by reference to a thermal history analysis) for quartz and ankerite; temperatures also were inferred from oxygen isotope data from ankerite (similarly converted to dates; Macaulay *et al.*, 1992, 1993a, b) and illite was dated directly using K–Ar methods (Emery *et al.*, 1993). All of these diagenetic ages converge and also relate to the inferred timing of oil filling (Fig. 2). Petrographic evidence suggests that kaolin growth also was at least partly synchronous with quartz cementation (Fig. 5a), although illite growth may have been slightly later than the kaolin (Fig. 5c). Thus quartz, ankerite, kaolin and illite cementation occurred over much the same time interval and broadly at the same time as oil emplacement. The evidence for the coincidence of the cementation processes and oil emplacement also has been presented previously (Emery *et al.*, 1993; Barclay & Worden, 2000b; Barclay *et al.*, 2000) using oil-fluid-inclusion abundance data and carbon stable-isotope data (suggesting that ankerite precipitated at the same time as an influx of source-rock-derived CO_2).

The relative timing of kaolin cement is more complex than the other cements because there was a (relatively) early transient telogenetic event at the crest of the field directly beneath the erosive unconformity that led to replacement of K-feldspar by kaolin (Emery *et al.*, 1990). However, this source of kaolin occurs only down to a few metres below the crest of the field according to Emery *et al.* (1990). Such material at the crest of the field was not sampled in this study so that this generation of kaolin will not be present in this study. Moreover, if this generation of kaolin was responsible for the overall pattern of kaolin abundance then the result would be an increase in abundance upward (Emery *et al.*, 1990) and not the upwards decrease in abundance as reported in Fig. 6a. The SEM evidence (Fig. 5c) showed that the majority of the kaolin grew after (or during) quartz cementation, placing the clay mineral within the same time frame as the quartz, ankerite and illite (i.e. during late burial and at least partly synchronous with oil emplacement).

Depositional facies: not a significant control on kaolin and illite distribution

Differences in the clay populations in a suite of sedimentary rocks that have undergone significant porosity loss and cementation might be explained by variations in depositional facies. However, in this study, we have compared rocks from the same depositional facies (massive structureless sands of the Bouma-A facies), so differences in the overall amounts of illite and kaolin are not the result of primary differences in facies.

In some depositional environments, there can be local differences in detrital mineralogy. For example, in some circumstances deltaic sandstones can have locally variable clay-mineral populations dependent on whether the sand was deposited in fresh water, normal salinity sea water or in brackish water. The Magnus Sandstone Member is a turbidite sandstone. It was deposited in an approximately uniform depth of sea water with no obvious conditions that could cause local differences in initial sediment mineralogy. Moreover, turbidity currents would have homogenized the initial Magnus sediment on the continental slope during transport and removed any possibility of the difference in clay mineralogy being the result of primary, intrafacies variations in

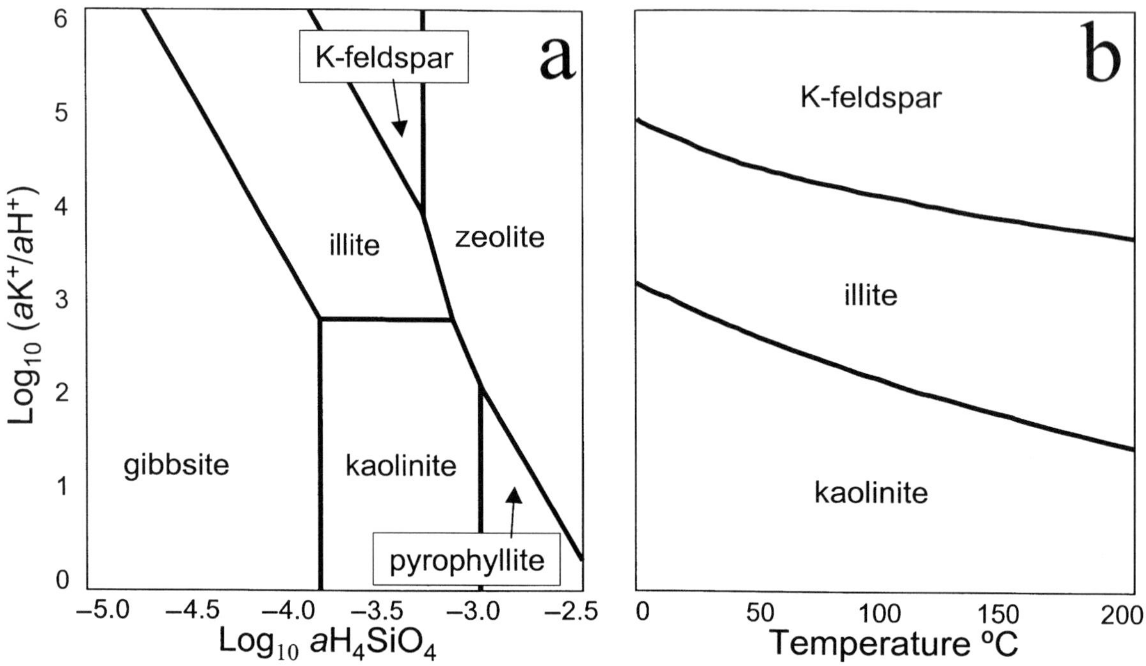

Fig. 7 Phase diagrams (generated using Geochemist's Workbench™) showing: (a) the aqueous geochemical conditions (aK^+/aH^+, $aSiO_2$) under which illite and kaolin can form at constant temperature (100°C) as a function of the silica content of the water and the ratio of the activities of potassium and acidity; (b) the effect of temperature change on the illite–kaolin system at quartz saturation showing that, if all else (e.g. aK^+/aH^+) remains constant and at equilibrium, kaolin will react to form illite with increasing temperature.

initial mineralogy. Thus, facies or primary depositional factors for the Magnus oil field cannot have been a control on the observed clay-mineral distribution pattern.

Possible aqueous geochemical links between illite and kaolin

It is likely that kaolin and illite cements originate from the diagenetic alteration of detrital feldspars because (i) the primary sandstones were subarkosic, (ii) there is abundant secondary porosity formed as a result of K-feldspar dissolution (Fig. 5a), and (iii) the abundant ankerite cement grew as a result of CO_2 influx, resulting in reduced pH of formation waters (see below), which could have initiated K-feldspar alteration. Simple reactions can be written describing the decay of K-feldspar to both illite and kaolin. The simplest geochemical reactions to describe these processes are:

$$\underset{\text{K-feldspar}}{3KAlSi_3O_8} + 2H^+ \Leftrightarrow \underset{\text{illite}}{KAl_3Si_3O_{10}(OH)_2} + \underset{\text{quartz}}{6SiO_2} + 2K^+ \quad (1)$$

$$\underset{\text{K-feldspar}}{2KAlSi_3O_8} + 2H^+ + H_2O \Leftrightarrow \underset{\text{kaolin}}{Al_2Si_2O_5(OH)_4} + \underset{\text{quartz}}{4SiO_2} + 2K^+ \quad (2)$$

The key difference between these reactions is the amount of potassium released to aqueous solution from a given amount of reactant K-feldspar. The illite-producing reaction leads to a smaller loss of potassium per unit of reactant than the kaolin-producing reaction. The geochemical system has been modelled using Geochemist's Workbench™ (Fig. 7a & b), showing that the maximum permissible value of aK^+/aH^+ in the sandstone formation is the main factor that controls the occurrence of reactions 1 or 2. Note that kaolin and K-feldspar appear to be unable to coexist under equilibrium thermodynamic conditions according to the phase diagram.

There is thus another geochemical reaction that can relate these clay minerals. A reaction between K-feldspar and kaolin can produce illite plus quartz:

$$\underset{\text{kaolin}}{Al_2Si_2O_5(OH)_4} + \underset{\text{K-feldspar}}{KAlSi_3O_8}$$

$$\Leftrightarrow \underset{\text{illite}}{KAl_3Si_3O_{10}(OH)_2} + \underset{\text{quartz}}{2SiO_2} + H_2O \quad (3)$$

There is SEM evidence that suggests that reaction 3 may have indeed happened, because illite can be seen growing on top of (i.e. cannibalizing) kaolin (Fig. 5d).

Other reactions also must occur to supply the protons needed for reactions 1 and 2 to proceed. It is most likely that this involved the influx of isotopically depleted CO_2, described earlier, dissolution and dissociation of the gas and then reaction to produce carbonate minerals (Barclay & Worden, 2000b). Note that ankerite is a common late diagenetic mineral as well as the clay minerals and quartz. These CO_2-related reactions thus may be written

$$CO_2 + H_2O \Leftrightarrow HCO_3^- + H^+ \quad (4)$$

$$0.8Mg^{2+} + Ca^{2+} + 0.2Fe^{2+} + 2HCO_3^-$$

$$\Leftrightarrow \underset{\text{ankerite}}{Ca(Mg_{0.8}Fe_{0.2})(CO_3)_2} + 2H^+ \quad (5)$$

Reaction 4 will be encouraged to move to the right-hand side because protons are consumed by reactions such as 1 and 2. The aqueous carbonate concentration will thus increase, encouraging reaction 5 also to move to the right-hand side of the equation.

Illite growth results from three moles of K-feldspar producing two moles of aqueous potassium. Kaolin growth results from two moles of K-feldspar producing two moles of aqueous potassium. Thus illite growth occurs either in situations where the aqueous potassium is kept at an elevated level or where the potassium is not allowed to move away from the reaction zone. It is noteworthy that illite is most concentrated in the oil zone and kaolin is most concentrated in the water zone. The implication from the clay-mineral distribution pattern and the geochemical reactions above is that potassium has been held within the sandstone in the oil zone but allowed to escape from the sandstone in the water zone.

The bulk geochemical data seem to corroborate the implications from the clay-mineral pattern drawn out above. There is more K_2O and a higher $K_2O/(K_2O + Al_2O_3)$ ratio above the transition zone than below and where there is a higher illite/(illite + kaolinite) value. The presence of illite is thus associated with a greater quantity of potassium in the rock. There is no reason why the rocks were different initially, so that the difference in potassium content and the differences in clay mineralogy must be the result of diagenetic alteration.

There are two possible ways to explain the differences in potassium content in the Magnus Sandstone. Either more potassium has been added to the oil zone than the water zone or more potassium has been lost from the water zone than the oil zone. Although the mobility of aluminium in sedimentary basins has been proposed (e.g. see Wilkinson *et al.*, this volume, pp. 147–160; Milliken, this volume, pp. 343–360), there are strong geochemical arguments supporting the fact that potassium is likely to be many orders of magnitude *more* mobile than aluminium:

1 the relative solubility of the two elements (with aluminium scarcely soluble and potassium highly soluble in water);

2 the commonplace illitization of smectite in mudstones, which typically is thought to require a supply of potassium (e.g. from sandstones);

3 the aqueous flux of potassium (by flow or diffusion) will be inhibited in the oil zone relative to the water zone (see later), providing us with a rationale for the potassium distribution pattern.

Thus in simple terms of relative mobility, it is more likely that potassium has been lost preferentially from the water zone and not preferentially gained in the oil zone.

If, as is likely, feldpars are ultimately responsible for the clay minerals, it is likely that the

Table 1 Definition of terms.

a	(Archie) constant
J_o	Diffusive flux in water in the water zone of a reservoir (mol cm^{-2} s^{-1})
J_t	Diffusive flux in water in the oil zone of a reservoir (mol cm^{-2} s^{-1})
k	Intrinsic permeability of a rock (mD)
k_{eff}	Effective permeability (mD)
k_r	Relative permeability factor
m	Cementation factor
n	Saturation exponent
$\varnothing$	Fractional porosity
R_t	Bulk resistivity of oil-bearing porous medium (ohm)
R_w	Resistivity of free water (ohm)
S_w	Fractional water saturation in oil field sandstone
S_{wi}	Irreducible water saturation (water that cannot be replaced by oil)

water zone has lost more potassium as a result of reaction 2 than the oil zone has by reaction 1. The main question remaining concerns the reason for the differential loss of potassium from the water zone in comparison with the oil zone.

Significance of illite and kaolin distribution: petrophysics

The occurrence of predominant illite or kaolinite seems to depend on the ability of the local sandstone to lose potassium. The movement of potassium will occur within the aqueous part of the oil–water–rock system (because potassium conventionally is not thought to be soluble in oil). Thus we need to assess the controls on the movement of aqueous species in water associated with oil fields.

With potassium being lost from the solid portion of the system (Fig. 3), there must be a net flux away from the Magnus Sandstone in the water zone. Otherwise, the potassium concentration would build up until the aK^+/aH^+ increased such that the system entered the illite stability field on Fig. 7a & b. In the presence of oil, the transport of aqueous species in water is relatively retarded. Transport can occur by two end-member mechanisms: diffusion and advection. We are concerned here with the *relative* rates of transport in the oil zone versus the water zone to try to explain the distribution pattern of illite and kaolin.

Although absolute diffusion rates in deeply buried sandstones are naturally of fundamental interest; we are most interested in the relative rates of diffusion in the oil and water zones. Worden *et al.* (1998) demonstrated that the ratio of the rates of diffusion in the oil and water zones (J_t/J_o) is a function of the water saturation (S_w) and the saturation exponent (n):

$$J_t/J_o = S_w^n \tag{6}$$

All terms are defined in Table 1. The saturation exponent (n) is broadly a function of wettability. As reservoirs become more oil-wet (increasing preference for oil rather than water) the saturation exponent increases. Reservoirs tend to become more oil-wet as oil saturation increases (Søndenå *et al.*, 1989), suggesting that oil fields become more oil wet as they fill and that the rate of diffusion will be slowed down to greater degrees as the volume of oil in a given portion of the pore system increases. Equation 6 illustrates that aqueous diffusion in the oil zone will be slowed relative to aqueous diffusion in the water zone. The physical explanation for this slowing is the increased path length (tortuosity), which results from the majority of the pores being filled with oil and the remaining water forming a thin network of films that cling to mineral surfaces. The value of the saturation exponent is specific to each reservoir but a default value of 2 often is selected (e.g. see Rider, 1986). Thus, at typical irreducible water saturations in sandstone reservoirs (0.2) and for a saturation exponent of 2, the rate of solute transport by diffusion in the oil zone is 4% (0.2^2) of that in the water zone. The wettability of the sandstone will have a marked influence upon the relative degree of diffusion retardation (Barclay & Worden, 2000a). The rate of diffusive transport of potassium in solution therefore will decrease significantly in oil-filled sandstones compared with water-filled ones. This phenomenon will increase in magnitude with decreasing water saturation and increasing oil wetting.

The relative diffusion rate of potassium in the oil and water zones can be computed using equation 6 and an assumption about the evolution of the wettability (and thus the saturation exponent) during reservoir filling. In the first instance, a suite of curves can be constructed for diffusion at variable water saturations and a range of constant values of the saturation exponent (Fig. 8). The next step is to recreate the evolution of wettability during oil filling. For high water saturations, a value of 1.75 was selected for the saturation exponent. This was increased progressively through to 2.5 for increasing water saturations (filled circles on the diffusion curves on Fig. 8). These points

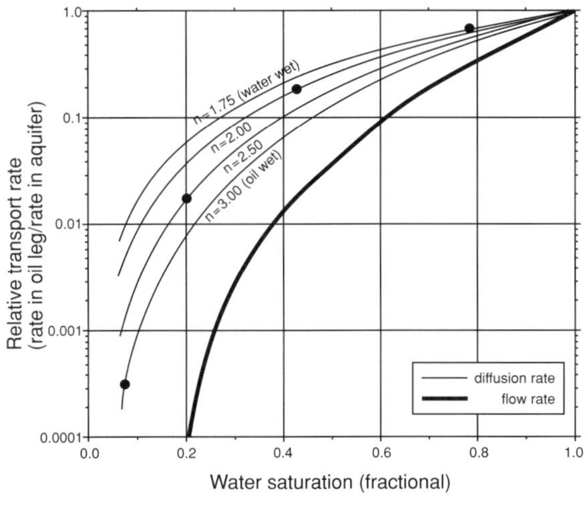

Fig. 8 Modelled relative effect of oil emplacement on (i) aqueous flow rate and (ii) aqueous diffusion rate. Aqueous transport by both methods is severely retarded in the presence of oil.

then can be joined up to simulate the relative diffusion-rate change during progressive oil filling (Fig. 8). At low oil saturations (≤ 20%), the diffusion of potassium from K-feldspar reaction sites will be little affected. At high oil saturations (≥ 80%), the diffusion rate of potassium in the oil zone will be ≥ 10^2 times slower than in the water zone. At very high oil saturations and in predominantly oil-wet sandstones, the diffusion rate of aqueous species in water in an oil zone effectively will be zero.

The approach using equation 6 makes a profound assumption regarding the nature of the thin water films on water coated grains in oil fields. It assumes that the mechanical properties of the thin water film are the same as those of free water. It is possible that at high capillary pressures the residual water film is extremely thin and possibly consisting only of the Gouy–Chapman double layer and the Stearn–Graham double layer (Melrose, 1984; Lindsay Kaye, personal communication, 1995). The density of hydrated cations in such a film is probably high and short-range forces of attraction and repulsion may influence the diffusive properties of the water. In this setting, the residual water may behave more like a solid than a free fluid in terms of diffusion. The diffusion coefficient may be very much lower in the residual water film than in free water. The diffusion coefficient of potassium in a solid is many orders of magnitude lower than the diffusion coefficient of the potassium–water system. This effect would tend to reduce the value of J_t/J_o beyond the influence of water saturation and saturation exponent values. The exact magnitude of this effect is difficult to quantify but may reduce the diffusive flux of potassium in the oil zone by many orders of magnitude relative to the water zone, especially at maximum oil saturation.

Advection of water is the other transport mechanism we need to consider. The *flow* of water from the reaction sites in the oil zone is complicated by the fact that two-phase flow, rather than single-phase flow, must be considered. Under conditions of uniform pressure gradient, the flow of water in the oil zone is dictated by its relative permeability (k_r):

$$k_{\text{eff}} = k_r \cdot k \tag{7}$$

where k is the intrinsic permeability of the rock. The relative permeability term is a dimensionless factor that reflects the permeability of porous media to two-phase flow. The relative permeability of water in the oil zone varies with water saturation (Archer & Wall, 1992). At typical values of oil-zone irreducible water saturation, the relative permeability to water is negligible. One empirical model for relative permeability has been defined by Archer & Wall (1992) as:

$$k_r = S_w^4 \cdot \left(\frac{S_w - S_{wi}}{1 - S_{wi}} \right)^{0.5} \tag{8}$$

At a water saturation of 0.3, for a reservoir with an irreducible water saturation of 0.2, the relative permeability of water is more than four orders of magnitude lower in the oil zone than in the water zone (Fig. 8). At close to the irreducible water saturation in the oil zone the relative permeability, and therefore the effective permeability, of water tends towards zero. Thus the flow of water and the advective transport of dissolved potassium by fluid flow will be slowed down by $> 10^3$ times in the oil zone in sandstone reservoirs relative to the water zone. At very high oil saturations, the flow of water and any consequent inorganic diagenetic phenomena will be stopped entirely (Worden *et al.*, 1998; Worden & Morad, 2000).

The relative degrees by which water flow and aqueous diffusion are slowed by oil emplacement may be compared to show the relative effects of oil on these transport mechanisms (Fig. 8). Both diffusion and advection of potassium from the oil zone will be slowed down relative to the water zone by at least several orders of magnitude. Therefore it is not necessary to deduce the mechanism of potassium transport that was operating in Magnus; because both mechanisms are slowed down to a great extent.

Overview of the controls on the diagenetic system

The description of the distribution pattern of clay minerals in the Magnus Sandstone, the fact that potassium was largely retained in the oil zone but lost from the water zone, and the analysis of the effect of oil emplacement on transport rate has led us to conclude that the addition of oil to the sandstone has resulted in quite different diagenetic reactions in the oil and water zones. In the oil zone, illite has grown, possibly partly at the expense of kaolin because potassium could not be transported out of the sandstone. In the water zone, potassium was able to move from the sandstone resulting in preferential growth, or retention, of kaolin. However, several questions remain.

The relative timing of clay mineral growth is uncertain. For example, did kaolin and illite grow at the same time in the oil and water zones or did they grow at different times? One piece of evidence we have suggests that kaolin may have grown throughout the sandstone in the first instance. Figure 5c shows illite growing on pre-existing kaolin booklets. From this image, it is apparent that the illite grew after the kaolin. It is still unclear whether all the illite grew after the kaolin or whether they were partly contemporary.

The preferential growth of kaolin occurred at the same time as loss of potassium from the sandstone. The predominant clay mineral in the finest grained (muddy) sediments in the turbidite sequence is illite with vanishingly little smectite in such rocks (McHardy *et al.*, 1982). It is possible that potassium was exported from the sandstone to the mudstone at localities where kaolin grew. Note that there is a growing body of evidence supporting this as a common pattern in sedimentary basins (Land & Milliken, 2000; Wilkinson *et al.*, this volume pp. 147–160). In cases where illite grew, there was less potassium lost from the sandstone. This leads to a hypothesis that the interlayered turbidite mudstones in the oil zone may have been less illitized than those in the water zone. Unfortunately, we did not collect samples of mudstone facies to test this idea.

The distribution of illite and kaolin broadly conforms to the oil–water transition zone, but there is not an exact correlation (Figs 4 & 5). Small amounts of illite grew in the water zone and kaolin grew, or remains, in the water zone. Given the rationale outlined above, using local transport rates of potassium as the most likely explanation for the clay mineral pattern, it is possible that local variations in water saturation, relative permeability, tortuosity, wettability and intrinsic permeability could have caused local variations in transport rate. It is difficult to test how local textural and mineralogical heterogeneity control clay-mineral diagenesis. This is especially difficult because some of the controls, such as wettability, can vary with time.

SUMMARY

1 There is a higher illite/(illite + kaolin) ratio in the oil zone than in the water zone of the Upper Jurassic Magnus Sandstone in the North Sea.
2 Illite grew at a similar time as, or slightly later than, kaolin cements.
3 Both illite and kaolin grew during deep burial diagenesis at about the same time as oil emplacement, quartz cementation and ankerite cementation.
4 There is a higher $K_2O/(K_2O + Al_2O_3)$ ratio in the oil zone than the water zone. It is most likely that potassium has been lost preferentially from the water leg rather than preferentially gained within the oil zone.
5 The clay-mineral distribution pattern cannot be the result of:
(i) primary facies variations, because all samples were collected from the same facies;
(ii) primary depositional differences, because turbidite sand bodies have broadly uniform character and early diagenesis;
(iii) differences in an early weathering event, because this led to minor kaolin growth at the crest of the field (the opposite to the pattern seen in this study);
(iv) differences in depth of burial and thus temperature, because this would have led to more illite in the deeper water zone than the oil zone (the opposite to that seen).
6 The clay mineral pattern is likely to be the result of different abilities of the oil and water zone to lose potassium from the local sandstone.
7 The oil zone has much (orders of magnitude) slower aqueous transport by both advection and diffusion than the water zone. This seems to have resulted in local retention of potassium in the sandstone and the consequent growth of potassium-rich clay (illite) instead of potassium-free clay.
8 The model and observations imply that kaolin may have grown in the oil zone in the first instance and was then illitized as the potassium activity of the water increased. Illite also may have grown directly in the oil zone.
9 The model and observations also imply that inorganic diagenesis is not necessarily stopped in zones of elevated oil saturation. However, it is apparent that the physicochemical environment in the oil zone is quite different to that in an aquifer and this has major implications for geochemical processes.

ACKNOWLEDGEMENTS

We offer our thanks to reviewers Warren Dickinson and Knut Bjørlykke for their valuable insights and critical comments. We also would like to thank editor Sadoon Morad for seeing the paper through and offering his own discerning comments. The oil company BP, John Rowse in particular, is acknowledged for allowing us to sample core, providing us with supporting oil field data and for giving us permission to publish this work. The Northern Ireland Department of Education is here graciously thanked for funding the research. We would like to thank all our most excellent colleagues from the School of Geosciences at Queen's University in Belfast for providing such a wonderfully intellectually nourishing environment. Professor Whalley is especially picked out in this respect.

REFERENCES

Archer, J.S. & Wall, P.G. (1992) *Petroleum Engineering. Principles and Practice*. Graham and Trotman, London, 362 pp.

Barclay, S.A. & Worden, R.H. (2000a) Effects of reservoir wettability on quartz cementation in oil fields. In: *Quartz Cementation in Sandstones*, International Association of Sedimentologists Special Publication 29 (Ed. by R.H. Worden and S. Morad), Blackwell Science, Oxford, pp. 103–117.

Barclay, S.A. & Worden, R.H. (2000b) Geochemical modelling of diagenetic reactions in a sub-arkosic sandstone. *Clay Mineral.*, **35**, 61–71.

Barclay, S.A., Worden, R.H., Parnell, J., Hall, D.H. & Sterner, M. (2000) Assessment of fluid contacts and compartmentalisation in sandstone reservoirs using fluid inclusions: an example from the Magnus oil field, North Sea. *AAPG Bull.*, **84**, 489–504.

Bethke, C.M. (1994) *The Geochemists Workbench, version 2.4, A Users Guide to Rxn, Act2, Tact, React and Gtplot.* Hydrogeology Program, University of Illinois, Urbana-Champaign.

Bish, D.L. & Post, J.E. (1993) Quantitative mineralogical analysis using the Rietveld full-pattern fitting method. *Am. Mineral.*, **78**, 932–940.

Darby, D., Wilkinson, M., Fallick, A.E. & Haszeldine, R.S. (1997) Illite dates record deep fluid movements in petroleum basins. *Petrol. Geosci.*, **3**, 133–140.

De'Ath, N.G. & Schuyleman, S.F. (1981) The geology of the Magnus oilfield. In: *Petroleum Geology of the Continental Shelf of North-West Europe* (Ed. by L.V. Illing and G.D. Hobson), Institute of Petroleum, London, pp. 342–351.

Delaney, J.M. & Lundeen, S.R. (1990) *The LLNL Thermochemical Database.* Lawrence Livermore National Laboratory Report UCRL-21658.

Ehrenberg, S.N. (1990) Relationship between diagenesis and reservoir quality in the Garn Formation, Haltenbanken, Mid-Norwegian continental shelf. *AAPG Bull.*, **74**, 1538–1558.

Emery, D., Myers, K.J. & Young, R. (1990) Ancient subaerial exposure and freshwater leaching in sandstones. *Geology*, **18**, 1178–1181.

Emery, D., Smalley, P.C. & Oxtoby, N.H. (1993) Synchronous oil migration and cementation in sandstone reservoirs demonstrated by quantitative description of diagenesis. *Philos. Trans. Roy. Soc.*, **A344**, 115–125.

Fein, J. (2000) Experimental and field constraints on the role of silica-organic complexation and silica-microbial interactions during sediment diagenesis. In: *Quartz Cementation in Sandstones*, International Association of Sedimentologists Special Publication 29 (Ed. by R.H. Worden and S. Morad), Blackwell Science, Oxford, pp. 119–127.

Hogg, A.J.C. & Pearson, M.J. (1995) An integrated thermal and isotopic study of the diagenesis of the Brent Group, Alwyn South, UK North Sea. *Appl. Geochem.*, **10**, 531–546.

Land, L.S. & Milliken, K.L. (2000) Regional loss of SiO_2 and $CaCO_3$ and gain of K_2O during burial diagenesis of Gulf Coast mudrocks, USA. In: *Quartz Cementation in Sandstones*, International Association of Sedimentologists Special Publication 29 (Ed. by R.H. Worden and S. Morad), Blackwell Science, Oxford, pp. 183–197.

Macaulay, C.I., Haszeldine, R.S. & Fallick, A.E. (1992) Diagenetic pore waters stratified for at least 35 million years: Magnus oil field, North Sea. *AAPG Bull.*, **76**, 1625–1634.

Macaulay, C.I., Haszeldine, R.S. & Fallick, A.E. (1993a) Distribution, chemistry, isotopic composition and origin of diagenetic carbonates: Magnus sandstone, North Sea. *J. Sed. Petrol.*, **63**, 33–43.

Macaulay, C.I., Fallick, A.E. & Haszeldine, R.S. (1993b) Textural and isotopic variation in diagenetic kaolinite from the Magnus oil field sandstones. *Clay Mineral.*, **28**, 625–639.

McHardy, W.J., Wilson, M.J. & Tait, J.M. (1982) Electron microscope and X-ray diffraction studies of filamentous illitic clay from sandstones of the Magnus Field. *Clay Mineral.*, **17**, 23–29.

Melrose, J.C. (1984) Physicochemical analysis of the reservoir wetting problem. *J. Am. Chem. Soc.*, **61**, 688–698.

Milliken, K.L. (2003) Microscale distribution of kaolinite in Breathitt Formation sandstones (middle Pennsylvanian): implications for mass balance. In: *Clay Mineral Cementation in Sandstones*, International Association of Sedimentologists Special Publication 34 (Ed. by R.H. Worden and S. Morad), Blackwell Publishing, Oxford, pp. 343–360.

Morad, S., Bergan, M., Knarud, R. & Nysteun, J.P. (1990) Albitisation of detrital plagioclase in Triassic reservoir sandstones from the Snorre Field, Norwegian North Sea. *J. Sed. Petrol.*, **60**, 411–425.

Rider, M.H. (1986) *The Geological Interpretation of Well Logs.* Blackie, Glasgow, 171 pp.

Robinson, A. & Gluyas, J. (1992) Duration of quartz cementation in sandstones, North Sea and Haltenbanken Basins. *Mar. Petrol. Geol.*, **9**, 324–327.

Shepherd, M., Kearney, C.J. & Milne, J.H. (1990) Magnus Field. In: *Atlas of Oil and Gas Fields: Structural Traps II* (Ed. by E.A. Beaumont and N.H. Foster), American Association of Petroleum Geologists, Tulsa, OK, pp. 95–125.

Søndenä, E., Bratteli, F., Kolltveit, K. & Normann, H.P. (1989) A comparison between capillary pressure data and saturation exponent obtained at ambient conditions and at reservoir conditions. In: *Proceedings of the 64th Annual Technical Conference and Exhibition of the Society of Petroleum Engineers*, San Antonio, W. SPE 19592, pp. 213–225.

Thomas, M. (1986) Diagenetic sequences and K/Ar dating in Jurassic sandstones, Central Viking Graben: effects on reservoir properties. *Clay Mineral.*, **21**, 695–710.

Walderhaug, O. (1994) Precipitation rates for quartz

cement in sandstones determined by fluid inclusion microthermometry and temperature history modelling. *J. Sed. Res.*, **A64**, 324–333.

Wilkinson, M., Haszeldine, R.S. & Milliken, K.L. (2003) Cross-formational flux of aluminium and potassium in Gulf Coast (USA) sediments. In: *Clay Mineral Cementation in Sandstones*, International Association of Sedimentologists Special Publication 34 (Ed. by R.H. Worden and S. Morad), Blackwell Publishing, Oxford, pp. 147–160.

Worden, R.H. & Barclay, S.A. (1999) Internally-sourced quartz cementation caused by externally-derived CO_2 in subarkosic sandstones of the Northern North Sea, UKCS. In: *Abstracts of the 1999 American Association of Petroleum Geologists Annual Convention, San Antonio*, 11–14 April, pp. A153–A154.

Worden, R.H. & Barclay, S.A. (2000) Internally-sourced quartz cement due to externally-derived CO_2 in subarkosic sandstones, Northern North Sea, UKCS. *J. Geochem. Explor.*, **69–70**, 649–653.

Worden, R.H. & Heasley, E.C. (2001) The effect of oil emplacement on diagenesis in carbonate reservoirs. *Bull. Soc. Geol. Francais*, **171**, 607–620.

Worden, R.H. & Morad, S. (2000) Quartz cement in oil field sandstones: a review of the critical problems. In: *Quartz Cementation in Sandstones*, International Association of Sedimentologists Special Publication 29 (Ed. by R.H. Worden and S. Morad), Blackwell Science, Oxford, pp. 1–20.

Worden, R.H., Oxtoby, N.H. & Smalley, P.C. (1998) Can oil emplacement stop quartz cementation in sandstones? *Petrol. Geosci.*, **4**, 129–138.

Glauconite case study

Int. Assoc. Sedimentol. Spec. Publ. (2003) **34**, 473–488

Application of glauconite morphology in geosteering and for on-site reservoir quality assessment in very fine-grained sandstones: Carnarvon Basin, Australia

J.P. SCHULZ-ROJAHN*, D.A. SEEBURGER† *and* G.J. BEACHER‡
West Australian Petroleum Pty Ltd. (WAPET), Box S1580 GPO, Perth, Western Australia 6001, Australia

ABSTRACT

This paper describes the innovative application of drill cuttings during the drilling of a horizontal development well in the Mardie Greensand of Early Cretaceous age, Saladin Field, Australian North West Shelf (Carnarvon Basin). Well results show that the morphology of glauconite pellets, as observed in drill cuttings under the binocular microscope, can be useful for making a first-pass assessment of reservoir quality when drilling with oil-based mud (OBM) in this very fine to fine-grained sandstone. The shape of the glauconite pellets is controlled by the degree of rock compaction, which, in turn, is controlled by the degree of framework grain support that drives reservoir quality. At Saladin, good-quality Mardie reservoir rock (up to 100 mD at least) capable of producing initial oil rates of up to 4000 BOPD (barrels of oil per day) from individual horizontal wells is dominated by more or less well-rounded ('marshmallow-shaped') glauconite pellets with gently curved flat tops and bottoms, consistent with very minor compaction owing to abundant framework grain support. In contrast, relatively poor-quality Mardie Greensand reservoir (≤ 10 mD) is dominated by thin glauconite pellets with long flat tops and bottoms, and non-reservoir Mardie Greensand lithologies contain disk- or plate-shaped glauconite pellets that are estimated to have compacted by as much as 80% or more. The description of the clay pellet style rather than just a conventional description of cuttings is the key to a reliable lithology indicator because the bulk of the very fine to fine sand grains tends to be lost at the shakers, whereas the larger glauconite pellets are retained preferentially. This simple recognition could have important implications for pay detection in low-resistivity (greensand) exploration settings, although unproven for non-OBM borehole conditions. Further, despite lag time, the simple but effective method can complement measurement-while-drilling (MWD) controlled geosteering, as exemplified by the Saladin Field, where along-hole variations in the shape of glauconite pellets provided key stratigraphical clues. In the event

*Present address: Shell International Exploration and Production B.V. (SIEP), c/o Brunei Shell Petroleum, EPE/21, KB 3534, Seria, Brunei Darussalam, e-mail: jorg.schulz-rojahn@shell.com
† Present address: Chevron Petroleum Technology Company (CPTC), 6001 Bollinger Canyon Road, San Ramon, CA 94583, USA, e-mail: dase@chevrontexaco.com
‡ Present address: Kalaro Pty Ltd., 40 Campbell Drive, Hillarys, Western Australia 6025, Australia, e-mail: gravi@ca.com.au

that exclusively disk- or plate-shaped glauconite pellets are encountered along supposed reservoir target sections, an early well-site warning system is provided that could accelerate the decision to plug and abandon. Finally, in horizontal or high-angle deviated wells the technique has potential applications for static reservoir modelling because it may allow the reservoir geologist to gain an improved understanding of lateral facies transitions and sand geometries in subtle greensand lithologies.

INTRODUCTION

Knowledge about clay minerals and diagenetic processes in sandstone reservoirs is to a very large extent based on core and outcrop data, however, valuable insights also can be gained from drill cuttings. Researchers in their reservoir studies, for a number of good reasons, traditionally have ignored drill cuttings: sample disintegration, sample contamination, ambiguity in depth derivation, and lack of accompanying sedimentological data. However, with recent advances in drilling technology, the time may have come to re-evaluate the usefulness of drill cuttings in reservoir studies because it is clear that cuttings can have certain advantages over core data, supplementing petrophysical evaluation. In horizontal or high-angle deviated wells where individual reservoir sections can be followed for a kilometre or more, drill cuttings can provide a more comprehensive two-dimensional appreciation of lateral reservoir variation than normally would be possible from core material, typically cut over relatively short depth intervals for cost efficiency reasons. With increases in the number of horizontal or deviated wells in a hydrocarbon asset, the potential is given for a three-dimensional approximation of lateral lithology, textural and diagenetic variations, improving static reservoir visualization.

In this paper an example is provided of the unexpected application of drill cuttings in the geological interpretation of an unconventional reservoir characterized by low-resistivity pay. The case study is from the Lower Cretaceous Mardie Greensand in the Saladin Field of Australia's North West Shelf (Fig. 1), operated by West Australian Petroleum Pty Ltd. (WAPET).

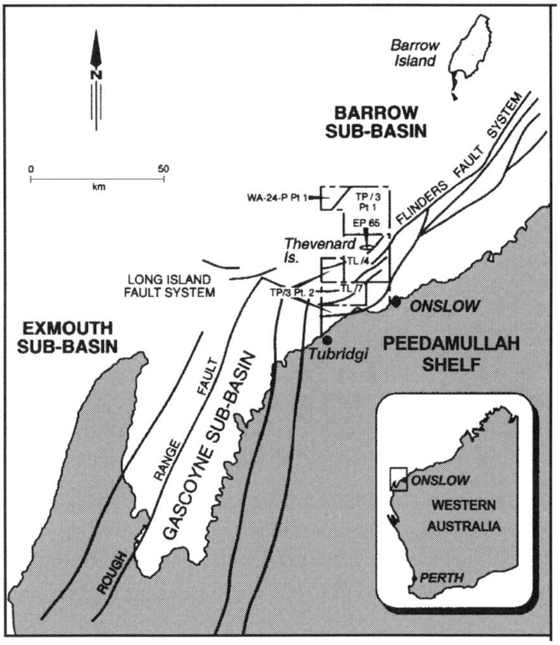

Fig. 1 Location map showing Thevenard Island and TL/4 permit in offshore Western Australia where the Saladin Field is situated (Barrow Sub-basin).

In this field, the surprising discovery was made that the shape of glauconite pellets can assist geosteering (directional drilling using real-time geological information). Furthermore, the study shows that the shape of glauconite pellets can be useful for first-pass reservoir quality assessment in very fine to fine-grained sandstones at the well site. The case study demonstrates that drill cuttings can yield a new perspective on the geological characterization of unconventional reservoirs.

Table 1 Dinoflagellate zones and approximate ages for Early Cretaceous sediments of Australia's North West Shelf (after Helby *et al.*, 1987).

Dinoflagellate zones	Age (Ma)
M. australis	120–126
M. testudinaria	126–130
P. burgeri	130–132.5
S. tabulata	132.5–135
S. areolata	135–137

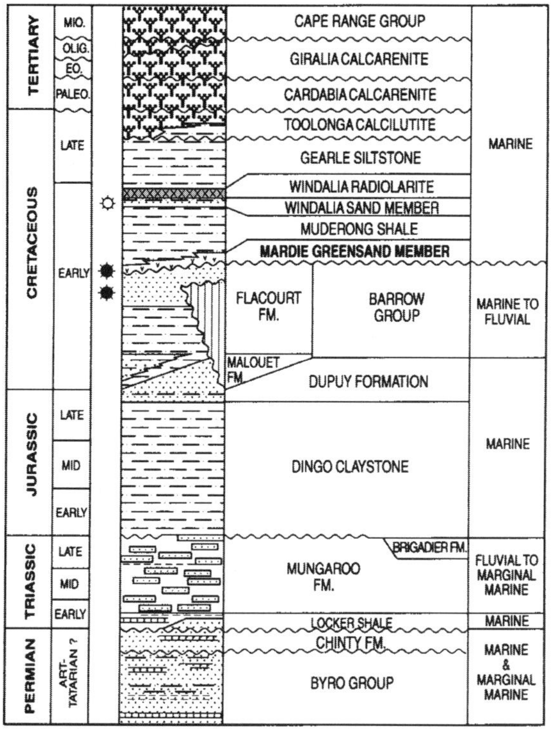

Fig. 2 Stratigraphical table for the Barrow Sub-basin of the Carnarvon Basin. Note the diachronous nature of the transgressive Lower Cretaceous Mardie Greensand that unconformably overlies the lowstand Barrow Group, and that grades upwards into the deep-water marine Muderong Shale of the same age.

REGIONAL FRAMEWORK

The Lower Cretaceous Mardie Greensand Member of the Muderong Shale (Thomas, 1978) and stratigraphical equivalents extend over much of the Carnarvon Basin, occupying an area of at least 30 000 km^2. Approximate depth to the top of the greensand varies between less than 100 m in onshore regions and more than 2 km in offshore regions. The thickness of the greensand ranges between about 15 and 25 m, although locally a thickness of up to 35–40 m is attained (Schulz-Rojahn, 1996).

The Mardie Greensand is a dark-green to green-grey glauconitic, heavily bioturbated siltstone or very fine to fine-grained sandstone interbedded with dark-grey siltstone and shale (Thomas, 1978; Seeburger *et al.*, 1998). The stratigraphic interval spans several dinoflagellate zones of Valanginian to Barremian age, including *Senoniasphaera aerolata*, *S. tabulata*, *Phoberocysta burgeri*, *Muderongia testudinaria* and *M. australis*, which together represent an age of about 17 Ma (Helby *et al.*, 1987) (Table 1). The diachronous Mardie Greensand constitutes a transgressive systems tract that unconformably overlies the lowstand progradational complex of the Barrow Group (Fig. 2). The Mardie Greensand grades upwards into the deep-water marine argillaceous facies of the Muderong Shale, interpreted to have been deposited during maximum marine flooding and sea-level highstand (Fig. 2) (Seeburger *et al.*, 1998).

Despite its name, the Mardie Greensand is a clay- (glauconite) dominated stratigraphic interval with only localized development of good reservoir quality. The lithology is interpreted to have been deposited slowly in a lower shoreface to outer marine shelf setting with very restricted influx of detrital sediments at an accumulation rate of between 0.5 m and 1 m per 100 000 yr (Seeburger *et al.*, 1998).

Based on 1368 ambient core plug measurements from 42 regional offshore wells, mean core porosity of the Mardie Greensand is 22 ± 6%, with about 85% of core plugs having permeabilities of less than 10 mD (Schulz-Rojahn, 1996). From an exploration perspective, the dominantly low-permeability character of the Mardie Greensand on the regional scale is made worse by a low-resistivity 'shaly' log response, including good-quality reservoir sections containing hydrocarbons (Seeburger *et al.*, 1998).

In many wells, this petrophysical challenge (Zhang *et al.*, 1996, 1997, 2000) is amplified by poor hole conditions owing to extensive wash-outs over reservoir and non-reservoir sections alike (Seeburger *et al.*, 1998). In cores, the reservoir zones are also very subtle and easily can be missed by the inexperienced geologist or newcomer to the Mardie Greensand, especially if initial core descriptions are carried out without availability of measured core plug or profile permeametry data (Schulz-Rojahn, 1996). Thus, log-derived pay recognition and prediction of reservoir quality trends represent two key challenges for the successful opening of the Mardie Greensand as a frontier exploration play.

The Mardie Greensand also lacks any form of seismic expression, rendering trap delineation difficult using conventional methods. Further, all hitherto known reservoir zones are relatively thin (maximum 11 m) as based on regional core and log data (Schulz-Rojahn, 1996), which makes it uncertain if a sufficiently large hydrocarbon accumulation can ever be found to justify a stand-alone offshore field development in the Mardie Greensand. Thus, despite hundreds of regional well-penetration points, only about two dozen drill-stem tests were attempted over the Mardie Greensand interval by different operators, with a disappointing outcome in almost every case, for a variety of reasons (Schulz-Rojahn, 1996). These facts have frustrated sporadic exploration efforts in the Mardie Greensand by a variety of different oil companies over the years. At best, the Mardie Greensand has been viewed as a secondary objective in the quest for richer pickings in the underlying prolific, high-quality Barrow Group sandstones. At worst, drilling programmes were designed without any Mardie Greensand evaluations in mind (Seeburger *et al.*, 1998).

Despite these frustrations and key challenges, three fields presently produce from the Mardie Greensand in the general study area (Fig. 1). The Barrow Island field produces from a relatively low-quality *P. burgeri/M. testudinaria* zone, whereas both the onshore Tubridgi gas field and the offshore Saladin oil field produce from high-quality lower *M. australis* reservoir intervals (Schulz-Rojahn, 1996). At Tubridgi, the producing lower (to middle?) *M. australis* zone consists of an unusually quartz-rich unit of medium-grey to light-green colour, with exceptionally low glauconite content. The occurrence of this unit at this biostratigraphical level is consistent with regional reservoir trends, which show that relatively high-permeability (quartz-rich) Mardie Greensand intervals occur chiefly in the Lower *M. australis* zone, or at the uppermost *M. testudinaria*/lower *M. australis* boundary (Schulz-Rojahn, 1996; Seeburger *et al.*, 1998).

These regional observations point towards a primary eustatic control on reservoir development in the Mardie Greensand (Schulz-Rojahn, 1996; Seeburger *et al.*, 1998). The occurrence of the relatively poor-quality *P. burgeri/M. testudinaria* reservoir at Barrow Island could be the result of additional eustatic events, however, no associated reservoir trends have been identified to date (Schulz-Rojahn, 1996; Seeburger *et al.*, 1998). Importantly, significant regional variations in reservoir quality at the uppermost *M. testudinaria*/lower *M. australis* boundary, over relatively short distances (1–2 km), show that important secondary controls are at work. These secondary controls are thought to be related to winnowing and sorting within the wavebase, along subtle syndepositional highs (Schulz-Rojahn, 1996; Seeburger *et al.*, 1998).

SALADIN FIELD

Saladin Field is located about 500–1000 m offshore from the eastern end of Thevenard Island in TL/4, Western Australia (Fig. 1). The field is a fault-bounded structure in which oil production initially was established only from the high-quality Barrow Group sandstones (core porosity 25–35%, permeability 1–20 darcy) (Tippet & Beacher, 2000).

In 1993, Saladin-1 tested the overlying Mardie Greensand, demonstrating the productive nature of this interval, despite a typical

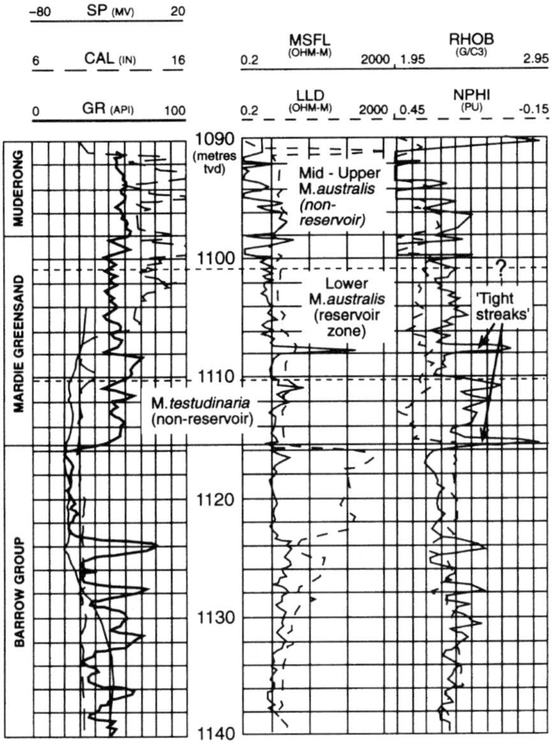

Fig. 3 Saladin-1 composite log signature over the Mardie Greensand, including approximate biostratigraphical zonation. Note the 'shaly' gamma-ray log response, the low-resistivity invasion profile and the density-neutron separation, rendering conventional log interpretation difficult over this oil-bearing reservoir. The high-density *M. testudinaria* zone is a proven seal on the production time-scale whereas the lower *M. australis* zone represents the actual producing reservoir.

'shaly' log response over the pay zone (Fig. 3). Conventional log interpretation indicates water saturations of 60–70% or greater, with resistivity logs showing only about 0.5 ohm-m difference between porous water-bearing sands and hydrocarbon-bearing intervals (Seeburger *et al.*, 1988). Despite the high apparent S_w, all early Mardie Greensand tests produced clean oil with no water at Saladin. Initial oil rates of up to 4000 BOPD have been achieved from individual horizontal wells in the Mardie Greensand. As of December 1997, Mardie Greensand production was about 12 000 BOPD, with less than 100 BWPD (barrels of water per day) from seven producers in Saladin Field (Seeburger *et al.*, 1998). With nearing depletion of Barrow Group reservoirs, the Mardie Greensand now supplies more than half of total Saladin oil production.

Along the Saladin structure, top Mardie reservoir depth varies between about 1050 and 1120 m subsea. Total thickness of the Mardie Greensand is a relatively constant 18 m but reservoir sands are found only in the lower middle 10 m of section (Fig. 3) (Seeburger *et al.*, 1998). This productive interval corresponds to the lower *M. australis* zone and is underlain by a 3–5 m thick, variably siderite and pyrite cemented member that is of dominantly *M. testudinaria* age, with a possible local influence of *P. burgeri* (Schulz-Rojahn, 1996). This basal cemented member, which includes a 15–60 cm thick pyrite-cemented conglomeratic unit directly overlying the Barrow Group sandstones, has been demonstrated to be a seal, at least on the production time-scale (Beacher, 1998). The sealing capacity stems from the intense siderite and/or pyrite cementation, and especially the poor to very poor sorting with resultant lack of residual interstitial porosity between isolated quartz grains that occur sprinkled between strongly compacted glauconite pellets (Fig. 4).

The lower *M. australis* reservoir interval consists of a succession of glauconitic sandstones, each between 2 and 3 m thick, which are separated by siderite-cemented zones or 'tight streaks' that in turn vary in thickness between 30 cm and less than 2 m (Schulz-Rojahn, 1996; Seeburger *et al.*, 1998) (Fig. 3). These siderite-cemented zones are interpreted to be of a relatively early diagenetic (methanogenic) origin on the basis of their cyclic stratigraphy, their correlatable nature across the field, isotope and petrographic evidence (Stuart *et al.*, 1995; Schulz-Rojahn, 1996). Siderite cementation is thought to have commenced at relatively shallow depth within the seafloor sediment, below the influence of erosive processes (Stuart *et al.*, 1995; Seeburger *et al.*, 1996). Although these siderite-cemented layers typically have low permeability (< 2 mD) they are not perceived to be seals on the geological time-scale, and a

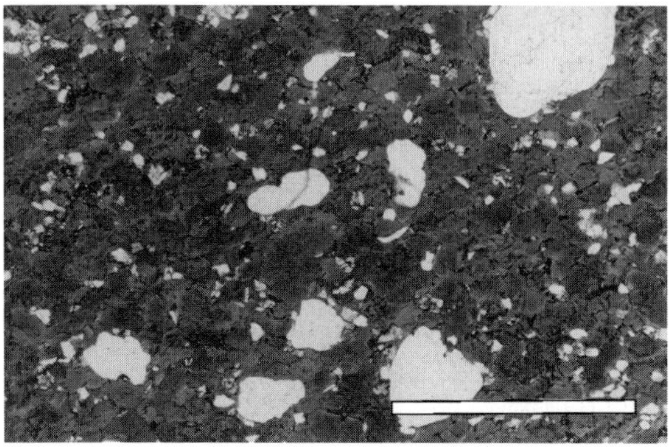

Fig. 4 Thin-section micrograph (plain polarized light) of the *M. testudinaria* greensand, about 1 m above the basal conglomerate. Individual quartz grains do not touch each other but occur interspersed between strongly compacted, dark-green glauconite pellets. The lack of a rigid grain framework accounts for the general lack of interstitial macroporosity in this biostratigraphical zone. Scale bar ≈ 2 mm.

(a)

(b)

Fig. 5 (a) Hand specimen view of the very fine-grained productive lower *M. australis* greensand characterized by a subtle wispy texture and intense bioturbation. The dots with regular spacing represent areas where profile permeametry measurements were undertaken. Saladin-2, depth 1099 mKB (metres below Kelly Bushing, a standard reference point for drilled depth). (b) Thin-section micrograph (plain polarized light) of the lower *M. australis* reservoir. Note the relative abundance of detrital quartz (white) producing a good-quality reservoir dominated by primary intergranular porosity. Further note that the glauconite pellets generally are larger than the detrital quartz grains. The sample contains minor siderite microspar that has helped to resist the effects of compaction in this sample. Ambient core porosity is around 27%, permeability 30–35 mD. Scale bar = 2 mm.

common oil–water contact is assumed for the various lower *M. australis* reservoir sands at Saladin.

In hand specimens, the oil-producing lower *M. australis* reservoir sands are characterized by a dark-green greyish colour, very fine to fine grain size, intense bioturbation and a subtle wispy texture (Fig. 5a). In thin-section these same reservoir sands are distinguished on the basis of their generally well-sorted nature and high textural maturity, an abundance of dark-green pelloidal glauconite (19–57%) and varying amounts of finely dispersed siderite microspar cement (0–10%). The glauconite pellets display no evidence of the primary substrate of glauconitization. Significantly, these glauconite pellets (average 200–286 μm) are larger than the associated detrital quartz grains (average 100–138 μm) (Fig. 5b).

Virtually all the visible thin-section porosity is of primary intergranular nature (Fig. 5b), with some minor contributions from feldspar dissolution. Abundant microporosity is associated with the glauconite pellets, as evident from a poor correlation between measured core porosity and visible thin-section porosity (Stuart *et al.*, 1995; Schulz-Rojahn, 1996) and also the lack of a good porosity–permeability correlation (Seeburger *et al.*, 1998). Although measured core porosity falls within the 21–29% range, core permeability varies between 6.5 and 106 mD at least over the lower *M. australis* reservoir zone (excluding hard streaks). Marginal reservoir facies are characterized by core permeabilities in the order of a few millidarcies to 10 mD. Typically, more than half of the measured core porosity is intrapelletal porosity (microporosity) associated with the glauconite. On the basis of high-pressure mercury porosimetry data and using a cutoff of 0.5 μm for distinguishing between productive and non-productive porosity, Seeburger *et al.* (1998) estimate that the Saladin productive porosity is in the 8–13% range.

Based on regional core data, including from adjacent areas, the relatively quartz-rich lower *M. australis* reservoir zone at Saladin grades upwards into an extremely quartz-poor greensand facies that is at least 3–4 m thick. The gradational nature of this contact explains why the best reservoir sands are found in the lower middle portion of the lower *M. australis* interval at Saladin (Seeburger *et al.*, 1998). In thin-section, the lack of any framework grain support is clearly evident in this very glauconite-rich greensand facies (Fig. 6) that lacks any reservoir potential (ambient core permeabilities of 1 mD or less). This facies becomes increasingly muddy (light-green greyish to light-grey) up-sequence until all glauconite has disappeared, marking the approximate transition point into the overlying deep-water grey marine Muderong Shale of late lower to middle *M. australis* age in the study area.

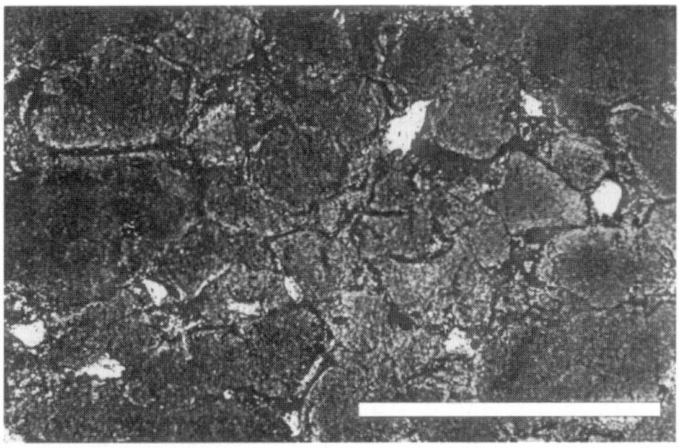

Fig. 6 Thin-section micrograph (plain polarized light) of the Mardie Greensand transition zone into the Muderong Shale. Although this micrograph is derived from a Barrow Island well (Barrow F35M, 837.6 mKB) the same feature is observed throughout the general study area (not actually cored at Saladin). Note that in this lithofacies, characterized by a lack of framework grain support, compaction has destroyed all visual porosity. The thin-section was cut broadly parallel to bedding plane and glauconite pellets appear round to subrounded in this view but are, in fact, flat. Detrital kaolin matrix (dark) surrounds the detrital glauconite pellets as based on X-ray diffraction results. Ambient core porosity is 21%, permeability 0.449 mD. All the measured porosity is microporosity associated with the clays. Scale bar = 0.5 mm.

A detailed knowledge of reservoir stratigraphy has proven to be an important prerequisite for the successful exploitation of the Mardie Greensand at Saladin. The Saladin field development concept is centred on a series of horizontal, relatively low-cost development wells from nearby Thevenard Island (Fig. 1). The horizontal wells are designed to maximize oil drainage from individual lower *M. australis* reservoirs through low-sinuosity well trajectories (Fig. 7). The siderite-cemented layers and intervening sands are hereby used to guide geosteering through application of measurement-while-drilling (MWD) logging technology, notably a gamma-ray and resistivity tool assembly (Fig. 7). By counting up or down the siderite-cemented layers during drilling, the operation geologist normally has a good idea as to which sands are being penetrated by the drill bit and can guide the drillers accordingly.

However, interpretative problems for geosteering can arise when drilling inclination is modified substantially, in poor borehole conditions, and/or if lateral variations in the intensity of siderite cementation are encountered. These factors can reduce drastically the confidence with which specific 'tight streaks' are identified. The MWD tool also can fail, as indeed occurred over parts of the reservoir section during the drilling of Saladin-12 (see below). During these times, uncertainty surrounds the precise stratigraphical position of the drill bit, with risk of suboptimal well path development and costly delays, possibly even total well failure. For example, important sands could be missed for oil development, or the drill bit could accidentally break through the basal 3–5 m of the Mardie Greensand (production time seal) into the Barrow Group sandstones.

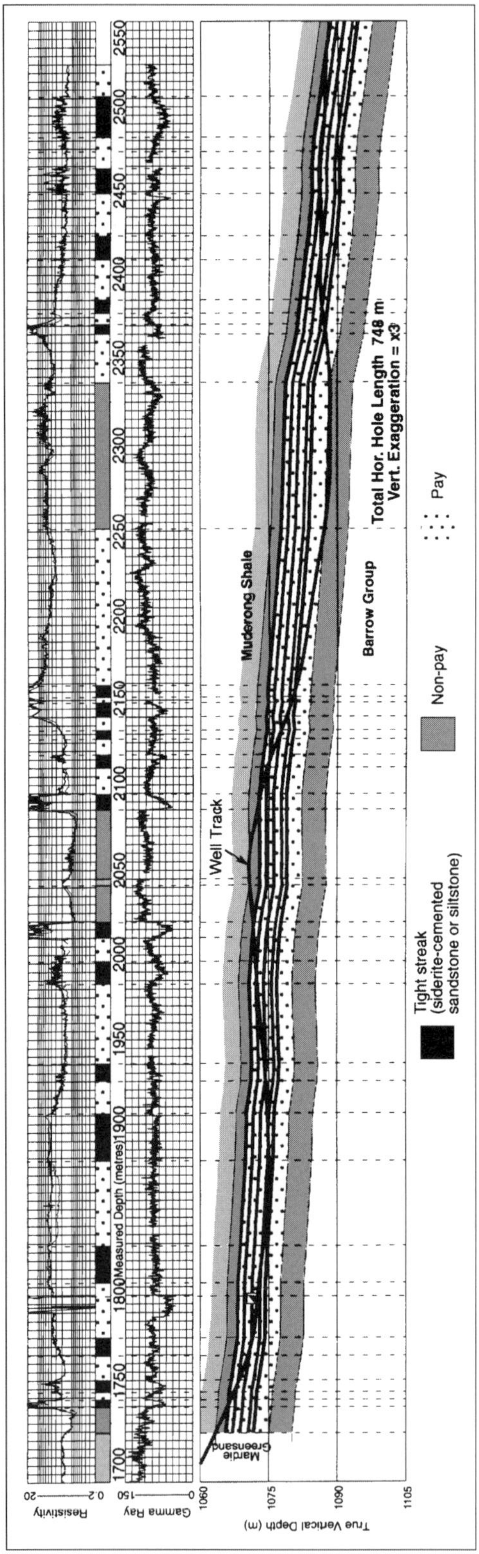

Fig. 7 Horizontal well path for Saladin-12 at Mardie Greensand level showing measurement-while-drilling (MWD) (gamma ray and resistivity) log response and the application of field-wide correlation of siderite-cemented layers and intervening sands to guide geosteering. The challenge for geosteering is to intersect (drain) as many individual Mardie Greensand reservoir sands as possible, without breaking into the underlying high-permeability Barrow Group, and preferably without grazing the base of the overlying Muderong Shale, for reasons of well cost and production optimization.

The latter would have disastrous consequences for Mardie Greensand production in view of the superior reservoir performance of the Barrow Group sandstones. These considerations are highlighted as they place the ensuing observations into their operational context.

WELL-SITE OBSERVATIONS

Saladin-12 was spudded from Thevenard Island in July 1996, with the sole objective of increasing oil production from the lower *M. australis* interval in the Mardie Greensand. The horizontal well was drilled using oil-based mud (OBM). Figure 7 illustrates the well path through the Mardie Greensand.

Drill cuttings were collected from the shakers by the mudlogger about every 10 m of along-hole section drilled using a 250-μm mesh (i.e. borderline fine to medium grain size). The samples were then gently washed in fresh water containing small amounts of detergent to remove as much as possible of the oil-based mud from the drill cuttings before inspection and description of the samples under the binocular microscope. Usually by the time this process was completed (against the backdrop of additional tasks) the next batch of drill cuttings was due for collection. Calculated travel time of the cuttings from reservoir depth to surface was no more than about 20 min on average.

On-site examination of the fresh drill cuttings under the binocular microscope showed that the samples derived from the Mardie Greensand are 90–95% dominated by clay, with only about 5–10% detrital quartz (Fig. 8A–C). The clay principally consists of dark-green pelloidal glauconite, but also brown (non-glauconitic) matrix clay in some samples (Fig. 8A). The very low proportion of detrital quartz grains over supposed reservoir sections prompted the mudlogging contractor to comment 'are you certain there is any reservoir?'

All observed quartz grains, typically located at the coarse end of the fine-grained spectrum, are substantially smaller than associated glauconite pellets. Typically, the glauconite pellets

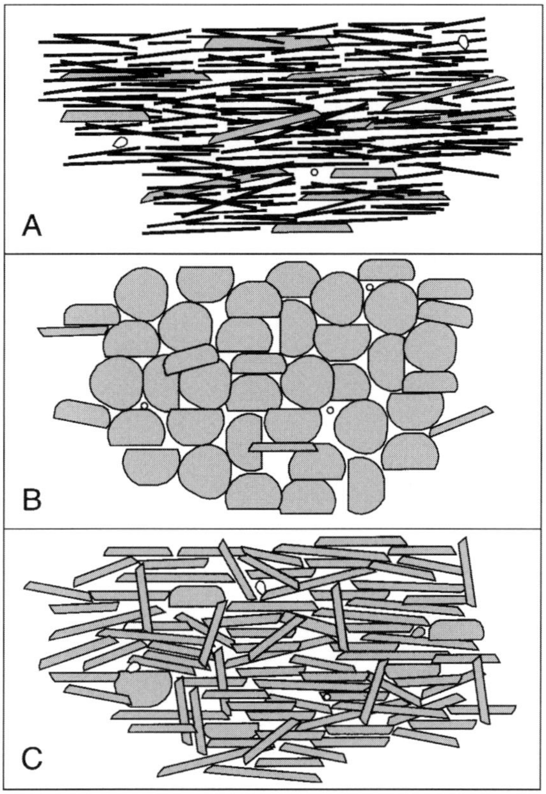

Fig. 8 Schematic representation of the various types of oil-based mud drill cuttings typical of the Mardie Greensand as observed in sample trays under the binocular microscope, fresh from the shakers at the Saladin-12 well site. Note the overall low content in detrital quartz (white) and the high proportion of clay in each basic sample type. (A) Cuttings style A is dominated by dark-brown (non-glauconitic) mud and disk- or plate-shaped glauconite pellets (grey) with rare quartz grains. (B) Cuttings style B is dominated by round to semi-oval glauconite pellets with flat tops and bases, without any brown mud, containing few quartz grains. The quartz grains are typically well rounded and of similar size (very fine to fine grained). (C) Cuttings style C is characterized by a dominance of disk- or plate-shaped glauconite pellets, again without any brown mud, and contains rare quartz grains of variable roundness but similar size (very fine to fine grained). Note that the size of the glauconite pellets typically exceeds the size of the quartz grains in all sample types.

have round to subround outlines but display a wide range of different height ratios in side view (Fig. 8A–C). Systematic description of these different glauconite morphologies during

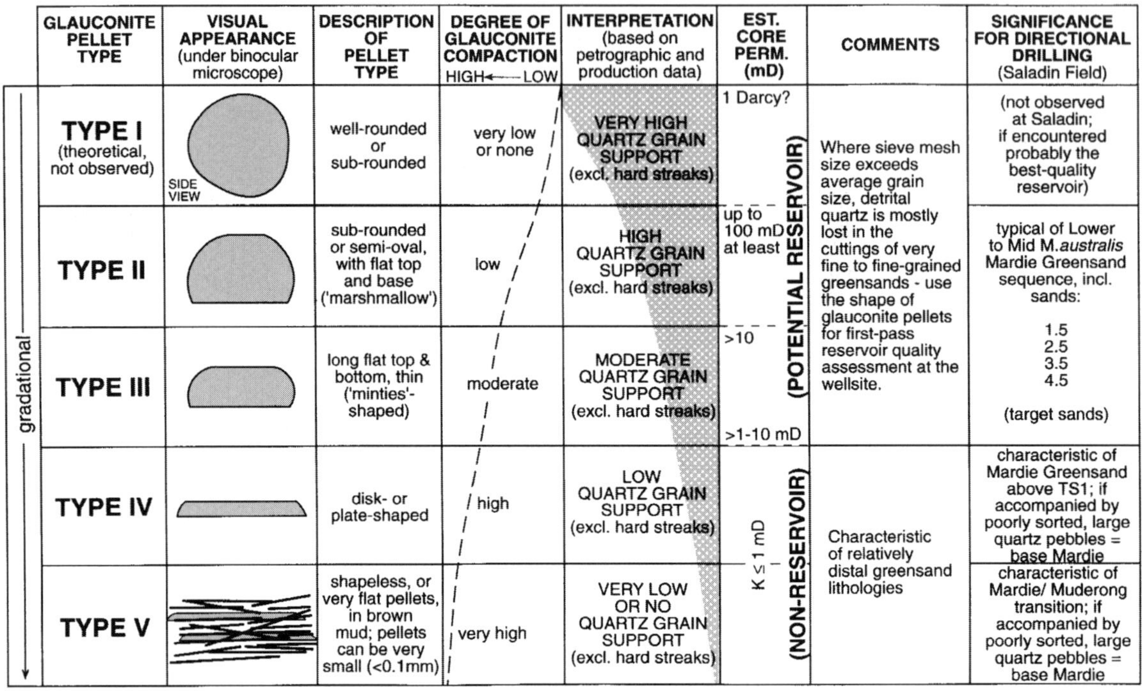

Fig. 9 Summary diagram showing the different types of glauconite pellets and their application for inferring subsurface reservoir quality in silt-size to very fine-grained Mardie Greensand lithologies (Saladin Field). The simple method based on drill cuttings may be applicable only under oil-based mud (OBM) borehole conditions, and should not be used in conjunction with shallow greensand reservoirs (≤ 500 m) where differential sediment compaction may be insufficiently pronounced. TS = tight streak.

drilling showed that many samples are dominated by more or less well-rounded or 'marshmallow'-shaped glauconite pellets (Fig. 8B) characterized by gently curved flat tops and bottoms, consistent with very minor compaction (type I/type II pellets; Fig. 9). Other cutting samples contain predominantly thin glauconite pellets with long flat tops and bottoms ('minties-shaped') indicative of about 40–50% compaction assuming an originally spherical shape (type III pellets). Still other cutting samples are dominated by disk- or plate-shaped dimensions (Fig. 8C), where compaction is estimated to be 80% or more (type IV pellets; Fig. 9). Last, a few drill samples were found exclusively to contain extremely flat glauconite pellets within a sticky mass of dark-brown (non-glauconitic) clay (Fig. 8A) consistent with very strong compaction (type V pellets) (Fig. 9).

Immediately following awareness of the existence of this range in glauconite styles, the possibility was examined that sampling or washing procedures could have influenced glauconite morphologies. This hypothesis, however, proved to be invalid. No difference in the dominant shape of glauconite pellets was observed during any of the different sample collection stages, although it was discovered that a small (≤ 1–5%) subpopulation of glauconitic pellets exists in most (possibly all) samples. This subpopulation of glauconitic pellets is distinguished by a light-green colour and the fact that these pellets easily disintegrate on gentle touch with a pair of tweezers but only after exposure to fresh water, presumably owing to smectite interstratification, as reported by Hatcher *et al.* (1996) and Baker *et al.* (1997). Apart from these characteristics, this subpopulation of light-green glauconitic pellets is indistinguish-

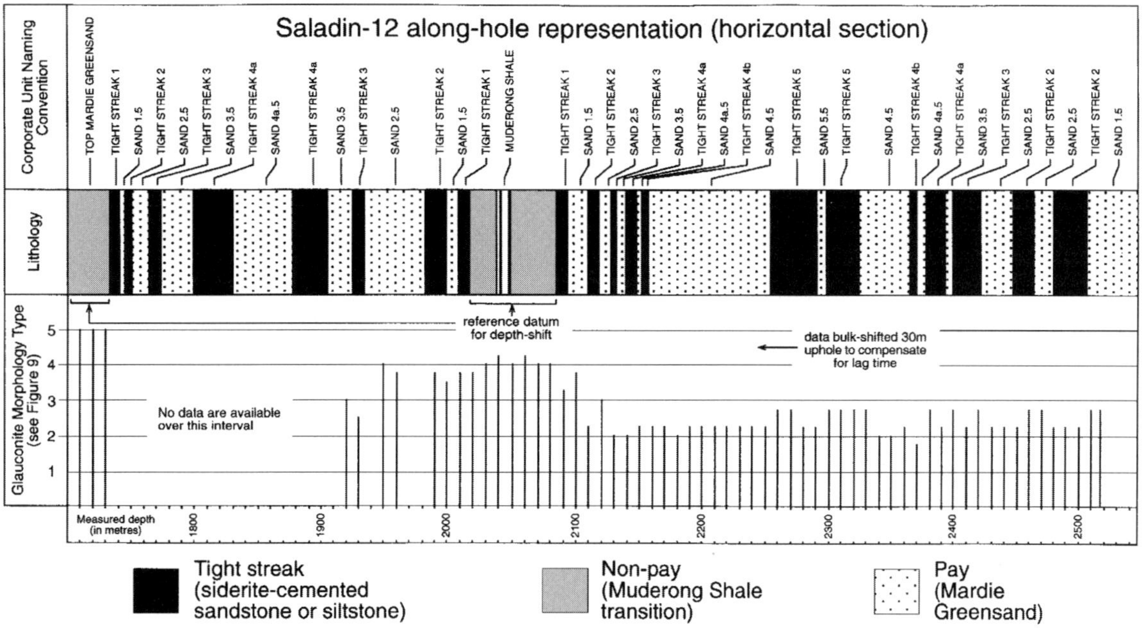

Fig. 10 Demonstration of the use of glauconite morphologies in the facilitation of horizontal geosteering. Although type II and type III glauconite morphologies were observed over actual lower *M. australis* reservoir sections, accidental grazing of the base Muderong Shale and top Mardie Greensand transition (2019–2085 m along-hole) produced a return to type IV and V glauconite morphologies. The same glauconite morphologies had been observed during initial penetration of the top Mardie Greensand (1700–1730 m along-hole). See text for full explanation.

able from the dominant (dark-green) glauconite variety under the binocular microscope: both glauconite types display the same range of morphologies.

Taking into account a 30 m lag in sample depth annotation, a clear correlation could be demonstrated between the dominant shape of the glauconite pellets and the known reservoir stratigraphy during well-site operations (Figs 7 & 10). Although the drill bit was within the lower *M. australis* reservoir zone, which subsequently produced oil at a rate of 3050 BOPD during the first production test, exclusively type II and type III glauconite morphologies were observed in the drill cuttings (Fig. 10). However, at a depth of between 2019 and 2085 m along-hole, the top of the (non-productive) Mardie Greensand was accidentally grazed, including the transition zone into the lowermost Muderong Shale (Figs 7 & 10). The cuttings that were recovered over this interval were found to be dominated by type IV and type V glauconite morphologies—the same types as had been encountered during initial top Mardie Greensand penetration between 1700 and 1730 m along-hole (Figs 7 & 10). The drilling of this suboptimal well path coincided with erratic MWD behaviour, including temporary MWD tool failure over the interval 1803 to 2547 m along-hole. As it happened, the drill cuttings in this case provided the first indication that the drill bit had penetrated non-reservoir section before the MWD logging tool yielded clear results again. Upon re-entering the reservoir section, a return to type II and type III glauconite morphologies was observed in the cuttings until total depth was reached at 2547 m along-hole (Figs 7 & 10).

DISCUSSION

The clear implication of the above observations is that during the sampling process the majority of the quartz grains that constitute the reservoir zones were being lost at the shakers. Monitoring of the quartz content in the drill cuttings provided no reliable indication as to whether a reservoir or non-reservoir zone was being penetrated by the drill bit. The mesh size used (250 μm) was simply too coarse for the generally very fine to fine-grained Mardie Greensand lithologies (62–170 μm). Although this sample bias can be easily remedied by selection of a finer mesh size, such simple modification would require more time by the mudlogger to wash, describe and bag samples at the rig site. The consequence of this action would be either a wider sample spacing (at least every 20 to 30 m) or the requirement for additional manpower to keep pace with a relatively fast rate of penetration. Although the former is clearly not desirable from a data acquisition perspective, the latter is equally not desirable from a cost leadership perspective. Thus, description of glauconite morphology has a small but viable role in downhole reservoir characterization based on drill cuttings in the Mardie Greensand, and possibly other greensands around the world.

At Saladin, the observed variety in glauconite morphologies is attributed to differential compaction of the pellets, which must be related to varying rock compositions at different stratigraphic levels within the Mardie Greensand. Principally the varying proportions of detrital quartz and to a lesser extent feldspars and other compaction-resistant framework grains (including minor cements) are thought to control glauconite morphology in the subsurface. In a quartz-rich greensand lithology, each individual glauconite pellet only has to support the weight of the detrital grains in its immediate vicinity because the bulk of the overburden weight is absorbed by a plethora of framework grains (types I–III glauconite morphologies). In contrast, glauconite pellets in lithologies that lack framework grain support can provide little mechanical resistance to the increasing overburden weight and, consequently, are squashed during sediment burial (types IV and V glauconite morphologies, Fig. 9). Thus, the dominant shape of glauconite pellets as observed in drill cuttings can be used to broadly estimate the level of framework grain support, and thereby indirectly reservoir quality in the Mardie Greensand, at least under oil-based mud conditions (Fig. 9).

This conclusion is based on the assumption that the glauconite pellets were more or less spherical before major sediment compaction was initiated. Various workers have shown that the shape of glauconite grains can depend upon the shape of the primary substrate of glauconitization, including quartz grains, mica flakes, microfossil tests, faecal pellets, shell fragments, etc. (Odin & Matter, 1981; Van Houten & Purucker, 1984; Odin & Fullagar, 1988). In the present study, no primary substrate is evident in any of the glauconite pellets as seen in thin-section, which would appear to be a widespread feature based on a regional petrographic data base for the Mardie Greensand (see Stuart *et al.*, 1995; Rezaee & Tingate, 1996; Schulz-Rojahn, 1996). Hatcher *et al.* (1996) demonstrate both highly evolved, dark-green glauconite and a small population of expandable pale-green nascent glaucony (glauconitic smectite) in the Mardie Greensand of the Saladin Field. Baker *et al.* (1997) also report abundant dark-green, low-expandability glauconite pellets characterized by an average K_2O content of 8.5% in broadly time-equivalent glauconitic sandstones of the northern Carnarvon Basin. Baker *et al.* (1997) again show that these dark-green pellets coexist with very low amounts of light-green, pelloidal glauconitic smectite that is prone to swelling and disintegration when exposed to fresh water.

These data, coupled with the chiefly dark-green colour of the glauconitic pellets and the complete dissolution of the substrate (see Amorosi, 1995, 1997) underline the dominance of an evolved or even highly evolved form of glaucony (glauconite) in the Mardie Greensand, including at Saladin. Thus, it is exceedingly unlikely that the shape of the glauconite pellets was inherited from the primary substrates of glauconitization, in particular because the pellets tend to have round to subround outlines and differ only in height ratios, which is explained most readily by differential compaction. Baker *et al.* (1997) also noted the role of compaction in influencing peloid shape. It seems reasonable to assume that the glauconite pellets that experienced the least degree of compaction (types I–III) provide a good approximation of how type IV and V glauconite pellets also must have looked like in Early Cretaceous times, on the seafloor, prior to the onset of major subsidence.

Petrographic evidence for this hypothesis stems from the abundant occurrence of microcrystalline quartz cement envelopes around spherical dissolution pores that locally still contain residual glauconite pellets in distal greensand facies of the Mardie Greensand, within a few kilometres of the Saladin Field (Rezaee & Tingate, 1996; Schulz-Rojahn, 1996). The shape of the dissolution pores vividly demonstrates that the network of microcrystalline quartz cement envelopes must have formed around virtually uncompacted (near-spherical) glauconitic pellets in the shallow subsurface prior to dissolution of the pellets (Rezaee & Tingate, 1996; Schulz-Rojahn, 1996). This zone of diagenetic alteration, characterized by an almost complete lack of framework grains, grades laterally and vertically into an unaltered greensand facies of the same type (see Fig. 6) dominated by dark-green type IV and V glauconite pellets (Schulz-Rojahn, 1996). Thus, the evidence available strongly suggests that compaction principally is responsible for the present-day shape of the peloidal grains rather than changes in spatial distribution of the substrates of glauconitization in the Mardie Greensand. Certainly the observation that type I–III glauconite morphologies dominate in reservoir sections (as proven by actual production data) whereas type IV and V glauconite morphologies occur in non-reservoir sections is consistent with this interpretation.

This simple recognition leads to a variety of possible industry or research applications. The Saladin-12 case study demonstrates that classification of glauconite morphology rather than just a conventional description of cuttings is the key to a reliable lithology indicator in the Mardie Greensand, unless a dedicated effort is made to capture the very fine grain fraction at the shakers, desander and desilter. Certainly under OBM conditions, descriptive focus on glauconite morphologies is required in order to identify reservoir zones that otherwise could be mistaken for clay-dominated lithologies in drill cuttings. Especially in low-resistivity (greensand) exploration settings where pay recognition can be a key subsurface challenge, the description of glauconite morphologies could provide an imperfect but low-cost insurance against the risk of pay zones being bypassed. However, it remains to be seen if the technique also can be used under non-OBM conditions, which tend to be preferred by explorationists because reservoir oil/condensates and oil-based drilling mud generally cannot be differentiated under the UV fluoroscope at the well site.

Further, the present case study indicates that glauconite morphologies can provide a broad qualitative indication of reservoir quality, well ahead of routine petrophysical evaluation, flow tests and/or core analysis. Subject to the rate of penetration, circulation, hole length and various other variables, cuttings from the Mardie Greensand are received on surface within a few tens of minutes at most. At Saladin, type II and III glauconite morphologies were observed to have been derived from reservoir zones capable of producing initial oil rates of up to 4000 BOPD, consistent with permeability up to about 100 mD at least (Seeburger *et al.*, 1998). In contrast, type IV and V glauconite morphologies encountered over non-reservoir Mardie sections are thought to be representative of glauconite-rich, quartz-poor greensand facies (Fig. 6) characterized by core permeability of 1 mD or less (Schulz-Rojahn, 1996). Thus, a broad framework is established that assigns a range of different permeability values to different glauconite morphologies (Fig. 9), which perhaps could be refined with growing experience in the future. This recognition is useful because in the event that exclusively type IV and/or type V glauconite morphologies are encountered along supposed reservoir target sections, an early well-site warning system is provided that could signal the selection of a poor drilling target owing to inadequacies in the geological model (or lack thereof). In such a case, the decision could be accelerated to plug and abandon, thereby reducing rig-based cost exposure. By the same token, systematic monitoring of glauconite morphologies along the well path could act as a basic cross-check on petrophysical evaluations in low-confidence interpretation situations (e.g. hole rugosity). In the same context, a hypothetical situation is conceivable whereby this type of independent cross-check has an impact on the selection of completion intervals in multilayered greensand reservoirs of variable reservoir quality and with known hydrocarbon fill(s).

Glauconite morphology as seen in drill cuttings also can provide a good indication of the stratigraphical level that is being penetrated by the drill bit, provided the stratigraphy and petrological characteristics of the target formation are well understood. This type of data can be helpful for geosteering, especially in cases where drilling rates are relatively slow, or MWD data are temporarily not available, as indeed was the case at Saladin-12. The present case study provides convincing evidence that monitoring of glauconite morphology can be helpful to the well-site geologist, who must decide whether the drill bit is too high or too low in the stratigraphical sequence, or indeed on track, during critical well-site operation stages.

Intriguingly, with a growing field data base, the potential is even there for indirect incorporation of differential glauconite compaction in the static reservoir model to reduce uncertainties in the three-dimensional porosity and permeability realization. In a formation that lacks seismic expression and where prediction of reservoir quality trends remains a key challenge, certainly on the regional scale, monitoring of drill cuttings could provide crucial clues in subtle reservoir lithologies such as the Mardie Greensand. On the field-scale, systematic description of glauconite style, both in the lateral and vertical sense, could provide the framework for a refined reservoir model (reservoir geometries, facies transitions) facilitating long-term production and facility optimization.

CONCLUSIONS

1 Study results show that, seemingly against all odds, glauconite pellets can survive in a near-perfect state a journey of several kilometres along borehole to surface, under oil-based mud conditions. This fact vividly demonstrates the high level of geological detail that can be obtained from drill cuttings, providing a fresh new perspective on the characterization of the unconventional Mardie Greensand reservoir. Drill cuttings represent a primary information source about the subsurface that always should be considered by the reservoir geologist.

2 The dominant shape of glauconite pellets, as observed under the binocular microscope at the well site, can be a first-pass indicator of subsurface reservoir quality ahead of petrophysical evaluation, flow tests and/or core analysis. In the Mardie Greensand, the shape of individual glauconite pellets is controlled by differential compaction, which, in turn, is controlled by the degree of framework grain support that drives reservoir quality. Glauconite pellets with round, oval or semi-oval shapes (types I–III) reflect quartz-rich greensand reservoirs with permeabilities of up to 100 mD at least, capable of producing initial oil rates of up to 4000 BOPD from individual horizontal wells. In contrast, disk- or plate-shaped glauconite pellets (types IV and V) dominate in very quartz-poor, strongly compacted greensand facies without any reservoir potential.

3 This simple reservoir characterization technique based on glauconite morphology finds its exclusive application in silt-size to fine-grained greensand lithologies, where it can be problematic to obtain representative drill cuttings, especially at high rates of penetration. Under these circumstances, the bulk of the silt-size to fine sand grains tend to be lost at the shakers, which can lead to prolific reservoir zones being mistaken for clay-dominated lithologies in drill cuttings, especially in cases where the glauconite pellets are larger than the framework grains.

4 Monitoring of glauconite morphology in drill cuttings has several potential, or proven, industry applications.

(i) Potential pay recognition in low-resistivity (greensand) exploration or appraisal settings, although as yet unproven for non-OBM borehole conditions.

(ii) The technique can be used as an independent, rather crude and basic cross-check on reservoir quality in low-confidence petrophysical interpretation situations. In the same context, the technique possibly could be used to cross-check selection of completion intervals in multilayered greensand reservoirs of variable reservoir quality and with known hydrocarbon fill(s).

(iii) Despite lag time of cuttings to surface, the technique can facilitate MWD-controlled geosteering, as exemplified by the horizontal Saladin-12 well, where along-hole variations in the shape of glauconite pellets provided helpful stratigraphical clues.

(iv) In the event that exclusively disk- or plate-shaped glauconite pellets are encountered along supposed reservoir target sections, an early well-site warning system is provided that could accelerate the decision to plug and abandon.

(v) In horizontal or high-angle deviated wells, the technique has potential applications for static reservoir modelling because it may allow the reservoir geologist to gain an improved understanding of lateral facies transitions and sand geometries in subtle greensand lithologies.

5 The case study demonstrates that oil company staff (well owners) have an active role in the geological description of drill cuttings, including for self-awareness reasons, especially during the early phases of a drilling campaign (exploration, appraisal and/or early field development phases). Failure to do so may lead to erroneous or ambiguous perceptions of reservoir quality and continuity because the mud-logging contractors more often than not will not be familiar with the key reservoir uncertainties and the potential subtleties of the target lithologies, usually through no fault of their own. At Saladin, description of glauconite pellet styles rather than just a conventional description of cuttings proved to be the key to a reliable lithology indicator.

6 A data acquisition strategy based on cuttings ideally should be tailored to the target lithology, which can be particularly important in unconventional and/or fine-grained reservoir lithologies. Sieve size, sampling rate and manpower (mud-logging) resource requirements should be considered carefully in advance, especially in formations where prediction of regional reservoir quality trends constitutes a key subsurface challenge.

ACKNOWLEDGEMENTS

We are indebted to West Australian Petroleum Pty Ltd. (WAPET), the operator on behalf of Chevron Asiatic Ltd., Texaco Inc., Mobil Exploration and Producing Australia (MEPA) and Shell (Development) Australia Pty Ltd., for permission to publish this work. We further would like to thank IAS reviewers E. McBride (University of Texas at Austin) and A. Amorosi (University of Bologna) for their constructive criticisms of the draft manuscript. We also express our warm appreciation to our colleagues S. Morad (University of Uppsala), J. Meath, J. Popek (Chevron Overseas Petroleum), M. Rezaee (University of Tehran) and R. Helby for their support. All opinions expressed reflect those of the authors. This paper evolved out of a collaborative project between WAPET and the Australian Petroleum Cooperative Research Centre (APCRC) as represented by the National Centre for Petroleum Geology and Geophysics (NCPGG) and the Centre for Petroleum Engineering (CPE) following recruitment of the geological team leader (senior author) by WAPET in 1996.

REFERENCES

Amorosi, A. (1995) Glaucony and sequence stratigraphy: a conceptual framework of distribution in siliciclastic sequences. *J. Sed. Res.*, **B65** (4), 419–425.

Amorosi, A. (1997) Detecting compositional, spatial, and temporal attributes of glaucony: a tool for provenance research. *Sed. Geol.*, **109**, 135–153.

Baker, J.C., Uwins, P.J.R. & Hamilton, P.J. (1997) Freshwater sensitivity of glauconitic hydrocarbon reservoirs. *J. Petrol. Sci. Eng.*, **18**, 83–91.

Beacher, G.J. (1998) Pressure study of the Flacourt Formation aquifer in the Thevenard Island area of the Barrow Sub-basin. *J. Aust. Petrol. Prod. Explor. Assoc.*, **38** (1), 438–452.

Hatcher, G.B., Chen, H., Rahman, S.S. & Hogg, P.F. (1996) Evaluating formation damage risks in a glauconitic sandstone reservoir: a case history from the Offshore North West Shelf of Australia. *Soc. Petrol. Eng. Pap.*, **37014**, 1–15.

Helby, R., Morgan, R. & Partridge, A.D. (1987) A palynological zonation of the Australian Mesozoic. *Mem. Assoc. Aust. Paleontol.*, **4**, 1–94.

Odin, G.S. & Fullagar, P.D. (1988) Geological significance of the glaucony facies. In: *Green Marine Clays*, Developments in Sedimentology 45 (Ed. by G.S. Odin), Elsevier, Amsterdam, pp. 295–332.

Odin, G.S. & Matter, A. (1981) De glauconiarum origine. *Sedimentology*, **28**, 611–641.

Rezaee, M. & Tingate, P.R. (1996) *Regional Reservoir Prospectivity of the Mardie Greensand, Part II: Petrology and Quantitative Lithofacies Identification.* Unpublished report on behalf of the Australian Petroleum Cooperative Research Center (APCRC) for West Australian Petroleum Pty Ltd, Perth, 35 pp.

Schulz-Rojahn, J.P. (1996) *Regional Reservoir Prospectivity of the Mardie Greensand, Carnarvon Basin—Part 1: Stratigraphic Controls on Reservoir Development and Implications for Forward Modeling.* Unpublished report on behalf of the Australian Petroleum Cooperative Research Center (APCRC) for West Australian Petroleum Pty Ltd, Perth, 48 pp.

Seeburger, D.A., Miller, N.W.D., Beacher, G.J., Schulz-Rojahn, J.P. & Popek, J.P. (1998) An evaluation of the Mardie Greensand reservoir, Thevenard Island area, Carnarvon Basin. In: *The Sedimentary Basins of Western Australia 2: Proceedings of the Petroleum Exploration Society of Australia Symposium* (Ed. by P.G. and R.R. Purcall), Perth, pp. 491–502.

Stuart, W.J., Schulz-Rojahn, J.P. & Hayball, A. (1995) *Lithofacies Identification, Correlation and Reservoir Characterization of the Mardie Greensand, Barrow Field Area.* Unpublished report by the Australian Petroleum Cooperative Research Center (APCRC) for West Australian Petroleum Pty Ltd, Perth.

Thomas, B.M. (1978) Robe River—an onshore shallow oil accumulation. *J. Aust. Petrol. Explor. Assoc.*, **18** (1), 3–12.

Tippet, P.J. & Beacher, G.J. (2000) Integration of subsurface disciplines to optimize the development of the Mardie Greensand at Thevenard Island, Western Australia. SPE paper 59410. *2000 SPE Asia Pacific Conference on Integrated Modeling for Asset Management*, Yokohoma, 25–26 April 2000.

Van Houten, F.B. & Purucker, M.E. (1984) Glauconitic peloids and chamositic ooids—favourable factors, constraints, and problems. *Earth Sci. Rev.*, **20**, 211–243.

Zhang, Y.J., Lollback, P.A., Schulz-Rojahn, J.P., Salisch, H.A. & Stuart, W.J. (1996) A methodology for estimating permeability from well logs in a formation of complex lithology. SPE paper 37025. In: *1996 SPE Asia Pacific Oil and Gas Conference*, 28–31 October, Adelaide, pp. 561–566.

Zhang, Y.J., Lollback, P.A., Salisch, H.A. & Schulz-Rojahn, J.P. (1997) Determination of permeability transforms from geophysical logs using statistical pattern recognition: the Mardie Greensand. *Explor. Geophys.*, **28**, 181–184.

Zhang, Y.J., Salisch, H.A. & Arns, C. (2000) Permeability evaluation in a glauconite-rich formation in the Carnarvon Basin, Western Australia. *Geophysics*, **65** (1), 46–53.

Index

Page numbers in *italics* refer to figures; those in **bold** refer to tables.